Phosphorus in Environmental Technology

Integrated Environmental Technology Series

The *Integrated Environmental Technology Series* addresses key themes and issues in the field of environmental technology from a multidisciplinary and integrated perspective.

An integrated approach is potentially the most viable solution to the major pollution issues that face the globe in the 21st century.

World experts are brought together to contribute to each volume, presenting a comprehensive blend of fundamental principles and applied technologies for each topic. Current practices and the state-of-the-art are reviewed, new developments in analytics, science and biotechnology are presented and, crucially, the theme of each volume is presented in relation to adjacent scientific, social and economic fields to provide solutions from a truly integrated perspective.

The *Integrated Environmental Technology Series* will form an invaluable and definitive resource in this rapidly evolving discipline.

Series Editor
Dr Ir Piet Lens, Sub-department of Environmental Technology, The University of Wageningen, P.O. Box 8129, 6700 EV Wageningen, The Netherlands.
(piet.lens@ algemeen.mt.wag-ur.nl)

Published titles
Biofilms in Medicine, Industry and Environmental Biotechnology:
  Characteristics, analysis and control
Decentralised Sanitation and Reuse: *Concepts, systems and implementation*
Environmental Technologies to Treat Sulfur Pollution: *Principles and engineering*
Water Recycling and Resource Recovery in Industries: *Analysis, technologies and implementation*

Forthcoming titles
Pond Treatment Technology
Resource Recovery and Reuse in Organic Solid Waste Management

www.iwapublishing.com

Phosphorus in Environmental Technology

Principles and Applications

Edited by

Eugenia Valsami-Jones

Published by IWA Publishing, Alliance House, 12 Caxton Street, London SW1H 0QS, UK
Telephone: +44 (0) 20 7654 5500; Fax: +44 (0) 20 7654 5555; Email: publications@iwap.co.uk
Web: www.iwapublishing.com

First published 2004
© 2004 IWA Publishing

Printed by TJ International (Ltd), Padstow, Cornwall, UK
Typeset by Gray Publishing, Tunbridge Wells, UK

Apart from any fair dealing for the purposes of research or private study, or criticism or review, as permitted under the UK Copyright, Designs and Patents Act (1998), no part of this publication may be reproduced, stored or transmitted in any form or by any means, without the prior permission in writing of the publisher, or, in the case of photographic reproduction, in accordance with the terms of licences issued by the Copyright Licensing Agency in the UK, or in accordance with the terms of licenses issued by the appropriate reproduction rights organization outside the UK. Enquiries concerning reproduction outside the terms stated here should be sent to IWA Publishing at the address printed above.

The publisher makes no representation, express or implied, with regard to the accuracy of the information contained in this book and cannot accept any legal responsibility or liability for errors or omissions that may be made.

Disclaimer
The information provided and the opinions given in this publication are not necessarily those of IWA or of the editors, and should not be acted upon without independent consideration and professional advice. IWA and the editors will not accept responsibility for any loss or damage suffered by any person acting or refraining from acting upon any material contained in this publication.

British Library Cataloguing in Publication Data
A CIP catalogue record for this book is available from the British Library

Library of Congress Cataloging-in-Publication Data
A catalog record for this book is available from the Library of Congress

ISBN: 1 84339 001 9

Contents

Contributors	xiii
Preface	xviii

Part One Principles of phosphorus distribution: chemistry, geochemistry, mineralogy, biology 1

1. The chemistry of phosphorus 3
   D. Bryant
   1.1 Introduction 3
   1.2 Atomic properties 4
   1.3 Electronic structure and bonding 4
   1.4 Phosphorus Compounds 5
   References 17

2. The geochemistry and mineralogy of phosphorus 20
   E. Valsami-Jones
   2.1 Introduction 20
   2.2 Geochemical abundance and distribution of phosphorus 22
   2.3 Phosphate minerals 32
   References 44

3. The biology of phosphorus 51
   C. Dobrota
   3.1 Introduction 51
   3.2 Phosphate acquisition 51
   3.3 Transport and translocation 52
   3.4 Phosphate assimilation 62
   References 74

Part Two Phosphorus in the environment — 77

4. Background and elevated phosphorus release from terrestrial environments — 79
   P.M. Haygarth & L.M. Condron
   - 4.1 Introduction — 79
   - 4.2 Background release — 80
   - 4.3 Elevated release — 81
   - 4.4 Conclusion — 87
   - References — 87

5. Phosphorus and crop nutrition: principles and practice — 93
   T.D. Evans & A.E. Johnston
   - 5.1 Introduction — 93
   - 5.2 Phosphorus resources — 95
   - 5.3 Phosphorus fertilisers — 96
   - 5.4 Phosphorus in soils — 98
   - 5.5 Phosphorus nutrition of plants — 103
   - 5.6 Fertilisation practices — 110
   - 5.7 Phosphorus fertilisation and environmental issues — 113
   - 5.8 Concluding remarks — 115
   - References — 116

6. Transfer of phosphorus to surface waters; eutrophication — 120
   S. Burke, L. Heathwaite and N. Preedy
   - 6.1 Introduction — 120
   - 6.2 Evidence of sources and pathways of P delivery to water — 122
   - 6.3 Transfer mechanisms and control — 131
   - 6.4 Management minimisation — 138
   - References — 140

7. Environmental chemistry of phosphonic acids — 147
   B. Nowack
   - 7.1 Introduction — 147
   - 7.2 Analysis of phosphonates — 150
   - 7.3 Exchange reactions — 151
   - 7.4 Degradation — 157
   - 7.5 Speciation — 161
   - 7.6 Environmental behavior — 163
   - 7.7 Conclusions — 165
   - Acknowledgements — 165
   - References — 166

8. Phosphate pollution: a global overview of the problem — 174
   A.M. Farmer
   - 8.1 Introduction — 174
   - 8.2 The European Union — 176

Contents vii

	8.3	The river Rhine: case study	181
	8.4	United States	183
	8.5	Australia	185
	8.6	Japan	186
	8.7	South and East Asia	188
	8.8	Africa	189
	8.9	The Antarctic	190
	8.10	Conclusions	190
		References	190

Part Three Phosphorus removal from water and waste water: principles and technologies **193**

9. Principles of phosphate dissolution and precipitation — 195
   P.G. Koutsoukos & E. Valsami-Jones
   9.1 Introduction — 195
   9.2 Equilibrium and kinetics: theory, mechanisms and equations — 196
   9.3 Applications relevant to the precipitation of calcium phosphates — 208
   9.4 Applications relevant to the dissolution of calcium phosphates — 234
   9.5 Metal phosphates, dissolution and precipitation applications — 238
   9.6 Concluding remarks — 239
   References — 241

10. Waste water treatment principles — 249
    S.A. Parsons & T. Stephenson
    10.1 Introduction — 249
    10.2 Waste water treatment processes — 250
    10.3 Sludge treatment and disposal — 257
    References — 259

11. Chemical phosphorus removal — 260
    S.A. Parsons & T-A. Berry
    11.1 Introduction — 260
    11.2 Phosphorus removal with metal salts — 261
    11.3 Concluding remarks — 269
    References — 270

12. Biological phosphorus removal — 272
    J.W. McGrath & J.P. Quinn
    12.1 Introduction — 272
    12.2 Process design of biological phosphorus removal plants — 274
    12.3 Polyphosphate — 276
    12.4 Isolation and identification of polyphosphate-accumulating microorganisms from the EBPR process — 282
    12.5 New approaches to biological phosphate removal — 283
    Acknowledgements — 286
    References — 286

13. A review of solid phase adsorbents for the removal of phosphorus
    from natural and waste waters   291
    G.B. Douglas, M.S. Robb, D.N. Coad & P.W. Ford
    - 13.1 Introduction   291
    - 13.2 Adsorbent materials   293
    - 13.3 Selection of an appropriate phosphorus adsorbent   307
    - 13.4 Concluding remarks   311
    - References   311

14. Removing phosphorus from sewage effluent and agricultural
    runoff using recovered ochre   321
    K.V. Heal, K.A. Smith, P.L. Younger, H. McHaffie & L.C. Batty
    - 14.1 Introduction   321
    - 14.2 Formation and properties of ochre   322
    - 14.3 Capacity of ochre for phosphorus removal   325
    - 14.4 Future developments   330
    - 14.5 Conclusions   333
    - Acknowledgements   334
    - References   334

Part Four   Phosphorus recovery for reuse: principles, technologies, feasibility   337

15. Phosphorus recovery in the context of industrial use   339
    I. Steén
    - 15.1 Introduction   339
    - 15.2 Why recovery?   341
    - 15.3 Possible technical-industrial pathways   344
    - 15.4 Recycling by the phosphate industry   348
    - 15.5 Future prospects   353
    - References   354

16. Fluid dynamic concepts for a phosphate precipitation reactor design   358
    D. Mangin & J.P. Klein
    - 16.1 Introduction   358
    - 16.2 Fluid dynamics and mixing   359
    - 16.3 Interaction between mixing and primary nucleation   366
    - 16.4 Interaction between mixing and other precipitation mechanisms   373
    - 16.5 Stirred vessels   379
    - 16.6 Fluidised bed reactors   388
    - 16.7 Air mixing   392
    - 16.8 Modelling of precipitation reactors   394
    - 16.9 Conclusion and perspectives   397
    - 16.10 Notation   398
    - References   400

		Contents	ix

17. Phosphorus recovery via struvite production at Slough sewage treatment works, UK – a case study — 402
    Y. Jaffer & P. Pearce
    - 17.1 Introduction — 402
    - 17.2 Site description — 403
    - 17.3 Phosphorus forms through the sewage treatment process — 405
    - 17.4 Phosphorus mass balance — 407
    - 17.5 Bench-scale work — 411
    - 17.6 The reactor — 415
    - 17.7 Preliminary results — 419
    - 17.8 Quality of struvite — 419
    - 17.9 Further work — 424
    - 17.10 Proposed use of product — 425
    - References — 426
    - Appendix — 427

18. Phosphorus recovery trials in Treviso, Italy – theory, modelling and application — 428
    P. Battistoni
    - 18.1 Introduction — 428
    - 18.2 Italian sewer waste water and WWTPs — 429
    - 18.3 Improvement of BNR technologies and performance — 437
    - 18.4 Phosphorus recovery test — 439
    - 18.5 FBR and air stripping — 443
    - 18.6 Case study: Treviso city waste water treatment plant — 455
    - 18.7 Long term performance of the reactor in Treviso — 463
    - 18.8 Perspective and conclusions — 465
    - Acknowledgements — 467
    - References — 467

19. The case study of a phosphorus recovery sewage treatment plant at Geestmerambacht, Holland – design and operation — 470
    P.G. Piekema
    - 19.1 Introduction — 470
    - 19.2 Overview of the plant before upgrading — 471
    - 19.3 Key features of the plant after upgrading — 472
    - 19.4 Design philosophy of the P-recovery process — 475
    - 19.5 Design of the biological part of the side-stream — 478
    - 19.6 Design of Crystalactor® in the side-stream — 484
    - 19.7 Control of sludge settling — 490
    - 19.8 Control of nitrogen removal — 491
    - 19.9 Full scale results — 493
    - 19.10 Cost — 494
    - 19.11 New developments — 495
    - References — 495

20.	Full scale struvite recovery in Japan *Y. Ueno*		496
	20.1	Introduction	496
	20.2	Full scale struvite recovery in Japan	497
	20.3	Summary	506
		References	506
21.	Phosphorus recovery from unprocessed manure *P. Hobbs*		507
	21.1	Introduction	507
	21.2	Profile of phosphorus species in different livestock manure	508
	21.3	Extraction of phosphorus from wastes and manure	511
	21.4	Concluding remarks	518
		References	519
22.	Scenarios of phosphorus recovery from sewage for industrial recycling *A. Klapwijk & H. Temmink*		521
	22.1	Introduction	521
	22.2	Wastewater and sludge treatment	522
	22.3	P-recovery from the end product of sludge treatment	524
	22.4	P-extraction before sludge treatment	525
	22.5	Discussion	527
	22.6	Conclusions	527
		References	528
23.	Phosphate recycling: regulation and economic analysis *J. Köhler*		529
	23.1	Introduction	529
	23.2	Why recycle phosphates?	530
	23.3	Regulation and economics of WwTPs and phosphates	532
	23.4	A system approach	540
	23.5	Conclusions: economics and policies	543
		References	543

Part Five	**Novel biotechnologies**		**547**
24.	Bacterial precipitation of metal phosphates *L.E. Macaskie, P. Yong & M. Paterson-Beedle*		549
	24.1	Introduction	549
	24.2	Heavy metal bioremediation: why select a phosphate-based precipitation process?	551
	24.3	Case history: metal phosphate biomineralization by *Serratia* sp.	553
	24.4	A conceptual model for phosphate biomineralization	557
	24.5	A quantitative model for metal phosphate biomineralization	560
	24.6	Phosphate biomineralization for the removal of transuranic elements: the need for nucleation processes	561

	24.7	Co-precipitative metal removal	562
	24.8	Use of metal phosphate as an ion exchanger	564
	24.9	Use of alternative phosphate donors	566
	24.10	Use of inorganic phosphate as the phosphate donor	569
	24.11	Conclusions	573
		References	575
25.	Developments in the use of calcium phosphates as biomaterials		582
	R.L. Sammons, P.M. Marquis, L.E. Macaskie, P. Yong & C. Basner		
	25.1	Introduction	582
	25.2	The demand for bone substitute materials	582
	25.3	The ideal bone-substitute: what it has to do?	583
	25.4	Successful applications of calcium phosphates in biomaterials	586
	25.5	Chemistry and manufacture of biomedical calcium phosphates	590
	25.6	Reactions of biomedical phosphates in the body	597
	25.7	Future developments	601
	25.8	Concluding remarks	603
		References	604
26.	Agronomic-based technologies towards more ecological use of phosphorus in agriculture		610
	W.J. Horst & M. Kamh		
	26.1	Introduction	610
	26.2	Defining optimum soil-P levels	611
	26.3	Phosphorus dynamics in soils	611
	26.4	Phosphorus utilisation efficiency of plants	612
	26.5	Agronomic practices	614
	26.6	Conclusion	622
		References	622
27.	Biodegradation of organophosphate nerve agents		629
	M. Shimazu, W. Chen & A. Mulchandani		
	27.1	Introduction	629
	27.2	Microbial degradation	631
	27.3	Enzymatic detoxification of OP neurotoxins	632
	27.4	Whole cell detoxification of OP neurotoxins	634
	27.5	Modifications of specificity and activity	636
	27.6	Organophosphorus acid anhydrolase	638
	27.7	Conclusion	639
		Acknowledgements	639
		References	639
Index			643

Contributors

Christine Basner
DENTSPLY FRIADENT, 68229,
Steinzeugstrasse 50, Mannheim,
Germany.

Paolo Battistoni
Sanitary and Environmental
Engineering, Engineering Faculty,
Hydraulics Institute,
University of Ancona,
via Brecce Bianche, 60131,
Ancona, Italy.

Lesley Batty
School of Civil Engineering and
Geosciences, Cassie Building,
University of Newcastle upon Tyne,
Newcastle upon Tyne,
NE1 7RU, UK.

Terri-Ann Berry
Thames Water,
Spencer House,
Manor Farm Road,
Reading,
RG2 0JN, UK.

David Bryant
Department of Chemistry,
University of Leeds,
Leeds,
LS2 9JT, UK.

Sean Burke
Environment Agency,
Phoenix House,
Leeds, LS11 8PG, UK.

Wilfred Chen
Department of Chemical and
Environmental Engineering,
University of California,
Riverside,
CA 92521, USA.

Dean N. Coad
CSIRO Land and Water,
Centre for Environment and
Life Sciences,
Private Bag No. 5,
Wembley, WA, 6913,
Australia.

Leo Condron
Soil, Plant & Ecological Sciences Division,
PO Box 84, Lincoln University,
Canterbury 8150,
New Zealand.

Cristina Dobrota
Babes-Bolyai University,
Faculty of Biology–Geology,
Plant Biology Dept.,
1 Kogalniceanu St., 3400
Cluj-Napoca, Romania.

Grant Douglas
CSIRO Land and Water,
Centre for Environment and Life Sciences,
Private Bag No. 5,
Wembley, WA, 6913, Australia.

Tim D. Evans
TIM EVANS ENVIRONMENT,
Stonecroft, Park Lane, Ashtead,
Surrey, KT21 1EU, UK.

Andrew M. Farmer
Institute for European
Environmental Policy,
Dean Bradley House,
52 Horseferry Road, London,
SW1P 2AG, UK.

Phillip W. Ford
CSIRO Land and Water,
Black Mountain Laboratories, GPO
Box 166, Canberra, ACT, 2601,
Australia.

Phil Haygarth
Soil Science and Environmental
Quality Team,
Institute of Grassland and
Environmental Research (IGER),
North Wyke Research Station,
Okehampton, Devon,
EX20 2SB, UK.

Kate Heal
University of Edinburgh,
School of GeoSciences,
Darwin Building, Mayfield Road,
Edinburgh, EH9 3JU, UK.

Louise Heathwaite
Department of Geography,
University of Sheffield,
Sheffield, S10 2TN, UK.

Phil Hobbs
Manures and Farm Resources,
Institute of Grassland and
Environmental Research (IGER),
North Wyke Research Station,
Okehampton, Devon,
EX20 2SB, UK.

Walter Horst
Institute for Plant Nutrition,
Faculty of Horticulture,
University of Hannover,
Herrenhaeuser Str. 2, D 30419
Hannover, Germany.

A.E. Johnny Johnston
Agriculture and the Environment Division,
Rothamsted Research,
Harpenden, Herts.,
AL5 2JQ, UK.

Yasmin Jaffer
Thames Water,
Spencer House,
Manor Farm Road, Reading,
RG2 0JN, UK.

Mahmoud Kamh
Soil and Water Science Department,
Faculty of Agriculture,
University of Alexandria,
El-Shatby, Alexandria,
Egypt.

Contributors

Abraham Klapwijk
University of Wageningen & Research Center,
Subdepartment Environmental Technology,
Bomenweg 2, POB 8129, NL-6700 EV Wageningen,
Netherlands.

Jean Paul Kline
LAGEP (Laboratory of Automatic Control and Process Engineering),
UMR CNRS n°5007 – University of Lyon 1 – ESCPE Lyon,
Bâtiment 308G,
43 Boulevard du 11 Novembre 1918,
F-69622 Villeurbanne cedex,
France.
and
IUT A, Chemical Engineering Department,
43 Boulevard du 11 Novembre 1918,
F-69622 Villeurbanne cedex,
France.

Jonathan Köhler
Department of Applied Economics,
University of Cambridge,
Sidgwick Avenue,
Cambridge, CB3 9DE, UK.

Petros G. Koutsoukos
Institute of Chemical Engineering and High Temperature Chemical Processes,
P.O. Box 1414,
Department of Chemical Engineering,
University of Patras, GR-26500,
Patras, Greece.

Lynne E. Macaskie
School of Biosciences,
University of Birmingham,
Birmingham,
B15 2TT, UK.

Denis Mangin
LAGEP (Laboratory of Automatic Control and Process Engineering),
UMR CNRS n°5007 – University of Lyon 1 – ESCPE Lyon,
Bâtiment 308G,
43 Boulevard du 11 Novembre 1918,
F-69622 Villeurbanne cedex,
France.
and
IUT A, Chemical Engineering Department,
43 Boulevard du 11 Novembre 1918,
F-69622 Villeurbanne cedex,
France.

Peter Marquis
University of Birmingham School of Dentistry,
St Chad's Queensway, Birmingham,
B4 6NN, UK.

John W. McGrath
Queen's University Belfast,
School of Biology and Biochemistry,
Medical Biology Centre,
97 Lisburn Road,
Belfast, BT9 7BL, Northern Ireland.

Heather McHaffie
University of Edinburgh,
School of GeoSciences,
Darwin Building, Mayfield Road,
Edinburgh, EH9 3JU, UK.

Ashok Mulchandani
Department of Chemical and Environmental Engineering,
University of California, Riverside,
CA 92521, USA.

Bernd Nowack
Institute of Terrestrial Ecology (IT),
Swiss Federal Institute of Technology,
Zürich (ETH), CH-8952 Schlieren,
Switzerland.

Simon A. Parsons
School of Water Sciences,
Cranfield University,
Cranfield, Bedfordshire,
MK43 0AL, UK.

Marion Paterson-Beedle
School of Biosciences,
University of Birmingham,
Birmingham,
B15 2TT, UK

Peter Pearce
Thames Water,
Spencer House,
Manor Farm Road,
Reading,
RG2 0JN, UK.

Peter G. Piekema
DWR Watermanagement and
Sewerage Department,
Water Board Amstel,
Gooi and Vecht (AGV),
PO Box 94370,
1090 GJ Amsterdam,
The Netherlands.

Neil Preedy
Department of Geography,
University of Sheffield, Sheffield,
S10 2TN,UK.

Malcolm S. Robb
Water and Rivers Commission,
Hyatt Centre, 3 Plain St, East Perth,
WA, 6004, Australia.

John P. Quinn
Queen's University Belfast,
School of Biology and Biochemistry,
Medical Biology Centre,
97 Lisburn Road, Belfast, BT9 7BL,
Northern Ireland.

Rachel Sammons
University of Birmingham School of
Dentistry,
St Chad's Queensway,
Birmingham, B4 6NN, UK.

Mark Shimazu
Department of Chemical and
Environmental Engineering,
University of California,
Riverside,
CA 92521, USA.

Keith Smith
University of Edinburgh,
School of GeoSciences,
Darwin Building, Mayfield Road,
Edinburgh,
EH9 3JU, UK.

Ingrid Steén
Kemira GrowHow Oy,
Mechelininkatu 1a, (PL 900),
00108 Helsinki, Finland.

Tom Stephenson
School of Water Sciences,
Cranfield University,
Cranfield, Bedfordshire,
MK43 0AL, UK.

Hans Temmink
University of Wageningen & Research
Center,
Subdepartment Environmental
Technology,
Bomenweg 2, POB 8129,
NL-6700 EV Wageningen,
Netherlands.

Yasunori Ueno
Technical Dept. Eng. Div.,
Unitika Ltd, MN611-0021,
23 Ujikozakura, Uji,
Kyoto, Japan.

Eugenia Valsami-Jones
Department of Mineralogy,
The Natural History Museum,
Cromwell Road,
London, SW7 5BD, UK.

Ping Yong
School of Biosciences,
University of Birmingham,
Birmingham, B15 2TT, UK.

Paul Younger
School of Civil Engineering and
Geosciences,
Cassie Building,
University of Newcastle upon Tyne,
Newcastle upon Tyne, NE1 7RU, UK.

Preface

Phosphorus is both hero and villain in the history of Earth. It receives good publicity for being the backbone of the molecule of life, DNA, and bad for being a persistent pollutant that costs UK industry, for example, more than £150 million per year due to freshwater eutrophication[a]. In addition, phosphorus is at the centre of the structure of the basic energy-providing biological entity, ATP, and is a key component of the scaffold of all vertebrates, bone. But it also forms some of the most toxic chemicals known, organophosphates, which have been used in applications as diverse as pesticides and nerve agents, and which have left a legacy of severe contamination. In between these extremes lie a whole host of other applications of phosphorus, from fertilisers and foodstuffs to detergents, catalysts, biomaterials, medicines, fire retardants, stabilisers, radioactive element hosts and more. Knowledge of the chemistry of phosphorus witnessed a vast expansion in the 20th century, rivalled only by carbon, with known phosphorus compounds now exceeding 100,000[b].

In this context, planning a book entitled "Phosphorus in Environmental Technology" means that the choices of topics to cover are endless. To give the book a focus, three main ideas were set at the core of planning. Firstly, that the book should be self-contained and therefore include aspects of the fundamental science of phosphorus. Secondly, that it would aim to describe technological applications with global industrial or economic implications, presented in dedicated chapters. Thirdly, that the book would emphasise and prioritise novel and topical technological themes. From the latter, two sections stand out: one dedicated to phosphorus recovery for reuse, a topical theme for many different indus-

---
[a]Pretty et al., 2003. Env. Sci. & Technology, vol. 37, pp. 201–208.
[b]Corbridge, D.E.C., Phosphorus 2000, Chemistry, Biochemistry and Technology. Elsevier.

trial technologies that is especially relevant to phosphorus, and a second dedicated to novel biotechnologies, an area of active and diverse expansion.

The book is divided into five parts. Part I is concerned with the chemistry, geochemistry, mineralogy and biology of phosphorus. The diverse roles of phosphorus in nature mean that this range of fundamental sciences underpin knowledge of its use in environmental technologies, and will serve to support later chapters.

Part II looks into the distribution of phosphorus in the environment. Beginning with a review of the mechanisms of phosphorus release, this part goes on to consider the role of phosphorus in agriculture and its transfer to sensitive waters, and to investigate the environmental fate of a special group of organophosphorus compounds, the phosphonates, commonly used in industrial applications. This part of the book ends with a global review of phosphorus pollution.

Part III focuses on phosphorus removal technologies and starts with the fundamental principles of phosphate dissolution-precipitation, which control phosphorus removal in industrial applications. That chapter is succeeded by a chapter dedicated to waste water treatment principles, followed by others on chemical and biological phosphorus removal. Solid phase adsorbents of phosphorus are then discussed and a case study on phosphorus removal by recovered ochre presented.

Part IV addresses a theme which has only emerged in the past few years: phosphorus recovery. The development of recovery technologies makes sense not only in terms of good environmental practice, but also because global phosphorus reserves are finite. Given the importance of phosphorus to life on Earth, its conservation may ultimately be of great significance. Phosphorus recovery is first examined through the eyes of the industry in a dedicated chapter. If recovery is to be implemented, appropriate recovery reactor technologies need to be in place, and the fluid dynamic principles of such constructions need to be considered; one chapter reviews the fundamentals of such technology. This is then followed by case study chapters, after which this part closes by considering the regulation and economics of recovery. The message from this last chapter is that, at present, there is no financial incentive for phosphorus recycling. Such incentives may take long to materialise and environmental criteria should perhaps be the driver to encourage recycling technologies.

The final part of the book sees the presentation of four novel phosphorus biotechnologies; the first involves the use of bacteria to precipitate metal phosphates as a mechanism of immobilising/neutralising toxic/radioactive metals. The second covers biomaterial technologies for bone substitution, and the third is about technologically-based rationalisation of phosphorus use in agriculture. The fourth covers novel methods for biodegradation of organophosphate nerve agents.

The editor's thanks go to all the authors of the book for their enthusiasm and diligence. I have also a great many people to thank for advice, reviews, or other help. In alphabetical order they are: Wole Akinremi, Paolo Battistoni, Raffaella Boccadoro, Robin Cocks, D.E.C. Corbridge, Gordon Cressey, Aleksandra Drizo, Bill Dubbin, Karin Hing, Mark Hodson, Yasmin Jaffer, Johnny Johnston, Greg Jones, Sharron McEldowney, John McGrath, A. Mersmann, Simon Parsons, Gary

Pierzynski, John Quinn, Willem Schipper, Erwin Temminghoff, Paul Schofield, Alessandro Spagni, Chris Thornton and Jacqueline van der Houwen.

The encouragement of the series editor, Piet Lens, has proved invaluable, as has the tireless enthusiasm, support and humour of Alan Click and Alan Peterson of IWA Publishing.

Eugenia (Éva) Valsami-Jones
London, April 2004

PART ONE

Principles of phosphorus distribution: chemistry, geochemistry, mineralogy, biology

1.	The chemistry of phosphorus	3
2.	The geochemistry and mineralogy of phosphorus	20
3.	The biology of phosphorus	51

1
The chemistry of phosphorus

David Bryant

1.1 INTRODUCTION

PHOSPHORUS: chemical symbol P is named from the Greek *phosphoros* meaning 'light-bearing', an ancient name for the planet Venus when appearing before sunrise.

The discovery of phosphorus is generally attributed to Hennig Brandt in 1669, who prepared it by the distillation of urine. One hundred years later it was discovered that much higher concentrations could be found in teeth and bones (Emsley, 2000) though now it is obtained from ore deposits.

Phosphorus has atomic number 15 and an atomic weight of 30.97376. In the cosmos the relative abundance of phosphorus atoms compared with hydrogen is $3.2 \times 10^{-7} : 1$. In crustal rocks on Earth it is the 11th element in order of abundance. Phosphorus compounds play a crucial role in life processes, forming part of the 'stuff of life', DNA, and being involved in metabolic energy transfer as ATP and related compounds, yet are also among the most toxic substances known, used as chemical warfare agents and as pesticides.

© 2004 IWA Publishing. *Phosphorus in Environmental Technology: Principles and Applications.* Edited by Eugenia Valsami-Jones. ISBN: 1 84339 001 9

Table 1.1 Phosphorus isotopes.

Isotope	Lifetime	Decay mode	Decay energy (MeV)	Particle energies (MeV)	Particle intensities (%)	Nuclear spin (I)	Magnetic moment (μ)
$_{15}P^{28}$	0.28 s	β^+	13.8	10.6	50		
$_{15}P^{29}$	4.4 s	β^+	4.95	3.96			
$_{15}P^{30}$	2.5 min	β^+	4.24	3.27			
$_{15}P^{31}$	–					½	+1.1317
$_{15}P^{32}$	14.3 d	β^-	1.71	1.71	100	1	−0.2523
$_{15}P^{33}$	25 d	β^-	0.248	0.249	100		
$_{15}P^{34}$	12.4 s	β^-	5.1	5.1	75		
				3.2	25		

1.2 ATOMIC PROPERTIES

Phosphorus can exist in seven isotopic forms but in nature only ^{31}P is present due to the short half-lives of the remaining isotopes. The properties of the isotopes are shown in Table 1.1.

Iodine-131 (^{131}I) and Phosphorus-32 (^{32}P) are two radioisotopes used for internal radiotherapy. ^{32}P is used to control excess red blood cell production in bone marrow. It is also used to label phosphorylated biomolecules in the study of biochemical processes.

Phosphorus Nuclear Magnetic Resonance (^{31}PNMR) Spectroscopy is a commonly used technique giving information about the structure of molecules. The ^{31}P nucleus has a sensitivity of 6.63×10^{-2} at constant field compared to the proton. However the fact that it has a spin of ½ and a large magnetic moment allows for easy gathering of spectroscopic data. Spectra are referenced against 85% H_3PO_4 solution electronically, and do not need to be collected in deuterated solvents. There is a wide range of chemical shifts, the highest known being +450 ppm for P_4 phosphorus due to the high strain in the bonds.

1.3 ELECTRONIC STRUCTURE AND BONDING

The electronic configuration of ^{31}P is $1s^2, 2s^2, 2p^6, 3s^2, 3p^3$ which can also be considered as [Ne] $3s^2, 3p^3$. In order to achieve the electronic structure of the next noble gas (argon), phosphorus would need to accept three electrons. This requires a very large 1450 kJ mol^{-1} and so ionic compounds containing P^{3-} are uncommon. The vast majority of the compounds are covalent in nature, with the $3p^3$ electrons being shared and the $3s^2$ electrons existing as a lone pair in the trivalent compounds, or all five electrons involved in bonding for pentavalency. However this picture is an oversimplification since if three coordinate compounds were formed simply using p-orbitals, the substituents could be expected to be found at 90° to each other. A comprehensive study of the range of angles between substituents has been

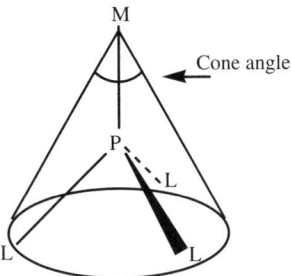

Figure 1.1   Tolman's cone angle.

carried out on a series of phosphines by Tolman (1977), who has termed them the 'cone-angles' (Figure 1.1). The values found range from 87° for phosphine itself to 212° for trimesityl phosphine.

The steric bulk of the substituents forces this increase in angle, and as a consequence the amount of s character in the bonds increases with a corresponding increase in p character for the orbital occupied by the lone pair. These changes manifest themselves in the chemical shift of the phosphorus nucleus. Infra-red studies reported in the same review show an increase in the P=O bond strength of phosphine oxides as the other substituents increase in size due to increased bonding orbital overlap with increasing p character. This bond is formed from the electrons, which existed as the lone pair in the corresponding phosphine. These same electrons form coordination compounds when phosphines act as ligands to metals. Phosphines are noted for their ability to stabilise late transition metals in low oxidation states. This is partly because of a phenomena known as back-bonding in which, apart from a conventional metal–phosphorus σ-bond, π-bonding occurs with electron density from the metal d-orbitals transferred into phosphorus orbitals which are hybrids of unoccupied phosphorus d-orbitals with anti-bonding $3p_x$ and $3p_y$ orbitals (Orpen and Connelly, 1990). Where the three substituents to phosphorus are nitrogen atoms, the lone pair can interact with the lone pair on nitrogen (Mitzel et al., 1996). Consequently, there is a loss of symmetry, since the phosphorus can only interact in this way with one of the three nitrogens and therefore one P–N bond is shorter than the other two.

1.4   PHOSPHORUS COMPOUNDS

There are now in excess of 100,000 known phosphorus compounds of which the majority contain linkages to oxygen, carbon, nitrogen and metals and a significant minority with linkages to boron, silicon, sulfur or halogens. Naturally occuring phosphorus is almost exclusively as phosphate containing P—O and P=O linkages. Biologically important phosphorus is either inorganic in nature (P_i) containing P—O and P=O linkages or is organic and contains one, two or three P—O—C linkages to form monoesters, diesters and triesters respectively.

1.4.1 Elemental phosphorus

Elemental phosphorus exists in four allotropic forms: white (or yellow), red, and black (or violet). White phosphorus has two modifications: α and β with a transition temperature at $-3.8°C$, a melting point of 44.1°C and boiling point of 280°C. The densities of the allotropes vary considerably; (white) sp. gr. 1.82, (red) 2.20, (black) 2.25–2.69.

Ordinary phosphorus is a waxy white solid, which is colourless and transparent when pure. It is insoluble in water but soluble in carbon disulfide. It catches fire spontaneously in air, burning to the pentoxide. It is very poisonous, an approximate fatal dose being 50 mg. The maximum recommended allowable concentration in air is $0.1\,mg/m^3$. White phosphorus should be kept under water as it is dangerously reactive in air, and it should be handled with forceps, as contact with the skin may cause severe burns. When exposed to sunlight or when heated in its own vapour to 250°C it is converted to the red variety, which does not luminesce in air, as does the white variety. This form does not ignite spontaneously and it is not as dangerous as white phosphorus. It should however, be handled with care as it can convert to the white form under certain conditions, can be ignited at 200°C, and it emits highly toxic fumes of the oxides of phosphorus when heated. The red allotrope is quite stable, subliming with a vapour pressure of 1 atmosphere at 417°C. It is used in the manufacture of safety matches, pyrotechnics, pesticides, incendiary shells, smoke bombs, tracer bullets, etc. Black phosphorus is formed by heating white phosphorus under high pressure or for 8 days in the presence of a mercury catalyst with a seed crystal of black P. It is the least reactive form of phosphorus and can only be ignited with difficulty. Black phosphorus is a semiconductor.

Perrin and Armerego (1997) describe the purification of the two common forms as follows:

Phosphorus (red) is boiled for 15 min. with distilled H_2O, allowed to settle and washed several times with boiling H_2O. It is transferred to a Büchner funnel, washed with hot H_2O until the washings are neutral, then dried at 100°C and stored in a desiccator.

Phosphorus (white) is purified by melting under dilute H_2SO_4/dichromate mixture and allowed to stand for several days in the dark at room temperature. It remains liquid, and the initial milky appearance due to insoluble, oxidisable material gradually disappears. The phosphorus can then be distilled under vacuum in the dark. Other methods include extraction with dry CS_2 followed by evaporation of the solvent, or washing with 6M HNO_3, then H_2O, and drying under vacuum.

White phosphorus exists as P_4 tetrahedra in all phases up to 800°C where dissociation to P_2 begins. Black phosphorus, in orthorhombic crystalline form, exists as a double layer of chains in which each atom is bonded to three neighbours, two in the chain and the third in the other layer. The structure of red phosphorus is complex and not well characterised. The high reactivity of white phosphorus is attributed to the strain involved in forming bond angles of 60° and in having all four lone electron pairs directed outwards. The difference in structures of the allotropes accounts for their marked differences in reactivity. A few organometallic compounds have been synthesised in which the P_4 unit has been retained and acts as a ligand. Examples include complexes where the P_4 unit is bound to the metal through one of its phosphorus atoms (Peruzzini *et al.*, 2000), through a P_4 tetrahedral face (Ginsberg *et al.*, 1971), as a bridge between two uranium atoms (Stephens *et al.*, 2000) and recently as an ionic species containing the $Ag(P_4)_2^+$ cation (Krossing, 2001). These exotic species are rare but do show that the reactivity of the unit can be moderated.

1.4.1.1 *Phosphorus and phosphorescence*

Although various forms of chemiluminescence and bioluminescence had been observed prior to the discovery of phosphorus, none of these persisted as long or gave so much light. Furthermore, the light emitted from phosphorus was not dependent on it having been previously exposed to sunlight (for in fact the light emitted by phosphorus is not phosphorescence despite the common origins of the words). These properties gave the new substance the name 'phosphorus mirabilis' (Harvey, 1957). The energy for the chemical luminescence is derived from the oxidation of phosphorus. The presence of water is necessary as no light emission occurs under rigorously dry conditions. Light emission is due to transitions of the excited states of P_2, HPO and principally PO, the exact balance depending on the concentrations of moisture and oxygen (Van Zee and Khan, 1976).

1.4.2 Phosphides

Phosphorus can be connected to a number of other partners, from one (phosphynes) to nine. The higher coordination numbers are almost invariably ionic lattice compounds formed with metals. These metal phosphides have the high densities and melting points associated with extended ionic lattice compounds. Almost all metals and boron form phosphides with binary phosphides in the composition range of M_3P to MP_3. Compounds of formula M_xP_y are usually divided into four categories for convenience.

   i. Metal rich $x > y$
   ii. Monophosphides $x = y$
   iii. Phosphorus rich $x < y$
   iv. Ionic phosphides $M_x^{n+}P_y^{n-}$

The metal rich phosphides and monophosphides are very chemically stable, generally resisting acids and bases. They are dense, hard, high melting and brittle, in common with isostructural carbides, borides and silicides.

Phosphorus rich phosphides of metals other than groups I and II contain phosphorus atoms linked to each other covalently in a wide variety of ways. They are less stable and on heating loose phosphorus to become monophosphides or metal rich phosphides.

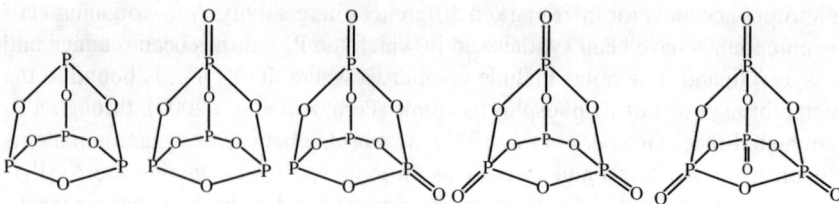

Figure 1.2 The oxides of phosphorus.

1.4.3 Oxides

A number of phosphorus oxides can be formed. If a P_4 tetrahedron has an oxygen inserted between each pair of three coordinate P atoms, then P_4O_6 the trioxide (P_2O_3) is obtained. Stepwise addition of a further oxygen to each phosphorus, to make it four coordinate, produces a series of five compounds (Figure 1.2), ending with phosphorus pentoxide: P_4O_{10} (P_2O_5). The P—O—P bond angles in phosphorus oxides are in the range 99–132°.

Phosphorus pentoxide is formed when phosphorus is burned in an excess of air. It is extremely hygroscopic, reacting exothermically with water to give orthophosphoric acid.

$$P_4O_{10} + 6H_2O \rightleftharpoons 4H_3PO_4 \qquad (1)$$

Consequently phosphorus pentoxide finds use as a powerful drying agent. There are a number of sub-oxides of phosphorus based on the P_4 tetrahedron, with oxygen at one or more vertices, or with oxygen bridging phosphorus atoms as in the series in Figure 1.2, but in each case also involving direct P–P bonding. The structures of these unstable compounds are inferred from infra red spectroscopic evidence (Mielke and Andrews, 1990).

1.4.4 Oxo-acids

There are more oxo-acids (Figure 1.3) of phosphorus than of any other element, though there is a simple series of principles to which they all adhere (Greenwood and Earnshaw, 1984).

i. All P atoms are four coordinate and have one double bond to oxygen (P=O).
ii. All P atoms have at least one ionizable P—OH group responsible for the acidity.
iii. Some species have one or more P—H group. These are not ionizable.
iv. Catenation is by P—O—P or direct P—P bonds in both linear and cyclic species.
v. Peroxo compounds contain either P—O—OH or P—O—O—P linkages.

The oxidation state for each phosphorus atom can be reckoned by counting +1 for each bond to O (+2 for a double bond) and −1 for each P—H bond and zero for P—P bonds.

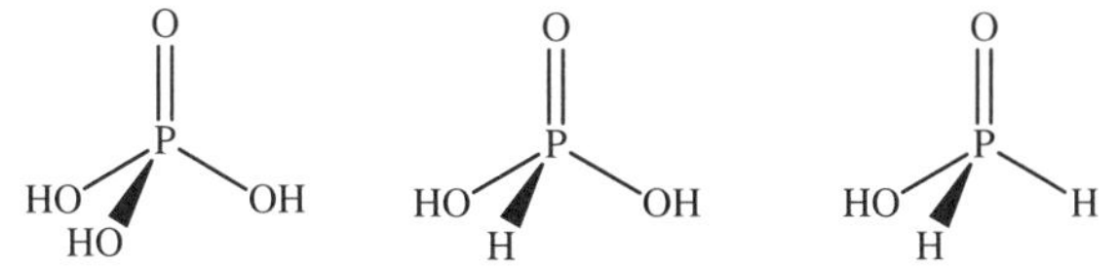

The chemistry of phosphorus

Figure 1.3 Phosphorus oxo-acids with one P atom.

Orthophosphoric acid — Phosphonic acid (phosphorous acid) — Phosphinic acid (hypophosphorous acid)

Figure 1.4 Tautomerism in phosphorous acid.

Dehydration of these single phosphorus oxo-acids leads to chains and cyclic forms. Polyphosphoric acid has the general formula $H_{n+2}P_nO_{3n+1}$ where members up to $n = 17$ are known. Thus, triphosphoric acid ($n = 3$) is $H_5P_3O_{10}$. Cyclic oxo-acid forms have the general formula $(HPO_3)_n$ and are known as metaphosphoric acids. In a similar fashion, phosphonic acid can be dimerised to diphosphonic acid (diphosphorous or pyrophosphorous acid) $H_4P_2O_5$.

A number of lower oxo-acids, containing direct P—P bonds, also exist as chains or rings.

Phosphorous acid and some related compounds undergo tautomerism where two structurally different forms with the same generic formula interchange (Figure 1.4). Where the rate of exchange is slow, it is sometimes possible to isolate each form.

1.4.5 Phosphates and polyphosphates

Phosphates and polyphosphates are the salts of the respective oxo-acids described above and therefore conform to the same range of structural types. By far the most common, naturally occurring forms of phosphorus are as the orthophosphate minerals known as apatites (see also Valsami-Jones, this volume). These have the general formula $Ca_5(PO_4)_3X$, where X = F for fluorapatite, X = Cl for chlorapatite and X = OH for hydroxylapatite. Hydroxylapatite forms about 60% by weight of animal bone and about 70% of teeth. These and other phosphate minerals are very insoluble, and consequently, in nature, the availability of phosphorus often controls proliferation of life forms, with major implications to agriculture. The intervention of man led to the development of monocalcium phosphate ($Ca(H_2PO_4)_2$) and ammonium phosphates as fertilisers, these being much more soluble. Phosphates also find uses as ingredients in toothpastes, flame retardants, baking powder and other food additives. Phosphates and phosphoric acid itself also have industrial uses, mainly in the field of metal surface treatment against corrosion.

The sodium salt of triphosphoric acid, $Na_5P_3O_{10}$ is a major component of many synthetic detergents due to its ability to sequester magnesium and calcium ions and thereby soften water. It is also a dispersing agent for clays in oil-well drilling and is used in cement technology. This widespread use of phosphates, has been linked to the problem of eutrophication (Burke et al., this volume).

1.4.6 Sulfides

Compounds containing only phosphorus and sulfur are based on the P_4 tetrahedron, yet only P_4S_{10} is a structural analogue of the oxide. However, there are similarities, as each sulfur atom bridges a pair of phosphorus atoms with P—S—P angles in the range of 95–111° and there are no S—S bonds, the remaining sulfur atoms forming exocyclic P=S bonds. All the compounds P_4S_3, P_4S_4, P_4S_5, P_4S_7, P_4S_9, P_4S_{10} are yellow. P_4S_3 is used in the manufacture of 'strike-anywhere' matches. Recently, the 'missing' member of the series, P_4S_6, has been synthesised (Blachnik et al., 1995).

1.4.7 Halides

The halides can be usefully grouped into trihalides (PX_3), tetrahalides (P_2X_4) and pentahalides (PX_5). Alongside the four binary trihalides are the following known mixed trihalides, PF_2Cl, $PFCl_2$, PF_2Br, $PFBr_2$ and PF_2I.

The trichloride is a convenient starting point for the synthesis of many organophosphorus compounds. In common with phosphine, all the trihalides have a pyramidal structure due to the lone pair of electrons.

Of the tetrahalides, the iodide is the most stable and easily made. The chloride is available only in low yield and decomposes slowly at room temperature. The fluoride hydrolyses to F_2POPF_2 and therefore is also unstable in the atmosphere. The bromide is only known as a coordinated compound (Hinke et al., 1981).

The four pentahalides are known but with some variation in their solid state structures. In the gas phase, they all exist as trigonal bipyramids; however, in the solid phase the chloride exists as $[PCl_4]^+[PCl_6]^-$ and the bromide and iodide as $[PX_4]^+[X]^-$; their form in solution depends on the nature of the solvent. The fluoride attracted interest as it led R.S. Berry (1960) to discover the phenomenon now known as the Berry pseudorotation. Figure 1.5 depicts the way in which the pentafluoride inverts from a trigonal bipyramid through a tetragonal pyramid back to a trigonal bipyramid with the substituents in new positions. The equivalence of all the fluorine atoms under examination by fluorine ^{19}F NMR provided the evidence for this discovery.

The anion PF_6^- is often used in organometallic chemistry due to its low coordinating ability. The range of three coordinate trivalent halides is complemented by the group of tetrahedral pentavalent oxohalides and thiohalides. The reactive lone pair has been replaced by the phosphorus–oxygen or phosphorus–sulfur double bond, yet the compounds remain quite reactive. The halogens are readily replaced forming phosphate esters with alcohols, e.g. $(RO)_3PO$, and with amines compounds of the form $(RNH)_3PO$.

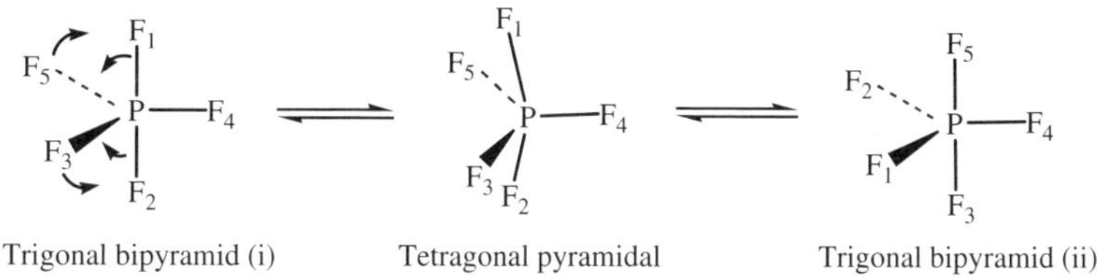

Figure 1.5 Berry pseudorotation leading to interchange of axial and equatorial fluorines in PF_5.

1.4.8 Hydrides

Phosphorus has two main hydrides, in common with nitrogen. Phosphine, PH_3 is an analogue of ammonia, and P_2H_4 is an analogue of hydrazine. Phosphine is a pyramidal molecule due to the lone pair. It undergoes a vibration known as inversion, often likened to an umbrella being blown inside out, in the same manner as ammonia but with a considerably higher energy barrier, 155 kJ mol^{-1} compared to 24 kJ mol^{-1} (Lehn and Munsch, 1969).

Unlike ammonia, phosphine is only sparingly water soluble, and forms neither acidic nor basic solutions. Diphosphine is a colourless liquid, unstable above room temperature.

1.4.9 Phosphines

The term phosphine is generally applied to three coordinate compounds of phosphorus where one or more of the substituents is an organic moiety. Strictly speaking these should be referred to as e.g. 'organophosphines' or 'chlorophosphines', to distinguish them from the phosphorus hydrides described above. There are an enormous number of compounds, which fit this description, making only a general overview including important examples possible here. Phosphines with alkyl substituents in particular, tend to react readily with atmospheric oxygen to form phosphine oxides and need to be handled in inert atmospheres. Aryl substituted phosphines are more stable and the ubiquitous triphenylphosphine is a fairly stable white solid. Triphenylphosphine is widely used in synthetic organic chemistry as a reagent for the Wittig reaction, where its affinity for oxygen is exploited (Wittig, 1980). An alkyl halide reacts with triphenylphosphine to form a phosphonium salt. A compound, known as an ylide, is formed from the salt by abstraction of a proton from the phosphonium salt; this, in turn, can transform a carbonyl compound into an alkene. This process is shown in the following reactions 2–4:

$$CH_3Br + Ph_3P \longrightarrow Ph_3P^+\text{---}CH_3\ Br^- \quad \text{phosphonium salt} \quad (2)$$

$$Ph_3P^+\text{---}CH_3\ Br^- + \text{base} \longrightarrow Ph_3P=CH_2 \quad \text{ylide} \quad (3)$$

$$Ph_3P=CH_2 + RR'C=O \longrightarrow RR'=CH_2 + Ph_3P=O \quad (4)$$

Phosphines are also widely used as ligands and are able to stabilize the late transition metals in low oxidation states by back donation of electron density into unfilled phosphorus orbitals as described. The significance of this is to facilitate the use of these metals as homogeneous catalysts. The electronic properties of the metal can be tailored by a suitable choice of phosphine. The hydrogenation of unsaturated compounds using homogeneous catalysts was made a practical possibility in the 1960s, when the team led by Geoffrey Wilkinson (Osborn et al., 1966) discovered the catalyst now named after him. The use of strong π-acceptor (π-acid) ligands such as phosphines was known to stabilise rhodium complexes against reduction to the metal and it was found that with an excess of triphenylphosphine, rhodium(III) chloride could be reduced to tris(triphenylphosphine)-chlororhodium(I).

In dilute solutions the complex is in equilibrium with the bis(triphenylphosphine)-chlororhodium(I) complex.

$$RhCl(PPh_3)_3 \longrightarrow RhCl(PPh_3)_2 + PPh_3 \qquad (5)$$

It is the vacant coordination sites, generated by the loss of ligand in reaction (5), that are the basis for the catalytic activity seen for this complex. The loss of ligand is thought to be driven by a combination of electronic effects (the *trans* effect) and relief of steric crowding of the bulky phosphine groups.

The hydrogenation catalyst ionic complex [Ir(COD)PCy$_3$(py)]PF$_6$, developed by Crabtree (1979), contains even bulkier tricyclohexylphosphine ligands and the hexafluorophosphate anion, and is particularly active in non-coordinating solvents such as dichloromethane. In this case the phosphines remain coordinated throughout the catalytic cycle and instead the cyclooctadiene (COD) is lost to give the vacant coordination site required to accommodate the substrate.

Homogeneous, catalytic hydrogenation generates the most commercial interest when applied to asymmetric systems. The ability to generate one enantiomer, in high excess, is exclusive to homogeneous systems and of great importance to the manufacture of biologically active molecules. Selectivity can be conferred on the catalyst by suitable design of the ligand system. Replacement of two of the PPh$_3$ ligands in Wilkinson's catalyst by an optically active, chelating diphosphine was an early approach.

An example of a chiral phosphine, now serving as a benchmark for future developments was reported by Kagan and Dang (1972). Starting with (−)-tartaric ester the phosphine known as DIOP was synthesised (Figure 1.6). This was allowed to react with [RhCl(cyclooctene)$_2$]$_2$ *in situ*, to form the active hydrogenation catalyst. Another important commonly used chiral phosphine is BINAP (Brown et al., 1984) (Figure 1.6).

More recently, considerable interest has been shown in the use of modified ferrocene based diphosphines as chiral ligands. Two different substituents on the same ferrocene ring confer chirality to the molecule (Togni, 1996). An important example of this, is one of the first industrial applications of this class of ligand, for the production of the herbicide Metolachlor (Figure 1.7). The list of chiral phosphines now available is extensive. The examples given are all chiral molecules but

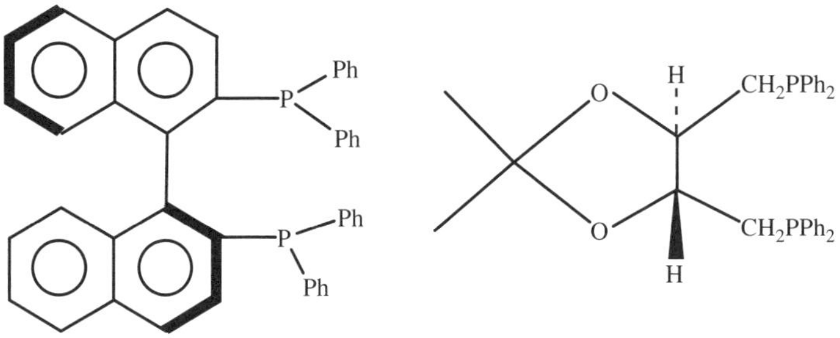

Figure 1.6 Two examples of commonly used chiral phosphines.

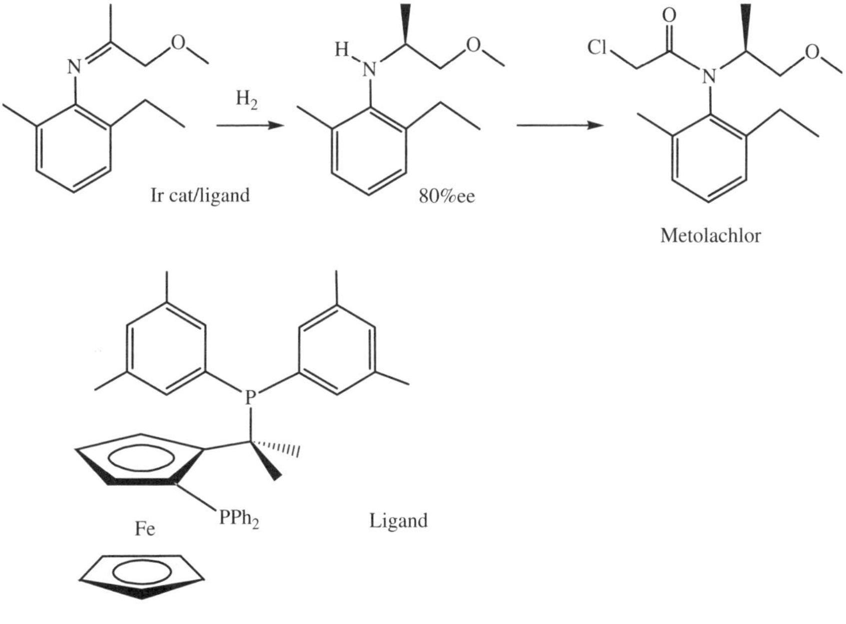

Figure 1.7 Industrial production of Metolachlor using a chiral ferrocene based phosphine ligand.

the chiral centre is not the phosphorus atom but the carbon based framework. Inversion at the phosphorus centre in the manner described for phosphine itself can occur despite the barrier to this inversion being higher, leading to a racemisation of any products from the reaction catalysed.

The chiral phosphines shown are special cases of the more general group of chelating diphosphines and triphosphines where the coordinating phosphorus atoms are linked by a supporting framework, in the simplest cases a hydrocarbon chain. The properties of the metal complex formed by coordination to these diphosphines are often different from those formed by two otherwise similar monophosphines. The metal is usually forced to adopt *cis* coordination and the angle between the two P–M bonds, known as the 'bite-angle' (Casey *et al.*, 1992) can influence the bonding at the metal and hence the product distribution of catalysis. Finally, phosphines have been incorporated into the structure of polymers to form polymeric ligands as a means of anchoring homogeneous catalysts with the aim of rendering easy their subsequent removal from a product solution (e.g. Evans *et al.*, 1974).

1.4.10 Phosphonates

Phosphonates are compounds where one of the P—O—H groups in phosphoric acid has been replaced by an organic group R to form RPO_3^{2-}. In many cases the R group acts as a link between PO_3^{2-} groups to form multi-phosphorus compounds. Phosphonates are extensively used in scale/corrosion inhibition, water treatment, metal finishing, ore recovery, oil drilling, industrial cleansing, pulp, paper and textile dyeing, and crop production. U.S. use of the industrial scale control and cleansing agent NTMP and the herbicide glyphosate, for example, both exceed four million kg/yr. Phosphonates have found favour as potential environmentally friendly fungicides, for example in the treatment of 'sudden oak death' where they can act as plant nutrient at the same time. They are found in medicine. Samarium-153 Ethylene Diamine Tetramethylene Phosphonate, known commercially as Quadramet, is used internally to reduce the pain associated with bony metastases of primary tumours of breast, prostate and some other cancers. Some of the more common phosphonates are listed below in Table 1.2 and their structures are depicted in Figure 1.8.

The most common route for the preparation of phosphonates is via the 'Arbusov (1964)' reaction (6) followed by acid hydrolysis of the ester groups (7).

$$P(OR)_3 + R'X \longrightarrow R'P(O)(OR)_2 + RX \qquad (6)$$

$$R'P(O)(OR)_2 + 2HCl \longrightarrow R'P(O)(OH)_2 + 2RCl \qquad (7)$$

Table 1.2 Important phosphonates.

Acronym	Other abbreviation	Name
MP	MPA	Methylphosphonic acid
HMP	HMPA	Hydroxymethyl-phosphonic acid
AMP	AMPA	Amino(methylene-phosphonic acid)
PBTC	Bayhibit AM	2-Phosphonobutane-1,2,4-tricarboxylic acid
HEDP	EHDP, HEBP, HEDPA, Dequest 2010	1-Hydroxyethan-1,1- diphosphonic acid
IDMP	IDP	Iminobis(methylenephosphonic acid)
NTMP	ATMP, NTP, AMP, NTPO, NTPH, Dequest 2000	Nitrilotris(methylenephosphonic acid)
EDTMP	EDTP, EDTPO, ENTMP, EDTMPO, EDTMPA, EDTPH, EDITEMPA, Dequest 2041	Ethylenedinitrilotetrakis(methylene-phosphonic acid)
DTPMP	DETPMPO, DETPMPA, DETPMP, DTPP, ETPP, Dequest 2060	Diethylenetrinitrilopentakis (methylenephosphonic acid)

1.4.11 Phosphorus–nitrogen compounds

Nitrogen is trivalent and can therefore form a triple bond with phosphorus to make P≡N. White phosphorus will react with nitrogen, although high temperatures and high pressures are required since PN is not stable in contrast to the very stable N_2 molecule. There is however an extensive chemistry of phosphorus–nitrogen compounds containing P—N single and P=N double bonds. There are analogues of phosphines and phosphorus halogen compounds containing P—NH_2 or P—NR_2 links and analogues of phosphine oxides containing P=NR links. The latter bonding arrangement provides the possibility of the formation of polymers using the third nitrogen valence leading to chains and cycles containing the P=N—P=N— backbone. There are also chains and rings containing only single bonds. The compounds containing the phosphorus–nitrogen double bond with 2-coordinate nitrogen and 4-coordinate phosphorus are known collectively as phosphazenes. These compounds are all very stable and contain only weakly basic nitrogen atoms. Phosphazene polymers constitute the broadest and most versatile of all known inorganic macromolecular systems. Despite the alternating P—N—P bonds being of equal length, the cyclic structures are not aromatic in the same way as benzene. Phosphorus can form compounds containing linkages to 3, 4, 5 and 6 nitrogens, though the simple $P(NH_2)_n$ does not exist for $n = 5, 6$. The direct reaction of PCl_3 with ammonia results in $P(NH_2)_3$ but the reaction of PCl_5 with ammonia leads to the salt $P(NH_2)_4^+Cl^-$. The tautomerism shown by phosphorous acid (Figure 1.4) has parallels in phosphorus–nitrogen compounds. The three coordinate PR(OR)(NHPh) is in equilibrium with the four coordinate PR(OR)(=NPh)H but in this case the equilibrium favours the three coordinate compound. Aminophosphines of a great variety can be synthesised by the general reaction (8). These aminophosphines will

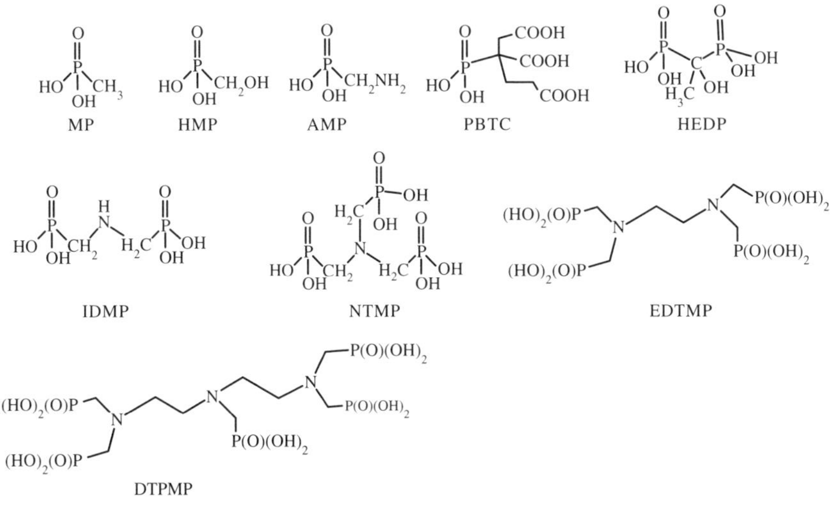

Figure 1.8 Phosphonate structures.

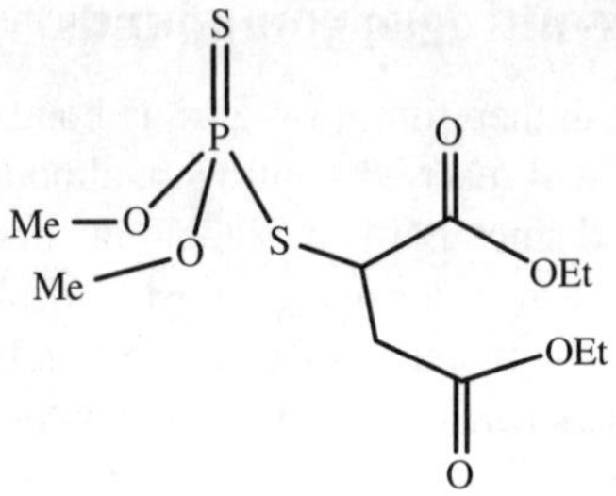

Figure 1.9 Structure of Malathion.

act as ligands to late transition metals in the same way as phosphines described above, with coordination to the metal through phosphorus and not nitrogen.

$$PR_2Cl + NR'_2H + Et_3N \longrightarrow R_2P\text{—}NR'_2 + Et_3NHCl \qquad (8)$$

1.4.12 Organophosphorus compounds

In addition to the phosphines and phosphonates mentioned earlier there are a very large number of other organophosphorus compounds. Phospha-alkyne HC≡P can be made in the gas phase at low temperatures from phosphine (Gier, 1961) and a few more stable derivatives exist where the H is replaced by an R group or there is coordination to a metal. There are a limited number of two coordinate compounds including phosphabenzene, the phosphorus analogue of pyridine which exhibits some aromatic character, and phospholes, the analogues of pyrroles. Figure 1.2 shows examples of three and four coordinate compounds, which are numerous. Five coordinate compounds are somewhat rarer, one example being PPh_5 made by Georg Wittig in 1948 (Wittig and Geissler, 1953).

Many organophosphorus compounds are very toxic including the notorious chemical weapons Tarin and Sabun. There are also many organophosphorus pesticides including the commonly used Malathion (Figure 1.9).

The World Health Organisation list 580 pesticides, and classifies them according to their hazard as class Ia (extremely hazardous), Ib (highly hazardous), class II (moderately hazardous) to class III (slightly hazardous), plus those which are unlikely to present a hazard (WHO 2001), based on rat LD_{50} figures. Half of each of the first two categories is made up by organophosphorus compounds.

The dialkyl monofluorophosphate esters have been found to inhibit cholinesterase in organisms and are therefore exceedingly toxic. Intake of organophosphorus compounds in general is revealed by the presence of organophosphate metabolites in the urine.

Organophosphates also find industrial uses, e.g. tributylphosphate is used in a kerosene solution to extract uranium during ore processing and, in common with other alkyl phosphates, as a plasticizer.

Included amongst organophosphorus compounds are the biologically occurring molecules of fundamental importance to life. All biomembranes have a structure

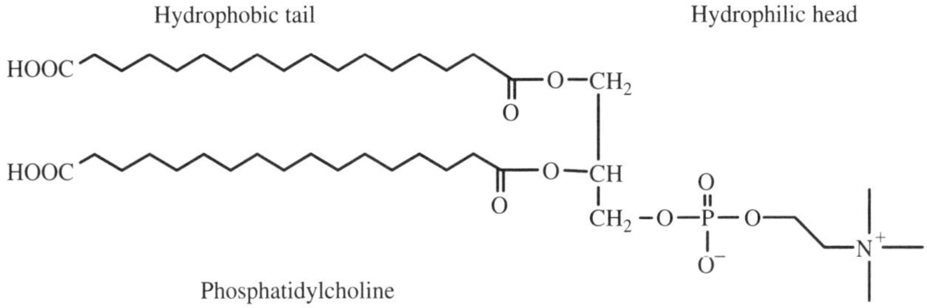

Figure 1.10 Phosphatidylcholine, a typical phospholipid.

based on a phospholipid bilayer (Lodish *et al.*, 2000). A phospholipid consists of hydrophobic fatty acid chains linked *via* an organophosphate group to a hydrophilic head as in phosphatidylcholine, depicted in Figure 1.10.

The transport of energy within living systems makes use of the phosphate esters ADP and ATP (see Dobrota, this volume) and the biomolecules RNA and DNA are polymers of nucleotides connected through phosphodiester linkages. Nowadays single strands of DNA up to about 100 nucleotides in length can be synthesised *in vitro*. Typically, the first nucleotide is bound to a glass support by the 3′-hydroxyl of its pentose sugar fragment. The second nucleotide in the sequence is derivatised by addition of 4′,4′-dimethoxytrityl (DMT) to protect the pentose 5′-hydroxyl group, and the 3′-hydroxyl is reacted with chloro-N, N-diisopropyl aminomethoxyphosphine, $(C_3H_7)_2NPCl(OCH_3)$, eliminating HCl (Beaucage and Carruthers, 1981). The nucleotides are mixed, in the presence of a weak acid the phosphorus–nitrogen bond is broken, and the two nucleotides couple with each other. The phosphorus is then oxidised to pentavalency and the protecting DMT group removed. This process is repeated for each subsequent nucleotide and finally the methyl groups are removed in alkali and the bond to the glass support cleaved. For a comprehensive overview of the chemistry of phosphorus see Corbridge (2000).

REFERENCES

Arbusov, B.A. (1964) The Michaelis-Arbusov and Perkow reactions. *Pure Applied Chem.* **9**, 307–312.

Beaucage, S.L. and Carruthers, M.H. (1981) Deoxynucleoside phosphoramidites: a new class of key intermediates for deoxypolynucleotide synthesis. *Tet. Lett.* **22**(20), 1859–1862.

Berry, R.S. (1960) Correlation of rates of intramolecular tunnelling processes, with application to some group V compounds. *J. Chem. Phys.* **32**, 933–938.

Blachnik, R., Peukert, U. and Czediwoda, A. (1995) Molecular composition of solidified phosphorus–sulfur melts and the crystal structure of βP4S6. *Z. Anorg. Allg. Chem.* **621**(10), 1637–1643.

Brown, K.J., Berry, M.S., Waterman, K.C., Lingenfelter, D. and Murdoch, J.R. (1984) Preparation and lithiation of optically active 2,2′-dihalo-1,1′binaphthyls. A general strategy for obtaining chiral, bidentate ligands for use in asymmetric synthesis. *J. Am. Chem. Soc.* **106**(17), 4717–4723.

Casey, C.P., Whiteker, G.P., Melville, M.G., Petrovitch, L.M., Gavney, J.A. and Powell, D.R. (1992) Diphosphines with natural bite angles near 120° increase selectivity for n-aldehyde formation in rhodium-catalyzed hydroformylation. *J. Am. Chem. Soc.* **114**, 5535–5543.

Corbridge, D.E.C. (2000) *Phosphorus 2000, Chemistry, Biochemistry and Technology.* Elsevier.

Crabtree, R. (1979) Iridium compounds in catalysis. *Acc. Chem. Res.* **12**, 331–337.

Emsley J. (2000) *The Shocking History of Phosphorus.* Macmillan.

Evans, G.O., Pittman, C.U., McMillan, R., Beach, R.T. and Jones, R. (1974) Synthetic and catalytic studies of polymer-bound metal carbonyls. *J. Organomet. Chem.* **67**, 295–314.

Gier, T.E. (1961) HCP a unique phosphorus compound. *J. Am. Chem. Soc.* **83**, 1769–1770.

Ginsberg, A.P. and Linsell, W.E. (1971) Rhodium complexes with the molecular P_4 as a ligand. *J. Am. Chem. Soc.* **93**(8), 2082–2084.

Greenwood, N.N. and Earnshaw, A. (1984) *Chemistry of the Elements.* Pergamon Press.

Harvey, E.N. (1957) *A history of luminescence from the earliest times to 1900.* The American Philosophical Society, Philadelphia.

Hinke, A., Kuchen, W. and Kutter, J. (1981) Stabilisation of diphosphorus tetrabromide as the bis (pentacarbonylchromium) complex. *Angew. Chem. Int. Ed. (Engl.)* **20**, 1060–1068.

Kagan, H.B. and Dang, T.P. (1972) Asymmetric catalytic reduction with transition metal complexes. I. Catalytic system of rhodium(I) with ($-$)-2,3-0-isopropylidene-2,3-dihydroxy-1, 4-bis(diphenylphosphino)butane, a new chiral diphosphine. *J. Am. Chem. Soc.* **94**(18), 6429–6433.

Krossing, I. (2001) $Ag(P_4)^{2+}$: The first homoleptic metal–phosphorus cation. *J. Am. Chem. Soc.* **123**(19), 4603–4604.

Lehn, J.M. and Munsch, B. (1969) Electronic structure and inversion barrier of phosphine. An *Ab Initio* SCF-LCAO-MO Study. *Chem. Comm.* 1327–1328.

Lodish, H., Berk, A., Zipursky, S.L., Matsudaira, P., Baltimore, D. and Darnell, J. (2000) *Molecular Cell Biology*, 4th edn. W.H. Freeman and Co.

Mielke, Z. and Andrews, L. (1990) Matrix infrared spectra of the products from photochemical reactions of tetraphosphorus with ozone and decomposition of tetraphosphorus hexoxide. *Inorganic Chem.* **29**, 2773–2779.

Mitzel, N.W., Smart, B.A., Dreihäupl, K-H., Rankin, D.W.H. and Schmidbaur, H. (1996) Low symmetry in P(NR2)3 skeletons and related fragments: an inherent phenomenon. *J. Am. Chem. Soc.* **118**, 12673–12682.

Orpen, A.P. and Connelly, N.G. (1990) Structural systematics: role of P-A σ* orbitals in metal–phosphorus π-bonding in redox related pairs of M-PA_3 complexes. *Organometallics* **9**, 1206–1210.

Osborn, J.A., Jardine, F.H., Young, J.F. and Wilkinson, G. (1966) Preparation and properties of tris(triphenylphosphine)halorhodium(I) and some reactions thereof including catalytic homogeneous hydrogenation of olefins and acetylenes and their derivatives. *J. Chem. Soc.* (A), 1711–1732.

Perrin, D.D. and Armarego, W.L.F. (1997) *Purification of Laboratory Chemicals*, 4th edn. Butterworth-Heinemann.

Peruzzini, Maurizio, Manas. S, Romerosa, A. and Vacca, A. (2000) Reaction of [(PPh3)3RuCl2] with white phosphorus: synthesis of the first RuII complex featuring a tetrahedro-tetraphosphorus ligand. *Mendeleev Commun.* **4**, 134–135.

Stephens, F.H., Arnold, P.L., Diaconescu, P.L. and Cummins, C.C. (2000) Novel uranium–phosphorus chemistry. *Abstr. Pap. Am. Chem. Soc.* 220th INOR-018.

Togni, A. (1996) Planar-chiral ferrocenes: synthetic methods and applications. *Angew. Chem. Intl. Ed. Engl.* **35**(13/14), 1475–1477.

Tolman, C.A. (1977) Steric effects of phosphorus ligands in organometallic chemistry and homogeneous catalysis. *Chem. Rev.* **77**(3), 313–348.

VanZee, R.J. and Khan, A. (1976) The phosphorescence of phosphorus. *J. Phys. Chem.* **80**(20), 2240–2242.
The WHO recommended classification of pesticides by hazard and guidelines to classification 2000–01. WHO/PCS/01.4
Wittig, G. and Geissler, G. (1953) Course of reactions of pentaphenylphosphorus and certain derivatives. *Annalen.* **580**, 44–57.
Wittig, G. (1980) From Diyls via Ylides to my Idyll (Nobel lecture). *Angew. Chem.* **92**(9), 671–675.

2
The geochemistry and mineralogy of phosphorus

E. Valsami-Jones

2.1 INTRODUCTION

This chapter presents an overview of the natural distribution of phosphorus (P), its occurrence and its most common forms. Phosphorus is widely present on the earth's surface, in rocks, soils, waters and in living organisms, although, having no gaseous phase under the conditions prevailing on Earth, it is not found in the atmosphere (Table 2.1). Under most conditions, phosphorus is exclusively combined with four oxygen molecules, forming the phosphate oxyanion. The phosphate electron configuration is such, that one of its 3s or 3p electrons can be transferred to the 3d orbital; this extra electron gives phosphate its chemical flexibility and reactivity that allows it to adopt to different structures and functions in living systems (Föllmi, 1996). A consequence of its chemical versatility is that

© 2004 IWA Publishing. *Phosphorus in Environmental Technology: Principles and Applications.* Edited by Eugenia Valsami-Jones. ISBN: 1 84339 001 9

Table 2.1 Phosphorus in the environment, and in natural and anthropogenic materials (wt% P, data from Corbridge, 2000).

Environment		Natural materials		Anthropogenic materials	
Air	0.00	Plants	0.05–1.0	Concrete	0.01–0.05
Sea water	0.0001–0.001	Body	1.0	Window glass	<0.01
Rain water	0–0.001	Blood	0.04	Wood ash	4.0–9.0
Igneous rocks	0.1	Bones	12.0	Wrought iron	0.1–0.2
Phosphate rock	10.5–15.0	Teeth	8.0	Steel	0.02–0.05
Soil	0.02–0.50	Brain	0.3	Sewage sludge (dried)	2.6
Meteorites	0.2	Milk (cow)	0.1		
		Brewer's yeast	1.8		

phosphorus represents the energy currency for organisms at cell level, and as such is essential for life (see Dobrota, this volume).

Although phosphorus is present in small quantities in all earth materials and life forms, it can be accumulated by certain biogeochemical processes. In geological time these processes create phosphate ore deposits, the primary bulk storage of phosphorus on Earth. Phosphorus is released in the environment via natural or human induced weathering. This is the only mechanism of introduction of new phosphorus to the Earth surface pool, and, since natural weathering processes are slow, represents a limit to its availability. As a result, phosphorus is often the nutrient limiting biological productivity, and as such, has a major influence on the carbon cycle, and, indirectly, on Earth's climate. Any phosphorus introduced in the environment is likely to undergo extensive recycling via chemical and/or biological processes.

Phosphate minerals are the main hosts of phosphorus during its environmental transformations. Whether in phosphate ores, rocks or soils, they act as the primary source of phosphorus. They can also form *in situ* in the environment, as transient pools of phosphorus during both bio- and geochemical processes. Their abundance, availability and reactivity are therefore key to the entire phosphorus cycle on Earth. Phosphate minerals have a unique role in biomineralization, by providing the inorganic component of all vertebrate's scaffold, the skeleton. Phosphates are also important in geochronology and material science and technology, including biomaterials. By including a variety of major and trace element components in their structures they can often control availability of elements other than phosphorus, with important environmental consequences in cases where these elements are toxic or radioactive (e.g. lead, cadmium, uranium).

This chapter will present a brief summary of the phosphorus cycle, and will discuss its distribution within individual pools in this cycle. It will then go on to describe the key features of phosphate minerals, in terms of chemistry, structures and applications relevant to the environmental role of phosphorus.

2.2 GEOCHEMICAL ABUNDANCE AND DISTRIBUTION OF PHOSPHORUS

2.2.1 Cosmic abundance

Phosphorus bearing compounds have been detected in space (e.g. MacKay and Charnley, 2001; Margules *et al.*, 2002). Phosphates and phosphides have been identified in meteorites; the former are mostly characteristic of stone meteorites and the latter of iron meteorites (Buchwald, 1984). Phosphorus is ubiquitous in meteorites, albeit present in small quantities; it is present in highest concentrations in iron rather than in stone meteorites (Buchwald, 1984). It has been shown that phosphate in Martian and carbonaceous chondrite meteorites is extractable, thus being available for would-be prebiotic synthesis and microbial nutrition (Mautner and Sinaj, 2002).

2.2.2 The global phosphorus cycle

The exogenic phosphorus cycle, which describes phosphorus movement on the Earth's surface, starts with its release from primary sources (phosphate minerals in rocks and soils), followed by cycling within soils, release and transport to rivers, as well as in the air in the form of particulates, and by rivers to lakes or the sea, and ends with its ultimate burial in sedimentary sequences. This phosphorus pool will then enter the endogenic phosphorus cycle, and will be recycled through global tectonic processes, via subduction, accretion and uplift (Figure 2.1).

Phosphorus, along with carbon (C) and nitrogen (N), regulate biological activity. Optimum productivity occurs when the composition of the three, approaches the Redfield ratio (Redfield, 1958; Redfield *et al.*, 1963), which represents the

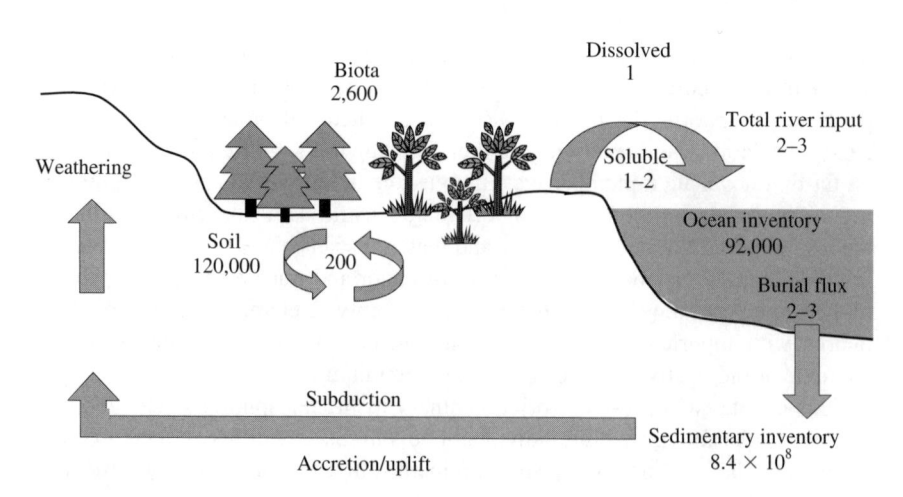

Figure 2.1 The dissolved phosphorus cycle prior to human influence (adapted from Filippelli, 2002). Reservoir capacities in Tg P, and fluxes (denoted by arrows) in Tg/year P.

optimum atomic proportions of these elements for phytoplankton growth, and is usually quoted as equal to $C:N:P = 106:16:1$, although a variety of modifications have been proposed (e.g. Anderson and Sarmiento, 1994; Geider and La Roche, 2002). Despite phosphorus requirement being the lowest of the three, the luck of an atmospheric pool for this element, unlike N and C, makes it the limiting nutrient, at least in terrestrial and freshwater environments, where N-fixing cyanobacteria counteract any limitation by nitrogen (Marino et al., 2002). It has been, however, argued that P-limitation may not be controlling in estuarine and coastal environments on short-term scales, and nitrogen limitation of net primary production may prevail (e.g. Howarth et al., 1995; Marino et al., 2002). It is unclear whether phosphorus or nitrogen limitation prevails in oceanic environments, generally phosphorus being considered the most likely limitation particularly in geological timescales (e.g. Bjerrum and Canfield, 2002; Compton et al., 2000; Howarth et al., 1995), even though it has been noted that phosphorus fluxes in the marine environment over the past two million years are too small to account for atmospheric carbon dioxide fluxes, whereas nitrogen fixation shows better correlation (Falkowski, 1997). Regardless of whether phosphorus or nitrogen is the ultimate limiting nutrient in any particular environment, they are both important in carbon fixing. However, there are presently many unknowns in assessing global fluxes for both elements reliably. For example, with regards to phosphorus, its rate of release from primary phosphate mineral dissolution is poorly constraint (Guidry and Mackenzie, 2000), resulting in large uncertainties when calculating fluxes.

Prior to human influence on global nutrient fluxes, the only significant source of phosphorus was weathering of primary phosphate minerals in rocks and soils on land, by far the most common of which is apatite (a generic name for a group of minerals, see later in this chapter). Other potential sources, such as hydrothermal activity and weathering of basaltic rocks, have been instead shown to be sinks due to scavenging by suspended iron oxides of hydrothermal origin (estimated to be 12% of the total annual phosphorus sink in the ocean) and by basalt weathering products (Froelich et al., 1982; Feely et al., 1990). The global P cycle has been

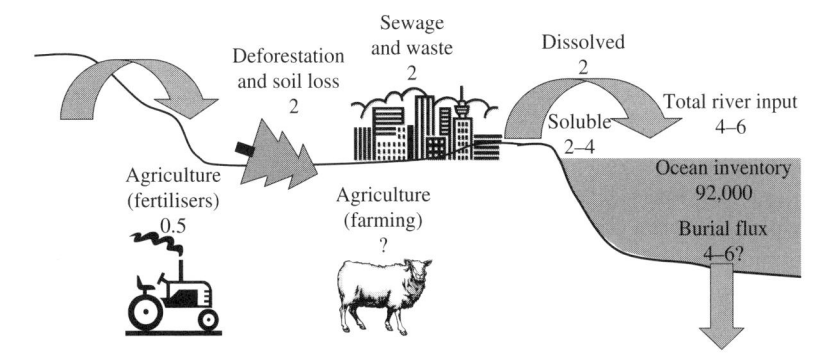

Figure 2.2 The anthropogenic dissolved P cycle (modified from Filippelli, 2002). Reservoirs are shown (in Tg P) and fluxes indicated by arrows (in Tg P/yr).

modified extensively with modern human (agricultural, urban and industrial) practices (compare Figure 2.1 with Figure 2.2) (e.g. Filippelli, 2002; Föllmi, 1996; Mckenzie *et al.*, 2002). As a result, fluxes of P, as well as C and N, have doubled during the past several centuries (Meybeck, 1993) and are expected to increase further (Tappin, 2002). Key to the global changes in phosphorus flux is the high rate of extraction of P from mineral deposits followed by widespread use in fertilisers and detergents, and production by extensive farming practices. Prediction of anthropogenic input to P cycle is that it represents 60% of its total flux (13.5×10^{12} g P/year) (Meybeck, 1993). Near industrialised areas, the net increase can be 10–100 times pre-anthropogenic levels (Caraco, 1993). The impact of such increased fluxes on the environment is widespread (e.g. Koelmans *et al.*, 2001; Sharpley *et al.*, 2000; Watson and Foy, 2001), and may extend as far as causing damage to coral reef ecosystems; mechanisms of P transport to coral reef environments include direct input during flood events, and re-release by reductive dissolution of precipitated ferric phosphates from within marine sediments (e.g. McCulloch *et al.*, 2003).

2.2.3 Phosphorus in soils and sediments

2.2.3.1 Soils

Being an essential nutrient, phosphorus is extensively recycled within soils, where its availability dominates biological processes. Soils effectively represent a temporary reservoir for phosphorus. Soil phosphate undergoes several biochemical transformations, including conversion to organic phosphate. Phosphorus is present in soils at relatively low concentrations in the range of 100–3,000 mg/kg (Sharpley, 2000). Its primary availability is controlled by sparingly soluble phosphate minerals. These conditions make it often the critical nutrient limiting plant growth. This applies particularly to unfertilised soils, since long-term fertilisation of soils in some parts of the world has had an effect on this balance (see Evans and Johnston, this volume).

Phosphorus in soils is present as a dissolved species in pore waters and in particulate form. In soil particles, phosphorus occurs in both inorganic (P_i) and organic (P_o) forms. Inorganic phosphorus, in turn, may be present as: (1) phosphate minerals (calcium, iron or aluminium phosphates) crystalline or amorphous, or (2) adsorbed on soil particles, mostly Fe and Al oxyhydroxides and clays (see Figure 2.3). In the former case P availability is controlled by dissolution precipitation reactions, in the latter by desorption/adsorption reactions. In reality, there is a continuum between P fixed by precipitation and P fixed by adsorption, and often the two processes are described by the bulk term 'sorption' (Frossard *et al.*, 1995).

From Figure 2.3, it is apparent that pH has a major control over phosphorus availability and that most bioavailable phosphorus occurs at pH 6–7. At higher pH, P is mostly fixed by calcium phosphates, mainly apatites (see later this chapter), whereas at pH < 6, P availability is considered to be controlled by adsorption on

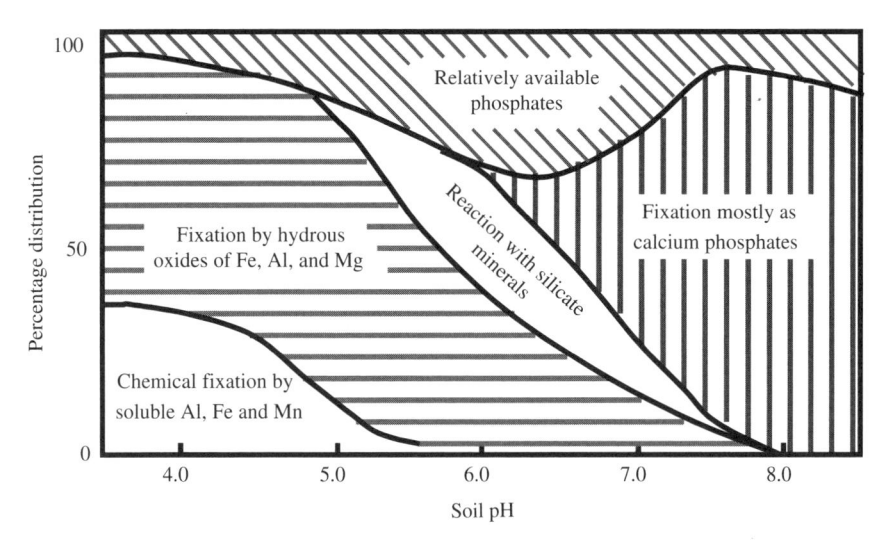

Figure 2.3 Speciation of phosphorus in the soil environment, showing the major phases controlling its availability as a function of pH (after Buckman and Brady, 1972).

and/or precipitation with Al and Fe phases (Buckman and Brady, 1972; Frossard et al., 1995). Availability of P in low pH soils is often modelled as being specifically controlled by the minerals strengite and variscite (Fe and Al phosphates respectively) and it is often assumed that amorphous Fe and Al phosphates will eventually transform into these two phases (e.g. Lindsay et al., 1989). However, well crystallized Fe and Al phosphates have not often been observed in soils (Frossard et al., 1995; Sharpley, 2000); it has also been suggested that crystallization of strengite and variscite require conditions which rarely occur in soils (e.g. Hsu, 1982a, b). Alternatively, it may be that Fe and Al control P availability through the formation of amorphous or other poorly crystalline phosphates. Phases that have been proposed in the past include sterrettite $((Al(OH)_2)_3HPO_4H_2PO_4$, van Riemsdjik et al., 1975), tinticite ($Fe_6(PO_4)_4(OH)_6 \cdot 7H_2O$, Jonasson et al. 1988) and griphite ($Fe_3Mn_2(PO_4)_3(OH)$, Martin et al., 1988). Other mixed Al-Fe-Si-P phases have also been observed in P-rich soils (Pierzynski et al., 1990), as well as mixed P mineral forms consisting of Fe, Al and Ca, indicating impregnation of primary Ca–P with Fe-oxyhydroxides, and interpreted as an intermediate stage in the transformation of primary P to secondary forms (Agbenin and Tiessen, 1994).

Sorption is mainly controlled by the most fine grained particle pool in the soil, i.e. the clay fraction, which represents the fraction with the highest and most reactive surface area of the soil, which is responsible for the bulk of reactions. The degree of reactivity will be a function of the surface charge of the mineral surfaces present in this fraction, and the overall soil pH. It is commonly recognized that Fe, Al and Mn oxyhydroxide particles will adsorb P in low pH environments, whereas as pH increases the negative surface charge of soil particles predominates and repels the phosphate oxyanions (e.g. Sharpley, 2000). Clay minerals

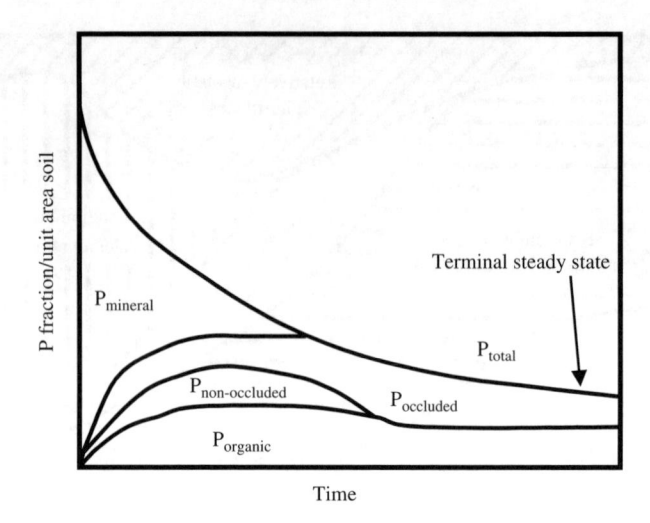

Figure 2.4 The effects of soil evolution on phosphate fraction (after Walker and Syers, 1976). The diagrams shows both overall loss of phosphate as a function of time, until reaching steady state, as well as gradual transformation from mineral to occluded (oxide-bound, unavailable) and organic forms.

sorb less phosphate than oxyhydroxides, and, notably, allophane-like phases have a particularly high sorption capacity for phosphate (Douglas et al., this volume; Frossard et al., 1995; Parfitt, 1989). Humic and fulvic acids are able to form metal complexes, which bind appreciable amounts of phosphorus in soils (e.g. Zalba and Peinemann, 2002).

Organic soil phosphorus compounds range from relatively available phospholipids and fulvic acids, to less available humic acids and inositols (Sharpley, 2000). A group of plant-derived organic phosphorus compounds, inositol phosphates, have been shown to accumulate in soils and potentially become the dominant class of organic P compounds; they have also been identified in high concentrations in aquatic environments and thus may play a role in eutrophication (Turner et al., 2002).

In most soils, the proportion of inorganic P ranges between 50–75%, but the actual range can vary between 10–90% (Sharpley, 2000). In young soils, P is mostly present as crystalline phosphate, whereas as soils age and weathering of the primary phases progresses, more P is present in the poorly crystalline and adsorbed fraction. This is shown schematically in Figure 2.4.

Whether organic, or inorganic, there is a gradient between bioavailable (labile) and unavailable (refractory) P in soils (Figure 2.5). Refractory forms include primary phosphate minerals, as well as secondary neoformed (precipitated *in-situ* in the soil) phosphate phases, or oxyhydroxides, where phosphorus has been incorporated into the structure (occluded). Non-occluded on the other hand represents the loosely surface bound, labile P fraction.

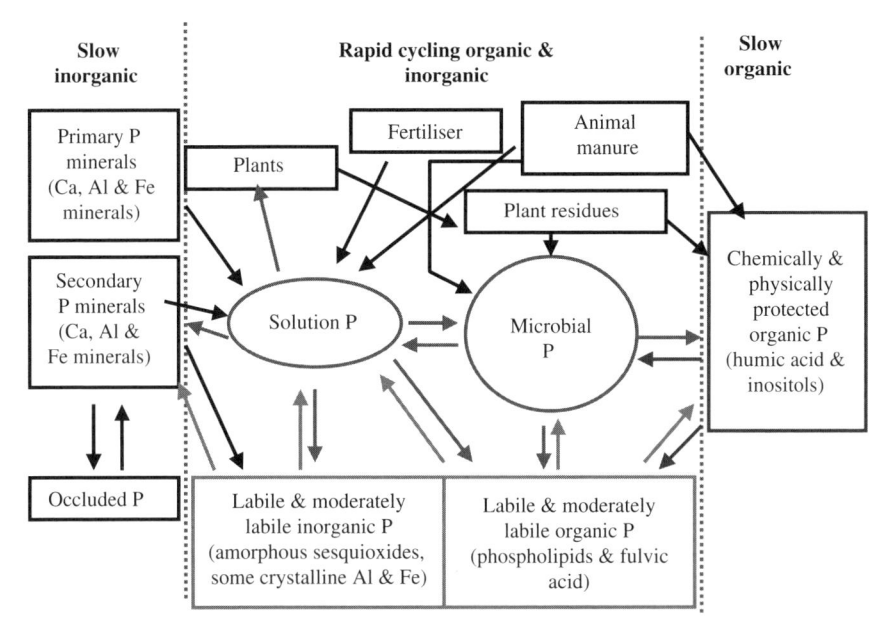

The geochemistry and mineralogy of phosphorus

Figure 2.5 The soil P-cycle, labile and fixed phosphorus pools and relative behaviour of organic and inorganic pools (adapted from Stewart and Sharpley, 1987).

Phosphorus accumulates in the top 20 centimetres of the soil, due to both chemical and biological processes responding fast to consume/immobilize any P additions (Sharpley, 2000). Inorganic phosphorus is the main source of P available to crops. However, transformation of labile organic to inorganic P (a process often described as 'mineralization' in the soil literature, even though the result of the process is soluble available P), mainly via microbial activity, has been shown to be important (e.g. Stewart and Tiesen, 1987).

Many organisms have developed strategies for P uptake, a classic example being plant strategies for P solubilization, via mycorrhizae. Mycorrhizal activity is through extension of the root system, which contributes to increased access and capacity for P adsorption, as well as through biochemical reactions enhanced by exudation of chelating compounds. Plants and soil microorganisms can also excrete phosphatase, an enzyme capable of releasing bioavailable phosphorus from soil organic matter (e.g. Kroehler et al., 1988; Vuorinen and Saharinen, 1996). Other organisms with a role in P transformations in soils include earthworms, which act mostly by making particulate phosphorus more reactive/available (by finely grinding through their digestive system) and increasing its surface area (Sharpley, 2000).

To differentiate between different pools of phosphorus in the soil, as shown in Figure 2.5, a variety of soil P tests have been developed (Table 2.2). These are extractions, involving a variety of acids and alkalis, each test dissolving an alternative P-pool. The true meaning of such extractions, in terms of P speciation and

Table 2.2 Soil P tests (from Westerman, 1990, and references therein).

Name	Composition
Citric acid	1% citric acid
Truog	0.001 M H_2SO_4 + $(NH_4)_2SO_4$
Morgan	0.54 M CH_3COOH + 0.7 M $NaC_2H_3O_2$ − pH4.8
Bray P_1	0.03 M NH_4F + 0.025 M HCl
Bray P_2	0.03 M NH_4F + 0.1 M HCl
Mehlich I	0.05 M HCl + 0.0125 M H_2SO_4
Olsen	0.5 M $NaHCO_3$ − pH8.5
Egner	0.01 M Ca lactate + 0.02 M HCl
Resin	Anion exchange resin
ISFEIP (Hunter)	0.25 M $NaHCO_3$ + 0.01 M NH_4F + 0.01 M EDTA − pH8.5
AB-DTPA	1 M NH_4HCO_3 + 0.005 M DTPA − pH7.5
Mehlich II	0.015 M NH_4F + 0.2 M CH_3COOH + 0.2 M NH_4Cl + 0.012 M HCl
Mehlich III	0.015 M NH_4F + 0.2 M CH_3COOH + 0.25 M NH_4NO_3 + 0.013 M HNO_3

availability, is still the subject of much debate (see, for example, Evans and Johnston, this volume, Neyroud and Lischer, 2003).

2.2.3.2 Sediments

Phosphate behaviour in sediments is similar to that in soils, in terms of geochemical controls on its distribution. Phosphate introduced in the marine environment, either via river discharge or airborne, will be in the form of both reactive and refractory fractions. The reactive, bioavailable fraction is consumed in the surface zone by biota and transformed into organic P. With time, phosphate is transferred to deeper oceanic zones, in inorganic particulate, organic particulate or, after oxidation of the latter, inorganic dissolved phosphate; this phosphate can be taken up again by deep upwelling water currents and be re-introduced into the photic zone and re-utilized (Föllmi, 1996).

Some of the particulate P delivered to the oceans is buried without transformation and is never solubilized or made biologically available. Detrital apatite eroded from land, for example, is unlikely to undergo substantial dissolution when reaching the sea (Froelich et al., 1982; Ruttenberg, 1993). Some particulate P however will be converted to dissolved forms. When considering the influence of P ocean inputs on primary productivity, it is desirable to know the proportion of total input that is potentially available to the biota. In the literature, this P pool − i.e. dissolved, as well as easily solubilized particulate, is often referred to as 'reactive' P (Froelich et al., 1982).

At the end of its oceanic cycle, a fraction of phosphate (estimated at 5% of reactive P by Föllmi et al., 1993) is removed from the water pool, into the sediments; there, phosphorus will occur as organic phosphate, authigenic inorganic phosphate or adsorbed inorganic phosphate (Berner et al., 1993).

Following similar methodologies to soils, sequential extractions are used to assess different P fractions in sediments. A widely recognised scheme for marine sediments (Ruttenberg, 1992) describes five P fractions: (1) loosely bound ($MgCl_2$ extractable); (2) Fe-bound (citrate-dithionite-bicarbonate); (3) authigenic apatite, calcite-bound phosphate and biogenic apatite (acetate buffer at pH4); (4) detrital apatite and other inorganic P (1 M HCl); and (5) organic P (ashing at 550°C + 1 M HCl). Refractive (non-reactive) fractions are the authigenic and detrital P. It should be noted that, in common with soils, extraction methods used to characterize different P fractions in sediments are rarely as specific as designed (e.g. Pardo *et al.*, 2003). Such uncertainties can be minimized if extractions are combined with better characterization of particulate fractions (e.g. Rao and Berner, 1995). In recent years, methods of characterization of nanosize particles (e.g. Hochella, 2002) are becoming widely available thus increasing the potential for better understanding of sediment (and soil) P mineralogy.

It is mostly refractory phosphate that becomes incorporated into marine sediments in the long term. A degree of mobilization of diagenetic phosphate has been established by observations suggesting that buried organic matter becomes depleted in P relative to C (Ingall and Van Cappellen, 1990). Estimates of the global average ratio of organic C/P is sediments vary (150–200: Filippelli, 2002; 250: Froelich, *et al.*, 1982; 496: Ramirez and Rose, 1992; 367: Williams *et al.*, 1976), but are distinct from the Redfield ratio of 106, the optimum ratio for organic matter (see above). It therefore appears that phosphorus is recycled preferentially to organic carbon, suggesting decoupling of these two elements in their long term cycles (Van Cappellen and Ingall, 1996), although the net differences may not be so dramatic when total P and C (rather than organic only) are considered (Filippelli, 2002).

The decrease in phosphorus concentration in marine sediments is a function of both depth and age. Sediments at depth contain less than 10% of the P of near-surface sediments, and this has been attributed to microbial release of organic phosphate in the sediment column (Filippelli and Delaney, 1996). Some of the released phosphate may reprecipitate as authigenic fluorapatite, the proportion of which shows an increase in sediments with age (Filippelli, 2002). Phosphorus concentration in marine sediments is also substantially influenced by the environment, with burial of only 1% of P in deep sea sediments compared to 50% in continental margin sediments (Ruttenberg and Berner, 1993). Margin sediments therefore contain 50–80% of the total P removed from the ocean (e.g. Ruttenberg, 1993; Filippelli, 2002).

Under certain conditions, phosphate-rich sediments, phosphorites, form in the marine environment. The main phosphate mineral present in these sediments is carbonate fluorapatite. Present day environments of phosphorite deposition are discussed in Föllmi (1996). They are areas characterized by high primary productivity rates in the surface waters, associated with upwelling of P-enriched waters, and low sedimentation rates, where phosphate enrichment is due to early diagenetic decomposition of organic matter and retention of phosphate in an oxidizing environment (Föllmi, 1996).

Lake sediments represent a significant P sink in continental settings. They are also environments where P transformations as a function of time, including anthropogenic influences, are captured in a well-defined sediment record. Lakes are particularly appropriate for the study of P mass balances on glacial/interglacial timescales (Filippelli, 2002). In many lakes, a glacial stage, dominated by P in primary mineral form, followed by an increase in organic and occluded P towards modern time, reflecting soil and forest development, have been well documented (e.g. Jackson Pond, Wilkins et al., 1991). This sequence is however not unique and will depend on the lake setting as this is influenced by climate and landscape history, which in turn will control the rate of replenishment and transformation of P in the lake sediment.

River sediments can also contain high concentrations of phosphorus, if they have been influenced by anthropogenic (e.g. waste water treatment) sources, and can store more than 30% of the river load throughput (House, 2003). Recent studies showed concentrations of several hundred to a few thousand μ mol g^{-1} P in the most impacted UK rivers (House and Denison, 2002; Owens and Walling, 2002). A combination of bio-physicochemical processes control phosphorus incorporation into river sediments. Mechanisms such as co-precipitation with calcite, and formation of vivianite in anoxic sediments, may control phosphorus availability in some rivers containing high P loads (House, 2003).

2.2.4 Phosphorus in waters

Transport of phosphorus from soils to surface waters takes place in the form of both chemical (dissolved) and physical (particulate) form, at both surface and subsurface, although the latter has been notoriously difficult to quantify, and only recent work has recognized its significance (see also Burke et al., this volume; Haygarth and Condron, this volume).

Dissolved phosphorus speciation is shown in Figure 2.6. The figure suggests that $H_2PO_4^-$ and HPO_4^{2-} are the dominant aqueous species. Dissolved P is usually considered a small proportion of total P transported and values of 10–25 μgl^{-1} have been reported (Meybeck, 1993); this is however the most reactive fraction of total P. Instead, phosphate is mostly transported in particulate form, adsorbed onto oxyhydroxides and clay particles, as phosphate mineral particles, or as mixed poorly crystalline phosphate phases (e.g. Fox, 1989; Howarth et al., 1995). The proportion of total P that is in particulate form is estimated at 95% (Föllmi, 1996). A large part (40%, Föllmi, 1996) of particulate P is associated with organic (living and dead) forms (Grobbelaar and House, 1995) according to some workers (but see also Salcedo and Madeiros (1995), who report examples of small organic P pool in river waters). Adsorbed P is mainly important in turbid waters with a lot of suspended material (Grobbelaar, 1983). The high proportion of particulate P makes global flux calculations imprecise, as particulate mass fluxes are disproportionately influenced by infrequent (and thus poorly sampled) high flow events (Tappin, 2002). Airborne transport of particulate P forms a significant part of its mass balance, probably of the order of 4–5% (e.g. Mackenzie et al., 1993), an estimated 0.95 Tg y^{-1} (Duce et al., 1991).

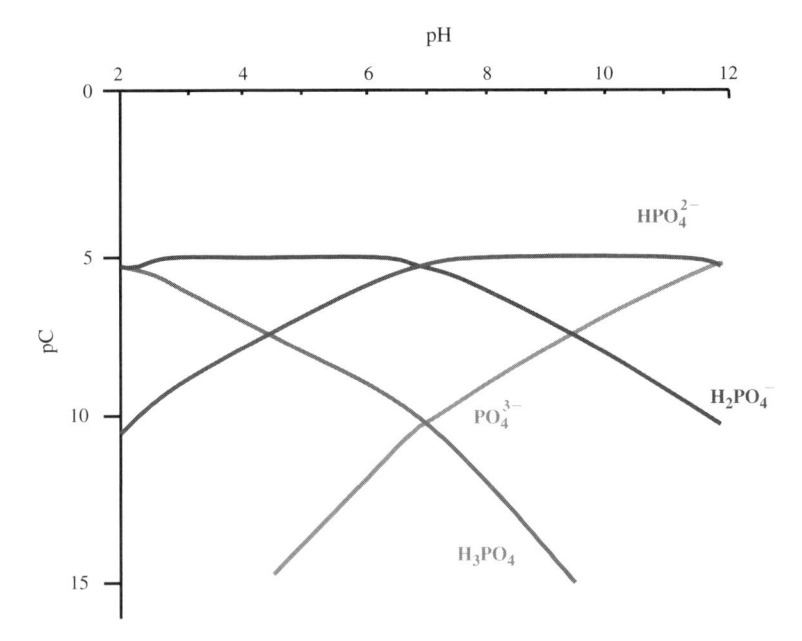

Figure 2.6 The aqueous speciation of phosphorus as a function of pH, based on a calculation using the HYDRAQL model (Papelis et al., 1988) for a total dissolved phosphate concentration of 10^{-5} M.

Some of the phosphorus transported in particulate form is likely to be non reactive, if present as sparingly soluble phosphate mineral particles, or if tightly bound onto other mineral particles. Also, some of the less reactive P transferred from continental environment to ocean, may be buried without further interactions (Meybeck, 1982; Froelich, 1988). However, in estuaries and river deltas substantial amounts of P may transform form particulate to reactive, due to redox and salinity changes (Howarth et al., 1995; Föllmi, 1996). It is estimated that 25% of total P eroded from continents enters oceans in bioavailable form (Froelich, 1988). In the marine environment, all reactive P is rapidly used up in photic zone. As a result, the P profile in the ocean shows depletions at the surface and enrichment at depth (Filippelli, 2002). The deep ocean water phosphate concentration increases with age of waters, and as a result the older Pacific deep waters contain more phosphate than the younger Atlantic (approximately 2.5 μM and 1.5 μM respectively, Broecker and Peng, 1982). An average dissolved phosphate concentration for seawater is 2 μM (Henderson, 1982).

The concentrations of dissolved phosphate in unpolluted river waters is estimated between 0.1–1 μM (0.003–0.03 mg l^{-1}, Meybeck, 1982) compared, for example, with a range of <0.01 to 7.85 mg l^{-1} for 98 rivers in the UK (Muscutt and Withers, 1996). This latter study showed seasonal variations of dissolved phosphate concentrations that reached maxima over the mid-summer to autumn period, and minima during mid-winter to spring. Dissolved organic phosphate can be significant in river waters: a compilation of values between 0.1–0.5 μM was presented by Salcedo and Medeiros (1995). Rainwater normally contains little P (0.16 μM, Meybeck, 1982) and therefore only contributes to river water load due to storm related erosion.

The abiotic regulation of soluble P can be both due to adsorption/desorption and dissolution/precipitation mechanisms, as discussed above (see also Frossard *et al.*, 1995), and some workers propose a two stage process where adsorption is followed by formation of a phosphate phase; interestingly such models have been able to predict phosphate concentrations in rivers (e.g. Fox, 1989; Salingar *et al.*, 1993).

Rivers, in turn, are the main source of P to the oceans, with estimates of the annual worldwide P flux between 21 and 39 Tg y^{-1} (Froelich *et al.*, 1982; Howarth *et al.*, 1995; Meybeck, 1982). Transport of phosphate occurs usually via estuaries, where a net release of phosphate is likely due to redox and salinity induced desorption, as described previously. This results in a net increase in concentrations of dissolved phosphate and values in the range 30–60 $\mu g \, l^{-1}$ (e.g. Billen *et al.*, 1991; Froelich, 1988; Lebo, 1991). The phosphate released is often used up by phytoplankton blooms or flocculated by humic complexes of Fe (Forsgren and Jansson, 1992; Salcedo and Medeiros, 1995). The fraction of phosphate that undergoes burial into sediments, as discussed earlier, will then become part of the endogenic P cycle (Figure 2.1).

2.3 PHOSPHATE MINERALS

The [PO_4] oxyanion combines with over 30 elements to form the reference structures of phosphate minerals, although virtually every known element has been found as a trace element in phosphates (Nriagu, 1984a). Some 308 phosphate minerals are now recognised (Clark, 1993). The importance of phosphate minerals is diverse, ranging from their role in biomineralization, to being hosts to a range of rare, including radioactive (U, Th), elements and their applications to geochronology, mineral engineering, agriculture, biomaterials and many other areas of science and technology.

Some general characteristics of phosphate minerals include mutual substitution of P by As, an element also occurring in oxyanionic form, and chemically akin to P. Solid solutions with silicates, also exist, although the substitution is of limited extent at ambient temperatures (Nriagu, 1984a). Phosphate in mineral structures can also be replaced by the oxyanions of B, V, Ge, Cr and S (Moore, 1984; Nriagu, 1984a).

The crystal chemistry of phosphate minerals has recently been presented in detail by Huminicki and Hawthorn (2002), who used cluster polymerization as the basis of the structural classification of the group.

Phosphate minerals occur as accessory phases in most rocks and are common in soils and sediments. Despite numerous phosphates described from igneous and metamorphic rocks, by far the most common mineral in such rocks is apatite; other reasonably common phosphates include monazite ((Ce,La,Y,Th)PO_4), xenotime (YPO_4) and whitlockite ($Ca_9Mg(PO_4)_6 \cdot (PO_3OH)$) (Nash, 1984). Apatite composition in igneous rocks reflects the conditions prevailing during its formation: for example enrichment in Si, Na, K, Sr, REE and F and depletions in Fe, Mg and Cl reflect increased magmatic differentiation (i.e. cooling and crystallization) (Nash, 1984). Structural changes may also occur as a result of differentiation, with apatite becoming more elongated with cooling (Piccoli and Candela, 2002). It has been suggested that at high supersaturation, surface controlled growth parallel to the c-axis results in acicular crystals, whereas low supersaturation, results in diffusion controlled growth and more equant crystals (Hoche et al., 2001). Apatites from metamorphic rocks have similar composition and properties to those of igneous rocks (Nash, 1984).

Apatite also dominates the phosphate mineral occurrences in sedimentary rocks, whereas other less common phosphates include crandallite ($CaAl_3(PO_4)_2(OH)_5 \cdot H_2O$), vivianite ($Fe_3(PO_4)_2 \cdot 8H_2O$), brushite ($CaHPO_4 \cdot 2H_2O$) and whitlockite (Cook, 1984). Although igneous phosphate deposits exist, by far the most common economic occurrences of phosphate rock are of sedimentary nature, with the biggest of those deposits occurring in Morocco and the United States (Cook, 1984). They are described as phosphorite deposits, and the mechanism of their formation has been briefly considered earlier in this chapter. Sedimentary apatites are substantially different from igneous apatites, with the key substitution of CO_3^{2-} for PO_4^{3-}; they are thus commonly referred to as carbonate fluorapatites (see later).

2.3.1 Calcium phosphates

There is a range of calcium phosphate minerals with [Ca]:[PO$_4$] ratios varying from 1.667 to 1; some contain, in addition to calcium and phosphate, hydroxyls (or F, Cl) or structural water in their reference formula; many other ions can enter their structures as substitutions for calcium and/or phosphate. Of the whole group, the most stable and common calcium phosphate is apatite, as discussed previously, whereas other members of the group are often considered as precursor phases to apatite, to which they recrystallize with time (see also Koutsoukos and Valsami-Jones, this volume). Some basic mineralogical characteristics of the most common naturally occurring calcium phosphates are shown in Table 2.3.

2.3.1.1 Apatite

The wide abundance of apatite is paralleled by its applications and its broad range of compositions. The formula commonly given for apatite, $Ca_{10}(PO_4)_6X_2$, where X=OH, F, Cl, reflects but a fraction of the known and possible substitutions. It is therefore more appropriate to refer to apatite as a group of minerals, whereas apatite, in the strict sense, can only be one of three minerals: hydroxylapatite [$Ca_{10}(PO_4)_6(OH)_2$], fluorapatite [$Ca_{10}(PO_4)_6F_2$] and chlorapatite [$Ca_{10}(PO_4)_6Cl_2$], or their solid solutions.

The atomic components of the ideal apatite structure combine to form hexagonal crystals, arranged in space group $P6_3/m$ (e.g. Hughes and Rakovan, 2002).

Table 2.3 Some mineralogical characteristics of selected calcium phosphate minerals (data from Nriagu, 1984a).

Mineral	Formula	Crystal system & lattice constants	Strongest diffraction lines
Hydroxylapatite	$Ca_{10}(PO_4)_6(OH)_2$	Hexagonal a = 9.418, c = 6.884 (Å)	2.814 (100), 2.778 (60), 2.720 (60)
Fluorapatite	$Ca_{10}(PO_4)_6F_2$	Hexagonal a = 9.36, c = 6.88 (Å)	2.800 (100), 2.702 (60), 2.772 (55)
Chlorapatite	$Ca_{10}(PO_4)_6Cl_2$	Hexagonal a = 9.64, c = 6.78 (Å) Monoclinic a = 19.21, b = 6.785, c = 9.605 (Å); β = 120°	2.78 (100), 2.86 (50–60)
Monetite	$CaHPO_4$	Triclinic a = 6.90, b = 7.00, c = 6.65 (Å); α = 96.35°, β = 91.27°, γ = 76.1°	2.96 (100), 3.35 (75), 3.37 (70)
Brushite	$CaHPO_4 \cdot 2H_2O$	Monoclinic a = 5.88, b = 15.15, c = 6.37 (Å); β = 117.47°	7.62 (100), 3.80 (30), 1.90 (10)
Whitlockite*	$Ca_9Mg(PO_4)_6 \cdot (PO_3OH)$	Hexagonal a = 10.32, c = 36.9 (Å) Rhombohedral a_{rh} = 13.67; a = 44.35°	2.837 (100), 2.572 (80), 1.701 (80)

* The pure calcium end member does not occur naturally.

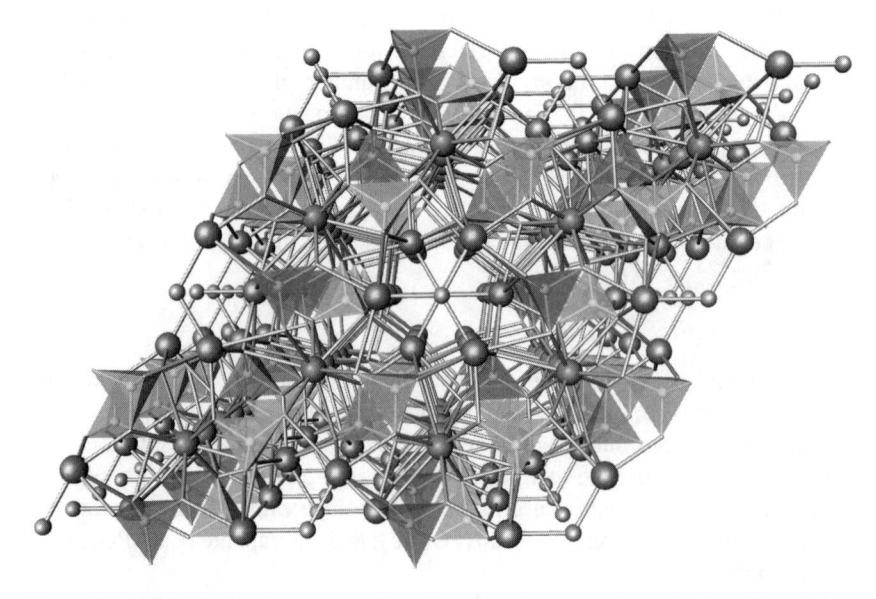

Figure 2.7 The idealized structure of apatite, viewed as a c-axis projection. Note the hexagonal structure, and the location of X ions (OH, F, Cl) along channels parallel to the c-axis. [This figure is also reproduced in colour in the plate section after page 336.]

More specifically, the structure consists of columns along the c-axis direction where the X ions reside (see Figure 2.7, in orange colour). These are surrounded by six calcium atoms (described as Ca(2) and seven coordinated) arranged in two triagonal mirror planes around the channel, displaced along the c direction and rotated by 120°, so as to give the impression of a hexagonal structure, when viewed in c-axis projection (Figure 2.7, in blue colour). The remaining structure is taken up by a second array of calcium atoms coordinated in the middle of the unit cell (described as Ca(1) and nine coordinated; Figure 2.7, in pink colour), and by the phosphate tetrahedra (Figure 2.7, in green colour).

The pure end member hydroxyl- and chlorapatite actually crystallize in the sub-symmetric monoclinic space group $P2_1/b$, due to the OH^- and Cl^- being too large to fit in the rigid triangular plane defined by the Ca^{2+} atoms along the c-axis, and having to be displaced above or below the plane, the larger Cl^- ion further than OH^- (~0.35Å and ~1.2Å respectively; Hughes and Rakovan, 2002). It is often the case that vacancies and substitutions (e.g. by F^-) allow the structure to retain a hexagonal symmetry, and natural hydroxylapatites are always hexagonal (Posner et al., 1984; Hughes and Rakovan, 2002). This can be achieved through the existence of reversal sites, where the hydroxyl orientation reverses within the column; such sites are centred around vacancies, or fluorine ions (Hughes and Rakovan, 2002). This structural destabilization may be the cause for the reluctance of low temperature hydroxylapatite to form large crystals and its tendency to stabilize with time by substituting OH^- for F^- (Skinner, 2000a). Furthermore the Ca—F bonds are stronger than Ca—OH bonds (Posner et al. 1984), which may explain the differences in solubility between the two minerals discussed elsewhere (see Koutsoukos and Valsami-Jones, this volume). Incorporation of the large Cl^- ion can only take place via reversals and obligatory vacancies, unless if adjustments occur elsewhere in the structure (Hughes and Rakovan, 2002).

The apatite group may be described by the generic formula: $M_{10}(ZO_4)_6X_2$, where $M = Ca^{2+}, Pb^{2+}, Sr^{2+}, Mg^{2+}, Fe^{2+}, Mn^{2+}, Cd^{2+}, Ba^{2+}, Co^{2+}, Ni^{2+}, Cu^{2+}, Zn^{2+}, Sn^{2+}, Eu^{2+}, Na^+, K^+, Li^+, Rb^+, NH_4^+, La^{3+}, Ce^{3+}, Sm^{3+}, Eu^{3+}, Y^{3+}, Bi^{3+}, Cr^{3+}, Th^{4+}, U^{4+}, U^{6+}, \square$; $Z = PO_4^{3-}, AsO_4^{3-}, SiO_4^{4-}, VO_4^{3-}, CrO_4^{3-}, CrO_4^{2-}, MnO_4^{3-}, SO_4^{2-}, SeO_4^{2-}, BeF_4^{2-}, GeO_4^{4-}, ReO_5^{3-}, SbO_3F^{4-}, SiO_3N^{5-}, BO_4^{5-}, BO_3^{3-}, CO_3^{2-}$ and $X = F^-, OH^-, Cl^-, O_2^-, O_3^-, NCO^-, BO_2^-, Br^-, I^-, NO_2^-, NO_3^-, CO_3^{2-}, O_2^{2-}, O^{2-}, S^{2-}, NCN^{2-}, NO_2^{2-}, \square$ (Pan and Fleet, 2002), where $\square$ represents vacancies or even vacancy clusters. There is ample evidence for the remarkable structural tolerance of apatite, allowing all these numerous substitutions, vacancies and distortions (see Elliott, 1994, Pan and Fleet, 2002 and references therein). Some of the substitutions listed above only exist in synthetically produced apatites, and do not occur in nature, but are indicative of the tolerance and versatility of the mineral structure. Many of the substitutions involve placement of an ion of different valency from that needed to preserve electrostatic neutrality, for example the M cations can be substituted by both monovalent (e.g. alkalis) and trivalent (e.g. lanthanides) ions. Such substitutions can be tolerated by a range of charge-compensating mechanisms, which involve structural vacancies (e.g. three Ca^{2+} atoms can be substituted by two trivalent ions and a vacancy) or coupled substitutions (e.g. two Ca^{2+} atoms can be substituted by one monovalent and one trivalent ion). Compensating mechanisms of similar nature are also involved in cases where substitutions involve larger or smaller ions. The existence of two non-equivalent Ca sites (Ca(1) and Ca(2)) in the apatite structure, described above, means that often the substituting ions show a preference for one or the other site, depending on their size or charge. For a detailed discussion on the range of possible structural substitutions, see Pan and Fleet (2002).

One of the more studied substitutions in the apatite structure is that of carbonate, usually to form carbonate fluorapatite (as for example the apatite in phosphorite deposits) and carbonate hydroxylapatite (a variety of which is the main component of bone and teeth of vertebrates, see later in this chapter). Older literature refers to these two apatites as 'francolite' and 'dahlite' respectively, although these are currently not considered valid mineral names (Knudsen and Gunter, 2002; Pan and Fleet, 2002). Carbonate can substitute two sites in the apatite structure, that of c-axis anions (X) as well as that of the phosphate tetrahedra (Z) (Elliott, 1994 and references therein). These are usually referred to as A-type and B-type carbonate apatites respectively (Bonel and Montel, 1964). The B-type substitution is limited not only by the charge difference between carbonate and phosphate, but also due to the effects of introducing the essentially planar carbonate ion in the place of the phosphate tetrahedron (Nathan, 1984). The carbonate ion probably lies in the position of one of the sloping faces of the replaced phosphate tetrahedron; for the A-type substitution, the carbonate ion plane is oriented parallel to the c-axis, to minimize strain due to the large size of the substituting carbonate ion (Elliott, 1994). The c-axis can also receive substitutions of organic molecules such as glycine (Rey et al., 1978) and acetate (Bacquet et al., 1981).

A trade-off for structural tolerance in some apatites can be a reduction in crystallinity. This is most apparent in the case of hydroxylapatite, which can occur as a well crystalline phase, usually the result of formation at high temperatures (igneous, hydrothermal or metamorphic conditions), but also as a low temperature very poorly crystalline phase with substantial vacancies, e.g. in bone; hydroxylapatites synthesized at low temperatures by precipitation, show a similar lack of crystallinity as, and are good laboratory analogues of, bioapatites (Figure 2.8); bioapatites will be discussed in more detail in the following section. Precipitated apatites are usually calcium and hydroxyl-deficient and contain HPO_4^{2-} substituting for PO_4^{3-} (Posner et al. 1984). Their Ca/P molar ratios usually range between 1.50 and 1.65 (e.g. Heughebaert et al., 1990), compared to the stoichiometric ratio of 1.667. It has been suggested that in OH-deficient apatites, it is Ca(2) ions, located around the c-axis channels that are missing (Posner and Perloff, 1957). A range of model formulas, of variable degrees of complexity, have been proposed to describe non-stoichiometric Ca-deficient apatites (see compilation in Elliott, 1994). A simple formula, proposed by Posner et al. (1960), which represents a good approximation of the unit cell content (Elliott, 2002) and assumes no carbonate present in the structure is:

$$Ca_{10-x}(HPO_4)_x(PO_4)_{6-x}(OH)_{2-x} \quad \text{where } 0 \leq x \leq 1 \tag{2.1}$$

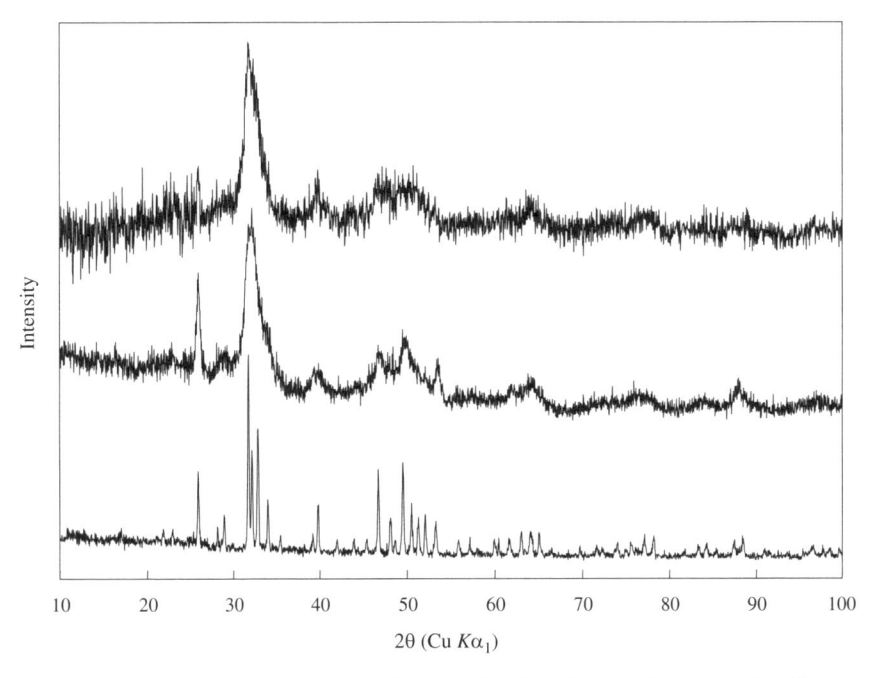

Figure 2.8 X-ray diffraction patterns of a natural hydroxylapatite from a type locality at Holly Springs, Georgia (NHM collection, BM 1958,313) (bottom), a synthetic (at 25°C) poorly crystalline hydroxylapatite (middle) and a bioapatite (sheep bone) (top). Note the increase in crystallinity (sharpness of diffraction peaks) from top to bottom. The patterns are courtesy of Gordon Cressey (natural hydroxylapatite and bone) and Jacqueline van der Houwen (synthetic apatite).

This model formula suggests that the presence of each HPO_4^{2-} is compensated by the loss of one Ca^{2+} and one OH^- ions. Model formulas to describe carbonated apatites, have also been proposed, for both A-type and B-type carbonated apatites, as follows (B-type originally by: Kühl and Nebergall, 1963; A-type originally by: Trombe, 1973; see also Elliott, 1994):

$Ca_{10}(PO_4)_6(OH)_{2-2x}(CO_3)_x$ where $0 \leq x \leq 1$ A-type   (2.2)

$Ca_{10-x+y}(CO_3)_x(PO_4)_{6-x}(OH)_{2-x+2y}$ where $0 \leq x \leq 2$ and $2y \leq x$ B-type   (2.3)

Consideration of combined formulas 2.1, 2.2. and 2.3 suggests that multiple substitutions in apatite for both X and Z ions would produce an apatite substantially deficient in calcium and particularly hydroxyl ions.

2.3.2 Biophosphates

Several phosphates of biological origin (biophosphates or biogenic phosphates) are known to exist. In a similar manner to non-biogenic phosphates, the group of biophosphates is dominated by the abundant bioapatite. Biominerals can be the product of normal physiological activity of the organism producing them, but can also form under pathological conditions. Pathologic minerals (also described as pathological calcifications) in humans form in specific sites: urinary system (kidney stones), oral cavity and cardiovascular system (plaque), pulmonary system (mineralization of trauma related scar tissue), joints (gout) and within tumours (LeGeros and LeGeros, 1984; Skinner, 2000b). It is interesting to note that the association of tumours, whether benign or cancerous, with mineral deposits allows their detection (Skinner, 2000b). Although the human body has developed mechanisms to dissolve ('resorb') minerals in normal mineralized tissue if necessary (see discussion on bone later in this chapter), pathologic mineral deposits occur in association with cells that are not programmed to resorb them, and thus accumulate rapidly with deleterious effects. A list of known biophosphates and their role in human biomineralization is presented in Table 2.4.

Table 2.4 Biogenic phosphates and their occurrence in human tissue (LeGeros and LeGeros, 1984; Elliott, 1994; Skinner, 2000b).

Name	Chemical formula	Occurrence
Bioapatite	$(Ca, Na)_5(HPO_4, PO_4, CO_3)_3(\square, OH, F)$	Bone, teeth, majority of pathological calcifications.
Monetite	$CaHPO_4$	Pathological calcifications, possibly bone.
Brushite	$CaHPO_4 \cdot 2H_2O$	Pathological calcifications (dental, urinary, joint), possibly teeth and bone, particularly fossil bone.
Whitlockite and tricalcium phosphate	$Ca_9Mg(PO_4)_6 \cdot (PO_3OH)$ $Ca_3(PO_4)_2$	Pathological calcifications (dental, pulmonary and cartilage), possibly teeth.
Octacalcium phosphate	$Ca_8H_2(PO_4)_6 \cdot 5H_2O$	Pathological calcifications (particularly dental and urinary), possibly bone and teeth.
Bobierrite	$Mg_3(PO_4)_2 \cdot 8H_2O$	Pathological calcifications.
Magnesium hydrogen phosphate	$MgHPO_4$	Pathological calcifications.
Struvite	$Mg(NH_4)(PO_4) \cdot 6H_2O$	Pathological calcifications (particularly urinary).
Diammonium calcium phosphate	$(NH_4)_2Ca_2(PO_4)_2$	Pathological calcifications.
Calcium pyrophosphate dihydrate	$Ca_2P_2O_7 \cdot 2H_2O$	Pathological calcifications (particularly joints).

$\square$ = vacancy

A phase described as 'amorphous calcium phosphate' has been previously described in biomineralized tissue, particularly as a precursor to bone apatite, although more recently its existence has been put to question (e.g. Elliott, 2002).

2.3.2.1 Bioapatites

The classic biophosphate is apatite, being the compound nature 'chose' to mineralize the bones and teeth of vertebrates. Bioapatite can also exist as a pathologic precipitate in most known pathological calcifications. The composition and structure of bioapatite varies depending on the type of mineralized tissue. There are three main types of bioapatite, found in enamel, dentine and bone.

Enamel, the thin top layer of the tooth, is the most highly mineralized tissue in the body containing up to 96% inorganic material by weight, whereas the remainder consists of around 3% water and 1% organic matter, proteins and lipids (Jones, 2001). Mature enamel is well crystalline, consisting of flattened hexagonal crystals of 70×25 nm wide, and from $0.1-5\,\mu$m or more long (Elliott, 2002). The cells that produce enamel disappear before eruption of the tooth, and the regeneration of this tissue is not possible (e.g. Skinner, 2000b).

Dentine, which represents the bulk of the tooth and forms concurrently with enamel, is a lot more porous, microcrystalline and organic-rich than enamel, containing around 21% by weight organic material (mainly collagen) (Jones, 2001). The bioapatite in dentine occurs in plates with dimensions 50×30 nm wide and 4 nm thick (Elliott, 2002). Odontoblasts, dentine-forming cells, retain their ability to form dentine throughout the life of the tooth (Jones, 2001), which, therefore, unlike enamel, can be regenerated.

Bone apatite, like dentine apatite, is held in an organic-rich matrix, and occurs as very small crystallites, with dimensions similar to those of dentine apatite (Elliott, 2002) (see Figure 2.9). Bone is a 50% by 50% inorganic–organic composite (Jones, 2001). The organic phase is mainly collagen, but mucopolysaccharides, cells and water are also present (Skinner, 2000b).

Bone formation starts with the synthesis of the triple stranded collagen, which forms an organic mesh, onto which bioapatites accumulate, their long dimension parallel with the collagen fibre direction, which is also the direction of the anatomical long axis of the bone (Skinner, 2000a; Elliott, 2002). Collagen fibres bundle together to form lamellae ($3-7\,\mu$m thick) located concentrically around the Haversian canal, where veins reside, and forming the osteon, which is the building block of bone (Gross and Berndt, 2002) (Figure 2.10). The cells that produce the organic matrix that mineralizes with bioapatite, called osteoblasts, and the cells that are responsible for the resorption of bioapatite, called osteoclasts, remain active and thus bone is remodelling throughout life (Skinner, 2000a; Jones, 2001). This gives bone the ability to grow with age, and to repair itself, when fractured.

Many questions remain with regards to the details of bioapatite formation. For example, it is still uncertain whether bioapatite precipitates directly, or via other calcium phosphate precursors, despite a large number or studies on this topic (see

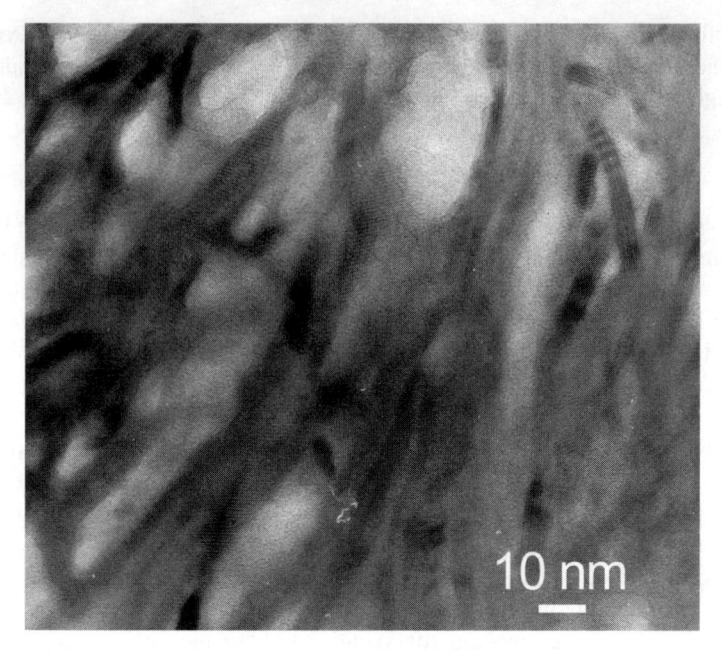

Figure 2.9 TEM image of modern sheep bone; note individual bioapatite crystallites, forming plates, which are viewed here in cross section. The plate width is up to 50 nm whereas their thickness is only 5 nm. Hexagonal lattice images can be seen where the plate has its c-axis parallel to the electron beam. For an XRD pattern of the same material, see Figure 2.8. Courtesy of: Barbara Cressey and Gordon Cressey.

for example Elliott, 2002). The purpose of the presence of almost up to 1% citrate in bone remains unclear (Dallemagne and Richelle, 1973; Elliott, 1994).

Substitutions and vacancies have an important role in the stability of bioapatites. It is important to consider for example that dental studies have shown that small quantities of fluorine reduce the solubility of bioapatites. A representative composition for mature human dental enamel (Elliott, 1997) is:

$$Ca_{8.86}Mg_{0.09}Na_{0.29}K_{0.01}(HPO_4)_{0.28}(CO_3)_{0.41}(PO_4)_{5.31}(OH)_{0.7}Cl_{0.08}(CO_3)_{0.05}$$

Note that in the above formula (based on published analysis), OH^- represents less than half of the X ions. It has been proposed that dentine contains even less OH^- and that bone apatite is totally OH-deficient (Pasteris *et al.*, 2004, and references therein) and, further, that the degree of hydroxylation may be the mechanism by which the body can control the function of its minerals. More specifically, non-hydroxylated bioapatites are the most soluble and are found in tissues where regular remodelling is required, whereas variable degrees of hydroxylation confer different resorbability to other bioapatites (none in enamel, but some in dentine) (Pasteris *et al.*, 2004).

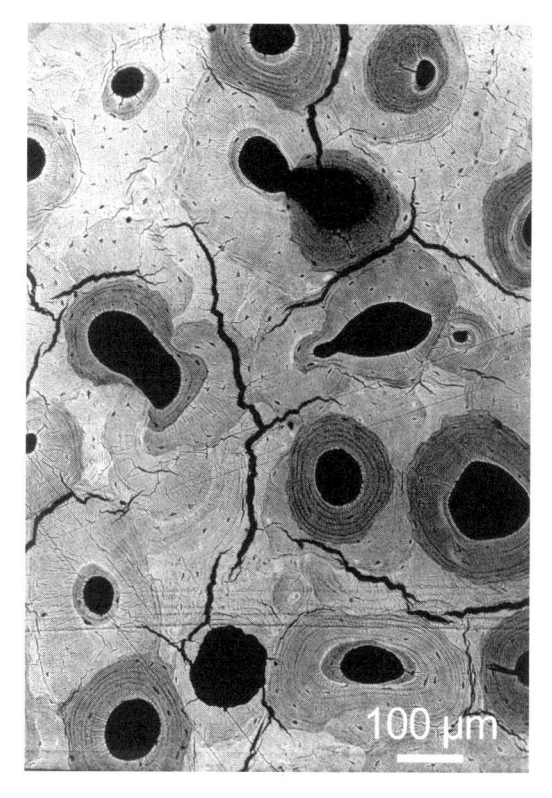

Figure 2.10 Backscatter electron image of 200-year-old human bone (source Spital Fields Cemetery, London). The image shows a section across the bone Haversian canals; age related recrystallization has occurred. Courtesy of: Theya Molleson, Barbara Cressey and Gordon Cressey.

2.3.3 Other phosphates of interest

2.3.3.1 Base metal phosphates

A key feature of many base metal phosphates is their low solubility, by far lower than their metal carbonate or oxide analogues (Nriagu, 1984b). The significance of this is that the availability of their component metal ions, which may be toxic at very low concentrations (e.g. Pb^{2+}), can be controlled at low levels, either naturally or by human intervention, by the presence (or induced formation) of metal phosphates. Low solubility uranium phosphates are also known (Nriagu, 1984b; Vieillard and Tardy, 1984), suggesting the potential for phosphates to form sparingly soluble matrices with aqueous radioactive waste. It has also been predicted that Ni and Cd should form phosphates with analogous compositions to calcium and zinc respectively, and that many calcium and aluminium phosphates contain significant amounts of these two metals (Nriagu, 1984b).

Table 2.5 Solubility constants for base metal phosphates; a selection of data from Nriagu (1974, 1984b).

Mineral name	Chemical formula	log K_{sp}
Plumbogummite	$PbAl_3(PO_4)_2(OH)_5 \cdot H_2O$	−99.3
Pyromorphite	$Pb_5(PO_4)_3Cl$	−84.4
Parsonsite	$Pb_2UO_2(PO_4)_2 \cdot 2H_2O$	−45.8
Tsumebite	$Pb_2Cu(PO_4)(OH)_3 \cdot 3H_2O$	−51.3
Dumontite	$Pb_2(UO_2)_2(PO_4)_2(OH)_4 \cdot 3H_2O$	−91.4
Corkite	$PbFe_3(PO_4)(OH)_6 \cdot SO_4$	−112.6
Hinsdalite	$PbAl_3(PO_4)(OH)_6 \cdot SO_4$	−99.1
Dewindtite	$Pb_2(UO_2)_4(PO_4)_2(OH)_4 \cdot 8H_2O$	−92.6
Renardite	$Pb_2(UO_2)_4(PO_4)_2(OH)_4 \cdot 7H_2O$	−93.7
Przhevalskite	$Pb_2(UO_2)_2(PO_4)_2 \cdot 4H_2O$	−47.4
Cornetite	$Cu_3(PO_4)(OH)_3$	−48.0
Libethenite	$Cu_2(PO_4)(OH)$	−28.0
Torbernite	$Cu(UO_2)_2(PO_4)_2 \cdot 10H_2O$	−41.0
Metatorbernite	$Cu(UO_2)_2(PO_4)_2 \cdot 8H_2O$	−41.3
Nissonite	$CuMg(PO_4)(OH) \cdot 2.5H_2O$	−23.6
Pseudomalachite	$Cu_5(PO_4)_2(OH)_4$	−75.8
Tagilite	$Cu_2(PO_4)(OH) \cdot H_2O$	−27.9
Pyromorphite	$Cu_5(PO_4)_3(OH)$	−65.6
Turquoise	$CuAl_6(PO_4)_4(OH)_8 \cdot 4H_2O$	−179.0
Chalcosiderite	$CuFe_6(PO_4)_4(OH)_8 \cdot 4H_2O$	−205.7
Veszelyite	$CuZn_2PO_4(OH)_3 \cdot 2H_2O$	−45.8
Hopeite	$Zn_3(PO_4)_2 \cdot 4H_2O$	−35.3
Phosphophyllite	$Zn_2Fe(PO_4)_2 \cdot 4H_2O$	−35.8
Scholzite	$Zn_2Ca(PO_4)_2 \cdot 2H_2O$	−34.1
Spencerite	$Zn_4(PO_4)_2(OH)_2 \cdot 3H_2O$	−52.8
Tarbuttite	$Zn_2(PO_4)(OH)$	−26.6
Zinc rockbridgeite	$ZnFe_4(PO_4)_3(OH)_5$	−138.6
Zinc pyromorphite	$Zn_5(PO_4)_3(OH)$	−63.1
Faustite	$ZnAl_6(PO_4)_4(OH)_8 \cdot 4H_2O$	−177.7

There is a surprising paucity of data on solubility of metal phosphates since the publication of the compilation by Nriagu (1984b), a selection of data from which are shown in Table 2.5.

2.3.3.2 Vivianite

Vivianite ($Fe_3(PO_4)_2 \cdot 8H_2O$), a ferrous iron phosphate, is a monoclinic mineral, consisting of single and double octahedral iron oxide and water groups bonded by phosphate groups to form sheets (Kostov, 1968). It is colourless when fresh, but darkening to shades of green and blue upon oxidation (Nriagu, 1984a). It forms authigenically in a variety of anoxic environments, such as river water sediments and waste waters (e.g. House, 2003), and is likely to be the phase that controls phosphorus availability in such environments. It is also found as secondary mineral in ore veins, as an alteration product of primary phosphates and as a primary phase in sedimentary deposits (Nriagu, 1984a). Vivianite is relatively insoluble with a reported solubility constant of log $K_{sp} = -35.767$. However, it oxidizes easily in air, and transforms to iron oxides releasing phosphate (Roldán et al., 2002). Its precipitation can be induced by iron reducing bacteria (e.g. Dong et al., 2000).

2.3.3.3 Struvite

Struvite ($Mg(NH_4)(PO_4)\cdot 6H_2O$), a magnesium ammonium phosphate, is an orthorhombic mineral, colourless, becoming white on dehydration (Nriagu, 1984a). Its main natural occurrence is in guano deposits, but it is a relatively common biomineral (see above) and is also found as a scale in waste water treatment systems. It can also precipitate via bacterial activity (e.g. Ben Omar et al., 1998). A solubility constant of $K_{sp} = 2.2 \times 10^{-13}$ has been reported (Aage et al., 1997). Despite numerous studies of experimental and actual struvite precipitates from either biological or engineered systems, there is a paucity of data regarding thermodynamics and kinetics of its formation. It has been established that the presence of aqueous calcium inhibits struvite crystal growth (LeGeros and LeGeros, 1984) and that its precipitation rates are dependent on solution supersaturation via a second-order equation, suggesting a surface diffusion control (Bouropoulos and Koutsoukos, 2000). Regarding struvite precipitation inhibition, EDTA has been found to be the most effective from a variety of chelating agents (Doyle et al., 2003). Struvite has been proposed as a potential candidate for recovery of phosphorus, via precipitation from waste waters (see Battistoni; Jaffer and Pearce; and Ueno, all this volume).

2.3.3.4 Monazites, xenotime and other REE (rare earth element) phosphates

A number of minerals exist with the structure $X(PO_4)$ where X may be a rare earth element (X = La, Ce, Nd, ...) to form the mineral monazite, or yttrium, to form the mineral xenotime. In general, anhydrous REE orthophosphates can have a monoclinic monazite structure, containing the lighter rare earths, or a tetragonal xenotime structure, containing the heavier rare earths (Boatner, 2002). Other ions of interest that may be incorporated in monazites are uranium and thorium (Boatner, 2002). A number of substitutions occur in both minerals, both in the cation site where lanthanides and actinides substitute, and the phosphate site, which can be occupied by arsenate, vanadate, or silicate. The monazite mineral structure involves nine-fold coordinated rare earth ions, linked by phosphate units to form chains, whereas the xenotime structure involves eight-fold coordinated rare earth ions linked with phosphates to form dodecahedra, which then link into chains (Huminicki and Hawthorn, 2002).

In terms of occurrences, both monazite and xenotime are accessory phases in igneous rocks, and monazite, being particularly resistant to weathering, can accumulate as a detrital mineral in placer deposits and beach sands (Nriagu, 1984a). A number of industrial applications for rare earth orthophosphates exist, including host minerals of radioactive waste (Ewing and Wang, 2002), scintillators for the detection of gamma rays, weak interfaces in ceramic composites and others (for a recent review of applications see Boatner, 2002).

REFERENCES

Aage, H.K., Andersen, B.L., Blom, A. and Jensen, I. (1997) The solubility of struvite. *J. Radioanal. Nucl. Ch.* **223**, 213–215.

Agbenin, J.O. and Tiessen, H. (1994) Phosphorus transformations in a toposequence of lithosols and cambisols from semi-arid northeastern brazil. *Geoderma* **62**, 345–362.

Alborno, A. and Tomson, M.B. (1994) The temperature-dependence of the solubility product constant of vivianite. *Geochim. Cosmochim. Acta* **58**, 5373–5378.

Anderson, L.A. and Sarmiento, J.L. (1994) Redfield ratios of remineralization determined by nutrient data-analysis. *Global Biogeochem. Cycles* **8**, 65–80.

Bacquet, G., Truong, V.Q., Vignoles, M. and Bonel, G. (1981) Electron-paramagnetic resonance detection of acetate ions trapping in B-type carbonated fluorapatites. *J. Solid State Chem.* **39**, 148–153.

Ben Omar, N., Gonzalez-Munoz, M.T. and Penalver, J.M.A. (1998) Struvite crystallization on *Myxococcus* cells. *Chemosphere* **36**, 75–481.

Berner, R.A., Ruttenberg, K.C., Ingall, E.D. and Rao, J.L. (1993) The nature of phosphorus barial in modern marine sediments. In *Interactions of C, N, P and S Biochemical Cycles and Global Change* (eds Wollast, R., Mackenzie, F.T. and Chou, L.), NATO ASI Series, 14, Springer, Berlin, pp. 365–378.

Billen, G., Lancelot, C., and Meybeck, M. (1991) N, P, and Si retention along the aquatic continuum from land to ocean. In *Ocean Margin Processes in Global Change* (eds Mantoura, R.F.C., Martin, J.M. and Wollast, R.), Wiley & Sons, Chichester. pp. 19–44.

Bjerrum, C.J. and Canfield, D.E. (2002) Ocean productivity before about 1.9 Gyr ago limited by phosphorus adsorption onto iron oxides. *Nature* **417**, 159–162.

Boatner, L.A. (2002) Synthesis, structure, properties of monazite, pretulite and xenotime. *Rev. Mineral. Geochem.* **48**, 87–121.

Bonel, G. and Montel, G. (1964) Sur une nouvelle apatite carbonatée synthétique. *Compt. Rend. Acad. Sci.* **258**, 923–926.

Bouropoulos, N.C. and Koutsoukos, P.G. (2000) Spontaneous precipitation of struvite from aqueous solutions. *J. Cryst. Growth* **213**, 381–388.

Broecker, W.S. and Peng, T.-H. (1982) *Tracers in the sea*. Eldigio Press, Palisades, New York.

Buchwald, V.F. (1984) Phosphate minerals in meteorites and lunar rocks. In *Phosphate Minerals* (eds Nriagu, J.O. and Moore, P.B.), Springer-Verlag, London, pp. 199–214.

Buckman, H.O. and Brady, N.C. (1972) *The Nature and Properties of Soils*, 7th edn. Published by Macmillan, New York.

Caraco, N.F. (1993) Disturbance of the phosphorus cycle: A case of indirect effects of human activity. *Trends in Ecol. and Evol.* **7**, 51–54.

Clark, A.M. (1993) Hey's mineral index. Mineral species, varieties and synonyms, 3rd edn. Chapman and Hall, London, 852 pp.

Compton, J., Mallinson, D., Glenn, C.R., Filippelli, G., Föllmi, K., Shields, G. and Zanin, Y. (2000) Variations in the global phosphorus cycle. In *Marine Authigenesis: From Global to Microbial. SEPM Special Publication* **66**, 21–33.

Cook, P.J. (1984) Spatial and temporal controls on the formation of phosphate deposits – a review. In *Phosphate Minerals* (eds Nriagu, J.O. and Moore, P.B.), Springer-Verlag, London, pp. 242–274.

Corbridge, D.E.C. (2000) *Phosphorus 2000, Chemistry, Biochemistry and Technology. Elsevier,* 1258 pp.

Dallemagne, M.J. and Richelle, L.J. (1973) Inorganic chemistry of bone. In *Biological Mineralization* (ed. Zipkin, I.), Wiley, New York, pp. 23–42.

Dong, H.L., Fredrickson, J.K., Kennedy, D.W., Zachara, J.M., Kukkadapu, R.K. and Onstott, T.C. (2000) Mineral transformation associated with the microbial reduction of magnetite. *Chem. Geol.* **169**, 299–318.

Doyle, J.D., Oldring, K., Churchley, J., Price, C. and Parsons, S. (2003) Chemical control of struvite precipitation. *J. Environ. Eng.-ASCE* **129**, 419–426.

Duce, R.A., Liss, P.S., Merrill, J.T., Atlans, E.L., Buat-Menard, P. Hicks, B.B., Miller, J.M., Prospero, J.M., Atimoto, R., Church, T.M., Ellis, W., Galloway, J.N., Hansen, L., Jickells, T.D., Knap, A.H., Reinhardt, K.H., Schneider, B., Soudine, A., Tokos, J.J., Tsunogai, S., Wollast, R. and Zhou, M. (1991) The atmospheric input of trace species to the world ocean. *Global Biogeochem. Cycles* **5**, 193–259.

Elliott, J.C. (1994) *Structure and chemistry of the apatites and other calcium orthophosphates.* Elsevier, Amsterdam, 389 pp.

Elliott, J.C. (1997) Structure, crystal chemistry and density of enamel apatites. In *Dental Enamel* (eds Chadwick, D. and Cardew G.), Wiley and Sons, Chichester, pp. 54–67.

Elliott, J.C. (2002) Calcium phosphate biominerals. *Rev. Mineral. Geochem.* **48**, 427–453.

Ewing, R.C. and Wang, L.M. (2002) Phosphates as nuclear waste forms. *Rev. Mineral. Geochem.* **48**, 673–699.

Falkowski, P.G. (1997) Evolution of the nitrogen cycle and its influence on the biological sequestration of CO_2 in the ocean. *Nature* **387**, 272–275.

Feely, R.A., Massoth, G.J., Baker, E.T., Cowen, J.P., Lamb, M.F. and Krogslund, K.A. (1990) The effect of hydrothermal processes on midwater phosphorus distributions in the Northeast Pacific. *Earth Planet. Sci. Lett.* **96**, 305–318.

Filippelli, G.M. (2002) The global phosphorus cycle. *Rev. Mineral. Geochem.* **48**, 391–425.

Filippelli, G.M. and Delaney, M.L. (1996) Phosphorus geochemistry of equatorial Pacific sediments. *Geochim. Cosmochim. Acta* **60**, 1479–1495.

Föllmi, K.B. (1996) The phosphorus cycle, phosphogenesis and marine phosphate-rich deposits. *Earth-Sci. Rev.* **40**, 55–124.

Föllmi, K.B., Weissert, H. and Lini, A. (1993) Nonlinearities in phosphogenesis and phosphorus-carbon coupling and their implications for global change. In *Interactions of C, N, P and S Biochemical Cycles and Global Change* (eds Wollast, R., Mackenzie, F.T. and Chou, L.), NATO ASI Series, 14, Springer, Berlin, pp. 447–474.

Forsgren, G. and Jansson, M. (1992) The turnover of river-transported iron, phosphorus, and organic carbon in the Ore estuary, northern Sweden. *Hydrobiologia* **235/236**, 585–596.

Fox, L.E. (1989) A model for inorganic control of phosphate concentrations in river waters. *Geochim. Cosmochim. Acta* **53**, 417–428.

Froelich, P.N. (1988) Kinetic control of dissolved phosphate in natural rivers and estuaries: a primer on the phosphate buffer mechanism. *Limnol. Oceanogr.* **33**, 649–668.

Froelich, P.N., Bender, M.L., Luedtke, N.A., Heath, G.R. and Devries, T. (1982) The marine phosphorus cycle. *Amer. J. Sci.* **282**, 474–511.

Frossard, E., Brossard, M., Hedley, M.J. and Metherell, A. (1995) Reactions controlling the cycling of P in soils. In *SCOPE 54 – Phosphorus in the Global Environment – Transfers, Cycles and Management* (ed. Tiessen, H.), Wiley, J. and Sons Ltd., Chichester, pp. 107–137.

Geider, R.J. and La Roche, J. (2002) Redfield revisited: variability of C:N:P in marine microalgae and its biochemical basis. *Eur. J. Phycol.* **37**, 1–17.

Grobbelaar, J.U. (1983) Availability to algae of N and P adsorbed on suspended solids in turbid waters of the Amazon River. *Arch. Hydrobiol.* **96**, 302–316.

Grobbelaar, J.U. and House, W.A. (1995) Phosphorus as a limiting resource in inland waters; interactions with nitrogen. In *SCOPE 54 – Phosphorus in the Global Environment – Transfers, Cycles and Management* (ed. Tiessen, H.), Wiley, J. and Sons Ltd., Chichester, pp. 255–273.

Gross, K.A. and Berndt, C.C. (2002) Biomedical applications of apatites. *Rev. Mineral. Geochem.* **48**, 631–672.

Guidry, M.W. and Mackenzie, F.T. (2000) Apatite weathering and the Phanerozoic phosphorus cycle. *Geology* **28**, 631–634.

Henderson, P. (1982) *Inorganic Geochemistry*. Pergamon Press, Oxford, 353 pp.

Heughebaert, J.C., Zawacki, S.J. and Nancollas, G.H. (1990) The growth of nonstoichiometric apatite from aqueous solution at 37°C. I. Methodology and growth at pH 7.4. *J. Colloid Interf. Sci.* **135**, 20–32.

Hoche, T., Moisescu, C., Avramov, I., Russel, C. and Heerdegen, W.D. (2001) Microstructure of SiO_2-Al_2O_3-CaO-P_2O_5-K_2O-F-glass ceramics. 1. Needlelike versus isometric morphology of apatite crystals. *Chem. Mater.* **13**, 1312–1319.

Hochella, M.F. (2002) Nanoscience and technology: the next revolution in the Earth sciences. *Earth Planet. Sci. Lett.* **203**, 593–605.

House, W.A. (2003) Geochemical cycling of phosphorus in rivers. *Appl. Geochem.* **18**, 739–748.

House, W.A. and Denison, F.H. (2002). Total phosphorus content of river sediments in relationship to calcium, iron and organic matter concentrations. *Sci. Total Environ.* **282–283**, 341–351.

Howarth, R.W., Jensen, H.S., Marino, R. and Postma, H. (1995) Transport to and processing of P in near-shore and oceanic waters. In *SCOPE 54 – Phosphorus in the Global Environment – Transfers, Cycles and Management* (ed. Tiessen, H.), Wiley, J. and Sons Ltd., Chichester, pp. 323–345.

Hsu, P.H. (1982a) Crystallization of variscite at room temperature. *Soil Sci.* **133**, 305–313.

Hsu, P.H. (1982b) Crystallization of iron(III) phosphate at room temperature. *Soil Sci. Soc. Am. J.* **46**, 928–932.

Hughes, J.M. and Rakovan, J. (2002) The crystal structure of apatite, $Ca_5(PO_4)_3(F,OH,Cl)$. *Rev. Mineral. Geochem.* **48**, 1–12.

Huminicki, D.M.C. and Hawthorn, F.C. (2002) The crystal chemistry of the phosphate minerals. *Rev. Mineral. Geochem.* **48**, 123–253.

Ingall, E.D. and Van Cappellen, P. (1990) Relation between sedimentation-rate and burial of organic phosphorus and organic-carbon in marine-sediments. *Geochim. Cosmochim. Acta* **54**, 373–386.

Jonasson, R.G., Martin, R.R., Giuliacci, M.E. and Tazaki, K. (1988) Surface reactions of goethite with phosphate. *J. Chem. Soc. Faraday Trans. 1.* **84**, 2311–2315.

Jones, F.H. (2001) Teeth and bones: applications of surface science to dental materials and related biomaterials. *Surface Sci. Reports* **42**, 75–205.

Knudsen, A.C. and Gunter, M.E. (2002) Sedimentary phosphorites – an example: Phosphoria Formation, Southeastern Idaho, U.S.A. *Rev. Mineral. Geochem.* **48**, 363–389.

Koelmans, A.A., Van der Heijde, A., Knijff, L.M. and Aalderink, R.H. (2001) Integrated modelling of eutrophication and organic contaminant fate & effects in aquatic ecosystems. A review. *Water Res.* **35**, 3517–3536.

Kostov, I. (1968) *Mineralogy*. Oliver and Boyd, Edinburgh.
Kroehler, C.J., Antibus, R.K. and Linkins, A.E. (1988) The effects of organic and inorganic phosphorus concentration on the acid-phosphatase activity of ectomycorrhizal fungi. *Can. J. Bot.* **66**, 750–756.
Kühl, G. von and Nebergall, W.H. (1963) Hydrogenphosphat- und Carbonatapatite. *Z. Anorg. Allg. Chem.* **324**, 313–320.
Lebo, M.E. (1991) Particle-bound phosphorus along an urbanized coastal plain estuary. *Mar. Chem.* **34**, 225–246.
LeGeros, R.Z. and LeGeros, J.P. (1984) Phosphate minerals in human tissues. In *Phosphate minerals* (eds Nriagu, J.O. and Moore, P.B.), Springer Verlag, Berlin Heidelberg, pp. 351–385.
Lindsay, W.L., Vlek, P.L.G. and Chien, S.H. (1989) Phosphate minerals. In *Minerals in soil environment* (eds Dixon, J.B. and Weed, S.B.), (2nd edn) SSSA Monograph no. Published by SSSA, Madison, Wisconsin, pp. 1089–1130.
MacKay, D.D.S. and Charnley, S.B. (2001) Phosphorus in circumstellar envelopes. *Mon. Not. R. Astron. Soc.* **325**, 545–549.
Mackenzie, F.T., Ver, L.M., Sabine, C., Lane, M. and Lerman, A. (1993) C, N, P, S global biogeochemical cycles and modelling of global change. In *Interactions of C, N, P and S biogeochemical cycles and global change* (eds Wollast, R., Mackenzie, F.T. and Chou, L.), Srpinger-Verlag, Berlin, pp. 1–62.
Mackenzie, F.T., Ver, L.M. and Lerman, A. (2002) Century-scale nitrogen and phosphorus controls of the carbon cycle. *Chem. Geol.* **190**, 13–32.
Margules, L., Herbst, E., Ahrens, V., Lewen, F., Winnewisser, G. and Muller, H.S.P. (2002) The phosphidogen radical, PH2: Terahertz spectrum and detectability in space. *J. Mol. Spectrosc.* **211**, 211–220.
Marino, R., Chan, F., Howarth, R.W., Pace, M. and Likens, G.E. (2002) Ecological and biogeochemical interactions constrain planktonic nitrogen fixation in estuaries. *Ecosystems* **5**, 719–725.
Martin, R.R., Smart, R.St.C. and Tazaki, K. (1988) Direct observation of phosphate precipitation in the goethite/phosphate system. *Soil Sci. Soc. Am. J.* **52**, 1492–1500.
Mautner, M.N. and Sinaj, S. (2002) Water-extractable and exchangeable phosphate in Martian and carbonaceous chondrite meteorites and in planetary soil analogs. *Geochim. Cosmochim. Acta* **66**, 3161–3174.
McCulloch, M., Pailles, C., Moody, P. and Martin, C.E. (2003) Tracing the source of sediment and phosphorus into the Great Barrier Reef lagoon. *Earth Planet. Sci. Lett.* **210**, 249–258.
Meybeck, M. (1982) Carbon, nitrogen, and phosphorus transport by world rivers. *Am. J. Sci.* **282**, 401–450.
Meybeck, M. (1993) C, N, P and S in rivers: from sources to global inputs. In *Interactions of C, N, P and S Biochemical Cycles and Global Change* (eds Wollast, R., Mackenzie, F.T. and Chou, L.), NATO ASI Series, 14, Springer, Berlin, pp. 163–193.
Moore, P.B. (1984) Crystallochemical aspects of the phosphate minerals. In *Phosphate Minerals* (eds Nriagu, J.O. and Moore, P.B.), Springer Verlag, Berlin Heidelberg, pp. 155–170.
Muscutt, A.D. and Wither, P.J.A. (1996) The phosphorus content of rivers in England and Wales. *Water Res.* **30**, 1258–1268.
Nash, W.P. (1984) Phosphate minerals in terrestrial igneous and metamorphic rocks. In *Phosphate Minerals* (eds Nriagu, J.O. and Moore, P.B.), Springer Verlag, Berlin Heidelberg, pp. 215–241.
Neyroud, J.A. and Lischer, P. (2003) Do different methods used to estimate soil phosphours availability accross Europe give comparable results? *J Plant Nutr. Soil Sc.* **166**(4), pp. 422–431

Nathan, Y. (1984) The mineralogy and geochemistry of phosphorites. In *Phosphate Minerals* (eds Nriagu, J.O. and Moore, P.B.), Springer Verlag, Berlin Heidelberg, pp. 275–291.

Nriagu, J.O. (1974) Lead Orthophosphates. IV – Formation and stability in the environment. *Geochim. Cosmochim. Acta* **38**, 887–898.

Nriagu, J.O. (1984a) Phosphate minerals: their properties and general modes of occurrence. In *Phosphate Minerals* (eds Nriagu, J.O. and Moore, P.B.), Springer Verlag, Berlin Heidelberg, pp. 1–136.

Nriagu, J.O. (1984b) Formation and stability of base metal phosphates in soils and sediments. In *Phosphate Minerals* (eds Nriagu, J.O. and Moore, P.B.), Springer Verlag, Berlin Heidelberg, pp. 318–329.

Nurnberg, G.K., Shaw, M., Dillon, P.J. and McQueen, D.J. (1986) Internal phosphorus load in an oligotrophic Precambrian shield lake with an anoxic hypolimnion. *Can. J. Fish Aquat. Sci.* **43**, 574–580.

Owens, P.N. and Walling, D.E. (2002) The phosphorus content of fluvial sediment in rural and industrialized river basins. *Water Res.* **36**, 685–701.

Pan, Y. and Fleet, M.E. (2002) Compositions of the apatite-group minerals: substitution mechanisms and controlling factors. *Rev. Mineral. Geochem.* **48**, 13–49.

Papelis, G., Hayes, K.F. and Leckie, J.O. (1988) HYDRAQL: A program for the computation of chemical equilibrium composition of aqueous batch system including surface complexation modeling of ion association at the oxide/solution interface. *Technical Report No. 306*, Stanford University, Stanford, CA.

Pardo, P., Rauret, G., Lopez-Sanchez, J.F. (2003) Analytical approaches to the determination of phosphorus partitioning patterns in sediments. *J. Environ. Monitor.* **5**, 312–318.

Parfitt, R.L. (1989) Phosphate reactions with natural allophane, ferrihydrite and goethite. *J. Soil Sci.* **40**, 359–369.

Pasteris, J.D., Wopenka, B., Freeman, J.J., Rogers, K., Valsami-Jones, E., van der Houwen, J.A.M. and Silva, M.J. (2004) Lack of OH in extremely nanocrystalline 'hyhdroxylapatite': implications for bone properties. *Biomaterials* **25**(2), 229–238.

Piccoli, P.M. and Candela, P.A. (2002) Apatite in igneous systems. *Rev. Mineral. Geochem.* **48**, 255–292.

Pierzynski, G.M., Logan, T.J., Traina, S.J. and Bigham, J.M. (1990) Phosphorus chemistry and mineralogy in excessively fertilized soils: descriptions of phosphorus-rich particles. *Soil Sci. Soc. Am. J.* **54**, 1583–1589.

Posner, A.S. and Perloff, A. (1957) Apatites deficient in divalent cations. *J. Res. Nat. Bur. Stds.* **58**, 279–286.

Posner, A.S., Blumenthal, N.C. and Betts, F. (1984) Chemistry and structure of precipitated hydroxyapatites. In *Phosphate Minerals* (eds Nriagu, J.O. and Moore, P.B.), Springer Verlag, Berlin Heidelberg, pp. 330–350.

Posner, A.S., Stutman, J.M. and Lippincott, E.R. (1960) Hydrogen-bonding in calcium-deficient hydroxyapatite. *Nature* **188**, 486–487.

Rao, J.L. and Berner, R.A. (1995) Development of an electron-microprobe method for the determination of phosphorus and associated elements in sediments. *Chem. Geol.* **125**, 169–183.

Ramirez, A.J. and Rose, A.W. (1992) Analytical geochemistry of organic phosphorus and its correlation with organic carbon in marine and fluvial sediments and soils. *Am. J. Sci.* **292**, 421–454.

Redfield, A.C. (1958) The biological control of chemical factors in the environment. *Am. Sci.* **46**, 205–221.

Redfield, A.C., Ketchum, B.H. and Richards, F.A. (1963) The influence of organisms on the composition of sea water. In *The Sea, II* (ed. Hill, M.N.), Interscience, New York, pp. 26–77.

Rey, C., Trombe, J.-C. and Montel, G. (1978) Sur la fixation de la glycine dans le réseau des phosphates à structure d'apatite. *J. Chem. Res.* **188**, 2401–2416.

Roldán, R. Barrón, V. and Torrent, J. (2002) Experimental alteration of vivianite to lepidocrocite in a calcareous medium. *Clay Min.* **37**, 709–718.

Ruttenberg, K.C. (1992) Development of a sequential extraction method for different forms of phosphorus in marine sediments. *Limnol. Oceanogr.* **37**, 1460–1482.

Ruttenberg, K.C. (1993) Reassessment of the oceanic residence time of phosphorus. *Chem. Geol.* **107**, 405–409.

Ruttenberg, K.C. and Berner, R.A. (1993) Authigenic apatite formation and burial in sediments from non-upwelling, continental margin environments. *Geochim. Cosmochim. Acta* **57**, 991–1007.

Salcedo, I.H. and Madeiros, C. (1995) Phosphorus transfer from tropical terrestrial to aquatic systems – mangroves. In *SCOPE 54 – Phosphorus in the Global Environment – Transfers, Cycles and Management,* (ed. Tiessen, H.), Wiley, J. and Sons Ltd., Chichester, pp. 347–362.

Salingar, Y., Geifman, Y. and Aronowich, M. (1993) Orthophosphate and calcium carbonate solubilities in the Upper Jordan watershed basin. *J. Environ. Qual.* **22**, 672–677.

Sharpley, A. (2000) Phosphorus availability. In *Handbook of Soil Science* (ed. Sumner, M.E.), CRC Press, D18–D38.

Sharpley, A., Foy, B. and Withers, P. (2000) Practical and innovative measures for the control of agricultural phosphorus losses to water: An overview. *J. Environ. Qual.* **29**, 1–9.

Skinner, H.C.W. (2000a) In praise of phosphates, or why vertebrates chose apatite to mineralise their skeletal elements. *Internat. Geol. Rev.* **42**, 232–240.

Skinner, H.C.W. (2000b) Minerals and human health. In *Environmental Mineralogy* (eds Vaughan, D.J. and Wogelius, R.A.), EMU Notes in Mineralogy, 2, Eötvös University Press, Budapest, pp. 383–412.

Stewart, J.W.B. and Sharpley, A.N. (1987) Controls on dynamics of soil and fertilizer phosphorus and sulfur. In *Soil fertility and organic matter as critical components of production* (eds Follet, R.F. *et al.*,), SSSA Spec. Pub. 19, Am. Soc. Agron., Madison, Wisconsin, pp. 101–121.

Stewart, J.W.B. and Tiesen, H. (1987) Dynamics of soil organic phosphorus. *Biogeochemistry* **4**, 41–60.

Tappin, A.D. (2002) An examination of the fluxes of nitrogen and phosphorus in temperate and tropical estuaries: Current estimates and uncertainties. *Estuar. Coast. Shelf. S.* **55**, 885–901.

Trombe, J.-C. (1973) Contribution à l'étude de la decomposition et de la réactivité de certaines apatites hydroxylées et carbonatées *Ann. Chim. (Paris) 14 series* **8**, 335–347.

Turner *et al.* (2002) Inositol phosphates in the envrironment. *Phil. Trans. R. Soc. Lond. B* **357**, 449–469.

Van Cappellen, P. and Ingall, E.D. (1996) Redox stabilization of the atmosphere and oceans by phosphorus-limited marine productivity. *Science* **271**, 493–496.

Van Riemsdjik, W.H., Weststrate, F.A. and Bolt, G.H. (1975) Evidence for a new aluminum phosphate phase from reaction rate of phosphate with aluminum hydroxide. *Nature* **257**, 473–474.

Vieillard, P. and Tardy, Y. (1984) Thermochemical properties of phosphates. In *Phosphate Minerals* (eds Nriagu, J.O. and Moore, P.B.), Springer Verlag, Berlin Heidelberg, pp. 171–198.

Vuorinen, A.H. and Saharinen, M.H. (1996) Effects of soil organic matter extracted from soil on acid phosphomonoesterase. *Soil Biol. Biochem.* **28**, 1477–1481.

Walker, T.W. and Syers, J.K. (1976) The fate of phosphorus during pedogenesis. *Geoderma* **15**, 1–19.

Watson, C.J. and Foy, R.H. (2001) Environmental impacts of nitrogen and phosphorus cycling in grassland systems. *Outlook Agr.* **30**, 117–127.

Westerman, R.L. (1990) *Soil testing and plant analysis*, 3rd edn. Soil Science Society of America book series No. 3. Madison, WI USA, 784 pp.

Wilkins, G.R., Delcourt, P.A., Delcourt, H.R., Harrison, F.W. and Turner, M.R. (1991) Paleoecology of central Kentucky since the last glacial maximum. *Quat. Res.* **36**, 224–239.

Williams, J.D.H., Murphy, T.P. and Meyer, T. (1976) Rates of accumulation of phosphorus forms in Lake Erie sediments. *J. Fisheries Res. Board Canada* **33**, 430–439.

Zalba, P. and Peinemann, N. (2002) Phosphorus content in soil in relation to fulvic acid carbon fraction. *Commun. Soil. Sci. Plan* **33**, 3737–3744.

3
The biology of phosphorus

C. Dobrota

3.1 INTRODUCTION

Living systems require a continuous flow of matter and energy. Matter is obtained from the environment and it is used to build new molecules, to replace old structures, to maintain appropriate concentrations of metabolites, to assemble molecules and to perform biological work. Energy is drawn from four physiological processes: photosynthesis, glycolysis, cellular respiration and fermentation. These are the biochemical and biophysical processes that initiate and maintain life.

Inorganic phosphate (P_i) is one of the key substrates in energy metabolism, biosynthesis of nucleic acids and membranes. Metabolic functions of inorganic phosphate include animal bone formation, buffering action in blood and urine, enzyme action, signal transduction, energy storage, muscle contraction and brain function. In most biochemical reactions organic phosphate esters are involved as energy carriers, as coenzymes or as intermediates.

3.2 PHOSPHATE ACQUISITION

The mechanisms of phosphorus (P) acquisition and its occurrence vary for different organisms. For plants, aqueous phosphate is taken up via the plant's root system.

© 2004 IWA Publishing. *Phosphorus in Environmental Technology: Principles and Applications.* Edited by Eugenia Valsami-Jones. ISBN: 1 84339 001 9

Plants commonly contain 0.05 to 0.5% of phosphorus on a dry matter basis during the vegetative stage of growth (Raghothama, 1999). It is incorporated into a variety of organic compounds, including sugar phosphates, phospholipids, phytin, coenzymes and nucleotides. Phosphorus may also be present, as inorganic orthophosphate, especially in the vacuoles.

Plants do not reduce phosphate and the ion enters into organic molecules largely unaltered. For example, phytin is the calcium–magnesium salt of phytic acid, the latter term being synonymous with inositol hexaphosphate ($C_6O_{24}P_6$, IP-6). Inositol phosphates are formed enzymatically and may have one to six phosphorus atoms per inositol unit. Inositol hexaphosphate is a naturally occurring component of plant fibre that may possess antioxidant, anticancer, and other beneficial properties.

Phosphorus is absorbed by animals from food as HPO_4^{2-} or $H_2PO_4^{-}$. From the small intestine it passes to the blood and then to the bones and tissues. The human dietary requirement for phosphorus is about 1.3 g per day (Corbridge, 2000). Large amounts of phosphorus are found in milk, cheese, meat, peanuts, brain, chocolate and eggs. Phosphorus deficiency in adult humans known as hypophosphatemia is extremely rare. However, excess of phosphorus in the diet can lead to a shortage of calcium producing hypocalcemia symptoms. In animals, the kidneys are controlling the excretion rate and are regulating phosphate homeostasis in the blood.

Unlike plants and animals, microorganisms are capable of acquiring phosphorus by releasing it from solid inorganic phases, in addition to their ability for acquisition from other, more easily available, aqueous sources. In microorganisms, most phosphorus can be found in the RNA, which accounts for one-third to half of all phosphorus of the cell. Large quantities, ranging around 15–25% of their total phosphorus content, are found in acid-soluble compounds, both inorganic and organic. The acid-soluble fraction of microbial protoplasm contains ortho- and metaphosphate, sugar phosphates, coenzymes and adenosine phosphates.

3.3 TRANSPORT AND TRANSLOCATION

3.3.1 Plants

Plants are seldom exposed to an unrestricted supply of phosphate in nature. Uptake of phosphorus by plants, under non-limiting conditions, appears to be regulated by the internal levels of phosphate.

In general, higher plants store P in the vacuole. The cytoplasmic phosphate is maintained at a relatively constant level during short-term nutrient deficiency, or variable supply, at the expense of the vacuolar phosphate, which may decrease significantly (Raghothama, 1999). Movement of P from and to the vacuole, and regulated plasma membrane efflux and influx, are the primary mechanisms that maintain P homeostasis.

The most common form of P transported into plant cells through the plasma membrane is the dihydrogen form of the orthophosphate ion ($H_2PO_4^{-}$)

(Raghothama, 1999). Vacuoles apparently play the dual role of sink and source for P in plant cells. In response to changing concentration, P transport across the tonoplast is activated in order to maintain homeostasis in the cytoplasm. This transport is performed via an $H^+/H_2PO_4^-$ symporter, requires ATP (adenosine triphosphate, a phosphate ester involved in biological energy transfer, see section 3.4.1), and is associated with cytoplasmic alkalisation (Sakano, 1990).

The excess protons entering the cytoplasm during phosphate transport may lead to activation of a biochemical pH mechanism during and after uptake. The inward and outward flux of P across the tonoplast occurs at high concentrations (5–20 mM) in either the vacuole, or the cytoplasm, or both (Raghothama, 1999). The tonoplast H^+-ATPase or pyrophosphatase provides the required energy to maintain an electrochemical gradient across the tonoplast that facilitates P transport.

At plasmalemma level, phosphate absorption is accompanied by H^+ influx, with a stoichiometry of 1/2 to 1/4 of $H^+/H_2PO_4^-$ transported (Sakano, 1990). The number of protons increases as the concentration of phosphate in the media decreases. The required protons are provided by the media and by the activated proton pumps in the membrane. The $H^+/H_2PO_4^-$ co-transport causes the pH of the cytoplasm to decrease transiently. A temporary depolarisation of the membrane occurs, caused by the plasma membrane H^+-ATPase, which pumps protons from the cytoplasm to maintain cellular pH and to provide the proton force to drive continued P uptake.

Phosphate transport in plants exhibits distinctive uptake kinetic properties at low (of the order of µM) and high (of the order of mM) concentrations in the external media. High-affinity P transporters are membrane-associated proteins, which translocate P from the external media, containing µM concentrations, to the cytoplasm. Phosphate transporter genes are encoded by a small family of genes in the genome (Muchhal *et al.*, 1996). The P transporters are membrane proteins consisting of 12 membrane-spanning regions, separated into two groups of six by a large hydrophilic charged region, a common feature shared by many proteins involved in transport of sugars, ions, antibiotics and amino acids (Henderson, 1993). Plant P transporters are members of the subfamily 9 of the Major Facilitator Superfamily (MFS) (Pao *et al.*, 1998). The MFS represents one of the largest families of single polypeptide-facilitated carriers capable of transporting small solutes utilising chemo-osmotic ion gradients.

P-efflux is another mechanism documented as regulating P concentration in the cytoplasm (Cogliati and Santa Maria, 1990). At higher external phosphate concentrations, increased P-efflux almost compensates for the P-influx, thus supporting the idea that under non-limiting P availability, P homeostasis is primarily controlled by P-efflux.

The cell wall chemistry and pH in the vicinity of the plasma membrane also has a strong influence on the concentration of P (Clarkson and Grignon, 1991). Owing to interactions of the anion with cell walls, the actual concentration of P at the plasma membrane may be ten times lower than that of the external solution.

The synthesis of starch and sucrose are competing processes occuring in different cellular compartments of the plant cell. Starch synthesis and starch deposits are localised in the chloroplast. Sucrose, the principal carbohydrate transported throughout most plants, is synthesised in the cytosol. The partitioning of triose phosphate ($C_3H_5O_6P$) between starch synthesis in the chloroplast and sucrose synthesis in the cytosol is determined by the orthophosphate concentration in these cellular compartments. When the cytosolic phosphate concentration is high, chloroplast triose phosphate is exported to the cytosol in exchange for phosphate, and sucrose is synthesised. When the cytosolic phosphate concentration is low, triose phosphate is retained within the chloroplast, and starch is synthesised. A balance between sucrose and starch synthesis can thus be achieved because phosphate is released into the cytosol whenever sucrose is synthesised.

Photosynthesis could be restricted when there is a lack of free orthophosphate in the chloroplast. If starch synthesis, which releases orthophosphate in the chloroplast, does not recycle phosphate fast enough, a deficiency of phosphate would result, and ATP synthesis and thus CO_2 fixation would decline. Under conditions of low demand by the sink, sucrose synthesis is usually reduced, and less phosphate is available for exchange with triose phosphate from the chloroplast (via the phosphate translocator).

Export of triose phosphate proceeds via the triose phosphate/phosphate translocator (TPT), located in the inner chloroplast membrane. The exported products are used for the biosynthesis of sucrose and amino acids, and the released phosphate is imported back into the chloroplasts via the TPT and used for the formation of ATP.

According to Wagner et al. (1989), TPT is a dimer composed of two identical subunits. As a substrate, TPT accepts inorganic phosphate or a phosphorylated triose. The exchange ratio is 1 : 1 under physiological conditions. Substrate is transported first across the membrane, and leaves the transport site before the second substrate can be bound and transported (Flügge, 1992). There is evidence that TPT has ion channel properties. It can behave as a voltage-dependent ion channel permeable to anions, as well as an antiporter (Flügge, 1999).

The shikimate pathway links metabolism of carbohydrates to biosynthesis of aromatic amino acids and many aromatic secondary metabolites. This pathway may be found only in microorganisms and plants, never in animals. The products of the shikimate pathways require phosphoenolpyruvate (PEP) as precursor, which is translocated by the PEP/phosphate translocator (PPT). PPT transports only PEP and inorganic phosphate, and accepts triose phosphates and 3-phosphoglycerate in very low amounts.

The import of hexose phosphates into non-green plastids can proceed via the glucose 6-phosphate/phosphate translocator (GPT). The phosphorylated hexoses imported from cytosol to plastids are used for starch and fatty acid biosynthesis. This translocator has a broad substrate specificity, transporting phosphate, phosphorylated triose and hexose phosphates (Heldt et al., 1991).

The overall similarities between the members of the TPT, PPT and GPT families are about 35% and are restricted in all translocator proteins to five regions, each of 15 to 30 amino acid residues in length. It may therefore be considered that there are three different classes of plastidic phosphate translocators (Kammerer et al., 1998).

Nucleotide phosphates and inorganic solutes containing phosphates and enzymes, such as protein kinases, are found in the phloem sap, the solution containing products of photosynthesis translocated via phloem. Phloem loading and unloading are important components of the internal movement of P in plants. There is genetic evidence for the involvement of P transporters in phloem loading of the nutrients.

Change in the cytosolic or vacuolar P can trigger a signal transduction pathway that activates P starvation rescue systems. At the organism level, the signalling mechanism involves the movement of P from old to young tissues, and from root to shoot and back to root.

Under phosphate deficiency, many biochemical changes occur. The induction of acid phosphatases in response to P starvation is a universal mechanism in higher plants (Duff et al., 1994). Phosphatases liberate P from organic materials. The levels of ATP and of the nucleotides are significantly reduced during starvation, and pyrophosphate, which is maintained at high levels, may function as an autonomous energy donor (Plaxton, 1996). Phosphorus limitation also results in activation of an alternate respiratory pathway (Rychter and Misulska, 1990) and in the decrease of the rate of photosynthesis and of stomatal conductance.

Transcriptional activation of P transporters in response to P starvation seems to be a major mechanism of regulation of P uptake. The spatial distribution of the P transporter gene expression indicates their function beyond the acquisition of P from soil. More specifically, some transporters may be involved in the intracellular movement or scavenging of P leaked into the apoplast. For example, tomato P transporter genes (LePT1 and LePT2) are expressed predominantly in the root epidermis and root hairs of P-deficient plants (Liu et al., 1998). The LePT1 message is also observed in the root cortical cells, palisade parenchyma and vascular bundles in the leaves.

3.3.2 Animals

In animals, inorganic ions, such as phosphorus, may serve as structural elements in the body, or provide a variety of functions such as nerve conduction, muscle contraction, act as cofactors in enzyme-catalysed reactions, etc. Phosphorus in the body of vertebrates may be found as calcium phosphate crystals in bones and teeth, forming the cement that contributes to the physical strength of these structures.

The main inorganic component of bone, dentine and tooth enamel is a form of the mineral apatite, often described as bioapatite (see also Valsami-Jones, this volume). The body of an average man weighing 70 kg contains about 780 g of P, of which about 700 g are present as bone apatite.

In bone, apatite occurs in the form of plate-shaped crystals associated with fibrous protein, collagen and bone cells, and embedded in the organic matrix. Other bone components containing P include: pyrophosphate, phosphoproteins, ATP and phospholipids. The Ca/P ratio in human bone is 1.50. Bone growth involves the crystallisation of apatite within the collagen fibres. A phosphoglycoprotein, named osteonectin is the site for the growth of apatite crystals, and for the anchorage to the collagen. The two organic molecules are linked via phosphate groups bound to the serine or tyrosine residues in the collagen or the osteonectin (Corbridge, 2000). Phospholipids are also found at the mineralisation front. Bone is in a constant state of remodeling, via the activity of osteoblasts, cells responsible for bone formation, and osteoclasts, cells causing bone resorption (removal).

Dentine contains about 72% apatite, 18% collagen and small quantities of phospholipids and other organic components. Unlike bone, dentine is not regenerated. Phosphoproteins called phosphoryns are present at the mineralisation front and have 90% phosphorylated serine residues. Dental enamel is harder and more rigid than dentine and serves to protect tooth. Enamel consists of 96% apatite and only a small proportion of organic residue (1%), which remains after the apatite crystals of the enamel are fully formed. A variety of phosphate minerals, such as octacalcium phosphate, brushite, whitlockite and struvite have been identified in pathological calcifications (dental calculus and urinary and goal stones; Corbridge, 2000; Elliott, 2002).

Muscle contains phosphoproteins, ATP and inorganic phosphate. Brain tissue contains phosphocreatine, ATP, sugar phosphates and phospholipids. Blood contains about 0.04% P (Corbridge, 2000). Red blood cells contain hemoglobin, ATP, 2,3-diphosphoglycerate and phospholipids but, unlike other body cells, have no DNA and RNA. The hemoglobin of the red blood cells contains a large amount of iron and P-containing compounds, ferritin and hemosiderin. Blood plasma contains most of the inorganic phosphate in blood as HPO_4^{2-} and $H_2PO_4^-$, which together with bicarbonate HCO_3^- act as buffering agents. The ratio $HPO_4^{2-}/H_2PO_4^-$ is 4:1. The concentration of both calcium and phosphate in the blood is controlled by the parathyroid hormone (PTH), which increases the absorption of calcium from the glomerular filtrate in the kidney and decreases the absorption of phosphate. PTH also increases the synthesis of 1,25-dihydroxycholecalciferol [1,25-$(OH)_2D_3$], the active form of vitamin D. This molecule stimulates release of calcium from bone and increases the active transport of calcium from the intestine to the blood. A rise in the extracellular calcium levels triggers the secretion of calcitonin by specialised cells found among the follicular cells of the thyroid gland. Calcitonin decreases the resorption of bone and increases the loss of both calcium and phosphate ions in the urine (Figure 3.1).

2,3-diphosphoglycerate (2,3-DPG) is an important regulator of the binding of oxygen to hemoglobin. It is the most abundant organic phosphate in the red blood cell, where it is found in concentrations roughly equivalent to that of hemoglobin. 2,3-DPG is synthesised from intermediates of glycolysis, and plays an important role in the normal transport function of hemoglobin. In the red blood cell, the presence of 2,3-DPG significantly reduces the affinity of hemoglobin for oxygen, which enables hemoglobin to release oxygen at the partial pressures needed by the tissues for oxidative metabolism. The level of 2,3-DPG increases in response to hypoxia such as that found in chronic anemia or in pulmonary emphysema. Elevated 2,3-DPG concentrations lower the oxygen affinity to hemoglobin, allowing greater unloading of oxygen in the capillaries of the tissues.

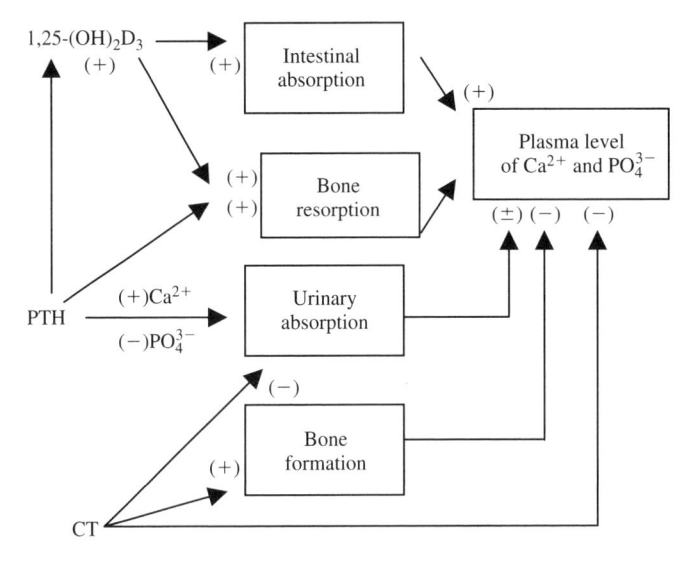

Figure 3.1 Homeostasis of PO_4^{3-} and Ca^{2+} in blood plasma. 1,25-$(OH)_2D_3$ – 1,25-dihydroxycholecalciferol, PTH – parathyroid hormone, CT – calcitonin, (+) – stimulation, (−) – inhibition.

3.3.3 Microorganisms

Phosphorus plays a fundamental role in microbial cell physiology and biochemistry, being a part of important constitutive biomolecules. This provides the opportunity for microorganisms to influence the phosphorus global cycle. More specifically, soil microorganisms are intimately involved in the cycling of soil P. They participate in the solubilization of inorganic P and the mineralization (i.e. transformation to inorganic) of organic P. They are important in the short-term biological immobilization of the available soil P, which protects some soil P from long-term fixation in minerals and may be beneficial to plants.

The main inorganic source of phosphorus on Earth is the mineral apatite (see also Valsami-Jones, this volume). Before phosphate from apatite can be made available to biological systems, this mineral must be dissolved, either chemically or via microbially induced mechanisms. Soluble phosphate in the soil and water is the principal source for autotrophic microorganisms, which fix it into organic molecules and pass it on to heterotrophs in this form.

The ability of microorganisms to solubilize inorganic phosphate can have major implications in agriculture and, for that reason, attempts have been made to isolate the specific bacteria responsible for solubilization, and to use such organisms as soil or seed inoculants. Most frequently isolated were species of *Pseudomonas* and *Bacillus*. In past years B. *megaterium* var. *phosphaticum* was widely used in a bacterial inoculant known as phosphobacterin (Paul and Clark, 1996). Thiobacilli are also known for their ability to solubilize insoluble rock phosphate in soil where there is a source of S, via the oxidation of S to H_2SO_4, which in turn partially dissolves the rock phosphate. It is considered that this method of solubilizing rock phosphate may be economically viable to use in wet tropical climates (Gianinazzi-Pearson and Diem, 1982). The activity of iron-reducing bacteria can also release P sorbed on iron oxides. It seems that the solubilization of P minerals may result from the synergistic action of a group of microflora rather than just one microorganism (Paul and Clark, 1996).

Some of the P polymers found in soils are believed to be of microbial origin. Some yeasts synthesize phosphorylated polymers of mannose and the cell walls of Gram positive bacteria contain both techoic acid (a polymer of ribitol phosphate and glycerophosphate), and 6-phosphate muramic acid. Of the total organic P in soil only about 1% can be identified as nucleic acids and their derivates.

Rhizosphere organisms are commonly high in P and often contain inositol polyphosphate granules. Away from roots, most bacterial activity is isolated in discrete micro-sites around organic residues. The non-water soluble P (mostly from diesters: nucleic acids, phospholipids, and phosphoproteins) remaining in the residues is used by fast growing saprophytic fungi, followed by slower growing fungi (Frossard et al., 1995).

Microbial phosphatases are produced by most organotrophic members of the soil microflora. These enzymes are involved in the mineralization of organic P, which otherwise is not available. Plant roots produce phosphatases able to hydrolyze a wide range of P and polyphosphates. The production of phosphatase either by roots or associated micro-organisms is an efficient strategy for the acquisition of P by plants.

ATP is very sensitive to phosphatases and does not persist in soil in a free state. During microbial growth, the C : ATP ratio can vary from 1000 : 1 to 40 : 1. In the resting state, the ATP : C : N : P : S ratios usually are 1 : 250 : 40 : 9 : 2.6. The biomass C content established by this ratio can be used as a measure of biological activity and is influenced by the soil P (Paul and Clark, 1996).

Mycorrhizal fungi increase the uptake of P and Zn from soils. More than 80% of the higher plants can enter into mycorrhizal associations. Mycorrhizal fungi are able to absorb P at low solution concentrations. The mechanism of P translocation within the mychorrhiza is believed to be due to cytoplasmic streaming. Polyphosphate makes up a significant portion of the P of mycelial strands, and can be translocated via streaming. It can be stored both in the arbuscular mycorrhiza hyphae, and in the sheath of ectomycorrhizae. Its insolubility maintains low internal concentrations, thus allowing for its continued uptake from the soil solution (Paul and Clark, 1996). Mycorrhizal infection also enhances the transfer of P between living roots and from dying roots to living roots. The high phosphate level of fertilized arable soils inhibits mycorrhizal development (Killham, 1994). Although the microbial biomass is less than that of the higher plants, its P content can be as much as 10 times higher. The microbial annual uptake of P often exceeds that of the higher plants in soils (Paul and Clark, 1996).

Microorganisms firstly consume the most bioavailable P forms in their environment, i.e. orthophosphate and phosphate esters containing phosphorus in its highest oxidation state. Bacteria have the ability to utilize more reduced organophosphorus compounds as phosphorus sources, in particular phosphonates.

Phosphonates belong to a family of biologically active organophosphorus compounds that contain C—P bonds (see Nowack, this volume). These compounds are chemically stable phosphate ester analogues, extremely resistant to chemical hydrolysis, thermal decomposition, and photolysis.

Biogenic phosphonates also include phosphonopyruvate and phosphonoacetate (PA). It is suggested (de Graaf et al., 1997) that phosphonates, as well as phosphites, emerged at an early stage of the Earth's evolution and could be prebiotic phosphorus carriers. The presence of methyl, ethyl, and other alkylphosphonates in a sample of the Murchison meteorite (Cooper et al., 1992) suggests that these compounds might also have occurred in high concentrations on the Earth when life first emerged.

Phosphonates are widespread among naturally occurring compounds in all kingdoms of life, but only procaryotic microorganisms are able to cleave the C—P bond via several pathways. Interest in microbial degradation of phosphonates has grown in recent years due to their toxicity. Organophosphorus compounds with the C—P bond have become common man-made chemical substances (xenobiotics), which uncontrollably enter and pollute the environment.

Phosphonates include herbicides, antibiotics synthesized by Streptomyces, insecticides, fungicides and chelate-additives to household detergents. Phosphonates are used as chemical additives to function as threshold antiscalants, corrosion inhibitors, sludge conditioners, deflocculants, dispersants, and crystal growth modifiers in various industrial water treatment processes.

There are numerous phosphonate degrading microorganisms in nature, including Gram positive and Gram negative bacteria, as well as some yeasts and fungi. However, the ability to degrade phosphonates is quite specialised to certain strains. These include various strains of *Arthrobacter* spp., *Rhizobium* spp., *Pseudomonas testosterone*, *Escherichia coli*, *Alcaligenes eutrophus*, *Agrobacterium radiobacter*, *Kluyvera ascorbata*, *Klebsiella pneumoniae*, *Bacillus megaterium* (Kononova, 2002).

In the mucosa of the stomach, the *Helicobacter pylori* has access to abundant organic and inorganic phosphates which it can use as nutrients. *H. pylori*, like some other bacteria, may utilize other phosphorus sources such as phosphonates and phosphites. Phosphonate breakdown is likely to be induced under conditions of inorganic phosphate limitation.

Three pathways exist for phosphonate breakdown in bacteria: the phosphonatase, the hydrolase, and the C—P lyase pathways, which differ with regard to their substrate specificity and their cleavage mechanisms.

The first reaction in the pathway of the C—P bond formation is intramolecular restructuring of phosphoenolpyruvate to phosphonopyruvate. The presence of enzymes of C—P bond biosynthesis suggested that under appropriate conditions in a cell they could also participate in decomposition of the C—P bond. It is suggested that the ability to synthesize phosphonates could give microorganisms advantages for survival in a potentially phosphate limited marine environment and other ecological niches, due to the presence of the chemically stable and phosphatase resistant C—P bond in these compounds (Kulaev and Kulakovskaya, 2000).

Phosphate may be so scarce in the environment that it severely limits growth, which explains why certain bacteria concentrate it intracellularly in metachromatic granules. When the environmental source of phosphate is depleted, the bacterial cell can gradually mobilize the stored source as required. Granules usually contain crystals of inorganic compounds and are not enclosed by membranes. They stain a contrasting colour (purple) in the presence of methylene blue.

Inorganic polyphosphate (polyP) is a linear polymer of many tens or hundreds of orthophosphate groups linked by high-energy, phosphoanhydride bonds. It is considered to be a precursor in prebiotic evolution, and a true 'molecule' (Kornberg et al., 1999).

The functionality of polyP has changed greatly during the evolution of living organisms. In prokaryotes, its main role has been as an energy source and a phosphate reserve. In eukaryotic microorganisms, regulatory functions predominate. As a result, a great difference can be observed between prokaryotes and eukaryotes in their polyP-metabolizing enzymes (Kulaev and Kulakovskaya, 2000).

In microbial cells, polyP plays a significant role in increasing cell resistance to unfavourable environmental conditions and in regulating different biochemical processes. The most important of its functions are: phosphate and energy reservation, cation sequestration and storage, membrane channel formation, participation in phosphate transport, involvement in cell envelope formation and function, gene activity control, regulation of enzyme activities, and a vital role in stress response and stationary-phase adaptation (Kulaev and Kulakovskaya, 2000). PolyP is involved in competence for bacterial transformation, ecological disposal of pollutant phosphate, and, of great interest, physiologic adjustments to growth, development, stress and deprivation.

Developmental changes in microorganisms-fruiting body and spore formation in Myxobacteria (e.g. *M. xanthus*), sporulation in bacteria (e.g. *Bacillus*) and fungi, and heterocyst formation in cyanobacteria (e.g. *Anabaena*) occur in response to starvation of one or another nutrient. In view of the involvement of polyP in the stationary stage of *E. coli*, polyP may well participate in certain instances of cellular adjustments to deprivation (Kornberg et al., 1999).

Another phosphorus compound, pyrophosphate ($P_2O_7^{4-}$), may play a crucial role in the survival of some parasites such as *Trypanosoma cruzi*, the causative agents of Chagas' disease, which is among the most widespread parasitic diseases in Latin America. Inorganic pyrophosphate can be found in acidocalcisomes, organelles with their own unique set of characteristics: acidic nature, high electron density, and a matrix consisting of pyrophosphate (PP_i), polyphosphate (polyP), calcium, magnesium, and other elements. Their chemical composition, in terms of high phosphorus content, indicates that one possible function of these organelles could be as an energy and phosphate reservoir.

Inorganic pyrophosphate was for a long time believed to be merely a byproduct of biosynthetic reactions (synthesis of nucleic acids, coenzymes, proteins, activation of fatty acids, and isoprenoid synthesis), and subject to immediate hydrolysis by inorganic pyrophosphatases. However at present this compound is considered to have an important bioenergetic and regulatory role. In plants and in some parasitic protozoa, pyrophosphate is used in the place of ATP as an energy donor in several reactions (Urbina et al., 1999).

2-aminoethylphosphonic acid (2 AEP) found in flagellates from cattle rumen was the first biogenic compound with a C—P bond to be identified. Further investigations established that 2 AEP is a constituent of certain lipids, known as phosphonolipids that have been found in protozoa, flagellates, coelenterates, mollusks, and lower fungi. Other than in lipids, 2 AEP was found to be a constituent of proteins and polysaccharides.

Phosphoglycans, containing glycosyl phosphate or oligoglycosyl phosphate repeat units, are immunologically active polymers of the cell wall or capsule of numerous microorganisms and often responsible for the microorganism infectivity and survival.

The phospholipid constituents of the cytoplasmic membrane have a unique role in allowing the formation of the bilayer structure and modulating its characteristics to ensure the continued function of the membrane under a variety of environmental conditions (Di Russo and Nystro, 1998). In the case of *Escherichia coli*, three major phospholipids are present in order of increasing abundances, shown in brackets: phosphatidylethanolamine (PE, 75%), phosphatidylglycerol (PG, 20%) and cardiolipin (diphosphatidylglycerol, CL, 1–5%). In addition to these major phospholipids, there are small amounts of phosphatidic acid (PA), phosphatidylserine (PS), lysophospholipids, and diacylglycerol. These minor constituents act mainly as biosynthetic intermediates or turnover products.

Microorganisms have some specific metabolic pathways in which phosphorylated compounds are involved. Bacteria that digest cellulose in the rumens of cattle are largely fermentative. After initially hydrolyzing cellulose to glucose, they ferment the glucose to organic acids, which are then absorbed as the animal's principal energy source. Certain bacteria follow a different pathway in carbohydrate catabolism: the phosphogluconate pathway (also called the hexose monophosphate shunt) that provides ways to anaerobically oxidize glucose and other hexoses, to release ATP, to produce large amounts of NADPH and to process pentoses. This pathway, common in heterolactic fermentative bacteria, yields various end products, including lactic acid, ethanol and carbon dioxide.

Oxygen releasing photosynthesis that occurs in plants, algae, and cyanobacteria is the dominant type on earth. Photosynthesis in some types of bacteria differs in several ways. Photosynthetic green and purple bacteria possess bacteriochlorophyll, which is more versatile in capturing light. They have only a cyclic photosystem I. This pathway generates a relatively small amount of ATP, and may not produce NADPH. Being photolithotrophs, these bacteria use H_2, H_2S or elemental sulfur rather than H_2O as a source of electrons and reducing power. As a consequence, they are anoxygenic and many are strict anaerobes (Talaro and Talaro, 1993).

The oxidative burst is a characteristic early feature of the plant host cell as a response to microbial infection attempts. A balance between phosphorylation and dephosphorylation events, termed phosphorylation poise, regulates the oxidative burst. Such a regulatory poise may be important for rapid induction and tight control of the magnitude and duration of the oxidative burst (Ridge, 2002).

3.4 PHOSPHATE ASSIMILATION

3.4.1 ATP biosynthesis

A large number of cellular energy transformations proceed by exchange of phosphate groups. Turnover of phosphate in organic compounds plays a fundamental role in metabolism and in many other cellular work processes. The main entry point of phosphate into assimilatory pathways occurs during the formation of ATP. In the overall reaction for this process, inorganic phosphate is added to the second phosphate group in adenosine diphosphate (ADP) to form a phosphate ester bond. ATP is a phosphate ester with a particularly high free energy of hydrolysis, compared to most other esters, the hydrolysis of which produces considerably less than half as much free energy as the hydrolysis of ATP (Figure 3.2).

ATP can be considered as the circulating currency of energy in living organisms. An active cell requires millions of molecules of ATP per second to drive its biochemical reactions. ATP is used for the synthesis of long-term energy storage compounds. Plant cells synthesise starch and fats and animal cells synthesise glycogen. For synthesis of peptide bonds, the cells use guanosine triphosphate (GTP) as energy donor.

In animal organisms, at physiological pH, ATP is highly negatively charged, having a total of three or four negative charges on its phosphates (Champ and Harvey, 1987). ATP forms stable complexes with Mg^{2+} and Mn^{2+}.

Phosphate-bond energy is utilised in biochemical reactions and represents the only form of energy that can be used by living cells. The standard free energy of hydrolysis, $\Delta G°$, of ATP is approximately -7300 cal/mole for each of the two terminal phosphate groups.

There are, however, compounds that contain phosphates with energy higher than that of ATP. These ultra high energy compounds include phosphoenolpyruvate, phosphocreatine and 1,3-bisphosphoglycerate, all of which have a standard free

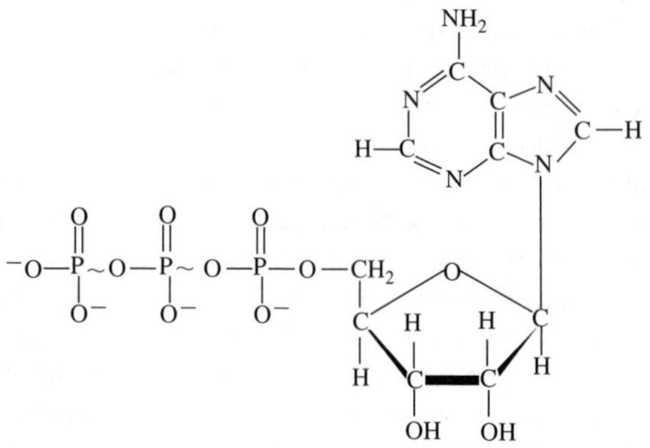

Figure 3.2 Adenosine triphosphate. The symbol ~ represents high-energy bonds.

energy of hydrolysis greater than −10,000 cal/mol. Other phosphate-containing compounds are low-energy phosphates, which have standard free energies of hydrolysis less than −4000 cal/mol. These include glucose 6-phosphate, AMP (adenosine monophosphate), and glycerol 3-phosphate. ATP lies in the middle of the phosphorylation scale. The adenylate system is therefore particularly suited to act as an intermediary energy cosubstrate between exergonic and endergonic areas of metabolism. ADP can serve as an acceptor of a phosphate group from cellular phosphate containing higher-energy compounds. ATP can donate phosphate groups forming phosphates of lower energy.

The molecules of ATP are bound by the sequences and structural motifs common to a large number of proteins, such as the P loop and the glycine-rich loop. The function of the P loop is to bind the phosphoryl group of ATP or GTP (also the phosphoryl group of ADP and GDP). Glycine residues provide an appropriate conformation for the loop, and, at the same time, leave space for the phosphoryl groups. Hydrogen bonds are often formed between NHs of the main chain within the loop and oxygen atoms of the phosphoryl groups. More specifically, the glycine residues within the sequence provide at least three functions: (1) they favour a loop conformation, (2) in the absence of steric hindrance between the phosphoryl group and the protein, provide hydrogen bonding with main chain NHs and (3) they assume torsion angles that would be energetically unfavourable for residues with side-chains (Matte and Delbaere, 2001).

Many different reactions in the cell can provide the energy to convert ADP to ATP. The most significant of these are the reactions of cellular respiration, in which the energy released from food molecules is trapped as the stored energy of ATP. Energy released in some spontaneous (exergonic) reactions may be captured in ATP for use at a later time or in another part of the cell.

The phosphorylation processes occurring in particular circumstances are referred to as substrate level phosphorylation, oxidative phosphorylation and photophosphorylation.

3.4.1.1 Substrate level phosphorylation

Substrate level phosphorylation implies the formation of ATP from ADP and involves the direct transfer of a phosphate moiety from a substrate molecule to ADP to form ATP as in the process of glycolysis.

Glycolysis is an example of an integrated modulatory system of metabolic feedback loops, which has the task of maintaining the concentration of adenylate system, and thus the phosphorylation potential in the cell, constant. The level of ATP hardly changes during the transition from aerobic respiration to anaerobic conditions, although the glycolytic flux changes are many-fold.

Glycolysis incorporates inorganic phosphate into 1,3-bisphosphoglyceric acid, forming a high-energy acyl phosphate group. This phosphate can be donated to ADP to form ATP. Once incorporated into ATP, the phosphate group may be transferred via many different reactions to form the various phosphorylated compounds found in cells (Taiz and Zeiger, 1998).

Phosphorylated sugar molecules penetrate cell membranes only with specific carriers, for example the irreversible phosphorylation reaction traps glucose in a form that does not diffuse out of the cell. The reaction is catalysed by two different enzymes: hexokinase, located in most of the animal and plant tissues, and glucokinase located in liver. Hexokinase phosphorylates glucose and several other hexoses; it works at low concentration of sugar and is inhibited by a high ATP/ADP ratio, and by the accumulation of glucose 6-phosphate.

Glucokinase phosphorylates only glucose. This enzyme, which functions only at high concentration of glucose, prevents large amounts of glucose from entering the circulation after consumption of a carbohydrate-rich meal (Champe and Harvey, 1987).

The second phosphorylation reaction of glycolysis consists of the irreversible phosphorylation of fructose 6-phosphate catalysed by phosphofructokinase I (PFK I), resulting in the formation of fructose 1,6-bisphosphate. This is considered to be the control point of the process because it is the rate-limiting step. The rate of PFK I is controlled by the concentration of fructose 6-phosphate and by ATP. The enzyme is activated by the concentration of AMP (which signals that the cell's energy stores are depleted), and inhibited by elevated levels of ATP (which signals that cell's energy stores are adequate).

In plant cells, the interconversion of fructose 6-phosphate and fructose 1,6-bisphosphate is made more complex by the presence of an additional enzyme, a pyrophosphate (PP_i)-dependent phosphofructokinase, which catalyses the following reversible reaction:

$$\text{fructose-6-P} + PP_i \rightleftharpoons \text{fructose-1,6-}P_2 + P_i \qquad (3.2)$$

where 'P_i' represents phosphate and 'P_2' bisphosphate. The enzyme is found in the cytosol of many plant tissues at levels higher than those of the ATP-dependent phosphofructokinase. Under some circumstances, operation of glycolytic (or gluconeogenic) pathway in plants differs from that in other organisms (Taiz and Zeiger, 1998).

Another compound, fructose 2,6-bisphosphate appears in animal cells in high amounts after a carbohydrate-rich meal and acts as an intracellular signal indicating that glucose is abundant. It is an activator of PFK I.

In a further step of the glycolysis process, the oxidation of the aldehyde group of glyceraldehyde 3-phosphate is coupled with the attachment of phosphate to a carboxyl group. The high-energy phosphate group at carbon 1 of 1,3-diphosphoglycerate conserves much of the free energy produced by the oxidation of glyceraldehyde 3-phosphate. The high-energy phosphate group of 1,3-diphosphoglycerate is used to synthesise ATP from ADP in a reaction catalysed by phosphoglycerate kinase.

The overall process of glycolysis can be represented by the reaction:

$$\text{glucose} + 2NAD^+ + 2ADP + 2P_i \longrightarrow 2\text{pyruvate} + 2NADH + 2ATP + 2H^+ + 2H_2O \qquad (3.3)$$

Nicotinamide adenine dinucleotide (NAD^+) serves as coenzyme in oxidation-reduction reactions in which the coenzyme undergoes reduction of the pyridine ring by accepting a hydride ion (hydrogen atoms plus one electron). Its phosphorylated derivative is nicotinamide adenine dinucleotide phosphate ($NADP^+$). The reduced forms of NAD^+ and $NADP^+$ are NADH and NADPH, respectively, which can be used to reduce substrates in other reactions (Champe and Harvey, 1987).

In the multi-step process of Equation 3.3, 2ATP molecules are consumed in total and 4ATP molecules are produced. On balance, 2ATP molecules are gained per molecule of glucose.

In case of O_2 deficiency, the reduction equivalents produced in glycolysis cannot be used to produce phosphorylation potential. Instead of water formation, other reactions accepting [H] take place producing the fermentation of the accumulating organic compounds. The anaerobic breakdown of the glucose to ethanol or lactate can be represented by the reactions:

$$\text{glucose} + 2ADP + 2P_i \longrightarrow 2\text{ethanol} + 2ATP + 2CO_2 \quad (3.4)$$

$$\text{glucose} + 2ADP + 2P_i \longrightarrow 2\text{lactate} + 2ATP \quad (3.5)$$

Reduction equivalents (NADH) released in glycolysis are consumed quantitatively for the formation of ethanol and lactate; thus, they do not appear in the equations. As the ATP yield of glycolysis is relatively low, both processes are strongly exergonic and take place with considerable release of heat. Many yeast cells can satisfy their total ATP requirements by anaerobic degradation of sugars to ethanol, which is excreted as a reduced waste product. Cells of higher plants are also able to produce alcohol by fermentation under O_2-deficient conditions. Lactic acid formation by fermentation is used in muscles to yield ATP anaerobically, in potato tubers and in the roots of Gramineae.

3.4.1.2 Oxidative phosphorylation

Oxidative phosphorylation consists of the formation of ATP from ADP, coupled to the electron transfer from NADH, or from another electron transferring cofactor, $FADH_2$, to oxygen. These processes take place in almost all types of cells under aerobic conditions.

In mitochondria, the energy for ATP synthesis derives from the oxidation of NADH. The Nobel Prize winner Peter Mitchell's hypothesis postulates that the electrochemical potential of a proton gradient drives phosphorylation through an ATP synthase situated anisotropically in the membrane (Taiz and Zeiger, 1998).

Therefore, the actual site of ATP formation on the mitochondrial inner membrane is the ATP synthase. In fact, ATP synthesis is not obligatorily linked to electron transport, if an alternative method of generating a proton gradient across the inner membrane exists. This can be experimentally demonstrated by inhibiting electron transport and subjecting the mitochondria to an artificially generated pH gradient.

By this mechanism the electrochemical potential of a proton gradient is converted into phosphorylation potential. ATP synthesis and ADP-driven proton pumps catalyse, in principle, the same reaction but work in opposite directions.

Once generated, the proton electrochemical gradient, referred to as proton motive force, can be utilized to produce chemical work (ATP synthesis). This gradient is coupled to the synthesis of ATP by an additional protein complex associated with the inner membrane, the F_0F_1-ATP synthase (also called complex V). This complex consists of two major components, F_1 and F_0. F_1 is a peripheral membrane protein complex that is composed of at least five different subunits and contains the catalytic site for converting ADP and phosphate to ATP. This complex is attached to the matrix side of the inner membrane. F_0 is an integral membrane protein complex that consists of at least three polypeptides that form the channel through which protons are able to cross the inner membrane. The passage of the protons through the channel is coupled to the catalytic cycle of the F_1 component of the ATP synthase, allowing the ongoing synthesis of ATP and simultaneously dissipating the proton electrochemical gradient. For their contributions to the elucidation of the mechanism of ATP synthesis, Paul Boyer and John Walker shared the Nobel Prize in medicine in 1997. They demonstrated the rotary motion of one unit of the enzyme (Taiz and Zeiger, 1998).

Although ATP is synthesised in the mitochondrial matrix, most of it is used outside the mitochondrion, thus an efficient mechanism is needed for moving ADP and ATP in and out of the organelle. This mechanism involves another inner membrane protein, the ADP/ATP transporter, which catalyses the exchange of ADP and ATP across the inner membrane. The electric potential gradient generated during electron transfer (negative inside) is such, that there will be a net movement of the more negatively charged ATP^{4-} out of the mitochondria in exchange for ADP^{3-}.

There are three sites of phosphorylation in the respiratory chain: site I in the region of the flavoproteins, site II between cytochrome b_{560} and cytochrome c_{552}, and site III between cytochrome a and a_3. At all three sites, one ATP is formed for every two transported electrons. The simplified reaction of respiratory phosphorylation can be written as follows:

$$\begin{cases} 3NADH + 3H^+ + 9ADP + 9P_i \\ FADH_2 + 2ADP + 2P_i \end{cases} + 2O_2 \longrightarrow \begin{cases} 9ATP + 3NAD^+ \\ 2ATP + FAD \end{cases} + 4H_2O$$

(3.6)

For the breakdown of pyruvate via the tricarboxylic acid cycle (TCA) and the respiratory chain, the following reaction can be written:

$$pyruvate + 2.5O_2 + 15ADP + 15P_i \longrightarrow 3CO_2 + 5H_2O + 15ATP \quad (3.7)$$

By including the reduction equivalents released in glycolysis and by decarboxylation of pyruvate, the oxidative breakdown of glucose via glycolysis, the TCA cycle and the respiratory chain, 38 molecules of ATP are theoretically produced in total.

3.4.1.3 Photophosphorylation

Photophosphorylation is the process through which photosynthetic organisms convert light energy to ATP. The chemical reactions culminating in ATP production are carried out by four membrane-associated protein complexes: photosystem II, the cytocrom b_6f complex, photosystem I and the ATP synthase. These complexes are located in the thylakoid membrane and are vectorially oriented so the water that is oxidised to O_2 in the thylakoid lumen $NADP^+$ is reduced to NADPH on the stromal side of the membrane, and ATP is released into the stroma by H^+ moving from the lumen to the stroma.

The energy of the captured light initiates an electron transfer and produces a charge separation across the membrane. Water oxidation and electron transfer between photosystems II and I generate a proton gradient. The protons accumulate on the lumenal side of the thylakoid membrane. The combination of a proton concentration and charge difference across the thylakoid membrane is termed the electrochemical proton gradient, and drives the synthesis of ATP from ADP and phosphate.

Under normal cellular conditions, photophosphorylation requires electron flow, although under some condition electron flow and photophosphorylation can take place independently of each other. It has been shown that the application of an artificial proton gradient ($\Delta pH = 3.5$) to thylakoids results in ATP synthesis in darkness. ATP can also be produced by an artificial transmembrane electrical potential. An increase of the membrane permeability to H^+, caused by 'uncouplers' such as gramicidin, FCCP and other protonophores, blocks phosphorylation without affecting electron transport. Electron flow without accompanying phosphorylation is referred to as uncoupled.

Photophosphorylation works via the chemiosmotic mechanism, which is applicable to phosphorylation during aerobic respiration in bacteria and mitochondria. Under conditions of steady state electron transport in chloroplasts, the membrane electric potential is quite small because of the inward movement across the membrane. The stoichiometry of protons translocated to ATP synthesised is four H^+ per ATP.

It was demonstrated that there is also a cyclic electron transport in the thylakoids, driven by one photosystem only. In this case the electrons are transferred from ferredoxin back to Chl a_1. No reduction equivalent can be produced, only phosphorylation potential (cyclic phosphorylation). The physiological significance of cyclic electron transport lies in the additional availability of ATP, especially in situations where reduction equivalents (NADPH) are available in excess. Cyclic phosphorylation supplies under anaerobic conditions ATP for light-dependent uptake of ions and non-electrolytes, for hexose–starch conversion and for N_2 fixation and photokinesis of photosynthesising prokaryotes (Mohr and Schopfer, 1995).

The proton electrochemical energy is used to produce ATP by a single thylakoid protein complex called ATP synthase. The uneven distribution of photosystems II and I and of ATP synthase on the thylakoid membrane poses some challenges for the formation of ATP. ATP synthase is found only in the stroma lamellae and at the edges of the grana stacks. The ATP is synthesised by a large (400 kDa) enzyme complex known by several names: ATP synthase, the coupling factor, and CF_0F_1-ATPase. The enzyme consists of two parts: a hydrophobic membrane-bound protein called CF_0 and a portion that sticks out into the stroma called CF_1. The latter is made up of five different proteins termed α, β, δ, ε and γ. Each interface of an $\alpha\beta$ dimer can bind ADP and phosphate and form ATP. Passage of protons through the CF_0 component is thought to change the conformation of one of the sites, causing ATP to be released. The model proposed by Noji *et al.* (1997) suggests that a large portion of the CF_1 rotates about a bearing consisting of the γ subunit, because of the passage of protons. A rapid interconversion of the nucleotide-binding sites occurs, followed by the release of ATP (Webber, 2001). The energy of the conformational movements is then translated into phosphoanhydride bond energy (Junge *et al.*, 1997). The structure and function of the plastidial ATP synthase is similar to that of the F_0F_1-ATPase in oxidative phosphorylation. Phosphorylation takes place only in intact thylakoids.

3.4.2 ATP breakdown

Many different enzymes can catalyse the breakdown of ATP, yielding adenosine diphosphate (ADP) and an inorganic phosphate ion. This breakdown is an exergonic reaction, yielding approximately 10 kcal of free energy per mole of ATP under biological conditions – sufficient to drive typical endergonic reactions in the cell. Simultaneous reactions, to which the energy could be transferred, are usually referred to as 'coupled reactions'. ADP, which possesses less free energy than ATP, can combine with a phosphate ion to make a new molecule of ATP, if enough energy is provided by some exergonic reactions.

In all cleavage reactions of ATP, a divalent metal ion is required, usually Mg^{2+}. There are three mechanisms for this catalytic reaction: (1) a single displacement, a direct transfer from MgATP to an acceptor where both substrates bind at adjacent sites; (2) a double displacement, with the formation of a covalent intermediate and the release of ADP or PP_i, followed by the second step, i.e. the group transfer to an acceptor; (3) an activation of MgATP to form a phosphorylated or adenylated product, with the release of ADP or PP_i, followed by the ligation with another substrate and release of phosphate or AMP (Cohn, 2001).

3.4.3 ATP functions

ATP is utilised in the processes related to (1) energy transduction, (2) metabolic regulation and (3) biosynthesis (Figure 3.3).

3.4.3.1 Energy transduction

The free energy of ATP hydrolysis provides the energy for the active transport of ions and molecules across membranes, for the mechanical work produced by contractile elements (muscle, fibres and flagella), and for bioluminescence. Because of its function as a transport molecule for phosphorylation potential, ATP has an enormously large turnover in the cell: in the human body 40 kg of ATP are produced and reused each day.

```
                              ┌ Active transport
         Energy transduction ─┤ Mechanical work
        ╱                     └ Bioluminescence
       ╱                      ┌ Enzyme regulation
[ATP] ──── Metabolic regulation ─┤
       ╲                      └ Signal transduction
        ╲                     ┌ Sugar phosphates
         Biosynthesis ────────┤ Protein phosphates
                              ┤ Phospholipids
                              ┤ Nucleic acids
                              └ Vitamins
```

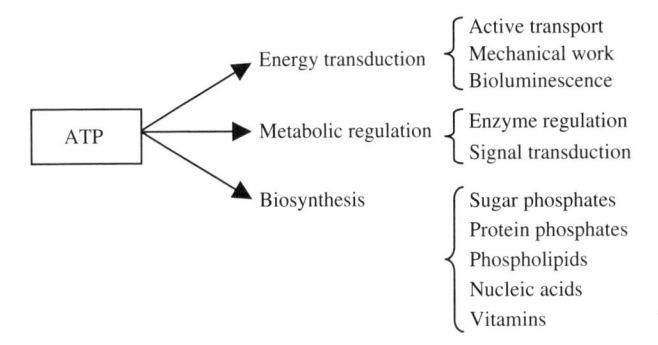

Figure 3.3 ATP utilisation.

Active transport of ions and non-electrolytes is supplied with energy by hydrolysis of ATP. The membrane proteins, which carry out primary active transport, are called pumps. Membrane ATPases, functioning as proton pumps, play a central role in transformation of the phosphorylation potential into electrochemical potential. ATPases consist of a cytoplasmic, catalytic domain where ATP binds and is hydrolysed and a membrane domain responsible for ion transport.

There are two types of ion motive pumps: the P-type, which is phosphorylated by ATP during its cycle, and the V-type, which is not phosphorylated (Guillain and Mintz, 2001).

The earliest ATPases to evolve were probably the multisubunit F_1F_0-ATPases of bacteria, transporting H^+ in one direction for ATP synthesis or in the other at the expense of ATP. The likely subsequent symbiosis of such bacteria with the ancestor of eukaryotic cells resulted in the presence of F_1F_0-ATPases in the inner membranes of mitochondria and chloroplast. This class of pump is unique in that it is biologically reversible, i.e. under certain conditions it can operate in the reverse direction, hydrolysing ATP and using this energy to pump protons across the membrane (Sachs and Keeling, 2001).

V-ATPases are present in all eukaryotic cells, in organelles such as synaptic vesicles, vacuoles, lysosomes, Golgi complexes, and in certain circumstances, are also present in the plasma membrane of the cell where they acidify the extracellular medium. All V-ATPases extrude H^+ out of the cytoplasm at the expense of ATP hydrolysis, H^+ being accumulated outside the cell or in the internal space of the organelle. Thus V-ATPases are electrogenic and create a membrane potential as well as a pH gradient. They are involved in the adjustment of the body's overall acid–base balance by transporting acid to the urine or to the blood, and have a role in the breakdown of bone osteoclasts and in sperm maturation when reduced pH is needed.

Unlike the F_1F_0-ATPases, V-ATPases cannot act in a reversible manner and are only capable of working in a hydrolytic, proton-pumping mode. The stoichiometry for V-ATPases corresponds to two protons transported for each ATP hydrolysed.

In the plasma membranes of plants, fungi and bacteria, as well as in plant tonoplasts and other plant and animal endomembranes, H^+ is the principal ion that is electrogenically pumped across the membrane. The plasma membrane H^+-ATPase creates the gradient of electrochemical potential of H^+ in the plasma membranes, while the vacuolar H^+-ATPase and the H^+-pyrophosphatase electrogenically pump protons in the lumen of the vacuole and the Golgi cisternae. When protons are extruded from the cytosol by electrogenic H^+-ATPases, both at the plasma membrane and at the vacuole membrane, a membrane potential and a pH gradient are created at the expense of ATP hydrolysis.

P-ATPases are the largest family of ATPases in prokaryotic and eukaryotic cells and essentially comprise one polypeptide (α chain) responsible for both ion transport and ATP hydrolysis. Some have a β chain, which forms an $\alpha\beta$ complex responsible for the insertion in the membrane. P-ATPases catalyse ion transport in a completely different manner using successive conformational changes induced by phosphorylation and dephosphorylation of their catalytic subunits, in order to pick up an ion on one side of the membrane and release it on the other. They are able to transport a wide range of ions. The mechanism also allows countertransport. However, only one ion can be pumped for each molecule of ATP hydrolysed.

The common feature of all types of ATPases is that they catalyse an electrogenic proton transport and thus polarise the membrane. They are therefore always coupled to other reactions, which depend on membrane and proton potential.

3.4.3.2 Metabolic regulation

Protein phosphorylation is a general mechanism for regulating cellular metabolism and development. Enzymes are proteins with a high catalytic efficiency, highly specific for the substrate. Enzymes are effective in extremely small concentrations, which can increase reaction rates by as much as 10^4–10^9 times. Enzyme activity can be regulated-activated or inhibited by binding allosteric modifiers or by covalent modification of the enzyme. Covalent modification occurs by the addition or removal of phosphate groups from specific residues of the enzyme.

Depending on the specific enzyme, the phosphorylated form may be more or less active than the unphosphorylated form. For example, the addition of phosphate to glycogen synthase results in a less active enzyme, whereas the addition of a phosphate group to glycogen phosphorylase increases the activity (Champe and Harvey, 1987).

Enzymes, which catalyze the phosphorylation of a protein, are referred to as protein kinases. Some of them are 'broad spectrum' and may assist in phosphorylation of several different proteins. Kinase-catalysed reactions are usually not reversible and it appears that different kinases may phosphorylate a given protein at different sites. The phosphorylation of an enzyme causes a conformational change that affects the active site, which can either increase the catalytic activity of some enzymes or decrease that of others. Effectors, which regulate protein kinases, are, for example, free Ca^{2+}, polyamines, etc. In plants more than 30 protein kinases were identified and 10 of them are dependent on Ca^{2+}.

Some of the kinases are AMP-dependent; cAMP activates a protein kinase which is a tetramer, having two regulatory subunits and two catalytic subunits. When cAMP is removed, the inactive tetramer is formed again.

The phosphorylated enzyme can be inactivated by hydrolytic removal of its phosphate by protein phosphatase. There is a balance between the activity of kinases and phosphatases. If the kinases have greater activity than the phosphatases, more of the regulated enzymes will be phosphorylated, and *vice versa*.

The key enzymes in the plant cell cytoplasm that are involved in the regulation of sucrose synthesis are sucrose phosphate synthase (SPS) and fructose 1,6-bisphosphatase. The key chloroplast enzyme involved in the regulation of starch synthesis is ADP-glucose pyrophosphorylase. The activities of these enzymes are controlled by the concentrations of key metabolites. SPS activity is related to sucrose export in the presence of light. SPS catalyses the formation of sucrose phosphate, which is subsequently hydrolysed to sucrose and inorganic phosphate. Glucose 6-phosphate stimulates the activity of SPS.

In animals, creatine phosphokinase (CPK) is commonly determined in the diagnosis of myocardial infarction. Fructokinase, found in the liver and the kidneys, converts fructose to fructose 1-phosphate using ATP as the phosphate donor.

There is evidence for a potential role for phosphorylation in the phytochrome action related to the red-light regulation of protein phosphorylation and phosphorylation-dependent binding of transcription factors to the promoters of phytochrome-regulated genes. Some highly purified phytochromes seem to have kinase activity, possessing the capacity to transfer phosphate groups from ATP to amino acids, such as serine and tyrosine, either on themselves or on other proteins. Kinases are often found in signal transduction pathways in which the addition or the removal of phosphate groups regulates the enzyme activity.

The basic mechanisms that enable organisms to sense and to respond to their environment are common to all cell sensory systems, and include stimulus detection, signal amplification, and the appropriate output response.

Cell growth, migration, shape, differentiation and apoptosis are cellular events that are regulated by signals that the cells receive from their environment, either through direct contact with other cells, interaction with matrix molecules, or via stimulation by soluble signalling factors (Heldin, 2001). Many signalling pathways have been shown to consist of modular units called transmitters and receivers. The signal is passed from transmitter domain to receiver domain via protein phosphorylation. Transmitter domains have the ability to phosphorylate themselves, using ATP, on a specific histidine residue near the amino terminus (Parkinson, 1993). For this reason, sensor proteins containing transmitter domains are called autophosphorylating histidine kinases. The receiver domain of the response contains a conserved aspartate. The transmitter phosphorylates itself as its conserved histidine, and transfers the phosphate to the aspartate of the response regulator. The response regulator then undergoes a conformational change leading to the response.

Many hormones, neurotransmitters etc. act to produce signals at the cell surface. The signals are then transmitted across the membrane and produce transient intracellular chemical messengers that greatly amplify the original signal. These are called 'second messengers' and trigger various biochemical pathways to produce the cell's response.

Some common second messengers are:

- cyclic adenosine monophosphate (cAMP), formed by the cyclization reaction of ATP catalysed by adenyl cyclase, which acts as a second messenger in the glycogen metabolism;
- cyclic guanosine monophosphate (cGMP), involved in the retinal visual processes;
- inositol trisphosphate (IP_3), which acts as a second messenger in the intracellular signalling systems.

In animal cells, certain hormones can induce a transient Ca^{2+} concentration, released from intracellular compartments by the opening of intracellular calcium channels. The coupling of hormone binding with the opening of intracellular calcium channels is mediated by inositol trisphosphate, which works as a second messenger. Phosphatidylinositol (PI) is a minor phospholipid component of the membranes, which can be converted to the phosphoinositide PI phosphate (PIP) and to PI bisphosphate (PIP_2) by kinases:

$$PI \xrightarrow[\text{PI kinase}]{ATP \quad ADP} PIP \xrightarrow[\text{PIP kinase}]{ATP \quad ADP} PIP_2 \begin{cases} \text{Activates protein kinase C} \\ \text{Releases } Ca^{2+} \text{ from endoplasmic reticulum (ER) and vacuole} \end{cases} \quad (3.8)$$

PIP_2 plays a central role in signal transduction. In animal cells, binding of a hormone, such as vasopressin, to its receptor leads to the activation of a G protein. The α subunit dissociates from G protein and activates a phosphoinositide specific phospholipase (PLC). The activated PLC rapidly hydrolyses PIP_2, generating inositol trisphosphate and diacylglycerol, each of these molecules playing an important role in cell signalling. IP_3 is water-soluble and diffuses through the cytosol, until it encounters IP_3-binding sites on the ER. These binding sites are IP_3 gated Ca^{2+} channels that open when they bind IP_3.

3.4.3.3 Biosynthesis

The monosaccharides found in living systems are mostly mono- and di-phosphate esters. Of greatest importance in plant and in animal metabolism are: glucose 1-phosphate, glucose 6-phosphate and fructose 6-phosphate. Glucose monophosphate has 5 possible isomeric forms with -OP (O) $(OH)_2$ groups being attached to each carbon atom except C_5. Glucose diphosphate has 10 possible isomers with the phosphate groups attached to $1:2, 1:3, 1:4, 1:6, 2:3, 2:4, 2:6, 3:4, 3:6$, and $4:6$ carbon atoms.

Ribose 5-phosphate is one of the important nucleotides and an intermediate of the Calvin-Benson cycle, and of the pentose phosphate pathway.

Polysaccharide phosphate esters can be found in bacterial and in plant cells, where the phosphate groups serve to link the saccharide rings to lipids or other units as in teichoic acids.

There are more than one hundred phosphoproteins in plant and animal cells. Phosphorylation of proteins occurs in serine, threonine, tyrosine, histidine and lysine residues. Phosphorylation of serine, threonine and tyrosine residues replaces -OH with -OP (O) (OH)$_2$, places a negative charge, and modifies the secondary and the tertiary structure of the protein.

Protein phosphorylation is involved in many biochemical processes such as gene transcription and translation, cell division, metabolic regulation, membrane transport, hormone response, intracellular signalling, muscle contraction, etc.

The main phosphoproteins, which store iron, are: ferritin, hemosiderin, phosvitin, found in egg yolk, and ovalbumin, found in egg white (Corbridge, 2000).

The lipids which contain at least one polar phosphate group, are named phospholipids. They are the major components of the cell membranes and may be found in bacteria, plant and animal cells. They are involved in enzyme activity, electron transport and in oxidative phosphorylation.

Phospholipids of the animal body are concentrated in brain, liver and kidneys and constitute more than 5% of the total weight of the nervous system. Phospholipids form about 75% of pulmonary surfactants found in alveolar spaces, which lower surface tension preventing lung collapse during exhalation. The main structural lipids in plant cell membranes are the polar glycerolipids. There are two categories of polar glycerolipids: glyceroglucolipids, in which sugars form the head group, and glycerophospholipids, in which the head group contains phosphate. The main glycerophospholipid components of the mitochondrion and the endoplasmic reticulum membranes are: phosphatidylglycerol (PG), phosphatidylcholine (PC), phosphatidylethanolamine (PE) and phosphatidylinositol (PI).

The first steps of glycerolipid synthesis are two acylation reactions that transfer fatty acids from acyl-ACP or acyl CoA to glycerol 3-phosphate to form phosphatidic acid. Diacylglycerol (DAG) is produced from phosphatidic acid by a specific phosphatase. Phosphatidic acid can be converted to PI or PG while DAG can give rise to PE or PC.

The extent of membrane unsaturation or the presence of particular lipids, such as disaturated PG can affect the response of plants to low temperature (Ishizaki-Nishizawa *et al.*, 1996).

Nucleic acids are high molecular weight polymers consisting of fundamental repeating units named mononucleotides. They are used for the storage and the expression of genetic information. There are two chemically distinct types of nucleic acids: deoxyribonucleic acid (DNA) and ribonucleic acid (RNA). Nucleotides are built from three main components: a nitrogenous base, a monosaccharide (ribose or deoxyribose) and one, two or three phosphate groups. Nucleotides are phosphate esters of nucleosides. The phosphate group is usually attached by an ester linkage to the 5th hydroxyl of the pentose. If one phosphate group is attached to the 5th carbon of the pentose, the structure forms a nucleoside monophosphate (NMP), like adenine dinucleotide monophosphate (AMP) or cytidine monophosphate (CMP). If a second or third phosphate is added to the nucleotide, the result is a nucleoside diphosphate or triphosphate, respectively.

The polynucleotide chains of DNA contain mononucleotides covalently linked to each other by 3',5'-phosphodiester bonds. These bonds join the 5th hydroxyl group of the pentose of one nucleotide to the 3rd hydroxyl group of the pentose of another nucleotide through a phosphate group.

Phosphodiester linkages between nucleotides can be cleaved hydrolytically by chemicals, or hydrolysed enzymatically by a family of nucleases, deoxyribonucleases or ribonucleases.

Water-soluble vitamins form parts of phosphate ester coenzymes. Vitamin B_1 is thiamin, which is often encountered in coenzyme form as thiamin pyrophosphate. It is obtained by phosphorylation of thiamin with ATP. Vitamin B_2 is riboflavin, which is utilised for synthesis of the coenzymes flavin mononucleotide (FMN) and flavin adenine dinucleotide (FAD). Vitamin B_6 can be found as phosphate esters of pyridoxal, pyridoxine and pyridoxamine. Pyridoxyl phosphate is a coenzyme, which participates in decarboxylation and transamination reactions (Corbridge, 2000).

REFERENCES

Boyer, P.D. (1997) The ATP synthase: a splendid molecular machine. *Annu. Rev. Biochem.* **66**, 717–749.
Champe, P.C. and Harvey, R.A. (1987) *Biochemistry*. J.B. Lippincott, Philadelphia, 441 pp.
Cogliati, D.H. and Santa Maria, G.E. (1990) Influx and efflux of phosphorus in roots of wheat plants in non-growth-limiting concentrations of phosphorus. *J. Exp. Bot.* **41**, 601–607.
Cohn, M. (2001) Adenosine Triphosphate. In *Encyclopedia of Life Science*, Nature Publishing Group.
Cooper, G.W., Onwo, W.M. and Cronin, J.R. (1992) Alkyl phosphonic acids and sulfonic acids in the Murchison meteorite *Geochim. Cosmochim. Acta* **56**, 4109–4115.
Corbridge, D.E.C. (2000) *Phosphorus 2000, Chemistry, Biochemistry & Technology*. Elsevier Science, Amsterdam, 1258 pp.
De Graaf, R.M., Visscher, J. and Schwartz, A.W. (1997) Reactive phosphonic acids as prebiotic carriers of phosphorus. *Mol. Evol.* **44**, 237–241.
Di Russo, C.C. and Nystro, M.T. (1998) The fats of *Escherichia coli* during infancy and old age: regulation by global regulators, alarmones and lipid intermediates. *Mol. Microbiol.* **27**, 1–8.
Duff, S.M.G., Sarath, G. and Plaxton, W.C. (1994) The role of acid phosphatase in plant phosphorus metabolism. *Physiol. Plant* **90**, 791–800.
Elliott, J.C. (2002) Calcium phosphate biominerals. In *Phosphates: Geochemical, Geobiological and Materials Importance*, (eds Kohn, M.J., Rakovan, J. and Hughes, J.M.), pp. 427–453.
Flügge, U.I. (1992) Reaction mechanism and asymmetric orientation of the reconstituted chloroplast phosphate translocator. *Biochim. Biophys. Acta* **1110**, 112–119.

Flügge, U.I. (1999) Phosphate translocators in plastids. *Annu. Rev. Plant Physiol. Plant Mol. Biol.* **50**, 27–47.
Frossard, E., Brossard, M., Hedley, M.J. and Metherell, A. (1995) Reactions controlling the cycling of P in soils. In *Phosphorus in the Global Environment*, pp. 107–138.
Gianinazzi-Pearson, V. and Diem, H.G. (1982) Endomycorrhizae in the tropics. In *Microbiology of Tropical Soils and Plant Productivity*, pp. 209–252.
Guillain, F. and Mintz, E. (2001) ATPases-Ion Motive. In *Encyclopedia of Life Science*, Nature Publishing Group.
Heldin, C.H. (2001) Signal transduction:overview. In *Encyclopedia of Life Science*, Nature Publishing Group.
Heldt, H.W., Flügge, U.I. and Borchert, S. (1991) Diversity of specificity and function of phosphate translocators in various plastids. *Plant Physiol.* **95**, 341–343.
Henderson, P.J.F. (1993) The 12-trans-membrane helix transporters. *Curr. Opin. Cell Biol.* **5**, 708–721.
Ishizaki-Nishizawa, O., Fujii, T., Azuma, M., Sekiguchi, K., Murata, N., Ohtani, T. and Toguri, T. (1996) Low-temperature resistance of higher plants is significantly enhanced by a nonspecific cyanobacterial desaturase. *Nature Biotechnol.* **14**, 1003–1006.
Junge, W., Lill, H. and Engelbrecht, S. (1997) ATP synthase: an electrochemical transducer with rotary mechanism. *TIBS* **22**, 420–423.
Kammerer, B., Fischer, K., Hilpert, B., Schubert, S. and Gutensohn, M. (1998) Molecular characterization of a carbon transporter in plastids from heterotrophic tissues: the glucose 6-phosphate/phosphate antiporter. *Plant Cell* **10**, 105–117.
Killham, K. (1994) *Soil Ecology*. Cambridge University Press, 242 pp.
Kononova, S.V. and Nesmeyanova, M.A. (2002) Phosphonates and their degradation by microorganisms. In *Biochemistry (Moscow)*, **67**, 184–195.
Kornberg, A., Rao, N.N. and Ault-Riché, D. (1999) Inorganic polyphosphate: a molecule of many functions. *Annu. Rev. Biochem.* **68**, 89–125.
Kulaev, I. and Kulakovskaya, T. (2000) Polyphosphate and phosphate pump. *Rev. Microbiol.* **54**, 709–734.
Liu, C., Muchhal, U.S., Mukatira, U., Konowicz, A.K. and Raghothama, K.G. (1998) Tomato phosphate transporter genes are differentially regulated in plant tissues by phosphorus. *Plant Physiol.* **116**, 91–99.
Matte, A. and Delbaere, L.T.J. (2001) ATP-binding motifs. In *Encyclopedia of Life Science*, Nature Publishing Group.
Mohr, H. and Schopfer, P. (1995) *Plant Physiology*. Springer-Verlag, Heidelberg, 678 pp.
Muchhal, U.S., Pardo, J.M. and Raghothama, K.G. (1996) Phosphate transporters from the higher plant *Arabidopsis thaliana*. *Proc. Natl. Acad. Sci. USA* **93**, 10519–10523.
Noji, H., Yasuda, R., Yoshida, M. and Kinosita, K. (1997) Direct observation of the rotation of F1-ATPase. *Nature* **386**, 299–302.
Pao, S.S., Paulsen, I.T. and Saier, M.H. (1998) Major facilitator superfamily. *Microbiol. Mol. Biol. Rev.* **62**, 1–34.
Parkinson, J.S. (1993) Signal transduction schemes of bacteria. *Cell* **73**, 857–871.
Paul, E.A. and Clark, F.E. (1996) *Soil Microbiology and Biochemistry*. Academic Press, Inc., 340 pp.
Plaxton, W.C. (1996) The organization and regulation of plant glycolysis. *Annu. Rev. Plant Physiol. Mol. Biol.* **47**, 185–214.
Raghothama, K.G. (1999) Phosphate acquisition. *Annu. Rev. Plant. Physiol. Plant Mol. Biol.* **50**, 665–693.
Ridge, I. (2002) *Plants*. Oxford University Press, 345 pp.
Rychter, A.M. and Misulska, M. (1990) The relation between phosphate status and cyanide-resistant respiration in bean roots. *Physiol. Plant* **79**, 663–667.
Sachs, G. and Keeling, D. (2001) Ion Motive ATPases: V- and P-type ATPases. In *Encyclopedia of Life Science*, Nature Publishing Group.

Sakano, K. (1990) Proton/phosphate stoichiometry in uptake of inorganic phosphate by cultured cells of *Catharanthus roseus* (L.). *Plant Physiol.* **93**, 479–483.

Taiz, L. and Zeiger, E. (1998) *Plant Physiology.* Sinauer Assoc. Inc., Sunderland, MA, 792 pp.

Talaro, K. and Talaro, A. (1993) *Foundation in Microbiology.* W.C. Brown Publishers, 685 pp.

Urbina, J.A., Moreno, B., Vierkotter, S., Oldfield, E., Payares, G., Sanoja, C., Bailey, B.N., Wen, Y., Scott, D.A., Moreno, S.N.J. and Docampo, R. (1999) *Trypanosoma cruzi* contains major pyrophosphate stores, and its growth *in vitro* and *in vivo* is blocked by pyrophosphate analogs. *J. Biol. Chem.* **274**, 33609–33615.

Wagner, R., Apley, E.C., Gross, A. and Flügge, U.I. (1989) The rotational diffusion of chloroplast phosphate translocator and of lipid molecules in bilayer membranes. *Eur. J. Biochem.* **182**, 165–173.

Webber, A.N. (2001) Photophosphorylation. In *Encyclopedia of Life Science*, Nature Publishing Group.

PART TWO

Phosphorus in the environment

4.	Background and elevated phosphorus release from terrestrial environments	79
5.	Phosphorus and crop nutrition: principles and practice	93
6.	Transfer of phosphorus to surface waters; eutrophication	120
7.	Environmental chemistry of phosphonic acids	147
8.	Phosphate pollution: a global overview of the problem	174

4
Background and elevated phosphorus release from terrestrial environments

P.M. Haygarth and L.M. Condron

4.1 INTRODUCTION

Release of phosphorus (P) from terrestrial environments can undermine water quality by contributing to eutrophication in lakes and rivers (Daniel *et al.*, 1998; de Clercq *et al.*, 2001; Watson and Foy, 2001; Withers *et al.*, 2000). Eutrophication arises from a complex interrelationship between nutrient status and ecological circumstances (Haygarth *et al.*, 2000a) that results in accelerated growth of algae or water plants (Pierzynski *et al.*, 2000). There is no internationally accepted 'critical' P concentration for water, but the Environment Agency for England and Wales proposed standards of 85 µg total P (TP) l^{-1} (annual geometric mean) for standing fresh waters and 200 µg soluble reactive P (SRP) l^{-1} (annual mean) for running fresh waters (Environment Agency, 1998). Presence of algae causes significant limitations on water use for drinking and fishing, as well as for industrial and recreational uses (Carpenter *et al.*, 1998). In Europe, 55% of river stations reported

© 2004 IWA Publishing. *Phosphorus in Environmental Technology: Principles and Applications.*
Edited by Eugenia Valsami-Jones. ISBN: 1 84339 001 9

annual average dissolved P concentrations in excess of 50 μg P l^{-1} over the period 1992–96 (Crouzet et al., 1999). A report on the state of New Zealand's environment revealed that approximately 10% of shallow lakes were classified as eutrophic (20–50 μg total P l^{-1}) or hyper-eutrophic (>50 μg total P l^{-1}) (Cameron et al., 2002). Accelerated eutrophication has also been linked with large-scale fish kills in some estuaries caused by increased populations of the dinoflagellate *Pfiesteria piscicida*, which in turn has the potential to adversely affect human health (Pierzynski et al., 2000).

The protection of water quality is, therefore, a major international environmental problem. In recent years, the level of focus and attention on point sources has diminished, because of the relative ease by which they have been identified and subsequently controlled (see also Farmer, this volume). In contrast, and partly due to the success with point sources, the perceived and actual importance of diffuse pollution has increased. In the European Union, the Water Framework Directive (2000/60/EC) is a key legislative driver behind this, and aims to restore all waters to 'good ecological status' by 2015, which will require specific action to reduce and control diffuse pollution from terrestrial environments. In this chapter we aim to evaluate the release of P from the terrestrial environments that contribute to these problems. In the first instance we shall consider the concept of background release, before a more in depth assessment of elevated P release from agricultural systems.

4.2 BACKGROUND RELEASE

We have chosen to conceptually separate background from elevated P release in order to emphasise that (1) some release occurs as a natural geological process (background) and (2) some release occurs as a result of long term anthropogenic interference and addition of P (elevated). Despite this conceptual separation, the mechanisms of release *per se* in the two circumstances are not necessarily different. The concept of background can be inferred and extrapolated from various studies (Letkeman et al., 1996; Walker and Syers, 1976) and is necessary so that we can set standards by which to compare types of release and its mitigation. There are many avenues to explore here and there are considerable opportunities for studies of (i) contemporary pristine environments, for example those conducted at Glacier Bay, Alaska (Chapin et al., 1994), and (ii) historically (long term) pristine environments that have since been perturbed, using retrospective palaeo-ecological techniques (Foster and Lees, 1999; Foster and Walling, 1994). Theoretically, background transfers will depend on a combination of factors including the nature of the (i) parent material, and (ii) the nature of climax vegetation. We postulate that, in the long term, background transfers will attain quasi-equilibrium, with changes only occurring as a result of the progression of ecological succession.

4.3 ELEVATED RELEASE

Elevated release exceeds the rate of background transfers because imported P is added to the system and thus the ability of soils to buffer and retain P is exceeded. The term 'elevated' is partially derived from underlying agricultural transfers, described previously by Haygarth et al. (2000c). The most obvious examples include additions of P via fertiliser or feed concentrate (via animal–excreta/manure) to agricultural soils (Frossard et al., 2000; Haygarth et al., 1998c). In Figure 4.1 we illustrate the background and elevated rates transfer showing them in context with one another over the long-term time-scale.

The P release from agricultural soils is generally small compared with the amounts of P added to soil as mineral fertiliser and organic manure. These can exceed $25\,kg\,P\,ha^{-1}\,yr^{-1}$ in many agroecosystems (Cameron et al., 2002; de Clercq et al., 2001; Haygarth et al., 1998a; Sibbesen and Runge-Metzger, 1995). Data from a wide range of field and catchment studies have shown that in most cases annual total P transfer from soil is less than $1\,kg\,ha^{-1}$, although higher rates of transfer (2–$6\,kg\,P\,ha^{-1}\,yr^{-1}$, up to $17\,kg\,P\,ha^{-1}\,yr^{-1}$) have been recorded from soil under intensive pastoral or arable farming, especially when animal manure is applied (Gillingham and Thorrold, 2000; Haygarth and Jarvis, 1999; Hooda et al., 2000; McDowell et al., 2001; Nash et al., 2000; Turner and Haygarth, 2000).

Phosphorus transfer from agricultural soil to water has attracted increased attention in the last 10 years, but it has often been difficult to achieve a balanced perspective on the important issues. Some approaches have highlighted the importance of soil erosion and physical transfer of P with soil particles from land to water. These mechanisms are important on arable soils in, for example, the midwestern USA (Sharpley, 1985; Sharpley and Smith, 1990). Other approaches have recognised the importance of fertiliser inputs and soil P levels, with benchmark work on English arable land by Heckrath et al. (1995), and other researchers noting the importance of manure inputs on grassland (Greatz and Nair, 1995; Lanyon, 1994). To help achieve a balanced perspective on release from agriculture, we have

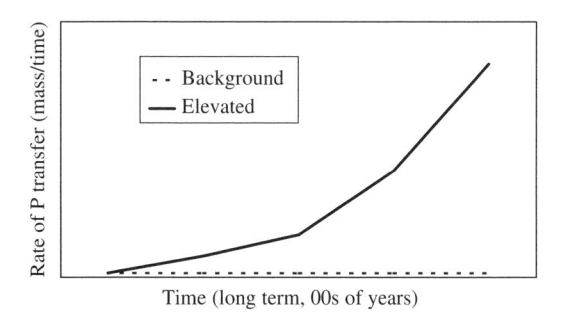

Figure 4.1 Conceptual and theoretical comparison for background and elevated phosphorus release in a long term time-scale, showing increased release following anthropogenic modification in elevated circumstances.

arranged the following subsections into that of sources, mobilisation (by solubilisation and detachment) and transport.

4.3.1 Sources

Sources reflect the input of P to the agricultural and soil reservoirs that represent the long term potential for transfer to the wider environment. Phosphorus transfer from land to water cannot occur without 'sources' of P, which come from (i) indigenous soil P, which relates to soil parent material including some atmospheric deposition (i.e. background), (ii) fertilisers, and (iii) imported livestock feed concentrates, which are returned to the land via direct excretion during grazing or as spread manure.

Continued inputs of P have resulted in significant accumulation of P in topsoil. Haygarth *et al.* (1998a) compiled comprehensive P budgets for representative intensive dairy and extensive upland sheep farming systems in the UK and determined an annual accumulation rate of 26 kg P ha^{-1} in dairy farms compared with only 0.28 kg P ha^{-1} with upland sheep. Withers *et al.* (2001) determined an average P surplus on arable and grassland farms in the UK of 1000 kg P ha^{-1} over the 65 years from 1935 to 2000 (15 kg P ha^{-1} yr^{-1}). High levels of P accumulation in soil have also been reported under intensive farming systems in other parts of Europe and North America (de Clercq *et al.*, 2001; Sibbesen and Runge-Metzger 1995; Sims *et al.*, 2000), together with consequent increases in plant available P (Tunney *et al.*, 1997). The accumulation of P in soil from imported feed is particularly important in areas of intensive livestock production (e.g. pigs, poultry, dairy) where large quantities of manure are applied to land (Sharpley and Tunney, 2000).

4.3.2 Mobilisation

Mobilisation describes the start of the transport process, embracing the way in which P molecules become separated from their source and is defined at the soil profile scale. This includes chemical, biological and physical processes that we group as 'solubilisation' and 'detachment'.

4.3.2.1 Solubilisation

The term 'solubilisation' replaces the term 'dissolution', which can be ambiguous or confused with the specific chemical process, and was first defined in this conceptual sense with regard to P transfer by Haygarth and Jarvis (1999). It refers to the release of molecules or macromolecules of P from soil surfaces and soil biota into soil water for potential transfer away from the point of source. Levels of P in soil leachate after solubilisation have been shown to be sufficiently high to contribute to eutrophication if transferred to watercourses. The magnitude with which solubilisation occurs will be affected by long-term management history and there is strong evidence that solubilisation potential increases with increased soil P status.

Solubilised soil P has been classified into various potentially mobile P forms (Haygarth and Jarvis, 1999), and this is mostly quantified through agronomic soil P tests. Various acid and alkali extractants have been developed as agronomic tests to determine quantities of potentially plant available P in soil and thereby assess P fertiliser requirements (Kamprath and Watson, 1980). These include the Olsen P (0.5 M $NaHCO_3$ = pH 8.5), Bray-1 P (0.025 M HCl + 0.03 M NH_4F), Mehlich-3 P (0.2 M CH_3COOH + 0.25 M NH_4NO_3 + 0.015 M NH_4F + 0.013 M HNO_3 + 0.001 M EDTA) and Morgan P (0.54 M CH_3COOH + 0.7 M CH_3COONa = pH 4.8) tests. Sharpley and Tunney (2000) have described environmental threshold values for P tests that are commonly two to four times greater than the corresponding values for optimum crop production (agronomic thresholds). For example, in the USA, the recommended environmental threshold levels of Bray-1 P are 75 mg kg^{-1} in Michigan and Wisconsin, compared with 150 mg kg^{-1} in Ohio. The corresponding Mehlich-3 P values are 130 and 150 mg kg^{-1} in Oklahoma and Arkansas, respectively. Land application of pig manure in Ireland is prohibited when the soil P test (Morgan P) value is greater than 15 mg kg^{-1} (equivalent to an Olsen P of 60 mg kg^{-1}) (Sharpley and Tunney, 2000). However, caution must be used when interpreting soil P test values within an environmental context as they comprise only a small percent of the total soil P reservoir and do not account for potential detachment. They therefore only provide a measure of potential and not actual P transfer.

Relationships between P transfer in overland or subsurface flow and P inputs and soil P status have also been examined extensively. This has been prompted by the desire to establish appropriate management thresholds for the mitigation of P transfer in agricultural watersheds (Gburek et al., 2000; Sims et al., 2000). This is a difficult task, given the complex combination of properties and processes that control P transfer. Heckrath et al. (1995) observed that concentrations of P in drainage water from long established arable field plots at Rothamsted (UK) were related to topsoil P status. They found that significant P transfer in drainage water ($>150\,\mu g\,P\,l^{-1}$) occurred during winter only when levels of plant available (Olsen) P in topsoil exceeded 60 mg P kg^{-1}. This critical level of soil P was termed the 'change point', and is a physico-chemical phenomenon that describes the relationship between the quantity of P held on adsorption sites and its release to soil solution. Change point is closely related to the degree of P saturation, which can also be used to assess the potential for P transfer from soil in subsurface flow (McDowell and Condron, 2000; Schoumans and Groenendijk, 2000). A number of studies have examined the change point and P saturation phenomena in different soils and agroecosystems (Blake et al., 2002; Daly et al., 2001; Hesketh and Brookes 2000; Hughes et al., 2000; Leinweber et al., 1997; Maguire and Sims, 2002; McDowell and Sharpley, 2001a, b). McDowell and Sharpley (2001a) examined relationships between extractable soil P (water, calcium chloride (0.01 M), soil tests (Olsen P, Mehlich-3 P)) and P loss by overland flow (surface runoff) and subsurface drainage from a range of soils. Comparison of water or calcium chloride extractable P with soil test P revealed change points which were closely related to P loss. Daly et al. (2001) examined P sorption and desorption dynamics for a range of soils in Ireland and confirmed that accelerated P loss was likely to be greater from peat and high organic matter soils. These findings indicate that soil test P levels and the degree of P saturation can be used to predict the risk of P transfer by overland or subsurface flow.

It is important to recognise the influence of physical and biological properties and processes of solubilisation. In particular, preferential flow through root channels and earthworm burrows etc. may effectively bypass a significant proportion of P sorption surfaces in the soil and thereby facilitate rapid transfer of P from the soil surface and topsoil (Haygarth et al., 2000b; Haygarth and Sharpley, 2000; Heathwaite and Dils, 2000). Several studies have also demonstrated that biological processes play an important role in determining the amounts and forms of P transfer from soil. For example, Turner and Haygarth (2001) found that the process of wetting and drying resulted in accelerated release of soluble organic P from soil biomass, which indicates that P loss may be influenced by environmental conditions that affect biological activity in the soil (Figure 4.2). This is confirmed by findings from several studies which revealed that a significant proportion of soluble P, and P in subsurface and overland flow from grassland soils is present as organic P (Haygarth and Jarvis, 1997; Heathwaite and Dils, 2000; MacLaren 1996; Turner and Haygarth, 2000; Turner et al., 2002).

Organic P is becoming increasingly acknowledged to have an important role in determining P solubilisation (Chardon et al., 1997; Hannapel et al., 1963a; Haygarth et al., 1998b) and may be especially important with soils with low fertiliser applications history (Ron Vaz et al., 1993). A recent study by Espinosa et al. (1999) successfully separated organic P compounds in soil leachate using liquid chromatography, and inositol hexaphosphate was found to be the dominant component. This form of P is derived from plant residues, excreta and soil organisms: critical reviews of the processes involved are available (Harrison, 1987; Magid et al., 1996; Stewart and Tiessen, 1987).

Soil biological activity plays a role in organic P mineralisation through ingestion/excretion and immobilising inorganic forms by retention within the biomass

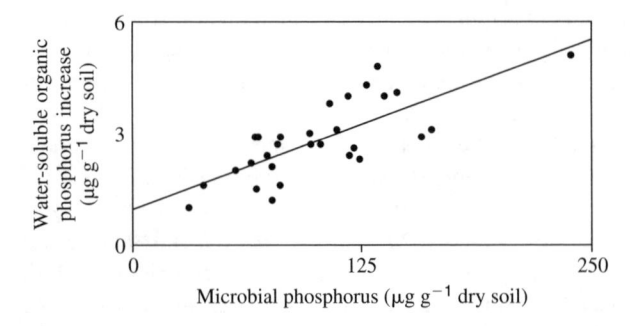

Figure 4.2 Release of water-soluble organic phosphorus after soil drying in relation to soil microbial phosphorus, for a range of permanent grassland soils from England and Wales (adapted from Turner and Haygarth, 2001).

(Brookes *et al.*, 1982; Cole *et al.*, 1978; Patra *et al.*, 1990). The balance between these two processes can influence the size of the potentially soluble P pool (Stewart and Tiessen, 1987). Mineralisation of soil organic P decreases with depth, with an associated increased proportion of organic P and decreased inorganic P (Cole *et al.*, 1977). This is associated with greater soil microbial activity in the surface layers and, since the potential for transfer to water is greatest in the upper few centimetres of soil (Ahuja, 1986; Haygarth *et al.*, 1998b), this may have a significant influence on solubilisation and release in runoff. Hydrolysis of organic P occurs because plant roots and soil microbes can exude extra-cellular acid and alkaline phosphatase enzymes (Brookes *et al.*, 1984; Leprince and Quiquampoix, 1996) which catalyze the hydrolysis of both esters and anhydrides of hydrogen phosphate (Page *et al.*, 1982). The increase in soluble P concentration in runoff with increased storm interval has been found to be related to the phosphatase activity of soils (Sharpley, 1980). Although there have been many references to the role of enzymes in soil ecology (Joner and Jakobsen, 1995; Joner *et al.*, 1995), their precise role in organic P mineralisation in the soil environment remains unclear.

4.3.2.2 Detachment

Detachment of soil particles containing P, often associated with soil erosion, is a physical mechanism for mobilising P from soil into waters (Kronvang, 1990; Sharpley and Smith, 1990). Soil erosion *per se* has been described many times (Burnham and Pitman, 1986; Elliot *et al.*, 1991; Evans 1990; Heathwaite and Burt, 1992; MAFF 1997; Morgan, 1980; Quinton 1997), as has the role of particle transfer in P loss (Catt *et al.*, 1994; Kronvang 1990; Sharpley and Smith, 1990; Zobisch *et al.*, 1994).

Various size thresholds have been used to operationally define detachment, with most examples using $>0.45\,\mu m$, although in some cases $>0.4\,\mu m$ or even $>0.2\,\mu m$ have been used for the threshold between 'dissolved' and 'particulate' P (Haygarth *et al.*, 1997). However, these filter sizes are somewhat arbitrary as P can occur in a continuum of particulate and colloidal sizes down to near molecular sizes (Aiken and Leenheer, 1993; Buffle *et al.*, 1978; De Haan *et al.*, 1984; Fox and Kemprath, 1971; Hannapel *et al.*, 1963b; Mayer and Jarrell, 1995; Nanny *et al.*, 1994). This has practical relevance because Fordham and Schwertmann (1977) have shown that deep penetration of P from liquid cattle manure into a sandy soil was best explained by transport of finely divided soil particles of inorganic P suspended in large volumes of liquid. More recent studies have demonstrated that soil solution and river water samples contain a continuum of $<0.45\,\mu m$ sized solids and macromolecules, often smaller than 10,000 Dalton molecular weight (Haygarth *et al.*, 1997; Matthews *et al.*, 1998). Categorisation of P into dissolved or particulate forms is therefore an analytical convenience and, apart from implications for downstream bioavailability, may have limited environmental or mechanistic relevance. A major challenge for understanding P detachment mechanisms is to untangle the importance of high volume, short-timescale 'soil erosion' events (perhaps associated with larger sized particles) from discrete long-term and more continuous detachment of soil colloids (perhaps mostly smaller particles and macromolecules). Much of the evidence and mechanistic understanding of detachment at the landscape scale is, at best, anecdotal and many fundamental hypotheses remain untested.

4.3.3 Transport

Transport is complex because it embraces a range of spatial and temporal scales that expand from the point of mobilisation, which are confounded by potential changes in P forms. It involves a multiplicity of hydrological conditions and pathways at both the hillslope and the catchment scale. Adding further difficulties to understanding transport mechanisms is the deposition and entrainment that can change P form in space and time. Because of this complexity, there is far less known about transport of P once it has been mobilised. Transport can be separated into transport from hillslope to stream and subsequently in-stream to point of impact.

Transfers from the hillslope to the stream are mainly controlled by a combination of hydrological factors that include the intensity and duration of rainfall (or irrigation) events, together with the spatial variables of scale and pathways (Haygarth et al., 2000b). Slope (topography) and drainage (substrate permeability) mainly influence P transfer pathways at the field scale. Phosphorus transfer to surface water occurs via overland flow and/or subsurface flow (interflow, drainflow) as base-flow and storm-flow, while transfer to groundwater occurs via a combination of throughflow (percolating water) and preferential flow.

Critical to transport potential from hillslope to stream are so called 'incidental transfers' that occur when applications of manure or fertiliser coincide with discharges (Haygarth and Jarvis, 1999), resulting in direct transfers to watercourses without the opportunity for the P molecule to be incorporated into the soil matrix (Preedy et al., 2001). The interaction between hydrology and agricultural management practice is clearly important with incidental transfer mechanisms (Haygarth et al., 2000b). Despite the proliferation of anecdotal evidence, there are only a few robust scientific studies providing evidence for incidental transfers (e.g. Harris et al., 1995; Haygarth and Jarvis, 1997) and an obvious opportunity exists for the development of some scientific criteria upon which to define and then subsequently manage these occurrences.

In streams, rivers and estuaries, the extent, nature and dynamics of interactions between soluble and particulate P in water and sediments play a critical role in the delivery of P to sites of potential impact. Baldwin et al. (2002) highlighted the importance of P-sediment interactions in determining the fate and environmental impact of P transfer from soil to water, which in turn are determined by a complex combination of chemical, physical and biological properties and processes. McDowell et al. (2001) demonstrated that base-flow concentrations of dissolved reactive P and total P in stream water were influenced by the sediment P sorption properties in an agricultural watershed. It is therefore important to examine the nature and associated chemical and biological processes that influence P retention and release in sediments.

Clearly, for all transport, 'forms' of P can be varied both in water and in soil and it is important when approaching the problem that they are tightly defined. Forms in water include inorganic and organic species, which may be defined operationally, mostly using molybdenum–blue chemistry (Murphy and Riley, 1962), or chemically (Espinosa *et al.*, 1999). The classification of forms of P in soil and water is a fundamental need and is defined and discussed in relation to P transfer by Haygarth and Sharpley (1999).

4.4 CONCLUSIONS

We have to accept that some P release from the terrestrial surface is inevitable through background transfers. There is a definite need for more fundamental studies here, as these transfers seem to have been overlooked in relation to elevated agricultural transfers. Issues such as the natural buffering capacities of different parent materials need to be examined, and data from existing literature re-examined. Theoretically, background P transfer rates will be primarily influenced by a combination of (i) parent material, (ii) effective rainfall, and (iii) drainage. Soil age and disturbance may also be issues that require focus. Until we have established a model framework for background transfers, we have no effective reference point against which to calibrate the extent of P transfers from agricultural soils.

In assessing P transfer from terrestrial surfaces, it is essential to consider an interdisciplinary perspective. This requires appreciation of physical and biochemical processes, within the framework sources-mobilisation transport. There is therefore no simple solution to the problem, but in the long term, elevated transfers from agricultural land are unacceptable and reflect the build up of soil P reserves and associated potential for increased losses (Heckrath *et al.*, 1995; Pote *et al.*, 1996). Accordingly, while it takes a number of years to build up P reserves, if the need arises in sensitive catchments to deplete these, this will also take several years to achieve.

REFERENCES

Ahuja, L.R. (1986) Characterization and modelling of chemical transfer to runoff. *Adv. Soil Sci.* **4**, 149–188.

Aiken, G. and Leenheer, J. (1993) Isolation and chemical characterization of dissolved and colloidal organic matter. *Chem. Ecol.* **8**, 135–151.

Baldwin, D.S., Mitchell, M.A. and Olley, J.M. (2002) Pollutant-sediment interactions: sorption, reactivity and transport of phosphorus. In *Agriculture, Hydrology and Water Quality* (eds Haygarth, P.M. and Jarvis, S.C.), pp. 265–280. CAB International, Wallingford.

Blake, L., Hesketh, N., Fortune, S. and Brookes, P.C. (2002) Assessing phosphorus 'change-points' and leaching potential by isotopic exchange and sequential fractionation. *Soil Use and Manage.* **18**, 199–207.

Brookes, P.C., Powlson, D.S. and Jenkinson, D.S. (1982) Measurement of microbial biomass phosphorus in soil. *Soil Biol. Biochem.* **14**, 319–329.

Brookes, P.C., Powlson, D.S. and Jenkinson, D.S. (1984) Phosphorus in the soil microbial biomass. *Soil Biol. Biochem.* **16**, 169–175.

Buffle, J., Deladoey, P. and Haerdi, W. (1978) The use of ultrafiltration for the separation and fractionation of organic ligands in freshwaters. *Analytica Chimica Acta* **101**, 339–357.

Burnham, C.P. and Pitman, J.I. (1986) *Soil Erosion*. The Ashford Press, Ashford, Kent.

Cameron, K.C., Di, H.J. and Condron, L.M. (2002) Nutrient and pesticide transfer from agricultural soils to water in New Zealand. In *Agriculture, Hydrology and Water Quality* (ed. P.M. Haygarth and S.C. Jarvis), pp. 373–393. CAB International, Wallingford.

Carpenter, S.R., Caraco, N.F., Correll, D.L., Howarth, R.W., Sharpley, A.N. and Smith, V.H. (1998) Nonpoint pollution of surface waters with phosphorus and nitrogen. *Ecol. Appl.* **8**, 559–568.

Catt, J.A., Quinton, J.N., Rickson, R.J. and Styles, P. (1994) Nutrient losses and crop yields in the Woburn erosion reference experiment. In *Conserving Soil Resources: European Perspectives* (ed. R.J. Rickson). CAB International, Wallingford.

Chapin, F.S., Walker, L.R., Fastie, C.L. and Sharman, L.C. (1994) Mechanisms of primary succession following deglaciation at Glacier Bay, Alaska. *Ecol. Monogr.* **64**, 149–175.

Chardon, W.J., Oenema, O., del Castilho, P., Vriesema, R., Japenga, J. and Blaauw, D. (1997) Organic phosphorus solutions and leachates from soils treated with animal slurries. *J. Environ. Qual.* **26**, 372–378.

Cole, C.V., Elliot, E.T., Hunt, H.W. and Coleman, D.C. (1978) Trophic interactions in soils as they affectenergy and nutrient dynamics. V. Phosphorus transformations. *Microb. Ecol.* **4**, 381–387.

Cole, C.V., Innis, G.S. and Stewart, J.W.B. (1977) Simulation of phosphorus cycling in semi-arid grassland. *Ecology* **58**, 1–15.

Crouzet, P., Leonard, J., Nixon, S., Rees, Y., Parr, W., Laffon, L., Bøgestrand, J., Kristensen, P., Lallana, C., Izzo, G., Bokn, T. and Bak, J., eds (1999) *Nutrients in European Ecosystems*, pp. 1–155. European Environment Agency, Copenhagen.

Daly, K., Jeffrey, D. and Tunney, H. (2001) The effect of soil type on phosphorus sorption capacity and desorption dynamics in Irish grassland soils. *Soil Use and Manage.* **17**, 12–20.

Daniel, T.C., Sharpley, A.N. and Lemunyon, J.L. (1998) Agricultural phosphorus and eutrophication: A symposium overview. *J. Environ. Qual.* **27**, 251–257.

De Clercq, P., Gertsis, A.C., Hofman, G., Jarvis, S.C., Neeteson, J.J. and Sinabell, F. (2001) *Nutrient Management Legislation in European Countries*, Wageningen.

De Haan, H., De Boer, T., Voerman, J., Kramer, A.J. and Moed, J.R. (1984) Size classes of 'dissolved' nutrients in shallow, alkaline, humic and eutrophic Tjeukemeer, The Netherlands, as fractionated by ultrafiltration. *Verh. Internat. Verein. Limnol.* **22**, 876–881.

Elliot, W.J., Foster, G.R. and Elliot, A.V. (1991) Soil erosion: processes, impacts and prediction. In *Soil Management for Sustainability* (ed. R. Lal and F.J. Pierce), pp. 25–34. Soil and Water Conservation Society, Alberta.

Environment Agency (1998) *Aquatic Eutrophication in England and Wales: A Proposed Management Strategy, Consultative Report*. Environment Agency, Bristol.

Espinosa, M., Turner, B.L. and Haygarth, P.M. (1999) Pre-concentration and separation of trace phosphorus compounds in soil leachate. *J. Environ. Qual.* **28**, 1497–1504.

Evans, R. (1990) Soils at risk of accelerated erosion in England and Wales. *Soil Use and Manage.* **6**, 125–131.

Fordham, A.W. and Schwertmann, U. (1977) Composition and reactions of liquid manure (Gulle), with particular reference to phosphate: 1. Analytical composition and reaction with poorly crystalline iron oxide (ferrihydrite). *J. Environ. Qual.* **6**, 133–136.

Foster, I.D.L. and Lees, J.A. (1999) Changing headwater suspended sediment yields in the LOIS catchments over the last century: a palaeolimnological approach. *Hydrol. Processes* **13**, 1137–1153.

Foster, I.D.L. and Walling, D.E. (1994) Using reservoir deposits to reconstruct changing sediment yields and sources in the catchment of the Old Mill Reservoir, South Devon, UK, over the past 50 years. *Hydrol. Sci.* **39**, 347–368.

Fox, R.L. and Kemprath, E.J. (1971) Adsorption and leaching of P in acid organic soils and high organic matter sand. *Soil Sci. Soc. Am. J.* **35**, 154–158.

Frossard, E., Condron, L.M., Oberson, A., Sinaj, S. and Fardeau, J.C. (2000) Processes governing phosphorus availability in temperate soils. *J. Environ. Qual.* **29**, 15–23.

Gburek, W.J., Sharpley, A.N., Heathwaite, A.L. and Folmar, G.J. (2000) Phosphorus management at the watershed scale: a modification of the phosphorus index. *J. Environ. Qual.* **29**, 130–144.

Gillingham, A.G. and Thorrold, B.S. (2000) A review of New Zealand research measuring phosphorus in runoff from pasture. *J. Environ. Qual.* **29**, 88–96.

Greatz, D.A. and Nair, V.D. (1995) Fate of phosphorus in Florida spodisols contaminated with cattle manure. *Ecol. Eng.* **5**, 163–181.

Hannapel, R.J., Fuller, W.H., Bosma, S. and Bullock, J.S. (1963a) Phosphorus movement in a calcareous soil: I. Predominance of organic forms of phosphorus in phosphorus movement. *Soil Sci.* **97**, 350–357.

Hannapel, R.J., Fuller, W.H. and Fox, R.H. (1963b) Phosphorus movement in a calcareous soil: II. Soil microbial activity and organic phosphorus movement. *Soil Sci.* **97**, 421–427.

Harris, R.A., Heathwaite, A.L. and Haygarth, P.M. (1995) High temporal resolution sampling of P exported from grassland soil during a storm, and the impact of slurry additions. In *Proceedings of the International Workshop on Phosphorus Loss to Water from Agriculture* (ed. A.E. Johnston), pp. 25–26. TEAGASC, Johnstown Castle, Wexford, Ireland.

Harrison, A.F. (1987) *Soil Organic Phosphorus – A Review of World Literature*, pp. 1–257. CAB International, Wallingford.

Haygarth, P.M., Chapman, P.J., Jarvis, S.C. and Smith, R.V. (1998a) Phosphorus budgets for two contrasting farming systems in the UK. *Soil Use and Manage.* **14**, 160–167.

Haygarth, P., Dils, R. and Leaf, S. (2000a) Phosphorus. In *Diffuse Pollution Impacts* (ed. D'Arcy, B.J., Ellis, J.B., Ferrier, R.C., Jenkins, A. and Dils, R.), pp. 73–84. Terence Dalton Publishers, The Lavenham Press, Lavenham, Suffolk.

Haygarth, P.M., Heathwaite, A.L., Jarvis, S.C. and Harrod, T.R. (2000b) Hydrological factors for phosphorus transfer from agricultural soils. *Adv. Agron.* **69**, 153–178.

Haygarth, P.M., Hepworth, L. and Jarvis, S.C. (1998b) Forms of phosphorus transfer in hydrological pathways from soil under grazed grassland. *E. J. Soil Sci.* **49**, 65–72.

Haygarth, P.M. and Jarvis, S.C. (1997) Soil derived phosphorus in surface runoff from grazed grassland lysimeters. *Water Res.* **31**, 140–148.

Haygarth, P.M. and Jarvis, S.C. (1999) Transfer of phosphorus from agricultural soils. *Adv. Agron.* **66**, 195–249.

Haygarth, P.M., Jarvis, S.C., Chapman, P. and Smith, R.V. (1998c) Phosphorus budgets for two contrasting grassland farming systems in the UK. *Soil Use and Manage.* **14**, 160–167.

Haygarth, P.M. and Sharpley, A.N. (2000) Terminology for phosphorus transfer. *J. Environ. Qual.* **29**, 10–15.

Haygarth, P.M., Turner, B.L., Fraser, A.I., Jarvis, S.C., Harrod, T.R., Nash, D.M. and Halliwell, D.J. (2000c) Prioritising mitigation of soil phosphorus transfer in relation to water flows. In *Grassland Farming: Balancing Environmental and Economic Demands* (ed. Soegaard, K., Ohlsson, C., Sehested, J., Hutchings, N.J. and Kristensen, T.), Vol. **5**. European Grassland Federation, Holstebro.

Haygarth, P.M., Warwick, M.S. and House, W.A. (1997) Size distribution of colloidal molybdate reactive phosphorus in river waters and soil solution. *Water Res.* **31**, 439–442.

Heathwaite, A.L. and Burt, T. (1992) The evidence for past and present erosion in the Slapton Catchment. In *Soil Erosion, Past and Present: Archeological and Geographical Perspectives* (ed. Bell, M. and Boardman, J.), Vol. **22**, pp. 89–100. Oxbow Monograph.

Heathwaite, A.L. and Dils, R.M. (2000) Characterising phosphorus loss in surface and subsurface hydrological pathways. *Sci. Total Environ.* **251/252**, 523–538.

Heckrath, G., Brookes, P.C., Poulton, P.R. and Goulding, K.W.T. (1995) Phosphorus leaching from soils containing different phosphorus concentrations in the Broadbalk Experiment. *J. Environ. Qual.* **24**, 904–910.

Hesketh, N. and Brookes, P.C. (2000) Development of an indicator for risk of phosphorus leaching. *J. Environ. Qual.* **29**, 105–110.

Hooda, P.S., Edwards, A.C., Anderson, H.A. and Miller, A. (2000) A review of water quality concerns in livestock farming areas. *Sci. Total Environ.* **250**, 143–167.

Hughes, S., Reynolds, B., Bell, S.A. and Gardner, C. (2000) Simple phosphorus saturation index to estimate risk of dissolved P in runoff from arable soils. *Soil Use and Manage.* **16**, 206–210.

Joner, E.J. and Jakobsen, I. (1995) Growth and extracellular phosphatase activity of arbuscular mycorrhizal hyphae as influenced by soil organic matter. *Soil Biol. Biochem.* **27**, 1153–1159.

Joner, E.J., Magid, J., Gahoonia, T.S. and Jakobsen, I. (1995) Phosphorus depletion and activity if phosphatases in the rhizosphere of mycorrhizal and non-mycorrhizal cucumber (*Cucumis sativus* L.). *Soil Biol. Biochem.* **27**, 1145–1151.

Kamprath, E.J. and Watson, M.E. (1980) Conventional soil and tisssue tests for assessing the phosphorus ststus of soils. In *The Role of Phosphorus in Agriculture* (ed. Khasawneh, F.E., Sample, E.C. and Kamprath, E.J.), pp. 433–469. ASA-CSSA-SSSA, Madison.

Kronvang, B. (1990) Sediment associated phosphorus transport from two intensively farmed catchment areas. In *Soil Erosion on Agricultural Land* (ed. J. Boardman, L.D.L. Foster and J.A. Dearing), pp. 131–330. John Wiley and Sons, Chichester.

Lanyon, L.E. (1994) Dairy manure and plant nutrient management issues affecting water quality and the dairy industry. *J. Dairy Sci.* **77**, 1999–2007.

Leinweber, P., Lunsmann, F. and Eckhardt, K.U. (1997) Phosphorus sorption capacities and saturation of soils in two regions with different livestock densities in northwest Germany. *Soil Use Manage.* **13**, 82–89.

Leprince, F. and Quiquampoix, H. (1996) Extracellular enzyme activity in soil: effect of pH and ionic strength on the interaction with montmorillonite of two acid phosphatases secreted by the ectomycorrhizal fungus *Hebeloma cylindrosporum*. *E. J. Soil Sci.* **47**, 511–522.

Letkeman, L.P., Tiessen, H. and Campbell, C.A. (1996) Phosphorus transformations and redistribution during pedogenesis of western Canadian soils. *Geoderma* **71**, 201–218.

MacLaren, J.P. (1996) *Environmental effects of planted forests in New Zealand*, New Zealand Forest Research Institute, Rotorua, New Zealand.

MAFF (1997) *Controlling Soil Erosion*. Her Majesty's Stationary Office, London.

Magid, J., Tiessen, H. and Condron, L.M. (1996) Dynamics of organic phosphorus in soils under natural and agricultural ecosystems. In *Humic Substances in Terrestrial Ecosystems* (ed. Piccolo, A.), pp. 429–466. Elsevier Science.

Maguire, R.O. and Sims, J.T. (2002) Soil testing to predict phosphorus leaching. *J. Environ. Qual.* **31**, 1601–1609.

Matthews, R.A., Preedy, N., Heathwaite, A.L. and Haygarth, P.M. (1998) Transfer of colloidal forms of phosphorus from grassland soils. In *3rd International Conference on Diffuse Pollution* (ed. SEPA), pp. 11–16. IAWQ, Edinburgh.

Mayer, T.D. and Jarrell, W.M. (1995) Assessing colloidal forms of phosphorus in the Tualatin River Basin. *J. Environ. Qual.* **24**, 1117–1124.

McDowell, R.W. and Condron, L.M. (2000) Chemical nature and potential mobility of phosphorus in fertilized grassland soils. *Nutrient Cycling in Agroecosystems* **57**, 225–233.

McDowell, R.W. and Sharpley, A.N. (2001a) Approximating phosphorus release from soils to surface runoff and subsurface drainage. *J. Environ. Qual.* **30**, 508–520.

McDowell, R.W. and Sharpley, A.N. (2001b) Phosphorus losses in subsurface flow before and after manure application to intensively farmed land. *Sci. Total Environ.* **278**, 113–125.

McDowell, R.W., Sharpley, A.N. and Folmer, G. (2001) Phosphorus export from an agricultural watershed. *J. Environ. Qual.* **30**, 1587–1595.

Morgan, R.P.C. (1980). Soil erosion and conservation in Britain. *Prog. Phys. Geogr.* **4**, 24–47.

Murphy, J. and Riley, J.P. (1962) A modified single solution method for the determination of phosphate in natural waters. *Analytica Chimica Acta* **27**, 31–36.

Nanny, M., Kim, S., Gadomski, J.E. and Minear, R.A. (1994) Aquatic soluble unreactive phosphorus: concentration by ultrafiltration and reverse osmosis membranes. *Water Res.* **6**, 1355–1365.

Nash, D., Hannah, M., Halliwell, D.J. and Murdoch, C. (2000) Factors affecting phosphorus export from a pasture-based grazing system. *J. Environ. Qual.* **29**, 1160–1166.

Page, A.L., Miller, R.H. and Kenney, D.R. (1982) *Methods of Soil Analysis. Part 2. Chemical and Microbiological Properties*, 2nd edn. American Society of Agronomy, Inc. and Soil Science Society of America, Inc., Madison, Wisconsin.

Patra, D.D., Brookes, P.C., Coleman, K. and Jenkinson, D.S. (1990) Seasonal changes in soil microbial biomass in an arable and a grassland soil which have been under uniform management for many years. *Soil Biol. Biochem.* **22**, 739–742.

Pierzynski, G.M., Sims, J.T. and Vance, G.F. (2000) Soil phosphorus and environmental quality. In *Soils and Environmental Quality*, pp. 155–207. CRC Press, Boca Raton.

Pote, D.H., Daniel, T.C., Sharpley, A.N., Moore, P.A., Edwards, D.R.J. and Nichols, D.J. (1996) Relating extractable soil phosphorus to phosphorus losses in runoff. *Soil Sci. Soc. Am. J.* **60**, 855–859.

Preedy, N., McTiernan, K.B., Matthews, R., Heathwaite, L. and Haygarth, P.M. (2001) Rapid incidental phosphorus transfers from grassland. *J. Environ. Qual.* **30**, 2105–2112.

Quinton, J.N. (1997) Reducing predictive uncertainty in model simulations: a comparison of two methods using the European Soil Erosion Model (EUROSEM). *Catena* **30**, 101–117.

Ron Vaz, M.D., Edwards, A.C., Shand, C.A. and Cresser, M.S. (1993) Phosphorus fractions in soil solution: Influence of soil acidity and fertilizer additions. *Plant Soil* **148**, 179–183.

Schoumans, O.F. and Groenendijk, P. (2000) Modeling soil phosphorus levels and phosphorus leaching from agricultural land in the Netherlands. *J. Environ. Qual.* **29**, 111–116.

Sharpley, A.N. (1980) The effect of storm interval on the transport of soluble phosphorus in runoff. *J. Environ. Quality* **9**, 575–578.

Sharpley, A.N. (1985) The selective erosion of plant nutrients in runoff. *Soil Sci. Soc. Am. J.* **49**, 1527–1534.

Sharpley, A.N. and Smith, S.J. (1990) Phosphorus transport in agricultural runoff: The role of soil erosion. In *Soil Erosion on Agricultural Land* (ed. Boardman, J., Foster, L.D.L. and Dearing, J.A.), pp. 351–366. John Wiley and Sons, Chichester, New York.

Sharpley, A.N. and Tunney, H. (2000) Phosphorus research strategies to meet agricultural and environmental challenges of the 21st century. *J. Environ. Qual.* **29**, 176–181.

Sibbesen, E. and Runge-Metzger, A. (1995) Phosphorus balance in European agriculture. In *Phosphorus cycling in terrestrial and aquatic ecosystems: Global perspective* (ed. Tiessen, H.), pp. 43–58. John Wiley, New York.

Sims, J.T., Edwards, A.C., Schoumans, O.F. and Simard, R. (2000) Integrating soil phosphorus testing into environmentally based agricultural management practices. *J. Environ. Qual.* **29**, 60–71.

Stewart, J.W.B. and Tiessen, H. (1987) Dynamics of soil organic phosphorus. *Biogeochemistry* **4**, 41–60.
Tunney, H., Breeuwsma, A., Withers, P.J.A. and Ehlert, P.A.I. (1997) Phosphorus fertilizer strategies: Present and future. In *Phosphorus Loss from Soil to Water* (ed. Tunney, H., Carton, O.T., Brookes, P.C. and Johnston, A.E.), pp. 177–204. CAB International, Wallingford.
Turner, B.L. and Haygarth, P.M. (2000) Phosphorus forms and concentrations in leachate under four grassland soil types. *Soil Sci. Soc. Am. J.* **64**, 1090–1097.
Turner, B.L. and Haygarth, P.M. (2001) Phosphorus solubilisation in rewetted soils. *Nature* **411**, 258.
Turner, B.L., McKelvie, I.D. and Haygarth, P.M. (2002) Characterization of water-extractable soil organic phosphorus by phosphatase hydrolysis. *Soil Biol. Biochem.* **34**, 27–35.
Walker, T.W. and Syers, J.K. (1976) The fate of phosphorus during pedogenesis. *Geoderma* **15**, 1–19.
Watson, C.J. and Foy, R.H. (2001) Environmental impacts of nitrogen and phosphorus cycling in grassland systems. *Outlook on Agr.* **30**, 117–127.
Withers, P.J.A., Davidson, I.A. and Foy, R.H. (2000) Prospects for controlling nonpoint phosphorus loss to water: a UK perspective. *J. Environ. Qual.* **29**, 167–175.
Withers, P.J.A., Edwards, A.C. and Foy, R.H. (2001) Phosphorus cycling in UK agriculture and implications for phophorus loss. *Soil Use and Manage.* **17**, 139–149.
Zobisch, M.A., Richter, C., Heiligtag, B. and Schlott, R. (1994) Nutrient losses from cropland in the central highlands of Kenya due to surface runoff and soil erosion. *Soil Tillage Res.* **33**, 109–116.

5
Phosphorus and crop nutrition: principles and practice

T.D. Evans and A.E. Johnston

5.1 INTRODUCTION

Plants need an adequate supply of light, to produce sugars by photosynthesis in their green leaves, carbon dioxide and water, to supply carbon, hydrogen and oxygen, and a number of other nutrients, including phosphorus, which they acquire from the soil. The rooting medium in which they grow must give support and provide a balanced supply of water and air because the roots respire and need oxygen. This balance is especially important for terrestrial plants because prolonged water logging can kill the roots.

Figure 5.1 is a representation of the 'Law of the Minimum' enunciated by Justus von Liebig, the famous German chemist, building on the earlier (1830–1840) work of his compatriot, Carl Sprengel. The law is illustrated using three nutrients, nitrogen (N), phosphorus (P) and potassium (K) and a stylised plant. In the first case (1) there is sufficient N and K available to the crop but too little P. Applying P corrects this deficiency (2), and so K becomes limiting. When K deficiency is corrected the optimum growth is obtained (3). In a similar way, once a deficiency of any other

© 2004 IWA Publishing. *Phosphorus in Environmental Technology: Principles and Applications.*
Edited by Eugenia Valsami-Jones. ISBN: 1 84339 001 9

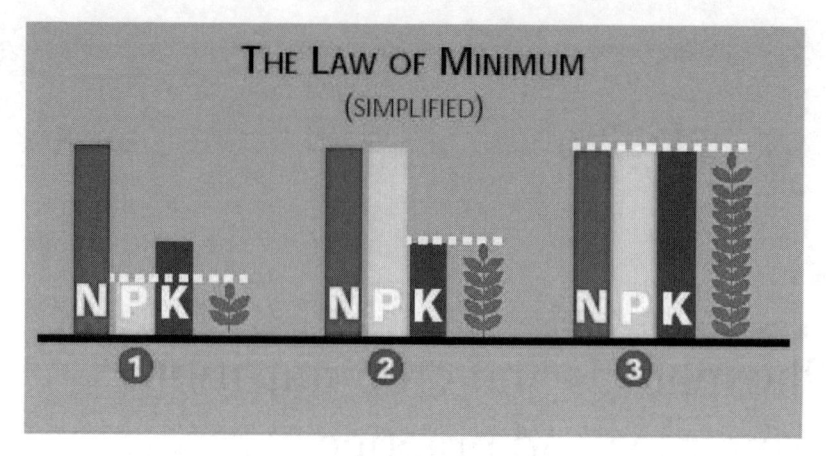

Figure 5.1 Simplified schematic of the law of minimum.

nutrient has been identified it must be corrected for the plant to achieve the maximum benefit from all the other nutrients. In some cases other factors, like soil structure or water supply, can limit yields irrespective of how much nutrients are applied.

The plant nutrients required from soil in the largest amounts (the major nutrients) are N, P, K, calcium (Ca), magnesium (Mg) and sulphur (S). In addition, other elements are needed in small or trace amounts (the minor elements). These include boron, chlorine, copper, iron, manganese, molybdenum, zinc and silicon. Other elements have been found important for some plants, symbiotic bacteria and/or for adequate animal and human nutrition (e.g. cobalt, iodine, selenium and sodium). For a more detailed account of the function of these elements see Mengel and Kirkby (1986). The availability of all nutrients to plants may be affected by both their quantity and distribution in soil, by soil pH and redox status (oxidation/reduction potential) and the amount of soil organic matter. If any nutrient is severely deficient, the effect of its application on plant growth can be dramatic. For example, in recent years, following controls on atmospheric emissions, the deposition of S to soil has been greatly reduced, sulphur deficiency in crops is becoming widespread and an application of S can dramatically increase growth (McGrath *et al*., 2002).

The composition of any soil is the product of the effects of a range of factors, including the parent material and the extent of weathering and aerial deposition. In many soils only a small proportion of the total quantity of any nutrient is readily available for uptake by plant roots. If the parent material is deficient in an element, then the soil will also be deficient. The amount that is available depends, amongst other factors, on the weathering of the parent material, nutrient cycling through vegetation, release from and exchange with reserves within the soil, and additions through manures and fertilisers. Besides by plant uptake, nutrients may be lost from the rooting zone within the soil either in solution in the drainage water (leaching) or attached to particles of soil (erosion by water or wind), or in the case of N by gaseous losses to the atmosphere.

This chapter discusses P in crop nutrition. Many soils are inherently deficient in P and where there is a long history of agriculture, as in Europe, yields until comparatively recently were frequently limited by this deficiency. This was because recycling of P through waste materials, like farmyard manure, was insufficient to meet demand. In Europe, this situation began to change in the early 1800s first with the importation of guano, and then from the mid-1800s with the increasing availability of superphosphate. British imports of guano exceeded 25,000 t in 1846 and in the late 1850s reached 300,000 t annually (Parliamentary Papers of the period). Phosphorus deficiency still limits yields in many poor developing countries.

Lack of plant available P in soils limits yields because P is essential to all known life forms. It is a key element in many physiological and biochemical processes and in these roles it cannot be replaced by any other element. It is found in every cell in all living organisms, for example it is in DNA and RNA molecules and is also an essential component of the energy transport systems in all cells (see Dobrota this volume).

5.2 PHOSPHORUS RESOURCES

Phosphorus is present in nature as phosphates, which can be inorganic or organic and range from the simple to the very complex, it does not occur as elemental P. Phosphorus is eleventh in order of abundance in the earth's crust, it is not a rare element, but the concentration in many rocks is very small. There are both sedimentary and igneous deposits that are sufficiently rich in P that extraction is commercially viable. These sources are found throughout the world and over 30 countries are currently producing phosphates for domestic use and/or for international trade. The main commercial deposits being exploited at present are in the United States, Morocco, China, West Africa, the Former Soviet Union, Israel and Jordan. The first three of these countries currently produce about two thirds of the global P requirement.

Because P is essential for all living processes, there is concern that the exploitation of this non-renewable resource to meet current demands is not sustainable. It is very important, however, to make a distinction between reserves, i.e. deposits that can be currently exploited in an economically viable way, and resources (or potential reserves), i.e. deposits that could be used subject to advances in processing technology, or the value of the finished product. Estimating reserves and resources is not easy, not least because a country or company may consider its estimates commercially sensitive. Estimates of current reserves that can be exploited vary from 250 years to as little as 60 years. One authoritative estimate puts economic reserves at about 100 years at the current rate of extraction (Driver *et al.*, 1999). If known potential reserves are taken into account, forecasts range from 600–1000 years. These estimates do not include the possible discovery of 'new' P resources. Yet, however large or small the reserves/resources may be, they are finite and must be used efficiently and in a sustainable way.

Two major opportunities exist for increasing the life expectancy of the world's P resources: recycling by recovery from municipal and other waste materials and in the efficient use of P in agriculture, both as fertilisers and organic manures. At present in Western Europe, of the total P use some 79% goes to make fertilisers, about 11% for feed grade additives for animal feeds and 7% for detergents (Johnston and Steén, 2000).

5.3 PHOSPHORUS FERTILISERS

In Britain in the early 1800s it was observed that bone trimmings, from the manufacture of bone handles for steel cutlery, when applied to land increased the yield of crops, especially turnips. This lead to a considerable trade in bones that was specific to the UK. Thompson (1968) estimated that during 1837–1842 the average annual British domestic supply was 27,000 t whilst imports averaged 46,000 t annually. But bones were not effective on all soils and many people started to treat bones with acid, mainly sulphuric acid. J. B. Lawes was among the first to do this and as a result of many experiments at Rothamsted, his small agricultural estate at Harpenden, some 40 km north of London, he established the optimum ratio of bones to sulphuric acid and the acid strength. Lawes' product was a solid cake that could be crushed to a powder and applied to land by machine. He called it superphosphate of lime (single superphosphate) and took out a manufacturing patent in 1842. In 1843 he had a factory in commercial production in London (Johnston, 1994) and is now often credited with founding the 'artificial' fertiliser industry. In 1843, with J. H. Gilbert, a chemist, he founded the Rothamsted Experimental Station (now Rothamsted Research), the oldest agricultural research station in the world. Bones were soon replaced first by coprolites then by phosphate rock most probably imported from North Africa. The active ingredient of single superphosphate was water-soluble monocalcium phosphate (MCP), $Ca(H_2PO_4)_2$.

Much of the P in bones and in P bearing minerals (phosphate rock, PR) is present as species of apatite; in bones a microcrystalline calcium phosphate, akin to hydroxylapatite, in PR principally calcium apatites related to fluorapatite $Ca_{10}(PO_4)_6F_2$. However there is considerable substitution of F^- by Cl^-, OH^- and CO_3^{2-}; of Ca^{2+} by Ba^{2+}, Mg^{2+}, Mn^{2+}, and Sr^{2+} and also in sedimentary rocks of PO_4^{3-} by $(CO_3^{2-} + F^-)$ and Ca^{2+} by Na^+ and Mg^{2+}. Non-nutrient and potentially toxic elements, like cadmium and lead, also substitute. In the initial stages of manufacture, as much as possible of the non-P bearing material is removed after crushing (beneficiation) to give a more uniform product for acid treatment. The phosphate rock can be treated with sulphuric acid to make single superphosphate or phosphoric acid, or with nitric acid to give nitrophosphates, or with phosphoric acid to give triple superphosphate (TSP). Monoammonium phosphate (MAP) or diammonium phosphate (DAP) are obtained by adding ammonia to phosphoric acid. TSP, MAP and DAP are the principal phosphatic fertilisers traded today.

In 1900, the United States was the world's largest producer of PR, accounting for 48% of world production of 3.15 million tonnes (M t). By that time, Florida had

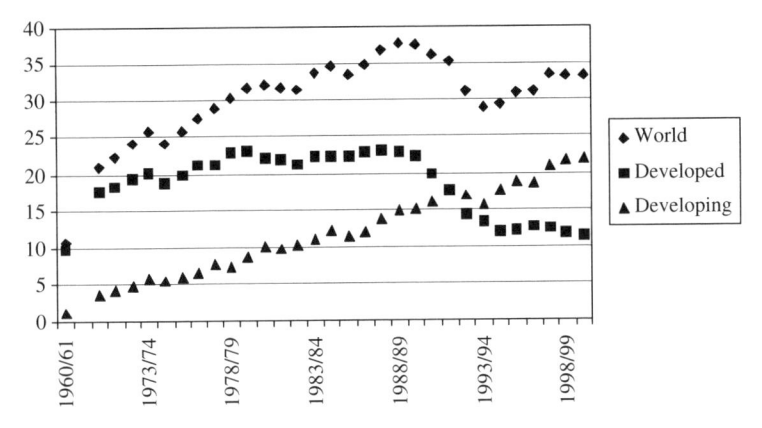

Figure 5.2 Phosphate fertiliser consumption – millions tonnes P_2O_5 (source http:// www.fertilizer.org/ifa/statistics).

become the dominant producing State surpassing South Carolina, where phosphate rock was first discovered and mined in the United States in 1867. In 1928, the flotation process (for beneficiation) was patented enabling companies to recover up to 95% of the phosphate that was previously discarded. In 2000, the United States was still the world's largest producer of PR, accounting for nearly 29% of total world production of 120 M t. However, in the early 1990s the United States producers shifted emphasis to exporting higher value fertiliser materials (see Jasinski, 2000 for detailed world production statistics).

Figure 5.2 shows the pattern of phosphate fertiliser use, expressed as P_2O_5, from 1960/61 to 1999/2000. Use in the developed world declined in the last decade of the twentieth century, but increased in the developing world so, after a period of stable or even declining world use, the trend is likely to be one of increasing global use.

Fertiliser regulations in many countries require the declaration of the phosphorus content, expressed as P_2O_5, as a warranty of quality. However, some countries declare the phosphate content of fertilisers as P rather than P_2O_5, which can be confusing. In the UK, the legal requirement for this dates from 1895 and Lawes was influential here too because he wanted to protect the reputation of his superphosphate from inferior products produced with insufficient sulphuric acid or diluted with inert material. In the EU and many other countries, for fertilisers with more than 2% water soluble P_2O_5, the P_2O_5 soluble in neutral ammonium citrate and water must be declared. For fertilisers containing less than 2% water soluble P_2O_5 the amount soluble in other reagents has to be given, for example, 2% formic acid for soft ground rock phosphates.

Low solubility sources of P were not effective at Rothamsted because the soils there are near neutral to alkaline. However, low solubility sources of P can be effective on acid soils. Thus basic slag from iron foundries was applied on acidic, upland soils in the UK to good effect. More recently much research has centred on using ground reactive PRs to increase the fertility of acid upland soils in the humid tropics (see the papers in Johnston and Syers, 1999). In both cases soil acidity gradually releases P from the material. To be effective, the material must be finely ground, which makes application difficult especially in windy weather. Un-reactive PRs, which are often hard crystalline apatites, are very insoluble and almost valueless. Heating PRs to a high temperature (calcination) can sometimes increase the availability of the P they contain and this might be economic where energy is cheap and sulphuric acid is expensive. The advantage of using ground reactive PRs on acid soils, especially strongly leached, free draining soils with low sorption capacity, is that the P, which is released gradually, is available for crop uptake.

5.4 PHOSPHORUS IN SOILS

Plants roots take up P as ortho-phosphate anions (i.e. negatively charged ions, HPO_4^{2-} or $H_2PO_4^-$), and crop growth will be restricted unless an adequate supply of these two ions can be maintained in the soil solution. However, most of the P in soils is unavailable to plants, and research on P nutrition has sought to define what is an adequate amount of readily plant available P in soil and how can this be maintained.

The chemistry of soil P is complex because it exists both as inorganic phosphates and as P within complex organic molecules, organic P, in soil organic matter. In soils cropped predominately with arable crops, inorganic P will be as much as 70% of the total P, and the annual net mineralisation of the organic P is invariably too little to meet the needs of large yielding crops (Chater and Mattingly, 1980). Conversely, the annual flux of P through soil organic matter in permanent grassland soils can be sufficient to meet the needs of the grass to sustain large yields (Brooks et al., 1984).

Before the 1950s much evidence, mainly from field experiments, suggested that the residue of P applied to a crop and not used by it, was unavailable to future crops. From the 1950s much work was done on the chemistry of soil P (for a general review see Sikora and Giordano, 1995). Early in this period, thermodynamic models predicted, and some experimental evidence using pure minerals supported, the formation of water insoluble (and by inference plant unavailable) P compounds when water soluble P (known to be plant available) was added to soil. These compounds included variscite ($AlPO_4 \cdot 2H_2O$) and strengite ($FePO_4 \cdot 2H_2O$) at low pH and a range of calcium phosphates at high pH (Lindsay, 1979). However, such simple reaction products are unlikely to be produced in the heterogeneous environment of the soil (see for example Barrow, 1983).

It is now generally considered that the behaviour of inorganic P in soil can be best explained by adsorption–desorption reactions (Barrow, 1980). It is suggested that added inorganic P is initially sorbed either weakly (physisorption, electrostatic forces) or strongly (chemisorption) on to variable charged surfaces associated with aluminium, iron or calcium compounds depending on soil type and pH (Syers and Curtin, 1988; for calcareous soils see Holford and Mattingly, 1975a, b). After the initial adsorption the P can become less labile, perhaps by diffusive penetration (absorption) of adsorbed phosphate ions into soil components (Barrow and Shaw, 1975; Evans and Syers, 1971). This absorption, which is gradual, makes this P less available for plant uptake, at least in the short term. This absorption could be as crystalline intermediate calcium phosphate compounds in neutral to alkaline soils, or ill defined amorphous Fe–P or Al–P compounds in acid soils (Pierzynski, 1991).

These more recent concepts of electrostatic and chemical adsorption, followed by absorption into the solid phase of the soil, make it easier to understand the time scale of P reactions in soil. They also begin to explain why different reagents used to extract P from soil in routine methods of soil analysis extract different amounts of P, yet, the amount extracted often correlates well with crop responses in the field.

These ideas on the behaviour of P in soil can be conceptualised and related to the availability of soil P to crops as shown in Figure 5.3. Soil P is considered to be in different pools or categories. When water-soluble P is added to soil in fertilisers or manures, it goes first into the soil solution and then distributes itself between the readily and less readily available pools. The most important feature in Figure 5.3 is the representation of the transfer of P in both directions between the soil solution, the readily available and the less readily available pools of P. Routine soil analysis to assess the P status of soils in relation to crop response to added P usually measures the P in the soil solution and the readily available pools.

Evidence for the reversible transfer of P between pools and the movement of added water soluble P to the less available pool comes from many long-term experiments on the silty clay loam soil at Rothamsted. Figure 5.4 (Johnston, 2001) shows that only about 13% of the P balance (P applied *minus* P removed) in many experiments has remained as Olsen P, i.e. P extracted by Olsen's reagent that correlates well with crop response to soil P and is a measure of water soluble/readily available

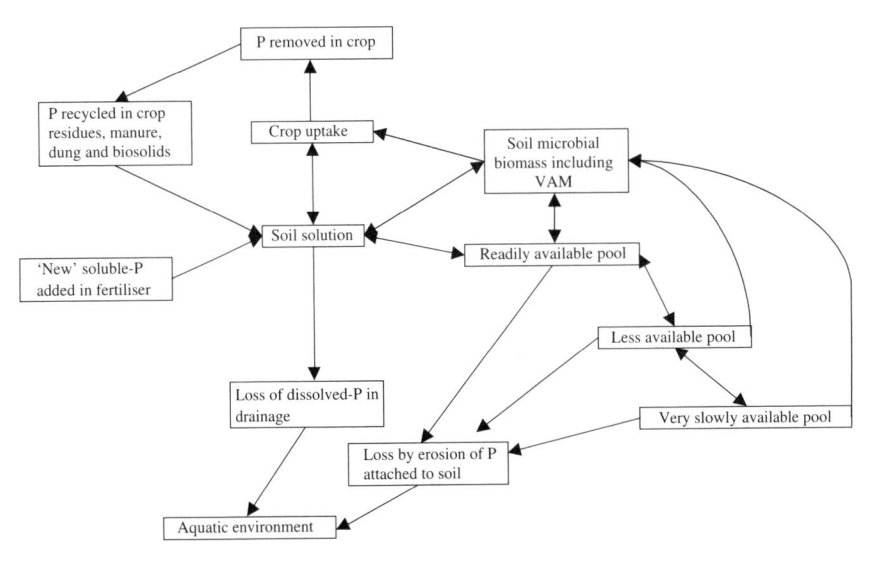

Figure 5.3 Simplified schematic representation of the phosphorus cycle in crops and soils showing the different pools of soil phosphorus.

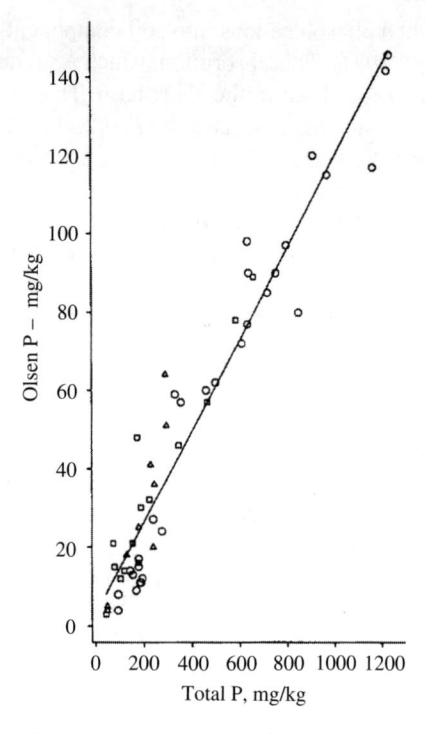

Figure 5.4 The relationship between the increase in Olsen P and the increase in total soil phosphorus in long-term experiments on a silty clay loam (circles), a sandy loam (squares) and a sandy clay loam (triangles) (Johnston, 2001).

soil P (for the significance of this test, see also section 5.5.2 and Westerman, 1990). The data in Table 5.1 is taken from an experiment in which from 1856 to 1901, P was added in both fertiliser and farmyard manure, first to winter wheat and then to potatoes and the P balance was positive. However, the increase in Olsen P was much smaller than the P balance indicating that much of the P had transferred to the less readily available pool. Then from 1901 onwards no further P was applied, either as fertiliser or manure, and cereals, mainly spring barley, were grown. The P balance was negative, but the decline in Olsen P was much less than the P removed in the harvested crops indicating that P in the less readily available pool had become available for crop uptake.

Further evidence that P is not irreversibly fixed in many soils and can become available for crop uptake is shown in Table 5.2. When a sandy clay loam soil with different levels of Olsen P was cropped for 16 years without P addition, not more than half the P offtake could be accounted for by the decline in Olsen P even on the soils with most Olsen P, i.e., some of the P in the crop had come from the less available pool (Johnston and Poulton, 1992). The effects of phosphate sorptivity on the long-term recovery of P accumulated in soil as reserves from past applications of

Table 5.1 Changes in Olsen P in the 0–23 cm topsoil during 1856–1903 and 1903–58 and the P balance, net gains or losses of P due to cropping and manuring, Exhaustion Land, Rothamsted (from Johnston and Poulton, 1977).

Treatment	Period*	P balance kg/ha	Change in Olsen P kg/ha	Change in Olsen P as a % of the P balance
None	1856–1903	−80	−5	6
FYM	1856–1903	+1030	+170	17
PK	1856–1903	+1220	+170	14
None	1903–1958	−190	−5	3
FYMr	1903–1958	−380	−130	34
PKr	1903–1958	−340	−140	41

*Fertilisers added 1856–1901, FYM added 1876–1901: no P, K or FYM added after 1901.

Table 5.2 Relationship between phosphorus balance and the decline in Olsen P in a sandy clay loam, Saxmundham, 1969–82 (adapted from Johnston and Poulton, 1992).

Olsen P, mg/kg, 1969	3	7	21	28	39	44	54	67
P removed in crops, kg/ha, 1969–82	94	153	217	237	253	256	263	263
Decrease in Olsen P, kg/ha, 1969–82	8	12	27	50	65	78	87	120
Change in Olsen P as % of crop uptake	8	8	12	21	26	30	33	46

P in fertilisers and manures and the effectiveness of fertiliser P added to soil has been reviewed by Holford (1982).

In addition to the evidence from crop responses that added P can transfer between the different soil P pools, further confirmation has come from research at Rothamsted funded by the World Phosphate Institute (IMPHOS) (Blake et al., 2003). A seven-step technique for sequentially extracting soil P was developed based on existing published methods. In order, the extractants were: anion exchange resin, 0.5 M $NaHCO_3$, 0.1 M NaOH, 1 M NaOH, 0.5 M H_2SO_4, concentrated HCl and finally residual P determined by fusing the soil residue with Na_2CO_3. This technique does not uniquely identify any one soil P fraction or compound but appears to sequentially remove P held with increasing bonding strengths and from a continuum of changing P compounds. Soil samples taken from the long-term Rothamsted experiments before and after long periods with and without P additions, were sequentially extracted with the reagents listed above. The results showed that more P was extracted by all of the first five extractants at the end of a long period of a positive P balance than at the beginning. Subsequently, after the soils had been cropped without P addition over many years and the P balance was negative, P declined in those five fractions where previously there had been an increase (Blake et al., 2003).

These measurable changes in the amount of P in the different pools could only be made over long time periods because the annual changes are too small to be determined accurately. This highlights the fact that many soil parameters change continuously and that these changes happen over different time-scales. Annual changes in soil organic matter, for example, are very small during the annual cycle of growth and decay, but the changes are much larger over a longer time scale when the management of a soil changes, e.g. from grassland to arable farming. The mineral fraction is also changing; for example, iron as non-crystalline hydrous oxide gels, which have a high sorption capacity, may crystallise over time to non-reactive forms that play little part in soil chemical reactions. Conversely, reactive gels may be replaced by the weathering of crystalline forms. So soil is not stable or inert, changes in the mineral components together with the activity of the living biological component affect many soil processes including the availability of P for plant growth.

Many other methods have been used to study P in soils. Scanning electron microscopy (SEM) with energy dispersive X-ray analysis (EDX) can be used to show the spatial distribution of elements within small units of soil. For example, an SEM image of anaerobically digested wastewater biosolids (sewage sludge), which had been thermally dried, showed, using EDX, P in several locations including a major concentration associated with Ca (Smith *et al.*, 2002).

Smith *et al.* (2002) also found that the availability of P in soils incubated with thermally dried biosolids was much reduced compared with the same biosolids that had not been through the drier. They concluded that a low-solubility Ca–P was formed within the drier. Studies using sequential extraction also suggest that a significant proportion of P in organic manures is associated with Ca, but perhaps drying encourages crystallisation. Drying Fe-rich biosolids had a similar effect in decreasing the plant availability of the P. If this decrease in P availability is permanent or at best long-term, then thermal drying of biosolids appears to reduce their value as a source of P.

The P sorption capacities of soils and minerals have been studied by reacting soils with solutions of differing strength and constructing sorption and desorption curves. It has been observed that such experiments produce curves that are not coincident, indicating fixing of some P; however, this gives little or no information that relates to plant uptake of P. More informative are studies using ^{32}P, which have estimated the lability of P in soil by isotopic exchange.

Solubility-product studies involve measuring the composition of equilibrium aqueous extracts over a range of pH and plotting the data on a diagram where the stability of different P mineral phases is mapped. If the data are close to the line for a particular mineral or minerals, then it is concluded that that mineral(s) is the dominant form of P in that soil. However, ion activity product-phase diagrams cannot account for non-crystalline or substituted inorganic forms or for organic forms of P in soil.

None of these techniques provides an absolute answer to the forms or behaviour of P in soil but if their limitations are accepted they are useful tools in helping to understand what is happening to soil P.

5.5 PHOSPHORUS NUTRITION OF PLANTS

It has been noted previously that many soils are inherently P deficient and this can lead to very distinct visual symptoms in plants grown on such soils. These symptoms tend to be seen more in crops with a short growing season and/or those with a poorly developed root system. Root systems develop more slowly, and P deficiency symptoms may be seen more often, when soils are too wet, too dry or too cold for rapid growth. The symptoms frequently disappear as soil conditions improve and root systems proliferate into the soil mass.

In many plants P deficiency shows itself as a purplish colouration in the older tissues due to the formation of anthocyanins. Typical P deficiency symptoms in spring barley, maize and tomato are shown in Figure 5.5. Symptoms are seen less frequently in winter wheat because it develops a sufficiently large root system in autumn when the soil is warm. Tomatoes require a high concentration of soil P for seedling growth and deficiency causes stunted growth with the stems becoming slender, fibrous and hard. Growth is slowed and maturity delayed. Leaves become dark green with purple areas between the veins on the underside of the leaf. Phosphorus deficiency symptoms are not readily recognised in potatoes unless the

Figure 5.5 Phosphorus deficiency symptoms in (a) barley (left), maize (right), (b) maize and (c) tomato. [This figure is also reproduced in colour in the plate section after page 336.]

deficiency is acute when the leaves roll and curl upwards revealing a grey-green lower surface.

Plant roots take up P from the soil solution as ortho-phosphate ions, $H_2PO_4^-$ and HPO_4^{2-}. However the concentration of these two ions in the soil solution is usually very small and is invariably less than the plant requires when it is growing vigorously. During active growth many plants maintain between 0.3 and 0.5% P in dry matter. To achieve this the plant needs to take up from the soil solution between 0.5 and 1.0 kg P, as phosphate ions, per ha each day. But at any one time, even on soils well supplied with P, the total quantity of P in the soil solution may be only between 0.3 and 3.0 kg/ha, and not all of this P is available to the crop. This is because the roots are only in contact with a limited volume of the soil solution and the diffusive flow of P to the root is very slow. Thus P in the soil solution must be replenished frequently, often several times each day, from the pool of readily available P in the soil (Marschner, 1995).

Figure 5.6 shows the daily rate of P uptake throughout the growth of a crop of spring barley grown in an experiment where there were soils with two levels of readily available soil P, estimated as Olsen P. These values were 100 and 5 mg/kg, respectively, the larger concentration being much more than sufficient for all arable crops. The maximum daily rate of P uptake was 0.6 kg/ha on the soil well supplied with P but only 0.2 kg/ha on the soil with only 5 mg/kg Olsen P. This difference in the availability of soil P and its effect on P uptake was reflected in the grain yields at harvest, 6.4 and 2.9 t/ha with and without an adequate supply of P. To ensure optimum economic yields and the efficient use of P means that the soil must contain sufficient readily available P to allow a crop to achieve the optimum daily uptake rate appropriate to the stage of growth.

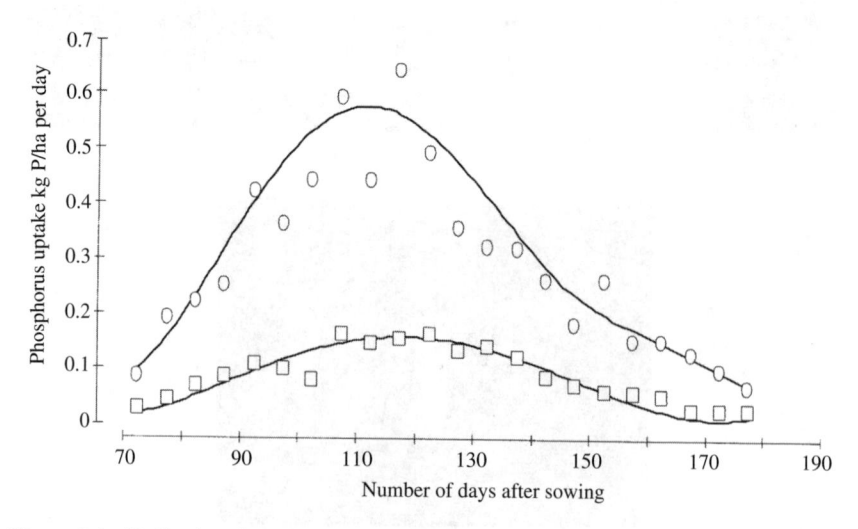

Figure 5.6 Daily phosphorus uptake rate, kg P/ha, by spring barley grown on a soil with adequate P reserves (circles) and too little readily available soil P (squares) (Johnston, 2000).

It is interesting that many plants have developed a symbiotic association with vesicular arbuscular mycorrhizal (VAM) fungi. The spores, which are found in many soils, develop hyphae that penetrate the root, take their food from it and grow out into the soil immediately surrounding the root (the rhizosphere). This effectively extends the root surface for nutrient acquisition, especially of P (Tinker, 1984). The fungal association is usually not well developed in soils rich in plant available P. Some plants do not have mycorrhizal fungi, notable amongst these are the brassicas such as sugar beet. Johnston *et al.* (1986) noted that spring barley grown following sugar beet on low P soils yielded less than barley following potatoes, because there were less VAM spores available to infect the barley roots in the soil after growing sugar beet.

5.5.1 Critical levels of phosphorus in plants

For annual arable crops it is well known that as a plant grows and accumulates dry matter, consisting largely of carbon compounds, the %P in the above ground biomass declines when expressed on a dry matter basis. In consequence it is difficult to define a critical concentration that would indicate whether or not a plant was deficient in P. In an attempt to overcome this problem it is usual to sample a specific plant part at a given growth stage. Some values taken from the literature and summarised from IFA (1992) are in Table 5.3. Although there is a broad similarity in %P across a wide spectrum of crops, there are differences not only between crops, but, where a range of values are given, within the range for each crop. Where

Table 5.3 The minimum percentage phosphorus in different plant parts.

Crop	Plant part	Minimum %P in dry matter
Cereals		
Maize	Ear leaf	0.25–0.30
Rice	Leaf blade	0.1
Wheat	Whole plant	0.20–0.29
Root and tuber crops		
Potato	Foliage	0.16–0.25
Cassava	4/5th leaf 4 months after planting	0.3
Sweet potato	Tissue	0.12
Pulses		
Beans	2/3rd leaf from apex	0.19
Peas	2/3rd leaf from apex	0.19
Chickpea	Whole shoot	0.24
Oilseeds		
Soybean	Fully developed leaf	<0.26
Rapeseed	Fully expanded leaf	<0.35
Sunflower	Fully expanded leaf	<0.30

Adapted from IFA 1992.

a single value is given, it is not possible to decide if that value is the average from a range of values in the literature or not. It would therefore be preferable to know whether the different values for any one crop were genuine and if so why, because it might be expected that there should be only one value for that crop part at that stage of growth. One explanation for the differences could be that although nominally the same part of the plant was sampled and analysed, the physiological age of the tissue was not the same in all studies. Another factor could be that the rate of growth, which depends on other factors like nitrogen and water supply, differed between the crops sampled. If attempts are to be made to increase the efficient use of P in crop production, it is essential that there are well recognised and agreed critical P concentrations in plants to give guidance as to whether too little P is being applied to achieve optimum yields.

Leigh and Johnston (1986), as part of an investigation examining the usefulness of expressing plant nutrient concentrations on a tissue water basis, rather than in dry matter, compared the concentration of P in spring barley from emergence to harvest when it was expressed both in dry matter and in tissue water. Whereas %P in dry matter declined during growth, P concentrations in tissue water decreased a little initially but then increased a little, until the onset of senescence when the increase was rapid as the crop began to lose water as it ripened. Nitrogen and water supply induced differences in %P in dry matter but there was less effect on the P concentrations in tissue water. Like %P in dry matter, P in tissue water was smaller in crops grown on a P deficient soil compared to those grown on a soil with adequate P. Leigh and Johnston (1986) considered that P concentrations in tissue water could perhaps provide a better method of assessing the P status of a crop that has some advantages over %P in dry matter but this needs further investigation.

5.5.2   Soil tests and fertiliser recommendations

There is no one universally accepted or applicable method to determine plant available P in all soils. As early as 1872, H. von Liebig, son of Justus von Liebig the famous German chemist, analysed soils from the Broadbalk experiment started some 30 years earlier in 1843 at Rothamsted. When he extracted the soils with dilute mineral acids, the extracts from soils that had received P annually for 30 years contained more P than those from soils that had not received P during the experiment (Liebig, 1872). Dyer (1902) subsequently used 1% citric acid as an extractant arguing that this level of acidity was near to that of cell sap and thinking that root exudates might increase the availability of soil P. Using this extractant, Dyer could clearly differentiate between soils with and without P over long-periods in the Rothamsted experiments. Subsequent research, however, showed that a 1% solution of citric acid extracted so much P from soil that it was not possible to reliably discriminate between soils where there were differences in crop yield but only very small differences in extracted P.

Many different extractants have been tested subsequently. Most are based on pragmatic experimentation that showed an acceptable relationship between crop yield and the amount of P extracted. Fixen and Grove (1990) listed many of the methods that are used. Four methods that indicate the range of acidic and alkaline reagents and the range of strengths used are: Bray-1, 0.03 M NH_4F + 0.025 M HCl (Bray and Kurtz, 1945); DL, 0.02 M Ca-lactate + 0.02 M HCl (Egner and Riehm 1955); Olsen, 0.5 M $NaHCO_3$ − pH 8.5 (Olsen *et al.*, 1954); and Morgan, 0.54 M CH_3COOH + 0.7 M $NaC_2H_3O_3$ (Morgan, 1941). Most methods seek to extract P that is weakly-bonded to soil constituents or P in those chemical combinations or compounds thought to predominate in different soil types, i.e. acidic extractants for acid soils and alkaline/neutral extractants for alkaline soils. Many of the reagents have been chosen because they will also extract plant available potassium (K). The use of anion exchange resins to extract P from the soil was proposed as a method that mimicked extraction by plants (Amer *et al.*, 1955; Hislop and Cooke, 1968). This method has not been widely adopted but it appears to have a greater range of applicability than many of the chemical methods that are currently used.

There are two approaches to using soil analysis as a basis for fertiliser recommendations, but both require an acceptable/reliable method of analysis. The soil samples should be representative of the area from which they are taken, and the depth of soil from which the plants are likely to acquire most of the P they require. This is usually the plough layer for arable crops and the top 15 cm for grassland. If there is obvious soil variability each soil unit should be sampled separately. Often data are presented showing variation in analytical values for a single field either within or between years (see for example Collins and Budden, 1998). It is difficult to know whether these are genuine changes or if they arise from an inherent lack of uniformity in nutrient status within a field and differences in sampling technique that has not been allowed for.

The first and well-tried approach for developing fertiliser recommendations uses the results of field experiments sited on soils with a uniform P level. Rates of applied P, usually at least six amounts, are tested on plots where all other growth-controlling factors are uniform. The response curve is fitted statistically to the relationship between yield and applied P. The data can be interpreted in two ways: (i) the amount of P required to get 95% of the maximum yield is estimated from the response curve and this would be the recommendation for the crop grown on that soil type with its level of available P; (ii) the economic application can be estimated from the response curve. This is the amount of P where the cost of adding one additional kg of P is more than the value of the additional crop produced. The economic application will change with the cost of the P source and the value of the crop. However, if the response curve has been determined with precision then different costs and values can be used as appropriate.

Recommendations based on soil analysis and field trials have to be made over a number of years, to minimise seasonal effects, and on a range of soil types with different levels of readily available P. This is time consuming and expensive but gradually a database is accumulated and general advice can be given with some confidence as in England and Wales with Bulletin RB 209 (MAFF, 2000). However these are general guidelines and can be modified on the basis of detailed local knowledge and experience.

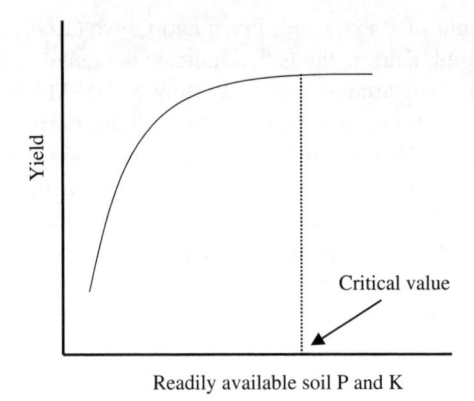

Figure 5.7 A schematic representation of the relationship between crop yield and readily available soil P and K.

Recently a second method of using soil analysis has been proposed to give advice on P additions. On soils were P accumulates as plant available reserves, it is appropriate to determine the extent to which such reserves should be built-up. The concept of plant response to nutrients, including P, is based on the Law of Diminishing Returns as illustrated by a Mitscherlich type response curve. This implies that as the readily available P in soil increases then yield will increase rapidly at first and then more slowly, to reach an asymptote (Figure 5.7). The Olsen P at which the yield reaches the asymptote can be considered the critical value for that crop and farming system. Below the critical value the loss of yield is a financial loss to the farmer. Above the critical value, there is no justification to further increase the available P because this is an unnecessary expense. There is also an increasing risk of loss of P from soil to water when soils are excessively enriched with P and this can lead to the unacceptable consequences of eutrophication in some water bodies.

Figure 5.8 shows examples of the yield/Olsen P relationship for various crops grown in Rothamsted experiments. Although yields of each crop often differed appreciably between years, mostly due to seasonal variations in weather or to the applications of different amounts of nitrogen fertiliser, the critical Olsen P changed little. Frequently it was smaller in the years with larger yields probably because soil conditions were favourable for roots to explore the soil for the nutrients the plant required. Olsen P invariably accounted for more than 60% and often more than 80% of the variance in yield in these experiments.

There are many reasons why farmers should have a good knowledge of the nutrient status of their fields as part of their soil fertility management practices. However, soil analysis appears to be less used than it should be. There are probably a number of reasons for this, principle amongst them being farmers' uncertainty about the reliability and interpretation of the data.

In many pot experiments in the glasshouse there is a very good relationship between yield and Olsen P. This suggests that there is little fundamentally wrong

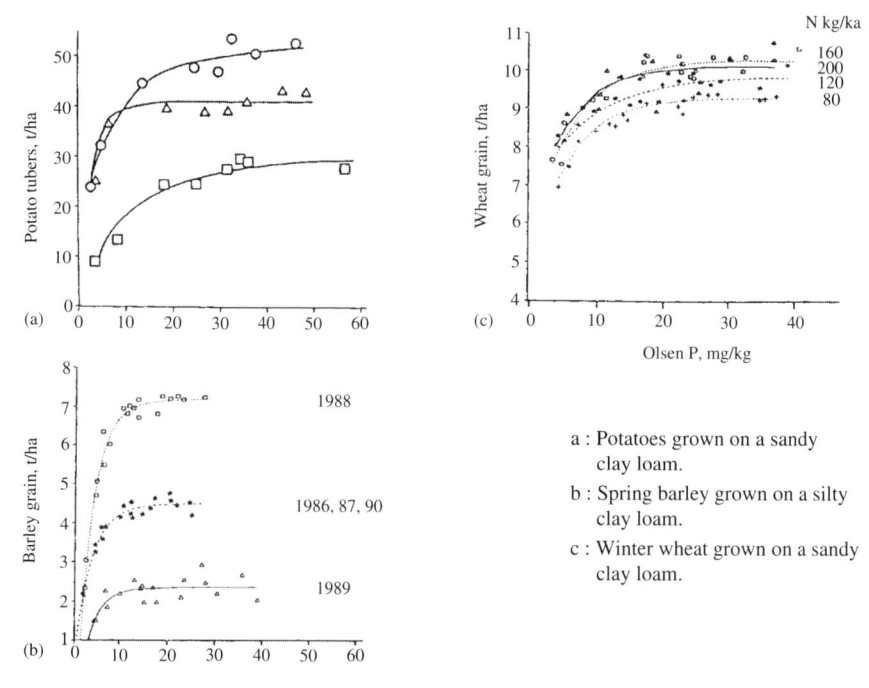

Figure 5.8 The relationship between the yields of three arable crops and Olsen P (adapted from Johnston, 2001).

a : Potatoes grown on a sandy clay loam.
b : Spring barley grown on a silty clay loam.
c : Winter wheat grown on a sandy clay loam.

with soil analysis provided that appropriate, well-tried methods are used. Figure 5.8 also shows that in field experiments where plots have been established with a range of Olsen P there is a good relationship between it and yield. Why then is soil analysis often less satisfactory when used to assess the P status of farmers' fields? There are two major reasons. The first is how representative was the sample of the whole field? Samples should consist of 16 to 25 individual cores, bulked and thoroughly mixed. The second reason is that the sample is air-dried and ground to pass a <2 mm sieve for analysis. When interpreting data, it is essential to remember that because a <2 mm sample was analysed the data are only applicable to that amount of this sized fraction of soil that the roots can explore to find the nutrients they require. The amount of fine soil will depend on the depth and stoniness of the soil, and how extensive is the root system will depend on the soil structure. Also the crop's response to applied fertiliser will depend on the uniformity with which the fertiliser was mixed with the soil in which the roots grow. Other factors like soil moisture and the control of weeds, pests and diseases all affect crop growth and yield.

Table 5.4 shows the % variance accounted for by the relationship between yield and Olsen P for three arable crops, and how much more Olsen P was required to achieve optimum yield in a poorly structured soil with little soil organic matter. When soil samples from all plots were cropped in the glasshouse with ryegrass under controlled conditions, the relationship between yield and Olsen P showed the

Table 5.4 The effect of soil organic matter (SOM) on the response of spring barley, potatoes and sugar from sugar beet to Olsen P in a field experiment and the relationship between the cumulative yield of five harvests of ryegrass and Olsen P when grown in uniform conditions in the greenhouse (Adapted from Johnston et al., 2001).

Crop	% SOM	Yield at 95% of the asymptote	Olsen P* mg/kg	% variance accounted for
Spring barley, grain	2.4	5.00	16	83
t/ha	1.5	4.45	45	46
Potatoes, tubers	2.4	44.7	16	89
t/ha	1.5	44.1	61	72
Sugar from beet	2.4	6.59	18	87
t/ha	1.5	6.56	32	61
Grass D.M.	2.4	6.46	23	96
g/pot	1.5	6.51	25	82

*Olsen P at which the yield was 95% of the asymptote.

same critical value for Olsen P irrespective of the level of soil organic matter. This strongly suggests that the different critical Olsen P values observed for the field-grown crops were due to differences in soil structure.

Thus there are a number of factors that can affect the relationship between yield and soil analysis data and these should be acknowledged. If soil analysis does not appear to work as well as expected, it is better to ask why rather than discard a valuable management tool in assessing the need for P additions.

More recently there has been a need to attempt to classify soils according to the risk of P loss from soil to water. Provided soils are not excessively enriched, in relation to the need to be able to supply sufficient P to meet the crop's requirement during growth, the risk of loss of P appears to be related to factors like soil type, topography and rainfall. The largest source of loss of P from soils is that of particulate P (see e.g. Haygarth and Condron this volume). Excessive plant-available P can be associated with increased risk of leaching, but the overall risk of transfer of P from soil to water is very often related to the soil's vulnerability to erosion. All this implies that not all soils in a catchment are likely to be a source of the P that can be lost to rivers, and any attempt to limit losses will need to recognise this fact.

5.6 FERTILISATION PRACTICES

Many phosphatic fertilisers are now available and farmers can make choices. Their first consideration must be whether the P in the fertiliser will benefit the growth and yield of the crops they wish to grow on their soil. The second and increasingly important consideration is cost. The criterion of suitability comes from experience and advice but there is often a need for guidance about quality. There is the statutory requirement to declare the composition of the fertiliser but an important aspect of quality of granular fertilisers is the strength of the granule for both transport and spreading. The Fertiliser Manufacturers' Association in the UK is seeking ways to set standards for this aspect of fertiliser quality.

The principal P fertilisers in use today are triple superphosphate (TSP) 47% P_2O_5, diammonium phosphate (DAP) 18% N, 46% P_2O_5, and monoammonium phosphate (MAP) 12% N, 52% P_2O_5. In these three fertilisers most of the P is water-soluble, an important aspect that Lawes sought to achieve initially when he treated phosphate rock with sulphuric acid. Research subsequent to Lawes' studies has shown that even on neutral and slightly calcareous soils nitrophosphates, with much less than all the phosphate in water-soluble forms, and in some cases dicalcium phosphate, with no water-soluble P, can give crop yields equal to those given by superphosphate (Johnston, 1993).

These fertilisers can be applied singly or combined with others to supply other nutrients, often K and S but increasingly with microelements like boron (B) and zinc (Zn), to provide crops with their nutritional requirements. There are 'compound' or 'complex' fertilisers, produced as granules, each of which contains all the nutrients stated to be in the fertiliser. For example, a 15:15:15 compound fertiliser will contain 15% N, 15% P_2O_5 and 15% K_2O in each granule. There are also blended fertilisers that are a physical mixture of individual fertilisers in proportions to provide the required amount of each nutrient. Overall P fertiliser use in Europe peaked in the 1970s and has been in gradual decline since, the decrease in the USA started later (Johnston *et al.*, 2001a). The reasons appear to relate to the falling prices for agricultural produce, concern to avoid environmental damage through over application and better advice on fertiliser use.

Other sources of P inputs to agriculture are organic manures, like farmyard manure and slurry, biosolids (sewage sludge), and recovered phosphates, like those produced from treating waste water streams in sewage treatment works (Johnston and Richards, 2003). Organic manures recycled to agricultural land supply valuable quantities of plant nutrients and organic matter that can help meet crop P requirements and maintain soil fertility (Smith and Chambers, 1995). However, land-applied manures can be a major potential source of diffuse pollution of P to surface waters (Smith and Chambers, 1995). Recent survey data show that manures are applied annually to about 15% of arable land and 43% of grassland in England and Wales, with farmyard manure (FYM) and biosolids as the most and least frequently used forms respectively (Chalmers *et al.*, 1999). The estimated average annual P loading to soils receiving manure is 69 kg/ha P_2O_5 (Smith and Chambers, 1995). Supplementary information from the British Survey of Fertiliser Practice found that although the majority of farmers said that they made some allowance for the nitrogen content of the manure they applied they made no or little allowance for the P (and K) content of the manure (Anon, 1995). This flies in the face of good practice and could lead to the over enrichment of some soils with P and the potential for the loss of P from soil to water.

Currently in Western Europe the average annual application of P in mineral fertiliser to arable land is 43 kg P_2O_5/ha, range 20 kg/ha in Denmark and Sweden to 52 and 55 kg/ha in the UK and Switzerland, respectively. The average input to the utilised agricultural area (UAA) that includes grassland and permanent crops is somewhat less, 27 kg P_2O_5/ha (Johnston and Steén, 2000). The P status of European soils is estimated, by routine soil analysis and for many countries, some 25% (5–55%) of soils test as very low and low in readily available P. On such soils it would be justified to apply more P than that which is removed in the harvested crops to increase soil P reserves and thus raise their fertility. For many countries some 40% of soils test as high and very high in readily available P, and on such soils it would be justified to apply less fertiliser P unless very P responsive crops are being grown. On soils where the P status is about the critical level it should be only necessary to apply the amount of P removed in the harvested crops (Steén, 1997).

It is widely accepted that all parts of a field are not equally productive. In some fields the most productive area is not the same in every year. For example, soil physical conditions and drainage might be more influential in some years than others, whereas plant nutrient content might be limiting in some areas and not in others. Techniques are being developed to use all fertilisers, including P fertilisers, more precisely (precision farming) including varying the amount applied both spatially and in time. One approach has been soil mapping on a grid (50 m × 50 m or 25 m × 25 m) to produce fertility contours and then applying fertiliser only on those areas where there is too little P. This grid size is usually too large to produce sufficient accuracy and even at this scale is very expensive in sample collection and analysis.

Another approach, made possible by using Global Positioning Satellite signals and the facility to record and map yields as a crop is harvested, is to produce yield maps of the field. Areas of small and large yields can then be examined in the field for physical differences, like soil compaction and drainage, and the soil can be sampled for chemical analysis in the laboratory to determine nutrient deficiencies or other factors likely to affect yield. Appropriate action can then be taken to rectify any factor that could be decreasing yield. One such approach is the use of digital yield maps to compute P offtakes and then vary, with on-tractor control, the application of P fertiliser (Fenton, 1998). Remote sensing can also be used to detect the onset and areal extent of disease or weeds, and attempts are in progress to use it to estimate a crop's yield potential and state of nutrition. These tools are enabling variable rate application of inputs to be targeted where they will have most benefit, and the complement of this is that the risk of wastage and loss to the wider environment is minimised.

Where plant available reserves of P can be accumulated in soil and where soils are at or about the critical level, applications of fertilisers and manures should aim to maintain this amount of plant available soil P by replacing the amount of P removed in the harvested crop. The success of this approach can be checked by soil sampling at intervals, say 3–4 years. On soils were the P is below the critical level it would be justified to apply more P than that removed in the crops until the soils have reached the critical level. As noted earlier it is never justified to apply P to soils were the available P is much above the critical value.

On many soils manures are often applied in autumn before ploughing and fertilisers are applied broadcast over the soil surface prior to seedbed preparation. For soils with only small amounts of readily available P, where large amounts of P would be required to get acceptable yields, the P fertiliser can be placed close to the rows of seeds, this is called 'banding' or placement of the fertiliser. Also for soils with a capacity to quickly immobilise P, the fertiliser can be banded. In both cases this results in larger local concentrations of soluble P in the bands leaving more of the P available for crop uptake.

5.7 PHOSPHORUS FERTILISATION AND ENVIRONMENTAL ISSUES

5.7.1 Phosphorus and the aquatic environment

Within many aquatic ecosystems there is a very complex, finely balanced, ecological structure often of great biodiversity. Not only productivity, but also the balance between species depends on the availability of nutrients that in most undisturbed systems have come from the land. Since the early 1970s there have been reports of undesirable biological changes in a number of aquatic ecosystems and these have often been associated with increasing concentrations of nutrients due to increased inputs. Very frequently it has been shown that P is the limiting nutrient for plant and algal growth, the cause of the undesirable changes. Initially much of this P was thought to come from sewage treatment works, but as the larger works have been required to remove P from the effluent they discharge to rivers, the problems associated with eutrophication have persisted. In some cases this has highlighted the possible loss of P from agricultural soils, discussed in more detail in other chapters, and supports the point made in this chapter that soils should not be over enriched with P but should be maintained at about the critical level of available P to ensure that yields are not jeopardised.

5.7.2 Phosphorus fertilisers and the terrestrial environment

5.7.2.1 Cadmium in phosphatic fertilisers

There is an issue around human health and cadmium (Cd) that has to be addressed because Cd is toxic and carcinogenic. Johnston and Jones (1995) using analyses of soils archived from the 1850s onwards from long-term experiments at Rothamsted, showed that in England, and probably in many countries with a long history of industrialisation, Cd had been added to agricultural soils by aerial deposition, in sewage sludge (biosolids) and in phosphatic fertilisers. The possibility of Cd accumulating in soil from any of these three sources is of concern because plant based foodstuffs are the largest source of Cd intake for non-smokers[1] (Johnston and Jones, 1995; Syers and Gochfeld, 2000 and references in both). The Cd inputs to agricultural soils from the first two sources have been declining. In Europe atmospheric

---

[1] Smokers receive high doses of Cd through its accumulation in tobacco plants treated with fertilisers high in Cd.

emissions have declined appreciably since the mid-1960s (Johnston and Jones, 1995). In 1996–1997 the median level of Cd in sewage sludge applied to agricultural soils in the UK was about 60 mg Cd/kg P_2O_5 (Gendebien *et al*., 1999). This is a considerable decrease compared with former years, and has been achieved through active control of Cd at source, backed up by legislation. The Cd content in sludge is likely to have continued to decline since that survey for the same reason, and because of increased extraction of P from waste waters. In Europe, as a consequence of the decrease in the inputs from the first two sources, inputs of Cd in phosphatic fertilisers have become relatively more important, but not necessarily greater in amount than previously. The importance of Cd inputs with P fertilisers has always been the case for countries where aerial deposition has never been a problem and little sewage sludge has been applied to agricultural soils.

Phosphate rocks contain very variable amounts of Cd, from less than 10 to more than 50 mg/kg of rock. The average Cd content of sedimentary rocks is appreciably larger than that of igneous rocks (van Kauwenbergh, 1997). Almost all of the Cd carries through into the processed fertiliser in the manufacture of TSP, MAP and DAP. World wide, to limit Cd inputs from phosphatic fertilisers, controls in some countries are already in place, in others they are being proposed or being considered. The European Union is currently seeking to harmonise the maximum permissible levels of Cd in P fertilisers to be sold within European countries. The level will probably be set on the basis of mg Cd/kg P_2O_5; figures in the range 20–60 mg Cd/kg P_2O_5 have been discussed. There are concerns that the level may be set such that many sedimentary rock deposits could not be used to produce fertilisers until methodologies are available to remove the Cd at some stage in the manufacturing process. This will undoubtedly increase the cost of the fertiliser.

The uptake of Cd by crops, i.e. the availability of soil Cd, is related to soil pH. Availability is increased in acid soils with pH (in water) less than 6.5 (Johnston and Jones, 1995). This can result in something of a dilemma. It has been shown that the most cost effective way of raising the productivity of many acid soils farmed by resource-poor farmers is to apply finely ground, reactive phosphate rock. If these rocks contain Cd it will become available for uptake by crops grown on the acid soil. So the alternatives are either to increase productivity with some risk of Cd contamination or condemn the farmers and their families to a life of poverty.

Factors controlling plant Cd levels include the following:

i. Direct aerial deposits if the Cd can be adsorbed through leaves; there is some evidence for this but aerial concentrations are declining.
ii. Dilution effects when the rate of Cd uptake is less than the rate of dry matter accumulation as a result of photosynthesis.
iii. Plant control over the transport and storage of Cd within the plant; there is some evidence that translocation into cereal grain is in part controlled.
iv. The addition of P fertilisers, these encourage root growth and this may increase Cd uptake, but root growth is essential for the plant to acquire nutrients.
v. The Cd in fertilisers may be more plant available than the Cd in soil until it has reacted with soil constituents that decrease its availability.

The issues about Cd and human health are complex and their resolution requires much more understanding of the soil-plant-animal-human Cd cycle, especially the bioavailability of Cd, its uptake by plants and its transfer through the food chain. One of the major problems is that 'natural' (geogenic) soil Cd levels can vary greatly. In England and Wales, most soils are in the range 0.2 to 1.5 mg Cd/kg topsoil, with less than 5% with >2 mg Cd/kg topsoil (McGrath and Loveland, 1992), but whether bioavailability increases proportionally is not known.

5.7.2.2 *Radioactivity and phosphatic fertilisers*

Radioactive elements, like uranium (U) and radium (Ra), are normal constituents of the earth's crust (Scholten and Timmermans, 1996). In some cases, the concentration of these elements has been increased in phosphatic rocks by geological processes and a proportion of these elements is carried through into processed P fertilisers. For some people the use of such fertilisers in agriculture has become an issue. This is because the radioactive elements they contain will be added to the soils on which food crops are grown and thus the elements could enter the food chain. Equally there is concern about the environmental impact and handling of the co-product phosphogypsum, which will also contain a proportion of the radioactive elements present in the original rock (Rutherford *et al.*, 1994) as well as some of the Cd. Currently, there is a discussion within the EU about the possible introduction of regulations relating to radiological protection from naturally occurring radioactive materials (NORM).

Soils from the Rothamsted long-term experiments, started between 1843 and 1856, were sampled in 1976 and analysed for U. The superphosphate, applied annually in each experiment, added 33 kg P and 15 g U/ha per year, giving a total addition of about 1300 g/ha U in more than 120 years. The soil is a silty clay loam and in 1976 most of the added U was found in the top 23 cm of soil where arable crops were grown and this was the depth of ploughing, or in the organic surface layer of permanent grassland. In an experiment in New Zealand most of the 330 g Cd/ha added in superphosphate to permanent grassland was also found in the surface layer. There is evidence that U (like P) can be lost by leaching from very sandy soils or can be transported to rivers and lakes with eroded soil (Rothbaum *et al.*, 1979).

There was no evidence of increased U uptake in the cereal grain grown on these experiments even where 1300 g U/ ha had been applied in superphosphate over a 120 year period (Smith, 1960). This supports the generally accepted view that there is little transfer of radionuclides from soil to vegetation (Rutherford *et al.*, 1994).

5.8 CONCLUDING REMARKS

Phosphorus is essential to all known life forms being a key, irreplaceable element in many physiological and biochemical processes. Many unamended soils are inherently deficient in P and while recycling P to them in organic waste products was the only way of adding P their yield potential was limited. This situation changed when the first P fertilisers became available less than 200 years ago. Now in Western Europe about 90% of the P used today is used in agriculture, about 80% as fertilisers and 10% as feed additives for animal feeds. Rightly there are current issues about using this P more efficiently. This is because it has been recognised that global resources of P, as phosphate rock, are finite, and because the loss of P from soil to water is a contributory cause of the undesirable disturbance of the biological balance in some surface fresh waters. The first and probably the second of these two issues can be addressed by ascertaining the critical level of plant available P in soil, i.e. the level of P below which there is a serious loss of yield and above which yield is not further increased by adding more P. Once the soil has been raised to its critical level, the P removed in the harvested crop should be replaced. To check that this replacement technique will suffice to maintain the appropriate soil P status, the soil should be sampled and analysed periodically. Conserving soil P resources suggests that as much P as possible should be recycled to minimise the extraction of phosphate rock for agricultural and domestic use. Husbanding soil so as to reduce the risk of erosion and consequent transfer to water of particulate P is also important for reducing eutrophication risk.

In addition to husbanding the reserves of phosphate rock and avoiding P enrichment of waters there has been concern about non-nutrient elements that are added to soils adventitiously with fertiliser. Principle amongst these is Cd. Many countries already have limits on the permissible level of Cd in P fertilisers and the EU is considering harmonising these levels amongst its member states. The EU is also considering regulations relating to human protection from naturally occurring radioactive materials. Because such elements are found in phosphate rocks, any regulations will apply to the handling of the rock, the manufactured products, if the radioactive elements carry through onto them, and any residual material.

REFERENCES

Amer, F., Bouldin, D.R., Black, C.A. and Duke, F.R. (1955) Characterisation of soil phosphorus by anion exchange resin adsorption and ^{32}P equilibrium. *Plant Soil* **6**, 391–408.

Anon (1995) Use of organic manure. Supplementary report. British Survey of Fertiliser Pratice. Edinburgh University Data Library, Edinburgh, UK.

Barrow, N.J. (1980) Evaluation and utilization of residual phosphorus in soils. In *The Role of Phosphorus in Agriculture*, American Society of Agronomy, Madison, Wisconsin. 333–360pp.

Barrow, N.J. (1983) A mechanistic model for describing the sorption and desorption of phosphate by soil. *J. Soil Sci.* **34**, 733–750.

Barrow, N.J. and Shaw, T.C. (1975) The slow reactions between soil and anions. 2. The effects of time and temperature on the decrease in isotopically exchangeable phosphorus. *Soil Sci.* **119**, 190–197.

Blake, L., Johnston, A.E., Poulton, P.R. and Goulding, K.W.T. (2003) Changes in soil phosphorus fractions following positive and negative phosphorus balances for long periods. *Plant Soil* (in press).

Bray, R.H. and Kurtz, T.L. (1945) Determination of total, organic and available forms of phosphorus in soils. *Soil Sci.* **59**, 39–45.

Brookes, P.C., Powlson, D.S. and Jenkinson, D.S. (1984) Phosphorus in the soil microbial biomass. *Soil Biol. Biochem.* **16**, 169–175.

Chater, M. and Mattingly, G.E.G. (1980) Changes in organic phosphorus contents of soils from long continued experiments at Rothamsted and Saxmundham. *Rothamsted Experimental Station Report for 1979*, Part 2, 41–61.

Chalmers, A.G., Renwick, A.W., Johnston, A.E. and Dawson, C.J. (1999) Design, development and use of a national survey of fertiliser applications. Proceedings No. 437, The International Fertiliser Society, York, UK. 42 pp.

Collins, C. and Budden, A.L. (1998) Soil analysis techniques – the need to combine precision and accuracy. Proceedings No. 418. The International Fertiliser Society, York, UK. 20 pp.

Driver, J., Lijmbach, D. and Steén, I. (1999) Why recover phosphorus for recycling and how? *Environ. Technol.* **20**, 651–662.

Dyer, B. (1902) *Results of Investigations on the Rothamsted Soils.* US Department of Agriculture Office of Experimental Stations, Bulletin 106, Washington, DC. 180 pp.

Egner, H. and Riehm, H. (1955) Lactatmethode zur bestimmung der bodenphosphorsaure. In: *Die Untersuchung von Boden*. Neumann, Radebeul. 177.

Evans, T.D. and Syers, J.K. (1971) Application of autoradiography to study the fate of 33P labelled orthophosphate added to soil crumbs. *Soil Sci. Soc. Am. Proc.* **35**, 906–909.

Fenton, J.P. (1998) On-farm experience of precision farming. Proceedings No. 426. The International Fertiliser Society, York, UK. 32 pp.

Fixen, P.E. and Grove, J.H. (1999) Testing soil for phosphorus. In *Soil Testing and Plant Analysis* (ed. Westerman, R.L.), 3rd edn. Soil Science Society of America, Book Series 3, Madison, Wisconsin, USA. 141–180.

Gendebien, A., Carlton-Smith, C., Izzo, M. and Hall, J.E. (1999) *UK Sewage Sludge Survey*. Environmental Agency, Bristol, UK. 71 pp.

Hislop, J. and Cooke, I.J. (1968) Anion exchange resin as a means of assessing soil phosphate status: a laboratory technique. *Soil Sci.* **105**, 8–11.

Holford, I.C.R. (1982) Effects of phosphate sorptivity on long-term plant recovery and effectiveness of fertilizer phosphate in soils. *Plant Soil* **64**, 225–236.

Holford, I.C.R. and Mattingly G.E.G. (1975a) the high- and low-energy phosphate adsorbing surfaces in calcareous soils. *J. Soil Sci.* **26**, 407–417.

Holford, I.C.R. and Mattingly, G.E.G. (1975b) Phosphate sorption by Jurassic oolitic limestones. *Geoderma* **13**, 257–264.

IFA (1992) *IFA World Fertilizer Use Manual*. International Fertilizer Industry Association, Paris. 632 pp.

Jasinski, S.M. (2000) Phosphate Rock. *US Geological Survey Minerals Yearbook – 2000*. http://minerals.usgs.gov/minerals/pubs/commodity/phosphate-rock/

Johnston, A.E. (1993) Nitrophosphates and their use in agriculture – a historical review. In *Nitric Acid-based Fertilizers and the Environment* (ed. Lee, R.G.), International Fertilizer Development Centre, Muscle Shoals, USA. 37–48.

Johnston, A.E. (1994) The Rothamsted Classical Experiments. In *Long-term Experiments in Agricultural and Ecological Sciences* (eds Leigh, R.A. and Johnston, A.E.), CAB International, Wallingford, UK. 428 pp.

Johnston, A.E. (2000) *Soil and Plant Phosphate*. International Fertilizer Industry Association, Paris. 46 pp.

Johnston, A.E. (2001) Principles of crop nutrition for sustainable food production. Proceedings No. 459, The International Fertiliser Society, York, UK. 39 pp.

Johnston, A.E. and Jones, K.C. (1995) The origin and fate of cadmium in soil. Proceedings No. 366, The International Fertiliser Society, York, UK. 31 pp.

Johnston, A.E. and Poulton, P.R. (1992) The role of phosphorus in crop production and soil fertility: 150 years of field experiments at Rothamsted, United Kingdom. In *Phosphorus,*

Life and Environment, from Research to Application. World Phosphate Institute, Casablanca. 539–572.

Johnston, A.E. and Richards, I.R. (2003) Effectiveness of the water-insoluble component of triple superphosphate for yield and phosphorus uptake by plants. *J. Agri. Sci. Cambridge* **140**, (in press).

Johnston, A.E. and Steén, I. (2000) *Understanding Phosphorus and Its Use in Agriculture.* European Fertilizer Manufacturers Association, Brussels. 36 pp.

Johnston, A.E. and Syers, J.K. (1998) *Nutrient Management for Sustainable Crop Production in Asia.* CAB International, Wallingford, UK. 394 pp.

Johnston, A.E., Goulding, K.W.T., Poulton, P.R. and Chalmers, A.G. (2001) Reducing fertiliser inputs: endangering arable soil fertility. Proceedings No. 487, The International Fertiliser Society, York, UK. 44 pp.

Johnston, A.E., Lane, P.W., Mattingly, G.E.G., Poulton, P.R. and Hewitt, M.V. (1986) Effects of soil and fertilizer P on yields of potatoes, sugar beet, barley and winter wheat on a sandt clay loam soil at Saxmundham, Suffolk. *J. Agri. Sci. Cambridge* **106**, 155–167.

Leigh, R.A. and Johnston, A.E. (1986) An investigation of the usefulness of phosphorus concentrations in tissue water as indicators of the phosphorus status of field grown spring barley. *J. Agri. Sci. Cambridge* **107**, 329–333.

Liebig, H., von (1872) Soil statics and soil analysis. *Z. landw. Ver.* Abstract in *J. Chem. Soc.* **25**, 318 and 837.

Lindsay, W.L. (1979) *Chemical Equilibrium in Soils.* John Wiley and Sons Inc., New York, USA.

MAFF (2000) *Fertilizer Recommendations for Agricultural and Horticultural Crops.* RB 209, 7th edn. Ministry of Agriculture, Fisheries and Food. The Stationary Office, London. 178 pp.

Marschner, H. (1995) *Mineral Nutrition of Higher Plants*, 2nd edn. Academic Press, London. 889 pp.

Mengel, K. and Kirkby, E.A. (1987) *Principles of Plant Nutrition*, 4th edn. International Potash Institute, Basel, Switzerland. 687 pp.

McGrath, S.P. and Loveland, P.J. (1992) *The Soil Geochemical Atlas of England and Wales.* Blackie, Glasgow, 101 pp.

McGrath, S.P., Zhao, F.J. and Blake-Kalff, M.M.A. (2002) Sulphur in soils: processes, behaviour and measurements. Proceedings No. 499, The International Fertilizer Society, York, UK. 28 pp.

Morgan, M.F. (1941) *Chemical soil diagnosis by the Universal Soils Testing System.* Connecticut Agricultural Experimental Station, Bulletin 450.

Olsen, S.R., Cole, C.V., Watanabe, F. and Dean, L.A. (1954) Estimation of available phosphorus in soils by extraction with sodium bicarbonate. *US Department of Agriculture, Circular No. 939.* US Government Printing Office, Washington, DC.

Pierzynski, G.M. (1991) The chemistry and biology of phosphorus in excessively fertilized soils. *Crit. Rev. Environ. Con.* **21**, 265–295.

Rothbaum, H.P., McGaveston, D.A., Wall, T., Johnston, A.E. and Mattingly, G.E.G. (1979) Uranium accumulation in soils from long continued applications of superphosphate. *J. Soil Sci.* **30**, 147–153.

Rutherford, P.J., Dudas, M.J. and Samek, R.A. (1994) Environmental impacts of phosphogypsum. *Sci. Total Environ.* **149**, 1–38.

Scholten, L.C. and Timmermans, C.W.M. (1996) Natural radioactivity in phosphate fertilizers. *Fertil. Res.* **43**, 103–107.

Sikora, F.J. and Giordano, P.M. (1995) Future directions for agricultural phosphorus research. *Fertil. Res.* **41**, 167–178.

Smith, G.H. (1960) Uranium in crops dressed frequently with super phosphate. *Rothamsted Experimental Station Report for 1959*, 58.

Smith, K.A. and Chambers, B.J. (1995) Muck: from waste to resource: Utilisation: the impacts and implications. *Agri. Eng.* **50**, 33–38.

Smith, S.R., Bellett-Travers, D.M., Morris, R. and Bell, J.R.B. (2002) Fertilizer value of enhanced treated and conventional biosolids products. *Biosolids: the Risks and Benefits CIWEM Conference*, London. January 2002.

Steén, I. (1997) A European fertilizer industry view on phosphorus retention and loss from agricultural soils. In *Phosphorus Loss from Soil to Water* (eds Tunney, H., Carton, O.T., Brookes, P.C. and Johnston, A.E.), CAB International, Wallingford, UK. 311–328.

Syers, J.K. and Curtin, D. (1988) Inorganic reactions controlling phosphorus cycling. In *Phosphorus Cycles in Terrestrial and Aquatic Ecosystems. I Europe* (ed. Tiessen, H.), SCOPE, UNDP. Saskatchewan Institute of Pedology, Saskatoon, Canada.

Syers, J.K. and Gochfeld, M. (eds) (2001) *Environmental Cadmium in the Food Chain: Sources, Pathways and Risks*. SCOPE, Paris, France. 204 pp.

Tinker, P.B. (1984) The role of microorganisms in mediating and facilitating uptake of plant nutrients from soil. *Plant Soil* **76**, 77–91.

Thompson, F.M.L. (1968) The Second Agricultural Revolution 1815–1880. *Economic History Review* 2nd Series XXI, 62–77.

Van Kauwenbergh, S.J. (1997) Cadmium and other minor elements in world resources of phosphate rock. Proceedings No. 400, The International Fertiliser Society, York, UK. 40 pp.

Westerman, R.L. (1990) *Soil Testing and Plant Analysis*, 3rd edn. Soil Science Society of America book series No. 3. Madison, WI USA.

6
Transfer of phosphorus to surface waters; eutrophication

S. Burke, L. Heathwaite and N. Preedy

6.1 INTRODUCTION

Phosphorus (P) is an essential element for plant growth and considered necessary for modern agricultural techniques. However, P is one of the key nutrients which have the potential to contribute to eutrophication in surface waters (Cooper *et al.*, 2002; Heathwaite *et al.*, 1996). While many sewage treatment works now have P removal technology in operation, P inputs into agricultural systems continue to supplement ever-increasing global food production. This will exacerbate the potential for increased eutrophication by the transfer of P from agricultural systems to surface waters (Daniel *et al.*, 1998; Edwards and Withers 1998; Heckrath *et al.*, 1995; Sharpley *et al.*, 1994, 1996; Withers and Sharpley 1995), to the extent that, on a global scale, less than 10% of surface waters may now be classified as pristine (Heathwaite *et al.*, 1996).

Eutrophication is defined by Pierzynski *et al.* (2000) as 'an increase in the fertility status of natural waters that causes accelerated growth of algae or water plants'. Changes in the trophic status of waterbodies are characterised by a slow shift from

© 2004 IWA Publishing. *Phosphorus in Environmental Technology: Principles and Applications.*
Edited by Eugenia Valsami-Jones. ISBN: 1 84339 001 9

in-lake biological production driven by external loading of nutrients, to biological production driven by in-lake processes (Rast and Thornton, 1996). The shift in trophic status from oligotrophic through to eutrophic and hypertrophic conditions is typically accompanied by changes in the composition of lake flora and fauna. However, this naturally slow shift can be greatly accelerated by human intervention. Over the last 60 years, the natural biogeochemical cycling of nutrients in agricultural catchments has been influenced by the input of nutrients in the form of fertilisers and animal feeds. These inputs have led to increased soil nutrient levels which have augmented the external loading of nutrients to aquatic systems. Indeed, the term eutrophication is now widely used to describe anthropogenically driven increases in the trophic status of lakes, streams and rivers alike, and there is strong evidence to implicate diffuse loading of P and nitrogen (N) on surface waters as the causal factor (Carpenter *et al.*, 1998). Some features of accelerated eutrophication are excessive algal and rooted plant growth, extensive de-oxygenation of the water, and reduced species diversity of aquatic flora and fauna. The degradation in water quality caused by accelerated eutrophication can create conditions that are detrimental to continued use of the water body for socio-economic purposes. In many areas of the world, the concentration of soil P is in excess of crop requirements (Sims *et al.*, 2000a) and it is feared that continued P amendments may cause soils to become P 'saturated' whereby they lose their capacity to retain P. Such levels of P 'saturation' would increase soil solution P levels and may result in newly applied P amendments being lost from the system (Haygarth and Jarvis 1999; Heckrath *et al.*, 1995; Sims *et al.*, 2000a) resulting in eutrophication and therefore deterioration of surface water quality.

Eutrophication and the associated increase in algae production of inland and coastal waters is a major environmental problem for much of northeast US, Western Europe and parts of Australia. The seriousness of water quality deterioration as a consequence of the export of nutrients in agricultural runoff has been documented recently for both the UK (Environment Agency, 2000) and Australia (Banens and Davis, 1998). The continued influx of P has had long term effects on water quality in some areas of the UK, e.g. the Norfolk Broads (Briggs and Smithson, 1992), however, in many areas algal blooms are quite often localised and short-lived occurrences (Withers, 1993). Yet, many areas of the world have persistent algal blooms such as the Great Lakes in North America (Haygarth, 1997). The ecological implications of increased algae production in rivers and lakes, such as changes in dominant biota, reduced diversity of flora and fauna and summer fish kills, are well documented (Foy and Withers, 1995; OECD, 1982), as are the effects on municipal resources, including reduced amenity and recreation value of fresh waters and increased costs of water treatment (Mason, 1991). Also, the growth of more toxic microbial species can be hazardous and pose considerable threat to livestock and human health. One such incident was the outbreak of *Pfiesteria piscicida* in inland waters of the Chesapeake Bay area of eastern USA (Burkholder and Glasgow, 1997; Satchell, 1997). Indeed, toxic algae have recently been described as one of the most serious global environmental problems (McCarthey, 1999). Such incidents have heightened public awareness and, in tandem with the financial implications of eutrophication, there is increasing political pressure to identify and ameliorate causative nutrient sources (Haygarth and Sharpley, 2000; Sharpley et al., 2000).

The number of rivers, canals and lakes designated as sensitive to eutrophication by the Environment Agency in England and Wales increased from 33 in 1994 to 61 in 1998 (Withers et al., 2000). In Northern Ireland, the eutrophic status of both Lough Neagh and Lough Erne has continued to worsen despite strict compliance with directives to reduce point source P inputs (Zhou et al., 2000). The increasing incidence of eutrophication, both globally and in the UK, has highlighted the contribution of non-point sources of P, especially from agricultural land (Sharpley et al., 2000).

Phosphorus is typically the key limiting nutrient (i.e. that which does or could control biological productivity) in freshwater ecosystems (Hession and Storm, 2000; Hudson et al., 2000). However, N can play an important role in hypertrophic situations, particularly in still waters with a large excess of P. In such cases, freshwater cyanobacteria can fix atmospheric N. In estuarine ecosystems, P tends to be the limiting nutrient at the freshwater extreme, grading through to N limitation at the seaward end, although other factors (light, turbidity, residence time) may limit algal growth. The ability of aquatic algae to efficiently absorb and utilise P in the production of biomass makes fresh water environments sensitive even to small increases in P supply, often manifested in algal-blooms. There are no definite criteria for P concentrations beyond which algal growth will proliferate, but concentrations in excess of $10\,\mu g\,P\,l^{-1}$ (as reactive P, defined as the $<0.45\,\mu m$ fraction) no longer limit the growth of many algae species (Reynolds, 1984). However, the concentrations of P that may be considered as eutrophic will vary for different water bodies. For example, nutrient rich lowland lakes may be able to assimilate certain levels of P influx with no adverse effects that similar nutrient poor upland lakes would not. Therefore, Vollenweider and Krekes (1982) proposed a banded classification of trophic status on the basis of P concentrations, which described lakes with concentrations of total, i.e. dissolved and particulate P (TP), between $35\text{--}100\,\mu g\,l^{-1}$ as eutrophic, with concentrations in excess of $100\,\mu g\,l^{-1}$ as hyper-eutrophic. The situation is somewhat different for running waters that require greater concentrations for similar problems to occur (Edwards et al., 1998). Much attention has been given to reactive (inorganic) $P<0.45\,\mu m$ in runoff, because of its perceived immediate bio-availability to aquatic algae (Sharpley and Smith, 1989). However, organic P forms can contribute to algal growth after the release of inorganic P by phosphatase enzymes (Jansson et al., 1988). Therefore, a measure of total P in runoff is perhaps the best environmental indicator for P.

6.2 EVIDENCE OF SOURCES AND PATHWAYS OF P DELIVERY TO WATER

Eutrophication as a consequence of excessive P concentrations in receiving waters may result from the export of P from point and/or non-point sources. Traditionally a point source is a clearly defined entry point such as a sewage treatment works, while a non-point source does not have a discrete discharge point, for example, agricultural runoff (O'Shea, 2002). However, it is important to recognise that within non-point sources, such as agricultural fields, the contribution to P delivery is not equal: some areas of a field may have greater connectivity with receiving waters and other areas may contain 'point' sources of P such as feeding and defecation areas.

Water quality in lakes and rivers reflects the dominant land use in their catchments (Arbuckle and Downing, 2001; Carpenter et al., 1998; Crosbie and Chow-Fraser, 1999). Harris (1999) discusses the relationships between changes in land use and the rates of export and predominant forms of carbon (C), N and P. Catchment exports of N and P from diffuse sources are a strong function of land use, population densities, agricultural practices and urban development (Carpenter et al., 1998). Forested catchments export mostly dissolved organic carbon and little N or P (Figure 6.1); whereas agricultural and urban catchments show increased exports of C, N and P with more particulate organic C (from eroded A horizons), particulate P from gullying, and more dissolved inorganic N and P from fertilisers and wastewater inflows.

Sewage treatment works release P in soluble reactive form to surface waters. At times of high flow in receiving waters, the P loadings can be modest with minimal environmental impact. However in times of low flow, the P discharged to surface water course can be significant; concentrations up to $20\,\mathrm{mg\,l^{-1}}$ dissolved reactive phosphorus ($<0.45\,\mu m$) have been recorded by the authors (unpublished data) in a small river.

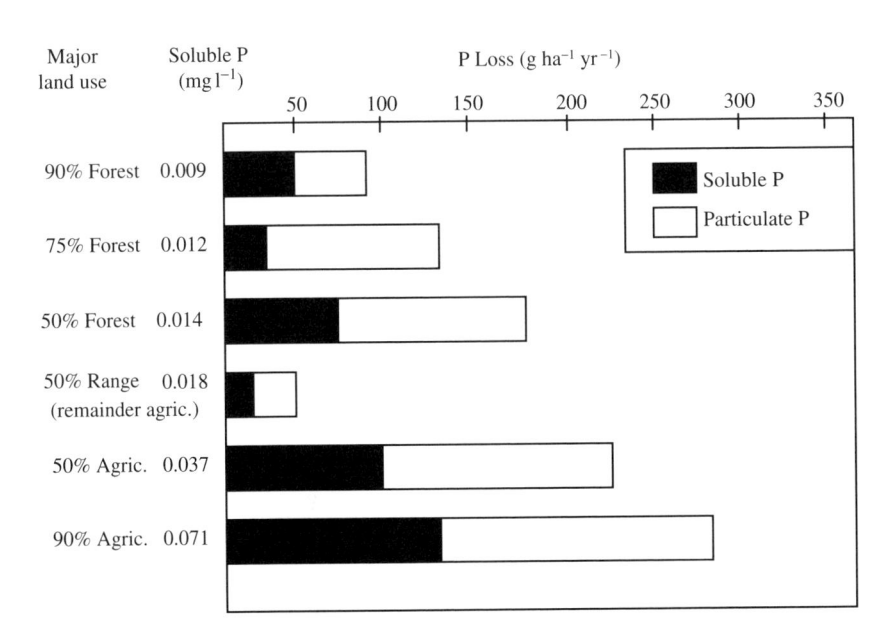

Figure 6.1. Effect of land use on the loss of soluble and particulate P from soils (adapted from Pierzynski et al., 2000).

Crop P requirements for agricultural land is often met by the routine application of inorganic P fertilisers, livestock manures or increasingly, sewage sludge (Withers et al., 2001). This may lead to regional and national imbalances in agricultural inputs and outputs of P (Heathwaite et al., 2000). The current view is that agriculture may be a major source of elevated nutrient concentrations in ground and surface waters around the world (Heathwaite et al., 1996; Sharpley et al., 1994) and often the source of eutrophication in surface waters. The build-up of soil P is of considerable importance in determining the cumulative balance between inputs and outputs since losses range significantly. Withers and Lord (2002) report the UK average annual P surplus ranges around $10\,kg\,ha^{-1}$, with estimated values in the US averaging $26\,kg\,ha^{-1}$ per annum (Gburek et al., 2000).

Total P levels in soils range from 50 to $1500\,mg\,kg^{-1}$, with 50 to 70% found in the inorganic form in mineral soils (Pierzynski et al., 2000). However the concentration of P in soil depends on its source and this varies considerably between fertilisers. Table 6.1 shows the major sources of P for crop production and the typical P and P_2O_5 levels.

However high soil P concentrations do not in themselves indicate areas that are susceptible to P transfer. The association between the source and its pathway is more important (Gburek et al., 2000; Heathwaite and Dils, 2000; Pionke et al., 1997). Hydrology provides both the energy and carrier for P transfer (Haygarth and Jarvis, 1999). Values for soil test P (STP) alone can only be used to indicate

Table 6.1. Major sources of P for crop production (adapted from Pierzynski et al., 2000).

P Source and chemical composition	P (%)	P_2O_5 (%)
Commercial fertilisers		
Ordinary super phosphate	7–10	16–23
Triple super phosphate	19–23	44–52
Monoammonium phosphate	26	61
Diammonium phosphate	23	53
Urea-ammonium phosphate	12	28
Ammonium polyphosphate	15	34
Rock phosphates		
US	14	33
Brazil	15	35
Morocco	14	33
Former U.S.S.R	17	39
Organic P Sources		
Beef cattle manure	0.9	2.1
Dairy cattle manure	0.6	1.4
Poultry manure	1.8	4.1
Pig manure	1.5	3.5
Aerobically digested sludge	3.3	7.6
Anaerobically digested sludge	3.6	8.3
Composted sludge	1.3	3.0

areas of high soil P as the vulnerability of inland waters to P loading is primarily a function of hydrological connectivity from P source areas to a watercourse (Heathwaite and Dils, 2000). Therefore, to mitigate P transfer necessitates a holistic understanding of interactions between: soil P concentrations and/or sources of P at the soil surface, the propensity of different hydrological pathways for P transfer and hydrological connectivity between P sources and surface waters. This concept was exemplified by Haygarth et al. (1998a) who found that for a poorly drained grassland soil with a low soil P status (7.5 mg kg^{-1} Olsen P), TP concentrations in overland flow and drainflow during storm runoff were still consistently in excess of the 100 µg l^{-1} required for algal growth. For the same soil, annual P exports were estimated at between 0.5 and 2 kg TP ha^{-1} (Haygarth and Jarvis, 1996; Haygarth et al., 1998a). This situation may be similar for many grassland areas in the UK, many of which are located on impermeable soils in upland areas that receive high rainfall.

Consequently, it is necessary to obtain a clear understanding of hillslope hydrological processes, and how the temporal and spatial complexities inherent within the hydrological regime can influence P transfer, pathways and subsequent eutrophication. Pionke et al. (1997) consider a catchment to be a collection of P sources, stores and sinks (Figure 6.2) that are knitted together on the hillslope by a flow framework. The hydrological connectivity of this framework at any one time is critical when considering the magnitude of P transfer during and following rainfall (Figure 6.2).

Perhaps the most important spatial concept in hillslope hydrology in relation to P transport is that of the variable source area (VSA) which, given the right hydrological conditions, can account for a significant proportion of a catchment. Variable source areas are dynamic areas on a hillslope that expand and contract in response to rainfall (Dunne, 1978). They occur where convergent subsurface flows meet at slope bases, slope hollows, slope concavities or areas of shallow soil (Anderson and Burt, 1990). Such areas generate rapid surface and subsurface runoff that can dominate catchment response during rainfall. In temperate climates surface runoff occurs as saturation excess overland flow when precipitation falls directly onto VSAs, but also as return flow generated during and after rainfall. In addition to the hydrological importance of VSAs, researchers investigating P transfer have reported VSAs to be the key driver for the majority of annual P transfer in some catchments (Pionke et al., 1996, 1997). The coincidence of P sources, stores and sinks in a catchment with that of VSAs has been used to define Critical Source Areas (CSA) (Gburek et al., 2000) for P transfer within catchments (Figure 6.3). Critical source areas of diffuse pollution occur where land with high nutrient status coincides with areas of significant hydrological connectivity with receiving waters (Table 6.2). As VSAs (or CSAs) expand during rainfall, the proportion of a catchment susceptible to P transfer may increase. In theory, maximum transfer will occur when there is complete connectivity between all source areas and the stream channel. If the connectivity to the stream channel is incomplete, P may be retained in temporary stores to be re-entrained when the hydrological framework is more complete (Figure 6.4). Although we know that subsurface flows are extremely important for generating VSAs, to date, the concept of CSAs for P transfer has been largely

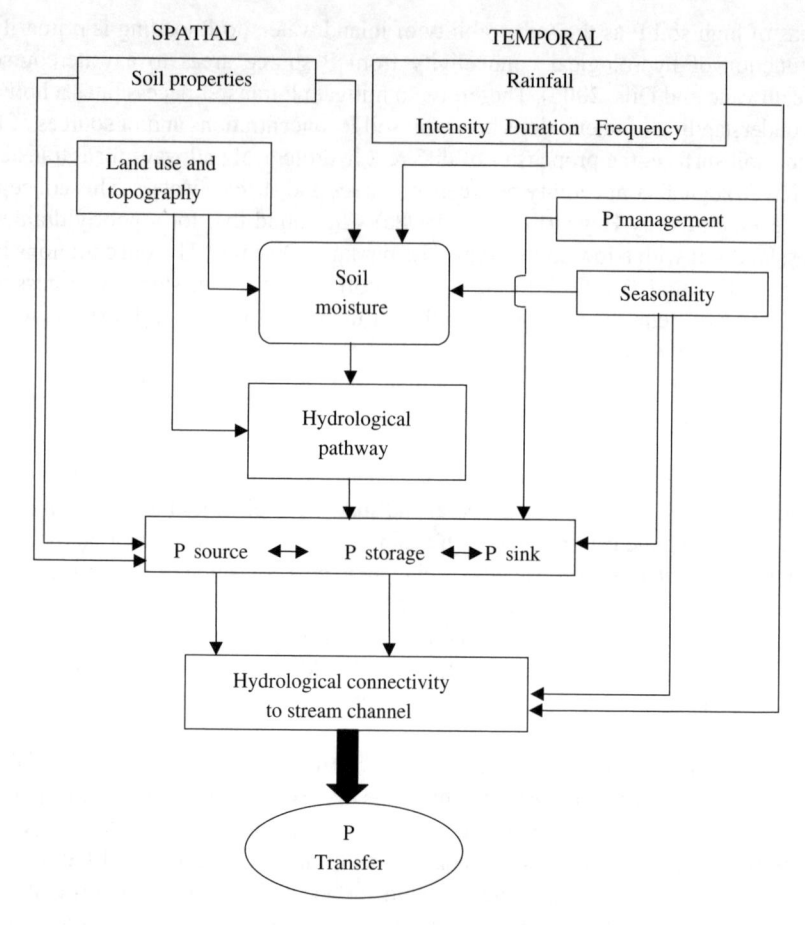

Figure 6.2 Temporal and spatial factors affecting hydrological connectivity, P transfer and subsequent eutrophication (Preedy *et al.*, 2000).

Figure 6.3 Critical source area concept (Gburek *et al.*, 2000).

used to examine the inter-connectivity between areas of surface runoff (Gburek *et al.*, 2000; Pionke *et al.*, 1997). The role of catchment-wide subsurface flow pathways and their contribution to CSAs remains more uncertain (Heathwaite *et al.*, 2000).

Table 6.2 Overview of temporal and spatial factors influencing hydrology and P transfer in soils (definitions of hydrology and P transfer terminology are summarised in Table 6.3).

Temporal	Hydrology	P mobilisation
Rainfall intensity	Low intensity rainfall readily infiltrated into the soil; high intensity rainfall leads to the initiation of overland flow.	Rain impact during intense rainfall encourages dispersal of soil aggregates at the soil surface making P associated with fine sediment more susceptible to erosion.
Rainfall duration	Soil moisture increases (toward saturation) with continued rainfall. When a soil is saturated any additional rainfall initiates saturation excess overland flow (even at low rainfall intensities).	The amount of loose pre-storm colloidal material available for erosion decreases with increasing rainfall duration.
Return period	As antecedent dry time increases large soil pores empty under gravity and the hydrological connectivity between soil pores is reduced. Conversely, the soils potential for infiltration and water storage increases.	Dry soils have low adhesion forces binding sediments together. Therefore, lengthy antecedent dry times encourage increasingly dry soils and increase the amount of potentially erodable colloidal and clay sized material.
Seasonality	Wet soils encourage a more rapid runoff response to rainfall whereas when soils drain beyond field capacity the connectivity between soil pore decreases and runoff response is reduced. Dry soils also have greater potential for infiltration and storage of rainfall. Evapotranspiration rates are also much higher during the summer.	Soil temperature and moisture conditions encourage an increase in microbial activity in the spring and autumn, and can raise microbial populations that increase the pool of organic P and the turnover of inorganic P.
Antecedent soil moisture	Soil moisture can effect runoff response to rainfall and determine the contribution of different hydrological pathways to stormflow response.	Saturated soils encourage reducing conditions that increase the potential for P solubilisation.
Infiltration rate	Sandy, dry or cultivated soils have a higher capacity for infiltration thus reducing overland flow potential.	
P management	In the short-term, the addition of slurry to the soil surface can increase moisture content. In the long-term, continued additions of organic amendments can increase the moisture retention capacity of the soil.	Continued P inputs raise soil P levels and increase the potential for P solubilisation. Also, amendments made prior to rainfall are vulnerable to direct transfer from the soil surface.
Soil P		Most P is strongly 'fixed' in the top few cm of the soil. Only a small proportion of soil P at any one time exists as soluble

(*continued*)

Table 6.2 (continued)

Temporal	Hydrology	P mobilisation
		phosphate in soil solution, instead up to 90% of soil solution P is organic P. Equilibrium soil solution P concentrations are increasing in agricultural soils because of continued P inputs and increasing soil P levels.
Soil texture	Soils with a high clay content are prone to cracking and the development of preferential flow pathways. Sandy soils allow more efficient movement of water through the soil matrix due to the interconnectivity between large soil pores.	Clay soils offer a large specific surface area for P adsorption and these soils can retain high soil P levels. Overland flow has a selective nature to erode P rich clay size particles. In contrast, sandy soils have a lower P retention capacity and also encourage rapid infiltration of rainwater.
Soil structure	Cracks and fissures encourage preferential flow. Also, low hydraulic conductivity and slowly impermeable soil horizons can encourage lateral subsurface flow and development of perched water tables.	
Soil pH		An increase in pH can release sediment bound P by increasing the charge of Fe and Al hydrous oxides and therefore increasing the competition between hydroxide and phosphate anions for sorption sites.
Drainage	Tile drains reduce waterlogging and overland flow generation and also offer a rapid route for the transfer of subsurface water.	Reduces the potential for overland flow and P transfer at the soil surface. However, land drains are also considered to be effective for P transfer.
Topography and slope	In temperate climes variable source area (VSA) hydrology can dominate catchment response to rainfall. VSAs are prone to saturation and the generation of overland flow. VSAs develop at slope bases, slope concavities or areas of shallow soil. Any increase in the angle of the slope can increase the erosive potential of overland flow.	
Stocking density	High stocking densities increase poaching at the soil surface. This encourages waterlogging and overland flow.	Livestock excreta (high in P) is a P source at the soil surface that is susceptible to transfer in surface and subsurface pathways.
Crop cover	Vegetation cover can increase infiltration. Bare soils are prone to capping and the generation of overland flow.	Bare soils encourage the loss of P rich sediment in sheet and rill erosion.

Table 6.3 Hydrological terminology for P transfer (adapted from Preedy *et al.*, 2000; Haygarth and Sharpley, 2000).

Term	Scale	Flow direction	Definition
Antecedent soil moisture			Moisture status of the soil preceding rainfall.
Baseflow	Slope/catchment		Typically describes background groundwater flow and low magnitude flow conditions.
Bypass flow-unsaturated subsurface flow	Soil/slope		Rapid subsurface flow of water through preferential flow pathways that enables subsurface water to pass through unsaturated areas.
Capping			Formation of crusts on bare soils due to dispersion of surface aggregates as a results of rainfall.
Convergent flow	Slope	Lateral	Subsurface flows that converge in slope hollows and concavities.
Critical source area	Slope/catchment		Descriptor for a hydrologically active area that is significant in contributing P to stream channels during storm flows.
Field capacity			Soil moisture status after gravitational drainage of soil pores.
Groundwater flow	Slope/catchment	Lateral	Specific pathway that describes the passage of deep groundwater.
Hortonian flow-infiltration-excess overland flow	Slope	Lateral	Descriptive terms for overland flow that is generated when rainfall intensity is in excess of infiltration capacity i.e., on capped bare soils.
Lagtime			Time delay from peak rainfall to peak discharge in the stream channel.
Land drains-Tile drains	Slope	Lateral	Specific pathway. Subsurface 'mole' and 'tile' drains artificially installed to aid water movement
Leaching	Soil	Lateral or vertical	Chemical mechanism describing the elluviation of chemicals through the soil profile.
Macropore flow-Preferential flow	Soil	Lateral or vertical	Specific pathway. Movement of water through large connected pores, wormholes, cracks and fissures.
Matrix flow	Soil	Lateral	Specific pathway referring to the movement of water through the smaller pore spaces in the bulk of the soil.
Overland flow-'Surface runoff'	Slope	Lateral	Movement of water exclusively over the soil surface.

(continued)

Table 6.3 (continued)

Term	Scale	Flow direction	Definition
Partial area	Slope		Catchment pathways such as roads, animal tracks etc. which have hydrological connectivity with streams during storms.
Percolation-Throughflow	Soil	Lateral or vertical	General non-specific term describing subsurface water movement.
Pipe flow	Slope	Lateral	Specific pathway for the rapid subsurface transfer of water through large soil pipes i.e. root channels and animal burrows.
P source/store/sink	Soil/slope		Area of soil that actively contributes to P transfer source areas/temporarily stores of P on the hillslope/area that retains P.
Rainfall return			Time interval between rainfall events.
Return flow	Slope	Lateral	Emergence of subsurface flow path at the soil surface.
Runoff	Slope	Lateral	General term referring to the composite of surface and subsurface flows influencing stream flow response.
Saturated overland flow	Slope	Lateral	Descriptive term for overland flow generated either when rain falls onto a saturated area or when convergent subsurface flows reach the soil surface as return flow.
Saturated subsurface flow	Soil	Lateral or vertical	Composite of matrix and macropore flow when the soil is uniformly saturated.
Soil moisture deficit			Soil moisture status after cessation of gravitational drainage (field capacity) when evapotranspiration is dominant.
Soil solution	Soil		General term referring to water held within the matrix of the soil profile.
Stormflow	Slope	Lateral	General term used to describe the rapid surface and shallow subsurface flows during and after rainfall that contribute to the storm hydrograph response of a catchment i.e. above baseflow.
Subsurface flow	Slope	Lateral	Non-specific term referring to a composite of shallow subsurface pathways.
Translatory flow	Slope	Lateral	Prestorm water stored in the soil that is displaced by rainfall to become a component of subsurface storm flow.
Variable source area	Slope		Temporally and spatially dynamic saturated area that can contribute to overland flows as return flow or by direct precipitation onto the saturated area.

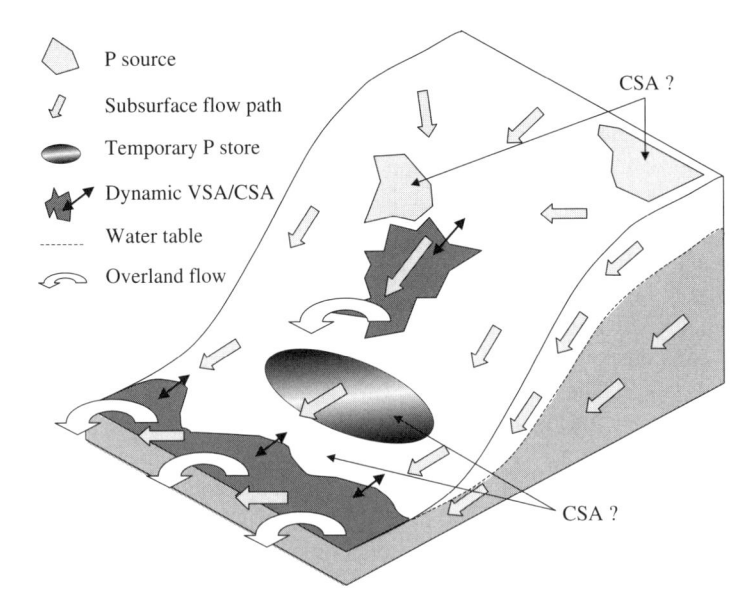

Figure 6.4 Hillslope connectivity: variable-source areas (VSA), critical source areas (CSA), temporary P stores and pathways for P transfer.

The relationship between imbalances in the agricultural P cycle and losses of P to surface and groundwaters is not straightforward. As discussed above, factors such as climate, soil type, land use, drainage and fertiliser timing and application rate are influencing the magnitude of P losses from agricultural land. Many studies have now been able to demonstrate a general relationship exists between land use/cover and river P concentration at the catchment and sub-catchment scale (Haygarth, 1997; Heathwaite and Johnes, 1996; Sharpley *et al.*, 1994). The link exists because cultivation modifies soil structure, releasing potentially transportable P. Evidence in support of this link is obtained from P concentrations and losses measured at plot, field and watershed scales which have shown that the concentrations of limnological significance (<0.1 mg P l^{-1}) can frequently occur in drainage water (Sharpley *et al.*, 2000).

6.3 TRANSFER MECHANISMS AND CONTROL

Phosphorus in the soil can be subdivided into inorganic and organic forms. Inorganic P exists in a range of mineral forms (e.g. aluminium (Al), iron (Fe) and calcium (Ca) phosphates) that are stable in the soil and also as orthophosphate adsorbed onto soil particles, particularly those with high surface areas such as clays, or the organic matter coatings of soil particles (Frossard *et al.*, 1995; Ryden *et al.*, 1973). Inorganic P, as soluble phosphate, has been identified as the most immediately plant and algal-available P form. As a result the transfer to water of inorganic P forms (<0.45 μm) that have been perceived as 'soluble' P are well researched (Ahuja et al., 1981; Heckrath et al., 1995; Sharpley, 1985; Sharpley and Smith, 1989).

In contrast, organic P forms are poorly understood in terms of the chemical forms in which they exist, their cycling, plant availability and transfer to water (Turner, 2000). A substantial proportion of the total P in soils exists in an organic form (Harrison, 1987), and organic P has been shown to account for up to 90% of total P in soil solution for a range of grassland soils (Chapman et al., 1997; Ron Vaz et al., 1993; Shand and Smith, 1997). Also, many forms of organic P are considered to be highly mobile in the soil (Chardon et al., 1997; Turner, 2000). Therefore, organic P may actually provide an important P source in surface waters where it can contribute to algal growth after the release of inorganic P by phosphatase enzymes (Jansson et al., 1988; Turner, 2000).

There are many mechanisms involved in P transfer. However the initial mobilisation of P may be considered to be the limiting step. Haygarth and Jarvis (1999) have defined three mechanisms by which P is initially mobilised. These are solubilisation, physical detachment and direct (incidental) transfer of recent P amendments. Solubilisation accounts for the dissolution and subsequent displacement of soil solution P, commonly referred to as leaching. Solubilisation is initiated through the weathering of Fe, Al and Ca phosphates and the subsequent release of labile inorganic P (Black, 1967). Solubilisation occurs from the organic P pool following the hydrolysis of organic P by phosphatase enzymes (Turner and Haygarth, 2002). The pool of soluble phosphate accounts for only a small proportion of total P in the soil system. In natural systems the concentration of phosphate in solution remains in constant equilibrium with plant requirements and P levels in the surrounding soil. However, continued P applications to agricultural systems have led to increasing soil P concentrations. This has had the dual effect of increasing soil solution P concentrations whilst 'saturating' available sorption sites, thus decreasing the ability of the soil to fix P in less mobile forms (Sims et al., 2000b).

There has been considerable research relating soil test 'dissolved' P to soil P levels (Heckrath et al., 1995; Pionke and Kunishi, 1992; Sharpley and Smith, 1989; Sharpley et al., 1996; Sims et al., 2000a). However, in these studies filtration through a 0.45 μm filter has often been used to define the difference between 'dissolved' and 'particulate' P. In fact much P (<0.45 μm) is not actually 'dissolved' but found in association with a continuum of colloids and particles >1000 MW (0.01 μm) (Haygarth and Sharpley, 2000; Haygarth et al., 1997) (Figure 6.5). The arbitrary analytical approach attaches considerable ambiguity to the definition of 'dissolved' or 'soluble' P when interpreting published literature. Despite this, the literature does offer some useful indicators for the factors affecting solubilisation. For example, Ron Vaz et al. (1993) found that inorganic P (<0.45 μm) accounted for only 12% of P in soil solution under a zero fertiliser treatment compared to 56% at a high P input site. Research in the USA has also shown clear relationships between the build up of soil P and associated increases in easily desorbed P and soluble P (<0.45 μm). These methods have been used to assess the degree of soil P saturation (DPS) whereby values of soluble P (<0.45 μm) above 35% were considered to be excessive to crop requirements (Pautler and Sims, 2000). The

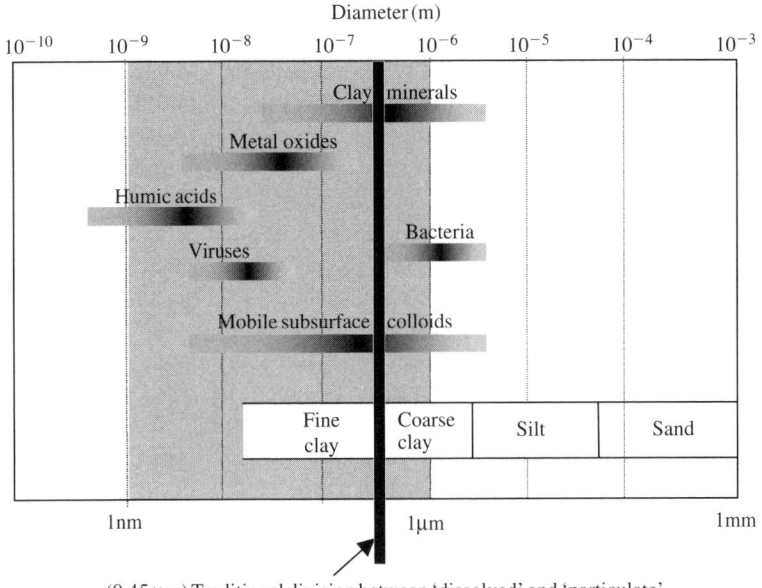

Figure 6.5 Size range of colloidal particles in subsurface environments. The classic size fractions commonly used in soil science, and P transfers are shown for comparison. The bold vertical line is showing the division between dissolved and particulate. After Kretzschmar et al. (1999) and Matthews et al. (1998).

authors have alluded to concerns that more solubilised P is becoming available to transfer in subsurface pathways.

Changing oxidation-reduction conditions caused by water table fluctuations can also affect P sorption-desorption kinetics in the soil profile. Vadas and Sims (1998) have shown that reduced conditions can increase levels of soluble and chemically extractable P. Concurrent increases in soluble Fe in solution suggested that this was a function of P dissolution from Fe compounds. They also showed that reduced conditions also decreased the soil's sorption capacity and had the effect of increasing equilibrium P concentrations in soil solution. Sallade and Sims (1997) found that soils with DPS values in excess of 40% had the greatest potential for P solubilisation under anoxic conditions. The contribution of inorganic solid phase P to soluble phase P is also influenced by pH and the considerable presence of other strongly sorbing species. An increase in pH can release sediment bound P by increasing the negative charge of Fe and Al hydrous oxides and therefore increasing the competition between hydroxide (OH^-) and phosphate ($H_2PO_4^-$) anions for sorption sites (Ron Vaz et al., 1993; Sosa, 1996). The presence of other strongly sorbing anions can have a similar effect by competing with phosphate anions for sorption sites.

Physical transfers occur when water carries non-dissolved P. It is well established that surface runoff erodes particles and colloids from the soil surface and that P is transferred in association with this material. Further to this, Kretzschmar et al. (1999) suggest that colloids can also be highly mobile in subsurface pathways (Figure 6.5) and that this may provide a vehicle for strongly sorbing contaminants such as P. To be significant for P transfer, these colloids would need to be transferred in sufficient quantities. However, at present the size fractions and loads of colloids transferred along different pathways are not well established. Colloidal organic P (bacteria, Figure 6.5) frequently forms the dominant fraction of the total P that is held in soil solution. This has been observed in a range of grassland soils with varying P input (Chapman et al., 1997; Ron Vaz et al., 1993; Shand and Smith, 1997). Apart from being an important source for P solubilisation, these colloids might also be important for physical transfers of P under grassland soils (He et al., 1995; Turner and Haygarth, 1999). Unlike inorganic P that has an affinity to be rapidly adsorbed at or near to the soil surface, organic P has a tendency to be more mobile when held in soil solution. Evidence of this P mobility has been illustrated by a number of workers who found that the ratio of organic P to inorganic P increases considerably with depth in the soil (Chardon et al., 1997; Dils and Heathwaite, 1996; Ron Vaz et al., 1993). In contrast to the soil solution P studies, both Chapman et al. (1997) and Turner and Haygarth (1999) found that 'dissolved' ($<0.45\,\mu m$) inorganic P accounted for up to 70% of total P in leachate from lysimeters. However, it is possible that this is a reflection of the experimental set up, whereby lysimeters can encourage vertical preferential flow pathways that facilitate the rapid transfer of 'dissolved' ($<0.45\,\mu m$) inorganic P from near the soil surface. Turner and Haygarth (1999) and Turner (2000) observed that colloidal organic P in leachate was greater in the springtime and this was coupled to elevated biological activity in the soil. Indeed, organic P ($>0.45\,\mu m$) accounted for a large proportion of total P transferred at this time, which might possibly consist of microbial cells and cell debris (Hannepel et al., 1963).

Much inorganic P $<0.45\,\mu m$ is not actually 'soluble' but is instead stable by its integration within inorganic complexes, such as Fe, Al and Ca phases or adsorbed to organic complexes that coat the surface of soil particles (Ryden et al., 1973). Indeed, the majority of inorganic P in many soils exists primarily in association with Fe and Al compounds, either as surface bound forms or within the matrix of such components (Ryden et al., 1973). These stable associations satisfy the tendency for P to be strongly adsorbed, however, if this material is easily eroded it may also enable P to be mobile in the soil (Kretzschmar et al., 1999). Of all the inorganic soil components with which P can be associated, it is the Fe and Al compounds in particular that have the potential to be mobile in subsurface environments, especially in acid soils. Indeed Wilson et al. (1991) found that the concentration of Fe and Al ($<0.45\,\mu m$) increased rapidly in a stream channel with the initiation of subsurface storm flow. Furthermore, although studies allude to the efficiency of macropore or preferential flow pathways for the transfer of 'dissolved' ($<0.45\,\mu m$) inorganic P (Haygarth et al., 1998a; Jensen et al., 1998; Simard et al., 2000), it is highly probable that much of this P is not actually soluble but associated with colloidal material. Although phosphate complexation limits the availability of P, upon reaching surface waters, it is inorganic P associated with Fe and Al oxyhydroxides that is considered to regulate net P mobilisation in these waters (Koski-Vahala *et al.*, 2001).

Direct transfers of fertiliser or manure (incidental losses) can occur when applications are coincident with the onset of rainfall. In reality, incidental losses will include solubilisation and physical detachment. However, Haygarth and Jarvis (1999) have argued that, conceptually, incidental transfer should be kept separate because of the unique circumstances leading up to its occurrence (Haygarth *et al.*, 2000).

Circumstances for incidental transfer are not uncommon in the UK. The frequency of rainfall can offer few opportunities to apply fertiliser or manure when soil and weather conditions are suitable. Furthermore, farmers are often under pressure to apply slurry (even to wet soils) to reduce pressure on limited storage facilities (Edwards and Withers, 1998; Withers *et al.*, 2000). Some scientific evidence of incidental P transfer has been observed in overland flow following the application of fertiliser (Haygarth and Jarvis, 1997; Nash *et al.*, 2000; Withers *et al.*, 2001) and different forms of manure (Misselbrook *et al.*, 1995; Wang *et al.*, 1996). In subsurface pathways, Sharpley and Syers (1979) observed an increase in drainflow P concentrations following the introduction of livestock onto pasture. This phenomena is elucidated in work by Chardon *et al.* (1997) who discovered that the amount of organic P increased significantly at depth following the application of manure to the surface of soil columns. Donald *et al.* (1993) suggested that increased levels of mobile colloidal C following slurry applications could be a mechanism for the transfer and redistribution of P in some soils. Indeed, increasing attention is being given to P transfer under slurry based systems (Barkle *et al.*, 1999; Chardon *et al.*, 1997). If C colloids do facilitate P transfer through the soil, subsurface pathways could be particularly important for P transfer under this kind of management.

Gburek and Sharpley (1998) state that the key to understanding P transfer is a detailed understanding of the pathways of flow within a catchment. Figure 6.6 illustrates the potential P pathways. Different pathways vary in their propensity for

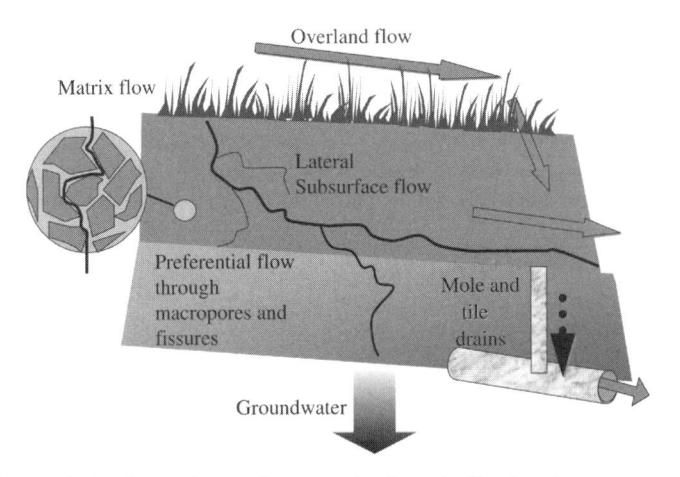

Figure 6.6 Hydrological pathways for P transfer in agricultural systems.

P transfer, and the load and chemical form of P entering a watercourse is entirely a product of a particular rainfall-soil-pathway interaction on a hillslope (Dils and Heathwaite, 1996). The pathways for water transfer (Figure 6.6), whether laterally along a hillslope or vertically through the soil, influence the amount and rate of P reaching a stream channel. The rapid transfer of water in lateral surface and shallow subsurface hillslope pathways is referred to as stormflow. Although the time in which stormflow occurs is proportionately small, it is considered to account for the majority of annual P transfer in agricultural catchments (Pionke and DeWalle, 1994; Pionke et al., 1996). Alternatively, vertical flows have been shown to be effective at leaching P in clay (Turner and Haygarth, 1999) and sandy textured soils (Chardon et al., 1997). The translocation of P down the soil profile may be important for P transfer in soils or topographical areas that encourage vertically percolating water. Also, vertical flows act to recharge groundwater stores that become the principal component of stream discharge during baseflow periods. However, more so than other transfer pathways, the contribution of groundwater flows to P transfer, although suspected to be minimal, is still poorly understood.

The link between high rainfall and large P transfer events has targeted research on the mechanisms of P mobilisation that occur during stormflow periods. However, because of the perceived immobility of P in subsurface soil and the tendency of P to rapidly adsorb on to particles the top few cm of the soil profile (especially in uncultivated grassland soils) (Haygarth et al., 1998b), most P transfer research has focussed on surface runoff (Burwell et al., 1975, 1977; He et al., 1995; Pionke et al., 1996; Sharpley, 1980; Sharpley and Smith, 1989). Surface runoff is effective at eroding P adsorbed to clay size particles and organic colloids at, or near to, the soil surface (Quinton et al., 2001; Ryden et al., 1973). Additionally, surface runoff can be effective at removing soluble P from the thin layer of soil (the effective depth of interaction) at the soil-surface and water interface (Ahuja et al., 1981; Sharpley, 1985).

Surface runoff rises as an exponential function of rainfall and runoff and erosion events are spatially limited and confined to high rainfall events (Heathwaite and Dils, 2000; McIvor et al., 1995). In many regions, storm flows carry a very large fraction of the total N and P loads in only a small portion of the year (Cullen and O'Loughlin, 1982; Olive and Walker, 1982) so that there are significant methodological problems in measuring the exports from catchments (Cullen et al., 1978). In Australia, for example, much of the water and nutrient runoff from catchments comes from relatively small areas and in relatively short time periods in most cases (Cullen et al., 1978; Cullen and O'Loughlin, 1982; Dillon and Molot, 1997; Olive and Walker, 1982; Prosser et al., 2001; Wallbrink et al., 1998).

Flow at the soil surface can be generated as 'Horton' (infiltration-excess) overland flow (Horton, 1945) or saturation-excess overland flow (Emmett, 1978). In natural systems, Hortonian flow is often localised in occurrence (Anderson and Burt, 1990), although newly tilled bare soils can be susceptible to surface sealing that reduces infiltration and encourages the generation of infiltration excess overland flow (Heathwaite, 1990). However, in established grassland systems overland flow is usually generated by saturation-excess mechanisms according to the theories of VSA hydrology (Anderson and Burt, 1990; Emmett, 1978). Surface runoff has long been considered as the principle pathway for P transfer, and in some areas of the world this is certainly the case. For example, in the semi-arid climate of south Australia very occasional, intense rainfall accounts for over 90% of annual P transfer (Fleming *et al.*, 1997, 2001). However, in the temperate climate of the UK, Quinton *et al.* (2001) observed that high-frequency, low-intensity rainfall accounted for a greater proportion of sediment bound P lost over a six-year period than large, infrequent events. They also observed temporal fluctuations in the transfer susceptibility of sediment bound P. Indeed, although concentrations generally increased with increasing peak discharge, the amount of available P transferred decreased with increasing event duration and shorter return period between rainfall events.

Much of the world's grassland agriculture is located in temperate climes and in these regions it is subsurface flows that are fundamental for the generation VSAs and surface runoff: in the UK the occurrences of 'intense' rainfall ($>10\,\text{mm hr}^{-1}$) are rare and the majority of annual rainfall occurs as high-frequency, low intensity rainfall (typically between $1-5\,\text{mm hr}^{-1}$) (Fraser, 2000). The frequency of rainfall renders the soil in excess of field capacity for long periods of the year (Armstrong *et al.*, 1984). The high antecedent moisture conditions invariably result in a 'flashy' response to rainfall ($>2-3\,\text{mm hr}^{-1}$) as surface and subsurface stormflow pathways become activated. As in other areas of the world, it is these periods of high runoff that are perceived to account for the majority of annual P transfer (Foy and Withers, 1995; Hawkins and Scholefield, 1996; Haygarth, 1997; Haygarth *et al.*, 2000; Withers, 1993). However, the slow drainage of soils toward field capacity that occurs between rainfalls, may account for a high proportion of annual runoff and requires investigation for its role in P transfer (Haygarth *et al.*, 2000).

The subsurface pathways that constitute stormflow, in both saturated and unsaturated conditions, are macropore flow (Bouma and Dekker, 1981), preferential flow (Armstrong *et al.*, 1999) pipe flow (especially important in uncultivated permanent grassland (Jones, 1997) and land drains (Skaggs *et al.*, 1994). These pathways bypass the slower diffusion of water in the soil matrix and offer a rapid route for the transfer of stormflow water (Chapman *et al.*, 1993; Nieber and Warner, 1991; Skaggs *et al.*, 1994). These pathways are an important transfer mechanism for P because, although they account for only a small portion of the whole soil volume, they can be responsible for the majority of subsurface water transfer (Stamm *et al.*, 1998). Indeed, the perception that the connectivity of surface runoff can dominate rapid hydrograph response during high rainfall may be a fallacy for many catchments. This has been shown in stormflow studies of agricultural catchments in the USA. The contribution of total overland flow and total subsurface flow to storm events were consistently recorded at around 10% and 40% respectively, for storm events of various rainfall intensities and antecedent conditions (Bazemore *et al.*, 1994; DeWalle and Pionke, 1994; Pionke *et al.*, 1993). Also, studies in upland areas of the UK have reported similar results, with subsurface pathways contributing greater than a half of total stormflow (Jones, 1997; Muscutt *et al.*, 1993; Nieber and Warner, 1991). The studies in the USA have also shown the importance of translatory flow (pre-event soil water) in contributing to storm flow. This soil solution water can account for between 25% and 65% of subsurface stormflow water (Bazemore et al., 1994; Mulholland, 1993; Pionke et al., 1993). Overall, the contribution of subsurface hydrological pathways is extremely important to channel discharge, both during and between stormflow periods. Yet, despite the hydrological importance of lateral subsurface pathways, they have traditionally been overlooked for their role in P transfer.

Recent P transfer research is, however, challenging the premise that surface runoff is the dominant pathway for P loss from soil to water. Firstly, investigations of P concentrations in subsurface pathways have revealed that they can be consistently in excess of those required for eutrophication (Dils and Heathwaite, 1996; Haygarth et al., 1998a, 2000; Heathwaite and Dils, 2000; Matthews et al., 1998; Simard et al., 2000; Turner and Haygarth, 1999). Secondly, increasing attention is being given to the transfer of P attached to, or integrated within, chemically stable colloidal material that may be highly mobile through subsurface pathways (Chardon et al., 1997; Haygarth and Jarvis, 1997; Kretzschmar et al., 1999; Matthews et al., 1998; Toran and Palumbo, 1992). Indeed, rapid flows of water in soil pipes, land drains and preferential flow pathways are not dissimilar to overland flow in that they have potential to erode colloidal forms of P (Jacobsen et al., 1997). Also, Jensen et al. (1998) have shown that soluble P can be efficiently transported in larger macropores and avoid adsorption to the bulk soil, either when the macropore walls are P saturated, or when rapid flowing water at the very centre of macropore channels avoids contact with the soil. Indeed, macropores may provide a route for the transfer of displaced soil solution P (translatory water) during stormflow (Ryden et al., 1973).

6.4 MANAGEMENT MINIMISATION

The diminution of eutrophication in surface waters requires a reductions in water P concentrations to a ecologically acceptable level. Reductions in point P loads from sewage treatment works to sensitive catchments are now being implemented. This has focused attention on the role of diffuse inputs from agriculture (Withers and Lord, 2002). The European Community Urban Waste Water Treatment Directive (UWWTD) has raised the profile of eutrophication and associated costs (Withers et al., 2000). The initial controls of P loss can be implemented at the source. This includes how the fertiliser and manures are stored and the level of P applied. Land management, including the farming methods used for a specific land use, e.g., direction and depth of cultivation, or the inclusion of short term cover crops in arable rotations are transport concerns that can also be addressed to reduce P loss (Withers and Lord, 2002).

Increased background losses of P due to surplus P accumulation in soils are difficult to rectify quickly and cheaply because soil P levels respond slowly to P uptake by crops (Sharpley et al., 2000). However, controls over nutrient and land

Transfer of phosphorus to surface waters; eutrophication 139

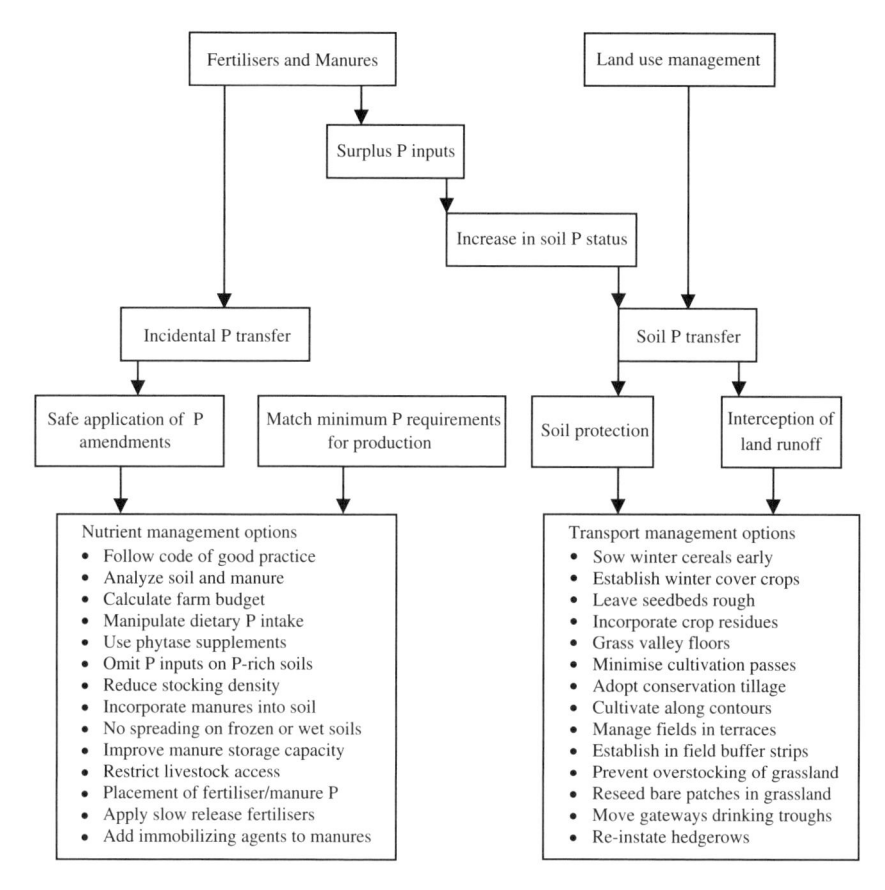

Figure 6.7 Management mechanisms, strategies and options for control of nonpoint P loss from agriculture (Withers *et al.*, 2000).

```
                    Fertilisers and Manures              Land use management
                              |                                  |
                              v                                  |
                         Surplus P inputs <---------------------+
                              |
                              v
                         Increase in soil P status
           |                                                      |
           v                                                      v
    Incidental P transfer                                    Soil P transfer
      |           |                                          |              |
      v           v                                          v              v
Safe application  Match minimum P                     Soil protection    Interception of
of P amendments   requirements for production                            land runoff
```

Nutrient management options
- Follow code of good practice
- Analyze soil and manure
- Calculate farm budget
- Manipulate dietary P intake
- Use phytase supplements
- Omit P inputs on P-rich soils
- Reduce stocking density
- Incorporate manures into soil
- No spreading on frozen or wet soils
- Improve manure storage capacity
- Restrict livestock access
- Placement of fertiliser/manure P
- Apply slow release fertilisers
- Add immobilizing agents to manures

Transport management options
- Sow winter cereals early
- Establish winter cover crops
- Leave seedbeds rough
- Incorporate crop residues
- Grass valley floors
- Minimise cultivation passes
- Adopt conservation tillage
- Cultivate along contours
- Manage fields in terraces
- Establish in field buffer strips
- Prevent overstocking of grassland
- Reseed bare patches in grassland
- Move gateways drinking troughs
- Re-instate hedgerows

management practices causing P loss are likely to have more measurable impacts on P export (Withers *et al.*, 2000). Therefore, to minimise P losses to surface waters there is a need to integrate both the source and transport issues and this should incorporate the interactions between soil P and surface runoff (Gburek *et al.*, 2000).

Soil P tests are used to indicate how much potentially plant P available is present in the soil and therefore determine how much P is to be applied to obtain the maximum crop yield. This is a well-established agricultural practice recognised as a cost-effective way to identify the P status of soils and avoid building soil P levels well above unnecessary current agronomic critical levels (Sims *et al.*, 2000a). However, many farmers continue to apply P to soils that may have excessive soil P levels leading to elevated P levels in surface runoff. Although several studies have shown that the loss of dissolved P in surface runoff is dependent on the P content of surface soil, that relationship is influenced by soil type, tillage and crop management. Sharpley *et al.* (2000) suggest that P loss via surface runoff and erosion may be reduced by crop residue management, conservation tillage, buffer strips, riparian zones, terracing, contour tillage, cover crops, constructed wetlands and impoundments. These are well-established practices and can be summarised by three main options, which exist for preventative management of agricultural P to reduce concentrations in receiving waters:

1. changes in land use (e.g., conversion of arable to low intensity grass as implemented in the 'set-aside' scheme in the UK),
2. changes in inputs
3. changes in management (e.g., timing of fertiliser applications).

The key proposed changes in farming practice contained in the various agricultural management schemes are: accurate determination of crop P requirements, P fertiliser application in late spring in order to maximise crop uptake, split fertiliser applications rather than one or two major dressings, use of slow-release fertilisers, use of autumn cover crops to reduce the area of bare ground and provision of buffer zones between arable land and receiving waters. Figure 6.7 identifies further management mechanisms and strategies for the control of non-point P loss.

With the identification of the problems of eutrophication many strategies for P management have been developed and implemented at the field and watershed scale. However Gburek *et al.* (2000) suggests that the prevention of P loss from agricultural watersheds must focus on defining, targeting and remediating the CSAs of P loss. Although many catchments exhibit differing hydrological processes CSAs can be identified within a catchment and preventative P loss from these areas can be implemented by the management methods shown in Figure 6.7.

REFERENCES

Ahuja, L.R., Sharpley, A.N., Yamamoto, M. and Menzel, R.G. (1981) The depth of rainfall–runoff–soil interaction as determined by P-32. *Water Resour. Res.* **17**(4), 969–974.

Anderson, M.G. and Burt, T.P. (1990) Process studies in hillslope hydrology: an overview. In *Process Studies in Hillslope Hydrology* (eds Anderson, M.G. and Burt, T.P.), John Wiley and Sons, pp. 1–8.

Arbuckle, K.E. and Downing, J.A. (2001) The influence of watershed land use on lake N: P in a predominantly agricultural landscape. *Limnol. Oceanogr.* **46**, 970–975.

Armstrong, A.C., Atkinson, J.L. and Garwood, E.A. (1984) *Grassland and Economics Experiment*, North Wyke, Devon. RD/FE/23, ADAS.

Armstrong, A.C., Leeds-Harrison, P.B., Harris, G.L. and Catt, J.A. (1999) Measurement of solute fluxes in macroporous soils: techniques, problems and precision. *Soil Use Manage.* **15**, 240–246.

Banens, R.J. and Davis, J.R. (1998) Comprehensive approaches to eutrophication management: the australian example. *Water Sci. Technol.* **37**(3), 217–225.

Barkle, G.F., Brown, T.N. and Painter, D.J. (1999) Leaching of particulate organic carbon from land-applied dairy farm effluent. *Soil Sci.* **164**(4), 252–263.

Bazemore, D.E., Eshleman, K.N. and Hollenbeck, K.J. (1994) The role of soil water in stormflow generation in a forested headwater catchment: synthesis of natural tracer and hydrometric evidence. *J. Hydrol.* **162**, 47–75.

Black, C.A. (1967) *Soil–Plant Relationships*. John Wiley and Sons Inc., New York, pp. 792.

Bouma, J. and Dekker, L.W. (1981) A method for measuring the vertical and horizontal Ksat of clay soils with macropores. *Soil Sci. Soc. Am. J.* **45**, 662–663.

Briggs, D. and Smithson, P. (1992) *Fundamentals of Physical Geography*. Unwin Hyman, London, pp. 558.

Burkholder, J.M. and Glasgow, H.B. (1997) *Pfiesteria piscicida* and other *Pfiesteria*-dinoflagellates behaviours, impacts, and environmental controls. *Limnol. Oceanogr.* **42**, 1052–1075.

Burwell, R.E., Schuman, G.E., Heinemann, H.G. and Spooner, R.G. (1977) Nitrogen and phosphorus movement from agricultural watersheds. *J. Soil Water Conserv.* **32**, 226–230.

Burwell, R.E., Timmons, D.R. and Holt, R.F. (1975) Nutrient transport in surface runoff as influenced by soil cover and seasonal periods. *Soil Sci. Soc. Am. Pro.* **39**, 523–528.

Carpenter, S.R., Caraco, N.F., Correll, D.L., Howarth, R.W., Sharpley, A.N., Smith, V.N. (1998) Nonpoint pollution of surface waters with phosphorus and nitrogen. *Ecol. Appl.* **8**(3), 559–568.

Chapman, P.J., Edwards, A.C. and Shand, C.A. (1997) The phosphorus composition of soil solutions and soil leachates: influence of soil : solution ratio. *Eur. J. Soil Sci.* **48**, 703–710.

Chapman, P.J., Reynolds, B. and Wheater, H.S. (1993) Hydrochemical changes along storm flow pathways in a small moorland headwater catchment in mid-wales, UK. *J. Hydrol.* **151**(2–4), 241–265.

Chardon, W.J., Oenema, O., delCastilho, P., Vriesema, R., Japenga, J., Blaauw, D. (1997) Organic phosphorus solutions and leachates from soils treated with animal slurries. *J. Environ. Qual.* **26**, 372–378.

Cooper, D.M., House, W.A., Reynolds, B., Hughes, S., May, L., Gannon, B. (2002) The phosphorus budget of the thame catchment, Oxfordshire: 2. modelling. *Sci. Total Environ.* **282**, 435–457.

Crosbie, B. and Chow-Fraser, P. (1999) Percentage land use in the watershed determines the water and sediment quality of 22 marshes in the great lakes basin. *Can. J. Fish. Aquat. Sci.* **56**, 1781–1791.

Cullen, P. and O'Loughlin, E.M. (1982) Non-point sources of pollution. pp. 437–453 In: (eds E.M. O'Loughlin and P. Cullen), *Prediction in Water Quality*, Australian Academy of Science, Canberra.

Cullen, P., Rosich, R. and Bek, P. (1978) A phosphorus budget for Lake Burley Griffin and management implication for urban lakes. Australian Water Resources Council, Technical Paper 31, AGPS, Canberra.

Daniel, T.C., Sharpley, A.N. and Lemunyon, J.L. (1998) Agricultural phosphorus and eutrophication: a symposium overview. *J. Environ. Qual.* **27**, 251–257.

DeWalle, D.R. and Pionke, H.B. (1994) Streamflow generation on a small agricultural catchment during autumn recharge. II stormflow periods. *J. Hydrol.* **163**, 23–42.

Dillon, P.J. and Molot, L.A. (1997) Effect of landscape form on export of dissolved organic carbon, iron and phosphorus from forested stream catchments. *Water Resour. Res.* **33**, 2591–2600.

Dils, R. and Heathwaite, A.L. (1996) Phosphorus fractionation in hillslope hydrological pathways contributing to agricultural runoff. In *Advances in Hillslope Processes* (eds Anderson, M.G. and Brooks, S.M.), John Wiley and Sons, Chichester, pp. 229–282.

Donald, R.G., Anderson, D.W. and Stewart, J.W.B. (1993) Potential role of dissolved organic carbon in phosphorus transport in forested soils. *Soil Sci. Soc. Am. J.* **57**, 1611–1618.

Dunne, T. (1978) Field studies of hillslope flow processes. In *Hillslope Hydrology* (ed. Kirkby, M.J.), John Wiley and Sons, Chichester, UK, pp. 227–289.

Edwards, A.C., Twist, H. and Codd, G.A. (1998) Novel techniques for assessing impacts of phosphorus transfer to running waters. In *Practical and Innovative Measures for the Control of Agricultural Phosphorus Losses to Water* (eds Foy, R.H. and Dils, R.), OECD, Antrim, Northern Ireland, pp. 25.

Edwards, A.C. and Withers, P.J.A. (1998) Soil phosphorus management and water quality: a UK perspective. *Soil Use Manage.* **14**, 124–130.

Emmett, W.W. (1978) Overland Flow. In *Hillslope Hydrology* (ed. Kirkby, M.J.), John Wiley and Sons, Chichester, UK, pp. 145–175.

Environment Agency (2000) *Aquatic entrophication in England and Wales: a management strategy*. National Centre for Ecotoxicology and Hazardous Substances, Wallingford, UK.

Fleming, N.K., Cox, J.W. and Chittleborough, D.J. (1997) *Optimising phosphorus fertilizer on dairy farms (catchment study) south Australia*. Dairy Research and Development Cooperation Project DAS053.

Fleming, N.K., Cox, J.W., Chittleborough, D.J. and Dyson, C.B. (2001) An analysis of chemical loads and forms in overland flow from dairy pasture in South Australia. *Hydrol. Process.* **15**(3), 393–405.

Foy, R.H. and Withers, P.J.A. (1995) The contribution of agricultural phosphorus to eutrophication. The fertiliser society, The Fertiliser Society in London, pp. 1–32.

Fraser, A.F. (2000) Hydrological pathways of phosphorus transfer from UK agricultural catchments. PhD Thesis, Cranfield University, Silsoe, UK.

Frossard, E., Brossard, M., Hedley, M.J. and Metherell, A. (1995) Reactions controlling the cycling of P in soils. In *Phosphorus in the Global Environment* (ed. Tiessen, H.), John Wiley and Sons, New York, pp. 107–138.

Gburek, W.J. and Sharpley, A.N. (1998) Hydrologic controls on phosphorus loss from upland agricultural watersheds. *J. Environ. Qual.* **27**, 267–277.

Gburek, W.J., Sharpley, A.N., Heathwaite, A.L. and Folmar, G.J. (2000) Phosphorus management at the watershed scale: a modification of the phosphorus index. *J. Environ. Qual.* **29**(1), 130–144.

Hannepel, R.J., Fuller, W.H., Bosma, S. and Bullock, J.S. (1963) Phosphorus movement in a calcareous soil: II soil microbial activity and organic phosphorus movement. *Soil Sci.* **97**, 421–427.

Harris, G.P. (1999) Comparison of the biogeochemistry of lakes and estuaries: ecosystem processes, functional groups, hysteresis effects and interactions between macro- and microbiology. *Mar. Fresh Water Res.* **50**, 791–811.

Harrison, A.F. (1987) Soil organic phosphorus – a review of world literature. CAB International, Wallingford, Oxford, UK.

Hawkins, J.M.B. and Scholefield, D. (1996) Molybdate reactive phosphorus losses in surface and drainage waters from permanent grassland. *J. Environ. Qual.* **25**(4), 727–732.

Haygarth, P.M. (1997) Agriculture as a source of phosphorus transfer to water: sources and pathways. 21, Scientific Committee on Phosphorus in Europe.

Haygarth, P.M., Chapman, P.J., Jarvis, S.C. and Smith, R.V. (1998a) Phosphorus budgets for two contrasting grassland farming systems in the UK. *Soil Use Manage.* **14**, 160–167.

Haygarth, P.M., Heathwaite, A.L., Jarvis, S.C. and Harrod, T.R. (2000) Hydrological factors for phosphorus transfer from agricultural soils. *Advan. Agron.* **69**, 153–178.

Haygarth, P.M., Hepworth, L. and Jarvis, S.C. (1998b) Forms of phosphorus transfer in hydrological pathways from soil under grazed grassland. *Eur. J. Soil Sci.* **49**(1), 65–72.

Haygarth, P.M. and Jarvis, S.C. (1996) Pathways and forms of phosphorus losses from grazed grassland hillslopes. In *Advances in Hillslope Processes* (eds Anderson, M.G. and Brooks, S.M.), John Wiley and Sons Ltd., Chichester, pp. 283–294.

Haygarth, P.M. and Jarvis, S.C. (1997) Soil derived phosphorus in surface runoff from grazed grassland lysimeters. *Wat. Res.* **31**(1), 140–148.

Haygarth, P.M. and Jarvis, S.C. (1999) Transfer of phosphorus from agricultural soils. *Advan. Agron.* **66**, 195–249.

Haygarth, P.M. and Sharpley, A.N. (2000) Terminology for P transfer. *J. Environ. Qual.* **29**(1), 10–15.

Haygarth, P.M., Warwick, M. and House, W.A. (1997) Size distribution of colloidal molybdate reactive phosphorus in river waters and soil solution. *Water Res.* **31**, 439–448.

He, Z.L., Wilson, M.J., Campbell, C.O., Edwards, A.C. and Chapman, S.J. (1995) Distribution of phosphorus in soil aggregate fractions and its significance with regard to phosphorus transport in agricultural runoff. *Water Air Soil Pollut.* **83**, 69–84.

Heathwaite, A.L. (1990) The effect of drainage on nutrient release from fen peat and its implications for water-quality – a laboratory simulation. *Water Air Soil Pollut.* **49**(1–2), 159–173.

Heathwaite, A.L. and Dils, R.M. (2000) Characterising phosphorus loss in surface and subsurface hydrological pathways. *Sci. Total Environ.* **251**, 523–538.

Heathwaite, A.L. and Johnes, P.J. (1996) Contribution of nitrogen species and phosphorus fractions to stream water quality in agricultural catchments. *Hydrol. Process.* **10**(7), 971–983.

Heathwaite, A.L., Johnes, P.J. and Peters, N.E. (1996) Trends in nutrients. *Hydrol. Process.* **10**, 263–293.

Heathwaite, L., Sharpley, A. and Gburek, W. (2000) A conceptual approach for integrating phosphorus and nitrogen management at watershed scales. *J. Environ. Qual.* **29**(1), 158–166.

Heckrath, G., Brookes, P.C., Poulton, P.R. and Goulding, K.W.T. (1995) Phosphorus leaching from soils containing different phosphorus concentrations in the broadbalk experiment. *J. Environ. Qual.* **24**(5), 904–910.

Hession, W.C. and Storm, D.E. (2000) Watershed level uncertainties: implications for phosphorus management and eutrophication. *J. Environ. Qual.* **29**, 1172–1179.

Horton, R.E. (1945) Erosional development of streams and their drainage basins: hydrophysical approach to quantitative morphology. *Bull. Ecol. Soc. Am.* **56**, 275–330.

Hudson, J.J., Taylor, W.D. and Schindler, D.W. (2000) Phosphate concentrations in lakes. *Nature* **406**, 54–56.

Jacobsen, O.H., Moldrup, P., Larsen, C., Konnerup, L. and Petersen, L.W. (1997) Particle transport in macropores of undisturbed soil columns. *J. Hydrol.* **196**, 185–203.

Jansson, M., Olsson, H. and Petterson, K. (1988) Phosphatase: origin, characterisation and function in lakes. *Hydrobiologia* **170**, 157–175.

Jensen, M.B., Jorgensen, P.R., Hansen, H.C.B. and Nielsen, N.E. (1998) Biopore mediated subsurface transport of dissolved orthophosphate. *J. Environ. Qual.* **27**, 1130–1137.

Jones, J.A.A. (1997) Pipeflow contributing areas and runoff repsonse. *Hydrol. Process.* **11**(1), 35–41.

Koski-Vahala, J., Hartikainen, H. and Tallberg, P. (2001) Phosphorus mobilization from various sediment pools in response to increased pH and silicate concentration. *J. Environ. Qual.* **30**, 546–552.

Kretzschmar, R., Borkovec, M., Grolimund, D. and Elimelech, M. (1999) Obile subsurface colloids and their role in contaminant transport. *Advan. Agron.* **66**, 121–193.

Mason, C.F. (1991) *Biology of Freshwater Pollution*. Longman Scientific and Technical, Harlow, UK.

Matthews, R.A., Preedy, N., Heathwaite, A.L. and Haygarth, P.M. (1998) Transfer of colloidal forms of phosphorus in subsurface hydrological pathways. In *Practical and Innovative Measures for the Control of Phosphorus Losses to Water* (eds Foy, R.H. and Dils, R.), OECD, Greemount College of Agriculture and Horticulture, Northern Ireland, pp. 162–163.

McCarthey, M. (1999) UN report warns of earth's unsustainable future. The Independent, London, UK.

McIvor, J.G., Williams, J. and Gardener, C.J. (1995) Pasture management influences runoff and soil movement in the semi-arid tropics. *Aust. J. Exp. Agr.* **35**, 55–65.

Misselbrook, T.H., Pain, B.F., Stone, A.C. and Scolefield, D. (1995) Nutrient runoff following application of livestock wastes to grassland. *Environ. Pollut.* **88**, 51–56.

Mulholland, P.J. (1993) Hydrometric and stream chemistry evidence of three storm flowpaths in walker branch watershed. *J. Hydrol.* **151**, 291–316.

Muscutt, A.D., Reynolds, B. and Wheater, H.S. (1993) Sources and controls of aluminiun in storm runoff from a headwater catchment in mid-wales. *J. Hydrol.* **142**(1–4), 409–425.

Nash, D., Murray, H., Halliwell, D. and Murdoch, C. (2000) Factors affecting phosphorus export from a pasture-based grazing system. *J. Eniron. Qual.* **29**, 1160–1166.

Nieber, J.L. and Warner, G.S. (1991) Soil pipe contribution to steady subsurface storm flow. *Hydrol. Process.* **5**(4), 329–344.

OECD (1982) *Eutrophication of Waters*. Monitoring, assessment and control. OECD, Paris.

Olive, L.J. and P.H. Walker (1982) Processes in overland flow – erosion and production of suspended material. In *'Prediction in Water Quality'* (eds O'Loughlin, E.M. and Cullen, P.), Australian Academy of Science, pp. 87–119. Canberra.

O'Shea, L. (2002) An economic approach to reducing water pollution: point and diffuse sources. *Sci. Total Environ.* **282**, 49–63.

Pautler, M.C. and Sims, J.T. (2000) Relationships between soil test phosphorus, soluble P and P saturation in delaware soils. *Soil Sci. Soc. Am. J.* **64**, 765–773.

Pierzynski, G.M., Vance, G.F and Sims, J.T. (2000) *Soils and Environmental Quality*. CRC Press.

Pionke, H.B. and DeWalle, D.R. (1994) Streamflow generation on a small agricultural catchment during autumn recharge: II. Stormflow periods. *J. Hydrol.* **163**, 23–42.

Pionke, H.B., Gburek, W.J. and Folmar, G.J. (1993) Quantifying stormflow components in a Pennsylvania watershed when O-18 input and storm conditions vary. *J. Hydrol.* **148**, 169–187.

Pionke, H.B., Gburek, W.J., Sharpley, A.N. and Schnabel, R.R. (1996) Flow and nutrient export patterns for an agricultural hill-land watershed. *Water Resour. Res.* **32**(6), 1795–1804.

Pionke, H.B., Gburek, W.J., Sharpley, A.N. and Zollweg, J.A. (1997) Hydrological and chemical controls on phosphorus loss from catchments. In *Phosphorus Loss from Soil to Water* (eds Tunney, H., Carton, O.T., Brookes, P.C. and Johnston, A.E.). CAB International, pp. 225–242.

Pionke, H.B. and Kunishi, H.M. (1992) Phosphorus status and the content of suspended sediment in a Pennsylvania watershed. *Soil Sci.* **153**(6), 452–462.

Preedy, N., Matthews, R., Heathwaite, L. and Haygarth, P. (2000) Simplified classification of hydrological terminology for phosphorus transfer. *Proceedings of the International Symposium on Impact of Landuse Change on Nutrient Loads from Diffuse Sources* (ed. Heathwaite, L.) IAHS publication 257, pp. 3–11.

Preedy, N., Mctirnan, K., Matthews, R., Heathwaite, L., Haygarth, P. (2001) Rapid incidental phosphorus transfers from grassland. *J. Environ. Qual.* **30**, 2105–2112.

Prosser, I.P., Rutherfurd, I.D., Olley, J.M., Young, W.J., Wallbrink, P.J. and Moran, C.J. (2001) Large scale patterns of erosion and sediment transport in river networks, with examples from Australia. *Mar. Fresh Water Res.* **52**, 81–99.

Quinton, J.N., Carr, J.A. and Hess, T.M. (2001) The selective removal of phosphorus from soil: is event size important? *J. Environ. Qual.* **30**, 538–545.

Rast, W. and Thornton, J.A. (1996) Trends in eutrophication research and control. *Hydrol. Process.* **10**(2), 295–313.

Reynolds, C.S. (1984) *The Ecology of Freshwater Phytoplankton*. Cambridge, UK.

Ron Vaz, M.D., Edwards, A.C., Shand, C.A. and Cresser, M.S. (1993) Phosphorus fractions in soil solution: influence of soil acidity and fertiliser additions. *Plant Soil* **148**, 179–183.

Ryden, R.C., Syers, J.K. and Harris, R.F. (1973) Phosphorus in runoff and streams. *Advan. Agron.* **25**, 1–47.

Sallade, Y.E. and Sims, J.T. (1997) Phosphorus transformations in the sediments of Delaware's agricultural drainageways: effect of reducing conditions on phosphorus release. *J. Environ. Qual.* **26**, 1579–1588.

Satchell, M. (1997) The cell from hell. *US News and World Report* **123**, 26–28.

Shand, C.A. and Smith, S. (1997) Enzymatic release of phosphate from model substrates and P compounds in soil solution from a peaty podzol. *Biol. Fert. Soils* **24**, 183–187.

Sharpley, A.N. (1980) The enrichment of phosphorus in runoff sediments. *J. Environ. Qual.* **9**(3), 521–525.
Sharpley, A.N. (1985) Depth of surface soil-runoff interaction as affected by rainfall, soil slope, and management. *Soil Sci. Soc. Am. J.* **49**, 1010–1015.
Sharpley, A.N., Chapra, S.C., Wedepohl, R., Sims, J.T., Daniel, T.C., Reddy, K.R. (1994) Managing agricultural phosphorus for protection of surface waters – issues and options. *J. Environ. Qual.* **23**(3), 437–451.
Sharpley, A.N., Daniel, T.C., Sims, J.T. and Pote, D.H. (1996) Determining environmentally sound soil phosphorus levels. *J. Soil Water Conserv.* **51**, 160–166.
Sharpley, A., Foy, B. and Withers, P. (2000) Practical and inovative measures for the control of agricultural phosphorus to water: an overview. *J. Environ. Qual.* **29**(1), 1–9.
Sharpley, A.N. and Smith, S.J. (1989) Prediction of soluble phosphorus transport in agricultural runoff. *J. Environ. Qual.* **18**, 313–316.
Sharpley, A.N. and Syers, J.K. (1979) Loss of nitrogen and phosphorus in tile drainage as influenced by urea application and grazing animals. *New Zeal. J. Agr. Res.* **22**, 127–131.
Simard, R.R., Beauchemin, S. and Haygarth, P.M. (2000) Potential for preferential pathways of phosphorus transport. *J. Environ. Qual.* **29**(1), 97–104.
Sims, J.T., Edwards, A.C., Schoumans, O.F. and Simard, R.R. (2000a) Integrating soil phosphorus testing into environmentally based agricultural management practices. *J. Environ. Qual.* **29**, 60–71.
Sims, J.T., Pautler, M.C., Gartley, K.L., Vadas, P.A., Maguire, R.O., Leytem, A.B., Lauahun, M.F., Luka, N.J., and Eaton, R.A. (2000b) Environmental aspects of soil phosphorus chemistry in the U.S Atlantic coastal plain. *International Conference on Agricultural Effects on Ground and surface waters*, IAHS, Wageningen, Netherlands.
Skaggs, R.W., Breve, M.A. and Gilliam, J.W. (1994) Hydrologic and water quality impacts of agricultural drainage. *Crit. Rev. Environ. Sci. Technol.* **24**(1), 1–32.
Sosa, V. (1996) The influence of lime application to the acid soil on bioavailability of phosphorus in runoff. *Water Sci. Technol.* **33**(4–5), 297–301.
Stamm, C., Fluhler, H., Leuenberger, J. and Wunderli, H. (1998) Preferential transport of phosphorus in drained grassland soils. *J. Environ. Qual.* **27**, 515–522.
Toran, L. and Palumbo, A.V. (1992) Colloid transport through fractured and unfractured laboratory sand columns. *J. Contam. Hydrol.* **9**, 289–303.
Turner, B.L. (2000) *Biological Controls on the Solubilisation of Phosphorus in Grassland Soils*. PhD Thesis, Royal holloway, London, pp. 344.
Turner, B.L. and Haygarth, P.M. (2000) Phosphorus forms and concentrations in leachate under four grassland soil types. *Soil Sci. Soc. Am. J.* **64**, 1090–1097.
Turner, B.L. and Haygarth, P.M. (1999) Phosphorus leaching under cut grassland. *Water Sci. Technol.* **39**, 63–67.
Turner, B.L., Haygarth, P.M. (2002) Influence of soil processes on solubilisation of P forms: A review of experimental data. Phosphorus Losses from agricultural soils: Processes at the Field Scale, (ed. Chardon, W. and Schoumans, O.). ALTERRA, Wageningen, The Netherlands. pp. 66–72.
Vadas, P.A. and Sims, J.T. (1998) Redox status, pultry litter, and phosphorus solubility in Atlantic coastal plain soils. *Soil Sci. Soc. Am. J.* **62**, 1025–1034.
Vollenweider, R.A. and Krekes, J.J. (1982) Eutrophication of Waters: Monitoring, Assessment and Control, Organisation for Economic Co-operation and Development, Paris.
Wang, Y., Edwards, D.R., Daniel, T.C. and Scott, H.D. (1996) Simulation of runoff transport of animal manure constituents. *T. Am. Soc. Agr. Eng.* **39**, 1367–1378.
Wallbrink, P.J., Murray, A.S. and Olley, J.M. (1998) Determining sources and transit times of suspended sediment in the Murrumbidgee River, New South Wales, Australia using ^{137}Cs and ^{210}Pb. *Water Resour. Res.* **34**, 879–887.
Wilson, G.V., Jardine, P.M., Luxmoore, R.J., Zelazny, L.W., Todd, D.E., Lietzke, D.A. (1991) Hydrogeochemical processes controlling subsurface transport from an upper

subcatchment of Walker Branch watershed during storm events. 2. Solute transport processes. *J. Hydrol.* **123**, 317–336.

Withers, P.J.A. (1993) The significance of agriculture as a source of phosphorus pollution in inland and coastal waters in the United Kingdom. Ministry of Agriculture, Fisheries and Food, London.

Withers, P.J.A., Clay, S.D. and Breeze, V.G. (2001) Phosphorus transfer in runoff following application of fertilizer, manure and sewage sludge. *J. Environ. Qual.* **30**, 180–188.

Withers, P.J.A., Davidson, I.A. and Foy, R.H. (2000) Prospects for controlling phosphorus loss to water: A UK perspective. *J. Environ. Qual.* **29**(1), 167–175.

Withers, P.J.A. and Lord, E.I. (2002) Agricultural nutrient inputs to rivers and groundwaters in the UK: policy, environmental management and research needs. *Sci. Total Environ.* **282**, 9–24.

Withers, P.J.A. and Sharpley, A.N. (eds), (1995) Phosphorus management in sustainable agriculture. *Soil Manage. Sustain. Agr.* Wye College Press, Ashford, Kent, pp. 201–297.

Zhou, O.X., Gibson, C.E. and Foy, R.H. (2000) Long term changes of nitrogen and phosphorus loadings to a large lake in north-west Ireland. *Water Res.* **34**, 922–926.

7
Environmental chemistry of phosphonic acids

Bernd Nowack

7.1 INTRODUCTION

7.1.1 Chelating agents in the environment

Phosphonic acids are a group of both synthetic and biogenic organophosphorous compounds characterized by a stable, covalent, carbon to phosphorous bond. Phosphonate compounds, especially the ones containing more than one phosphonate group, are effective chelating agents. The most common phosphonates are structural analogues of the well-known aminopolycarboxylates EDTA, NTA and DTPA.

Chelating agents have the potential to perturb the natural speciation of metals and to influence metal bioavailability (Anderson et al., 1985). They also have the potential to remobilize metals from sediments and aquifers, posing a potential risk to groundwater and drinking water (Hering, 1995; Müller and Förstner, 1976). Strong chelating agents occur in natural waters always in the form of metal complexes. Therefore, a discussion of the fate of a chelating agent always has to address the

© 2004 IWA Publishing. *Phosphorus in Environmental Technology: Principles and Applications.* Edited by Eugenia Valsami-Jones. ISBN: 1 84339 001 9

presence of metals and how they interact with the chelating agent. There are many studies emphasizing the importance of chelation on metal bioavailability, uptake, toxicity, transport, adsorption, distribution and fate. In a reciprocal manner, chelating agents are also influenced by the presence of metals. The different reactivities of different metal-chelating agent complexes have to be considered when assessing the reactions and ultimate fate of chelating agents in the environment.

The environmental fate of aminopolycarboxylate chelating agents such as EDTA and NTA has received considerable attention (Bucheli-Witschel and Egli, 2001, Means and Alexander, 1981; Nowack, 2002a; Sillanpää, 1997; Wolf and Gilbert, 1992). Much less is known about the corresponding phosphonates (Gledhill and Feijtel, 1992; Jaworska *et al.*, 2002). The aim of this chapter is to provide an overview of the current knowledge about the environmental chemistry of polyphosphonates (such as HEDP) and aminopolyphosphonates (such as NTMP, EDTMP, and DTPMP). Table 7.1 lists the names and structures of the chelating agents discussed

Table 7.1 Names and abbreviations of the phosphonates discussed in this chapter together with other abbreviation used for the same compound in the literature.

Abbreviation	Other abbreviations also in use	Name	Structure
HEDP	EHDP, HEDPA, HEBP	1-Hydroxyethane (1,1-diylbis phosphonic acid) *also called* 1-Hydroxyethane (1,1-diphosphonic acid)	$PO(OH)_2$, OH, $PO(OH)_2$
NTMP	ATMP, NTP, AMP, NTPH, NTPO, TPMA	Nitrilotris (methylene phosphonic acid)	$(OH)_2OP$–$N(PO(OH)_2)_2$
EDTMP	EDTP, EDTPO, ENTMP, EDTMPO, EDTMPA, EDITEMPA, EDTPH, TePMEDA	1,2-Diamino ethanetetrakis (methylene phosphonic acid) *also called* Ethylenediaminetetrakis (methylene phosphonic acid)	$(OH)_2OP$, $(OH)_2OP$–N–N–$PO(OH)_2$, $PO(OH)$
DTPMP	DETPMP, DTPP, DETPMPA, DTPPH, DETPMPO, ETPP	Diethylenetriaminepentakis (methylene phosphonic acid)	$(OH)_2OP$, $(OH)_2OP$–N–N–N–$PO(OH)_2$, $PO(OH)_2$, $PO(OH)_2$

in this chapter together with other abbreviation used for the same phosphonates. These compounds are known under many different abbreviations, that vary between the disciplines and countries and have changed with time. The chemistry of the phosphonates is compared to that of the corresponding polycarboxylates, e.g. EDTA. There is a lot more information about the environmental behaviour of the latter compounds than of the phosphonates. The similarities and dissimilarities of the two groups will be discussed.

7.1.2 Use and properties of phosphonates

Phosphonic acids are organic compounds containing one or more C—PO(OH)$_2$ groups. Bisphosphonates were first synthesized in 1897 by von Baeyer and Hofmann (1897). An example of such a bisphosphonate is HEDP (see Table 7.1). The work of Schwarzenbach et al. (1949) established phosphonic acids as effective complexing agents. The introduction of an amine group into the molecule to obtain —NH$_2$—C—PO(OH)$_2$ increases the metal binding abilities of the phosphonate. Examples for such compounds are NTMP, EDTMP and DTPMP (see Table 7.1). The herbicide glyphosate (N-phosphonomethylglycine) contains a phosphonate, an amine and a carboxylate functional group. This compound has received considerable attention as herbicide (Atkinson, 1985; Franz et al., 1997; Stalikas and Konidari, 2001), but is not discussed in detail in this chapter. 2-aminoethylphosphonic acid was the first phosphonate identified in an organism (Horiguchi and Kandatsu, 1959) and occurs in plants and many animals, mostly in membranes. Certain phosphonates are quite common among different organisms, from prokaryotes to eubacteria, fungi, mollusks and insects. The biological role of the natural phosphonates is still poorly understood (Wanner, 1994). Until now no bis- or polyphosphonates have been found to occur naturally.

Phosphonates have three main properties: they are effective chelating agents for di- and trivalent metal ions, they are inhibiting crystal growth and scale formation and they are quite stable at high temperature, low and high pH and in the presence of oxidants. An important industrial use of phosphonates is in cooling waters, desalination systems, and in oil fields to inhibit scale formation. In pulp and paper manufacturing and in the textile industry they are used as peroxide bleach stabilizers, acting as chelating agents for metals that could inactivate the peroxide. In detergents they are used as a combination of chelating agent, scale inhibitor and bleach stabilizer (May et al., 1986). Phosphonates are also used increasingly in medicine to treat various bone and calcium metabolism diseases (Fleisch, 1989; Francis and Centner, 1978) and as carriers for radionuclides in bone cancer treatments (de Klerk et al., 1992; de Witt et al., 1996).

In 1998 the consumption of aminopolycarboxylate chelating agents (excluding NTA) was 155,000 tons in the US, Europe and Japan; the consumption of phosphonates was 56,000 tons worldwide – 40,000 tons in the US, 15,000 tons in Europe and less than 800 tons in Japan (Davenport et al., 2000). The demand for phosphonates has been growing steadily at 3% annually.

7.2 ANALYSIS OF PHOSPHONATES

7.2.1 Analytical techniques

The standard method for the determination of phosphonates is ion-chromatography followed by post-column reaction with Fe(III) and detection of the Fe(III)-complexes at 300–330 nm (Felber *et al.*, 1995; Tschäbunin *et al.*, 1989a, b; Weiss and Hägele, 1987). This method has a detection limit of about 2–10 μM. Other methods have been developed based on post-column oxidation of the phosphonate to phosphate and detection of phosphate with the molybdenum blue method (Vaeth *et al.*, 1987; Waldhoff and Sladek, 1985). These include ion-chromatography with pulsed amperometric detection of amine-containing phosphonates (Mahabir-Jagessar *et al.*, 1997) and with indirect photometric detection (Thompson *et al.*, 1994), and capillary electrophoresis with indirect photometric detection (Shamsi and Danielson, 1995). These methods have all high detection limits of 1 μM or more and are therefore not suitable for natural systems.

A very promising method is the derivatization of the phosphonic acid group with diazomethane and separation and detection of the derivatives by HPLC-MS (Klinger *et al.*, 1997). This method is, however, not applicable to natural waters due to interference by the major cations and anions of the water matrix. The only method currently available by which phosphonates can be measured at low concentration in natural samples, is an ion-pair HPLC method with precolumn formation of the Fe(III)-complexes (Nowack, 1997). The phosphonates can be measured down to a detection limit of 0.05 μM in natural waters and waste waters (Figure 7.1). Using this method, phosphonate behaviour during waste water treatment has been investigated (Nowack, 1998). The method, however, is not able to quantify bisphosphonic acids such as HEDP.

The breakdown products of the Mn(II)-catalyzed degradation of NTMP, IDMP (iminodimethylenephosphonic acid) and FIDMP (formyl-iminodimethylenephosphonic acid) (Nowack and Stone, 2000), can be detected after derivatization of the

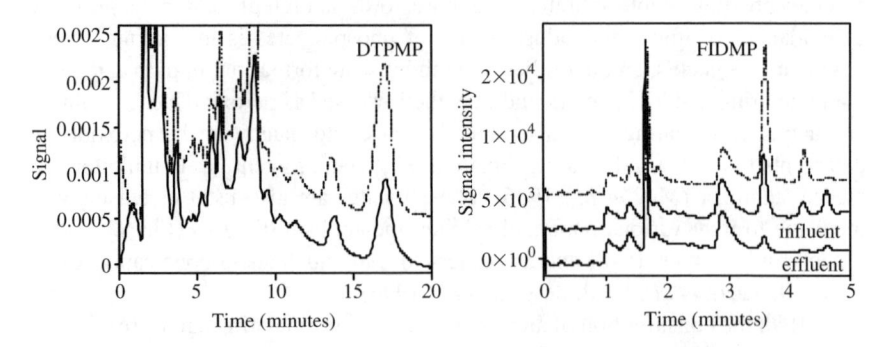

Figure 7.1 Determination of DTPMP in the influent (left) and FIDMP (right) in the influent and effluent of two waste water, treatment plants. The chromatograms are redrawn from Nowack (1997, 2002c). Chromatograms are also shown for samples containing added standard compounds to the influent (dotted line).

aldehyde group in FIDMP by 2,4-dinitrophenylhydrazine (DNPH) and of the imine-group in IDMP by 9-fluorenyl methylchloroformate (FMOC) (Nowack, 2002c). A detection limit of 0.01 μM FIDMP and 0.02 μM IDMP has been achieved. Using this method, the two breakdown products have been detected in waste water (Figure 7.1).

Preconcentration of phosphonates from natural water samples using different adsorbents has been tested (Frigge and Jackwerth, 1991). It was found that the investigated phosphonates HEDP, NTMP, and EDTMP differed so much in their chemical behaviour that a simultaneous enrichment from natural samples cannot be achieved. Successful preconcentration of the phosphonates NTMP, EDTMP and DTPMP from waters (e.g. waste water) was achieved using freshly precipitated $CaCO_3$ (Nowack, 1997). Recoveries at the 1 μM level were 95% or above for an influent sample of a waste water treatment plant.

7.2.2 Concentrations in the environment

No measurements of phosphonates in natural samples have been reported and only data for waste waters are available. This is mainly due to the fact that most analytical methods are not able to quantify phosphonates in natural waters at low concentrations. Using the HPLC method of Nowack (1997), phosphonates have been measured in Swiss waste water treatment plants (WWTP) (Nowack, 1998). The influent concentrations of NTMP were between <0.05 μM and 0.85 μM, of EDTMP between <0.05 μM and 0.15 μM, and of DTPMP between <0.05 μM and 1.7 μM. The highest concentration of DTPMP was found in a WWTP influenced by textile industry. Effluent samples from all investigated WWTP were, with the exception of one case, always below the detection limit. Another WWTP influenced by textile industry contained NTMP concentrations in the influent between 0.2 μM and 1.1 μM (Nowack, 2002b).

The oxidative breakdown products of NTMP, IDMP and FIDMP, have been detected in two WWTP receiving water from textile industry at concentrations of 0.08 μM and 0.015 μM FIDMP and 0.49 μM and 0.3 μM IDMP in the influent (Nowack, 2002c).

7.3 EXCHANGE REACTIONS

7.3.1 Adsorption

Phosphonates have a strong tendency to adsorb onto almost all types of mineral surfaces. Examples include calcite (Xyla *et al.*, 1992), clays (Fischer, 1991; Morillo *et al.*, 1997), aluminum oxides (Gerbino, 1996; Laiti *et al.*, 1995; Laiti and Öhmann, 1996; Liu *et al.*, 2000), iron oxides (Barja *et al.*, 1999; Day *et al.*, 1997; Nowack and Stone, 1999a, b), cassiterite (tin oxide) (Kuys and Roberts, 1987), zinc oxide (Nowack and Stone, 1999b), hydroxylapatite (Amjad, 1987; Chirby *et al.*, 1988; Jung *et al.*, 1973; Rawls *et al.*, 1982) and barite (Black *et al.*, 1991). Adsorption is also very strong on natural adsorbents such as sewage sludge (Fischer, 1992; Horstmann and Grohmann, 1988; Nowack, 2002; Steber and Wierich, 1986, 1987), sediments (Fischer, 1992) and soils (Held, 1989). Most of these studies, however, have not considered that metal ions might significantly alter the adsorption of the chelating agent. EDTA, for example, is very weakly adsorbed in the presence of Co(III), moderately in the presence of divalent metals like Cu and Zn, and very strongly in the presence of Pd(II) or La(III) (Nowack and Sigg, 1996; Nowack et al., 1996a). Fe(III) reduces its adsorption at low pH, and increases its adsorption at high pH. Contrary to EDTA adsorption, however, no influence of Fe(III), Zn, and Cu(II) on phosphonate adsorption onto goethite was observed (Nowack and Stone, 1999b). This was explained by the much stronger adsorption of phosphonates than EDTA, which resulted in a dissociation of the complex at the surface and separate adsorption of the metal and the phosphonate onto different surface sites. Calcium, however, has a very strong positive effect on phosphonate adsorption. In the presence of millimolar Ca concentrations, phosphonates were completely adsorbed up to pH of 12. The maximum surface concentration of phosphonates was also greatly enhanced in the presence of Ca. When evaluating the adsorptive capacity of a surface towards phosphonates in a natural system, it is therefore necessary to conduct the adsorption experiments under natural Ca concentrations.

Chelating agents are able to considerably alter the adsorption behaviour of a metal onto oxide surfaces. In general, while a metal shows increasing adsorption with increasing pH, a metal-chelating agent complex shows decreasing adsorption with increasing pH. The chelating agent is therefore mobilizing metals at high pH and immobilizing them at low pH. This effect of complexation on adsorption of metals is of outmost importance to the fate of a metal in the environment (Schindler, 1990).

The aminocarboxylate EDTA has a very strong influence on metal adsorption (Figure 7.2). For example, the adsorption edge of Cu onto the iron oxide goethite is

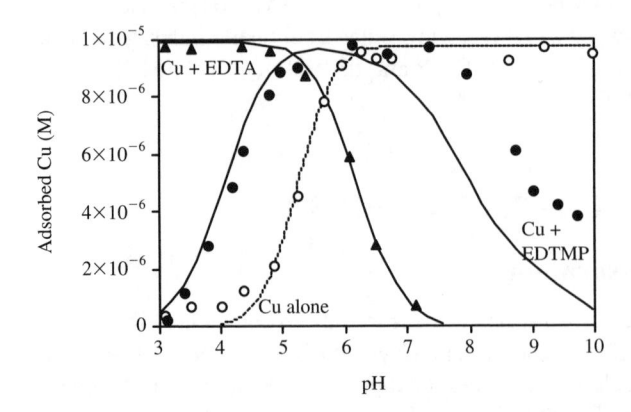

Figure 7.2 Influence of EDTA and EDTMP on Cu adsorption onto goethite. The lines represent model calculations. The data were taken from Nowack (1996) and Nowack and Stone (1999b). Conditions: 10 μM Cu and NTMP or EDTA, 0.42 g/l goethite, 0.01 M $NaNO_3$.

completely reversed. While Cu is increasingly adsorbed with increasing pH in the absence of EDTA, the presence of EDTA results in strong Cu adsorption at low pH and a decreasing adsorption at high pH. This effect can be explained by the formation of ternary Cu-EDTA-surface complexes. Cu forms surface complexes on goethite of the type (where "≡FeOH" stands for a surface site):

$$Cu^{2+} + \equiv FeOH \rightleftharpoons \equiv FeO-Cu^+ + H^+$$

In the presence of EDTA, the following ternary, ligand-like surface complex is formed (Nowack and Sigg, 1996):

$$CuEDTA^{2-} + \equiv FeOH + H^+ \rightleftharpoons \equiv Fe-EDTA-Cu^- + H_2O$$

This ternary surface complex is not formed in a system with Cu and polyphosphonates (Figure 7.2), although the complex formation in solution is of comparable strength. In the presence of the goethite surface, the Cu-phosphonate complex dissociates and Cu and phosphonate are adsorbed onto different surface sites (Nowack and Stone, 1999b). This is shown with CuNTMP at pH 7 as an example:

$$CuNTMP^{4-} + 2\equiv FeOH + H^+ \rightleftharpoons \equiv FeO-Cu^+ + \equiv Fe-NTMPH^{4-} + H_2O$$

Figure 7.2 shows that there is a slight increase in Cu adsorption at low pH, which is caused by electrostatic effects and that at high pH there is a mobilization of Cu due to the formation of dissolved CuEDTMP complexes. Overall, the influence of phosphonates on metal adsorption in the natural pH range from 4 to 8 is weak compared to EDTA.

Although not observed with the aminopolyphosphonates, ternary surface complexes may form in the presence of other phosphonates. Glyphosate forms a ternary surface complex with Cu on gibbsite (Dubbin *et al.*, 2000). The phosphonate moiety is adsorbed to the surface, and the Cu is bound to the amine and the carboxylate functional groups.

7.3.2 Dissolution of minerals

Dissolution of a mineral phase by chelating agents can be explained in terms of a ligand exchange process. Rate laws for ligand-promoted dissolution can be derived, that are related to the surface binding of ligands. The ligands stabilize the metal–oxygen bonds on the surface and enhance the release of metal ions from the surface into solution (Stumm, 1997). For aminocarboxylates, reactions with iron oxides are of great importance to the speciation of the ligand in solution. Dissolution of iron oxides by EDTA at pH <6 results in the formation of Fe(III)EDTA and the liberation of the complexed metal ion (Nowack and Sigg, 1997):

$$MeEDTA + FeOOH(s) \longrightarrow Fe(III)EDTA + Me(adsorbed)$$

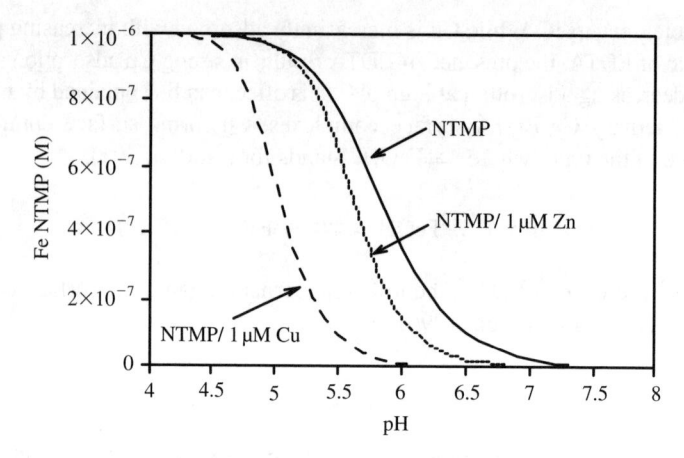

Figure 7.3 Speciation of 1 μM NTMP in contact with amorphous iron hydroxide in the presence and absence of 1 μM Cu or Zn.

Reactions like this have actually been observed in subsurface systems and have a pronounced influence on the mobility of heavy metals (Davis et al., 1993; Jardine et al., 1993).

Very little is known about the dissolution of iron oxides by phosphonates. Figure 7.3 shows the equilibrium speciation calculation for NTMP in contact with amorphous iron hydroxide (HFO), calculated with the NTMP-Fe(III) stability constants from Lacour et al. (1998). The formation of Fe(III)NTMP is possible at pH values below 6.5 in the absence of other metal ions. In the presence of Zn and Cu the pH region for FeNTMP formation is shifted to lower pH values. We can therefore conclude that dissolution reactions can occur at low pH. It was observed by Müller et al. (1984) that HEDP was significantly mobilizing Fe from natural sediments but no information was given about the pH value of the experiments. Bordas and Bourg (1998) on the other hand did not observe enhanced solubilization of Fe from river sediment at pH 3 by 0.01 M NTMP.

7.3.3 Remobilization of metals

In a similar manner to the dissolution of a solid phase, metals adsorbed onto a surface can be solubilized by chelating agents. This process has always been considered as one of the most adverse effects of elevated chelating agent concentrations in the environment (e.g. Hering, 1995; Müller and Förstner, 1976). Aminocarboxylates, such as EDTA, have mainly been studied with respect to remobilization of heavy metals. In all remobilization studies (e.g. Bordas and Bourg, 1998; Gonsior et al., 1997; Kedziorek et al., 1998; Müller and Förstner, 1976), complexation with dissolved metals, dissolution of minerals, and remobilization of adsorbed metals occurred simultaneously.

A closer look at Figure 7.2 shows that in the pH range from 5–8, EDTA has a pronounced influence on Cu adsorption, resulting in the enhanced solubilization of the metal. In the presence of EDTMP, however, there is no such influence, or even enhanced adsorption of Cu in the presence of the phosphonate at low pH. Only at pH >8 mobilization of Cu occurs. We can therefore expect that the polyphosphonates discussed in this review have only a slight influence on metal remobilization in natural systems. This was actually observed during the study of metal mobilization from river sediments by the phosphonate HEDP (Müller *et al.*, 1984). The only metal to be remobilized was Fe whereas Zn, Cr, Ni, Cu, Pb and Cd were not increased compared to a blank sample. Only dissolution of iron oxides was therefore observed. Bordas and Bourg (1998) did only observe a remobilization of Cu, Cd, and Pb from river sediment at NTMP concentrations above 0.1 mM.

7.3.4 Precipitation

In many instances, phosphonates are added to high salinity waters to prevent the formation of scale. However, due to the insolubility of some complexes they can potentially precipitate. This phenomenon often occurs in oil field applications when phosphonates are injected into the subsurface and are left to interact with calcium-containing formation waters. This precipitation mechanism is, in fact, often advantageous since the precipitation/dissolution process can enhance the amount of phosphonate placed during a treatment as well as the release characteristics of the phosphonates onto the fluids during production (Browning and Fogler, 1995; Gerbino, 1996). The fast precipitation forms a pool of phosphonates in the subsurface that slowly dissolves and thereby results in a steady, low concentration of phosphonates in the brine.

Whereas the metal-aminocarboxylate complexes are very soluble, there are several metal-phosphonate complexes that are insoluble. The solubility of precipitates of NTMP with divalent metals decreases in the order Ca > Ba > Sr > Mg (Samakaev *et al.*, 1984). The insoluble Ca precipitates of DTPMP (Kan *et al.*, 1994; Oddo and Tomson, 1990), NTMP (Pairat *et al.*, 1997), and HEDP (Browning and Fogler, 1995, 1996) and the precipitates of NTMP with Fe(II) (Friedfeld *et al.*, 1998) and Fe(III) (Zholnin *et al.*, 1990) have been investigated in detail. Insoluble products of HEDP are also formed with heavy metals such as Pb and Cd (Deluchat *et al.*, 1995).

Each phosphonate forms several precipitates with any metal cation, each with different stoichiometry and solubility. Commonly, an amorphous phase is formed first, and then transformed into a more thermodynamically stable and crystalline phase with time (Gerbino, 1996; Tomson *et al.*, 1995). For DTPMP for example two phases of the form Ca_3H_4DTPMP have been described (Kan *et al.*, 1994).

These precipitates are important in oil field applications or in engineered systems where high phosphonate and high ion concentration occur simultaneously. In natural waters or waste waters, the phosphonate or Ca concentrations are far too low to exert any influence on phosphonate concentrations. Figure 7.4 shows the solubility of NTMP in the presence of 1 and 5 mM Ca. The dissolved NTMP concentrations are always above 200 μM. In natural waters precipitation reactions are therefore not important.

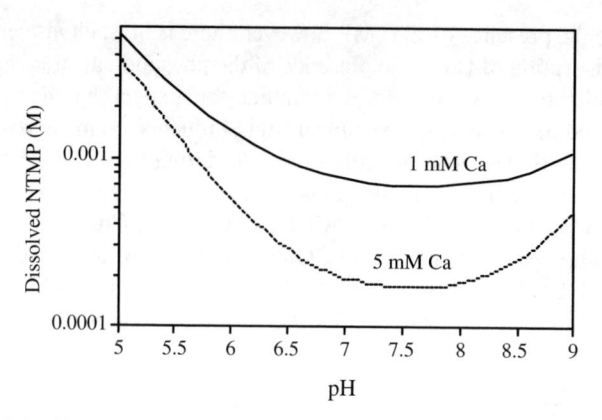

Figure 7.4 Solubility of NTMP in the presence of 1 mM and 5 mM Ca as a function of pH. The diagram was calculated for a system with $Ca_{2.5}HNTMP$ as solid phase and the solubility product from Gerbino (1996) adjusted for 25°C and 0.01 M ionic strength.

7.3.5 Inhibition of dissolution and precipitation

Scale formation (e.g. precipitation of calcium carbonate or calcium sulfate) is a significant problem in commercial water treatment processes including cooling water technology, desalination and oil field applications. This scale formation can be alleviated by the use of chemical water treatment additives, known as 'threshold inhibitors'. Phosphonic acids are among the most potent scale inhibitors next to the polyphosphates. These compounds inhibit crystal growth at concentrations far below those required to chelate stoichiometric amounts of the reactive cations. Models for this retardation include inhibition of nucleation, adsorption onto growth sites, distortion of the crystal lattice, changes in surface charge, and association with precursors of crystal formation (Gal et al., 1996; Gratz and Hillner, 1993).

The morphology of crystals formed in the presence of phosphonates is markedly different from those in the absence of phosphonates (Benton et al., 1993). Phosphonates limit the size of the growing crystals and produce a lag phase in which crystal growth is greatly reduced (Davis et al., 1995). It was found that the ability of different phosphonates to inhibit crystal growth can be interpreted in terms of the Langmuir adsorption model, with the strongest inhibitory effect caused by the compounds that adsorb the strongest (Zieba et al., 1996).

Due to their inhibitory effect on crystal growth, it has been argued that phosphonates may have an adverse effect on phosphate elimination by precipitation with iron or aluminum salts during waste water treatment (Horstmann and Grohmann, 1984, 1986). It was, however, found that although phosphonates had a negative influence on flocculation, it was possible to compensate by increased addition of flocculating agent. The resulting particulate precipitation products were stabilized by the dispersing action of the phosphonates and not retained in the sand filter. Another study, however, found no influence of HEDP on phosphate elimination (Müller et al., 1984).

7.4 DEGRADATION

7.4.1 Biodegradation

Phosphonates are similar to phosphate esters except that phosphonates have a carbon-phosphorous (C—P) bond in place of the more familiar carbon–oxygen–phosphorous linkage. Because this bond is less labile than the O—P, N—P or S—P linkage, biodegradation of phosphonate-containing compounds is limited and mainly restricted to phosphonates containing only one functional group (Huang and Chen, 2000; Kononova and Nesmeyanova, 2002; Ternan et al., 1998). Phosphonate biotransformation can occur on the C moiety or can result in breakage of the C—P bond. The former is specific to the C moiety; the latter may be quite nonspecific. In nature bacteria play a major role in phosphonate biodegradation. Apparently due to the presence of natural phosphonates in the environment, bacteria have evolved the ability to metabolize phosphonates as nutrient sources. Those bacteria capable of cleaving the C—P bond are able to use phosphonates as a phosphorous source for growth. Three pathways appear to exist for the use of phosphonates as a P source: (i) the phosphonatase, (ii) the C—P lyase and (iii) the phosphonoacetate hydrolase pathways (Wanner, 1994). Aminophosphonates can also be used as sole nitrogen source by some bacteria (McMullan and Quinn, 1993).

The polyphosphonate chelating agents discussed here differ significantly from natural phosphonates, such as 2-aminoethylphosphonic acid, because they are much larger, carry a high negative charge and are complexed with metals. However, bacterial strains capable of degrading aminopolyphosphonates and HEDP, under P-limited conditions, have been isolated from soils, lakes, waste water, activated sludge, and compost (Schowanek and Verstraete, 1990a). Also the phosphonate phosphonobutane-tricarboxylic acid (PBTC) was rapidly degraded by microbial enrichment cultures from a variety of ecosystems under conditions of low phosphate availability (Raschke et al., 1994).

The effects of other more accessible P sources on phosphonate uptake and degradation are of great environmental importance. Many environments such as activated sludge, sediments and soils that act as a sink for phosphonates are not characterized by a lack of P most of the time. Because phosphonates are utilized almost exclusively as P-source, little biodegradation can be expected under these conditions. It was demonstrated, however, that simultaneous phosphate and phosphonate utilization by bacteria can occur (Schowanek and Verstraete, 1990b).

Biodegradation tests with HEDP and NTMP showed no indication for any degradation (Huber, 1975). In standard biodegradation tests with sludge from municipal sewage treatment plants, little or no biodegradation of HEDP (Steber and Wierich, 1986) and NTMP (Steber and Wierich, 1987) was observed. An investigation of HEDP, NTMP, EDTMP and DTPMP in standard biodegradation tests also failed to identify any biodegradation (Horstmann and Grohmann, 1988). It was noted, however, that in some tests due to the high sludge to phosphonate ratio, removal of the test substance from solution was observed. This was attributed to adsorption rather than biodegradation because no concomitant increase in CO_2 was observed. Nowack and Baumann (1998) have shown that the photolysis products of Fe(III)phosphonates have no better biodegradability than the parent compounds.

7.4.2 Photodegradation

The Fe(III) complexes of EDTA are susceptible to rapid photodegradation (Carey and Langford, 1973). Fe(III) and Cu(II) complexes of other chelating agents such as NTA, DTPA and EDDS (ethylenediamine-disuccinic acid) are also rapidly photodegraded (Langford et al., 1973; Metsärinne et al., 2001; Svenson et al., 1989).

The Fe(III) complexes of phosphonates are photodegraded in an analogous reaction. HEDP is photolyzed to acetate and phosphate within few days (Steber and Wierich, 1986). In distilled water and in the presence of Ca no photodegradation of HEDP was observed but the addition of Fe(III) and Cu(II) resulted in rapid photodegradation (>98% removal after 40 h irradiation) (Fischer, 1993). The mechanism of Fe(III)EDTMP photodegradation was investigated in detail (Matthijs et al., 1989) and shown to be equivalent to the degradation of Fe(III)EDTA (Lockhard and Blakeley, 1975). Fe(III)EDTMP is degraded in a stepwise process from the parent compound via the ethylenediaminetrimethylene phosphonate and ethylenediaminedimethylene phosphonate to ethylenediaminemonomethylene phosphonate which is stable in the presence of Fe(III) and light. HEDP was photodegraded in the presence of titanium dioxide, in the absence of metal ions (Sabin et al., 1992). For EDTA it has been shown that photodegradation of the Fe(III)complexes is the major elimination pathway in natural waters (Kari and Giger, 1995). We can therefore expect that photodegradation is also very important for the fate of phosphonates in surface waters. Whereas the photodegradation products of Fe(III)EDTA are readily biodegradable, this is not the case for the phosphonates (Nowack and Baumann, 1998).

7.4.3 Chemical degradation

Synthetic chelating agents are selected so that corresponding metal ion complexes are stable during the technical operation under consideration. Breakdown typically requires longer timescales and more severe chemical conditions. Free aminocarboxylates do not break down until a temperature of 175°C is achieved, at which point hydrolytic C–N fission takes place (Motekaitis et al., 1982; Venetzki and Moniz, 1969). Complexation by divalent metal ions substantially lowers breakdown rates (Booy and Swaddle, 1977). Under these circumstances temperatures of 300°C are necessary for degradation, and rates of breakdown decrease in the order Mg > Ca > Zn > Fe(II) > Ni, reflecting increasing percentages of chelating agents complexed by the added metal ion. However, metal ions capable of existing in two or more oxidation states may induce oxidation–reduction pathways for chelating agent breakdown at temperatures of only 100°C (Motekaitis et al., 1980). Although phosphonates are more resistant to decomposition than corresponding aminocarboxylates, decomposition still occurs if temperatures are high enough. At temperatures above 200°C free NTMP decomposes to various breakdown products (Kaslina *et al.*, 1985; Martell *et al.*, 1975).

These studies are important for understanding the fate of the chelating agents in engineered systems at elevated temperatures (e.g. in cooling waters of power plants). For natural systems, these conditions are not relevant. One study performed at room temperature within the pH range of 2–10 reported that over a several month period, EDTMP hydrolyzed under formation of phosphate, phosphite and hydroxymethylphosphonate (Tschäbunin *et al.*, 1989c). Other phosphonate-containing breakdown products were present, but were not identified. No information on the kinetics or the percentage degraded was given.

In natural waters chelating agents always occur in the form of metal complexes. Preliminary studies of metal ion effects on phosphonate decomposition have been reported. Degradation of the amine linkage-containing phosphonates NTMP, EDTMP, and DTPMP was negligible in metal-ion free oxygenated solutions, but Ca, Mg, and Fe(II) brought about conversion to free phosphate at a rate of approximately 1% per day (Schowanek and Verstraete, 1991). Although the degradation was classified as hydrolysis, the conversion rate dropped to negligible levels in the absence of dissolved $O_2(g)$, indicating that redox reactions play a role. HEDP, which does not contain an amine linkage, degrades approximately 20 times more slowly.

A loss of NTMP in different natural waters (river waters, groundwaters) and appearance of the degradation products have been observed (Steber and Wierich, 1987). The conversion of NTMP into iminodimethylenephosphonate (IDMP) and hydroxymethylphosphonate (HMP) was attributed to abiotic hydrolysis, whereas the subsequent conversion to aminomethylphosphonate (AMP) and CO_2 to microbial degradation. The authors performed a follow-up study in an abiotic medium, that contained mM levels of Ca, Mg, K, and Na and trace levels (<1 µM) of Fe(III), Cu(II), Mn(II), and Zn. Complete conversion of NTMP to IDMP, HMP and AMP occurred within 32 hours. Because multiple metal ions were present in these investigations (Schowanek and Verstraete, 1991; Steber and Wierich, 1987), it was not possible to identify the catalytic agent.

A systematic study of the influence of metal ions on phosphonate breakdown has been conducted by Nowack and Stone (2000). No breakdown of NTMP was observed in metal-free systems and in the presence of Ca, Mg, Zn, Cu(II) and Fe(III). This stands in disagreement to the results of Steber and Wierich (1987) who found degradation of NTMP in the absence of metals and in the presence of Ca or Mg. Very rapid degradation of aminopolyphosphonates occurred in the presence of Mn(II) and molecular oxygen (Nowack and Stone, 2000). The half-life for the reaction of NTMP in the presence of equimolar Mn(II) and in equilibrium with 0.21 atm O_2 is 10 minutes at pH 6.5. The reaction occurs more slowly at pH below and above 6.5. The presence of other cations such as Ca(II), Zn(II) and Cu(II) can considerably slow down the reaction by competing with Mn(II) for NTMP. Catalytic Mn(II) is regenerated in cyclic fashion as the reaction takes place (Figure 7.5). Formate, orthophosphate, IDMP and N-formyl-iminodimethylenephosphonic acid (FIDMP) breakdown products have been identified. Breakdown also occurs in

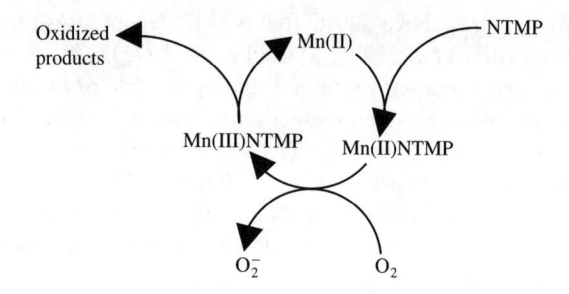

Figure 7.5 The catalytic cycle in the system NTMP-Mn(II)-O_2 resulting in the oxidation of NTMP by molecular oxygen.

oxygen-free suspension of the Mn(III) containing the mineral manganite (MnOOH, Nowack and Stone, 2002). Reaction of Mn(III)OOH with aminocarboxylates yields reductive dissolution products. The Mn(III)EDTA complexes decompose rather rapidly in aqueous solution with a pH-dependent rate (Schroeder and Hamm, 1964; Klewicki and Morgan, 1998). It is also known that Mn(III)EDTMP undergoes an intramolecular redox reaction, resulting in the appearance of Mn(II) and disappearance of EDTMP (Kurochkina, 1978). Very rapid degradation of NTMP is also observed with manganite in the presence of oxygen (Nowack and Stone, 2002). EDTMP and DTPMP are also degraded in the presence of Mn(II) and oxygen, although at a slower rate, but not the amine-free HEDP (Nowack and Stone, 2000). Using two derivative HPLC methods, two of the breakdown products of NTMP, IDMP and FIDMP, have been detected in waste water (Nowack, 2002c). This work indicates that manganese-catalyzed oxidation of aminopolyphosphonate is likely to be an important degradation mechanism in natural waters.

7.4.4 Degradation during oxidation processes

Ozonation of NTMP, EDTMP and DTPMP results in the rapid disappearance of the parent compound in less than a minute (Klinger et al., 1998a). Around 60–70% of the degraded phosphonate was found as phosphate; aminomethylphosphonic acid (AMP) and phosphonoformic acid were also detected. The amine-free HEDP was degraded much slower with only 15% degradation after 30 min. The reaction pathway of EDTMP during ozonation is equivalent to that of EDTA (Gilbert and Hoffmann-Glewe, 1990). The herbicide glyphosate (containing a phosphonic acid group) was formed during the reaction with concentration of up to 10 nM; also AMP was formed at high concentrations as an intermediate (Klinger et al., 1998b). The environmental fate, behaviour and analysis of both AMP and glyphosate has received considerable attention (Stalikas and Konidari, 2001) and the detection of these compounds during ozonation of an aminopolyphosphonate may change the risk analysis of these compounds considerably.

No information is available about the behaviour of phosphonates during chlorination.

7.5 SPECIATION

7.5.1 Stability constants

A recent IUPAC Technical Report (Popov *et al.*, 2001) critically evaluates the available experimental data on stability constants of proton and metal complexes for phosphonic acids, including the four compounds discussed in this chapter. It presents a set of high-quality data as 'recommended' or 'provisional' constants while for example all constants for DTPMP have been rejected due to insufficient purity of the parent compound. This review will be of great use for all future speciation calculations and should be the sole source of stability constants whenever possible.

7.5.2 Speciation calculations

The speciation of chelating agents in the environment can be calculated based on the known stability constants of the metal–ligand complexes and the measured total concentrations of metals and chelating agents. This approach has been used to predict the speciation of EDTMP in water from the Rhine (Gledhill and Feijtel, 1992). The simulated speciation was dominated by CuEDTMP and ZnEDTMP. HEDP was predicted to be mainly complexed with Ca, NTMP with Cu and Zn (Deluchat *et al.*, 1997, 2002; Lacour *et al.*, 1999). However, the accuracy of such calculation may be questionable. There are several points to consider: first, it is assumed that equilibrium has been reached in the system; this is not always the case. It has been found, for example, that Fe(III)EDTA is not in equilibrium with other metals in river water due to slow exchange kinetics of Fe(III)EDTA (Kari *et al.*, 1995). The concentration of Fe(III)EDTA in a water sample can therefore never be calculated, it is solely accessible by analytical determination. Other EDTA complexes, such as NiEDTA, have also very slow exchange kinetics and equilibrium is not achieved within the time scale of river flow. The concentration of these species is determined by their speciation at the source (e.g. industrial effluent). Almost nothing is known about the exchange kinetics of metal–phosphonate complexes and therefore all equilibrium calculations have to be treated with care.

Secondly, most calculations do not consider that besides the chelating agent of interest, other chelating agents and natural ligands are present in the water and compete for available metals. The interaction between phosphonates and fulvic acids is weak (Deluchat *et al.*, 1998), but Sigg and Xue (1994) have shown that considering the natural ligands for Cu and Zn is critical for obtaining an accurate speciation of EDTA and NTA.

A speciation model for three phosphonates has been developed, based on a river water sample from Switzerland with well known composition of metals and anthropogenic and natural ligands (Nowack *et al.*, 1997) (Table 7.2). This river water contains not only EDTA and NTA but also strong natural ligands for metals like Cu (Xue *et al.*, 1996), Zn (Xue and Sigg, 1994), and Ni (Xue *et al.*, 2001). While the chemical identity of these ligands is still unclear (Wu and Tanoue, 2001), they can

Table 7.2 Calculated species distribution of HEDP, NTMP, and DTPMP in river water under different conditions (20 nM phosphonates, 29.4 nM EDTA, 8.6 nM NTA; natural ligands for Cu, Zn and Ni according to Nowack et al., 1997 and Xue et al., 2001).

	% of total phosphonate			
	Ca	Mg	Zn	Cu
HEDP				
only HEDP	88	12	0	0.1
with EDTA/NTA	88	12	0	0
with EDTA/NTA/natural ligands	88	12	0	0
NTMP				
only NTMP	33	25	11	28
with EDTA/NTA	38	29	10	21
with EDTA/NTA/natural ligands	55	42	0	0
DTPMP				
only DTPMP	0	0	24	76
with EDTA/NTA	0	0	28	71
with EDTA/NTA/natural ligands	41	21	35	2

clearly be linked to biological activity (Xue and Sigg, 1993). These ligands compete with EDTA, NTA, and the phosphonates for the same metals and have to be included in the speciation calculation.

The concentration of the phosphonates in the calculations was set to 20 nM, comparable to EDTA at that location. The speciation was calculated for HEDP and NTMP with the constants from Popov et al. (2001) and for DTPMP with the constants from Rizkalla (1983). If only total metals and the phosphonates are taken into consideration, speciation is dominated by Cu for DTPMP, Ca for HEDP, and Ca, Mg, Zn and Cu for NTMP. The speciation does not change significantly when EDTA and NTA are included; however, as soon as the natural ligands for Cu and Zn are considered, the calculated speciation for NTMP and DTPMP changes drastically. For NTMP the Cu and Zn complexes disappear now totally due to the very strong binding of Cu to the natural ligands and CaNTMP and MgNTMP are dominant. For DTPMP the Ca and Mg complexes also become very important with more than 60% of the DTPMP complexed by these metals. CuDTPMP is only a minor species under these conditions. For HEDP the alkaline earth metals Ca and Mg are the major bound metals under all conditions. The fraction of other metal complexes is never above 0.1%. It can be concluded that phosphonates are most probably complexed to alkaline earth metals in natural waters. This calculation shows that considering the natural ligands is crucial for obtaining an accurate result for phosphonate speciation.

7.5.3 Determination of the speciation

Analytical methods have been developed to determine directly the speciation of aminocarboxylate chelating agents. In principle these methods should also be applicable to phosphonates. It is possible to detect the EDTA complexes of Cd, Co(II), Cu(II), Pb and Zn in wastewater with a detection limit of less than 0.1 µM (Bedsworth and Sedlak, 2001). Fe(III)EDTA can be determined in water samples by illumination of a sample, which results in the complete and selective photo-degradation of only Fe(III)EDTA (Kari and Giger, 1995). NiEDTA can be quantified by methods involving pre-column formation of Fe(III)EDTA and using the slow exchange kinetics of NiEDTA (Bedsworth and Sedlak, 1999; Nirel *et al.*, 1998; Nowack *et al.*, 1996a). With such methods it should also be possible to determine the speciation of phosphonates, especially the distinction between Fe(III)phosphonates and the photo-stable complexes.

A recent very promising method uses anion-exchange chromatography coupled to ICP-MS for the separation of metal-chelating agent complexes (Ammann 2002a, 2002b). The method is also applicable to phosphonates; it has been shown that the CuEDTMP complex can be determined.

7.6 ENVIRONMENTAL BEHAVIOUR

7.6.1 Behaviour during waste water treatment

Studies about the behaviour of phosphonates during waste water treatment can be divided into two groups: field studies involving the addition of elevated concentrations of phosphonates to the influent, and investigations at ambient concentrations. Müller *et al.* (1984) added 9.7 µM HEDP to a WWTP influent during a field study. The elimination during sedimentation was about 60%, whereas the elimination in the biological step with simultaneous $FeCl_3$ precipitation was 90–97.5%. Another study involved the addition of 5–10 µM HEDP and 3–7 µM NTMP to a WWTP without iron-addition; this study found removal rates of about 50–60% (Metzner, 1990). A field experiment involving the addition of 4.5–12 µM DTPMP was also carried out (Nowack, 2002b). The DTPMP concentration after the biological step was between 0.1 and 0.4 µM. Hence, the removal efficiency after this step was 95%. After the precipitation step with aluminum sulfate the concentration of dissolved DTPMP was between 0.1 and 0.3 µM. About 97% of the added DTPMP had been removed after this step. The concentration after the flocculation step with polyacrylamide was around 0.1 µM. This investigation has shown that even without simultaneous addition of iron or aluminum salts, very good removal in the biological step can be achieved.

The second group of studies investigated the fate of phosphonates that are already present in the influent of the WWTP. Nowack (1998) has reported after a 13-day field study that a total amount of 117 moles DTPMP in the influent of the WWTP compared to an effluent load of 17 moles, meaning that the removal efficiency was 85%. Figure 7.6 shows the load of DTPMP in the influent and effluent found in this study.

Elimination of NTMP and EDTMP from another WWTP was at least 80 and 70%, respectively (Nowack, 1998). Because the concentration in the effluent was below the detection limit, this removal efficiency is the lower limit. The fate of

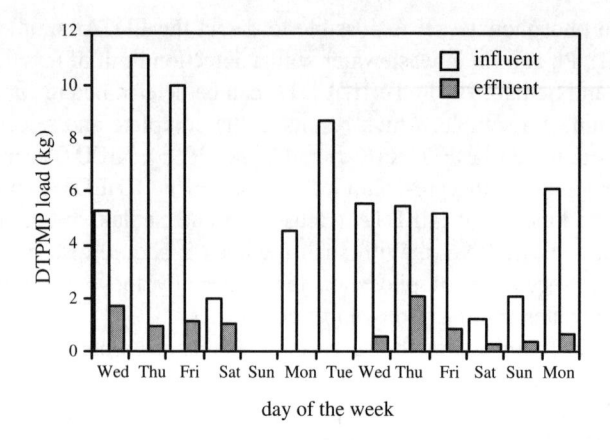

Figure 7.6 Load of DTPMP in the influent and effluent of a WWTP receiving wastewater from textile industry. Data from Nowack (1998).

NTMP was followed in another WWTP receiving waste water from textile industry (Nowack, 2002b). The load of NTMP in the influent was 324 moles during the 2-week period. The maximum effluent load for the two-week period can be calculated to be 23 moles, taking the detection limit of 0.05 μM as the upper limit of NTMP concentration. The removal efficiency of the WWTP was therefore at least 93%. Also the two breakdown products of NTMP, IDMP and FIDMP, were present at much higher concentrations in the influent than in the effluent (Nowack, 2002c), with a removal of 87% FIDMP and 96% IDMP.

The results from field studies and field measurements have shown that phosphonates are removed very efficiently in most WWTP and pose only little risk to the receiving waters.

7.6.2 Toxicology

Peer-reviewed data on the environmental toxicology of phosphonate chelating agents are very rare (Jaworska et al., 2002), except for glyphosate (Atkinson, 1985), which, however, is not discussed in this review. Huber (1975) has reported 48 h LC_{50} values for fish for NTMP and HEDP at concentrations of 0.88 mM and 1.08 mM, respectively. Similar values for fish toxicity were given by Schöberl and Huber (1988) at 0.1 mM for HEDP, 0.8 mM for NTMP and 0.36 mM for EDTMP. The EC_{50} to Daphnia was 1 mM for HEDP and NTMP and 0.9 mM for EDTMP. Van Hullebusch et al. (2002) have presented some results about effects of HEDP and NTMP on bacterial Cu toxicity. At concentrations higher than 1 μM phosphonate the Cu-toxicity was reduced due to the decrease in the free Cu^{2+} concentration. At a concentration of 0.1 μM phosphonate, however, an enhancement of bacterial Cu-toxicity was observed. No mechanistic interpretation of this effect was given.

A study of the bioconcentration of 0.25 μM HEDP and 0.005 μM NTMP in fish found a bioconcentration factor of 20 and 24, which decreased rapidly in phosphonate-free water (Steber and Wierich, 1986, 1987). No chronic toxicity from EDTMP was observed in rats, even at a dose of 0.2 mM kg^{-1} day^{-1} (Calvin et al., 1988) and no genotoxicity was observed at that dose. EDTMP was poorly absorbed in the gastro-intestinal tract, and most of the absorbed dose was rapidly excreted by the kidneys or sequestered in bone. A significant fraction of the administered EDTMP was found after a few hours in the bones (Jarvis et al., 1995). This effect is due to the strong adsorption of phosphonates onto the bone mineral hydroxylapatite (Amjad, 1987; Chirby et al., 1988; Jung et al., 1973; Rawls et al., 1982). At very high doses, the phosphonates interfere with the calcium metabolism. With a dose of 0.3 mM HEDP kg^{-1} day^{-1} the skeletal mineralization of rats was inhibited (Trechsel et al., 1977). However, if orally administered, almost no effect of EDTMP was observed in rats at a dose of 0.8 mM EDTMP kg^{-1} day^{-1}, but oral administration of 1.7 mM HEDP kg^{-1} day^{-1} had profound effects on calcium absorption and bone growth (Miller et al., 1985). Human toxicity is also low, which can be seen in the fact that phosphonates are used to treat various diseases (Fleisch, 1989; Francis and Centner, 1978). The toxicology studies show that at concentrations found in wastewater (0.01–1 μM), no toxic effect on aquatic organisms and humans is expected.

7.7 CONCLUSIONS

Although the use of phosphonates is almost as extensive as that of the aminocarboxylates, a lot less is known about their behaviour. Some of the properties of the two groups are very similar, however there are characteristics of the phosphonates that are substantially different and that affect significantly their environmental behaviour. Firstly, phosphonates interact strongly with surfaces, which results in a significant removal in natural systems. Secondly, aminopolyphosphonates are rapidly oxidized in the presence of Mn(II) and stable breakdown products are formed that have been detected in waste water. Analysis of the parent phosphonate alone is therefore not sufficient. The lack of information about phosphonates is certainly linked to analytical problems of their determination at trace concentrations in natural waters. Further method development is thus urgently needed in this area. Finally, there is no information available on the exchange kinetics of metal-phosphonate complexes. Slow exchange kinetics have been found to be particularly important to the environmental speciation of aminocarboxylates.

ACKNOWLEDGEMENTS

The author is indebted to Jean-Claude Bollinger and Véronique Deluchat for their fruitful comments to earlier versions of this manuscript. This chapter was prepared in part during a stay at the University of Limoges, France.

REFERENCES

Amjad, Z. (1987) The influence of polyphosphates, phosphonates, and polycarboxylic acids on the crystal growth of hydroxyapatite. *Langmuir* **3**, 1063–1069.

Ammann, A. (2002a) Determination of strong binding chelators and their metal complexes by anion-exchange chromatography and inductively couples plasma mass spectrometry. *J. Chromatogr. A* **947**, 205–216.

Ammann, A. (2002b) Speciation of heavy metals in environmental water by ion chromatography coupled to ICP-MS. *Anal. Bioanal. Chem.* **372**, 448–452.

Anderson, R.L., Bishop, W.E. and Campbell, R.L. (1985) A review of the environmental and mammalian toxicology of nitrilotriacetic acid. *CRC Crit. Rev. Toxicol.* **15**, 1–102.

Atkinson, D. (1985) Toxicological properties of glyphosate – a summary. pp. 127–133. In *The Herbicide Glyphosate* (ed. Grossbard, E. and Atkinson, D.) Butterworths, London.

Barja, B.C., Tejedor-Tejedor, M.I. and Anderson, M.A. (1999) Complexation of methylphosphonic acid with the surface of goethite particles in aqueous suspension. *Langmuir* **15**, 2316–2321.

Bedsworth, W.W. and Sedlak, D.L. (2001) Determination of metal complexes of ethylenediaminetetraacetate in the presence of organic matter by high-performance liquid chromatography. *J. Chromatogr. A* **905**, 157–162.

Bedsworth, W.W. and Sedlak, D.L. (1999) Sources and environmental fate of strongly complexed nickel in estuarine waters: the role of ethylenediaminetetraacetate. *Environ. Sci. Technol.* **33**, 926–931.

Benton, W.J., Collins, I.R., Grimsey, I.M., Parkinson, G.M. and Rodger, S.A. (1993) Nucleation, growth and inhibition of barium sulfate-controlled modification with organic and inorganic additives. *Faraday Discuss* **95**, 281–297.

Black, S.N., Bromley, L.A., Cottier, D., Davey, R.J., Dobbs, B. and Rout, J.E. (1991) Interactions of the organic/inorganic interface: binding motifs for phosphonates at the surface of barite crystals. *J. Chem. Soc. Faraday Trans.* **87**, 3409–3414.

Booy, M. and Swaddle, T.W. (1977) Chelating agents in high temperature aqueous chemistry. 2. The thermal decomposition of some transition metal complexes of nitrilotriacetic acid (NTA). *Can. J. Chem.* **55**, 1770–1776.

Bordas, F. and Bourg, A.C.M. (1998) Effect of complexing agents (EDTA and ATMP) on the remobilization of heavy metals from a polluted river sediment. *Aquat. Geochem.* **4**, 201–214.

Browning, F.H. and Fogler, H.S. (1996) Effect of precipitating conditions on the formation of calcium-HEDP precipitates. *Langmuir* **12**, 5231–5238.

Browning, F.H. and Fogler, H.S. (1995) Effect of synthesis parameters on the properties of calcium phosphonate precipitates. *Langmuir* **11**, 4143–4152.

Bucheli-Witschel, M. and Egli, T. (2001) Environmental fate and microbial degradation of aminopolycarboxylic acids. *FEMS Microbiol. Rev.* **25**, 69–106.

Calvin, G., Long, P.H., Stitzel, K.A., Anderson, R.L., Balmbra, R.R. and Bruce, R.D. (1988) Ethylenediaminetetra(methylenephosphonic acid): genotoxicity, biodistribution, and subchronic toxicity in rats. *Fd. Chem. Toxic.* **26**, 601–610.

Carey, J.H. and Langford, C.H. (1973) Photodecomposition of Fe(III) aminocarboxylates. *Can. J. Chem.* **51**, 3665–3670.

Chirby, D., Franck, S. and Troutner, D.E. (1988) Adsorption of ^{153}Sm-EDTMP on calcium hydroxyapatite. *Appl. Radiat. Isot.* **39**, 495–499.

Davenport, B., DeBoo, A., Dubois, F. and Kishi, A. (2000) CEH Report: Chelating agents. SRI Consulting, Menlo Park, Ca, USA.

Davis, R.V., Carter, P.W., Kamrath, M.A., Johnson, D.A. and Reed, P.E. (1995) The use of modern methods in the development of calcium carbonate inhibitors for cooling water systems. In *Mineral Scale Formation and Inhibition* (ed. Amjad, Z.), Plenum Press, New York, pp. 33–46.

Davis, J.A., Kent, D.B., Rea, B.A., Maest, A.S. and Garabedian, S.P. (1993) Influence of redox environment and aqueous speciation on metal transport in groundwater. In *Metals in Groundwater* (eds. Allen, H.E., Perdue, E.M. and Brown, D.S.), Lewis Publishers, pp. 223–273.

Day, G.M., Hart, B.T., McKelvie, I.D. and Beckett, R. (1997) Influence of natural organic matter on the sorption of biocides onto goethite. II. glyphosate. *Environ. Technol.* **18**, 781–794.

De Klerk, J.M.H., van Dijk, A., van het Schip, A.D., Zonnenberg, B.A. and van Rijk, P.P. (1992) Pharmacokinetics of rhenium-186 after administration of rhenium-186-HEDP to patients with bone metastases. *J. Nucl. Med.* **33**, 646–651.

Deluchat, V., Lacour, S., Serpaud, B. and Bollinger, J.C. (2002) Washing powders and the environment: has TAED any influence on the complexing behavior of phosphonic acids? *Water Res.* **36**, 4301–4306.

Deluchat-Antony, V., Bollinger, J.C., Serpaud, B. and Caullet, C. (1998) Influence of some washing powder components on copper(II) and calcium(II) complexation with fulvic acids. *Intern. J. Environ. Anal. Chem.* **68**, 123–135.

Deluchat, V., Bollinger, J.C., Serpaud, B. and Caullet, C. (1997) Divalent cations speciation with three phosphonate ligands in the pH-range of natural waters. *Talanta* **44**, 897–907.

Deluchat, V., Serpaud, B., Caullet, C. and Bollinger, J.C. (1995) Constantes de protonation et de complexation de l'acide 1-hydroxyethane-1,1'diphosphonique (HEDP) vis-à-vis de cations divalents: étude de complexes peu soluble de HEDP avec Pb(II) et Cd(II). *Phosphorous, Sulfur, and Silicon* **104**, 81–93.

De Witt, G.C., May, P.M., Webb, J. and Hefter, G. (1996) Biospeciation, by potentiometry and computer simulation of SmEDTMP, a bone tumor palliative agent. *BioMetals* **9**, 351–361.

Dubbin, W.E., Sposito, G. and Zavarin, M. (2000) X-ray absorption spectroscopic study of Cu-glyphosate adsorbed by microcrystalline gibbsite. *Soil Sci.* **165**, 699–707.

Felber, H., Hegetschweiler, K., Müller, M., Odermatt, R. and Wampfler, B. (1995) Quantitative Bestimmung von Phosphonaten in Waschmitteln. *Chimia* **49**, 179–181.

Fischer, K. (1991) Sorption of chelating agents (HEDP and NTA) onto mineral phases and sediments in aquatic model systems. Part I: sorption onto clay minerals. *Chemosphere* **22**, 15–27.

Fischer, K. (1992) Sorption of chelating agents (HEDP and NTA) onto mineral phases and sediments in aquatic model systems. Part II: sorption onto sediments and sewage sludges. *Chemosphere* **24**, 51–62.

Fischer, K. (1993) Distribution and elimination of HEDP in aquatic test systems. *Water Res.* **27**, 485–493.

Fleisch, H. (1989) Bisphosphonates: a new class of drugs in diseases of bone and calcium metabolisms. *Recent Results Cancer Res.* **116**, 1–28.

Francis, M.D. and Centner, R.L. (1978) The development of diphosphonates as significant health care products. *J. Chem. Educ.* **55**, 760–766.

Franz, J.E., Mao, M.K. and Sikorski, J.A. (1997) Glyphosate: A unique global herbicide. *ACS monograph 189*, American Chemical Society, Washington, DC. 653 p.

Friedfeld, S.J., He, S. and Tomson, M.B. (1998) The temperature and ionic strength dependence of the solubility product constant of ferrous phosphonate. *Langmuir* **14**, 3698–3703.

Frigge, E. and Jackwerth, E. (1991) Preconcentration and determination of organophosphonic acids: application to natural waters. *Anal. Chim. Acta* **254**, 65–73.

Gal, J.Y., Bollinger, J.C., Tolosa, H. and Gache, N. (1996) Calcium carbonate solubility: a reappraisal of scale formation and inhibition. *Talanta* **43**, 1497–1509.

Gerbino, A.J. (1996) Quantifying the retention and release of polyphosphonates in oil and gas producing formations using surface complexation and precipitation theory. Dissertation, Rice University, Texas.

Gilbert, E. and Hoffmann-Glewe, S. (1990) Ozonation of ethylenediaminetetraacetate (EDTA) in aqueous solution, influence of pH value and metal ions. *Water Res.* **24**, 39–44.

Gledhill, W.E. and Feijtel, T.C.J. (1992) Environmental properties and safety assessment of organic phosphonates used for detergent and water treatment applications. In *The Handbook of Environmental Chemistry.* (ed. Hutzinger O.), Springer Verlag Berlin, Heidelberg **3**, Part F, pp. 261–285.

Gonsior, S.J., Sorci, J.J., Zoellner, M.J. and Landenberger, B.D. (1997) The effects of EDTA on metal solubilization in river sediment/water systems. *J. Environ. Qual.* **26**, 957–966.

Gratz, A.J. and Hillner, P.E. (1993) Poisoning of calcite growth viewed in the atomic force microscope (AFM). *J. Cryst. Growth* **129**, 789–793.

Held, S. (1989) Zum Umweltverhalten von Komplexbildnern auf Phosphonsäurebasis. *Textilveredlung* **24**, 394–398.

Hering, J.G. (1995) Implications of complexation, sorption and dissolution kinetics for metal transport in soils. In *Metal Speciation and Contamination of Soils* (eds Allen, H.E., Huang, C.P., Bailey, G.W. and Bowers, A.R.), Lewis.

Horiguchi, M. and Kandatsu, M. (1959) Isolation of 2-aminoethane phosphonic acid from rumen protozoa. *Nature* **184**, 901–902.

Horstmann, B. and Grohmann, A. (1984) The influence of phosphonates upon phosphate elimination. 1. communication: 1-hydroxyethane-1,1-diphosphonic acid (HEDP). *Z. Wasser-Abwasser-Forsch.* **17**, 177–181.

Horstmann, B. and Grohmann, A. (1986) The influence of phosphonates upon phosphate elimination. Part 2: 2-phosphono-butane-1.2.4-tricarbonic acid, aminotri(methylenephosphonic acid); ethylenediaminetetra(methylenephosphonic acid). *Z. Wasser-Abwasser-Forsch.* **19**, 236–240.

Horstmann, B. and Grohmann, A. (1988) Investigations into the biodegradability of phosphonates. *Vom Wasser* **70**, 163–178.

Huang, J. and Chen, R. (2000) An overview of recent advances on the synthesis and biological activity of α-aminophosphonic acid derivatives. *Heteroatom Chem.* **11**, 480–492.

Huber, L. (1975) Untersuchungen über die biologische Abbaufähigkeit und Fischtoxizität von 2 organischen Komplexbildnern auf Phosphonsäurebasis (ATMP und HEDP). *Tenside Detergents* **12**, 316–322.

Jardine, P.M., Jacobs, G.K. and O'Dell, J.D. (1993) Unsaturated transport processes in undisturbed heterogeneous porous media: II. Co-contaminants. *Soil Sci. Soc. Am. J.* **57**, 954–962.

Jarvis, N.V., Wagener, J.M. and Jackson, G.E. (1995) Metal-ion speciation in blood plasma as a tool for elucidating the *in vivo* behavior of radiopharmaceuticals containing ^{153}Sm and ^{166}Ho. *J. Chem. Soc. Dalton Trans.* 1411–1415.

Jaworska, J., Van Genderen-Takken, H., Hanstveit, A., van de Plassche, E. and Feijtel, T. (2002) Environmental risk assessment of phosphonates, used in domestic laundry and cleaning agents in the Netherlands. *Chemosphere* in press.

Jung, A., Bisaz, S. and Fleisch, H. (1973) The binding of pyrophosphate and two diphosphonates by hydroxyapatite crystals. *Calc. Tiss. Res.* **11**, 269–280.

Kan, A.T., Oddo, J.E. and Tomson, M.B. (1994) Formation of two calcium diethylenetriaminepentakis (methylene phosphonic acid) precipitates and their physical chemical properties. *Langmuir* **10**, 1450–1455.

Kari, F.G. and Giger, W. (1995) Modeling the photochemical degradation of ethylenediaminetetraacetate in the river Glatt. *Environ. Sci. Technol.* **29**, 2814–2827.

Kari, F.G., Xue, H.B. and Sigg, L. (1995) Speciation of EDTA in natural waters: Exchange kinetics of Fe-EDTA in river water. *Environ. Sci. Technol.* **29**, 59–68.

Kaslina, N.A., Polyakova, I.A., Kessenikh, A.V., Zhadanov, B.V., Rudomino, M.V., Churilina, N.V. and Kabachnik, M.I. (1985) Thermal decomposition of nitrilotrimethylenephosphonic acid in aqueous solution. *J. Gen. Chem. USSR* **55**, 472–475.

Kedziorek, M.A.M., Dupuy, A., Bourg, A.C.M. and Compere, F. (1998) Leaching of Cd and Pb from a polluted soil during the percolation of EDTA: Laboratory column experiments modeled with a non-equilibrium solubilization step. *Environ. Sci. Technol.* **32**, 1609–1614.

Klewicki, J.K. and Morgan, J.J. (1998) Kinetic behavior of Mn(III) complexes of pyrophosphate, EDTA, and citrate. *Environ. Sci. Technol.* **32**, 2916–2922.

Klinger, J., Sacher, F., Brauch, H.J., Maier, D. and Worch, E. (1998a) Behavior of phosphonic acids during drinking water treatment. *Vom Wasser* **91**, 15–27.

Klinger, J., Lang, M., Sacher, F., Brauch, H.J., Maier, D. and Worch, E. (1998b) Formation of glyphosate and AMPA during ozonation of waters containing ethylenediaminetetra (methylenephosphonic acid). *Ozone Sci. Eng.* **20**, 99–110.

Klinger, J., Sacher, F., Brauch, H.J. and Maier, D. (1997) Determination of organic phosphonates in aqueous samples using liquid chromatography/particle-beam mass spectrometry. *Acta Hydrochim. Hydrobiol.* **25**, 79–86.

Kononova, S.V. and Nesmyanova, M.A. (2002) Phosphonates and their degradation by microorganisms. *Biochemistry (Moscow)* **67**, 184–195.

Kurochkina, L.V., Pechurova, N.I., Snezhko, N.I. and Spitsyn, V.I. (1978) Interaction of manganese(III) with certain phosphorous-containing complexones. *Russ. J. Inorg. Chem.* **23**, 1481–1484.

Kuys, K.J. and Roberts, N.K. (1987) In situ investigation of the adsorption of styrene phosphonic acid on cassiterite by FTIR-ATR spectroscopy. *Colloids Surf.* **24**, 1–17.

Lacour, S., Deluchat, V., Bollinger, J.C. and Serpaud, B. (1999) Influence of carbonate and calcium ions on the phosphonate complexation with Cu, Zn, Cd, and Ni in fresh waters: an evaluation of thermodynamic constants and a chemical model. *Environ. Technol.* **20**, 249–257.

Lacour, S., Deluchat, V., Bollinger, J.C. and Serpaud, B. (1998) Complexation of trivalent cations (Al(III), Cr(III), Fe(III)) with two phosphonic acids in the pH range of fresh waters. *Talanta* **46**, 999–1009.

Laiti, E. and Öhmann, L.O. (1996) Acid/base properties and phenylphosphonic acid complexation at the boehmite/water interface. *J. Colloid Interface Sci.* **183**, 441–452.

Laiti, E., Öhman, L.O., Nordin, J. and Sjöberg, S. (1995) Acid/base properties and phenylphosphonic acid complexation at the aged γ-Al_2O_3/water interface. *J. Colloid Interface Sci.* **175**, 230–238.

Langford, C.H., Wingham, M. and Sastri, V.S. (1973) Ligand photooxidation in copper(II)complexes of nitrilotriacetric acid. *Environ. Sci. Technol.* **7**, 820–822.

Liu, Y., Gao, L., Yu, L. and Guo, J. (2000) Adsorption of PBTCA on alumina surfaces and its influence on the fractal characteristics of sediments. *J. Colloid Interface Sci.* **227**, 164–170.

Lockhart, H.B. and Blakeley, R.V. (1975) Aerobic photodegradation of X(N) chelates of ethylenedinitrilotetraacetate (EDTA): implications for natural waters. *Environ. Lett.* **9**, 19–31.

Mahabir-Jagessar Tewari, K. and van Stroe-Bieze, S.A.M. (1997) Analysis of aminecontaining phosphonates in detergent powders by anion-exchange chromatography with pulsed amperometric detection. *J. Chromatogr. A* **771**, 155–161.

Martell, A.E., Motekaitis, R.J., Fried, A.R., Wilson, J.S. and MacMillan, D.T. (1975) Thermal decomposition of EDTA, NTA, and nitrilotrimethylenephosphonic acid in aqueous solution. *Can. J. Chem.* **53**, 3471–3476.

Matthijs, E., de Oude, N.T., Bolte, M. and Lemaire, J. (1989) Photodegradation of ferric ethylenediaminetetramethylene-phosphonic acid (EDTMP) in aqueous solution. *Water Res.* **23**, 845–851.

May, H.B., Nijs, H. and Godecharles, V. (1986) Phosphonates, multifunctional ingredients for laundry detergents. *Household and Personal Product Industry* **23**, 50–90.

McMullan, G. and Quinn, J.P. (1993) The utilization of aminoalkylphosphonic acids as sole nitrogen source by an environmental bacterial isolate. *Lett. Appl. Microbiol.* **17**, 135–138.

Means, J.L. and Alexander, C.A. (1981) The environmental biogeochemistry of chelating agents and recommendations for the disposal of chelated radioactive wastes. *Nucl. Chem. Waste Manage.* **2**, 183–196.

Metsärinne, S., Tuhkanen, T. and Aksela, R. (2001) Photodegradation of ethylenediamine tetraacetic acid (EDTA) and ethylenediamine disuccinic acid (EDDS) within natural UV radiation range. *Chemosphere* **45**, 949–955.

Metzner, G. (1990) Verhalten von organischen Komplexbildnern auf Phosphonsäurebasis in Kläranlagen. *Beiträge Abwasser- Fisch- Flussbiol.* **44**, 323–336.
Miller, S.C., Jee, W.S.S., Woodbury, D.M. and Kemp, J.W. (1985) Effects of N,N,N',N'-ethylenediaminetetramethylene phosphonic acid and 1-hydroxyethylidene-1,1-bisphosphonic acid on calcium absorption, plasma calcium, longitudinal bone growth, and bone histology in the growing rat. *Toxicol. Appl. Pharmacol.* **77**, 230–239.
Morillo, E., Undabeytia, T. and Maqueda, C. (1997) Adsorption of glyphosate on the clay mineral montmorillonite: effect of Cu(II) in solution and adsorbed on the mineral. *Environ. Sci. Technol.* **31**, 3588–3592.
Motekaitis, R.J., Cox, X.B., Taylor, P., Martell, A.E., Miles, B. and Tvedit, T.J. (1982) Thermal degradation of EDTA chelates in aqueous solution. *Can. J. Chem.* **60**, 1207–1213.
Motekaitis, R.J., Martell, A.E., Hayes, D. and Frenier, W.W. (1980) The iron(III)-catalyzed oxidation of EDTA in aqueous solution. *Can. J. Chem.* **58**, 1999–2005.
Müller, G., Steber, J. and Waldhoff, H. (1984) The effect of hydroxyethane diphosphonic acid on phosphate elimination with $FeCl_3$ and remobilization of heavy metals: results from laboratory experiments and field trials. *Vom Wasser* **63**, 63–78.
Müller, G. and Förstner, U. (1976) Experimental mobilization of copper and zinc from aquatic sediments by some polyphosphate substitutes in detergents. *Z. f. Wasser- und Abwasser-Forschung* **9**, 150.
Nirel, P.M., Pardo, P.E., Landry, J.C. and Revaclier, R. (1998) Method for EDTA speciation determination: application to sewage treatment plant effluents. *Water Res.* **32**, 3615–3620.
Nowack, B. (2002a) Environmental chemistry of aminopolycarboxylate chelating agents. *Environ. Sci. Technol.* **36**, 4009–4016.
Nowack, B. (2002b) Aminopolyphosphonate removal during wastewater treatment. *Water Res.* **36**, 4636–4642.
Nowack, B. (2002c) Determination of phosphonic acid breakdown products by high performance liquid chromatography after derivatization. *J. Chromatogr. A* **942**, 185–190.
Nowack, B. and Stone, A.T. (2002) Heterogeneous and homogeneous oxidation of nitrilotrismethylenephosphonate (NTMP) in the presence of oxygen and manganese (II, III). *J. Phys. Chem. B.* **106**, 6227–6233.
Nowack, B. and Stone, A.T. (2000) Degradation of Nitrilotris(methylenephosphonic acid) and related (amino)phosphonate chelating agents in the presence of manganese and molecular oxygen. *Environ. Sci. Technol.* **34**, 4759–4765.
Nowack, B. and Stone, A.T. (1999a) Adsorption of phosphonates onto the goethite-water interface. *J. Colloid Interface Sci.* **214**, 20–30.
Nowack, B. and Stone, A.T. (1999b) Influence of metals on the adsorption of phosphonates onto goethite. *Environ. Sci. Technol.* **33**, 3627–3633.
Nowack, B. (1998) Behavior of phosphonates in wastewater treatment plants of Switzerland. *Water Res.* **32**, 1271–1279.
Nowack, B. and Baumann, U. (1998) Biodegradation of the photolysis products of Fe(III)EDTA. *Acta Hydrochim. Hydrobiol.* **26**, 1–5.
Nowack, B., Xue, H.B. and Sigg, L. (1997) Influence of natural and anthropogenic ligands on metal transport during infiltration of river water to groundwater. *Environ. Sci. Technol.* **31**, 866–872.
Nowack, B. and Sigg, L. (1997) Dissolution of iron(III)hydroxides by metal–EDTA complexes. *Geochim. Cosmochim. Acta* **61**, 951–963.
Nowack, B. (1997) Determination of phosphonates in natural waters by ion-pair high performance liquid chromatography. *J. Chromatogr. A* **773**, 139–146.
Nowack, B. and Sigg, L. (1996) Adsorption of EDTA and metal–EDTA complexes to goethite. *J. Colloid Interface Sci.* **177**, 106–121.
Nowack, B., Kari, F.G., Hilger, S.U. and Sigg, L. (1996a) Determination of adsorbed and dissolved EDTA-species in natural waters. *Anal. Chem.* **68**, 561–566.
Nowack, B., Lützenkirchen, J., Behra, P. and Sigg, L. (1996b) Modeling the adsorption of metal-EDTA complexes onto oxides. *Environ. Sci. Technol.* **30**, 2397–2405.

Nowack, B. (1996) Behavior of EDTA in groundwater: a study of the surface reactions of metal–EDTA complexes. Diss ETH Nr. 11392, Zürich.
Oddo, J.E. and Tomson, M.B. (1990) The solubility and stoichiometry of calcium-diethylenetriamine-penta (methylenephosphonate) at 70°C in brine solution at 4.7 and 5.0 pH. *Appl. Geochem.* **5**, 527–532.
Pairat, R., Sumeath, C., Browning, F.H. and Fogler, H.S. (1997) Precipitation and dissolution of calcium-ATMP precipitates for the inhibition of scale formation in porous media. *Langmuir* **13**, 1791–1798.
Popov, K., Rönkkömäki, H. and Lajunen, L.H.J. (2001) Critical evaluation of stability constants of phosphonic acids. *Pure Appl. Chem.* **73**, 1641–1677.
Raschke, H., Rast, H.G., Kleinstück, R., Sicius, H. and Wischer, D. (1994) Utilization of 2-phosphonobutane-1,2,4-tricarboxylic acid by environmental isolates. *Chemosphere* **29**, 81–88.
Rawls, H.R., Bartels, T. and Arends, J. (1982) Binding of polyphosphonates at the water/hydroxyapatite interface. *J. Colloid Interface Sci.* **87**, 339–345.
Rizkalla, E.N. (1984) Metal chelates of phosphonate-containing ligands. *Rev. Inorg. Chem.* **5**, 223–304.
Sabin, F., Türk, T. and Vogler, A. (1992) Photo-oxidation of organic compounds in the presence of titanium dioxide: determination of the efficiency. *J. Photochem. Photobiol. A. Chem.* **63**, 99–106.
Samakaev, R.K., Dyatlova, N.M. and Dytyuk, L.T. (1984) The solubility in water of the nitrilotrimethylenephosphonates of the group II elements. *Russ. J. Inorg. Chem.* **29**, 1819–1820.
Schindler, P.W. (1990) Co-adsorption of metal ions and organic ligands: formation of ternary surface complexes. *Rev. Mineral.* **23**, 281–307.
Schöberl, P. and Huber, L. (1988) Ökologisch relevante Daten von nichttensidischen Inhaltsstoffen in Wasch- und Reinigungsmitteln. *Tenside Surfactants Detergents* **25**, 99–107.
Schowanek, D. and Verstraete, W. (1991) Hydrolysis and free radical mediated degradation of phosphonates. *J. Environ. Qual.* **20**, 769–776.
Schowanek, D. and Verstraete, W. (1990a) Phosphonate utilization by bacterial cultures and enrichments from environmental samples. *Appl. Environ. Microbiol.* **56**, 895–903.
Schowanek, D. and Verstraete, W. (1990b) Phosphonate utilization by bacteria in the presence of alternative phosphorous sources. *Biodegradation* **1**, 43–53.
Schroeder, K.A. and Hamm, R.E. (1964) Decomposition of the ethylenediaminetetraacetate complex of manganese(III). *Inorg. Chem.* **3**, 391–395.
Schwarzenbach, G., Ackermann, H. and Ruckstuhl, P. (1949) Neue Derivate der Iminodiessigsäure und ihre Erdalkalikomplexe. *Helv. Chim. Acta* **32**, 1175–1186.
Shamsi, S.A. and Danielson, N.D. (1995) Ribonucleotide electrolytes for capillary electrophoresis of polyphosphates and polyphosphonates with indirect photometric detection. *Anal. Chem.* **67**, 1845–1852.
Sigg, L. and Xue, H.B. (1994) Metal speciation: concepts, analysis and effects. In *Chemistry of Aquatic Systems: Local and Global Perspectives* (eds. Bidoglio, C. and Stumm, W.), ECSC, EEC, EAEC, Brussels and Luxembourg, pp. 153–181.
Sillanpää, M. (1997) Environmental fate of EDTA and DTPA. *Rev. Environ. Contam. Toxicol.* **152**, 85–111.
Stalikas, C.D. and Konidari, C.N. (2001) Analytical methods to determine phosphonic and amino acid group-containing pesticides. *J. Chromatogr. A* **907**, 1–19.
Steber, J. and Wierich, P. (1986) Properties of hydroxyethane diphosphonate affecting its environmental fate: degradability, sludge adsorption, mobility in soils, and bioconcentration. *Chemosphere* **15**, 929–945.
Steber, J. and Wierich, P. (1987) Properties of aminotris(methylenephosphonate) affecting its environmental fate: degradability, sludge adsorption, mobility in soils, and bioconcentration. *Chemosphere* **16**, 1323–1337.

Stumm, W. (1997) Reactivity at the mineral-water interface: dissolution and inhibition. *Colloids Surf. A* **120**, 143–166.
Svenson, A., Kaj, L. and Björndal, H. (1989) Aqueous photolysis of the iron(III) complexes of NTA, EDTA and DTPA. *Chemosphere* **18**, 1805–1808.
Ternan, N.G., McGrath, J.W., McMullan, G. and Quinn, J.P. (1998) Review: Organophosphonates: Occurrence, synthesis and biodegradation by microorganisms. *World J. Microbiol. Biotechnol.* **14**, 635–647.
Thompson, R., Grinberg, N., Perpall, H., Bicker, G. and Tway, P. (1994) Separation of organophosphonates by ion chromatography with indirect photometric detection. *J. Liquid Chromatogr.* **17**, 2511–2531.
Tomson, M., Kan, A.T., Oddo, J.E. and Gerbino, J. (1995) Solution and precipitation chemistry of phosphonate scale inhibitors. In *Mineral Scale Formation and Inhibition* (ed. Amjad, Z.), Plenum Press, New York, pp. 307–322.
Trechsel, U., Schenk, R., Bonjour, J.P., Russell, R.G.G. and Fleisch, H. (1977) Relation between bone mineralization, Ca absorption, and plasma Ca in phosphonate-treated rats. *Am. J. Physiol.* **232**, 298–305.
Tschäbunin, G., Fischer, P. and Schwedt, G. (1989a) On the analysis of polymethylenephosphonic acids. I. A systematic survey of the ion-chromatography of organophosphonic acids. *Fresenius J. Anal. Chem.* **333**, 111–116.
Tschäbunin, G., Fischer, P. and Schwedt, G. (1989b) On the analysis of polymethylenephosphonic acids. II. A systematic survey of the post-column derivatization in ion chromatography. *Fresenius J. Anal. Chem.* **333**, 117–122.
Tschäbunin, G., Fischer, P. and Schwedt, G. (1989c) On the analysis of polymethylenephosphonic acids. III. Stability of ethylenediaminetetramethylenephosphonic acid (EDTMP) in aqueous solution. *Fresenius J. Anal. Chem.* **333**, 123–128.
Vaeth, E., Sladek, P. and Kenar, K. (1987) Ion chromatography of polyphosphates and phosphonates. *Fresenius J. Anal. Chem.* **329**, 584–589.
Van Hullebusch, E., Chazal, P.M. and Deluchat, V. (2002) Influence of phosphonic acids and EDTA on bacterial copper toxicity. *Toxicol. Environ. Chem.* **82**, 75–91.
Venetzky, D.L. and Moniz, W.B. (1969) Nuclear magnetic resonance study of the thermal decomposition of ethylenedinitrilotetraacetic acid and its salts in aqueous solutions. *Anal. Chem.* **41**, 11–16.
Von Baeyer, H. and Hofmann, K.A. (1897) Acetodiphosphorige Säure. *Ber. d. D. Chem. Ges.* **30**, 1973–1978.
Waldhoff, H. and Sladek, P. (1985) Qualitative and quantitative determination of phosphonic acids with an autoanalyzer system. *Fresenuis J. Anal. Chem.* **320**, 163–168.
Wanner, B.L. (1994) Molecular genetics of carbon–phosphorous bond cleavage in bacteria. *Biodegradation* **5**, 175–184.
Weiss, J. and Hägele, G. (1987) Ion-chromatographic analysis of inorganic and organic complexing agents. *Fresenius J. Anal. Chem.* **328**, 46–50.
Wolf, K. and Gilbert, P.A. (1992) EDTA – Ethylenediaminetetraacetic acid. In *The Handbook of Environmental Chemistry* (ed. Hutzinger, O.) **3**, Part F, pp. 243–259.
Wu, F.C. and Tanoue, E. (2001) Isolation and partial characterization of dissolved copper-complexing ligands in streamwaters. *Environ. Sci. Technol.* **35**, 3646–3652.
Xue, H.B., Jansen, S., Prasch, A. and Sigg, L. (2001) Nickel speciation and complexation kinetics in freshwater by ligand exchange and DPCSV. *Environ. Sci. Technol.* **35**, 539–546.
Xue, H., Oestreich, A. and Sigg, L. (1996) Free cupric ion concentrations and Cu complexation in selected Swiss lake and rivers. *Aquat. Sci.* **58**, 69.
Xue, H.B. and Sigg, L. (1994) Zinc speciation in lake waters and its determination by ligand exchange with EDTA and differential pulse anodic stripping voltammetry. *Anal. Chim. Acta* **284**, 505–515.
Xue, H.B. and Sigg, L. (1993) Free cupric ion concentration and Cu(II) speciation in a eutrophic lake. *Limnol. Oceanogr.* **38**, 1200–1213.

Xyla, A.G., Mikroyannidis, J. and Koutsoukos, P.G. (1992) The inhibition of calcium carbonate precipitation in aqueous media by organophosphorus compounds. *J. Colloid Interface Sci.* **153**, 537–551.

Zholnin, A.V., Zudov, V.G., Leitsin, V.A., Nosova, R.L. and Smirnov, E.M. (1990) Preparation of the iron complexes of nitrilotrimethylenephosphonic acid from zinc-production waste. *J. Appl. Chem. USSR* **63**, 803–807.

Zieba, A., Sethuraman, G., Perez, F., Nancollas, G.H. and Cameron, D. (1996) Influence of organic phosphonates on hydroxyapatite crystal growth kinetics. *Langmuir* **12**, 2853–2858.

8

Phosphate pollution: a global overview of the problem

A.M. Farmer

8.1   INTRODUCTION

This chapter will provide an overview of phosphate pollution problems in different regions of the world. It will begin by focusing on the European Union, considering both waste water discharges and agricultural inputs. It will then look more widely at issues in North America, Asia and Australia and comment on other regions of the world. While the chapter will provide some basic assessment of pollution problems in waters and phosphate pollution sources, it will also, where possible, comment upon policy developments affecting these issues, not least to provide some guide to future developments.

The pollution of surface waters by phosphates from a wide variety of sources is a serious problem in many parts of the world. Phosphorus is a plant nutrient and excess quantities of it (alone or together with an excess of another main plant nutrient, nitrogen) can lead to extensive growth of phytoplankton, macroalgae and higher plants. In some cases, and along with other factors such as reduced flow rates resulting from dams or weirs or changes in the balance of food webs, this

© 2004 IWA Publishing. *Phosphorus in Environmental Technology: Principles and Applications.*
Edited by Eugenia Valsami-Jones. ISBN: 1 84339 001 9

can adversely affect invertebrate and fish populations and even result in the production of toxic substances from algal blooms (Farmer, 1997).

Plants require both phosphorus and nitrogen to grow and one or other of these nutrients is usually the limiting factor preventing excess growth. In freshwaters this tends to be phosphorus, while in many marine waters nitrogen is often limiting. Thus prevention of eutrophication symptoms in freshwaters usually emphasises targeted action against phosphorus and, for marine waters, action against nitrogen. However, there are instances where this pattern is different and action may often need to be targeted at both pollutants. Obtaining comparative data on the extent of the problem at a global level can be difficult. However, Table 8.1 provides an overview of the extent of eutrophication problems in standing freshwaters around the world. This demonstrates clearly that eutrophication is both widespread and significant.

In natural, unimpacted freshwaters total phosphorus concentrations are usually below 25 mg P/l. Unless otherwise demonstrated, it is generally assumed that concentrations above 50 mg P/l are the result of anthropogenic influences. Elevated phosphorus concentrations in surface waters can sometimes be of natural origin (bedrock), but are often the result of soil erosion, agricultural run-off and discharges of municipal and industrial waste waters. Sewage, for example, contains phosphates from human sources (about 2 g P/person/day), detergents, food waste, food additives and other products. In this case the quantities of phosphate discharged to receiving waters depend upon population size and levels of waste

Table 8.1 Percentage of lakes and reservoirs with eutrophication problems in different regions of the world.

Region	Percentage
Africa	28
Europe	53
North America	48
South America	41
South East Asia	54

Source: UNEP (1994).

Table 8.2 Total quantities of phosphorus fertiliser used in different regions around the world in 2000.

Continent	Phosphorus fertiliser use (Mt)
Africa	954,921
Asia	17,686,320
Europe	4,115,326
Latin America and Caribbean	3,912,686
North America	4,738,719
Oceania	1,562,006

Source: UNEP (2000).

water treatment. Agricultural inputs will depend upon tillage practices, crop types, climate, etc. However, a major element is the quantity of phosphate-based fertiliser used. This varies significantly around the world (Table 8.2), reflecting both population levels and the relative wealth of the economies.

8.2 THE EUROPEAN UNION

There are marked trends in phosphate concentrations in surface fresh waters across Europe (EEA, 1994, 1998). The lowest concentrations of phosphate are found in the streams and rivers of the Nordic countries. Here 91% of the monitoring stations report mean concentrations below 30 mg P/l and 50% below 4 mg P/l. This pattern results from a generally very low human population density and the presence of a slow-weathering and nutrient-poor bedrock. Higher phosphorus concentrations are found in a band stretching across central Europe, from southern England to Romania. While some monitoring sites in southern Europe also report high phosphate levels, the general pattern is for lower concentrations than central and eastern Europe, often because much of the municipal waste water produced is discharged directly to the sea.

A survey of rivers across Europe (EEA, 1998) found that a large proportion of 1000 monitoring stations had levels of total phosphate exceeding 50 mg P/l, many by a considerable margin. Indeed only about 10% of the monitoring stations reported mean total phosphorus concentrations below 50 mg P/l. Tables 8.3 and 8.4 provide recent data on phosphorus levels in rivers in selected EU Member States. A similar pattern of phosphate levels in lakes can also be found, with low concentrations in the Nordic and Alpine regions. However, in many parts of Europe lakes can be severely affected by phosphate eutrophication. This can be a particularly difficult problem to address as phosphates may accumulate in lake sediments. Even when the phosphate pollution source is controlled, these accumulated phosphates may be released (e.g. by climatic events or the action of fish) into the water column and maintain a eutrophic condition. Overcoming this problem requires expensive manipulation of the sediments and lake biota.

Table 8.3   Mean annual average orthophosphate concentrations (μg P/l) at aggregated river stations (number given in brackets) between 1990 and 1998 in selected western European countries.

Country	1990	1991	1992	1993	1994	1995	1996	1997	1998
Denmark (30)	133	106	93	79	84	78	87	75	71
Finland (52)	12	12	10	12	11	12	11	11	11
France (254)	141	121	105	102	84	95	100	91	78
Germany (89)	155	120	97	89	68	68	76	73	73
UK (89)	109	87	74	63	60	74	79	83	68

Source: EEA (1998).

Table 8.4 Number of aggregated river stations for each EU Member State, for the most recent year, and their total phosphorus (μg/l) concentration distribution.

Country	Date of most recent year	<25	25 to <50	50 to <125	125 to <250	250 to <500	>500
Austria	1998	56	53	93	28	7	3
Belgium	1995	0	0	0	1	4	7
Denmark	1998	0	1	15	17	3	1
Finland	1998	87	25	34	11	2	0
France	1998	1	9	100	181	81	35
Germany	1998	4	4	27	65	32	3
Greece	1998	10	4	4	8	6	11
Italy	1992	4	3	6	4	3	3
Netherlands	1998	0	0	7	18	8	2
Spain	1996	7	10	20	16	6	4
Sweden	1997	37	19	20	4	0	0
UK	1998	0	3	9	1	1	1

Source: EEA (2001).

Phosphate levels have shown a general decline across Europe during the late 1980s and 1990s. This trend occurs for annual average concentrations for both total and dissolved phosphorus, but it appears that high peak concentrations still occur even where improvements are being made. Improvements have been found in most regions and much of this is due to increased treatment of urban and industrial waste water and some decline in agricultural fertiliser use. There has also been an improvement in lake water phosphate concentrations, although many lakes are still seriously affected.

Phosphate concentrations in the marine environment are more variable, with high levels being found in the North Sea, Mediterranean, Baltic and Black Seas. Phosphate discharges have continued to increase in, for example, the North Sea, although in other areas they have largely remained stable.

8.2.1 Phosphorus pollution from municipal waste water

Phosphate pollution arises from a number of sources. Point sources (e.g. from municipal waste water treatment plants, industry and some agricultural) account for more than half of the phosphates discharged in Europe. The principle sources of phosphate in municipal waste water come from human waste and, in some countries, from the use of detergents. Emissions of phosphates from sewage into surface waters, across Europe, have typically fallen by about 30–60% since the mid-1980s (EEA, 1998). However, this decline has varied considerably across the continent. For example, declines in emissions from the Netherlands and Denmark have been of the order of 70–90%, largely as a result of highly improved waste water treatment.

Sewage treatment works collect waste water and may provide various levels of treatment, from simple removal of solids to complex treatment removing nutrients, etc. The changes in pollution from these sources is best viewed in relation to the objectives of the EU's urban waste water treatment Directive (Farmer, 2001). For nutrients this, inter alia, required nutrient removal by the end of 1998, for discharges to waters that are eutrophic or that may become eutrophic. Member States had two options. They could either designate individual waters as 'sensitive', in which case waste water discharged from sewage treatment works serving a population equivalent (pe – a standard measure of organic input) of more than 10,000 would require either 80% phosphorus removal and/or 70% nitrogen removal (depending on the potential impact) or a combined phosphorus and nitrogen removal of 75%. Alternatively, Member States could designate their entire territories as sensitive and meet a 75% reduction for both parameters for all waste water treatment plants. In November 2001, the European Commission published a short report detailing compliance with these requirements (COM(2001)685). The main findings of the European Commission demonstrated a wide variation in compliance with the Directive.

A number of Member States had either designated their whole territory as sensitive (e.g. Denmark, the Netherlands and Sweden), or most of the territory (e.g. Germany). In these countries action on phosphorus removal was largely complete (with occasional exceptions).

In contrast, some other Member States were considered to have undertaken very limited sensitive area designations (e.g. France, Ireland and Spain – see also Farmer et al., 2000). The Commission argued that further sensitive areas should be designated. Some designations were late (e.g. Greece and, partially, the UK), which has delayed implementing the Directive. Italy, which also had very limited designations, responded by producing a list of 187 new sensitive areas, none of which had agglomerations of more than 10,000 pe – a rather pointless exercise when significant eutrophication problems are occurring in the country. Overall, these delays (deliberate or otherwise) have postponed investment and restricted the scope for environmental improvements.

In conclusion, action to reduce phosphorus pollution is highly variable across the Member States. Some, especially in northern and central Europe, have taken significant action, with little additional phosphorus removal required. However, major additional investment will be required in southern Member States in particular, if phosphate pollution is to be controlled. However, it is also becoming apparent that the basis for the Commission's assessment is also being brought into question. For example, the assertion that the Humber Estuary in the UK should be a sensitive area is being challenged on the basis that the turbidity of the water prevents algal blooms and, therefore, the consequences of nutrient discharges.

The Directive is also now being applied across the Candidate Countries (mostly in central and eastern Europe), ten of which will (subject to referenda) join the EU in May 2004. A number of transition periods have been agreed with many of these countries on aspects of the Directive. Significant investments here will also result in major reductions in phosphorus discharges. However, the extent and timing will only become apparent in the next few years as investments are made and the European Commission assesses the degree of compliance.

8.2.2 Agricultural diffuse pollution

Diffuse phosphorus sources are also a significant problem in Europe. The degree of significance of such sources is highly variable across the continent. Control of such pollution is also related to other types of agricultural diffuse pollution (e.g. nitrates and pesticides) and coloured by the relative lack of EU wide measures on agricultural phosphorus pollution compared to point sources. It is, therefore, useful to examine this problem by considering four example Member States.

Denmark: 62% of the total area of Denmark is agricultural land. There are sizeable areas of cereals, particularly, barley grown for livestock, a substantial dairy herd and highly successful pig sector. Denmark has had extensive problems with agricultural nutrient problems in surface and ground waters. However, there has been some decline in agricultural pressures with a decline in total fertiliser consumption of 36% and in potash and phosphorus by 62% since the early 1980s. The trend has been facilitated by a reduction in the agricultural area of 13.5% between 1983/4 and 1998/9. The Danish government has achieved this through the adoption of several national reduction plans, which have been implemented in four–five years phases. They have become more stringent and progressive over the last ten years. However, early plans have focused on nitrates, although consideration is also being given to the development of a national phosphate reduction plan.

France: France is the EU's largest agricultural producer, responsible for one-quarter of the EU's total farm output; agriculture occupies approximately 55% of the land area. One third of farmland is given over to grazing (chiefly for cattle) and the other two thirds to arable crops. Intensification and modernisation have environmental impacts, which vary depending on the region and the type of farming operation. For example, livestock production has become a major source of water pollution affecting many areas of France, but is best epitomised in Britanny.

The government response to these concerns has not been as pro-active as in many other countries. The main approach has been to use voluntary national initiatives. One of the best known, called Ferti Mieux, is an education programme set up to reduce nitrate and phosphate pollution in 1991. Another, 'The Agricultural Pollution Control Programme' was set up in 1993 to contribute to the control of nutrient leaching from livestock housing.

Germany: Agriculture accounts for about 50% of land use in Germany. Arable and permanent crops cover two thirds of the total farmed area with the remainder dedicated to grassland. Specialised dairy farming is very important in grassland areas of the north German lowlands, in the central and southern uplands and in the Alps. Most of the eastern Länder are dominated by large scale arable farming. The most fertile regions in the western Länder (East Schleswig-Holstein) also tend to specialise in arable farming. Very intensive, specialised livestock production is found in coastal areas of the northwest. Mixed farms, however, are still an important part of the farming sector, especially in central and southern Germany.

Around 2% of Germany's agricultural land is farmed organically, which makes it one of the largest organic sectors in the EU.

German agriculture is not as intensive or industrialised as in, for example, the Netherlands but there are regions with very high input levels and livestock densities, particularly in the north. Reports from the German Joint Commission on Water Research indicate that phosphate pollution of many freshwater bodies has historically been at a relatively high level. The Federal Environment Agency (Schulz, 1999) records that between 1993 and 1997 agricultural run-off and ground water phosphate emissions into surface waters accounted for 66% of total point and diffuse phosphorus emissions.

There is a substantial body of national legislation concerned with input use, good agricultural practice and pollution control but it is notably less stringent than in Denmark. Much of the legislation is enacted at the federal and Länder level (e.g., water legislation and the German 'special agricultural legislation'), and implemented at Land level. German 'special agricultural legislation' constitutes a comprehensive legal framework for all Länder. It applies to all agricultural practices and covers environmentally-relevant regulations on the application of fertilisers (including inorganic fertilisers, sewage sludge and bio-wastes).

The Netherlands: The Netherlands is one of the most intensively farmed areas in Europe and despite its small land size, is ranked the third largest agricultural exporter in the world, after the US and France. The main areas of production are horticulture, pigs and poultry and dairy farming. Livestock farming is concentrated in the south and east of the country and accounts for about 60% of the Netherlands' agricultural production. Arable farming is concentrated in the north and southeastern regions of the country and is mostly given to the production of cereals, animal fodder, sugar beet, vegetables and flowers.

Due to the intensiveness of agricultural practice, and particularly livestock production, the Netherlands has suffered some of the worst pollution levels in Europe. The Dutch government has historically failed to take action in proportion to the scale of the problem and is under considerable pressure to curb the environmental impacts of intensive farming. The most significant concern has been the high nitrate and phosphate concentrations in freshwaters, which have been held attributable to the vast quantities of manure produced from intensive livestock farming. It has been estimated that the Netherlands produces more manure per hectare than in any other country in Europe, nearly 5 times the EU average.

This manure and slurry is spread on the fields and leaches into freshwater systems and some soils are saturated with phosphates. It is estimated that agriculture's overall contribution to surface water pollution (from the application of organic and inorganic mineral inputs) amounts to 40% of total phosphate pollution. A family of measures has been introduced to reduce water pollution, including an increasingly strict manure policy, stringent fertiliser accounting systems and a 20% cut to the country's pig population.

A number of studies have been undertaken to assess the proportion of phosphorus inputs to rivers in Europe that is derived from non-point sources. It is not

Table 8.5 Percentage contribution of non-point sources of phosphorus to different European rivers (from Macleod and Haygarth, 2003).

River	Year of study	Percentage of phosphorus derived from non-point sources
Po, Italy	Early 1990s	22–25
Rhine	Early 1990s	13–21
Elbe	Early 1990s	11–16
Rhine	1993–1997	42
Elbe	1993–1997	44
Danube	1996	44
Frome, UK	1998	60
Thames, UK	1996	15
Thames, UK	1999	36–53
Krka, Slovenia	1996–1997	41
Kennet, UK	1997	2
Kennet, UK	1998–1999	29–45
Avon, UK	2000–2001	24

possible to detail all of these in this chapter and the following case study of the Rhine provides just one example. However, Macleod and Haygarth (2003) have summarised the current state of knowledge (Table 8.5). The results show considerable variability across Europe and changes over time. For example, the percentage contribution from non-point sources has risen in the Elbe during the 1990s, largely as a result of a reduction in point source inputs. This highlights the growing need to adopt effective policies to tackle non-point sources.

8.3 THE RIVER RHINE: CASE STUDY

While it is useful to consider developments in point and diffuse phosphorus sources by individual countries, it is also beneficial to view these developments in the context of an individual river basin. This section will consider changes to phosphorus sources and concentrations in the River Rhine, one of the most important river basins in Europe, characterised by a long history of elevated population density, industrial production and agricultural activity.

The International Commission for the Protection of the Rhine has undertaken two full surveys of phosphorus inputs in the Rhine catchment in 1985 and 1996. The sources are divided into two general source types – point sources (such as a waste water treatment works, industrial discharges) and diffuse sources (such as agricultural and other land surface run off and drainage).

The data show (Table 8.6) that, in 1985, point sources accounted for about 75% of the total phosphorus input to the Rhine, with urban (i.e. sewage) discharges being about twice that of industrial inputs. By 1996, the relative importance of urban and industrial sources remained similar. However, by this time the relative contribution of point and diffuse sources was roughly equal. Between 1985 and 1996

Table 8.6 Sources of total phosphorus in 1985 and 1996 to the Rhine. All values are in tonnes per year.

Country	Diffuse		Domestic point		Industrial point		Natural		Total	
	1985	1996	1985	1996	1985	1996	1985	1996	1985	1996
Switzerland	448	449	2,300	900	150	35	98	138	2,996	1,522
Germany	8,987	6,452	25,970	4,925	3,370	590	625	605	38,952	12,572
France	2,190	1,527	3,520	830	1,280	410	108	108	7,098	2,875
Netherlands	5,430	4,229	6,749	2,071	11,989	3,000	524	524	24,692	9,824
Total	17,055	12,657	38,539	8,726	16,789	4,035	1,355	1,375	73,738	26,793

Source: Farmer and Braun (2003).

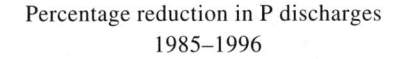

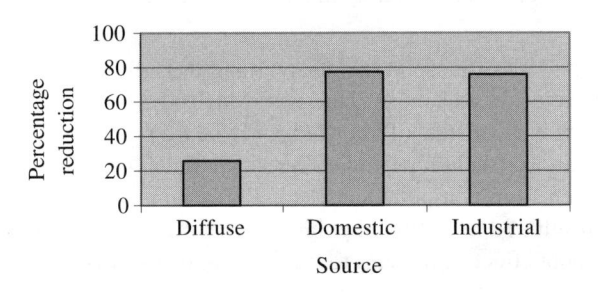

Figure 8.1 Percentage reduction in phosphorus discharges from anthropogenic sources to the Rhine, by source, between 1985 and 1996 (from Farmer and Braun, 2003).

the total input of phosphorus from human activity reduced from 72,400 t P/a to about 25,400 t P/a. This is a reduction of about 65% and was well above the target in the Rhine Action Programme of a 50% reduction by 1995. The decline was driven overwhelmingly by a 77% reduction from urban point sources and a 76% reduction from industrial point sources. Diffuse source inputs were reduced by 59%. These changes are illustrated in Figure 8.1.

In 1985 more than a third of all phosphorus inputs to the Rhine arose from urban pollution from Germany. The 81% decline in this source by 1996 is highly important in driving improvement in the river. Early investment in phosphorus removal was driven by a domestic political agenda (both to improve conditions in the Rhine and also for other water bodies). In later years the requirements of the EC 1991 urban waste water treatment Directive (see above) were also important. The 1998 deadline in this Directive suggests that improvements in point sources have continued beyond 1996.

Water quality has been monitored for many years at different locations along the length of the Rhine. Phosphorus concentrations have shown a dramatic decline

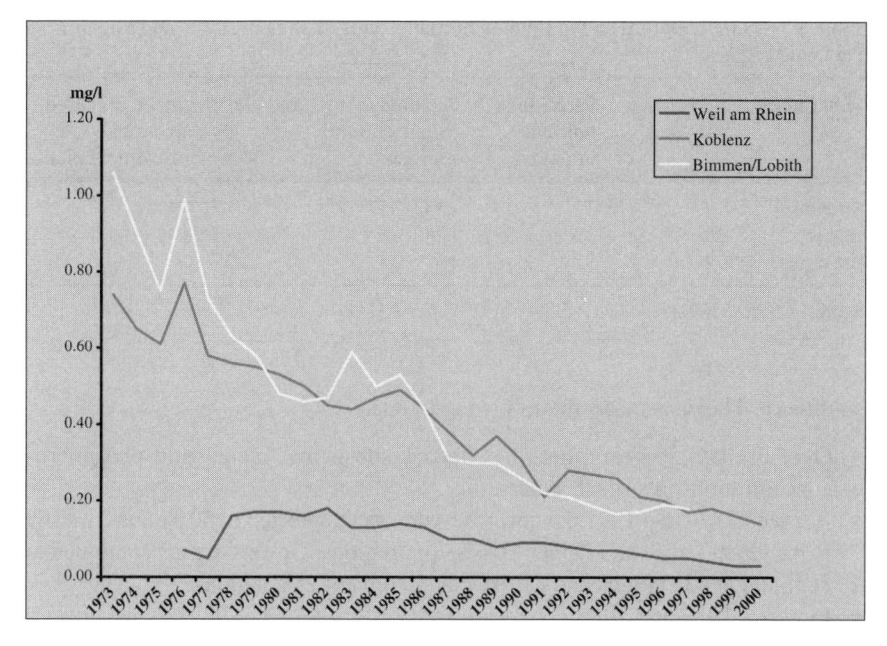

Figure 8.2 Phosphorus concentrations at three locations along the Rhine 1971–2000. The locations are Bimmen/Liboth (upstream), Koblenz (mid-Rhine) and Weil am Rhein (downstream).

(Figure 8.2), so that the target concentration has been met along much of the river. This trend has occurred since the mid-1970s in the mid/lower monitoring stations and since the early 1980s in the upper monitoring station (where phosphorus levels were already nearly at the target concentration). Phosphorus concentrations in the river have responded relatively well to changes in inputs, and the significant decline in discharges (see above) is reflected in the improved river water concentrations.

In conclusion the data on river phosphorus concentrations shows a clear response to the reduction in phosphorus sources to the catchment. This has been driven by reductions in industrial and domestic phosphorus discharges, although there has also been a decline in agricultural inputs. While river water quality objectives for phosphorus have largely been achieved, it is likely that further reductions, particularly in agricultural inputs, will be required in order to assist in achieving environmental objectives for the North Sea, to which the Rhine discharges.

8.4 UNITED STATES

8.4.1 Water quality

Nation-wide information on phosphorus concentrations in waters is incomplete. Litke (1999) reviews current information and while there has been an increase in monitoring stations across the country, there is still poor representation in the

Table 8.7 Summary of data for 1996 on nutrient status of streams, lakes and estuaries in the United States.

Water body	Percentage assessed	Percentage nutrient impaired	Percentage impaired by agricultural sources	Percentage impaired by municipal waste water treatment plants
Streams	19	14	25	5
Lakes	20	20	19	7
Estuaries	72	22	10	17

Source: USEPA (1997).

south-east. The data shows the following trends:

- Over the last 50 years there has been a substantial increase in phosphorus concentrations in surface waters.
- Between 1974 and 1981 downwards trends were detected in 50 stations, mostly in the Great Lakes and Upper Mississippi regions. Upward trends continued to be found at 43 stations and no detectable trend at 288 stations.
- Between 1982 and 1989 downward trends were found at 92 stations, upward trends at 19 stations and no trends at 299 stations.

However, while trends have important policy implications, the degree to which concentrations exceed ecological objectives drives practical decision-making. The USEPA has a recommended total phosphorus concentration of 0.1 mg/l. An examination of 410 stations between 1982 and 1989 showed that the number where the annual average total phosphorus concentrations exceeded this recommendation decreased from 54% to 42%. Data for 1990–1995 indicated that 32% of hydrological units had more than 50% of their observed total phosphorus concentrations exceeding this recommendation. An additional 44% of hydrological units had from 10–50% of observed concentrations exceeding the recommendation. These data clearly indicate that there is significant potential for adverse ecological consequences from existing phosphorus concentrations in surface waters across the United States. In the early 1970s the USEPA began its National Eutrophication Survey of streams and lakes. The results indicated that 10–20% of all lakes and reservoirs in the US were eutrophic and that phosphorus was the limiting nutrient in 67% of lakes studied. The survey has since been expanded and data for 1996 are given in Table 8.7.

8.4.2 Point source controls

Significant effort of the control of point source pollution to waters in the USA stems from the Federal Water Pollution Control Act. Between 1972 and 1991 this resulted in a spend of $354 billion on the construction and operation of waste water treatment plants. Much of this effort has been to upgrade existing facilities, e.g. from primary treatment to tertiary treatment.

Table 8.8 Summary of facilities discharging to water with phosphorus requirements derived from the USEPA Permit Compliance System.

Facility types	Total number of facilities	Number required to monitor phosphorus	Facilities with phosphorus concentration limits
Municipal	15,939	2,437	1,163
Industrial	50,599	2,379	877
Federal	1,119	110	50
Other	5,087	142	26
Total	72,744	5,068	2,116

Source: Litke (1999).

Table 8.8 summarises discharge information from the USEPA's permit monitoring database. This includes about 16,000 municipal waste water treatment plants, serving around 190 million people (70% of the population). Of these only 7.3% have phosphorus concentrations limits on their discharges. There is also a very uneven distribution in the location of these plants across the country. The largest concentration of plants with strict phosphorus limits is in those States on the Great Lakes and parts of the East Coast. Very few such plants occur west of the Mississippi.

A further means to examine changes in phosphorus levels in waters is to consider overall loading. These data are available for selected locations. For example, annual municipal phosphorus loads to Lake Erie have been reduced from 14,000 T in 1972 to 2000 T in 1990 and total point-source loads to Chesapeake Bay have been reduced from 5100 T in 1985 to 2500 T in 1996.

Phosphorus pollution, therefore, remains of concern in many parts of the United States. Significant improvements in discharges and surface water quality have taken place in recent years, indicating that the correct policy framework can deliver the necessary environmental objectives. However, these are not equally applied across all States and further action will be necessary. In particular, action is needed on agricultural phosphorus discharges in a number of locations, which require a more localised, integrated approach.

8.5 AUSTRALIA

Eutrophication of waters in Australia is important, but localised. This results in a range of problems, from the growth of sub-tropical invasive weeds (e.g. the water hyacynth) to toxic algal blooms. For example, between 1988 and 1994, 84% of algal blooms in Victoria were blue-green algae, with 75% of blooms containing potentially toxic species.

Point sources contribute between 5 and 35% of the total nutrients entering waterways in Australia, depending upon location. However, proportionately, their impact can be greater than this suggests, as their input is often continuous, can be concentrated, and diffuse source inputs may be minimal during dry weather when

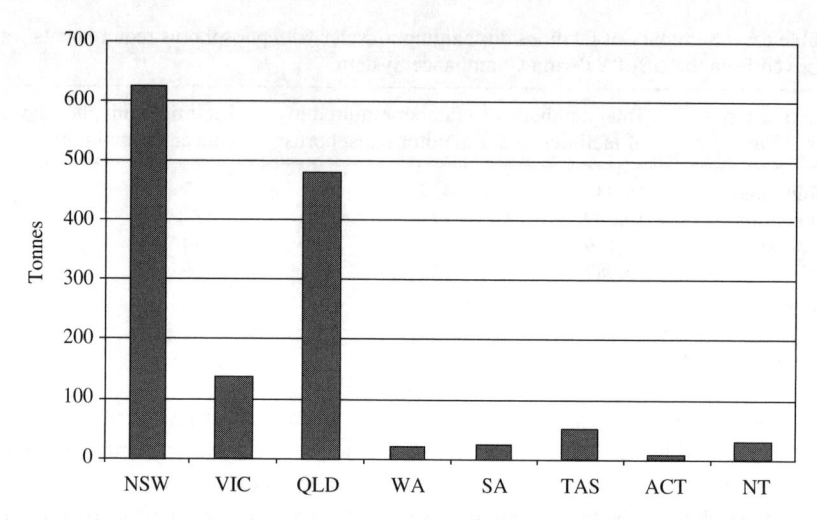

Figure 8.3 Discharge of phosphorus from inland sewage treatment plants in 2000. NSW (New South Wales), VIC (Victoria), QLD (Queensland), WA (Western Australia), SA (South Australia), TAS (Tasmania), ACT (Australian Capital Territory), NT (Northern Territory). Source: Environment Australia (2001).

low river flows may concentrate point source contribution. Diffuse source input includes inputs from the use of phosphorus-based fertilisers. However, the largest diffuse source is from soil erosion, made particularly acute in Australia by the specific climatic conditions found there, combined with extensive grazing of vegetation along the banks of water courses. Thus, approximately 1.6 billion tonnes of soil are lost annually to erosion, 60% of this occurring in native pasture lands (Ball, 2001).

Comprehensive data on point source phosphorus inputs is incomplete, but Figure 8.3 provides data for 2000 for municipal sewage treatment works. The relative input reflects population size, with New South Wales discharging the highest quantity of phosphorus. However, levels of treatment vary significantly. Thus while Queensland and Victoria discharge similar volumes of waste water, plants in Queensland discharge approximately three times the total quantity of phosphorus. In Victoria, about 75% of waste water treatment plants have tertiary phosphorus removal, but only 5% of plants in Queensland remove phosphorus (Ball, 2001). As a result, the four river systems receiving greater than 30 tonnes of phosphorus per year are the Murrumbidgee, Hawkesbury-Nepean, Namoi and Hunter catchments.

8.6 JAPAN

Japan has had a long history of eutrophication problems, particularly in lakes and in enclosed coastal areas, most famously the Inland Seto Sea.

Table 8.9 Total phosphorus concentrations in selected Japanese lakes and reservoirs in 1994 and target environmental quality standards (most to be met by 2000).

Lake	Sub-basin	1994 total P concentration	Target total P concentration
Kamahusa Dam Reservoir		0.017	0.013
Lake Kasumigaura	Nishiura	0.14	0.10
	Kitaura	0.086	0.086
	Hitachi-Tonegawa	0.090	0.078
Lake Inbanuma		0.15	0.098
Lake Teganuma		0.49	0.37
Lake Suwa		0.11	0.072
Lake Nojiri		0.004	0.005
Lake Biwa	Northern Lake	0.008	–
	Southern Lake	0.018	0.015
Lake Nakaumi		0.10	0.069
Lake Shinji		0.053	0.040
Lake Kojima		0.21	0.17

Source: Ministry of Environment (2001).

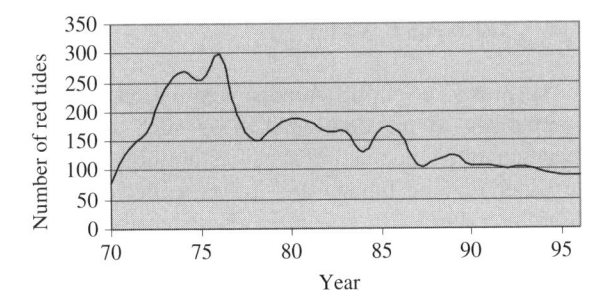

Figure 8.4 Number of red tides since 1970 in the Inland Seto Sea.

Various policy measures have been adopted to counter eutrophication. In the early 1980s environmental quality standards were established for phosphorus for the first time and these have been progressively applied to individual water bodies by the Director-General of the Environment Agency and by prefectural governors. These allow for planning of pollution control measures by comparing actual quality with the objective environmental quality standard. Examples are given in Table 8.9.

The Inland Seto Sea has been a particular source of concern. This area is semi-enclosed by the southern Islands of Japan and is surrounded by areas of high population density and agricultural activity. In the early 1970s eutrophication of the Sea began to result in an increase in the number of confirmed red tides (a specific type of algal bloom). These increased dramatically by the end of the decade, but have slowly declined as nutrient control measures have been put in place (Figure 8.4). The first general law to protect the Sea was enacted in 1973 and revised in 1978.

Table 8.10 Use of fertilisers in South China Sea countries.

Country	Padi field area (thousands of hectares)	Fertiliser use (t/y)
Cambodia	1,800	>40,000
China	3,400	3,640,000
Indonesia	5,000	>5,600,000
Philippines	1,200	181,000
Thailand	8,600	n.d.
Viet Nam	1,500	110,000

Source: Sien and Kirkman (2000).
n.d.: not determined.

Much of the phosphorus control focus has been on setting emission limits for waste water treatment. Uniform emission limits have been in place since 1985 and such limits have been applied for phosphorus for 1066 lakes and reservoirs. The use of emission limits for phosphorus for coastal waters was established in 1993. These are in place for 88 designated coastal areas, including Tokyo Bay, Ise Bay and Osaka Bay, which have had extensive summer algal blooms.

8.7 SOUTH AND EAST ASIA

South and East Asia present a very different context for the control of phosphorus pollution in comparison to the developed areas such as the EU, USA and Australia. In many countries population densities are high, but the collection of domestic waste water may be limited and, where collected, treatment is often basic at best. Agricultural production is often also intensive. However, the use of fertilisers varies significantly, depending upon the ability of farmers to purchase them. It is also important to note that rapid economic growth in a number of countries in this region is causing significant changes in all sectors, including growing pollution control measures.

Over the period of 1961 to 1997 the use of fertilisers in East and South East Asia rose by an average of 8.9% per annum. The average application rate is 147 kg/ha, compared with the developing world average of 90 kg/ha. However, this higher rate is due to the large use in China, as the average for the rest of the region is 93 kg/ha, with China consuming 73% of all fertilisers in the region. The rate of increase of application is expected to slow by 2030, resulting in an average consumption of 180 kg/ha in China and 106 kg/ha for the rest of the region (Dixon et al., 2001). Table 8.10 provides comparative information on fertiliser use in rice growing areas in countries of the region.

Across South Asia, the use of fertilisers has grown significantly. In 1970 the average input was 3 kg/ha, but this had risen to 79 kg/ha by the mid-1990s (Dixon et al., 2001). In India there are pressures on farmers in relation to the financing of fertiliser use, since the government deregulated the prices of phosphorus fertilisers. As a result, there is a continuous reduction in phosphorus levels in soils. This is

Table 8.11 Population and estimated BOD generation and removal in selected South China Sea countries.

Country	Population (thousand persons)	Percentage of population in cities	Population growth rate (%)	BOD generated (1000 t/y)	BOD removed by sewage treatment (1000 t/y)
Cambodia	1,985	89	2.7	36.2	No treatment
China	59,694	35	1.6	1,089.4	<109
Indonesia	105,217	48	2.9	1,920.2	364
Malaysia	10,336	15	3.3	188.6	53
Philippines	23,633	27	2.1	431.3	149
Thailand	37,142	?	1.4	677.8	89
Viet Nam	75,124	3	1.6	1,371.0	No treatment
Total	313,131	>27	1.4	5,714.5	655

Source: Sien and Kirkman (2000).

reducing long-term productivity as well as reducing some diffuse phosphorus sources to water.

Accurate data on point source phosphorus discharges in the region is difficult to obtain. In effect it is necessary to consider surrogate measures such as overall population levels and biological oxygen demand (BOD) in waste water discharges. Table 8.11 provides data for selected countries for areas discharging to the South China Sea. It can be seen that for most countries waste water treatment removes very little of the BOD. It can be assumed, therefore, that almost no phosphorus is removed through treatment. The table also indicates the current population growth rate for the area, which suggests that the pressure on surface waters from phosphorus in waste water discharges will increase significantly in the immediate future.

8.8 AFRICA

Eutrophication of surface waters in Africa is localised, but important in some areas. Although overall fertiliser use is low compared to other regions of the world (see above), agricultural phosphorus pollution can be important in some locations. More importantly the level of waste water collection and treatment is very poor. This means that catchments and coastal areas with large cities are particularly vulnerable to phosphorus pollution. There has been no comprehensive study of the problem in the continent. However, examples include (UNEP, 1998):

- In Mozambique, only the city of Maputo has a sewage collection system and, of this, only 50% receives any form of treatment.
- Eutrophication resulting from sewage discharge from Zanzibar has been identified as a cause of damage to coral reefs.
- In Tanzania, intensive use of fertilisers for rice production might be a cause of pollution in the Rufiji Delta.

- The use of septic tanks might prove an option in areas where collection is problematic, but ancestral customs in Madagascar prevent the storage of human excreta.

8.9 THE ANTARCTIC

Although phosphorus pollution in this continent is of a small scale compared to other parts of the world, it is interesting to note that even the Antarctic is not immune to the problem. There are a limited number of lakes in the coastal areas of Antarctica. A number of these are of high ecological importance, especially within the context of the overall value of the continent in global terms. Even the limited human activity on the continent has result in eutrophication problems. For example, Lake Glubokoye in the Schirmacher Oasis has received high quantities of phosphorus from waste water discharges from the Russian Station Novolazarevskaya, although the rate of pollution has decreased recently due to a decline in scientific activity (UNEP, 2000). It must be noted, however, that not all phosphorus pollution problems are man-made. Thus lakes in the South Orkney Islands have undergone rapid eutrophication in recent years due to seals transferring marine nutrients onto the lake catchments.

8.10 CONCLUSIONS

This chapter has attempted to provide an overview of aspects of phosphate pollution in some regions of the world, illustrating the different degree to which the problem occurs and measures taken to control it. Phosphorus pollution remains a significant issue in developed countries, in the EU, USA and Japan. Although control measures (focusing on environmental objectives and emission controls) have been put into place, these are either not always properly implemented or policies are not sufficiently comprehensive. Developing countries also have significant eutrophication problems and these are driven both by increasing fertiliser use, as well as very poor treatment of urban waste water from increasing populations. The relative balance between point and non-point sources varies and, as sewage discharges are gradually tackled, the relative importance of agriculture and soil erosion will increase. The lessons from Europe, North America and Japan are that managing point sources is not sufficient, but that non-point source management is difficult to achieve. Overcoming these problems will form a major challenge for investment, not just for ecosystem protection, but in tackling human health protection requirements.

REFERENCES

Ball, J. (2001) State of the Environment Australia 2001. Inland waters theme report. CSIRO Publishing, Canberra.

Dixon, J., Gulliver, A. and Gibbon, D. (2001) Farming systems and poverty. FAO and World Bank, Washington.

European Commission (1998) Implementation of council directive 91/271/EEC of 21 May 1991 concerning urban waste water treatment, as amended by commission directive 98/15/EC of 27 February 1998: summary of the measures implemented by the member states and assessment of the information received pursuant to articles 17 and 13 of the directive. COM(98)775.

EEA (1994) European rivers and lakes: assessment of their environmental state. EEA environmental monographs: 1. European environment agency, Copenhagen.

EEA (1998) Europe's environment: the second assessment. European environment agency, Copenhagen.

Farmer, A.M. (1997) *Managing Environmental Pollution*. Routledge, London.

Farmer, A.M. (2001) Reducing phosphate discharges: the role of the 1991 EU urban waste water treatment directive. *Water Sci. Technol.* **44**, 41–48.

Farmer, A.M. and Braun, M. (2003) Fifty years of the Rhine commission: a success story in nutrient reduction. SCOPE Newsletter No. 47. 16 pp. CEEP, Brussels.

Farmer, A.M., Precioso, B.L., Latorre, F., Tuddenham, M. and Thornton, C. (2000) The role of the 1991 EU urban waste water treatment directive in reducing phosphorus discharges in France and Spain. *Eur. Water Manage.* **3**, 35–43.

Litke, D.W. (1999) Review of phosphorus control measures in the United States and their effects on water quality. *Water Resources Investigations Report* 99-4007. US Geological Survey, Denver.

Macleod, C. and Haygarth, P. (2003) A review of the significance of non-point source agricultural phosphorus to surface water. CEEP, Brussels.

Sien, C.L. and Kirkman, H. (2000) Overview on land-based sources and activities affecting the marine environment in the East Asian Seas. UNEP regional seas reports and studies No. 173. United Nations Environment Programme, Nairobi.

UNEP (1994) The pollution of lakes and reservoirs. Environment library No. 12. United Nations Environment Programme, Nairobi.

UNEP (1998) Overview of land-based sources and activities affecting the marine, coastal and associated freshwater environment in the eastern african region. UNEP Regional Seas Reports and Studies No. 167. United Nations Environment Programme, Nairobi.

UNEP (2000) Global environmental outlook 2000. United Nations Environment Programme. Earthscan Publications, London.

USEPA (1997) National water quality inventory. Report EPA-841-R-97-008. United States Environmental Protection Agency, Washington.

PART THREE

Phosphorus removal from water and waste water: principles and techologies

9. Principles of phosphate dissolution and
   precipitation   195

10. Waste water treatment principles   249

11. Chemical phosphorus removal   260

12. Biological phosphorus removal   272

13. A review of solid phase adsorbents for
    the removal of phosphorus from natural and waste waters   291

14. Removing phosphorus from sewage effluent
    and agricultural runoff using recovered ochre   321

9
Principles of phosphate dissolution and precipitation

P.G. Koutsoukos and E. Valsami-Jones

9.1 INTRODUCTION

The thermodynamic properties of phosphates are less well known than other groups of minerals, such as the silicates and carbonates. However, recent years have seen an increased interest in the dissolution and precipitation processes of the sparingly soluble phosphates, stemming from their important applications. Phosphate phases control phosphorus availability in soils and surface waters, and often that of elements such as calcium, aluminium, iron, and other, potentially toxic, metals. Phosphates can precipitate in waste waters and thus provide a mechanism for phosphorus removal, and perhaps recovery. Also, importantly, phosphates can be used as biomaterials, fertilisers, food additives and in many other applications. Some of these applications are explored in other chapters of this book. Here we introduce the basic principles governing the dissolution and precipitation of phosphates, which, together with adsorption/desorption processes, control inorganic phosphorus mobility and availability in natural and engineered systems.

Many questions about the environmental role of phosphates remain. For example, although phosphates have been released into aquatic systems (rivers and lakes) near urban or agricultural areas at increasing rates, this has not always been followed by elevated phosphate concentrations, suggesting that some is removed, probably through precipitation processes (Nancollas, 1984; Brown, 1973); the solubility and availability of such precipitates in the long term is not known. Little is also known about the conversions from one phosphorus compound to another, including organic–inorganic transformations. Furthermore, and despite the calcium phosphate system having been extensively studied in recent years, there are still considerable uncertainties about individual phosphate phase stability. This is because of the ability of some phosphate minerals, notably apatite, to incorporate numerous impurities and vacancies, and due to the common formation of unstable, often non-stoichiometric phases, which are difficult to identify, but which may persist for long periods, instead of other, thermodynamically more stable ones. These uncertainties are to a large extent due to the influence of kinetic factors which often control which phase is most likely to persist.

Of all phosphates, calcium phosphate phases are the most intensely studied for a number of reasons. They represent the most abundant phosphates, and can be found naturally in all types of rocks from igneous to sedimentary and metamorphic. They are the main components of phosphorite deposits, which represent the only major primary source of phosphorus on Earth. They control phosphorus availability in neutral and alkaline soils. They form hard scale deposits in cooling systems, particularly when recycled and/or hard water is used. Finally, they are important biological minerals forming an integral part of the skeleton and teeth of vertebrates, as well as pathological calcifications.

9.2 EQUILIBRIUM AND KINETICS: THEORY, MECHANISMS AND EQUATIONS

Whether in the soil, air, water, or even within the body of an organism, inorganic solids (minerals, salts)* are in a constant state of remodelling. The presence of an aqueous phase facilitates the process, by providing a medium where dissolution or precipitation occurs, but is not a prerequisite; for example freshly exposed calcite surfaces have been shown to restructure in vacuum (Stipp, 1999). The reactivity of an inorganic solid will depend on its surface properties, and these will be, to some extent, controlled by the conditions prevailing during its formation.

In contact with an aqueous phase, inorganic solids will precipitate or dissolve, depending on the concentration of their constituent components in solution, and the prevailing conditions. The dissolution and precipitation reactions will change the properties of both the solid and the aqueous phase.

9.2.1 Precipitation mechanisms

Precipitation of a salt, in general, takes place in two stages: nucleation and crystal growth. During the first step, atomic size clusters, nuclei, of the precipitating phase are formed from the crystal building blocks, which grow subsequently to

---

*Minerals may be defined as naturally occurring salts.

Principles of phosphate dissolution and precipitation

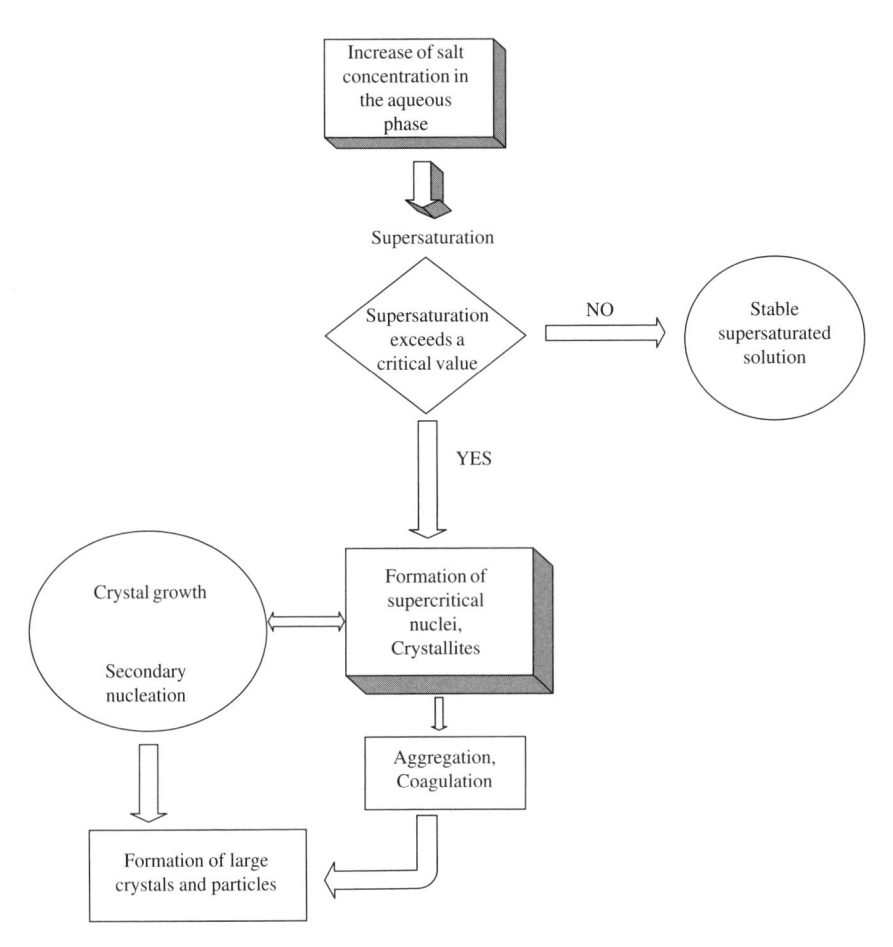

Figure 9.1 Schematic representation of the stages involved in the precipitation of solids from solutions.

macroscopic crystals; these may undergo secondary processes such as aggregation and secondary nucleation. The primary condition necessary for the precipitation process is the establishment of supersaturation, which may be attained by increasing the concentration of the aqueous components of the precipitating phase above equilibrium, by evaporation, addition of components etc. The overall process is described in Figure 9.1.

The formation of a salt in solution by supersaturation is also shown graphically in Figure 9.2. The solubility curve represents the points of equilibrium between the solid and the liquid phases. Below this line the solution is undersaturated with respect to the salt and above it, supersaturated. Supersaturated solutions may be stable for practically infinite time periods. There is however a threshold, marked by the dashed line in Figure 9.2, beyond which spontaneous precipitation occurs regardless of induction time required for the formation and further growth of supercritical

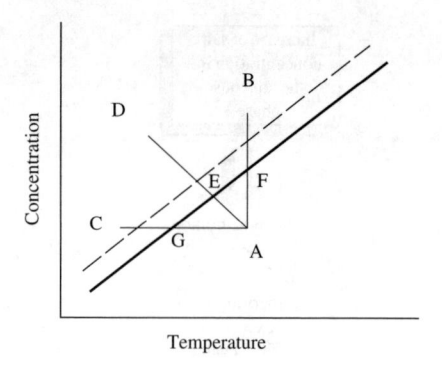

Figure 9.2 The solubility curve of a salt (solid bold line). Lines AGC, AED and AFB represent possible routes leading to the development of supersaturation in solutions starting from undersaturation at point A. Dashed line in the supersolubility curve.

nuclei. The range of supersaturations within which precipitation may be slow or even absent (between solid and dashed lines in Figure 9.2) defines the labile region; the dashed line is known as the supersolubility curve. Notably, the supersolubility curve is not well defined and depends on other factors such as the presence of foreign suspended particles, agitation, temperature, pH etc. The formation and subsequent precipitation of solids occurs only when the solution conditions correspond to the metastable or the labile region. Below the solubility curve, where solutions are undersaturated, precipitation cannot take place. Instead, the introduction of crystals of the salt in such solutions would result in dissolution.

9.2.2 Precipitation equations

The supersaturation that leads to precipitation can be reached in many ways including temperature or pH change, mixing of solutions, concentration increases by evaporation or solid addition etc. Although supersaturation is the driving force for the formation of a salt, the exact value at which precipitation occurs differs from salt to salt; as a rule, the degree of supersaturation needed for a sparingly soluble salt is orders of magnitude higher than the corresponding value for a readily soluble salt. Quantitatively, supersaturation may be expressed in several types of units (Cayes and Estrin, 1967; Mullin, 1993; Myerson, 1993), and is often expressed as concentration difference:

$$\Delta C = C - C_\infty \quad (1)$$

where C and C_∞ are the solute concentration in solution and at equilibrium respectively. The supersaturation ratio S is defined as:

$$S = \frac{C}{C_\infty} \quad (2)$$

being a number >1 for supersaturated solutions. A related expression, the relative supersaturation, σ, is defined as:

$$\sigma = \frac{C - C_\infty}{C_\infty} \quad (3)$$

and as can be seen from Equations 2 and 3:

$$\sigma = \frac{C}{C_\infty} - 1 = S - 1 \quad (4)$$

For sparingly soluble salts $M_{\nu+}A_{\nu-}$ the supersaturation ratio is defined as:

$$S = \left\{ \frac{(\alpha_{M^{m+}})_s^{\nu+} (\alpha_{A^{a-}})_s^{\nu-}}{(\alpha_{M^{m+}})_\infty^{\nu+} (\alpha_{A^{a-}})_\infty^{\nu-}} \right\}^{1/\nu} = \left(\frac{IP}{K_s^o} \right)^{1/\nu} \quad (5)$$

where subscripts s and ∞ refer to solution and equilibrium conditions respectively, α denotes the activities of the respective ions (M or A), m+ and α- are the positive and negative ionic charges and $\nu_+ + \nu_- = \nu$. IP and K_s^0 are the ion products in the supersaturated solution and at equilibrium respectively.

The fundamental driving force for the formation of a salt from a supersaturated solution is the difference in chemical potential of the solute in the supersaturated solution from the value at equilibrium:

$$\Delta \mu = \mu_\infty - \mu_s \quad (6)$$

Since the chemical potential per mole of the solute is expressed in terms of the standard potential and the activity, α, of the solute:

$$\mu = \mu^0 + RT \ln \alpha \quad (7)$$

where R and T are Boltzmann's constant and absolute temperature respectively. Substitution of Equation 7 to Equation 6 gives the driving force for precipitation (Garside and Davey, 1980):

$$\frac{\Delta \mu}{RT} = \ln \left(\frac{\alpha_s}{\alpha_\infty} \right) = \ln S \quad (8)$$

The calculation of the thermodynamic driving force involves the computation of the activities of the component free ions of the salts and consequently of the respective activity coefficients. Waters in both engineered and natural systems can often be of high ionic strength (0.6–3.0 M). For the computation of the activity coefficients in such media, it is not possible to use the Debye Hückel expressions, or even extended forms such as the Davies formulation (Davies, 1962). Semi empirical expressions for the activity coefficients have to be used in such cases (Pitzer, 1979; Ball and Nordstrom, 1991). It is now common practice for speciation computations to be performed using one of many computer software packages. The principles of speciation computation have been critically reviewed by

Nordstom et al. (1979; see also Bethke, 1996). The calculations proceed iteratively with the minimization of the Gibbs free energy using equilibrium constants for relevant equilibria regardless of mass balance constraints (Nancollas, 1966).

Nucleation takes place after supersaturation is established. Although there is no general agreement on a definition, primary nucleation is described as the formation of new crystal nuclei, which takes place in the absence of crystalline or any other type of suspended matter. When the new crystals are generated in the neighbourhood of existing suspended crystallites or particles, nucleation is termed as secondary. Primary nucleation may be further distinguished into homogeneous and heterogeneous to denote situations in which it starts spontaneously or is catalysed by the presence of foreign particles respectively. In primary nucleation, the corresponding rates depend strongly on supersaturation while in secondary nucleation the size of the crystallites present is important (Cayes and Estrin, 1967; Mullin, 1993). In theory, nucleation proceeds through an auto catalytic process of bimolecular reactions resulting in the formation of dimmers, trimers etc. of crystal building blocks which are the embryos of the crystals to be formed:

$$A + A \rightarrow A2$$
$$A2 + A \rightarrow A3$$
$$A3 + A \rightarrow A4$$
$$. \quad . \quad .$$
$$. \quad . \quad .$$
$$. \quad . \quad .$$
$$An - 1 + A \rightarrow An*$$

The cluster $An*$ has a critical size which allows the nucleus to grow further to a macroscopic crystallite. Embryos smaller than the critical size redissolve. The nucleation process may be influenced greatly by the presence of impurities in the solutions, which either suppress primary nucleation (Garside and Davey, 1980) or promote secondary nucleation (Mullin et al., 1970; Botsaris et al., 1972).

Very often a period of time elapses between the achievement of supersaturation and the detection of the formation of the first crystals. This time, defined as the induction time, τ, is considered to correspond to the time needed for the development of supercritical nuclei. The induction time is inversely proportional to the rate of nucleation and, according to classical nucleation theory, the following relationship exists (Malollari et al., 1995):

$$\log \tau = A + \frac{B\gamma_s^3}{(2.303kT)^3 \log S} \qquad (9)$$

Once stable, supercritical nuclei have been formed in a supersaturated solution, they eventually grow into crystals large enough to be visible. The rate of crystal growth may be defined as the displacement velocity of a crystal face relative to a fixed point of the crystal. This definition however cannot be easily applied to the formation of polycrystalline deposits such as in the case of the formation of phosphates. In this case, the rates of growth may be expressed in terms of the molar rate deposition by equation:

Principles of phosphate dissolution and precipitation

$$R_g = \frac{1}{A} \cdot \frac{dm}{dt} \tag{10}$$

Where m is the number of moles of the solid deposited on a substrate in contact with the supersaturated solution, and A, the surface area of the substrate. Linear rates, $\dot{r}$, are often used assuming the shape of the polycrystalline deposits is spherical, of equivalent mean radius $\bar{r}$:

$$\dot{r} = \frac{d\bar{r}}{dt} \tag{11}$$

The molar rate is related to the linear rate via Equation 12:

$$\dot{r} = R_g \frac{M}{\rho} \tag{12}$$

where M is the molecular weight and ρ the density of the crystalline deposit.

The rate laws used to express the dependence of the rates on solution supersaturation provide mechanistic information for the precipitation process. At a microscopic scale the sequence of steps followed for the growth of crystals are shown in Figure 9.3:

(i) Transport of building blocks to the surface on crystal terraces by convection or by diffusion (step 1).
(ii) Adsorption at a step representing the emergence of a lattice dislocation at the crystal surface, accompanied by partial dehydration (step 2).
(iii) Migration along the step, integration at a kink site on the step and further dehydration of the growth units (step 3).

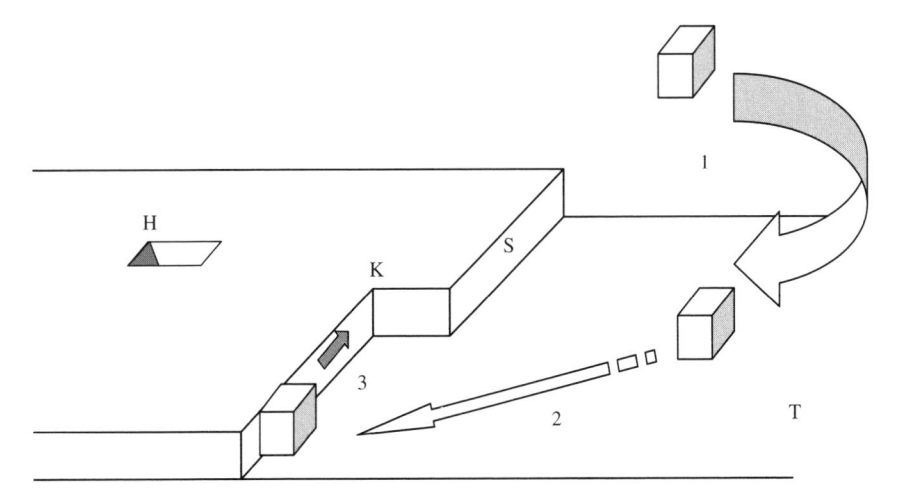

Figure 9.3 Model for the steps involved in the process of crystal growth of the supercritical nuclei. H: dissolution pit, K: kink, S: step, T: terrace

The rate of crystallization can be expressed in terms of the simple semi empirical kinetic equation:

$$R_g = k_g f(S) \sigma^n \tag{13}$$

where k_g is the rate constant for crystal growth, $f(S)$ a function of the total number of the available growth sites and n the apparent order of the crystal growth process. When mass transport (step 1) is the rate-determining step, the growth rate is given by:

$$R_d = k_d \sigma \tag{14}$$

where k_d is the diffusion rate constant which is given by:

$$k_d = \frac{Dv C_\infty}{\delta} \tag{15}$$

where D is the mean diffusion coefficient of the lattice ions in solution, v the molar volume of the crystalline material, C_∞ the solubility of the precipitating phase and δ the thickness of the diffusion layer at the crystal surface (Burton et al., 1951; Nielsen, 1984; Nancollas, 1989).

From a mechanistic point of view, it is possible to interpret kinetic data on the basis of theoretical models, the most important of which include adsorption and diffusion-reaction. The concept of crystal growth proceeding on the basis of an adsorbed monolayer of solute atoms, molecules or ion clusters was first suggested by Volmer (1939). Through this monolayer it is possible to exchange ions or molecules between the bulk solution and the crystal surface; the rate in this case is (Nielsen and Toft, 1984):

$$R'_g = k_{ad} \sigma \tag{16}$$

where the rate constant k_{ad} is given by:

$$k_{ad} = a v_{ad} v C_\infty \tag{17}$$

In Equation (17), a is the jump distance and v_{ad} the jump frequency of an ion into the adsorption layer.

Since as a rule the growing crystals exhibit the presence of steps on their surfaces, it may be assumed that these steps originate from imperfections or dislocations in the crystal lattice. The uniform deposition of growth units along a step results in the development of spirals, which allow for the continuous growth of the crystals, as long as the presence of active sites is maintained. The theory for spiral growth mechanism was developed by Burton, Cabrera and Frank (1951; BCF theory). According to this theory, the curvature of the spirals near their origin is related to the distance between the successive turns of the spirals and to the solution supersaturation. The growth rate, R_{BCF}, at all supersaturations is given by:

$$R_{BCF} = A'\sigma^2 \tanh\left(\frac{B'}{\sigma}\right) \quad (18)$$

where A' and B' are constants which depend on temperature and on the step spacings. At low supersaturations, $\tanh(B'/\sigma) \to 1$ and $R_{BCF} \propto \sigma^2$ while at high supersaturations, $\tanh(B'/\sigma) \to B'/\sigma$ and $R_{BCF} \propto \sigma$, i.e., the dependence of the rate of crystal growth on the solution supersaturation changes from a parabolic to a linear dependence as supersaturation increases. It should be noted that although the BCF theory was developed for vapours, it is also valid for liquids with different expressions for the constants A' and B' (Chernov, 1961). At relatively high supersaturations, it is possible that several two-dimensional nuclei form on the surface of the crystals and spread. In this case crystal growth can take place either by the completion of one step at a time (mononuclear mechanism) or by the development of numerous surface nuclei which form steps and kinks in which the incoming ions may be incorporated (polynucleation mechanisms) (O'Hara and Reid, 1973). The surface nuclei are for the most part of relatively large size, allowing for the development on their surface of new nuclei, i.e., deposition of islands upon islands. In this 'birth and spread' model the rate of crystal growth is:

$$R_p = k_p f(s) \exp(-k_p/\ln S) \quad (19)$$

where the polynucleation constant, k_p is:

$$k_p = \frac{\pi \gamma_s^2}{3kT^2} \quad (20)$$

In Equation 20, γ_s is the surface energy of the nucleating solid, and in Equation 19, $f(s)$ is:

$$f(s) = S^{7/6}(S-1)^{2/3}(\ln S)^{1/6} \quad (21)$$

Spiral and polynuclear controls occur in parallel, with the faster controlling the overall mechanism. However, mass transport to the crystal surface, adsorption of lattice ions and the mechanisms at the crystal surface are consecutive steps and the slowest is the rate determining step. Knowledge of the operative mechanism may be obtained by fitting experimental data into Equation (13). Supporting evidence may also be obtained by the investigation of the role of fluid dynamics on the rates of crystal growth (see Mangin and Kline, this volume). Dependence of the growth measured on the fluid velocity suggests mass transport control for the rates of deposition.

An estimate of the relative occurrence and of the potential for precipitation of salts in natural or engineered systems should be based on the combination of careful thermodynamic calculations with kinetics measurements from which useful mechanistic conclusions may be drawn. These should proceed in tandem with thorough characterisation of any newly formed precipitates, ideally using more than one methods.

9.2.3 Dissolution mechanisms

Mechanistically, the same three fundamental processes that govern precipitation, i.e., mass transport, diffusion and surface reaction, also control dissolution. Unlike readily soluble mineral salts (e.g. halite, gypsum), whose dissolution is controlled by the rate of mass transfer to the aqueous phase, the dissolution of sparingly soluble salts (such as most phosphates) is controlled by the kinetics of detachment of reacted components from their surface (e.g. Appelo and Postma, 1999).

According to transition state theory, the reactants and products of a dissolution reaction are separated by an energy maximum. In physical terms, this energy maximum represents the stage in dissolution where an activated complex exists at the salt surface, separating the reactants from the products, according to the equation (Aagaard and Helgeson, 1982; Langmuir, 1997; Lasaga, 1981):

$$A + B \rightleftharpoons C \rightleftharpoons AB \qquad (22)$$

where C is the activated complex, A, B the reactants and AB the product (dissolved species).

Dissolution is initiated at points on the dissolving surface, where the activation energy to derive C, and for the reaction to move forward, is lower. It is assumed that the number of such points, the 'active' sites of the dissolving surface is constant due to the continuous regeneration of these sites during dissolution.

For the overall dissolution reaction, the rate controlling step is the bulk rate of detachment, or desorption, of the activated complex from the surface (Stumm and Wollast, 1990). The activated complex forms when a reactant attaches to surface sites, to form active sites. A reactant is an aqueous species, i.e. H^+, OH^-, a ligand or a reductant/oxidant (the latter only for surfaces that may undergo redox reactions), capable of electrostatic attachment to the surface. Attachment will be followed by charge redistribution and destabilisation of the metal–oxygen bond in the mineral structure and ultimately detachment of the surface species. The dissolution reaction can then be described as a two-step process as follows (Stumm and Morgan, 1996):

$$\text{Surface sites} + \text{reactants (H}^+, \text{OH}^-, \text{ligands)} \xrightarrow{\text{fast}} \text{surface species} \qquad (23)$$

$$\text{Surface species} \xrightarrow{\text{slow}} \text{aqueous species} \qquad (24)$$

The slower of the two, Equation 24, is the rate limiting step. The rate law for the overall reaction will depend on the concentration of the surface species. For a given surface, the quantity of reactants attached to the surface will depend on the surface charge, which is a function of pH. It is thus possible to generalize that dissolution rates will increase both with increasing positive surface charge (i.e. decreasing pH)

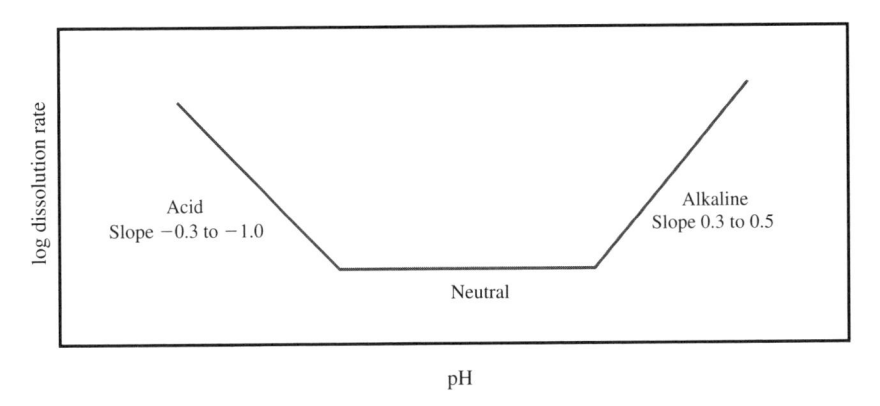

Figure 9.4 Relationship between pH and dissolution rate, derived from silicate minerals. Redrawn after Drever, 1994.

and increasing negative charge (i.e. increasing pH) and are likely to be minimum where the net surface charge is zero. This principle is shown in a simplified form in Figure 9.4.

A lot of what is known about sparingly soluble salt dissolution has been derived from studies of oxide and silicate minerals. In simple single oxides, dissolution requires the breaking of only one metal–oxygen bond, and therefore the dissolution rate is a function of the rate at which the surface species is likely to break this bond and become an aqueous species. However, in multioxide minerals (such as silicates and phosphates) where more than one metal–oxygen bonds are present, dissolution may require the breaking of more than one metal–oxygen bonds, some of which will break faster than others. Dissolution is therefore likely to be a multi-tier process, the overall rate of which will be defined by the slowest breaking bonds (Oelkers, 2001). The selective removal of the more weakly bonded metals of the structure will result in the formation of a 'leached layer' (e.g. Nesbitt and Muir, 1988; Hellmann *et al.*, 1990), i.e. a layer lacking some of the components of the mineral, practically, in silicate minerals, a silica layer. The net effect of variable rate of removal of different metals from a mineral structure is that at the early stages dissolution will be non-stoichiometric, until the leached layer is fully developed and dissolution reaches a steady state. This is likely to apply to phosphate minerals too, and non-stoichiometric early dissolution has been reported for apatite (e.g. Manecki *et al.*, 2000; van der Houwen, 2002), although the effect of non-stoichiometric release can be difficult to separate from stoichiometric release from a non-stoichiometric apatite. To date, there have been no attempts to perform long-term kinetic experiments, to assess when or how a steady state is reached in the dissolution of phosphates.

9.2.4 Dissolution equations

If, in Equation 2 $S < 1$, then the solution is undersaturated with respect to the salt considered, and this will dissolve. From Reaction 24 above, and by analogy to

Equation 16, which defines surface controlled precipitation rates, the dissolution rate of a mineral salt is proportional to the concentration of the activated surface species:

$$R_d = k_j C_j^p \qquad (25)$$

where R_d is the dissolution rate (in mol m^{-2} h^{-1}), k_j is a rate constant, C_j is the concentration of the surface species (in mol m^{-2}) and $p \geq 1$. C_i may be $<\equiv$MeOH$_2^+>^j$, in proton promoted dissolution, $<\equiv$MeO$^->^i$, in basic (hydroxide) promoted dissolution, $<\equiv$MeL$>$, in ligand promoted dissolution, or $<\equiv$MeR$>$ in oxidative/reductive dissolution (Casey and Ludwig, 1995; Stumm and Morgan, 1996). The symbol $\equiv$ denotes the mineral surface, and $<>$ surface concentrations, whereas Me is a metal, and j and i are exponents representing the metal ion charge.

In a closed system, a salt will dissolve until it reaches equilibrium. The concentration of its components in solution will therefore follow the pattern of Figure 9.5. An empirical assessment of the dissolution rate of a salt can be made based on a calculation of the mass released (dm), as a function of time (dt), representing the slope of the curve in the far from equilibrium part of Figure 9.5, normalised over surface area (A); this is the same formula as Equation 10 used earlier to describe precipitation rates:

$$R_d = \frac{1}{A}\frac{dm}{dt} \qquad (26)$$

Note that, as concentrations in solution approach equilibrium, dissolution will become dependent on these component concentrations. Dissolution rates therefore apply to the far from equilibrium region, which is defined as the part of the curve where affinity, Ă $\geq$ 15 kJ/mol. The affinity definition is based on transition

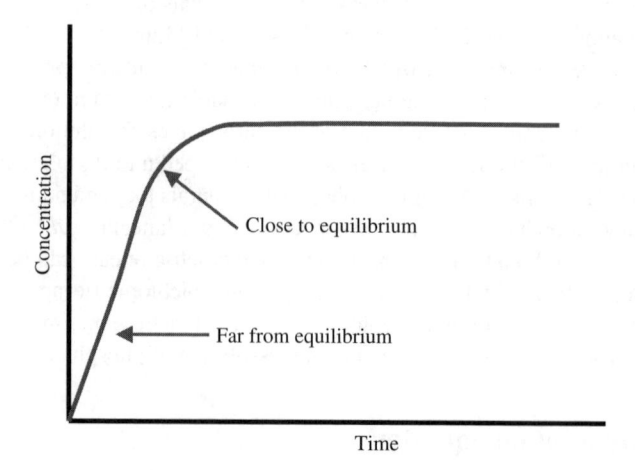

Figure 9.5 Concentration of the components of a dissolving salt, as a function of time, in a closed system.

state theory (Aagaard and Helgeson, 1982; Lasaga, 1981; Oelkers *et al.*, 1994), and can be derived from:

$$\breve{A} = RT \ln(K/Q) \tag{27}$$

where R (mJ/mol K) is the gas constant, T (K) is absolute temperature, K (mol/kg) is the equilibrium constant for the mineral hydrolysis reaction and Q (mol/kg) is the ion activity for the reaction.

Equation 26 shows that knowledge of the surface area of the dissolving mineral is essential. Surface area can be calculated, using geometric models (e.g. Oelkers, 2002), or measured by gas adsorption using the method developed by Brunauer, Emmet and Teller (1938), often referred to as BET surface area. The BET surface area can account for submicroscopic surface roughness and porosity, and thus appears to be more representative than simple geometric surface areas. However, its relevance to true, reactive mineral surfaces is still under debate (e.g. Brantley and Chen, 1995; Hodson, 1998; Oelkers, 2002).

As a measure of the time it takes for a sparingly soluble mineral particle to dissolve, Stumm and Morgan (1996) presented the following calculation: assuming there are about 10 functional groups on a mineral surface per nm^2, and for a mineral dissolving at a rate of 10^{-4} mol m^{-2} y^{-1} (a rate similar to that of apatite dissolving in neutral pH, see later), it would require 6 years for the detachments of a whole surface layer (the depth of a few Angstroms) from the mineral surface.

It is important to mention here that the dissolution laws described above apply to particles above a certain size (probably >1 μm). Nanoparticles do not obey standard thermodynamic laws and instead are a lot more soluble than larger particles. In a natural system, nanoparticles will have a very short lifetime, and will dissolve and recrystallize into larger grains. The smaller the particles, the more soluble according to the formula (Adamson, 1982):

$$S/S_0 = \exp[2\gamma V/RTr] \tag{28}$$

where S (mol/kg) is the solubility of (nano)particle of radius r (m), S_0 is the bulk solubility, γ (mJ/m^2) is the surface free energy, V (m^3/mol) is the molar volume, R (mJ/mol K) is the gas constant and T (K) is the absolute temperature. Figure 9.6 shows a plot of Equation 28, calculated for a silicate mineral (quartz). The figure suggest that effects on solubility begin to appear at particle sizes <100 nm. However, particles in the range 1–10 nm show a dramatic increase in their solubility. A similar effect was observed for a carbonate mineral (calcite, Morse and Mackenzie, 1990). This discussion has great relevance to phosphates, particularly biogenic phosphates, which are typically nanosized (see Valsami-Jones, this volume).

Another situation where solubility behaviour is affected occurs when foreign constituents are incorporated in the crystal lattice. This is again very relevant to phosphates, particularly apatites, which commonly contain a variety of substitutions. In situations where a minor component is substituting isomorphously into the structure of a solid phase (i.e. forming a solid solution), the solubility of the

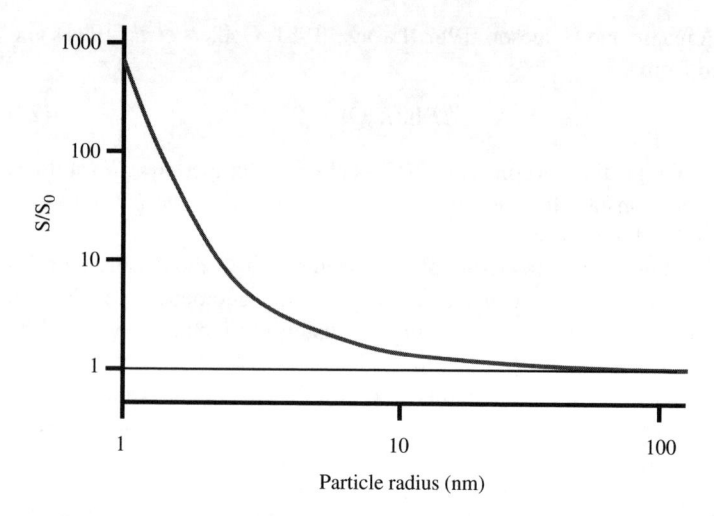

Figure 9.6 A schematic presentation of the deviation of solubility of nanoparticles as a function of size. Whereas at 100 nm solubility is indistinguishable from that of larger particles, at 1 nm the predicted solubility is almost three orders of magnitude higher. The calculations are based on quartz particles. Modified after Hochella (2002).

minor component is greatly reduced, if their solubility was higher than that of their host phase (Stumm and Morgan, 1996).

9.3 APPLICATIONS RELEVANT TO THE PRECIPITATION OF CALCIUM PHOSPHATES

Calcium phosphates represent the most studied group of phosphate minerals. This is both due to their being the most common phosphates and also their significance as biominerals, present in bone, teeth and pathologic mineral deposits (see also Valsami-Jones, this volume). The precipitation characteristics and properties of calcium phosphates are particularly well studied, often as a means of developing new, or improving existing biomaterials, and there is also a significant amount of literature covering issues of their dissolution. The most studied calcium phosphate is hydroxylapatite, being an analogue for bone/tooth biogenic apatite.

9.3.1 Calcium phosphate bulk precipitation

We have previously alluded to the fact that when the aqueous concentrations of phosphate and calcium are high, a number of calcium phosphates with variable stoichiometries of the constituent ions may form. A list of calcium phosphates forming in aqueous solutions, in the order of decreasing solubility, are shown in Table 9.1.

Principles of phosphate dissolution and precipitation

Table 9.1 Calcium phosphate crystalline phases, formulae and corresponding thermodynamic solubility products (Data from NIST, 2001, except defect apatite from van der Houwen, 2002). Mineral names: DCPD, brushite; DCPA, monetite; TCP, calcium end member of whitlockite group.

Solid Phase	Abbrev.	Formula	Thermodynamic Solubility Product
Dicalcium phosphate dihydrate	DCPD	$CaHPO_4 \cdot 2H_2O$	$1.87 \times 10^{-7} (mol\ l^{-1})^2$
Dicalcium phosphate anhydrous	DCPA	$CaHPO_4$	$9.2 \times 10^{-8} (mol\ l^{-1})^2$
β Tricalcium phosphate	TCP	$Ca_3(PO_4)_2$	$2.8 \times 10^{-9} (mol\ l^{-1})^{15}$
Octacalcium phosphate	OCP	$Ca_8H_2(PO_4)_6 \cdot 5H_2O$	$2.5 \times 10^{-99} (mol\ l^{-1})^{16}$
Defect apatite	DA	$Ca_{9.4}(HPO_4)_{0.6}(PO_4)_{5.4}(OH)_{1.4}$	$2.0 \times 10^{-104} (mol\ l^{-1})^{18}$
Hydroxylapatite	HAP	$Ca_{10}(PO_4)_6(OH)_2$	$5.5 \times 10^{-118} (mol\ l^{-1})^{18}$

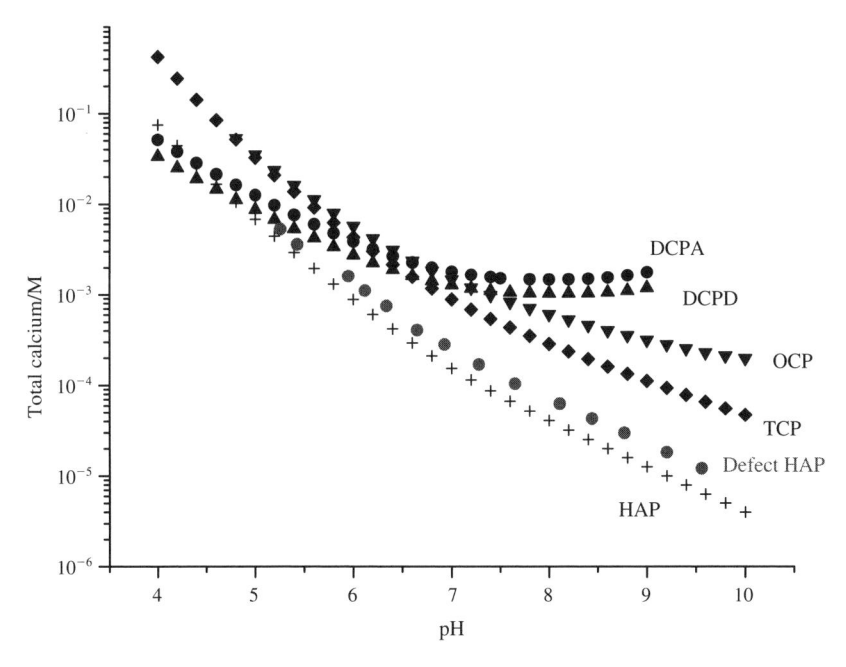

Figure 9.7 A plot of total calcium as a function of the solution pH showing calculated solubility isotherms for the calcium phosphate phases (for abbreviations, see Table 9.1). Solution conditions: 0.1 M NaCl, 25°C.

On the basis of the values of the solubility products given above, solubility isotherms calculated for the various calcium phosphate phases are shown in Figure 9.7. The tendency for a particular calcium phosphate phase to form in supersaturated aqueous media may be determined from the solubility phase diagrams such as the diagram shown in Figure 9.7.

At high aqueous concentrations, more than one calcium phosphate may be supersaturated. It is generally believed that in such cases a number of phases may initially form, depending on the solution pH. These phases will act as precursors, and ultimately transform into the thermodynamically more stable HAP. This is in accordance to Ostwald's rule of stages, which predicts that the least stable phase having the highest solubility is formed preferentially during a stepwise precipitation process. It is well established that kinetic factors may be more important in determining the nature and hence the characteristics of the solid deposits formed during the precipitation process than the respective equilibrium considerations. Complications may also arise from the formation of mixed solid phases resulting from the overgrowth of one crystalline phase over another.

In highly supersaturated solutions an unstable precursor phase has been reported to form, characterized by the absence of peaks in the powder X-ray diffraction pattern; this phase has been described as amorphous calcium phosphate (ACP). The composition of ACP appears to depend upon the precipitation conditions and is usually formed in supersaturated solutions at pH >7.0 (Eanes et al., 1965; LeGeros et al., 1975; Newesely, 1966). In slightly acidic calcium phosphate solutions the monoclinic DCPD forms (Betts and Posner, 1974; Brown and Lehr, 1959; Lehr et al., 1959). OCP is formed by the hydrolysis of DCPD in solutions of pH 5–6 and may also precipitate heterogeneously upon TCP (Brown et al., 1957; DeRooij et al., 1984).

HAP, the thermodynamically most stable calcium phosphate phase, can sustain many substitutions in its lattice (see also Valsami-Jones, this volume). Common substitutions include that of OH^- by F^- or Cl^- ions, of the phosphate by sulphate, carbonate and hydrogen phosphate and of the calcium by Sr^{2+}, Mg^{2+} and Na^+ ions (Elliott, 1994; Heughebaert et al., 1983; LeGeros and LeGeros, 1984; Moreno and Varughese, 1981; Nathan, 1984).

In previous studies of the relative stabilities of calcium phosphates, identification of precipitated phases has been based on the stoichiometric molar ratio of calcium to phosphate, calculated from their respective changes in the solutions. This ratio has been found in several cases to be 1.45 ± 0.05, which is considerably lower than the value of 1.67 corresponding to stoichiometric HAP. On the basis of such results, a number of different precursor phases have been postulated to form including TCP (Montel et al., 1981; Narasaraju and Phebe, 1996; Walton, 1967), OCP (Eanes and Posner, 1968; Posner, 1969) and DCPD (Furedi-Milhofer et al., 1976). However, it is now recognised that defect apatites can have a ratio of calcium to phosphate around 1.45, as a result of substitution and lattice vacancies (e.g. Elliott, 1994). Note also, how the solubility of a defect apatite approaches that of other calcium phosphates in Table 9.1. To conclude, although the presence of precursor phases cannot be dismissed, particularly since such phases may be of transient nature and recrystallize shortly after their precipitation and prior to identification, some previous interpretations require revision.

A further problem in spontaneous precipitation studies is that of changing saturation in solution, due to the precipitating phase lowering the ionic activities of the precipitating components (Figure 9.8); this in turn can lead to changes in the

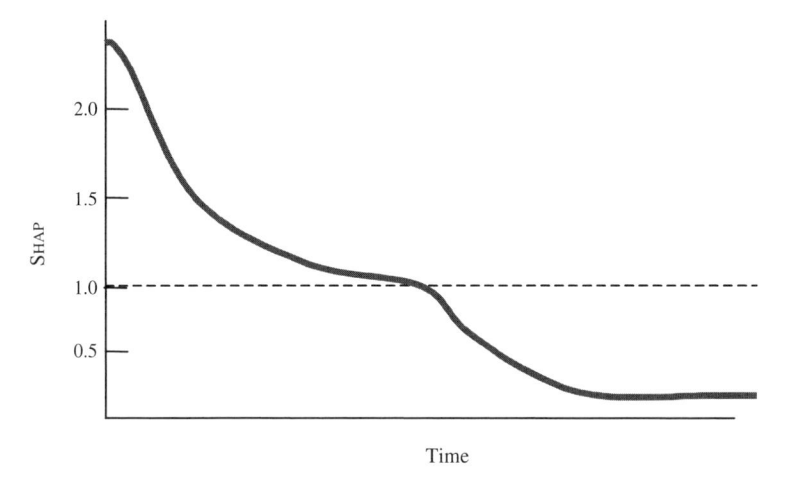

Figure 9.8 Change in solution saturation with respect to HAP as a function of time due to the reduction of ionic activities in solution during precipitation.

nature of the supersaturated phase as the experiment progresses. Furthermore, a phase may form and subsequently dissolve yielding finally a different calcium phosphate. Spontaneous precipitation experiments have thus been associated with reduced reproducibility and low accuracy of rates measured. Some of these problems can be overcome by using seeded crystal growth techniques, which can be performed at lower supersaturations.

However, even in seeded growth experiments at low supersaturation (at constant or variable pH) calcium and phosphate ion concentrations decrease appreciably as equilibrium concentration is approached, and a number of phases may form and redissolve adding to the uncertainty of the kinetics measured. These problems were overcome with the development of the constant composition technique in which the chemical potentials of the solution species were kept constant throughout the precipitation or crystal growth process (Tomson and Nancollas, 1978). The principle of this technique is that the lattice ion activities of the precipitating mineral are kept constant by the addition of titrant solutions at stoichiometric (for the precipitating phase) proportions (Dalpi et al., 1993; Heughebaert and Nancollas, 1984; Hohl et al., 1982). This is achieved by monitoring the pH, and triggering automatic dosing of the lattice ions in response to drop in pH (which corresponds to precipitation, if the precipitation reaction consumes hydroxyls, as is the case of HAP). Initiation of precipitation is commonly, but not necessarily, triggered by addition of seed crystals.

The application of this methodology to the investigation of the kinetics of crystal growth of calcium phosphates both on seed crystals and on diverse substrates, confirmed their formation through kinetic stabilization (Dalas et al., 1991; Koutsoukos, 1998; Koutsoukos and Nancollas, 1981a, 1981b, 1987; LeGeros et al.,

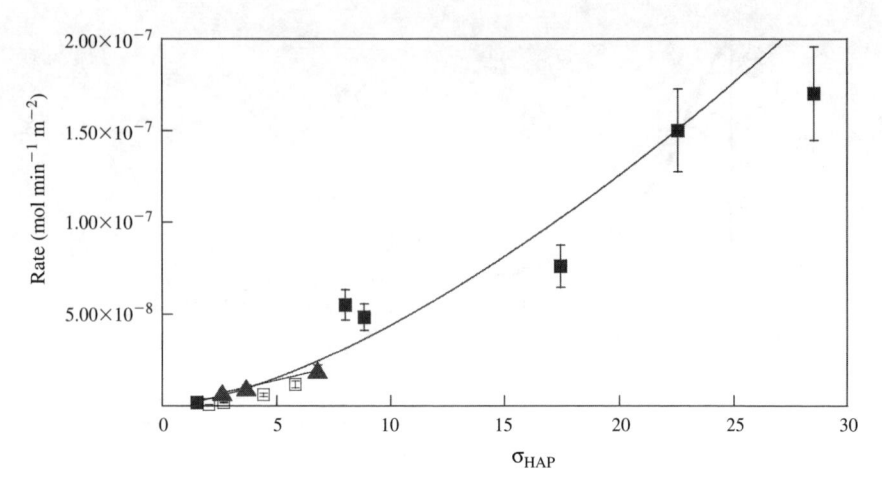

Figure 9.9 Dependence of the rate of crystal growth of HAP on HAP seed crystals at 37°C. Data from (■): Spanos et al., (2001); (▲) Kapolos and Koutsoukos, 1999; (▣) Dalpi et al., 1993.

1981). In the absence of any inhibitor, the growth rate as a function of the solution supersaturation was found to exhibit parabolic dependence, as may be seen in Figure 9.9. The second order dependence suggested that the mechanism of HAP crystal growth is controlled by surface diffusion of the growth units to the active growth sites.

9.3.2 Calcium phosphate nucleation at the atomic level

In the absence of a substrate, the first step in the formation of a new phase from a supersaturated solution is nucleation. Nucleation occurs via the formation of atomic clusters, which then grow to form particles, as discussed earlier. There are few studies investigating nucleation from solution supersaturated with respect to calcium phosphate. These studies, based mostly on theoretical calculations, predict that nucleation is likely to proceed via the formation of calcium phosphate clusters such as: $Ca_3(PO_4)_2$. It is currently uncertain whether single or multiple $Ca_3(PO_4)_2$ units give rise to a calcium phosphate proto-nuclei but *ab initio* calculations suggest the so-called Posner's cluster: $[Ca_3(PO_4)_2]_3$ is the energetically most stable configuration (Treboux et al., 2000). A Posner's cluster measures 0.815 nm and 0.87 nm in the α- and γ- directions respectively (Onuma and Ito, 1998). It is currently unknown how the transition from Posner's clusters (Ca : PO_4 ratio of 1.50) to calcium phosphates, and notably hydroxylapatite (Ca : PO_4 ratio of 1.67) occurs in solution, and how this may be affected/controlled by the presence of other ionic species. It is also not yet known whether the same cluster configuration leads to the formation of all calcium phosphates.

After nucleation has taken place, the conditions that lead to crystal growth are better understood and will be described next.

9.3.3 Calcium phosphate crystal growth and inhibition at the molecular level

At a molecular level, the growth of crystals may be considered as a sequence of growth unit additions onto a crystal step as shown earlier (see Figure 9.3). At the interface region, the growth unit transport is restricted to molecular diffusion through a diffusion boundary layer, regardless of the convection conditions away from the interface region (Rosenberger, 1986). The interface region is estimated to extend over a range of ca. 100 Å and plays an important role in crystal growth and secondary nucleation (Mullin, 1972). Following diffusion of the solute molecules from the bulk to the interface, they adsorb onto the surface and then diffuse across the surface. During this diffusion process, bonds between solute–solvent are broken, while new bonds between the solute and the surface ions are formed. This process is known as desolvation and, when the solvent is water, dehydration. Depending on the relative ease and speed of dehydration (desolvation) this step can be rate determining. The relative importance of bulk diffusion, desolvation and integration depends on the solid state properties and the solution properties. According to Bennema (1969), the various processes can be treated as a reaction pathway similar to homogeneous chemical reactions pathways. The volume diffusion step may be analysed in the usual manner, but the surface diffusion processes require consideration of the structure of the interface and the physical and chemical nature of adsorption and surface diffusion.

The explanation of the effectiveness of the presence of trace amounts of inhibitors to retard or terminate crystal growth processes may be described by taking into account the presence of key growth sites on the crystals (Davey, 1976). According to the Kossel model for crystals, a number of crystalline inhomogeneities are present on the crystal surfaces. The crystal interface contains distinct regions (Figure 9.3): flat surfaces or terraces which are atomically smooth, steps which separate terraces and kink sites formed from incomplete regions on steps. The sites with the lowest energy (highest binding energy) and therefore those in which the solute integration is most probable are the kinks. Steps are sites of intermediate energy and terraces of highest energy (lowest binding energies). Since the growth process consists of surface diffusion along a step to a kink with subsequent incorporation into the lattice, the concentration of steps and kinks on the surface of a crystal are important parameters of the crystal growth process. It has been shown experimentally that crystal faces which are atomically rough, i.e. contain numerous kink sites, grow more rapidly. The concentration of kinks may be quite high, reaching one kink per five molecules (Burton *et al.*, 1951). The experimental observation that crystals can grow at extremely low supersaturations led to the conclusion that sources of steps are almost always present on the crystal surfaces. According to Burton *et al.* (1951) a mechanism of continuously

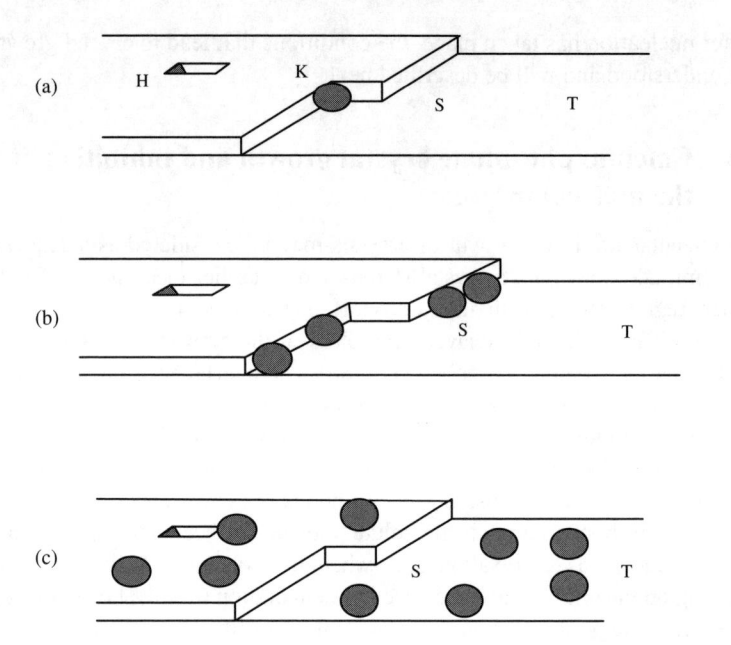

Figure 9.10 (a–c) Surface structures on an idealized crystal face: kinks (K); steps (S); terraces (T); and dissolution pits (H). Adsorbed impurities at each of these sites are illustrated.

producing steps is via surface dislocations. The steps move away from the source leading to the growth of crystals by layers. From this model it can be easily understood that the presence of traces of impurities is possible to poison the kinks and hence can have dramatic effects on the crystal growth kinetics. At higher solution supersaturation the number of dislocations increase and growth rates are higher at higher dislocation densities.

The adsorption of the impurities and their specific effect on growth processes depends on the chemical bonds between the impurity, the solvent and the surface. The forming bond controls whether the impurity will ultimately adsorb on terraces, steps or kinks, as shown in Figure 9.10.

The interaction of the impurity with the substrate is similar to the crystal growth process. More specifically the 'mobility' of an impurity depends on factors such as supersaturation, temperature and concentration. As a general rule the efficacy of an impurity is reduced with increasing temperature and saturation.

Surface adsorbed impurities reduce the rates of crystal growth by hindering the movement of the steps on the crystals. Two extreme cases may be distinguished: immobile impurities, as shown in Figure 9.11, which are fixed at the site where they first reach the crystal surface, and completely mobile impurities, which adsorb first and then diffuse two dimensionally on the surface. The strengths of bonds between the substrate and the impurity determine the type of interaction.

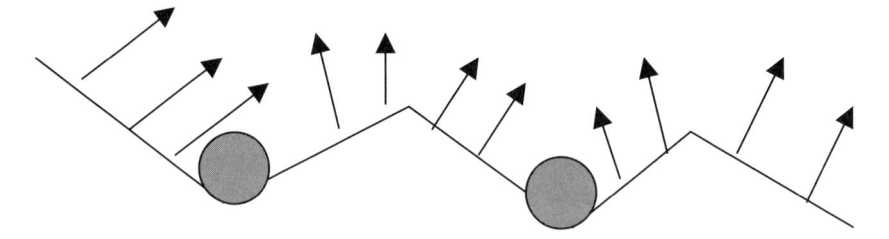

Figure 9.11 Pinning of an advancing step by immobile adsorbed impurities. Arrows show the direction of growth of the crystal faces.

The presence of mobile impurities as a rule does not cause morphological changes on the surface, as they are expelled from the surface with the advancement of steps. However, they may influence the relative velocities of growth of the steps on different crystal faces, and may also change the overall crystal morphology.

Inhibitory activity is manifested via the influence of the concentration of the impurities on the rates of crystal growth and may be related to: (i) ionic association in solution, thereby decreasing the driving force, i.e. the degree of supersaturation, and (ii) blocking of crystal growth sites by adsorption. The inhibition of the crystal growth of hydroxylapatite by amino-acids and proteins is a good example of this case (Moreno et al., 1979). HAP crystal growth was studied in the presence of L-serine, using the computer code HYDRAQL (Papelis et al., 1988) to calculate saturation and known stability constants for the HAP L-serine system. According to these calculations, at a concentration of serine of 0.01M only 0.6% of the total calcium contributes to the formation of an ion pair in solution. It is therefore evident that the presence of serine in the concentration range 1×10^{-3} to 1×10^{-2} M does not affect to any significant extent the concentration of free Ca^{2+} ions in solution and therefore the degree of the solution supersaturation with respect to HAP. Consequently, the inhibitory effect of L-serine may be ascribed to adsorption onto HAP surfaces and subsequent blocking of the active growth sites. This suggestion was tested by assuming Langmuir-type adsorption, according to which the rate of HAP crystal growth in the presence of the inhibitor, R_i, is given by (Bliznakov et al., 1971; Koutsoukos et al., 1980):

$$R_i = R_o(1 - \theta) \quad (29)$$

where R_o is the crystallization rate in the absence of inhibitors and θ is the fraction of the active growth sites occupied by the inhibitor adsorbed onto the HAP seed crystals ($0 < \theta < 1$). This can be described by a Langmuir isotherm:

$$1 - \theta = \frac{1}{1 + K_L C_i} \quad (30)$$

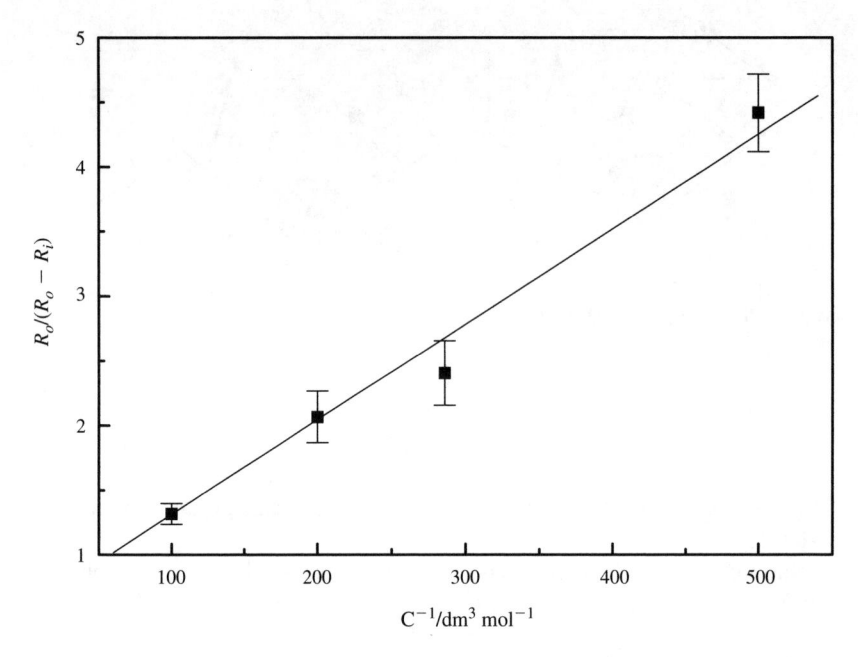

Figure 9.12 Kinetics of HAP crystal growth in the presence of various concentrations of L-serine, according to a Langmuir-type kinetic model. Experimental conditions: 37°C, pH 7.4. Modified from Spanos and Koutsoukos (2001).

In Equation 30 the adsorption constant K_L is the ratio of the rate constants for adsorption and desorption respectively, and may be considered as a measure of the affinity of the adsorbate for the adsorbent; C_i is the total equilibrium concentration of the adsorbate (L-serine). Combination of Equations 29 and 30 yields (Christoffersen and Christoffersen, 1981):

$$\frac{R_o}{R_o - R_i} = 1 + \frac{1}{K_L C_i} \tag{31}$$

Plots of the parameters of Equation 31 are shown in Figure 9.12. The linear fit of the kinetic data suggested that for the concentration range investigated, the inhibitory activity of L-serine may be explained by blocking of the active growth sites by a Langmuir-type adsorption.

From the slope of the straight line a value of 130 dm³/mol was obtained for the affinity constant. This value is much lower than those obtained kinetically for other inhibitors, reported in the literature (Koutsoukos et al., 1981; Amjad, 1987; Dalas and Koutsoukos, 1989b). The considerably lower affinity of HAP for L-serine is reflected by the significantly higher concentrations of L-serine needed for an appreciable inhibition (order of magnitude 10^{-3} M), as compared with the concentrations of other more effective inhibitors of HAP crystal growth (order of magnitude 10^{-6} or 10^{-5} M).

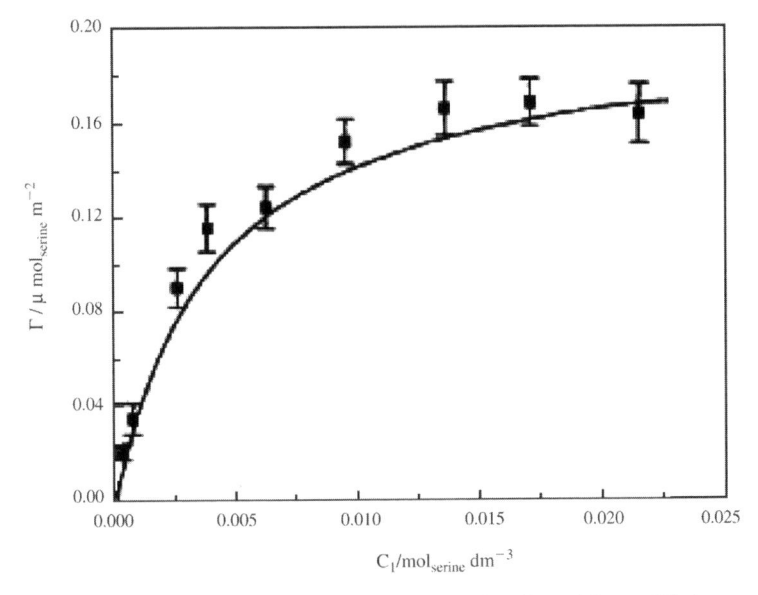

Principles of phosphate dissolution and precipitation 217

Figure 9.13 Uptake of L-serine adsorbed on HAP as a function of the equilibrium concentration; ■: experimental isotherm; Solid line: theoretical isotherm calculated on the basis of adsorption equilibria, reactions 32 and 33. Experimental conditions: pH = 7.4, T = 37°C, ionic strength 0.01 M NaNO$_3$. Modified from Spanos and Koutsoukos (2001).

The low affinity of HAP for L-serine may be corroborated from the analysis of the equilibrium adsorption results. Figure 9.13 illustrates the adsorption isotherm of L-serine onto HAP, obtained experimentally at pH = 7.4 ± 0.3. The L-type isotherm obtained suggests adsorption on distinct, energetically equivalent sites with no lateral interactions between the adsorbed species, i.e. Langmuir-type adsorption, in agreement with the adsorption-type on which the analysis of the kinetic results was based. Analysis of the adsorption results also allows the determination of the adsorption constant, K_L, and the saturated surface concentration (i.e. the surface concentration of adsorbate corresponding to monolayer surface coverage), Γ_m. The fraction of the active growth sites occupied by the adsorbate, θ, is related with Γ_m by: θ = Γ_m/Γ, where Γ is the total number of surface sites. The values 0.16 μmol m^{-2} and 546 dm^3 mol^{-1} are calculated for Γ_m and K_L respectively.

Despite the differences between the affinity constant obtained from the kinetics of the heterogeneous precipitation of HAP (on an HAP substrate), and that from the adsorption of serine onto HAP, they both are of the same order of magnitude. The discrepancy between the values of the affinity constant may be ascribed to the fact that the lower value was obtained from kinetics measurements (non-equilibrium), while the higher value was obtained from adsorption isotherms (equilibrium). Similar values of the affinity constant, calculated from adsorption isotherms, have been reported for other non-phosphorylated amino acids.

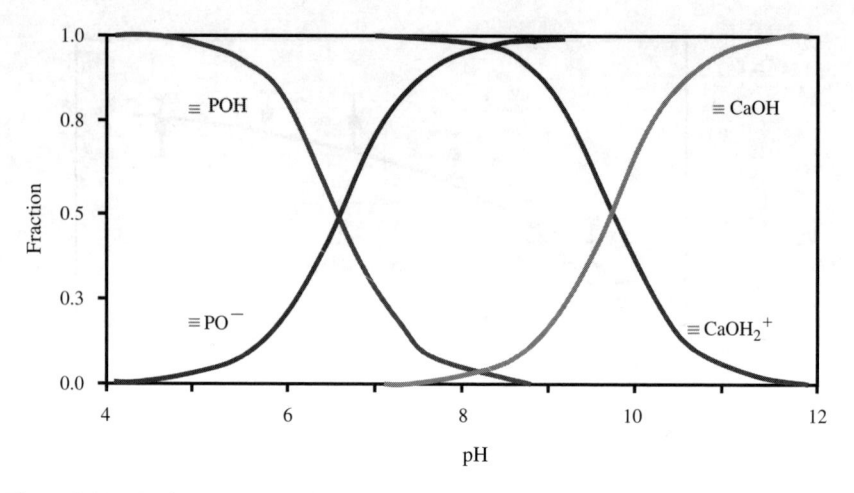

Figure 9.14 Surface speciation for HAP, as a function of pH (modified from Wu et al., 1991).

9.3.4 Investigation of the mechanism of adsorption on calcium phosphates

Given that both growth and dissolution are controlled by surface processes, and are triggered by the adsorption of solution components onto the surface of the phase under consideration, knowledge of surface speciation of the phase is essential, as this will reveal the likely tendency for it to react at any specific pH. Of all calcium phosphates, only HAP surface speciation is known, and this is shown in Figure 9.14. The figure suggests that, in the acidic to neutral pH range $\equiv POH^0$ and $\equiv CaOH_2^+$ predominate, whereas from neutral to alkaline $\equiv PO^-$, and $\equiv CaOH^0$ are the species present on HAP surfaces. The pH_{pzc} for HAP, where net surface charge is neutral, has been calculated as 8.15, in a carbonate free system (Wu et al., 1991, Figure 9.14), which shifts to 7.13 when carbon dioxide is present.

If we now return to the discussion of L-serine, and its adsorption on HAP, it is possible to evaluate the behaviour reported in previous adsorption work (Spanos et al., 2001). At pH = 10.0 the adsorption of L-serine on the surface of HAP was negligible. This may be explained by considering the speciation of both the HAP surface and that of L-serine. The pH dependent equilibria in which serine participates are the following:

$$HO-CH_2\underset{\underset{(H_2L^+)}{NH_3^+}}{CHCOOH} \xrightleftharpoons{K_1} HO-CH_2\underset{\underset{(HL)}{NH_3^+}}{CHCOO^-} + H^+ \quad (32)$$

and

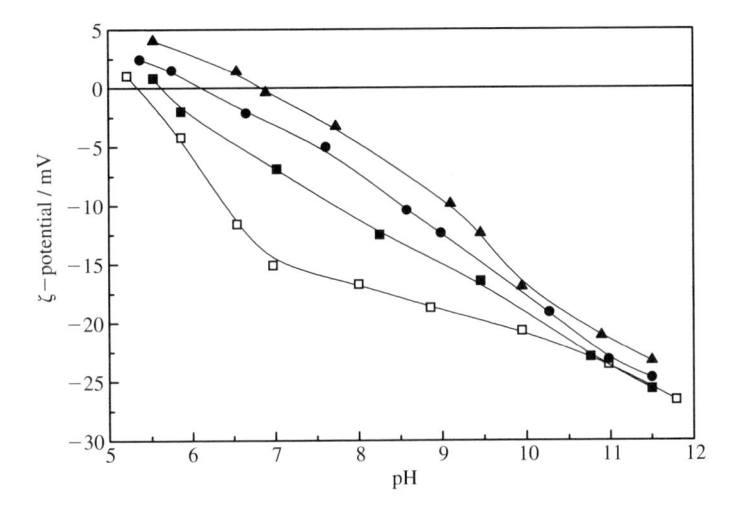

Figure 9.15 ζ-potential of HAP particles as a function of pH, in 0.01 M NaNO$_3$, 25°C: (□) in the absence of any additive; (■), (●) and (▲) in the presence of 3.5 × 10^{-3} M, 1 × 10^{-2} M and 2 × 10^{-2} M of L-serine, respectively. Modified from Spanos and Koutsoukos (2001).

$$\text{HO}-\text{CH}_2\text{CHCOOH}^- \xrightleftharpoons{K_2} \text{HO}-\text{CH}_2\text{CHCOO}^- + \text{H}^+ \\ \phantom{\text{HO}-\text{CH}_2\text{CH}}| \phantom{\xrightleftharpoons{K_2} \text{HO}-\text{CH}_2\text{CH}}| \\ \phantom{\text{HO}-\text{CH}_2\text{C}}\text{NH}_3^+ \phantom{\xrightleftharpoons{K_2} \text{HO}-\text{CH}_2\text{C}}\text{NH}_2 \\ \phantom{\text{HO}-\text{CH}_2\text{C}}(\text{HL}) \phantom{\xrightleftharpoons{K_2} \text{HO}-\text{CH}_2\text{CH}}(\text{L}^-)$$

(33)

The concentration of the three serine species at various pH values may be calculated by inserting equilibria 32 and 33 with their constants K_1 and K_2 (Martell and Smith, 1981), in the computer code HYDRAQL (Papelis *et al.*, 1988). According to these calculations the predominant species at pH 7.4 and 10.0 is HL (containing a deprotonated carboxyl group) and L$^-$ (deprotonated carboxyl and amino groups), respectively. It therefore appears that HL, predominant at pH 7.4, adsorbs substantially on the HAP surface, whereas the species L$^-$, predominant at pH 10.0, does not adsorb. Moreover, measurements of the ζ-potential, i.e. the potential at the shear plane of HAP particles in the absence of any additive (Figure 9.15), showed that the negative surface charge of HAP changes only gradually. So the only remarkable change in the system adsorbent-adsorbate is the deprotonation of the amino group, which takes place as pH increases from 7.4 to 10.0, suggesting that this group is mainly responsible for the adsorption of L-serine onto the HAP surface. Specifically, the protonated amino group ($-\text{NH}_3^+$) of the species HL, predominant at pH 7.4, interacts electrostatically with the negatively charged sites of the HAP surface, $\equiv\text{PO}^-$, forming ion pairs of complex adsorbed species. The fact that no calcium or phosphate ions were released from the solid HAP to the solution during the adsorption

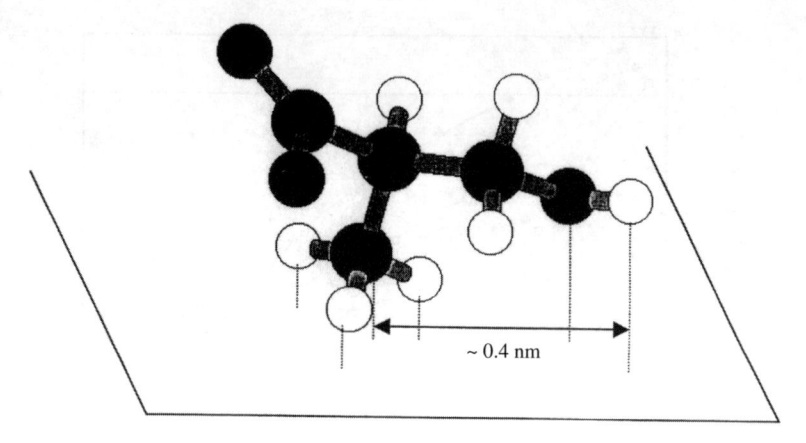

Figure 9.16 Model of L-serine molecule adsorbed on HAP based on minimum energy calculations.

isotherm experiments suggested that adsorption of serine is not a reactive process, i.e. ion exchange adsorption, but results from plain electrostatic surface attraction.

On the basis of the above discussion, a simplified model for the adsorption of L-serine on HAP can be developed. The protonated amino group may form complexes with the negatively charged HAP surface, whereas the deprotonated carboxyl group orients itself with the maximum possible distance from the surface because of electrostatic repulsion. The most likely position of the adsorbed serine molecule with respect to the surface is illustrated in Figure 9.16, in which L-serine molecule is drawn using calculations from commercially available molecular modelling software [CERIUS2 Molecular Simulations Inc.]. The projected molecular radius from the centre of the nitrogen atom is between 0.36 and 0.43 nm, and the mean effective area of a rotating molecule is about $0.48\,nm^2$. Assuming a density of 2.1 ions/nm^2 for the [$\equiv$ CaOH] groups and 3.1 ions/nm^2 for the phosphate [$\equiv$ PO$^-$] groups, the adsorption of one serine molecule on one phosphate surface group is accompanied by the simultaneous coverage of the adjacent hydroxyl group, being within a radius of 0.48 nm.

By inserting in the computer code HYDRAQL the above parameters and the total solution concentration of serine, the concentration of the adsorbed serine, HL$_{ads}$, may be calculated. It is therefore possible to calculate the uptake of serine at any serine concentration, pH and temperature. A calculated adsorption isotherm, has been compared with experimental data and found in perfect agreement (Figure 9.13; Spanos *et al.*, 2001).

The proposed adsorption model may also be tested by the results obtained from measurements of the ζ-potential of HAP particles in the absence and in the presence of L-serine. The adsorption of L-serine through electrostatic interactions exerted between the negative sites of the HAP surface and the positively charged protonated amino groups of the molecule of L-serine may cause a shift of ζ-potential to less negative values. The anionic charge introduced by the deprotonated carboxyl group, which according to the proposed model orients itself towards the layer of the counter ions (i.e. sodium ions in solution), is partially neutralized. The complexity of L-serine molecule favours the trapping of counter ions, which are drawn along with the moving particles, also shifting the ζ-potential to less negative values. The variation of ζ-potential as a function of pH in the presence of different concentrations of L-serine, showed the expected shift of the ζ-potential to less negative values (Figure 9.15). Increasing the concentration of L-serine in the solution, accompanied by increase of the adsorbed L-serine, caused a further ζ-potential shift to less negative values. Figure 9.15 also shows that the shift of ζ-potential to less negative values is more pronounced in the vicinity of pH 7.4 where adsorption of L-serine onto the HAP surface takes place. However, at higher pH, as adsorption becomes negligible, ζ-potential was not significantly affected by the presence of L-serine in the solution.

The example of the investigation on L-Serine – HAP interactions helps to demonstrate an integrated methodology for the study of inhibitors of crystal growth. To summarize, a detailed study involves the quantitative description of the interaction of the inhibitor/additive with the crystal growing from solution (adsorption isotherms and zeta potential measurements), measurement of the rates of crystallization in the presence and in the absence of the inhibitor and finally molecular modelling for the interaction of the inhibitor with the crystal surface.

9.3.5 Heterogeneous nucleation-epitaxial growth of phosphates on foreign substrates

Crystal growth of a crystalline phase may, in principle, take place on a crystalline substrate introduced in the corresponding metastable solution at conditions of equilibrium between the aqueous phase and the solid substrate. For the particular case in which the formation of crystalline overgrowth on a substrate is oriented, the process is termed epitaxial growth, from the composite Greek word επιταξια (επι + ταξις) meaning 'upon(=overgrowth) with orderliness'. Epitaxy involves the growth of one crystalline phase over another, provided that the atomic dimensions of one or more commonly occurring faces of each are similar. Epitaxy finds many applications; it is, for example, the trigger in artificial rain, caused by seeding clouds with silver iodide, the lattice of which resembles closely that of ice. The epitaxial growth of one calcium phosphate phase upon another or upon foreign (both organic and inorganic) substrates may be important steps during the crystallisation of apatites and in bio-mineralization. Thermodynamics, kinetics, and surface phenomena are all important factors in the epitaxial formation of a mineral salt on a substrate. In the following we review the considerable work done on the epitaxial growth of calcium phosphates not only on other calcium phosphate phases but also on foreign substrates both inorganic and organic. Thermodynamic and kinetics results will be presented and the importance of the substrate for the epitaxial growth will be investigated through published results.

As mentioned above, the nucleation process may be represented by a set of autocatalytic reactions, which lead to the formation of nuclei, which, upon reach-

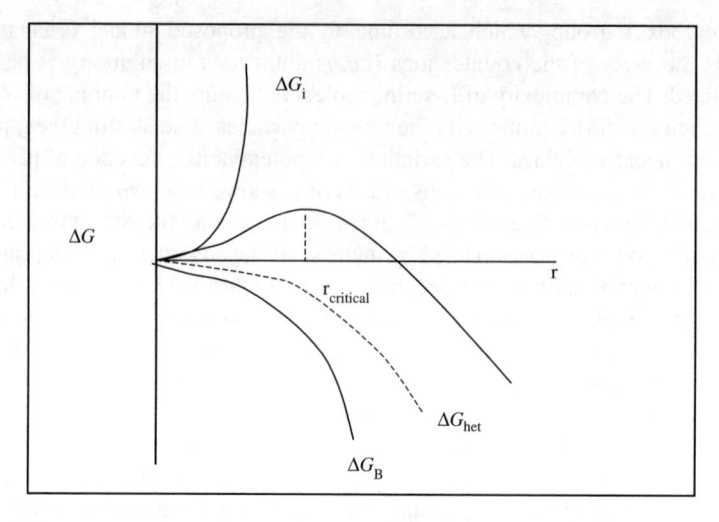

Figure 9.17 Free energy variation as a function of the cluster radius (r). Dashed line: heterogeneous nucleation ΔG_{het}. See text for definitions of ΔG_i and ΔG_B.

ing a critical size, grow macroscopically. Prerequisite to the formation of critical nuclei is exceeding an activation energy barrier. Nucleation requires that stable nuclei consisting of crystal building blocks are formed, in which the energy needed to form a new interface (ΔG_i) is less than the energy released in the formation of the bonds in the solid phase (ΔG_B). Nucleation is favoured at a critical nucleus size, which is proportional to the ratio $\Delta G_i/\Delta G_B$. The activation energy for nucleation, ΔG_N is given by Walton (1967):

$$\Delta G_N = \frac{16\pi(\Delta G_i)^3}{3(\Delta G_B)^2} \tag{34}$$

and

$$\Delta G_B = kT \ln S \tag{35}$$

In the absence of foreign substrates the energy barrier needed for nucleation is very high. The presence of foreign substrates facilitates nucleation in two ways: by influencing local supersaturation through ion binding of the substrate with the lattice ions of the mineral phase, and by lowering the activation energy barrier inducing heterogeneous nucleation (ΔG_{het}), as may be seen in Figure 9.17.

The presence of a substrate may even eliminate the free energy barrier and nucleation may take place instantaneously, upon the introduction of the foreign substrate. Beyond lowering this non-specific energy barrier however, a complementary factor is the matching of structural parameters between the substrate and the overgrowth. The latter effect is related to epitaxy, in which the lattice matching at the interface leads to oriented overgrowth of preferred crystal faces. Thus it has been shown that HAP crystals may grow on DCPD seed crystals introduced in calcium phosphate solutions supersaturated only with respect to the former phase (Koutsoukos and Nancollas, 1981a). A comparison between the (010) and (110) crystal planes of HAP and DCPD showed a minimum lattice mismatch which did not exceed the 15%. Synergism between calcium binding and structural matching was shown to favour the formation of HAP on synthetic polymers with phosphonic groups (Dalas et al., 1991). It is always difficult to prove oriented overgrowth in polycrystalline systems, especially in the absence of specialized facilities such as high resolution TEM or AFM. However the methodology of investigation of nucleation and growth of calcium phosphates at constant supersaturation offers a powerful tool for the experimental investigation of these processes since it allows for the precise measurement of very low rates of crystal growth (Dalpi et al., 1993; Koutsoukos et al., 1980; Tomson and Nancollas, 1978).

Epitaxial growth of a crystalline phase on a substrate takes place in solutions, provided that there is a chemical potential difference between the non-equilibrium state, $\mu_s = (\sum_i \mu_i)_s$ and the equilibrium, $\mu_\infty = (\sum_i \mu_i)_\infty$.

$$\Delta\mu = \mu_s - \mu_\infty \tag{36}$$

The non equilibrium state is created through the preparation of supersaturated solutions from the salt components (in the case of calcium phosphates from stock solutions of Ca^{2+}, PO_4^{3-} and OH^- ions) in such concentrations that the solutions are stable, i.e. no spontaneous precipitation occurs for practically infinitely long time periods. Regardless whether the process of epitaxial growth is considered macroscopically or at the atomic level, the system may be regarded as consisting of three components: the substrate, normally crystalline material, the bulk solution and an intermediate zone, adjacent to the crystalline substrate which transforms into the overgrown phase by the gradual increase of the orderliness of the growing phase. The molecular or ionic units in this intermediate zone are more disordered than the corresponding in the crystalline substrate, but less so than those of the fluid phase. All processes related to epitaxy take place precisely at this layer, which may be controlled by the conditions chosen.

Most often, in systems in which epitaxy is investigated, single crystals are selected as crystalline substrates. In the case of calcium phosphates however the majority of the reported investigations is concerned with polycrystalline materials in which it is difficult to distinguish whether there is true epitaxy, i.e. oriented overgrowth of a calcium phosphate phase over a substrate. The term epitaxy is however often used in crystallography and the use of this term may be justified in view of the selectivity of the substrates for the overgrown crystalline phase.

The role of the substrate is decisive for epitaxial growth, provided that the thermodynamic driving force in the solution is present. The key criterion for the relationship between the substrate and the overgrowth is the compatibility of their crystal lattices in geometrical terms. This implies favourable arrangement of the atomic units in the crystal lattices of the substrate and the overgrowth which may be attained either through the accommodation of structural defects in the layer or by strain associated with the relevant interfacial potential energy. A quantitative measure for the lattice compatibility between substrates and overgrown phases has been termed as lattice misfit by van der Merwe (1978). Assuming a rectangular symmetry for both the substrate and the overgrowth, the lattice misfit, δ_i, may be defined as:

$$\delta_i = \frac{\alpha_{oi} - \alpha_{si}}{\alpha_{si}} \qquad (37)$$

where α_{si}, α_{oi} are the lattice constants of the substrate and of the overgrowth respectively. Since the substrate surface however is not perfect the lattice misfit is expressed in relation to the fixed interatomic spacing.

The kinetics and energetics of solid/solid interfaces depend strongly on the spacing and structure of the lattices involved, as well as on their relative orientations. There are two extreme cases for the relationship between the substrate and the overgrowth: (a) complete coherence and (b) complete incoherence. In the first case both the substrate and the overgrowth have the same or very similar lattice parameters (Figure 9.18a). In this case, the crystal lattice shows continuity across the interface and it is only the chemical composition change that differentiates the overgrowth from the substrate. In the case of different materials the change in chemical composition may be sharp but in cases in which there is overgrowth of one calcium phosphate phase over another (e.g. HAP on DCPD), the composition change zone may be diffuse and spread over the surface of the substrate (Russel, 1980). The crystal lattice coherence may be limited to one orientation only as is the case for the close packed planes in adjoining hexagonal and face centred cubic crystals. It is usually impossible to achieve coherence between a substrate and an overgrowth when their crystal structures are entirely different. In most cases however, partial coherence is encountered (Figure 9.18b). The overgrowth of a crystalline

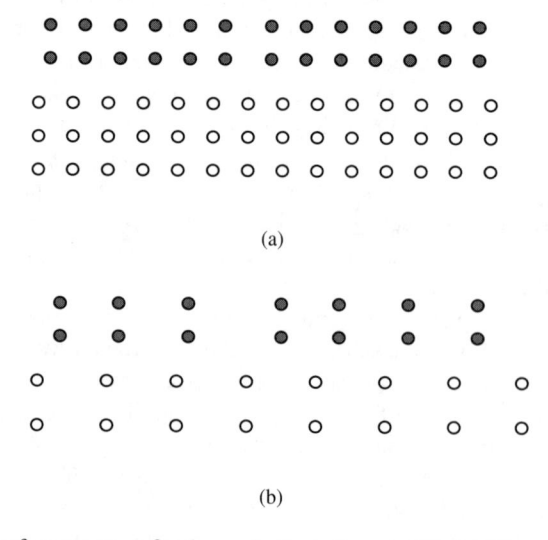

Figure 9.18 Interface structure for the nucleation of a crystalline solid on a crystalline substrate; (a) fully coherent interface (b) partially coherent interface.

Principles of phosphate dissolution and precipitation

layer on a substrate occurs in the direction of energy minimization. In incoherent interfaces, where there is little or no bonding, the energy is relatively higher in comparison with coherent interfaces.

In the majority of cases, however, interfaces are partially coherent and there are regions of high coherence separated by disordered regions. In the cases of formation of thin epilayers (1 or 2 atomic layers thick) the interatomic distances at equilibrium may differ by 5% and Equation 37 is used to express misfit. For thick epilayers however, the definition of lattice misfit is modified so that both the substrate and the overgrowth are treated equivalently (Herman, 1994):

$$\delta_i = \frac{2(\alpha_{si} - \alpha_{oi})}{\alpha_{si} + \alpha_{oi}} \tag{38}$$

When the lattice misfit between the substrate and the overgrowth is sufficiently small, the strain between the two lattices is minimum and as a result a coherent epilayer is formed on the substrate, which is a few atomic layers thick. Upon increase of the epilayer thickness dislocations develop in order to reduce the increasing strain energy. It should be noted that the criteria of lattice misfit posed by Equations 37 and 38 are mainly valid for substrate-overgrowth systems of the same crystallographic symmetry. If the heteroepitaxial systems differ crystallographically from the substrates, the interfacial translational symmetry should be taken into consideration in the comparisons of the characteristics of the bulk phases (Zur and McGill, 1984). Thus, in such cases, favourable lattice matching between two crystalline systems exists, provided that the interface translational symmetry is comparable with the symmetry on both sides of the interface.

The overgrowth of an epilayer over a substrate is preceded by the nucleation process in which the heteronuclei are forming on the substrate, according to the crystallographic limitations presented above. In the case of favourable lattice matching, the substrate may act as the catalyst of a series of stepwise bimolecular reactions leading to the formation of the supercritical nuclei.

The mechanism is the same as in homogeneous nucleation except for the fact that in the presence of a substrate the supersaturation needed for the phenomenon

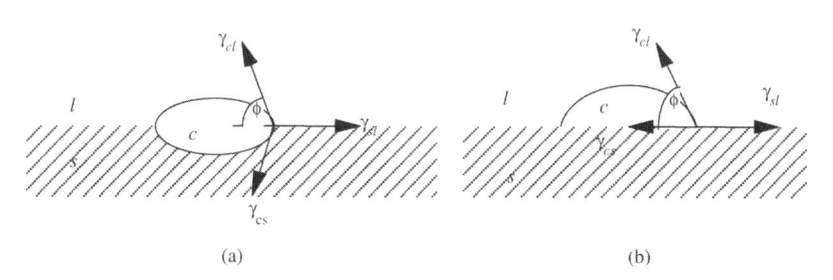

Figure 9.19 Nucleation of a crystalline solid, c, on a substrate, s; (a) incoherent nucleus; (b) fully coherent nucleus. l: liquid phase.

to occur is quite low. The relationship between the driving force for homogeneous nucleation, ΔG_{hom}, with the corresponding for heterogeneous, ΔG_{het}, is:

$$\Delta G_{het} = \Phi \Delta G_{hom} \qquad (39)$$

where Φ is a factor ($0 < \Phi < 1$) related to the contact angles, ϕ, between the substrate and the hetero-nucleus forming on the substrate (Nyvlt, 1971). As shown in Figures 9.19a and 9.19b, for the formation of coherent heteronuclei c on a phase s, it is necessary to have a favourable balance of the surface energies (Kern, 1978) (Figure 9.19b).

The contact angle between the overgrowth and the substrate in terms of the interfacial energies is given by Equation 40 (Venables and Price, 1975):

$$\cos \varphi = \frac{\gamma_{sl} - \gamma_{cs}}{\gamma_{cl}} \qquad (40)$$

and the relationship between the contact angle and the factor Φ is:

$$\Phi = \frac{(2 + \cos \varphi)(1 - \cos \varphi)^2}{4} \qquad (41)$$

In either homogeneous or heterogeneous nucleation, it is clear that the supersaturation is also very important, as it is represents the driving force. The ratio:

$$S = \left(\frac{(\Pi \alpha_i)_s}{K_s^0} \right)^{1/\nu} \qquad (42)$$

of the activity product α_i of the ν ions constituting the calcium phosphate phase forming, over the respective thermodynamic solubility product K_s^0 is the measure of departure from equilibrium, i.e. supersaturation. For the various calcium phosphates considered here the expressions for the activity products are:

DCPD: $(Ca^{2+})(HPO_4^{2-})$ $\qquad \nu = 2$
OCP: $(Ca^{2+})^4 PO_4^{3-})^3$ $\qquad \nu = 7$
β-TCP: $(Ca^{2+})^3 (PO_4^{3-})^2$ $\qquad \nu = 5$
HAP: $(Ca^{2+})^5 (PO_4^{3-})^3 (OH^-)$ $\qquad \nu = 9$

The rate, J, of formation of new nuclei on favourable substrates is given by Equation 43:

$$J = A \exp\left(-\frac{\varphi \Delta G}{kT}\right) \qquad (43)$$

where A is a pre-exponential factor related to the mechanism of incorporation of the growth units in the forming nucleus. Taking into consideration the Volmer-

Weber-Becker theory, the rate of precipitation as a function of the solution supersaturation S may be expressed as (Nyvlt et al., 1985):

$$J = A\exp\left(-\frac{\varphi\beta\sigma^3 v^2}{k^3 T^3}\frac{1}{(\ln S)^2}\right) \quad (44)$$

β, is a shape factor ($=16\pi/3$ for spherical nuclei), v the molecular volume of the growing phase and σ its surface energy. In most cases, the formation of heteronuclei is preceded by induction times, τ, inversely proportional to the solution supersaturation. The induction time is the period between establishment of supersaturation in the presence of the substrate, until the observation of the first changes in the solution properties, corresponding to the formation of calcium phosphates. This represents the time needed for the formation of critical nuclei, which may grow further, forming macroscopic crystalline overgrowth layers. Since the induction times are inversely proportional to the rates of growth of the new phase, a relationship similar to Equation 44 may be derived for the dependence of the induction time on supersaturation:

$$\tau = \exp\left(\frac{\Phi\beta\sigma^3 v^2}{(kT)^3}\frac{1}{(\ln S)^2}\right) \quad (45)$$

Experimentally, heterogeneous nucleation of a calcium phosphate phase on a substrate suspended in supersaturated solutions may be studied by monitoring the change of solution variables associated with the formation the new phase. Such solution variables for calcium phosphates are the pH, changes in total calcium, ΔCa_t, and total phosphate, ΔP_t (Moreno et al., 1977). The main problems associated with this approach include analytical errors, which render the identification of the precipitating phase based on calcium: phosphate molar ratio questionable, and the possible initial surges observed in heterogeneous precipitation processes. The surges introduce large uncertainties in the measurements of the rates of precipitation. These problems were discussed earlier, but are emphasized again here, as they are more pronounced in investigations of epitaxial growth, where the overgrowth may proceed very slowly, past relatively long induction times. The constant composition method serves again as the best means to ascertain well controlled experimental conditions (Koutsoukos et al., 1980; Tomson and Nancollas, 1978), and allows monitoring of overgrowth processes, which proceed at very low rates, and at the same time measuring accurately the precipitation rates from the traces of titrant solutions added as a function of time. Through the maintenance of constant supersaturation and of pseudo-steady state conditions, it is also possible to maintain the precipitation process for sufficiently long periods to grow enough material for the characterization, even at very low supersaturation (Koutsoukos et al., 1980). The invariance of the solution species activities throughout the crystal growth processes, which depend strongly and directly on the titrant solutions composition, is sometimes taken as a proof of the stoichiometry of the precipitating phase.

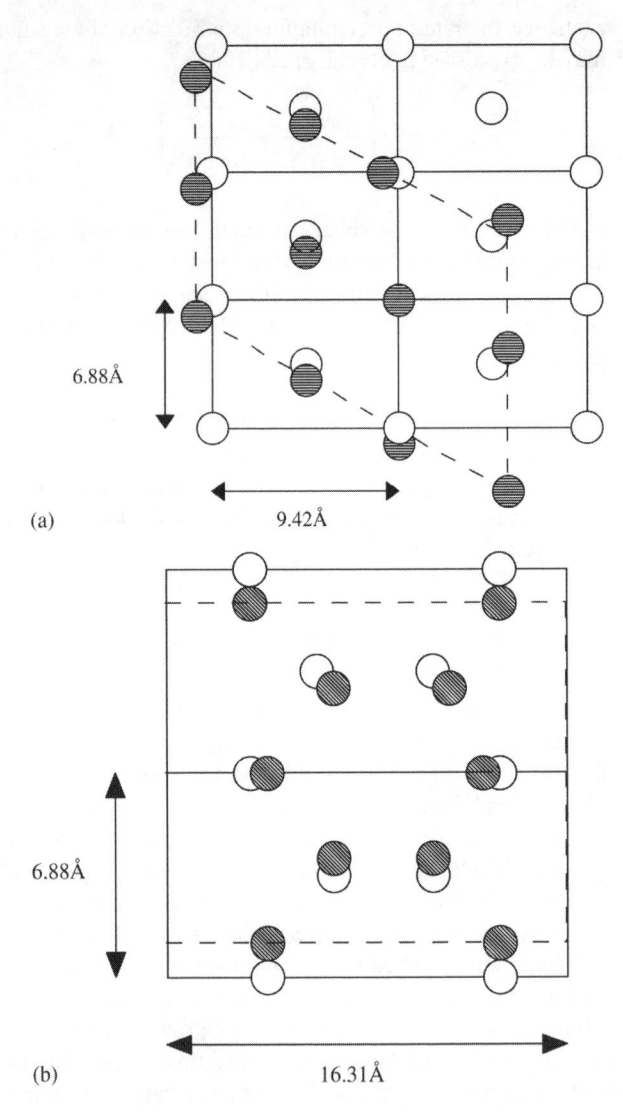

Figure 9.20 Calcium atoms in the unit cell at HAP (○) and DCPD (●). (a) (010) plane; (b) (110) plane.

DCPD is the calcium phosphate phase, which has been suggested to form at the initial stages of bone mineralization, being, according to some workers, the precursor of HAP (Francis and Webb, 1971; Kani et al., 1983; Neuman and Neuman, 1953; Neuman and Bareham, 1975). As may be seen from the solubility isotherms of the calcium phosphates shown in Figure 9.7, although HAP is the most stable phase, DCPD may exceed its stability if the solution is sufficiently acid. Upon increasing acidity, a more acid calcium phosphate may be expected to form on a calcium phosphate substrate. Seeded growth studies have shown that even at pH 7.40 the formation of DCPD is possible in aqueous solutions supersaturated with respect to all calcium phosphate phases (Koutsoukos et al., 1979). Moreover, DCPD together with HAP have been identified in embryonic chick bone (Roufosse et al., 1979). Experiments at lower pH (5.0–5.6) have confirmed that DCPD may be selectively grown on HAP seed crystals following the same kinetics as the growth of DCPD on DCPD seed crystals at similar conditions (Barone and Nancollas, 1977). The experimental data available corroborate the suggestion of epitaxial growth of HAP on DCPD (Koutsoukos and Nancollas, 1981a). Comparison of the (010) and (110) planes of the unit cells of the two phases showed that in these planes lattice matching is favourable. The compatibility between the unit cell lattice parameters for HAP and DCPD (less than 15% misfit) may also be seen in Figure 9.20.

However, the reasonable fit of the two lattices does not constitute, on its own, a necessary condition for epitaxial growth. In experimental solutions supersaturated with respect to TCP and HAP only, or saturated with respect to DCPD, and seeded with DCPD, it was observed that HAP was growing selectively on the seed crystals past the lapse of induction times inversely proportional to the solution supersaturation. Moreover, the rates of growth of HAP on the DCPD seed crystals increased with increasing supersaturation and were independent of the amount of the inoculating seed crystals. These two indications suggested that the growth of HAP on DCPD is selective and it is not secondary nucleation and growth (Brown et al., 1957).

The growth of HAP on CaF_2 crystals has also been reported (Koutsoukos and Nancollas, 1981a). In CaF_2, the calcium ions are found on a face centred cubic lattice. Each fluorine atom is surrounded by four calcium atoms and each calcium atom is surrounded by eight fluorine atoms (8,4 coordination). The lattice matching for the (111) and (1$\bar{1}$0) faces of CaF_2 with the corresponding faces for HAP is better than 8% (Koutsoukos, 1980).

The growth of DCPD on DCPA seed crystals has also been demonstrated at 60°C (Frèche, 1989). The rates obtained were higher compared to the rates of the growth of DCPD on DCPD at 37°C although in the latter case the supersaturation was higher (Salimi, 1985). The higher rates for the DCPD overgrowth on DCPA suggested a high activation energy. The linear dependence of the rates on the relative supersaturation suggested a surface controlled process. Moreover it should be noted that using the constant composition method, it was possible to demonstrate that at pH 5.0, heterogeneous nucleation of DCPD on OCP seed crystals was induced, past induction times and at a rate inversely proportional to the solution supersaturation (Feenstra and de Bruyn, 1979). Increase of pH to the domain where DCPD and OCP are not stable (pH 7.4–8.0) resulted in the hydrolysis of the substrate to non stoichiometric apatitic phases.

When considering the use of calcium phosphates as ceramic implants, it is important to know whether they have the capability of nucleation of apatite or apatitic minerals after sintering, or any other processing in which they are subjected, where phase changes may also occur (Dalas and Koutsoukos, 1989a; Osborn and

Newesely, 1980). HAP powders sintered at 900°C were converted into mixtures of TCP an HAP and introduced in solution supersaturated with respect to HAP and TCP only. It was, unexpectedly, observed that TCP was deposited on the calcined surfaces, showing direct formation of this, otherwise high temperature phase (Dalas and Koutsoukos, 1989a).

The fact that bone and teeth contain a fibrous protein, which forms the substrate on which calcium phosphate mineralization takes place has encouraged studies into the relationship between the organic matrix and the inorganic salts, mainly calcium phosphates, found in calcified tissue (Bills, *et al.*, 1982; Glimcher, 1976; Jenkins, 1978; Veis and Sabsay, 1983). Although the mediation of biochemical processes is a very important part of the biological mineralization, for the sake of simplicity we shall view here the mineralization process as heterogeneous nucleation. The substrate is provided by the biopolymers and biological fluids provide the supersaturation conditions, which are prerequisite for the formation of the mineral phase. The formation of enamel is an example of matrix directed mineralization where the reactions in the mineralization front, where HAP is formed by heterogeneous crystallization, are mediated by phosphophoryns. In bone, X-ray diffraction studies have revealed crystallographic relationship between small HAP crystallites (ca 5 nm × 100 nm) and the collagenous matrix, formed from a simple protein containing polypeptide chains of 1000 amino acid residues (Bills *et al.*, 1982).

The heterogeneous nucleation of calcium phosphates on polymeric substrates has been studied in order to assess the significance of the polymer structure in inducing nucleation of calcium phosphates (Dalas *et al.*, 1991). Polymers modified to contain phosphonic groups ($>(P=0)-O-R$) with various degrees of phosphorylation showed that increase of phosphorus content in the polymer resulted in higher rates of HAP deposition. Moreover, a favourable polymer conformation was found to be necessary for the enhancement of HAP deposition. It is interesting to note that in all polymers tested, in which phosphorus content varied between 0.5–2.0% w/w, the formation of HAP took place without any appreciable induction time.

Experiments using collagen type I showed also that HAP was formed at low supersaturation without the lapse of induction times (Koutsoukos and Nancollas, 1987). This result is in agreement with the results obtained from the phosphorylated polymers, since exposure of collagen to phosphate containing solutions resulted in the uptake of considerable amounts of phosphate on the collagenous matrix (Koutsoukos and Nancollas, 1986). It has been postulated that phosphorylation of the collagenous matrix and more specifically of the serine amino acid is a prerequisite for the formation of nucleation sites on collagen (Glimcher, 1959, 1960). Nucleation and growth of HAP on collagen type I from bovine Achilles tendon and on collagen obtained from demineralized bone occurred following the same mechanism, yet the rates of crystallization at constant supersaturation conditions were lower for the collagen type I substrate.

At higher supersaturations OCP could nucleate and grow past induction times both in collagen type I (Koutsoukos and Nancollas, 1987) and collagen from dem-

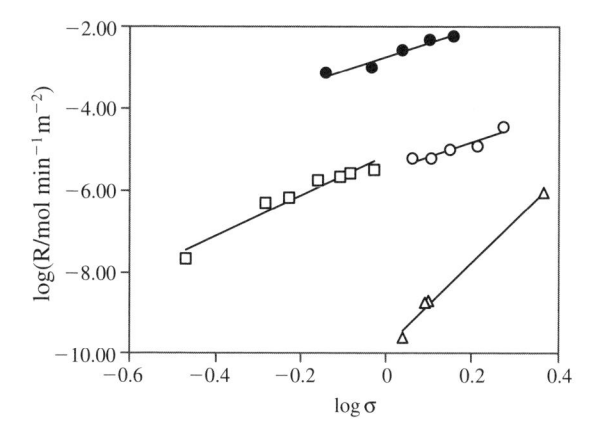

Figure 9.21 Logarithm of the crystal growth rates of OCP as a function of the logarithm of the relative supersaturation with respect to this phase at conditions of constant supersaturation and at 37°C. (●) OCP on heart valves (Kapolos et al., 1997); (□) OCP on OCP seed crystals (Heughebaert and Nancollas, 1984); (○) OCP on collagen from demineralized collagen (Combes, 1996); (△) OCP on Achilles tendon collagen type I (Koutsoukos and Nancollas, 1987).

ineralized bone (Combes, 1996). Experiments on glutaraldehyde treated porcine heart valves showed also the possibility of these surfaces to nucleate OCP without induction times (Kapolos et al., 1997); the rates of precipitation, compared with those of overgrowths of OCP on OCP seed crystals (Heughebaert and Nancollas, 1984), are shown in Figure 9.21.

It may be seen that the growth of OCP on collagen type I obtained from demineralized bone was closer to the growth of OCP on OCP seed crystals. Type I collagen from Achilles tendon yielded lower rates, a fact which may be ascribed both to the different type of collagen but also to the lower purity of this product, which was obtained commercially and may contain low molecular weight collagenous fractions which inhibit the formation of HAP. Heart valves yielded faster rates. It should be noted however in this case that the reported rates do not distinguish between deposition on the collagenous matrix and on the elastin rich matrix present in one of the two sides of the valve membranes. Elastin is a biopolymer highly cross-linked, found in arterial walls and ligaments. Experiments in which calcium phosphate supersaturated solutions were seeded with HAP showed that elastin could induce the formation of HAP, past induction times (Koutsopoulos et al., 1994). As anticipated for selective overgrowth, the induction times were inversely proportional whereas the rates of HAP crystallization at constant supersaturation were proportional to supersaturation. As in the case of collagen, significant amounts of orthophosphate were adsorbed from the phosphate containing solutions in contact with elastin, suggesting a similar mechanism for the initiation of the mineral deposition.

9.3.6 Competition/co-precipitation with carbonates

In the environment and/or biological fluids the presence of other ions in abundance, such as the carbonate ions, results in their incorporation into the apatitic lattice of calcium phosphates forming at low supersaturation. The presence of carbonate ions in supersaturated calcium phosphate solutions, in which the crystal growth of HAP seed crystals is occurring at pH 7.40, 37°C and 0.15 M NaCl, caused a reduction to the rates of HAP growth past a supersaturation threshold. Carbonate ions were incorporated into the HAP lattice substituting phosphate ions and also resulting in changes in the morphology of the HAP crystallites, which showed a plate-like habit. The incorporation of carbonate in the apatite lattice was further evidenced from the increase of the molar Ca:P ratio from 1.67 to 1.74 (Kapolos and Koutsoukos, 1999). Phosphate has long been used for water treatment and as an agent against calcium carbonate scale formation, and has been shown to inhibit calcite precipitation in many studies (e.g. Plant and House, 2002; Sawada, 1997). However, the extremely low solubility of calcium phosphates may also cause their direct precipitation in the presence of carbonate, in which case the carbonate ions will act as the inhibitors for phosphate precipitation. Mixed carbonate-phosphate phases may also form in cases where supersaturation with respect to both these two salts is exceeded. The calcium phosphate most likely to form, directly or through a precursor phase, is HAP, as discussed earlier. The stable calcium carbonate polymorphs in aqueous supersaturated solutions include in the order of increasing solubility calcite, aragonite and vaterite. It should be noted that the solubility product for the calcium carbonate is approximately fifty orders of magnitude higher than that of HAP. It is therefore anticipated that the presence of low phosphate and calcium concentrations (<0.5 mM of alkaline pH) may cause the spontaneous precipitation of HAP, while the precipitation of calcium carbonate is inhibited by the presence of the phosphate ions (Giannimaras and Koutsoukos, 1987; Plant and House, 2002); the threshold for spontaneous precipitation of calcium carbonate at these conditions is at considerably higher concentrations (Xyla *et al.*, 1991).

The presence however of HAP crystalline deposits in an aqueous medium supersaturated with respect to calcium carbonate may selectively induce the precipitation of one of the calcium carbonate polymorphs by providing the appropriate template for crystal growth. Work undertaken to investigate the possibility of selective formation of calcium carbonate onto HAP substrates, and the measurement of the kinetics of the preferred overgrowth, showed that the inoculation of calcium carbonate supersaturated solutions with HAP seed crystals resulted in the formation of calcite, 65 hours after the lapse of the induction period. The dependence of the induction times on the initial calcium concentration in the supersaturated solutions is shown in Figure 9.22.

Furthermore, the rates of $CaCO_3$ overgrowth on HAP seed crystals introduced in supersaturated solutions showed a second order dependence on the relative supersaturation, as may be seen in Figure 9.23, suggesting surface controlled process. The overgrowth was promoted by the lattice compatibility of HAP and $CaCO_3$.

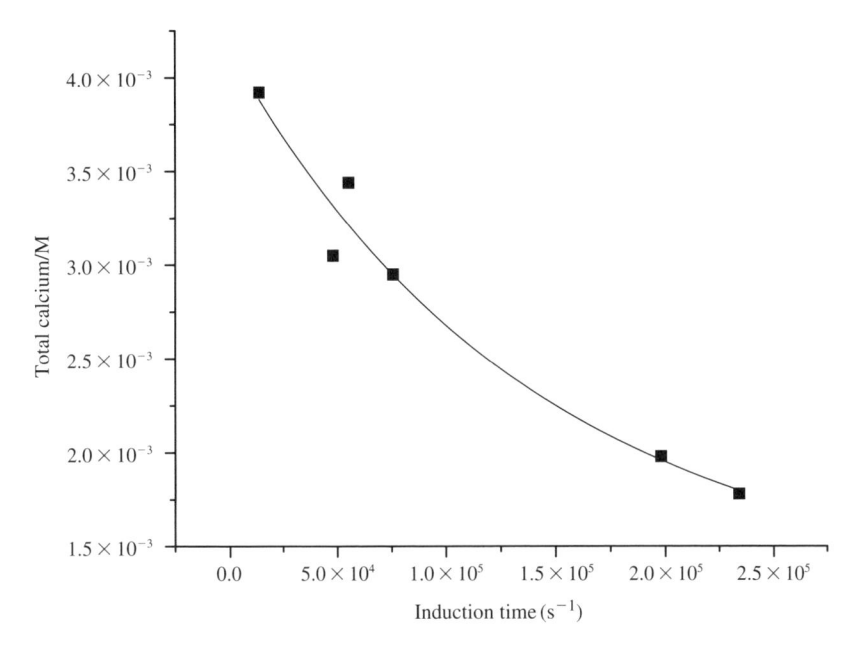

Figure 9.22 Dependence of the induction times on the initial calcium concentration in the supersaturated solutions for the crystallization of calcite on HAP seed crystals at constant supersaturation (pH 8.50, 25°C).

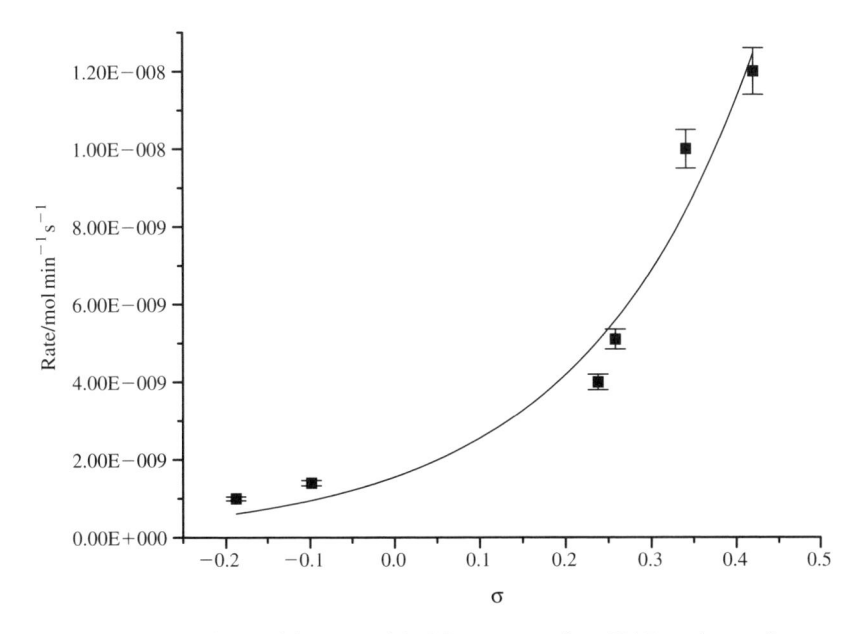

Figure 9.23 Dependence of the rates of $CaCO_3$ overgrowth on HAP seed crystals on saturation, at constant supersaturation (pH 8.50, 25°C).

The $(10\bar{1}0)$ $(10\bar{1}0)$ face of calcite, with net dimensions at 29.92Å (90°), gave a misfit with the $(10\bar{1}0)$ face of HAP (linear dimensions 28.26Å × 34.4Å (90°)) of 5.9% and 0.8% respectively (Lonsdale, 1968). Moreover comparison with the dimensions of $1\bar{1}0$ face of HAP 27.52Å × 32.62Å (90°) yielded a mismatch of 4.4% and 8.1% respectively. This lattice matching perhaps suggests that calcite may be a favourable substrate for HAP (Koutsoukos and Nancollas, 1981a) and vice versa. The selective deposition of calcite on the HAP substrate was demonstrated by experiments in which the amount of the seed crystals was varied. Using 100 mg of HAP instead of 50 mg to inoculate the supersaturated solutions, the induction times remained the same while the rates were doubled, suggesting a constant number of active sites. However, a difficulty presented in the experimental work which added to the uncertainty of the measured kinetics parameters, was the long induction times and the very slow rates of calcite overgrowth. Extension of the induction times and reduction of the rates was anticipated from earlier seeded crystal growth studies of calcite, which have shown that the presence of very low concentrations of orthophosphate had a strong inhibitory effect (Giannimaras and Koutsoukos, 1987).

9.4 APPLICATIONS RELEVANT TO THE DISSOLUTION OF CALCIUM PHOSPHATES

The literature on dissolution of calcium phosphates is currently dominated by applications related to bone loss and dental calculus, or the stability of calcium phosphate biomaterials. Some, mostly earlier, work has, however, addressed the fundamental dissolution properties of calcium phosphates.

Most phosphate minerals are known to be sparingly soluble (e.g. Vieillard and Tardy, 1984), although this rule does not apply to some members of the calcium phosphate group (see examples in Table 9.1). Figure 9.7 shows that by far the least soluble calcium phosphate is hydroxylapatite, which, for this reason, dominates the natural occurrences of calcium phosphates. At pH 7 and 25°C, the solubility order of calcium phosphates, from least to most soluble is: hydroxylapatite (HAP), tricalcium phosphate (TCP), dicalcium phosphate dihydrate (DCPD), octacalcium phosphate (OCP) and anhydrous dicalcium phosphate (DCPA). However, from Figure 9.7, it is apparent that the combined solubility of calcium phosphates is more complex, and this order is but a snapshot of their relative stabilities. In fact at low pH (<5) DCPD becomes the most stable calcium phosphate. For this reason, as well as due to precipitation and transformation kinetics (discussed earlier), all calcium phosphates can be found forming naturally, sometimes in extreme or special environments (see Valsami-Jones, this volume), or as short lived, transient phases, although by far the most common is HAP, to which other calcium phosphates tend to recrystallize. It should be noted that although the hydroxylated variety of apatite is the dominant form in environmental (earth surface) systems, fluorapatite is the commonest stable calcium phosphate in geological (subsurface) systems. This is an important observation, as it suggests that availability of phosphorus in environments where apatite is

Principles of phosphate dissolution and precipitation 235

derived from a rock source will be controlled by a less soluble mineral: fluorapatite is a few orders of magnitude less soluble than hydroxylapatite (their thermodynamic stability constants differ by 6–10 units, depending on the data source and equation configuration, see for example: Vieillard and Tardy, 1984; Valsami-Jones *et al.*, 1998). A further variety of apatite, chlorapatite is more soluble than both hydroxyl- and fluorapatite (e.g. Vieillard and Tardy, 1984).

There is no clear consensus in the literature, regarding the dissolution mechanisms or solubility constants for apatite, despite the substantial body of work over the last four decades. If we consider dissolution mechanisms first, both surface controlled (e.g. Christoffersen and Christoffersen, 1979; Margolis and Moreno, 1992; Valsami-Jones *et al.*, 1998) and diffusion controlled (e.g. Brown and Chow, 1981; Lower *et al.*, 1998) dissolution have been suggested. Similarly, the thermodynamic solubility products for HAP reported in the literature vary by several orders of magnitude (see for example the compilation in Rakovan (2002) where listed values range from $10^{-56.02}$–$10^{-88.5}$). There are many reasons for this variability: the choice of thermodynamic data (i.e. dissociation constants) used in calculations is one. Another reason is the variation in composition, crystallinity and surface properties of the material used. Apatites can contain a variety of trace components, which are known to affect their solubility. Magnesium and carbonate ions for example have been reported to reduce crystallinity and increase the solubility of apatites (e.g. LeGeros *et al.*, 1995). More importantly, apatites come in a variety of crystallinities, from well crystallized, stoichiometric (e.g. rock apatites) to poorly crystalline

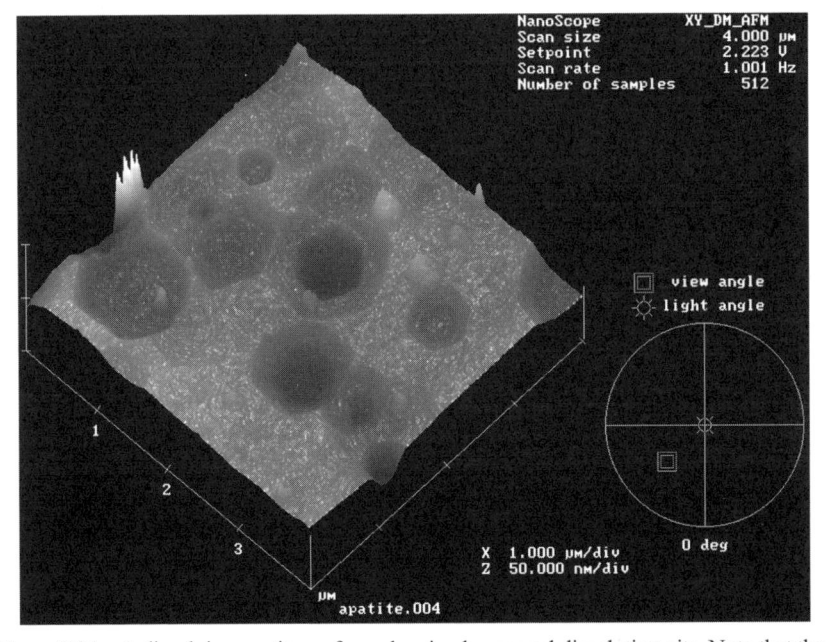

Figure 9.24 A dissolving apatite surface, showing hexagonal dissolution pits. Note that the vertical scale (Z) is 50 nanometers; X and Y scales are in μm. Data collected using Atomic Force Microscopy, courtesy of Andrew Putnis and Dirk Bosbach, Münster University. [This figure is also reproduced in colour in the plate section after page 336.]

phases with multiple vacancies (e.g. bone apatites) and these differences are likely to have a significant effect on structural stability, and consequently solubility. Many studies have used synthetic apatites to assess dissolution; these have formed in aqueous media containing organic molecules and/or other impurities, which may affect apatite surface properties; some synthetic apatites have been thermally treated, and this, again will have an effect on their behaviour. Such variability is not documented fully in the literature, resulting in the reported inconsistencies.

It is likely that well crystallized rock apatite dissolves via a surface controlled mechanism. Sparingly soluble phases in general, are considered to have surface controlled dissolution due to the fact that the detachment of the unstable surface species (see Equation 24) is likely to be a lot slower compared to aqueous diffusion. The formation of hexagonal etch pits during dissolution is a further indication. Figure 9.24 shows a (0001) growth surface of a well-crystallized natural fluorapatite, imaged with Atomic Force Microscopy (AFM) and displaying hexagonal etch pits with six-fold symmetry. These were formed after exposure of an atomically flat growth surface to a pH 3 solution (acidified with HNO_3). In contrast to this, a study by Lower et al. (1998), which also used AFM to image dissolving apatite surfaces, concluded that their dissolution was diffusion controlled, partly based on the lack of etch pits on the dissolving apatite surfaces. However, the apatite studied was a commercially available, poorly crystalline (or even potentially amorphous) non-stoichiometric hydroxylapatite.

The literature on the dissolution mechanism of other calcium phosphates is more limited. It has been proposed that TCP dissolution is diffusion controlled (Bohner et al., 1997). TCP does not form in aqueous systems and is usually synthesised at high temperatures (Elliott, 1994); for this reason, it tends to be unstable in aqueous media. Unlike apatite, the presence of magnesium reduces the solubility of TCP (Verbeek et al., 1986) due to a lattice stabilisation by the smaller magnesium ion

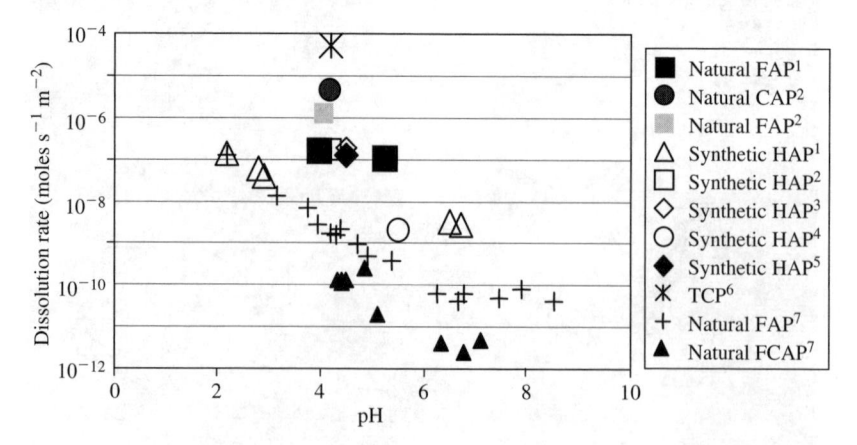

Figure 9.25 Dissolution rates of apatites as a function of pH. FAP: fluorapatite. CAP: chlorapatite. HAP: hydroxylapatite. CFAP: carbonated fluorapatite. TCP: tricalcium phosphate. Sources: [1]Valsami-Jones et al. (1998); [2]Manecki et al. (2000); [3]Christoffersen and Christoffersen 1979; [4]Tang et al. (2003); [5]Budz and Nancollas (1988); [6]Bohner et al. (1997); [7]Guidry and Mackenzie (2003).

(Elliott, 1994), a fact that is reflected by the far more common occurrence of the mineral whitlockite (a Ca–Mg phosphate with the structure of TCP) in natural systems. Both Verbeeck and Devenyns (1992) and Zhang and Nancollas (1992a) investigated the dissolution of octacalcium phosphate and proposed a surface control for this phase. Unlike apatite, OCP dissolution is not affected by the presence of carbonate, most likely because it is not a lattice component of OCP (Tung et al., 1988). For DCPD, both diffusion controlled (Nancollas and Marshall, 1971) and surface controlled (Zhang and Nancollas, 1992b) dissolution have been reported.

Dissolution kinetics data in the literature is only available for apatite with a few exceptions for other calcium phosphates (Figure 9.25). In some studies, rates are reported normalised to mass, rather than surface area, and such data have been excluded from the compilation shown on Figure 9.25. The dissolution rates of apatites are about an order of magnitude higher than the rates of common silicates and oxides (cf. Figure 9.25 with Figure 13.9 in Stumm and Morgan, 1996). It is difficult to interpret the combined data of Figure 9.25 in terms of a model such as that of Figure 9.4, but there appears to be a trend for slower kinetics, as pH increases. Variations caused by the source (natural vs. synthetic) and quality (preparation method, crystallinity, composition) of material, discussed earlier, also apply to kinetic experiments. Note, for example, that the rates of dissolution of a natural FAP and a synthetic HAP, reported in the same study (Valsami-Jones et al., 1998) and measured under identical experimental conditions vary by an order of magnitude, with FAP dissolving faster than HAP. Higher surface roughness and chemical heterogeneity of the natural phase were proposed as the factors causing this variation. Natural chlorapatite dissolves faster than fluorapatite (Manecki et al., 2000). Of all calcium phosphates, only data for TCP are available in the literature. Figure 9.25 shows that this phase dissolves much faster than apatite, in agreement with its diffusion controlled dissolution mechanism.

The presence of other dissolved species can have a major influence in the dissolution of calcium phosphates, in two opposite ways. It can enhance dissolution, by forming an unstable surface species on the mineral surface (according to Reactions 23 and 24) and facilitating detachment, or it can inhibit dissolution, by stably attaching to the surface and blocking surface sites. The same species can have a different effect on different calcium phosphates. For example, citric acid accelerates the dissolution of TCP, inhibits that of HAP and has no effect on DCPD, OCP and a carbonated apatite (Tang et al., 2003). HAP dissolution rates also increase in the presence of acetic, lactic, phosphoric and oxalic acid (Margolis and Moreno, 1992; Welch et al., 2002). Dissolution of apatite by organic acids (particularly lactic and acetic) is important, as it represents the process that leads to dental carries. It has been suggested that tooth apatite undergoes a constant dissolution/reprecipitation due to the presence of organic acids produced by bacteria; dental carries is the result of dissolution exceeding reprecipitation, whereas the presence of fluorine can decelerate dissolution. This occurs by replacing the existing more soluble magnesium carbonate apatite with a less soluble magnesium and carbonate poor and fluoride rich apatite (LeGeros, 1999). The dissolution of apatite by another group of organic ligands, humic and fulvic acids, is also an important process, as it models the conditions of apatite-controlled phosphorus availability in soils. Increased dissolution rates, as a function of pH and humic or fulvic acid concentration, and formation of P-humic acid complexes has been reported (Lobartini *et al.*, 1994).

Enhanced apatite dissolution has also been reported in the presence of microbial cultures in the lab (Welch *et al.*, 2002) and ectomycorrhizae in a field study (Blum *et al.*, 2002). Microorganisms are likely to gain a major advantage if they are able to solubilize phosphate minerals, and many studies have shown this to be the case; interestingly, it has been demonstrated that both fungi and bacteria can enhance the dissolution of highly insoluble metal phosphates (e.g. Di Simine *et al.*, 1998; Sayer *et al.*, 1999).

Anionic polymers are reported to inhibit apatite dissolution, via a mechanism that involves binding to calcium on the apatite surface by a phosphate or phosphonate group in the polymer (Schaad *et al.*, 1994).

Organic molecules, associated with bone formation, have been tested *in-vitro* for their effect on apatite and/or other calcium phosphate dissolution. Examples, from a rather extended literature, include evidence that osteoclasts (the organic molecules responsible for bone resorption) have little effect on apatite dissolution at physiologic pH, compared to a control, but become activated when pH is reduced (Kim *et al.*, 2001). A potential mechanism by which bone apatite dissolution may be controlled has been explained in a study demonstrating that proteins can be incorporated into the apatite crystal lattice, and decrease its solubility as well as increase its strength (Liu *et al.*, 2003). Proteins (osteocalcin, bovine serum album) have been shown to control calcium phosphate dissolution and precipitation by binding to surfaces and controlling growth in certain crystallographic directions (Fan *et al.*, 2000; Flade *et al.*, 2001).

9.5 METAL PHOSPHATES, DISSOLUTION AND PRECIPITATION APPLICATIONS

Phosphate minerals, other than calcium phosphates, are less abundant and, perhaps for this reason, have generally received less mention in the literature. It was not until Nriagu (1974, 1984) established solubility data for a range of metal phosphates and discussed their potential role in the dispersion of base metals in earth surface systems, that their significance became apparent. The composition and solubility constants of some of the most common metal phosphates has been presented elsewhere (Valsami-Jones, this volume); here we discuss examples of how the solubility of metal phosphates may be controlling a range of environmental processes.

Firstly, it should be emphasized that metal phosphates are highly insoluble, a lot more than their carbonate or sulphate counterparts, and can precipitate from aqueous media under earth surface conditions, unlike some metal silicates which often require high temperatures to stabilize. That makes them particularly good candidates to fix metals as phosphates in the environment. This process has often to be inferred rather than demonstrated, because very small concentrations suffice for such insoluble phases to form. However, once formed in small quantities, and within heterogeneous environments (e.g. soils), they are very difficult to detect. Recent studies providing evidence of their presence will be discussed next.

In terms of mechanisms, when dissolved phosphate is not available for metal phosphates to form, apatite may be the only source of phosphorus in soil or surface water environments. Many early experimental studies of the interaction between apatite and aqueous metals (Pb^{2+}, Cu^{2+}, Cd^{2+}, Zn^{2+}, Ni^{2+}, Co^{2+}, Mn^{2+}, Mg^{2+}, Ba^{2+}) invoked an ionic exchange by diffusion mechanism to explain the loss of metal from solution, coupled with an increase in calcium (e.g. Jeanjean et al., 1994; Suzuki et al., 1981, 1982; Takeuchi and Arai, 1990). Increasingly, evidence suggests that the mechanism is that of dissolution-precipitation, which fortuitously appears as an exchange due to rapid uptake of any phosphate released from apatite by the metal, followed by precipitation of metal phosphate. The removal of the phosphate from the system due to metal precipitation drives apatite dissolution forward and apatite continues to dissolve. In one study, it was shown that if sufficient metal (lead) is present in solution, the entire mass of apatite (a poorly crystalline synthetic phase) will dissolve, and all available phosphate will precipitate as lead phosphate (lead hydroxylapatite[1], Valsami-Jones et al., 1998). Another study observed some inhibition of apatite dissolution by the precipitating metal phosphate; the precipitating phase in that case was a pyromorphite (Xu and Schwartz, 1994). It is interesting to note that at least two studies have shown that the formation of lead phosphate, in systems where the dissolving apatite is the phosphate source, appears to happen epitaxially (Lower et al., 1998; Valsami-Jones et al., 1998), in other words the apatitic substrate forms a template and facilitates the precipitation of the metal phosphate (see also discussion in section 9.3.5 above). This mechanism is shown in Figures 9.26 a–c, where SEM and AFM images are combined to show the growth of a lead hydroxylapatite in two different apatitic substrates: a sheet like poorly crystallized hydroxylapatite (9.26a and b) and a crystalline fluorapatite (9.26c).

There is evidence that such processes take place in metal contaminated environments. For example Nriagu (1984) reported lead phosphate formation in roadside ecosystems and Cotter-Howells and co-workers identified metal phosphates in mine waste contaminated soils (Cotter-Howells and Thornton, 1991; Cotter-Howells et al., 1994). Jerden and Sinha (2003) reported fixing of uranium as a phosphate in a soil profile. Some recent work has also attempted to exploit this process and use as a means of locking-up metals, as insoluble phosphate, in contaminated soils and waters (Hodson et al., 2000, 2001; Ma et al., 1993; Seaman et al., 2001). For microbially mediated precipitation of metal phosphates see also Macaskie et al. (this volume).

Other metal phosphates of interest include strengite ($FePO_4 \cdot 2H_2O$) and variscite ($AlPO_4 \cdot 2H_2O$), which, probably in a poorly crystallized form, are the phases controlling phosphate availability in soils and surface waters at low pH (e.g. Gosh et al., 1996; Petticrew and Arocena, 2001; Zhang et al., 2001),

---

[1] Note that the names lead hydroxylapatite and hydroxypyromorphite are both used in the literature to describe the same phase: $Pb_5(PO_4)_3OH$.

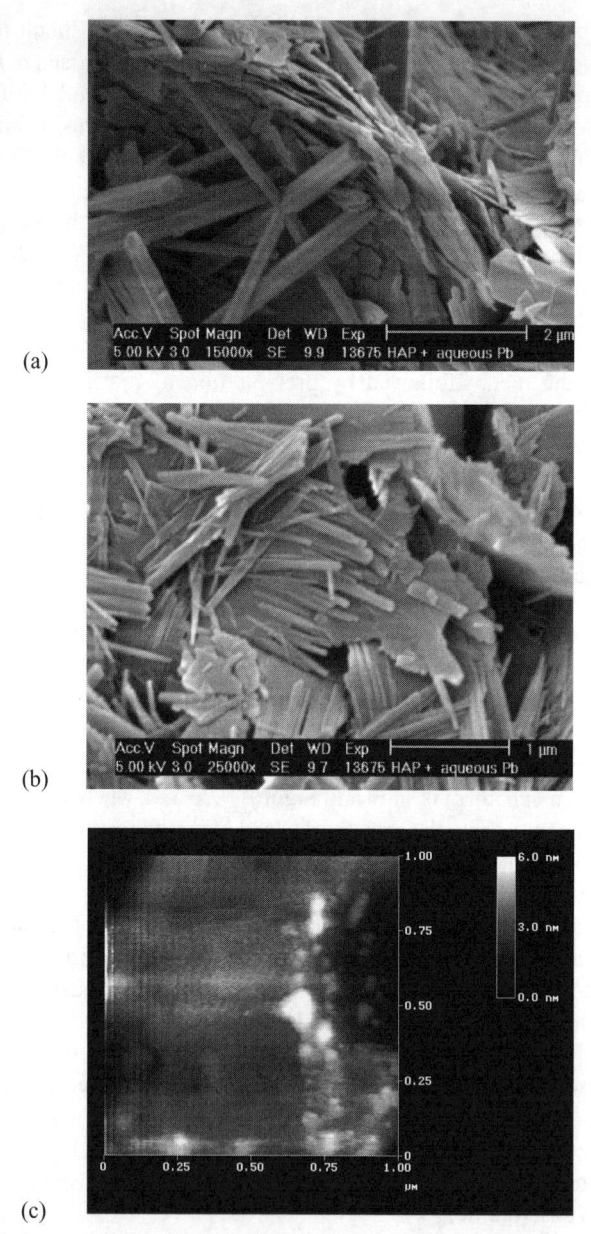

Figure 9.26 (a) SEM image of a Pb-phosphate (elongated hexagonal crystals) nucleating on synthetic HAP crystals (2-dimentional sheets). (b) As before; note nucleation of the lead phosphate needles along HAP sheets. (c) AFM image of Pb-hydroxylapatite (light areas) precipitating on the surface of a dissolving natural apatite, showing hexagonal dissolution pits (see also Figure 9.24); note the alignment of the precipitating hexagonal crystals along crystallographic orientations of the apatitic substrate. AFM image extracted from a figure in Valsami-Jones et al. (1998). [This figure is also reproduced in colour in the plate section after page 336.]

where apatite is less stable. In anoxic environments, vivianite ($Fe_3(PO_4)_2 \cdot 8H_2O$) has been found to control phosphorus availability; examples include waste water systems (Robertson *et al.*, 1998) and river sediments (House and Denison, 2002).

9.6 CONCLUDING REMARKS

Standard thermodynamic principles of precipitation and dissolution were presented in this chapter, and were put in the context of phosphate mineral behaviour. The studies summarized here were the result of research mostly instigated by engineering, environmental and biotechnical applications. Some of these applications are discussed further in forthcoming chapters of this book. Many of the studies presented here have mostly had a quantitative approach. As a result, mechanistic information of phosphate dissolution, and to some extent precipitation, is not well established, particularly when compared to other groups of minerals, i.e. oxides and silicates. However, recent advances in spectroscopic, microscopic and computational techniques, coupled with a need to qualify quantitative results for better process understanding, are likely to improve our mechanistic knowledge of phosphate formation, breakdown and stability in the near future.

REFERENCES

Aagaard, P. and Helgeson, H.C. (1982) Thermodynamic and kinetic constraints on reaction rates among minerals and aqueous solutions: I. Theoretical considerations. *Am. J. Sci.* **282**, 237–285.

Adamson, A.W. (1982) *Physical chemistry of surfaces*, 4th edn. Wiley, New York, 664 pp.

Amjad, Z. (1987) The influence of polyphosphates, phosphonates, and poly(carboxylic acids) on the crystal-growth of hydroxyapatite. *Langmuir* **3**, 1063–1069.

Amjad, Z. (1988) *Calcium Phosphates in Biological and Industrial Systems*. Kluwer Academic Publishers, London.

Amjad, Z., Koutsoukos, P.G. and Nancollas, G.H. (1984) The crystallization of hydroxyapatite and fluorapatite in the presence of magnesium-ions. *J. Colloid Interface Sci.* **101**, 250–256.

Appelo, C.A.J. and Postma, D. (1999) *Geochemistry, Groundwater and Pollution*. 4th corrected print, Balkema, Rotterdam, 536 pp.

Barone, J.P. and Nancollas, G.H. (1977) The seeded growth of calcium phosphates. The effect of solid/solution ratio in controlling the nature of the grown phase. *J. Coll. Interface Sci.* **62**, 421–429.

Ball, J.W. and Nordstrom, D.K. (1991) WATEQ4F-User's manual with revised thermodynamic data base and test cases for calculating speciation of major, trace and redox elements in natural waters. *U.S.Geological Survey Open File Report 90–129*, 185 pp.

Barone, J.P. and Nancollas, G.H. (1977) The seeded growth of calcium phosphates. The effect of solid/solution ratio in controlling the nature of the grown phase. *J. Colloid Interface Sci.* **62**, 421–429.

Bennema, P. (1969) The influence of surface diffusion for crystal growth from solution. *J. Cryst. Growth* **5**, 29–43.

Bethke, C.M. (1996) *Geochemical Reaction Modelling*. Oxford University Press, 397 pp.

Betts, F. and Posner, A.S. (1974) An X-ray radial distribution study of amorphous calcium phosphate. *Mater. Res. Bull.* **9**, 907–914.

Bills, P.M., Lewis, D. and Wheeler, E.J. (1982) Mineral-Collagen orientation relationships in bone. *J. Crystallographic Spectroscopic Res.* **12**, 51–63.
Bliznakov, G., Kirkova, E. and Nikolayeva, R. (1971) A study of the rate controlling stage of the process of crystal growth in solutions. *Kristall und technik* **6**, 33–38.
Blum, J.D., Klaue, A., Nezat, C.A., Driscoll, C.T., Johnson, C.E., Siccama, T.G., Eagar, C., Fahey, T.J. and Likens, G.E. (2002) Mycorrhizal weathering of apatite as an important calcium source in base-poor forest ecosystems. *Nature* **417**, 729–731.
Bohner, M., Lemaître, J. and Ring, T.A. (1997) Kinetics of dissolution of β-tricalcium phosphate. *J.Colloid Interf. Sci.* **190**, 37–48.
Borowitzka, M.A. (1989) Carbonate calcification in algae-initiation and control. In *Biomineralization* (eds Mann, S., Webb, J. and Williams, R.J.P.), VCH Weinheim, pp. 63–94.
Botsaris, G.D., Denk, E.G. and Chua, J. (1972) Nucleation in an impurity concentration gradient. A new mechanism of secondary nucleation. *AIChe. Symp. Ser. No. 121* **68**, 21–30.
Brantley, S.L. and Chen, Y. (1995) Chemical weathering rates of pyroxenes and amphiboles. *Rev. Mineral.* **31**, 119–172.
Brown W.E. (1973) Solubilities of phosphates and other sparingly soluble compounds. In *Environmental Phosphorus Handbook* (eds Griffith, E.J., Beeton, A., Spencer, J.M. and Mitchell, D.T.), Wiley and Sons, New York, pp. 203.
Brown, W.E. and Chow, L.C. (1981) Thermodynamics of apatite crystal-growth and dissolution. *J. Cryst. Growth* **53**, 31–41.
Brown, W.E. and Lehr, J.R. (1959) Application of phase rule to the chemical behavior of monocalcium phosphate monohydrate in soils. *Soil Sci. Am. Proc.* **23**, 7–12.
Brown, W.E., Lehr, J.R., Smith, J.P. and Frazier, A.W. (1957) Crystallography of octacalcium phosphate. *J. Am. Chem. Soc.* **79**, 5318–5319.
Brunauer, S., Emmet, P.H. and Teller, E. (1938) Adsorption of gases in multimolecular layers. *J. Am. Chem. Soc.* **60**, 309–319.
Budz, J.A. and Nancollas, G.H. (1988) The mechanism of dissolution of hydroxylapatite and carbonated apatite in acid solutions. *J. Cryst. Growth* **91**, 490–496.
Burton, W.K., Cabrera, N. and Frank, F.C. (1951) The growth of crystals and the equilibrium structure of their surfaces. *Phil. Trans. Royal Society (London)* **243**, 299.
Casey, W.H. and Ludwig, C. (1995) Silicate mineral dissolution as a ligand-exchange reaction. *Rev. Mineral.* **31**, 87–117.
Cayes, N.W. and Estrin, J. (1967) Secondary nucleation in agitated $MgSO_4$ solutions. *Ind. Engin. Chem. Fundamentals* **6**, 13–20.
Chernov, A.A. (1961) The spiral growth of crystals. *Soviet Physics Uspekhi* **4**, 116–148.
Christoffersen, J. and Christoffersen, M.R. (1979) Kinetics of dissolution of calcium hydroxyapatite. *J. Cryst. Growth* **47**, 671–679.
Christoffersen, J. and Christoffersen, M.R. (1981) Kinetics of dissolution of calcium hydroxyapatite. *J. Cryst. Growth* **53**, 42–54.
Combes, C. (1996) Croissance cristalline de phosphates de calcium sur des substrats d'interêt biologique: Le titane et le Collagene, Thèse doctorale, INPT, Toulouse, France, No 1189.
Cotter-Howells, J. and Thornton, I. (1991) Sources and pathways of environmental lead to children in a Derbyshire mining village. *Environ. Geochem. Hlth.* **13**, 127–135.
Cotter-Howells, J.D., Champness, P.E., Charnock, J.M. and Pattrick, R.A.D. (1994) Identification of PM in mine-waste contaminated soils by ATEM and EXAFS. *Eur. J. Soil. Sci.* **45**, 393–402.
Cowan, J.C. and Weintritt, D.J. (1976) *Water Formed Scale Deposits*. Gulf Publ. Co. Houston TX, pp. 204–206.
Dalas, E. and Koutsoukos, P.G. (1989a) The growth of calcium phosphate on ceramic surfaces. *J. Mater. Sci.* **24**, 999–1004.
Dalas, E. and Koutsoukos, P.G. (1989b) The effect of glucose on the crystallization of hydroxyapatite in aqueous-solutions. *J. Chem. Soc. Faraday Trans.* **85**(8), 2465–2472.
Dalas, E., Kallitsis, J. and Koutsoukos, P.G. (1991) Crystallization of hydroxyapatite on polymers. *Langmuir* **7**, 1822–1826.
Dalpi, M., Karayanni, E. and Koutsoukos, P.G. (1993) Inhibition of hydroxyapatite formation in aqueous solutions by zinc and 1,2-Dihydroxy-1,2-bis(dihydroxyphosphonyl)ethane. *J. Chem. Soc. Farad. Trans.* **89**, 965–969.
Davey, R.J. (1976) The effect of impurity adsorption on the kinetics of crystal growth from solution. *J. Cryst. Growth* **34**, 109–119.

Davies, C.W. (1962) *Ion Association*. Butterworths, London.
Davies, C.W. and Jones, A.L. (1949) The precipitation of silver chloride from aqueous solutions. Part I. *Discussions Farad. Soc.* **5**, 103–111.
DeRooij, J.F., Heughebaert, J.C. and Nancollas, G.H. (1984) A pH study of calcium-phosphate seeded precipitation. *J. Colloid Interf. Sci.* **100**, 350–358.
Di Simine, C.D., Sayer, J.A. and Gadd, G.M. (1998) Solubilization of zinc phosphate by a strain of Pseudomonas fluorescens isolated from a forest soil. *Biol. Fert. Soils* **28**, 87–94.
Drever, J.T. (1994) The effect of land plants on weathering rates of silicate minerals. *Geochim. Cosmochim. Acta* **58**, 2325–2332.
Eanes, E.D. and Posner, A.S. (1968) Alkaline earth intermediate phases in the basic solution preparation of phosphates. *Calcif. Tiss. Res.* **2**, 38–48.
Eanes, E.A., Gillessen, I.H. and Posner, A.S. (1965) Intermediate stages in the precipitation of hydroxyapatite. *Nature* **208**, 365–367.
Elliott, J.C. (1994) *Structure and Chemistry of the Apatites and Other Calcium Orthophosphates*. Elsevier, Amsterdam, 389 pp.
Fan, H.S., Qu, S.X. and Zhang, X.D. (2000) The precipitation of bone-like apatite on the surface of calcium phosphate-the effect of bovine serum album. *Key Eng. Mat.* **192–1**, 163–166.
Feenstra, T.P. and de Bruyn, P.L. (1979) Formation of calcium phosphates in moderately supersaturated solutions. *J. Phys. Chem.* **83**, 535–542.
Flade, K., Lau, C., Mertig, M. and Pompe, W. (2001) Osteocalcin-controlled dissolution-reprecipitation of calcium phosphate under biomimetic conditions. *Chem. Mater.* **13**, 3596–3602.
Francis, M.D. and Webb, N.C. (1971) Hydroxyapatite formation from a hydrated calcium monohydrogen phosphate precursor. *Calcif. Tissue Res.* **6**, 335–342.
Frèche, M. (1989) Contribution à l' étude des phosphates de calcium. Thèse, No. 162, INP Toulouse, Toulouse France.
Furedi-Milhofer, H., Brecevic, L. and Purgaric, B. (1976) Crystal growth and phase transformation in the precipitation of calcium phosphates. *Faraday Discussions Chem. Soc.* **61**, 184–193.
Garside, J. and Davey, R.J. (1980) Secondary contact nucleation: kinetics, growth and scale-up. *Chem. Eng. Commun.* **4**, 393–424.
Ghosh, G.K., Mohan, K.S. and Sarkar, A.K. (1996) Characterization of soil-fertilizer P reaction products and their evaluation as sources of P for gram (Cicer arietinum L). *Nutr. Cycl. Agroecosys.* **46**, 71–79.
Giannimaras, E.K. and Koutsoukos, P.G. (1987) The crystallization of calcite in the presence of orthophosphate *J. Colloid Interf. Sci.* **116**, 423–432.
Glimcher, M.J. (1959) Molecular biology of mineralized tissues with particular reference to bone. *Rev. Mod. Phys.* **31**, 359–303.
Glimcher, M.J. (1960) Specificity of the molecular structure of organic matrices in mineralization In (ed. Sognnaes, R.F.), Calcification in Biological Systems, *Am. Assoc. Adv. Sci.* Washington DC 1960 p. 421.
Glimcher, M.J. (1976) Composition, structure and organization of bone and other mineralized tissues and the mechanism of calcification. In *Handbook of Physiology vol. 7* (ed. Aurbach, G.D.), *Am. Physiol. Soc.* Washingtom DC, pp. 25–116.
Guidry, M.W. and Mackenzie, F.T. (2003) Experimental study of igneous and sedimentary apatite dissolution: Control of pH, distance from equilibrium, and temperature on dissolution rates. *Geochim. Cosmochim. Acta* **67**(16), 2949–2963.
Hellman, R., Eggleston, C.M., Hochella, M.F. and Crerar, D.A. (1990) The formation of leached layers on albite surfaces during dissolution under hydrothermal conditions. *Geochim. Cosmochim. Acta* **54**, 1267–1281.
Herman, M.A. (1994) Crystallization of epitaxial layers: fundamentals and basic problems. In *Elementary Crystal Growth* (ed. Sangwall, K.), Saan Publ. Lublin, pp. 377–407.
Heughebaert, J.C. and Nancollas, G.H. (1984) Kinetics of crystallization of octacalcium phosphate. *J. Phys. Chem.* **88**, 2478–2481.
Heughebaert, J.C., Zawacki, S.J. and Nancollas, G.H. (1983) The growth of octacalcium phosphate on beta-tricalcium phosphate. *J. Cryst. Growth* **63**, 83–90.
Hochella, M.F. Jr (2002) Nanoscience and technology: the next revolution in the earth sciences. *Earth Planet Sci. Lett.* **203**, 593–605.

Hodson, M.E. (1998) Micropore surface area variation with grain size in unweathered alkali feldspars: Implications for surface roughness and dissolution studies. *Geochim. Cosmochim. Acta* **62**, 3429–3435.

Hodson, M.E., Valsami-Jones, E. and Cotter-Howells, J.D. (2000) Bonemeal additions as a remediation treatment for metal contaminated soil. *Environ. Sci.Technol.* **36**, 3501–3507.

Hodson, M.E., Valsami-Jones, E., Cotter-Howells, J.D., Dubbin, W.E., Kemp, A.J., Thornton I. and Warren, A. (2001) Effects of bone meal (calcium phosphate) amendments on metal release from contaminated soils-a leaching column study. *Environ. Poll.* **112**, 233–243.

Hohl, H., Koutsoukos, P.G. and Nancollas, G.H. (1982) The crystallization of hydroxyapatite and dicalcium phosphate dihydrate. Representation of the growth curves. *J. Cryst. Growth* **57**, 325–335.

Hottenhuis, M.H.J. and Lucasius, C.B. (1989) The influence of internal crystal-structure on surface morphology – in situ observations of potassium hydrogen phthalate (010) *J. Cryst. Growth* **94**, 708–720.

House, W.A. and Denison, F.H. (2002) Total phosphorus content of river sediments in relationship to calcium, iron and organic matter concentrations. *Sci. Total Environ.* **282–283**, 341–351.

Jeanjean, J., Vincent, U. and Fedoroff, M. (1994) Structural modification of calcium hydroxyapatite induced by sorption of cadmium ions. *J. Solid State Chem.* **108**, 68–72.

Jenkins, G.N. (1978) The physiology and biochemicstry of the month, Blackwell Sci, 4th edn. Oxford, pp. 113–163.

Jerden, J.L. and Sinha, A.K. (2003) Phosphate based immobilization of uranium in an oxidizing bedrock aquifer. *Appl. Geochem.* **18**, 823–843.

Kani, T., Kani, M., Moriwaki, Y. and Doi, Y. (1983) X-ray diffraction analysis of dental calculus. *J. Dent. Res.* **62**, 92–95.

Kapolos, J. and Koutsoukos, P.G. (1999) Formation of calcium phosphates in aqueous solutions in the presence of carbonate ions. *Langmuir* **15**, 6557–6562.

Kapolos, J., Mavrilas, D., Missirlis, Y. and Koutsoukos, P.G. (1997) Model experimental system for investigation of heart valve calcification *in vitro*. *J. Biomed. Mater. Res. (Appl. Biomater)* **38**, 183–190.

Kazmierczak, T.F., Tomson, M.B. and Nancollas, G.H. (1982) Crystal growth of carbonate: a controlled composition kinetic study. *J. Phys. Chem.* **86**, 103–108.

Kern, R. (1978) Croissance épitaxique (aspects topologiques et structuraux). *Bull Mineral.* **101**, 202–233.

Kibalczyc, W., Melikhov, I.V. and Komarov, V.F. (1982) In *Industrial Crystallization* **81**. (eds Jancic, S.J. and de Jong, E.J.), North Holland, Amsterdam pp. 291.

Kim, H.M., Kim, Y.S., Woo, K.M., Park, S.J., Rey, C., Kim, Y., Kim, J.K.and Ko, J.S. (2001) Dissolution of poorly crystalline apatite crystals by osteoclasts determined on artificial thin-film apatite. *J. Biomed. Mater. Res.* **56**, 250–256.

Koutsopoulos, S., Paschalakis, P.C. and E. Dalas (1994) The calcification of elastin *in vitro*. *Langmuir* **10**, 2423–2428.

Koutsoukos, P.G. (1980) *Kinetics of Precipitation of Hydroxyapatite from Aqueous Solutions*, SUNY at Buffalo, NY, 308 pp.

Koutsoukos, P.G. (1998) Growth of calcium phosphates on different substrates: epitaxial considerations. In *Calcium Phosphates in Biological and Industrial Systems (ed. Amjad, Z.),* Kluwer Acad. Publ., Boston, pp. 41–66.

Koutsoukos, P.G. and Nancollas, G.H. (1981a) Crystal growth of calcium phosphates-epitaxial considerations. *J. Cryst. Growth* **53**, 10–19.

Koutsoukos, P.G. and Nancollas, G.H. (1981b) The influence of strontium ion on the crystallization of hydroxyapatite. *J. Phys. Chem.* **85**, 2403–2414.

Koutsoukos, P.G. and Nancollas, G.H. (1986) The adsorption of inorganic phosphate by collagen. *Colloids Surf.* **17**, 81–90.

Koutsoukos, P.G. and Nancollas, G.H. (1987) The mineralization of collagen *in vitro*. *Colloids Surf.* **28**, 95–108.

Koutsoukos, P., Amjad, Z. and Nancollas, G.H. (1979) The precipitation of hydroxyapatite at constant supersaturation. *J. Dent. Res.* **58A**, 167.

Koutsoukos, P.G., Amjad, Z. and Nancollas, G.H. (1981) The influence of phytate and phosphonate on the crystal-growth of fluorapatite and hydroxyapatite. *J. Colloid Interf. Sci.* **83**, 599–605.

Koutsoukos, P.G., Amjad, Z., Tomson, M.B. and Nancollas, G.H. (1980) Crystallization of calcium phosphates. A constant composition study. *J. Am. Chem. Soc.* **102**, 1553–1557.

Kukura, M., Bell, L.C., Posner, A.M. and Quirk, J.P. (1972)Radioisotope determination of the surface concentrations of calcium and phosphorus or hydroxyapatite in aqueous solutions. *J. Phys. Chem.* **76**, 900–904.

Langmuir, D. (1997) *Aqueous Environmental Geochemistry*. Prentice Hall, New Jersey, 600.

Lasaga, A.C. (1981) Transition state theory. In Kinetics of geochemical processes (eds Lasaga, A.C. *et al.*,), *Rev. Mineral.* **8**, 135–169.

Langmuir, D. (1997) *Aqueous Environmental Geochemistry*. Prentice Hall, New Jersey, 600 pp.

LeGeros R.Z. (1999) Calcium phosphates in demineralization/remineralization processes. *J. Clin. Dent.* **10**, 65–73.

LeGeros, R.Z. and LeGeros, J.P. (1984) Phosphate minerals in human tissues. In *Phosphate Minerals* (eds Nriagu, J.O. and Moore, P.B.), Springer Velag, Berlin Heidelberg, pp. 351–385.

LeGeros, R.Z., Kijkowska, R., Bautista, C. and LeGeros, J.P. (1995) Synergistic effects of magnesium and carbonate on properties of biological and synthetic apatites. *Connect. Tissue Res.* **32**, 525–531.

LeGeros, R.Z., Shirra, W.P., Mirawite, M.A. and LeGeros, J.P. (1975) In *Physico-Chimie et Cristallographie des Apatites d' Interêt Biologique*, Colloque Internationaux, Paris: CNRS 230, 105.

LeGeros, R.Z., Taheri, M.H., Quirologico, G.B. and LeGeros, J.P. (1981) Formation and stability of apatites: effects of some cationic substituents. In *Proceedings of the International Congress on Phosphorous Compounds*, IMPHOS, Rabat, pp. 89–103.

Lehr, J.R., Brown, W.E. and Brown, E.H. (1959) Chemical behavior of monocalcium phosphate monohydrate in soils. *Soil Sci. Soc. Am. Proc.* **23**, 3–7.

Liu, Y., Hunziker, E.B., Randall, N.X., de Groot, K. and Layrolle, P. (2003) Proteins incorporated into biomimetically prepared calcium phosphate coatings modulate their mechanical strength and dissolution rate. *Biomaterials* **24**, 65–70.

Lobartini, J.C., Tan, K.H. and Pape, C. (1994) The nature of humic acid-apatite interaction products and their availability to plant – growth. *Commun. Soil. Sci. Plant Anal.* **25**, 2355–2369.

Lonsdale, K. (1968) The solid state epitaxy as a growth factor urinary calculi and gallstones. *Nature* **217**, 56–58.

Lower, S.K., Maurice, P.A. and Traina, S.J. (1998) Simultaneous dissolution of hydroxylapatite and precipitation of hydroxypyromorphite: Direct evidence of homogeneous nucleation. *Geochim. Cosmochim. Acta* **62**, 1773–1780.

Ma, Q.Y., Traina, S.J. and Logan, T.J. (1993) In situ lead immobilization by apatite. *Environ. Sci. Technol.* **27**, 1803–1810.

Malollari, I.X., Klepetsanis, P.G. and Koutsoukos, P.G. (1995) Precipitation of strontium sulfate in aqueous solutions at 25°C. *J. Cryst. Growth* **155**, 240–246.

Manecki, M., Maurice, P.A. and Traina, S.J. (2000) Kinetics of aqueous Pb reaction with apatites. *Soil Sci.* **165**, 920–933.

Margolis, H.C. and Moreno, E.C. (1992) Kinetics of hydroxyapatite dissolution in acetic, lactic and phosphoric acid solutions. *Calcif. Tissue Int.* **50**, 137–143.

Martell, A.E. and Smith, R.M. (1981) *Critical Stability Constants* Vol. 1, Plenum Press, New York.

Meyer, J.L. and Eanes, E.D. (1978) A thermodynamic analysis of the amorphous to crystalline calcium phosphate. *Calcif. Tissue* **25**, 59–68.

Montel, G., Bonel, G., Heughebaert, J.C., Trombe J.C. and Rey, C. (1981) New concepts in the composition, crystallization and growth of the mineral component of calcified tissue. *J. Cryst. Growth* **53**, 74–99.

Moreno E.C. and Varughese K.J. (1981) Crystal growth of apatites from dilute solutions. *Cryst. Growth* **53**, 20–30.

Moreno, E.C., Varughese, K. and Hay, D.I. (1979) Effect of human salivary proteins on the precipitation kinetics of calcium phosphate. *Calsif. Tissue Int.* **28**, 7–16.

Moreno, E.C., Zahradnik, R.T., Glazman, A. and Hwu, R. (1977) Precipitaion of hydroxyapatite from dilute solutions upon seeding. *Calcif. Tissue Res.* **24**, 47–57.

Morse, J.W. and Mackenzie, F.J. (1990) *Geochemistry of Sedimentary Carbonates*. Elsevier, Amsterdam.

Mullin, J.W. (1972) *Crystallization*, 2nd edn. CRC Press, Bocca Raton Fl., 482 pp.
Mullin, J.W. (1993) *Crystallization*, 3rd edn. Butterworth-Heinemann, Oxford, pp. 118–122.
Mullin, J.W., Chakraborty, M. and Mehtak, (1970) Nucleation and growth of ammonium sulphate crystals. *J. Appl. Chem.* **13**, 423–429.
Myerson, A.S. (1993) Solutions and solution properties. In *Handbook of Industrial Crystallization* (ed. Myerson, A.S.), Bufferworth-Heinemann, Boston MA, pp. 1–31.
Nancollas, G.H. (1966) *Interactions in Electrolyte Solutions*. Elsevier, Amsterdam, pp. 73–90.
Nancollas, G.H. (1984) The nucleation and growth of phosphate minerals. In *Phosphate minerals* (eds Nriagu, J.O. and Moore, P.B.), Springer Verlag, Berlin Heidelberg, pp. 137–154.
Nancollas, G.H. (1989) In vitro studies of calcium phosphate crystallization. In *Biomineralization* (eds Mann, S., Weloto, J. and Williams, R.J.P.), VCH Publ., New York, pp. 157–187.
Nancollas, C.H. and Marshall, R.W. (1971) Kinetics of dissolution of dicalcium phosphate dihydrate. *J. Dent. Res.* **50**, 1268–1272.
Nancollas, G.H. and Mohan, M.S. (1970) The growth of hydroxyapatite crystals. *Archs. Oral. Biol.* **15**, 731–745.
Nancollas, G.N. and Purdie, N. (1964) The kinetics of crystal growth. *Quart. Rev. (London) Chem. Soc.* **18**, 1–20.
Nancollas, G.H. and Tomazic, B. (1974) Growth of calcium phosphate on hydroxyapatite crystals. Effect of supersaturation and ionic medium. *J. Phys. Chem.* **78**, 2218–2225.
Nancollas, G.H., Tomazic, B. and Tomson, M.B. (1976) The precipitation of calcium phosphates in the presence of magnesium. *Croat Chim. Acta* **48**, 431–438.
Narasaraju, T.S.B. and Phebe, D.E. (1996) Some physico-chemical aspects of hydroxyapatite. *J. Mat. Sci.* **31**, 1–21.
Nathan, Y. (1984) The mineralogy and geochemistry of phosphorites. In *Phosphate Minerals* (eds Nriagu, J.O. and Moore, P.B.), Springer Velag, Berlin Heidelberg, pp. 275–291.
Nesbitt, H.W. and Muir, I.J. (1988) SIMS depth profiles of weathered plagioclase and processes affecting dissolved Al and Si in some acidic soils. *Nature* **334**, 336–338.
Neuman, W.F. and Neuman, M.W. (1953) The nature of the mineral phase of bone. *Chem. Rev.* **53**, 1–45.
Neuman, W.F. and Bareham, B.J. (1975) Evidence for the presence of secondary calcium phosphate in bone and its stabilization by acid production. *Calcif. Tissue Res.* **18**, 161–172.
Newesely, H. (1966) Changes in Crystal types of low solubility calcium phosphates in of accompanying ions. *Arch. Oral Biol. Sp. Suppl.* **6**, 174.
Nielsen, A.E. (1984) Electrolyte crystal growth mechanisms. *J. Cryst. Growth* **67**, 289–303.
Nielsen, A.E. and Toft, J.M. (1984) Electrolyte crystal growth kinetics. *J. Cryst. Growth* **67**, 278–288.
NIST, 2001 Standard Reference Database 46, Version 6.0, NIST Standard reference Data, Gaithesburg, MD, USA.
Nordstom, D.K., Plummer, L.N., Wigley, T.M.L., Wolery, T.J., Ball, J.W., Jenne, E.A., Basset, R.L., Crerar, D.A., Florence, T.M., Fritz, B., Hoffman, M., Holdren, G.R., Lafon, G.M., Mattigod, S.V., McDuff, R.E., Morel, F., Reddy, M.M., Sposito, G. and Thrailkill, J. (1979) A comparison of computerized chemical models for equilibrium calculations in aqueous systems. In *Chemical Modelling in Aqueous Systems. Speciation, Solubility and Kinetics.* E.A. Jenne (ed.) ACS Symposium Series No. 93, ACS, Washington DC, pp. 857–892.
Nriagu, J.O. (1974) Lead Orthophosphates. IV – Formation and stability in the environment. *Geochim. Cosmochim. Acta* **38**, 887–898.
Nriagu, J.O. (1984) Formation and stability of base metal phosphates in soils and sediments. In *Phosphate Minerals* (eds Nriagu, J.O. and Moore, P.B.), Springer Verlag, Berlin Heidelberg, 318–329 pp.
Nyvlt, J. (1971) *Industrial Crystallization from Solutions*. Butterworths, London, pp. 45–50.
Nyvlt, J., Sohnel, O., Matuchova, M. and Broul, M. (1985) *The Kinetics of Industrial Crystallization*, Elsevier, Amsterdam, pp. 68, 284.
Oelkers, E.H. (2001) General kinetic description of multioxide silicate mineral and glass dissolution. *Geochim. Cosmochim. Acta* **65**, 3703–3719.
Oelkers, E.H. (2002) The surface areas of rocks and minerals. In *Proceedings of the Arezzo Seminar on Fluids Geochemistry* (eds Buccianti *et al.*,), Paninieditore, Pisa, pp. 18–30.

Oelkers, E.H., Schott, J. and Devidal, J.L. (1994) The effect of aluminium, pH, and chemical affinity on the rates of aluminosilicate dissolution reactions. *Geochim. Cosmochim. Acta* **58**(9), 2011–2024.
O'Hara, M. and Reid, R.C. (1973) Modelling crystal growth rates from solution, Prentice Hall, New York.
Onuma, K. and Ito, A. (1998) Cluster growth model for hydroxyapatite. *Chem. Mater.* **10**, 3346–3351.
Osborn, J.F. and Newesely, H. (1980) The material science of calcium phosphate ceramics. *Biomaterials* **1**, 108–111.
Pak, C.Y.C., Diller, E.C., Smith, G.W. and Howe, E.S. (1969) Renal stones of calcium phosphate: physicochemical basis for their formation. *Proc. Soc. Exp. Biol. Med.* **130**, 753–757.
Papelis, G., Hayes, K.F. and Leckie, J.O. (1988) HYDRAQL: A program for the computation of chemical equilibrium composition of aqueous batch system including surface complexation modeling of ion association at the oxide/solution interface. *Technical Report No. 306,* Stanford University, Stanford, CA.
Petticrew, E.L. and Arocena, J.M. (2001) Evaluation of iron-phosphate as a source of internal lake phosphorus loadings. *Sci. Total Environ.* **266**, 87–93.
Pitzer K.S. (1979) Theory-ion interaction approach. In *Activity Coefficients in Electrolyte Solutions* (ed. Pytkowitz, R.M.), CRC Press, Inc., Boca raton, Fla. USA, **1**, 157–208.
Plant, L.J. and House, W.A. (2002) Precipitation of calcite in the presence of inorganic phosphate. *Colloid Surf. A* **203**, 143–153.
Posner, A.S. (1969) Crystal chemistry of bone mineral. *Physiol. Rev.* **49**, 760–792.
Posner, A.S., Blumenthal, N.C. and Betts, F. (1984) Chemistry and structure of precipitated hydroxyapatites. In *Phosphate Minerals* (eds Nriagu, J.O. and Moore, P.B.), Springer Velag, Berlin Heidelberg, pp. 330–350.
Rakovan, J. (2002) Growth and surface properties of apatite. *Rev. Mineral. Geochem.* **48**, 51–86.
Robertson, W.D., Schiff, S.L. and Ptacek, C.J. (1998) Review of phosphate mobility and persistence in 10 septic system plumes. *Ground Water* **36**, 1000–1010.
Rosenberger, F. (1986) Inorganic and protein crystal growth – similarities and differences. *J. Cryst. Growth* **76**, 618–636.
Roufosse, A.H., Landis, W.J., Sabine, W.K. and Glimcher, M.J. (1979) Identification of brushite in newly deposited bone mineral from embryonic chick. *J. Ultrastruct. Res.* **68**, 235–255.
Russel, K.C. (1980) Nucleation in solids: the induction and steady state effects. *Adv. Coll. Interf. Sci.* **13**, 205–318.
Salimi, M.H. (1985) *The kinetics of growth of calcium phosphate*. Ph.D. Thesis, State Univ. New York, Buffalo, NY, USA.
Sawada, K. (1997) The mechanisms of crystallization and transformation of calcium carbonates. *Pure Appl. Chem.* **69**, 921–928.
Sayer, J.A., Cotter-Howells, J.D., Watson, C., Hillier, S. and Gadd, G.M. (1999) Lead mineral transformation by fungi. *Curr. Biol.* **9**, 691–694.
Schaad, P., Thomann, J.M., Voegel, J.C. and Gramain, P. (1994) Inhibition of dissolution of hydroxylapatite powder by adsorbed anionic polymers. *Coll. Surf. A* **83**, 285–292.
Seaman, J.C., Arey, J.S., Bertsch, P.M. (2001) Immobilization of nickel and other metals in contaminated sediments by hydroxyapatite addition. *J. Environ. Qual.* **30**, 460–469.
Spanos, N. and Koutsoukos, P.G. (2001) Model studies of the effect of orthophosphol-L-serine on biological mineralization. *Langmuir* **17**, 866–872.
Spanos, N., Klepetsanis, P.G. and Koutsoukos, P.G. (2001) Model studies on the interaction of amino acids with biominerals: the effect on L-serine at the hydroxyapatite-water interface. *J. Coll. Interf. Sci.* **236**, 260–265.
Stipp, S.L.S. (1999) Toward a conceptual model of the calcite surface: Hydration, hydrolysis, and surface potential. *Geochim. Cosmochim. Acta* **63**, 3121–3131.
Stumm, W. and Morgan, J.J. (1996) *Aquatic chemistry*, 3rd edn. Wiley, New York, 1022 pp.
Stumm, W. and Wollast, R. (1990) Coordination chemistry of weathering: kinetics of the surface-controlled dissolution of oxide minerals. *Rev. Geophys.* **28**, 53–69.

Suzuki, T., Hatsushika, T. and Hayakawa, Y. (1981) Synthetic hydroxyapatites employed as inorganic cation-exchangers. *J. Chem. Soc. Faraday Trans. 1*, **77**, 1059–1062.

Suzuki, T., Hatsushika, T. and Miyake, M. (1982) Synthetic hydroxyapatites as inorganic cation exchangers. Part 2. *J. Chem. Soc. Faraday Trans. 1*, **78**, 3605–3611.

Takeuchi, Y. and Arai, H. (1990) Removal of coexisting Pb^{2+}, Cu^{2+} and Cd^{2+} ions from water by addition of hydroxyapatite powder. *J. Chem. Eng. Japan* **23**, 75–80.

Tang, R., Henneman, Z.J. and Nancollas, G.H. (2003) Constant composition kinetics study of carbonated apatite dissolution. *J. Cryst. Growth* **249**, 614–624.

Tomson, M.B. and Nancollas, G.H. (1978) Mineralization kinetics: A constant composition approach. *Science* **200**, 1059–1060.

Treboux, G., Layrolle, P., Kanzaki, N., Onuma, K. and Ito, A. (2000) Existence of Posner's cluster in vacuum. *J. Phys. Chem. A* **104**, 5111–5114.

Tung, M.S., Eidelman, N., Sieck, B. and Brown, W.E. (1988) Octacalcium phosphate solubility product from 4 to 37°C. *J. Res. Natn. Bur. Stands* **93**, 613–624.

Valsami-Jones, E., Ragnarsottir, K.V., Putnis, A., Bosbach, D., Kemp, A.J. and Cressey, G. (1998) The dissolution of apatite in the presence of aqueous metal cations at pH 2–7. *Chem. Geol.* **151**, 215–233.

Van der Houwen, J.A.M. (2002) Chemical principles of calcium phosphate dissolution and precipitation. PhD Thesis, University of Reading, 323 pp.

Van der Houwen, J.A.M., Cressey, G., Cressey, B.A. and Valsami-Jones, E. (2003) The effect of organic ligands on the crystallinity of calcium phosphate. *J. Cryst. Growth* **249**, 572–583.

Van der Merwe, H. (1978) Misfitting monolayers and oriented overgrowth. *CRC Crit. Rev. Solid Stat. Mater. Sci.* **7**, 209–225.

Veis, A. and Sabsay, B. (1983) Bone and tooth formation. Insights into mineralization strategies. In *Biomineralization and Biological Metal Accumulation* (eds Westbroek, P. and de Jong, E.W.), D. Reidel Publ. Co., Dordrecht, pp. 273–284.

Venables, J.A. and Price, G.L. (1975) Nucleation of thin films. In *Epitaxial growth* (ed. Mathews, J.W.), Acad. Press. New York, pp. 381–436.

Verbeek, R.M.H., Debruyne, P.A.M., Driessens, F.C.M., Terpstra, R.A. and Verbeek, F. (1986) Solubility behavior of Mg-containing beta-$Ca_3(PO_4)_2$. *B. Soc. Chim. Belg.* **95**, 455–476.

Verbeek, R.M.H. and Devenyns, J.A.H. (1992) The kinetics of dissolution of octacalcium phosphate 2. The combined effects of pH and solution Ca/P ratio. *J. Cryst. Growth* **121**, 335–348.

Vieillard, P. and Tardy, Y. (1984) Thermochemical properties of phosphates. In *Phosphate Minerals* (eds Nriagu, J.O. and Moore, P.B.), Springer Verlag, Berlin Heidelberg, 171–198 pp.

Volmer, M. (1939) *Kinetic der Phasenbildung*, Steinkopff, Leipzig.

Walton, A.G. (1967) *The formation and Properties of Precipitates*. Interscience, New York.

Welch, S.A., Taunton, A.E. and Banfield, J.F. (2002) Effect of microorganisms and microbial metabolites on apatite dissolution. *Geomicrobiol. J.* **19**(3), 343–367.

Wu, L., Rorsling, W. and Schindler, P.W. (1991) Surface complexation of calcium minerals in aqueous solution. *J. Coll. Interf. Sci.* **147**, 178–185.

Xu, Y. and Schwartz, F.W. (1994) Lead immobilization by hydroxyapatite in aqueous solutions. *J. Contam. Hydrol.* **15**, 187–206.

Xyla, A.G., Giannimaras, E.K. and Koutsoukos, P.G. (1991) The precipitation of calcium carbonate in aqueous solutions. *Colloids Surf.* **53**, 241–255.

Zawacki, S.J., Koutsoukos, P.G., Salimi, M.H. and Nancollas, G.H. (1986) The growth of calcium phosphates. In *Geochemical Processes at Mineral Surfaces*. ACS Special Publication series No. 323 (eds Davis, J.A. and Hayes, K.F.), Washington DC, pp. 650–662.

Zhang, J. and Nancollas, H. (1992a) Kinetics and mechanisms of octacalcium phosphate dissolution at 37°C. *J. Phys. Chem.* **96**, 5478–5483.

Zhang, J. and Nancollas, H. (1992b) Interpretation of dissolution kinetics of dicalcium phosphate dihydrate. *J. Cryst. Growth* **125**, 251–269.

Zhang, M., Alva, A.K., Li, Y.C. and Calvert, D.V. (2001) Aluminum and iron fractions affecting phosphorus solubility and reactions in selected sandy soils. *Soil Sci.* **166**, 940–948.

Zur, A. and McGill, T.C. (1984) Lattice match: an application to heteroepitaxy. *J. Appl. Phys.* **55**, 378–386.

10
Waste water treatment principles

Simon A. Parsons and Tom Stephenson

10.1 INTRODUCTION

Domestic waste water varies from country to country depending on local diet, cleansing habits, sanitation systems and environmental conditions. In the western world, where water carriage is normal and grey water (water used for washing and laundering) is disposed of together with black water (that which contains faecal matter) the sewage strength is lower because of dilution. In some cases this is further exaggerated by the disposal of surface run-off (rain water) via the sewer system. A typical analysis of raw sewage is given in Table 10.1.

Discharges of waste water from municipal sewage treatment works, are normally controlled by a central government agency – in the UK this is the Environment Agency. The quality of discharges in respect of parameters such as BOD (Biological Oxygen Demand), are set by EU Directive (The Urban Wastewater Treatment Directive) which generally requires that 'background' levels in the receiving water are not exceeded. Traditionally the UK has worked to a discharge standard of 20/30 – 20 mg/l BOD and 30 mg/l suspended solids – but recent legislation has set standards of total nitrogen and phosphate that are dependent on the size of the treatment works. Typical limits are 15 mg/l N and 2 mg/l P for 10,000–100,000 population equivalent (p.e.) works and 10 mg/l N and 1 mg/l P for >100,000 p.e. works.

© 2004 IWA Publishing. *Phosphorus in Environmental Technology: Principles and Applications.*
Edited by Eugenia Valsami-Jones. ISBN: 1 84339 001 9

Table 10.1 Typical raw sewage analysis.

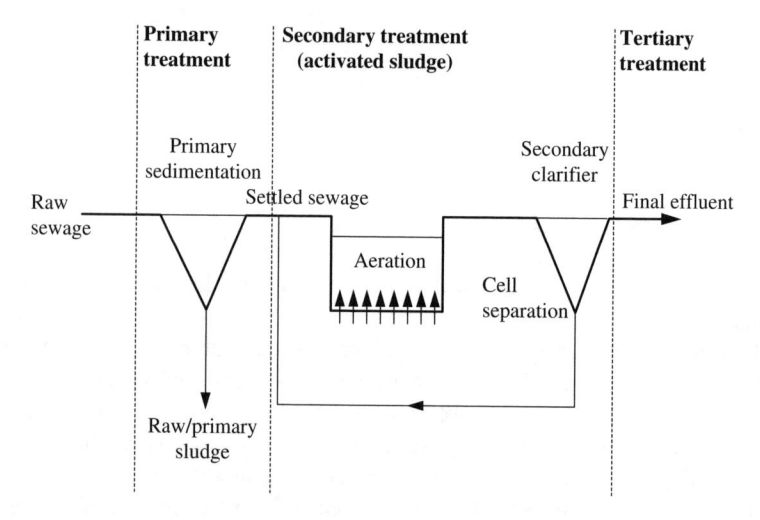

Parameter	Units	Concentration
Total dissolved solids	mg/l	250–850
Total suspended solids	mg/l	100–350
BOD_5	mg/l	110–400
COD	mg/l	250–1000
Ammonia	mg/l N	10–50
Nitrate	mg/l N	0–5
Phosphate	mg/l P	5–10
Chloride	mg/l Cl	30–100
Sulphate	mg/l SO_4	20–50
Alkalinity	mg/l $CaCO_3$	50–200
Fat, oil and grease	mg/l	50–150
VOCs	µg/l	50–500
Total coliform	cfu/ml	10^4–10^7

Figure 10.1 Levels of treatment used in municipal waste water treatment.

10.2 WASTE WATER TREATMENT PROCESSES

This chapter will focus on those processes typically used for the treatment of municipal waste water. The treatment flow sheet typically contains different stages of treatment (see Figure 10.1). These are:

- Preliminary treatments to remove gross solids.
- Primary treatment to remove biological solids and some organic load.
- Secondary treatment to remove organic carbon and nutrients.
- Tertiary treatment to remove fine solids and nutrients.

10.2.1 Preliminary treatment

Gross solids to be removed from waste waters by preliminary treatment can be placed in one of two categories:

- large floating and suspended solid objects, e.g. pieces of wood, vegetable debris, animal carcasses, plastic debris, paper and rags; and
- solids with high density, i.e. grit, silt and sand.

When removing gross solids it is necessary to prevent removal of organic matter, as the latter are treated in subsequent primary and secondary processes. Floating solids are usually removed by *screens* and grit removed by *sedimentation*.

Screens and grit removal are collectively termed preliminary processes when used for municipal waste water treatment. The main purpose for this application is to protect the downstream equipment and processes. Removal of gross solids can also protect sludge treatment processes. Grit removal is for small, dense particles, e.g. sand. Grit is always present where there are combined sewerage systems, i.e. the sewers collect surface water drainage in addition to waste waters from domestic and industrial premises. Screens and grit removal equipment are often supplied as factory-built proprietary processes; screens in particular. Therefore the design principles are also based on proprietary knowledge and 'rules of thumb'. Modern screening systems use travelling screen systems that rely upon the whole screen moving, an action that helps to prevent blinding and headloss build-up. Often these devices consist of a series of teeth, made of metal or nylon travelling along a continuous belt – effectively a screen using slots (see Figure 10.2). In addition to this type of travelling screen, continuous perforated band screens can provide effective fine screening – effectively a screen using holes.

10.2.2 Primary treatment

When talking about primary treatment processes in waste water treatment we typically mean the sedimentation of solids by gravity. Sedimentation is superficially a simple process of which there are many examples in nature. Sedimentary rocks are the result of geological solids–liquid separation processes and the bottom muds in rivers and lakes are also the result of this fundamental physical activity.

Sedimentation process can typically remove a major proportion of the settable solids from waste water, which make up 50–60% of the overall suspended solids. Performance can be predicted if you know the suspended solids size and density where for Reynolds numbers ($N_{Re} = \rho v l/\mu$, where ρ = density, v = velocity, l = characteristic length of the system and μ = viscosity) less than 0.5 solids settle according to Stokes Law.

$$v_s = \frac{g(\rho_s - \rho)d^2}{18\mu}$$

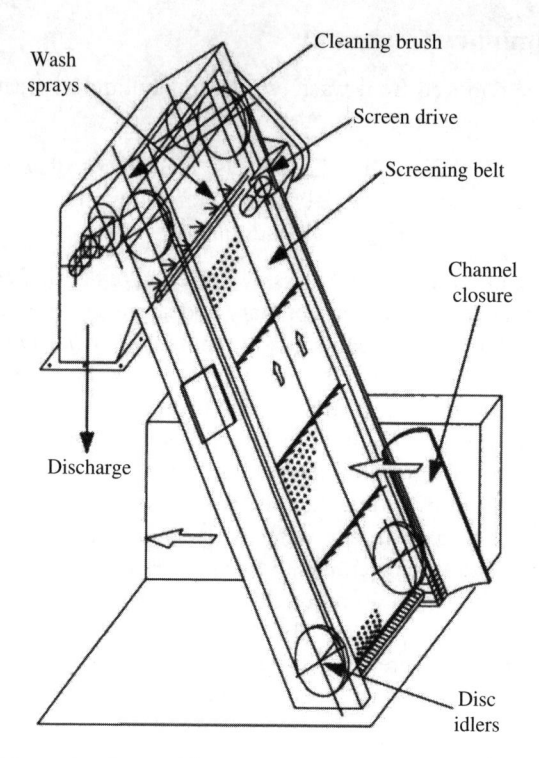

Figure 10.2 Schematic of a travelling band screen.

where v_s is terminal settling velocity, g is acceleration due to gravity, d is particle diameter, $(\rho_s - \rho)$ is density difference and μ is viscosity.

It will be clear from the above expression that the settling velocity will be affected by the size and density of the particle and also by the viscosity of the water that is itself a function of temperature. Primary sedimentation tanks are normally designed for a surface loading of 1–1.5 m/h with a depth of 4 m (Figure 10.3). More detailed design information can be found in standard textbooks (e.g. Metcalf and Eddy, 1991).

10.2.3 Secondary treatment

Conventional waste water treatment typically incorporates biological waste water treatment systems that are based on the simple concept of micro-organisms consuming soluble and colloidal organic compounds (measured as BOD) present in the waste water as food. The micro-organisms grow and are subsequently separated from water which has had the BOD removed. These micro-organisms derive energy and cellular material from the oxidation of this organic matter and can be aerobic or anaerobic. Biological waste water treatment is also capable of removing other waste water components besides organic matter, including suspended solids, ammonia, heavy metals and xenobiotics.

Figure 10.3 Radial sedimentation tank treating municipal waste water. [This figure is also reproduced in colour in the plate section after page 336.]

Advantages of biological processes are that:

- they are 'natural'
- wastes are converted to gases
- the sludge produced is treatable
- one of the waste products (methane gas) can be used as an energy source
- odours are usually reduced
- a high removal efficiency is possible
- variable loads can be tolerated
- constituents other than BOD are removed.

Disadvantages are that these processes are:

- susceptible to toxic chemicals
- slow compared to chemical processing
- generate solids that need to be disposed of
- produce noxious compounds
- produce aerosols
- energy consuming (if aerobic).

10.2.3.1 *Microbial groups in biological waste water treatment*

Many microorganisms have a role in biological treatment systems.

Bacteria
- conversion of soluble and particulate organic compounds into biomass and gaseous waste products (CO_2 and CH_4)

Table 10.2 Examples of bacterial genera in biological waste water treatment processes.

Genus	Function
Pseudomonas	Removal of carbohydrate, slime production, denitrification
Zoogloea	Slime production, floc formation
Bacillus	Protein and starch degradation
Arthrobacter	Carbohydrate degradation
Microthrix	Fat degradation, filamentous growth
Acinetobacter	Phosphorus uptake and release
Nitrosomonas	Nitrification (to nitrate)
Nitrobacter	Nitrification (to nitrate)
Achromobacter	Denitrification
Clostridium	Anaerobic hydrolysis of biochemicals
Desulfovibrio	Sulphate reduction
Methanothrix	Methane production

- conversion of ammonia to nitrate (nitrification)
- conversion of nitrate to nitrogen gas (denitrification)
- conversion of soluble phosphate into insoluble intracellular phosphate (permitting removal from dischargeable effluent).

Protozoa
- consume particulate organics (including bacteria)
- play an important role in the removal of suspended solids.

Fungi
- may assist bacteria in the removal of organics in trickling systems.

Algae
- role in nutrient uptake in specialised tertiary treatment systems.

Others
- larger biological species such as rotifers, nematode worms and insect larvae may contribute to the consumption of particulate organic matter, especially in trickling filter systems.

In most biological systems the bacteria play the main role in the primary reduction of contaminants. The genera responsible for significant processes within waste water treatment are shown in Table 10.2. Those bacteria responsible for BOD removal and denitrification are heterotrophs, i.e., organic carbon is required as an energy source and as the carbon source. Those responsible for nitrification, sulphate reduction and methanogenesis are autotrophs, i.e. inorganic chemical reactions provide the energy and inorganic carbon is needed as the carbon source.

10.2.3.2 Biological treatment processes

There are many types of biological process that are commonly used. Of these it should be noted that there are many generic and proprietary processes on the market but they can all be fitted into one or other of the categories listed. The two main process types are suspended growth (activated sludge) or fixed film (trickling filters). Suspended growth processes rely upon the microbes (biomass) being in free suspension. Contact with the waste water constituents will rely upon good mixing in such reactors. Fixed film reactors provide an inert support material on which the micro-organisms can grow. The microbial population is then in contact with the waste water as it is passed over the support matrix. It is important to note that with both types of systems a separation stage is needed to remove the biomass from the clean effluent. There are also a number of alternative biological treatment processes that are available, such as moving bed biofilm reactors (MBBR) and membrane bioreactors (MBR), that are a combination of suspended and fixed film processes or in the case of MBRs a combination of suspended growth and membrane separation processes.

10.2.3.3 Suspended growth processes

The activated sludge process is the best known of all suspended growth processes. This process was first operated at large scale in 1914 at the Davyhulme Works, in Manchester, with the first full-scale plant operated at Worcester in 1916.

Activated sludge may be defined as the flocculant microbial mass, which is produced when sewage is continuously aerated. It consists mainly of organisms which are able to metabolise and break down the principal contaminants of waste water.

This definition does not tell the whole story. An activated sludge plant consists of an aeration tank (Figure 10.4), to encourage the growth of BOD-consuming micro-organisms, followed by a sedimentation tank to separate flocculated micro-organisms from the clean water for final disposal. A means of returning a portion of the settled activated sludge flocs to the aeration tank is also required, along with some means of removing surplus activated sludge. The process thus relies upon developing a microbial population that will consume the organic waste as food in the presence of oxygen. The settlement and recycle means that a flocculent population will develop, as the sedimentation is a major selection pressure. Therefore the activated sludge process actually combines two unit operations – an aerobic suspended growth bioreactor and a sedimentation stage (Figure 10.1). Many designs of aeration units and system configurations are possible, optimising conditions for different goals such as nutrient removal.

10.2.3.4 Fixed film processes

The best known and most widely adopted fixed-film process for sewage treatment is trickling (or percolating) filtration. The trickling filter is the main alternative to activated sludge for the secondary treatment of settled domestic waste water for BOD and NH_3 reduction. The trickling filter can be designed to produce a 20/30 standard, fully nitrified effluent.

Trickling filtration gets its name from the nature of waste water flow through the system. Settled waste water is distributed mechanically over a bed of porous material.

Figure 10.4 Example of a surface aerated activated sludge plant treating municipal waste water.

The waste water trickles down through the bed, coming into close contact with filter particles and associated biofilm. Treatment is facilitated by the interaction of the waste water and the biofilm that coats the filter particles. The treated effluent flows via an underdrain to a sedimentation/humus tank to separate the sloughed biofilm prior to discharge of the treated, settled effluent to a watercourse.

Trickling filters at sewage treatment works may be circular or rectangular beds and are normally filled with mineral media (rock) (Figure 10.5). Circular beds (maximum 60 m diameter) have a rotary distribution arm that may be driven by the force of the exiting waste water or by a motor. Rectangular beds (maximum 10 m × 100 m) normally have motor-driven distributors, which move up and down the bed. The depth of mineral media filters is normally in the range 1.5–3 m (1.8 m on average). The 1983 British Standard Code of Practice for Small Sewage Works set a minimum of 1.2 m to maintain a reasonable retention time and to encourage nitrifying bacteria. More recently, plastic media has been developed for use in trickling filters. Plastic media is light allowing inexpensive filter towers to be constructed up to 12 m high.

10.2.4 Tertiary treatment processes

Tertiary treatment processes are used primarily to control the levels of nutrients such as phosphorus (P) and nitrogen (N) being discharged into the environment. P removal is typically achieved by precipitation with an iron or aluminium

Figure 10.5 Trickling filter treating municipal waste water. [This figure is also reproduced in colour in the plate section after page 336.]

coagulant followed by a separation stage such as a deep bed or a moving bed sand filter. This type of processes is widely used and can produce effluents with levels of P consistently below 1 mg/l. This will be covered in a later chapter (Parsons and Berry, this volume).

Nitrogen removal process selection will depend on whether ammonia or nitrate is being removed where nitrification processes oxidise ammonia to nitrate, whilst denitrification processes reduce nitrate to nitrogen gas. There are wide ranges of process options for nitrification/denitrification, such as biofilters, biological aerated filters and fluidised beds. In addition, there are a number of commercial denitrification processes that use deep bed granular sand filter. Here, the microorganisms develop on the fixed sand particles and, in order to accelerate the denitrification reaction, a carbon source (normally methanol) is added to the nitrified effluent to supply the energy required by the denitrifiers. Typical removal rates were ~ 0.9 kg NO_3 removed per m^3 per day (Upton, 1993).

10.3 SLUDGE TREATMENT AND DISPOSAL

The end product of most of the processes that occur at a typical waste water treatment plant is sludge, which can be defined as a solid–liquid suspension of organic materials in water. These suspensions can arise from primary and secondary sedimentation, biological effluent treatment and various anaerobic digester processes.

The water content is still very high, with often only 0.25 to 12% of the sludge weight consisting of solids. Sludges are often high in volume and require processing and a suitable method of disposal.

Primary sludge is one of the main sources of domestic sludge. This is the sludge from the primary settling tank. It is usually grey and is characteristic by its offensive (normally hydrogen sulphide) odour. However, this type of sludge can often be readily digested. Other organic sludges include activated (aerobic) and digested (mainly anaerobic) sludges.

The treatment of sludge typically follows three stages of treatment (i) thickening (ii) digestion and (iii) dewatering before disposal.

Thickening is a procedure used to increase the solids content of sludge by removing a portion of the liquid fraction. This serves to decrease the overall volume of sludge. For example, sludge content can change from 3% to 6% solids during the thickening process. This means an overall volume reduction of 50%. Thickening is generally achieved by physical means and the most common thickeners are gravity, dissolved-air flotation and centrifugation.

Anaerobic treatment of primary and co-settled sewage sludge is common practice and can be defined as the biological breakdown of solids, producing methane and carbon dioxide. This takes place in a sealed, airtight tank. Anaerobic sludge digestion has several purposes:

- To reduce the odour of sludge.
- A reduction in sludge solids (aim for 35–40% reduction).
- To reduce the numbers of pathogens for land spreading.
- To solubilise nitrogen for fertiliser action.
- To produce methane gas.

It is a common misconception that the main reason for the anaerobic treatment of sludge is to produce energy, as methane. However, it is the reduction in solids and pathogens that are most important. Mesophilic anaerobic digestion (20 days at 35°C) almost completely destroys pathogens of most concern in the U.K, i.e., *Salmonella* and beef tapeworm. Like many other biological processes, anaerobic digestion was used before its biochemistry and microbiology were understood. Prior to 1910 processes were operated without heating or mixing. Experiments undertaken in 1931 in Birmingham indicated that heating and mixing aided digestion. Sludge digesters are normally operated in the mesophilic temperature range, i.e. 35°C, although some digesters are designed to operate in the thermophilic range, i.e. 55–60°C.

Dewatering is a physical or mechanical unit operation used to reduce the moisture content, and hence the volume, of the sludge. There are a number of reasons why such a system is used:

- Reduction in trucking costs for the lower volumes of sludge.
- Dewatered sludge is easier to handle than a thickened or liquid sludge.
- Dewatering can produce an odourless sludge that does not putrefy.

Dewatering characteristics are difficult to predict. The degree of difficulty increases as the proportion of secondary sludge increases with activated sludge being more difficult than conventional humus sludge. Co-settled sludge is more difficult than new sludge and older sludges are worse than fresh sludges. A number of techniques are available for the dewatering process including filter and belt presses and centrifuges. Selection of these processes must take account of:

- The dry solids content of the cake.
- The chemical content of the cake.
- Solids capture efficiency.
- Changes in sludge characteristics.
- Labour and maintenance costs.
- Capital costs.
- The final disposal route.

REFERENCES

Upton, J. (1993) Denitrification in Deep Bed Filters. *Water Sci. Technol.* **27**, 381–390.
Metcalf & Eddy Inc. (1991) *Wastewater Engineering Treatment, Disposal and Reuse*, 3rd edn. McGraw-Hill Publishing Company, London.
Droste, R.L. (1997) *Theory and Practice of Water and Wastewater Treatment*. John Wiley & Sons Inc., Chichester.
Stephenson, T. and Stuetz, R. (eds) (2003) *Water and Wastewater Treatment Principles*. IWAP, London.

11
Chemical phosphorus removal

Simon A. Parsons and Terri-Ann Berry

11.1 INTRODUCTION

Phosphorus (P) is an essential element for the growth of plants and as such is known as a nutrient. Excess quantities of phosphorus, together with an excess of the other main plant nutrient, nitrogen, can lead to extensive growth of phytoplankton, macro algae and higher plants and result in water quality problems. The needs for phosphorus removal processes have increased in Europe due to the establishment of the Urban Wastewater Directive. This was proposed by the EC in 1991. The Directive set discharge limits for some of the established sanitary determinants, e.g. biological oxygen demand (BOD) and suspended solids, as well as the nutrients, for rivers and designated '*sensitive areas*' (House of Lords, 1991; EC, 1992).

Several technologies for phosphorus removal exist. The choice of the specific technology to be used will depend on the field situation (e.g. rural communities, or small farms are better off with constructed wetland systems or filter beds; large cities, that have treatment plants already installed, will probably use addition of metal salts). This chapter will focus on the use of metal salts, which can be employed to remove phosphorus from a variety of waste waters (municipal, industrial, agricultural).

© 2004 IWA Publishing. *Phosphorus in Environmental Technology: Principles and Applications.*
Edited by Eugenia Valsami-Jones. ISBN: 1 84339 001 9

11.2 PHOSPHORUS REMOVAL WITH METAL SALTS

The chemical precipitation of phosphorus is typically undertaken with the addition of the salts of di- and tri-valent metal ions or lime that form precipitates of sparingly soluble phosphates (Cooper *et al.*,1994; Koutsoukos and Valsami-Jones, this volume; IWEM, 1994). The three types of metal precipitants most commonly used for chemical phosphorus removal are iron (II), iron (III) and aluminium (III). Other precipitants, such as lime, have also been extensively and successfully used, although the high doses needed and the resulting excess sludge production mean they have little application in municipal waste water treatment today (Horan, 1992; Tchobanoglous and Burton, 1991). The quantity of information available in this subject area is extensive, and is therefore only covered briefly in this review. A list of recent reviews and their objectives is provided below to aid the interested reader find more detailed and specific information (Table 11.1).

Most conventional sewage treatment plants are not able to achieve high phosphate removal (Horan, 1992; Tchobanoglous and Burton, 1991). A typical sewage influent contains 12 mg P/l and is usually reduced just by 25% producing the effluent of 9 mg P/l (Lees, 2002). Additional treatment systems were developed in order to improve phosphorus removal from waste water effluents such are ion exchange, reverse osmosis, chemical coagulation, electrodialysis, adsorption using soil spreading, use of chemical adsorbents and biological nutrient removal (Horan, 1990; McGrath and Quinn this volume). However, adsorption using soil spreading requires large areas of land increasing the cost of operation. Electrodialysis is only capable of removing 25% of the total phosphorus loading. Ion exchange and reverse osmosis will typically remove 90–98% phosphorus but are extremely expensive and require skilled operators (Horan, 1992). Chemical adsorbents are discussed in more depth in a later chapter (Douglas *et al.*, this volume).

Iron chloride and sulphate are readily available at low cost as refined by-products of the steel and titanium industries, however they require pH adjustment with sodium hydroxide or lime. Aluminium salts are generally more expensive and

Table 11.1 Literature review of chemical phosphorus removal.

Coagulant/flocculant	Objective	Author
Pre-polymerised inorganic coagulants	Phosphorus	Jiang & Graham (1998)
Lime	Suspended solids, phosphorus, BOD	Vosser (1980), Horan (1992), Tchobanoglous and Burton (1991)
Ferric chloride & anionic polymer	Suspended solids, nitrogen, Phosphorus	Poon & Chu (1999)
Polyelectrolytes	Review of types and mechanisms of coagulation	Dentel (1991)

there are public health concerns regarding application of aluminium due to its possible toxicity (e.g. Pearce, 1998).

The principle of chemical P removal from waste water is to transfer dissolved ortho-phosphates into particulate form by producing chemical precipitates of low solubility from the addition of metal salts. The newly formed precipitates and particulate phosphates are usually removed by solids separation processes such as sedimentation, flotation or filtration commonly applied in waste water treatment (Tchobanoglous and Burton, 1991).

Polyphosphates and organic phosphorus may take part in adsorption reactions, but only to a limited extent. Recht and Ghassemi (1970) showed that polyphosphates might adsorb or precipitate in combination with Fe- or Al-salts, three decades ago. But with a required pH of 4.0 and 5.5 for Fe (III)- and Al(III)-salts, respectively, the effective pH range is very narrow and far below that of domestic waste water. It is therefore important to know the exact proportions of different phosphorus fractions, i.e. easily, versus partly removable phosphorus fractions. These fractions may vary along the scheme of a waste water treatment plant and are subject to strong diurnal variations. The following parameters highly influence process performance and achievable levels of residual phosphorus in the effluent:

- the raw water quality (pH, suspended solids, dissolved organics)
- type and dose of precipitant
- location of dose application
- chemical speciation
- mixing conditions
- process configuration
- the expected quality of the treated water according to targets.

In waste water treatment, one key variable is the location of dose application (Figure 11.1). The metal salts may be added at the primary clarifier (pre-precipitation), directly to the mixed liquor, either in the aeration basin or upstream of the secondary

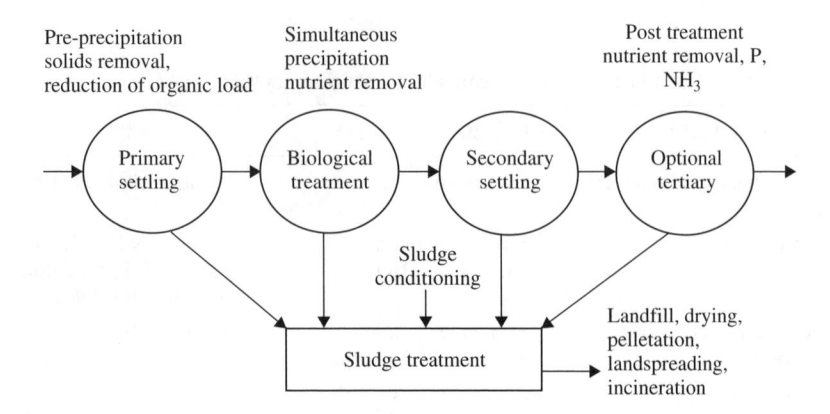

Figure 11.1 Options for chemical phosphorus removal in waste water treatment.

clarifier (simultaneous precipitation), or after the secondary portion of the plant in the feed to a tertiary solid–liquid separation process such as a filter or a reactor-clarifier (post-precipitation). Multiple point addition may also be employed.

The choice of operating mode (pre-precipitation, simultaneous precipitation or post-precipitation) depends on the physical configuration of the plant, chemical cost factors as well as effluent quality requirements.

11.2.1 Iron salts

Iron salts (Fe^{2+} and Fe^{3+}) are commonly used in the precipitation of phosphate from waste water. Chemical properties of the salt can cause corrosion, staining, iron carryover and coloured effluents. Trivalent salts usually have twice higher cost than divalent salts and they also have a lower metal content. The cost of divalent salts is so low that delivery distance is a major factor, with prices depending significantly on location.

Both ferrous (Fe^{2+}) and ferric (Fe^{3+}) ions can be used in the form of chloride or sulphate. A typical reaction between ferric chloride and phosphate is:

$$FeCl_3 + PO_4^{3-} \longrightarrow FePO_4 + 3Cl^-$$

The molar ratio of Fe:P is 1:1. 163.3 g of $FeCl_3$ will react with 95 g of PO_4^{3-} to form 150.8 g of $FePO_4$. The weight ratio of Fe:P is 1.8:1, while the weight ratio of $FeCl_3$:P is 5.2:1. The exact reaction mechanism is more complex than the above equation, and this also applies to alum (Brett et al., 1997). The reaction of ferrous salts and phosphate is approximated by:

$$3FeCl_2 + 2PO_4^{3-} \longrightarrow Fe_3(PO_4)_2 + 6Cl^-$$
$$3FeSO_4 + 2PO_4^{3-} \longrightarrow Fe_3(PO_4)_2 + 3SO_4^{2-}$$

The molar ratio of Fe:P is 3:2, while the weight ratio of the ferrous ion to phosphorus is 3.2:1 (Brett et al., 1997). Experimental results have indicated that the Fe^{3+} dose is usually considerably higher than predicted by $FePO_4$ stoichiometry. Competition between OH^- and phosphate for Fe^{3+} at the point of addition of ferric salt combined with a requirement for excess Fe^{3+} to destabilise $FePO_4$ and other colloids (e.g. dispersed microorganisms), probably account for the requirement for the stoichiometric excess. Reported optimum ratios for iron, have varied from less than 1 to as high as 7.5. The U.S. EPA (1987) suggests that a weight ratio of about 2.0 Al:P or 3.0 Fe:P is needed to achieve 95% removal of phosphorus from municipal waste water. Figure 11.2 shows an example of the correlation between %P removal and Fe dose. Typical performance data are presented in Table 11.2.

For the ferric iron the optimum pH range is 4.5–5.0. However, significant removal of phosphate can be achieved at a higher pH. For the ferrous ion, the optimum pH is around 8, and good removal rates can be achieved from 7 to 8. When an iron salt is added to waste water the metal ions will react with the abundance of water molecules and form hydrolysis products. If a hydrolysis product gets close to orthophosphate molecules, a reaction will occur. However, if the hydrolysis

Table 11.2 Plant performance data using metal salts for phosphorus removal.

Location/type	Chemical feed point	Chemical used	Metal ion: TP molar ratio	Chemical dose mg l^{-1}	Inf P	Eff P	Removal (%)	Reference
Lagoon	1 biol	Alum	3.2	–	4.8	0.5	90	Narasiah, 1991
Cranfield: AS	1 biol	Alum	3.0	–	6.8	2.0	71	Clark, 1999
Canada: modified AS	1 clarifier	Alum	–	–	14.5	0.9	95	Yeoman, 1988
USA: modified AS	1 clarifier	Alum	–	90	7.8	1.0	86	Yeoman, 1988
Letchworth: AS	2 biol	Ferric sulphate	0.89	–	–	1.0	80	Strickland, 1998
			1.11	–	–	0.51	93	
			1.61	–	–	0.36	96	
Sheboygan: AS	2 clarifier	Ferric chloride	0.89	10.2	6.38	0.9	85	EPA, 1987
Alton: TF	Plant influent	Ferric sulphate	1.61	–	5.6	0.8	86	Pearce, 1998
Marlborough: TF	Plant influent	Ferric sulphate	1.11	–	5.9	0.95	84	Pearce, 1998
			0.83	–	6.6	1.6	76	
Letchworth: AS	2 biol	Ferrous sulphate	0.72	–	–	2.3	63	Strickland, 1998
			1.28	–	–	0.72	86	
			1.56	–	–	0.62	93	
Appleton: AS	Plant influent	Ferrous chloride	0.89	16.8	10.5	0.8	92	EPA, 1987
Niles: AS	2 biol	Ferrous chloride	1.47	10.9	4.1	0.7	83	EPA, 1987
Port Washington: AS	1 clarifier	Ferrous chloride	0.8	8.5	5.9	1.0	83	EPA, 1987
Port Clinton: AS	2 biol	Ferrous chloride	1.09	10.2	5.2	0.5	90	EPA, 1987
Oberlin: AS	1 clarifier	Ferrous chloride	0.6	6.4	5.9	1.0	83	EPA, 1987

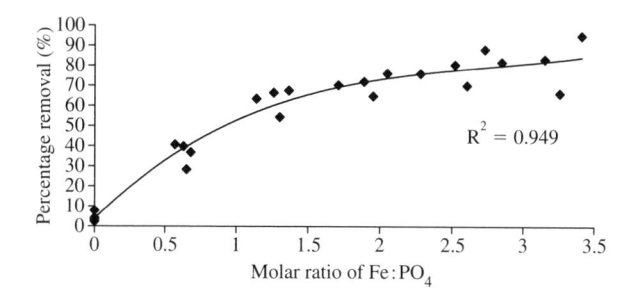

Figure 11.2 Percentage removal of orthophosphate by Fe(II), at increasing molar ratio of Fe : P.

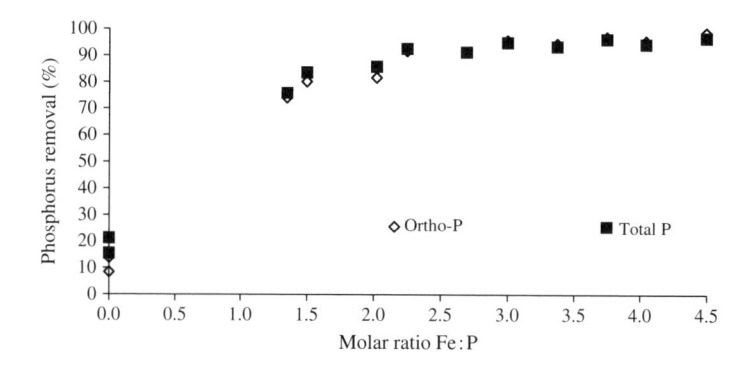

Figure 11.3 Phosphorus removal efficiency at varying doses of iron (III) chloride.

products do not come close to an orthophosphate ion, or other charged species, they will form metal hydroxides. The higher the pH the larger the probability that metal hydroxides will be formed through reactions with OH^- (Gilberg et al., 1996). The formation of metal hydroxides will affect the efficiency of phosphorus removal as demonstrated in Figures 11.3 and 11.4, by the higher than stoichiometric Fe : PO_4 molar ratios required for efficient removal. The reaction between iron (III) and orthophosphate is very rapid (completed in less than 1 s) and may be considered instantaneous (Recht and Ghassemi, 1971). However, due to the similarly rapid hydrolysis of the iron (III) when added to the waste water stream, it has been found that proper initial mixing conditions can be important to avoid excess chemical use (Thistleton et al., 2002).

The performance of ferrous ions requires the conversion of iron (II) to iron (III). Generally, it has been found that low dissolved oxygen (DO), low redox (E_h) and low pH resulted in poor conversions of iron (II) to iron (III) (Thistleton et al., 2001). Waste water with a very negative redox potential produced the lowest conversion averaging just 32.4% over 30 min. The highest conversions (72.3%) were observed with high DO concentrations, which (by nature of the relationship between DO and redox) coincided with redox potentials between 50–100 mV (Thisleton et al., 2002).

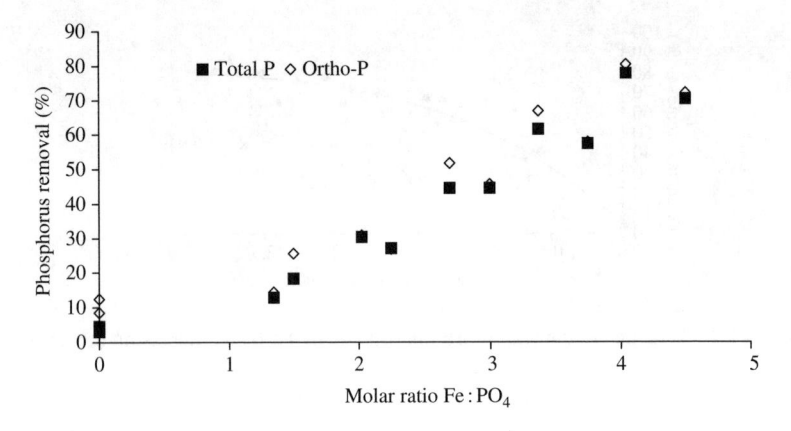

Figure 11.4 Phosphorus removal efficiency at varying doses of iron (III) hydroxide.

11.2.2 Aluminium salts

Aluminium salts behave similarly to iron (III) salts when dosed for phosphorus removal. They can be successfully applied for pre-, co- and post-precipitation although the optimal pH for efficient precipitation is slightly higher. As for iron (III) salts, the hydroxides can be formed easily resulting in excessive stoichiometry requirements to achieve full phosphorus removal.

There is far less information on the performances of aluminium dosing for P removal than for iron; however, on-site experiences have demonstrated that precipitation with iron salts can be adversely affected by incoming waste water quality, in particular a high content of complex organic components. The apparent complexation of iron with organic compounds has been observed to create a 'haze' in the effluent, resulting in slower settling which could contribute to increased phosphorus concentrations. On the contrary, addition of aluminium to the same waste water did not create this 'haze' effect and effluent phosphorus concentrations were improved. This effect was observed at both pilot and full scale systems, in particular on the sites with high industrial input phosphorus concentrations and especially when treating effluents from the malting industries. The adverse impact of iron on sludge volume index (SVI), compared to other chemicals (aluminium sulphate and ABP – an aluminium based polyelectrolyte) is shown in Figure 11.5.

One possible explanation of this interference could be gleaned from the relative positions of iron and aluminium in the periodic table. Iron is a transition (or d-block) element whereas aluminium is a main group (or p-block) element. The most significant feature of metals from the transition group is the presence of partially filled d-orbitals which may be of suitable energy, symmetry, directionality and occupancy to enter into very effective multiple bonding ($d\pi$–$p\pi$). Furthermore the iron–organic complexes are known to be very stable (for example, hemoglobin which remains unaffected even during the digestion process of the human stomach). However, it is generally accepted that d-orbitals do not play a major role in the bonding of p-block elements (such as aluminium) to carbon. These elements are

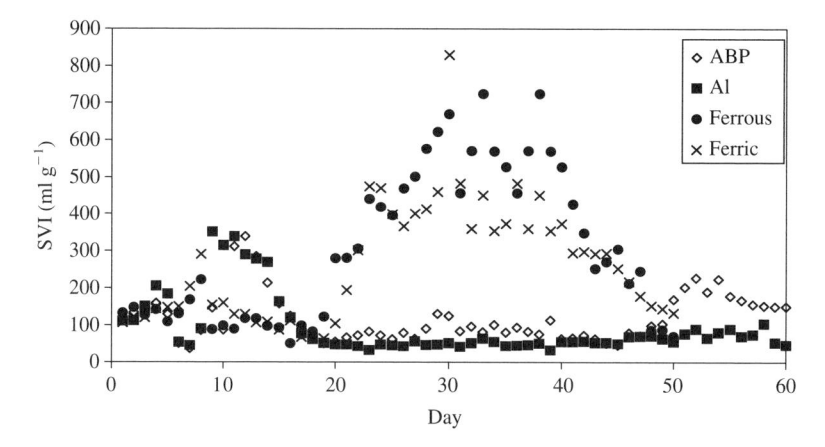

Figure 11.5 The impact of chemical addition on SVI (sludge volume index).

restricted to a limited number of sigma bonds and pπ–pπ bonds. Therefore complexation of various compounds in waste water with iron is achievable more easily and, once formed, they are also more stable. Even though iron remains an efficient precipitant for phosphorus, complexation with interference molecules (which are usually organic in nature) can occur and affect settlement to a great extent.

The literature on the efficacy of aluminium dosing for phosphorus removal is sparse, however some disadvantages of using aluminium salts instead of iron were reported. In general, it appears that although the stoichiometric requirements are the same for both elements, aluminium dosing often needs to be higher to achieve the same effluent quality (Table 11.2). In their laboratory experiments, Thistleton *et al.* (2002) demonstrated that P removal with pre-formed iron hydroxide is possible, but less efficient. This could be due to the ability of iron to form multiple bonds, according to the mechanism described above, i.e. using its outer d-orbitals resulting in iron hydroxyl-phosphate complexes. If conditions are not ideal for P removal, the addition of iron may result in higher effluent phosphorus concentrations compared to aluminium, due to the transitional nature of the former. This may explain why chemical addition for phosphorus removal in waste waters is termed 'site specific' as incoming waste water interferences must be taken into account before making a selection.

Apart from the overall effects of the chemical application when used for P removal, the effects on the biological system itself should be taken into consideration. They may be significant, particularly during co-precipitation. The effects on species diversity have been examined for iron and aluminium and this may influence the selection of the chemical for long term dosing to an aeration system (Clark *et al.*, 1999, 2000).

11.2.3 Downstream impacts

The addition of metal salts to municipal waste water has been shown to have a great potential for phosphorus removal and control. However, the impact of

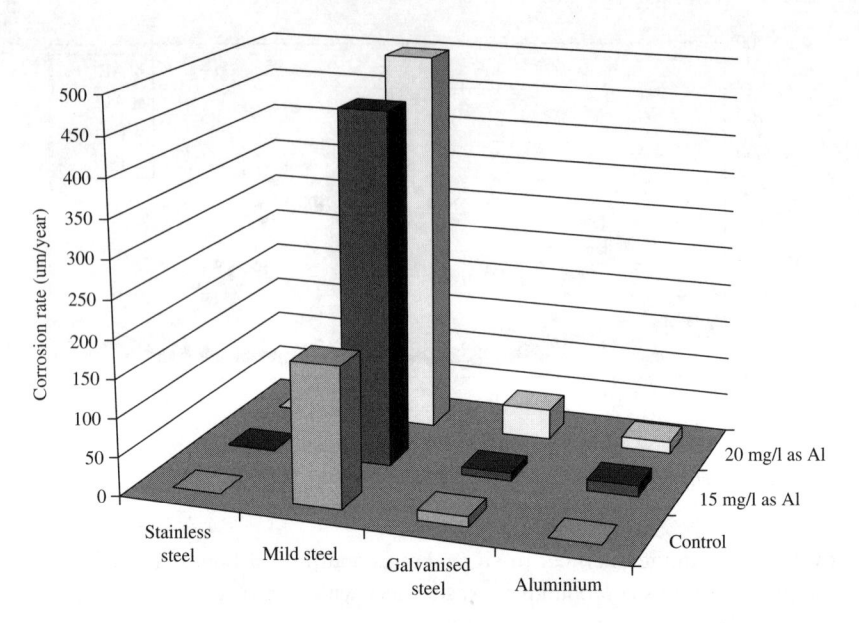

Figure 11.6 The effect of dosing with poly aluminium chloride on the corrosion rate of different material coupons housed in an RBC (Barugh, 1999).

adding significant quantities of metal ions into a waste water, through dosing, must be taken into consideration. For example, a number of studies conducted recently reported the effects of residual coagulant on nitrification, sludge production and dewaterability (Clark, 1999; Lees, 2002). Lees (2002) showed that dosing with $25\,\text{mg}\,l^{-1}$ of iron (III) chloride, iron (III) sulphate and aluminium sulphate significantly reduced the potential for an activated sludge pilot plant to remove ammonia. In the control experiment with no chemical addition, an influent ammonia concentration of $33\,\text{mg}\,l^{-1}$ was reduced to only $4\,\text{mg}\,l^{-1}$, compared to $10.5\,\text{mg}\,l^{-1}$ for iron (III) sulphate, $15.6\,\text{mg}\,l^{-1}$ for aluminium sulphate and $13.5\,\text{mg}\,l^{-1}$ for iron (III) chloride.

The corrodibility of the metal salts was another issue of concern (Clark, 1999; Barugh, 1999). Barugh (1999) showed that when dosing poly aluminium chloride in to a rotating biological contactor (RBC), this reduced P levels from $9\,\text{mg}\,l^{-1}$ to below $2\,\text{mg}\,l^{-1}$ but at the same time, the integrity of RBC might have been compromised by accelerated corrosion (Figure 11.6).

It is well known that metal salt dosing will lead to increased sludge production and that the metal ions also have impact on the thickening and dewatering of the sludge (Horan, 1992). Figure 11.7 shows the effect of iron dose on the sludge density that can be achieved by thickening. It is clear that as iron dose increases, the sludge thickness decreases, e.g. from 6% to below 2% at an iron (III) dose of $\sim 80\,\text{mg}\,l^{-1}$, although this sludge was much easier to dewater. At a dose of $20\,\text{mg}\,l^{-1}$ the sludge had a capillary suction time (CST) of $140\,\text{s}$ compared to $40\,\text{s}$ at $80\,\text{mg}\,l^{-1}$.

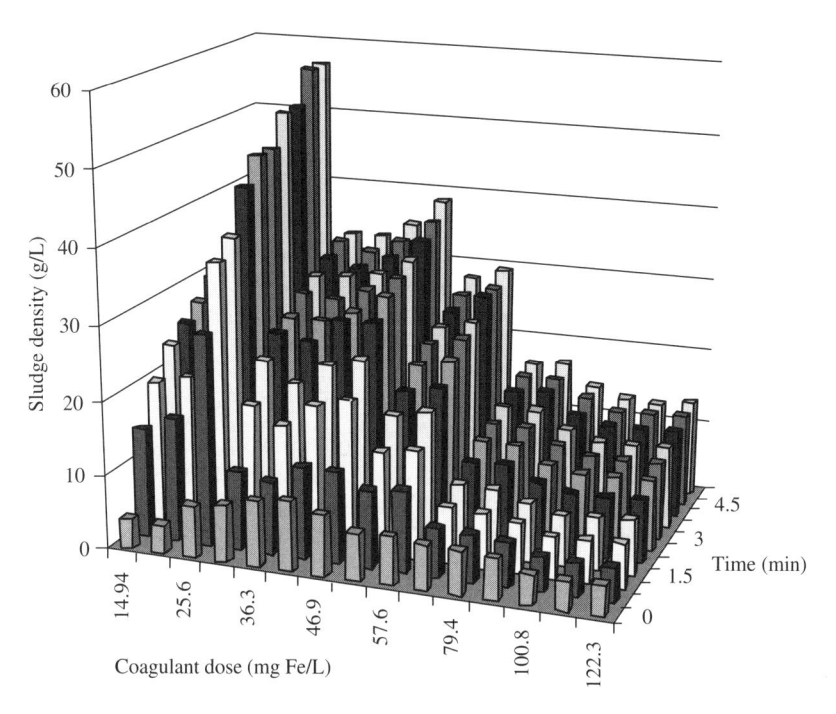

Figure 11.7 Effect of coagulant time on the thickening of a sludge produced from the pre-precipitation of municipal waste water with ferric sulphate.

11.3 CONCLUDING REMARKS

- The chemical precipitation of phosphorus is brought about by the addition of the salts of di- and tri-valent metal ions that form precipitates of sparingly soluble phosphates. The reduction of phosphorus levels in this way is widely practised and performance, whilst affected by water quality, can be controlled. Phosphorus removal in excess of 90% is achievable under optimised dose and pH conditions.
- The process is robust and flexible and given efficient dosing and mixing, chemical, residuals and cost savings can be made.
- The impact of chemical dosing on downstream biological and sludge treatment has recently gained much interest in terms of effect on biological oxidation processes, sludge production and material selection although further research is need to better understand the effects seen.
- Whilst there is considerable interest in biological phosphorus removal, there are waste water treatment sites where, due to e.g. size or waste water strength, chemical dosing is and will be the best available option for reducing phosphorus levels.

REFERENCES

Barugh, A. (1999) MSc Thesis. Chemical phosphorus removal from a RBC. Cranfield University, Bedford, UK.
Brett, S., Guy, J., Morse, G.K. and Lester, J.N. (1997) *Phosphorus Removal and Recovery Technologies.*, Selper Publications, London.
Clark, T. (1999) PhD Thesis: The impact of chemical addition for phosphorus removal on activated sludge treatment. Cranfield University, Bedford, UK.
Clark, T., Stephenson, T. and Arnold-Smith, A.K. (1999) The impact of aluminium based co-precipitants on the activated sludge process. *Trans Inst. Chem. Eng.* **77**, Part B, 31–36.
Clark, T., Burgess, J.E., Stephenson, T. and Arnold-Smith, A.K. (2000) The impact of iron based co-precipitants on the activated sludge process. *Trans Inst. Chem. Eng.* **78**, Part B, 405–410.
Cooper, P.F., Day, M. and Thomas, V. (1994) Process options for phosphorus and nitrogen removal from wastewater. *J. Inst. Water Environ. Manage.* **8**, 84–92.
Cooper, P.F., Job, G.D., Green, M.B. and Shutes, R.B.E. (1996) Reed beds and constructed wetlands for wastewater treatment. *Proceedings of the water Research Council/Severn Trent Water Workshop on Reed beds and constructed wetlands*, Kenilworth, Warwickshire, June 1986, WRc Ltd, Swindon.
Council of European Communities (1992) *Directive concerning urban waste water treatment* (91/271/EEC). Official Journal L135, 40–52.
Dentel, S.K. (1991) Coagulant control in water treatment. *Crit. Rev. in Environ. Control* **21**, 41–135.
Environmental Protection Agency (1987) Design manual: phosphorus removal. EPA/625.1-87/001. EPA, Cincinnati, Ohio.
Gilberg, L., Nilsson, D. and Akesson, M. (1996) The influence of pH when precipitating orthophosphate with aluminium and iron salts. *Proceedings of the 7th Gothenburg Symposium, Sep. 23–25, 1996, Edinburgh, Scotland*, 95–106.
Horan, N.J. (1992) Nutrient removal from wastewaters. *Water Wastewater Technol.* **2**, 16–17.
House of Lords (1991) Environment Agencies Bill. HMSO, London.
Institution of Water and Environmental Management (IWEM) (1994) *Tertiary Treatment*, 2nd edn. IWEM, London.
Jiang, J. and Graham, N.J.D. (1998) Pre-polymerised inorganic coagulants and phosphorus removal by coagulation – A review. *Water South Africa* **24**, 237–244.
Lees, E. (2002) PhD Thesis: The impact of chemically assisted sedimentation on downstream treatment processes. Cranfield University, Bedford, UK.
Narasiah K.S., Morasse, C. and Lemay, J. (1991) Nutrient removal from aerated lagoons using alum and ferric-chloride – a case-study. *Water Sci. Technol.* **23**, 1563–1572.
Pearce, P. (1998) Options for phosphorus removal on trickling filter plants. In *Chemical water and wastewater treatment V*, Springer-Verlag, Berlin-Heidelberg.
Poon, C.S. and Chu, C.W. (1999) The use of ferric chloride and anionic polymers in the chemically assisted primary sedimentation process. *Chemosphere* **39**, 1573–1582.
Recht, H.L., Ghassemi, M. and Kleber, E.V. (1970) Precipitation of phosphates from water and wastewaters using lanthanum salts. *Proc. Adv. Water Pollut. Res. 5th Intl. Conf.* **1**, 1–17.
Strickland, J. (1998) The development and application of phosphorus removal from wastewater using biological and metal precipitation techniques. *J. Inst. Water Environ. Manage.* **12**, 30–37.
Tchobanoglous, G. and Burton, F.I. (1991) *Wastewater Engineering*, 3rd edn. McGraw-Hill, Inc., New York.
Thistleton, J., Clark, T., Pearce, P. and Parsons, S.A. (2001) Mechanisms of chemical phosphorus removal I. Iron (II) salts. *Trans Inst. Chem. Eng.* **79**, Part B, 339–344.

Thistleton, J., Clark, T., Pearce, P. and Parsons, S.A. (2002) Mechanisms of chemical phosphorus removal II. Iron (III) salts. *Trans Inst. Chem. Eng.* **80**, Part B, 1–5.

Vosser, J.L. (1980) Chemically aided sedimentation of crude sewage. *Prog Water Technol.* **12**, 411–426.

Yeomen, S., Stephenson, T., Lester, J.N. and Perry, R. (1988) The removal of phosphorus during wastewater treatment – a review. *Environ. Pollut.* **49**, 183–233.

12
Biological phosphorus removal

John W. McGrath and John P. Quinn

12.1 INTRODUCTION

The recognition and systematic study of the phenomenon of enhanced biological phosphate removal by microorganisms dates back to 1955 when Greenberg et al. (1955) proposed that under certain circumstances activated sludge had the ability to accumulate phosphate in excess of that required for balanced microbial growth. Subsequently Greenberg's hypothesis was demonstrated by Srinath et al. (1959) who, while studying the feasibility of growing rice plants on the surface of an activated sludge plant, found that the plants suffered from the characteristic symptoms of phosphate deficiency exemplified by excessive vegetative growth and diminished grain formation. This lack of soluble phosphorus (P) was traced to its excessive accumulation by the sludge biomass. Levin and Shapiro (1965) reported similarly enhanced phosphate removal by activated sludge in the absence of added chemical precipitants; addition of 2,4-dinitrophenol, a biocide, to the sludge biomass inhibited phosphate uptake however, thus confirming a biological process. In later years this phenomenon was frequently referred to as 'luxury' phosphate uptake; its exploitation appeared to provide the basis of a possible biological alternative to chemical precipitation for phosphate removal from waste streams and

© 2004 IWA Publishing. *Phosphorus in Environmental Technology: Principles and Applications.*
Edited by Eugenia Valsami-Jones. ISBN: 1 84339 001 9

ultimately led to the development of the Enhanced Biological Phosphate Removal (EBPR) process.

The EBPR process is based upon the exposure of activated sludge to alternating periods in the absence and presence of oxygen – the anaerobic and aerobic phases. This is achieved by configuring the treatment system (see Parsons and Stephenson, this volume) such that an anaerobic zone is added upstream of the traditional aerobic phase; influent waste water is introduced into this anaerobic zone (reviewed in Brett *et al.*, 1997; Kortstee *et al.*, 1994; Mino *et al.*, 1998; van Loosdrecht *et al.*, 1997; Yeoman *et al.*, 1988).

In terms of the microbial biochemistry that underpins it, the EBPR process can be described as a tale of two biopolymers. In the initial anaerobic phase it appears that a specialized group (or groups) of microorganisms is able to gain a selective advantage through their ability to take up the short chain fatty acid molecules, such as acetate, that are present in the liquor as fermentation by-products, and to convert these to a carbon storage polymer. This is normally poly-β-hydroxybutyrate (PHB), although poly-β-hydroxyvalerate (PHV) has also been observed. The energy required for the uptake and synthesis of PHB comes at the expense of another biopolymer, polyphosphate (polyP), whose synthesis is a second defining characteristic of the microbial groups responsible for EBPR. PolyP consists of a linear chain of phosphate groups linked together by high-energy phosphoanhydride bonds and ranges in length from three to greater than 1000 orthophosphate groups (Kulaev, 1979). Hydrolysis of the intracellular polyP reserves stored by EBPR microorganisms provides the energy required for their accumulation of PHB under anaerobic conditions; as a result phosphate release to the extracellular medium is characteristic of this phase of the EBPR process.

In the subsequent, aerobic, stage of EBPR those microorganisms that contain stored PHB/PHV are able to replenish their internal polyP reserves; in doing so they remove not only the phosphate released during the anaerobic phase of the process but also almost all the available phosphate from the surrounding oxygen-rich environment. As much as 30% of the PHB/PHV previously formed during the anaerobic phase is thus consumed by EBPR microorganisms under aerobic conditions as they take up phosphate and convert it to polyP. The exact physiological or environmental factors that trigger this 'sacrifice' of stored PHB/PHV by the members of the EBPR microflora in order to replenish their polyP reserves is at present unknown (Kortstee *et al.*, 2000). However it seems clear that the catabolic and anabolic interplay between polyP and PHB is central to microbial phosphate removal during the EBPR process.

When operated under favourable conditions EBPR plants are often able to remove 80–90% of influent phosphate (Water Environment Federation, 1998), compared to the 20–40% removal typical of conventional activated sludge treatment (Streichan *et al.*, 1990). Ultimate elimination of phosphate from the system is then achieved by the wastage of P-rich excess sludge. It must be noted, however, that the absence of nitrate in the anaerobic zone is essential for polyP utilisation during PHB formation (and thus the effectiveness of the EBPR process); nitric oxide exhibits an inhibitory effect on the enzyme adenylate kinase, the function of which is described in Dobrota (this volume). Adenylate kinase plays a key role in energy generation from accumulated polyP (van Niel *et al.*, 1998). Anoxic denitrification stages are therefore often built into EBPR systems, thus ensuring nitrate removal prior to anaerobiosis.

12.2 PROCESS DESIGN OF BIOLOGICAL PHOSPHORUS REMOVAL PLANTS

An essential prerequisite for efficient EBPR is the presence of a substantial concentration of readily available organic carbon in the anaerobic phase. This permits the microorganisms responsible to accumulate reserves of PHB/PHV, at the expense of their stored polyP, and with the concomitant release of phosphate. Much of the work reported on EBPR has been conducted in North America and South Africa where stronger sewages (i.e. having BOD:P ratios greater than 20) facilitate the anaerobic formation of PHB/PHV (Upton *et al.*, 1996). In countries like the UK such BOD:P ratios are found infrequently, therefore several commercial processes have been developed to overcome these differing sewage characteristics. Such modified biological phosphate removal systems may take the form of either a sidestream process (treating only a percentage of the flow e.g. the Phostrip process) or a mainstream system (treating the entire flow e.g. the University of Cape Town process). Many other process configurations exist for EBPR including the Modified Bardenpho process, the Three Stage Phoredox process, the A/O process, the Rotanox process and the Modified University of Capetown process (reviewed in Brett *et al.*, 1997; Yeoman *et al.*, 1988). Irrespective of their individual features, however, the success of each is necessarily dependent on the ability of the microflora of the system to take up and store phosphate intracellularly in the form of polyP.

12.2.1 The Phostrip process

This is perhaps the most widely used sidestream process for phosphate removal from waste waters. The process incorporates both biological and chemical phosphorus removal, with a proportion of the return activated sludge from the aerobic stage of a conventional EBPR plant being channelled into an anaerobic phosphorus stripper tank (Figure 12.1). Anaerobiosis induces phosphate release from the sludge biomass, producing a phosphate-rich supernatant which is continuously fed into a chemical treatment zone for lime dosing. This raises the pH of the effluent to approximately 9.0, bringing about phosphate precipitation as calcium phosphate. Insoluble calcium phosphate is then either directly removed from the system or returned to the main stream and eliminated together with the primary sludge (Brett *et al.*, 1997).

As a sidestream process Phostrip technology has a number of advantages for phosphate removal. Most importantly, effluent concentrations of less than 1 mg/l total phosphorus can be readily achieved even at low BOD loadings, e.g. a full scale

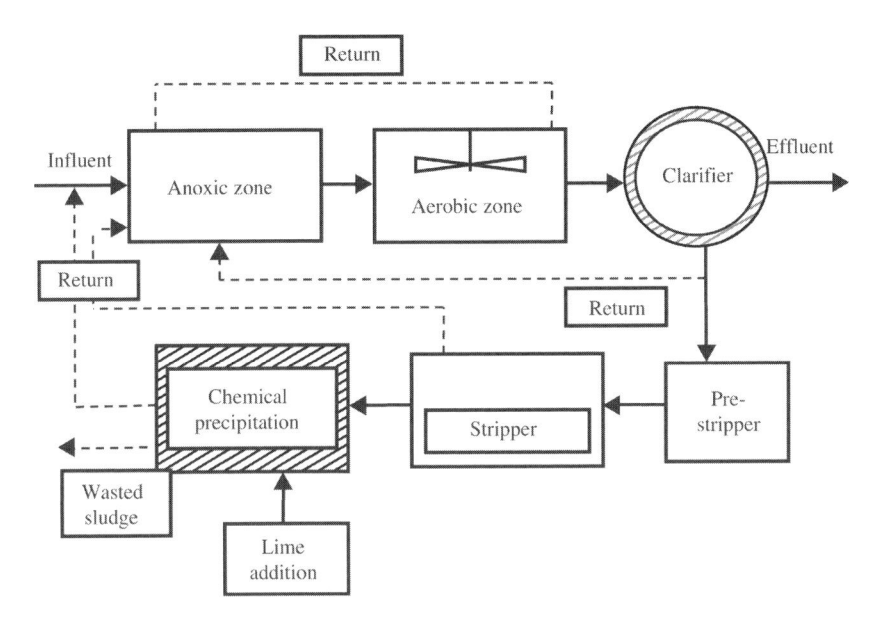

Figure 12.1 The Phostrip process.

Phostrip process operated by Severn Trent Water Ltd at Stratford-upon-Avon consistently achieved an effluent standard of 1 mg/l P (Upton *et al.*, 1996). In addition, phosphate is removed from the Phostrip system as a chemical sludge thus reducing its mobility and potential leakage from the biomass. Moreover the quantities of chemicals required (and thus the costs) of the Phostrip process are lower than those required for conventional chemical precipitation. However the Process does require greater operator skill and process control relative to traditional chemical dosing technology (Brett *et al.*, 1997).

12.2.2 The University of Capetown process (UCT)

The UCT process is just one example of a mainstream EBPR process and was developed in response to the need to prevent nitrate entering the anaerobic zone. As shown in Figure 12.2, return activated sludge is diverted to an anoxic zone instead of the traditional anaerobic phase. Nitrate generated under aerobic conditions is thereby denitrified in the anoxic phase producing a very low nitrate sludge, thus minimising the potential for nitrate to act as a terminal electron acceptor in the oxidation of BOD under anaerobiosis. In addition, recycling of the mixed liquor from the anoxic zone to the anaerobic stage provides optimum conditions for the formation of short chain fatty acid fermentation products. It is the production of these short chain fatty acids, in sufficiently high concentration, that is essential for a mainstream EBPR system to operate optimally (Brett *et al.*, 1997; Yeoman *et al.*, 1988). The requirement for readily biodegradable COD is demonstrated by a case study conducted by Severn Trent Water Ltd on their UCT-based system

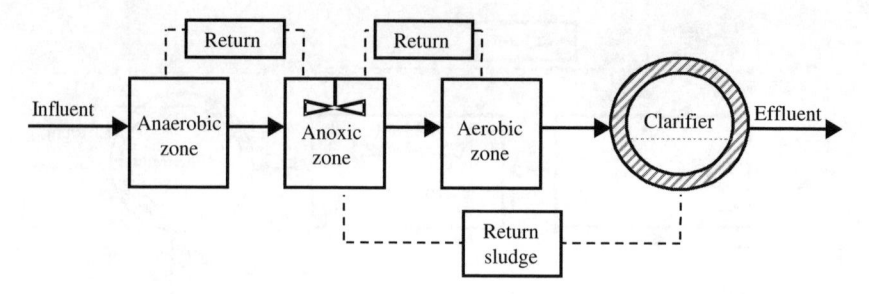

Figure 12.2 The University of Capetown (UCT) process.

commissioned for biological phosphate removal at Stratford-upon-Avon (Upton *et al.*, 1996). Despite receiving the discharge from a vegetable processing plant, and in consequence a strong sewage in which the BOD:P ratio varied between 25:1 and 28:1, an effluent standard of less than 2 mg/l P could be produced by the UCT system. Good EBPR, leading to effluents of below 1 mg/l P, could only be achieved after the addition of short chain fatty acids in the form of liquor from an on-site anaerobic digester (Upton *et al.*, 1996). This dependency of EBPR performance on waste water strength, set against the inherent variation in settled sewage characteristics, e.g. at night, during periods of inclement weather, or at weekends, when an extremely weak sewage may prevail, is one of the main disadvantages of the EBPR process as a biotechnological tool for phosphate removal (Upton *et al.*, 1996).

12.3 POLYPHOSPHATE

Over the past decade numerous reviews have been published on both the biochemistry and ecology of EBPR (Brett *et al.*, 1997; Kortstee *et al.*, 1994; Mino *et al.*, 1998; van Loosdrecht *et al.*, 1997; Yeoman *et al.*, 1988). In the absence of studies on the metabolic basis of the process in pure microbial cultures, many of these have focused on the biochemical models that have been produced to account for carbon turnover during the process, and in particular the formation of PHB/PHV in the course of its anaerobic phase. These models primarily seek to explain the source of the reducing power required for the conversion of acetyl-Coenzyme A (or propionyl-Coenzyme A) into PHB/PHV. For a comprehensive review of anaerobic carbon metabolism during EBPR the reader is referred to the papers of Mino *et al.* (1998), Kortstee *et al.* (2000) and Hesselmann *et al.* (2000).

The cornerstone of EBPR, however, is the microbial synthesis, intracellular storage and ultimate hydrolysis of polyP; without the ability of environmental micro-organisms to accumulate exogenous phosphate intracellularly in polymerized form, no biological process for the removal of phosphate would be possible. The remainder of this review will therefore concentrate on the microbial metabolism of this ubiquitous natural phosphate reserve material.

$$\left[\begin{array}{c} O^- \\ | \\ ^-O-P-O \\ \| \\ O^- \end{array} \begin{array}{c} O^- \\ | \\ -P-O \\ \| \\ O^- \end{array} \right]_n \begin{array}{c} O^- \\ | \\ -P-O^- \\ \| \\ O^- \end{array}$$

Figure 12.3 The structure of inorganic polyphosphate.

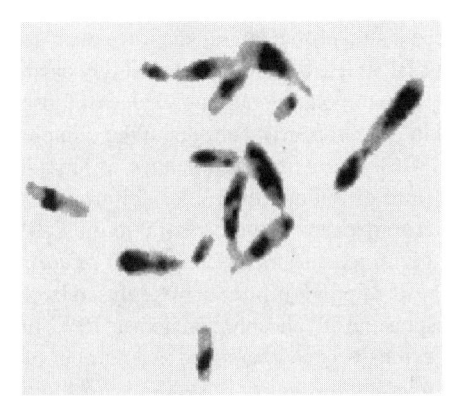

Figure 12.4 Stained cells of *Candida humicola* G-1 showing polyphosphate granules coloured black with Neisser stain (×1000).

Intracellular granules that stain metachromatically with basic dyes have been described in the microbiological literature since the early 1900s. It was not however until 1947 that Wiame (1947, 1948) demonstrated that these 'volutin' granules corresponded to deposits of polyP (Figure 12.3), a molecule that is today recognised as one of the most widely distributed natural biopolymers, having been detected in many bacteria, fungi, yeasts, plants and animals (Dawes and Senior, 1973; Kulaev and Vagabov, 1983; Kulaev et al., 1999). On Neisser staining (Gurr, 1965) intracellular polyP inclusions appear as dark granules (Figure 12.4). Under optimal conditions polyP may amount to 10–20% of the cellular dry weight and as such greatly exceeds the P requirements of the cell – which suggests that it may fulfil a metabolic role(s) other than simply that of a phosphate reserve (Pick et al., 1990). Reviews by Harold (1966), Dawes and Senior (1973), Kortstee et al. (1994), Kulaev et al. (1999) and Kornberg and Fraley (2000) detail many of the physical and chemical properties of polyP as well as its predicted role in prebiotic evolution.

12.3.1 Polyphosphate-metabolizing enzymes in the microorganisms of the EBPR process

Despite extensive research on the metabolic basis of the EBPR process, and its detailed biochemical modelling, the identities of those enzymes actually involved in polyP turnover *in situ*, and the basis of their regulation, remain unclear (Trelstad *et al.*, 1999). Moreover such studies must be qualified by the fact that the process described as enhanced 'biological' phosphate removal also involves chemical mechanisms such as phosphate precipitation or its adsorption at the outer membranes of microorganisms (Bark *et al.*, 1992, 1993); the relative contribution of such processes is difficult to quantify.

Given that the major component of EBPR is, however, biologically-mediated, it is clear that the majority of work on microbial polyP metabolism has involved strains that are unlikely to participate in the EBPR process. This is particularly true in the light of more recent studies that largely disprove the extensive involvement of *Acinetobacter* spp. in EBPR; it is becoming increasingly clear that the process cannot be defined by simply extrapolating our knowledge of those metabolic processes observed or assumed in *Acinetobacter* cultures to the complexity of the field situation (Weltin *et al.*, 1996). The wider significance of any physiological and biochemical studies, on pure or defined microbial cultures, is thus likely to remain uncertain until the microorganisms that are central to the EBPR phenomenon have been unambiguously identified and methods for their *in vitro* culture established.

Microbial biosynthesis of polyP is primarily catalysed by the enzyme polyphosphate kinase (polyphosphate: ADP phosphotransferase; PPK; EC 2.7.4.1) (Kornberg, 1995). PPK has been extensively characterised in a number of prokaryotes, including *Vibrio cholera, Pseudomonas aeruginosa* 8830, *Propionibacterium shermaii, Acinetobacter* sp. Strain ADP1, *Neisseria meningitidis, Arthrobacter atrocyaneus, Corynebacterium xerosis, Salmonella minnesota, Burkholderia cepacia* AM19, *Cryptococcus humicolus, Sulfolobus acidocaldarius* and, most extensively, in *Escherichia coli* (Kornberg and Fraley, 2000; Kulaev *et al.*, 1999; McGrath and Quinn, 2000; Mullan *et al.*, 2002). The enzyme catalyses the progressive synthesis of the polyP chain through the reversible transfer of the gamma phosphate from ATP to polyP (Figure 12.5) (Akiyama *et al.*, 1992; Geißdörfer *et al.*, 1998; Kato *et al.*, 1993; Tinsley and Gotschlich, 1995).

Despite the fact that PPK is the most extensively studied polyP-synthesising enzyme, for the reasons detailed above the true importance of even its involvement in the EBPR process is unclear. For example the evidence for its activity within the activated sludge microbial community during the aerobic phase of the EBPR process is far from conclusive. Significant levels of the enzyme were found in only one of six polyP-accumulating *Acinetobacter* strains isolated from EBPR sludge (van Groenestijn *et al.*, 1989) and in four of twenty-one similar isolates from two further studies (Bark *et al.*, 1993; Weltin *et al.*, 1996). Moreover, in a number of isolates production of the enzyme was found to be inducible in response to phosphate starvation but repressed under conditions of phosphate surplus (Trelstad *et al.*, 1999). In the light of this apparent paradox, the fact that it may be the 'reverse'

$$PolyP_n + ATP \rightleftharpoons PolyP_{n+1} + ADP$$

Figure 12.5 Polyphosphate synthesis reaction catalysed by polyphosphate kinase (PPK).

PPK activity (12.3.1.5; Figure 12.11) that is physiologically significant must be considered; alternatively the regulation of its activity by allosteric activation (or other non-genetic 'fine control' mechanisms) cannot be ruled out. An additional complicating factor is that the EBPR process aims to achieve residual phosphate target levels of some 50 μmoles l^{-1}; at these values phosphate starvation-inducible gene expression may well occur. Finally the possibility of alternative, as-yet-unrecognized routes of microbial polyP biosynthesis remains (Ogawa *et al.*, 2000; Zago *et al.*, 1999).

Equally important for the EBPR process, of course, is the possession, by the organisms responsible, of an efficient energy-generating system from polyP under anaerobic conditions. However numerous studies have been unable to conclusively attribute the degradation of polyP to any specific enzymatic activity. A number of different enzyme systems capable of polyP degradation have nevertheless been detected. These will be described in the following five sections.

12.3.1.1 Exopolyphosphatase (EC 3.6.1.11)

The processive hydrolytic cleavage of phosphate from the end of the polyP chain is catalysed by the enzyme exopolyphosphatase. This reaction may continue until only pyrophosphate remains (Kornberg *et al.*, 1999) (Figure 12.6).

Exopolyphosphatases has been identified from a number of bacterial and eukaryotic sources. Bacterial exopolyphosphatases commonly have low activity on short-chain polyP; however a specific tripolyphosphatase (EC 3.6.1.25) has been purified from *Methanobacterium thermoautotrophicum* (van Alebeek *et al.*, 1994); hydrolysis of tripolyphosphate by a purified exopolyphosphatase from *Saccharomyces cerevisiae* has also been reported (Kulakovskaya *et al.*, 1999).

$$PolyP_n + H_2O \longrightarrow PolyP_{n-1} + P_i$$

Figure 12.6 Polyphosphate degradation by exopolyphosphatase (PPX).

12.3.1.2 Endopolyphosphatase (EC 3.6.1.10)

Endopolyphosphatases catalyze the internal hydrolytic cleavage of polyP (Figure 12.7).

Endopolyphosphatase activity has been detected in eukaryotic cells but not in prokaryotes (Kulaev and Kulakovskaya, 2000), and the metal-dependent vacuolar enzyme from *Saccharomyces cerevisiae* has been purified (Kumble and Kornberg, 1996). It cleaves long chain polyP ($\sim$700 groups) to tri-polyP and in addition gives rise to an intermediate product of about 60 residues.

$$PolyP_n + H_2O \longrightarrow PolyP_{n-x} + PolyP_x$$

Figure 12.7 Endopolyphosphatase degradation of polyphosphate.

12.3.1.3 Polyphosphate: AMP phosphotransferase and adenylate kinase

PolyP: AMP phosphotransferase catalyzes attack at the terminus of the polyP chain by AMP to produce ADP (Figure 12.8).

The ADP formed can subsequently serve as a substrate for adenylate kinase which interconverts ADP and ATP (Figure 12.9).

ATP formation from AMP and polyP can thus occur through the combined action of the two enzymes, even in the absence of either an electron donor (organic carbon source) or electron acceptor (oxygen) (van Groenestijn *et al.*, 1987, 1989). However very recent studies dispute even the existence of such an enzyme activity in microbial cells and suggest that the polyP:AMP phosphotransferase activity observed in crude extracts of *E. coli* is catalyzed by an enzyme complex that is formed between PPK (polyP:ADP phosphotransferase) and adenylate kinase in the presence of polyP (Ishige and Noguchi, 2000). On the other hand, reports have described the purification of a polyP:AMP phosphotransferase like activity from *Acinetobacter* (Bonting *et al.*, 1991) and reported its identification in extracts of both *E. coli* and *Myxococcus xanthus* (Kornberg *et al.*, 1999).

$$PolyP_n + AMP \rightleftharpoons PolyP_{n-1} + ADP$$

Figure 12.8 Formation of ADP from AMP and polyphosphate by polyphosphate: AMP phosphotransferase.

$$2ADP \rightleftharpoons AMP + ATP$$

Figure 12.9 Adenylate kinase conversion of ADP into AMP and ATP.

12.3.1.4 Polyphosphate: glucose phosphotransferase (polyphosphate glucokinase; EC 2.7.1.63)

PolyP: glucose-6-phosphate phosphotransferase (glucokinase) (EC 2.7.1.63) catalyzes an attack by glucose at the end of the polyP chain (Figure 12.10).

Glucokinase activity has been identified in a wide variety of bacteria (Wood and Clark, 1988); its purification has been reported from *Mycobacterium tuberculosis*, *Mycobacterium phlei* (Hsieh *et al.*, 1993; Szymona and Ostrowski, 1964) and from *Propionibacterium shermanii* (Phillips *et al.*, 1993). The enzyme is specific to glucose and glucosamine and possesses two active centres, at one of which

$$PolyP_n + Glucose \longrightarrow PolyP_{n-1} + \text{Glucose-6-phosphate}$$

Figure 12.10 Polyphosphate: glucose-6-phosphate phosphotransferase catalyzed formation of glucose-6-phosphate.

ATP serves as phosphate donor. In the most phylogenetically-ancient organisms, the catalytic efficiency of polyP-glucokinase activity is greater than ATP-glucokinase activity (Hsieh *et al.*, 1996a, 1996b), suggesting that the enzyme evolved at a time when polyP played a more prominent role in bioenergetics. In more recent phylogenetic groups the polyP-dependent activity appears to be a 'fossil' reaction.

12.3.1.5 Polyphosphate: ADP phosphotransferase (reverse PPK activity)

PolyP: ADP phosphotransferase activity catalyses reverse polyphosphate kinase activity, i.e. formation of ATP from ADP and polyP (Figure 12.11).

Its existence in *Pseudomonas aeruginosa* as a discrete protein, distinct from the reverse reaction of PPK has very recently been proposed (Ishige and Noguchi, 2001). The enzyme has now been purified and appears to display an almost threefold higher specificity for GDP (guanosine triphosphate) than ADP: high levels of GTP formation may be required by this organism for exopolysaccharide production. This activity has been designated as PPK2 (Ishige and Noguchi, 2001; Ishige *et al.*, 2002).

The enzymes polyP: AMP phosphotransferase and adenylate kinase have been shown to form ATP from polyP in a number of *Acinetobacter* strains (Bonting *et al.*, 1991; Kampfer *et al.*, 1992; van Groenestijn *et al.*, 1987). Van Groenestijn *et al.* (1989) have extended this work by demonstrating that a positive correlation exists between the latter activity in cell extracts of a number of activated sludge samples and the percentage of P removed from waste water by those sludges.

By contrast, however, eleven polyP-accumulating activated sludge isolates (none identified as *Acinetobacter* spp.) lacked detectable polyP : AMP phosphotransferase activity, and only three showed (low) levels of PPK. All, however, contained adenylate kinase activity, while in four cases even higher levels of polyP : glucokinase activity were detected; three strains contained a glucose-6-phosphate-dependent nicotinamide adenine dinucleotide (NAD) kinase (Bark *et al.*, 1993). It was speculated that the latter two enzymes in concert could provide an effective polyP-dependent NADP regeneration system. It is also possible (van Veen *et al.*, 1994) that phosphate generated from polyP by celullar polyphosphatase activity could contribute to the transmembrane proton gradient and thus to partial conservation of the energy of the polyP phosphoanhydride bond.

The lack of a clear picture with regard to polyP metabolism may also be due to a shortage of suitable methods for the quantification of PPK and polyP. Current methods of enzymatic analysis of polyP synthesis are either (a) inexact, varying in product recovery yields obtained, (b) cumbersome, requiring extensive extraction and enrichment steps, or the use of radiolabelled phosphate, or (c) indirect, and

$$PolyP_n + ADP \rightleftharpoons PolyP_{n-1} + ATP$$

Figure 12.11 Action of polyphosphate: ADP phosphotransferase activity. This activity is the reverse of polyphosphate kinase activity (Figure 12.5).

therefore subject to interference by other cellular enzymes if crude cell-extracts are used. The development of assays which directly measure changes in polyP concentration should facilitate an increased understanding of those enzymes involved in polyP turnover during the EBPR process. Such a system, based on the well-established metachromatic reaction of toluidine blue with polyP has recently been shown to be effective in the characterisation of PPK activity in crude extracts of an environmental *Burkholderia cepacia* isolate (Mullan *et al.*, 2002).

12.4 ISOLATION AND IDENTIFICATION OF POLYPHOSPHATE-ACCUMULATING MICROORGANISMS FROM THE EBPR PROCESS

Early attempts to identify and study polyP-accumulating microorganisms ['PAOs' : (van Loosdrecht *et al.*, 1997)] relied on their enrichment from pilot plant or full-scale reactors in which EBPR had been successfully established, followed by their isolation and cultivation in the laboratory (Bond *et al.*, 1999; van Loosdrecht *et al.*, 1997). Such studies frequently indicated the numerical dominance of members of the *Acinetobacter-Moraxella* group, and other members of the *gamma* subclass of the class *Proteobacteria* in EBPR sludges (Deinema *et al.*, 1980; Fuhs and Chen, 1975; Streichan *et al.*, 1990; Wentzel *et al.*, 1988). Numerous subsequent studies on such isolates in pure culture have generally indicated that they are capable of polyP accumulation and release (Beacham *et al.*, 1992; Deinema *et al.*, 1980; Ohtake *et al.*, 1985; Vasiliadis *et al.*, 1990). Nevertheless in many instances the characteristics of the process are not consistent with those observed in EBPR sludges with a high phosphate removal capacity (Tandoi *et al.*, 1998; van Groenestijn *et al.*, 1989).

More recently, however, the development of a range of culture-independent methods that allow the *in situ* identification of the individual members of complex microbial communities, such as activated sludge, has caused a radical revision of our understanding of the microbial ecology of the EBPR process. Polyamines (Auling *et al.*, 1991), respiratory quinones (Hiraishi *et al.*, 1998), and fatty acid profiles (Melasniemi *et al.*, 1998) have been used as chemical biomarkers to distinguish between the global community structures of polyP-accumulating and non-accumulating sludges. More commonly, genetic approaches have also been applied to the issue of unambiguously determining the identities of PAOs. Such studies have frequently involved fluorescence *in situ* hybridization (FISH) using probes ('phylogenetic stains') from the domain, division, and sub-division levels; alternatively the construction and analysis of cloned 16S ribosomal DNA 'libraries' or the electrophoretic separation and analysis of PCR-amplified 16S rDNA fragments have been carried out (Bond *et al.*, 1995; Christensson *et al.*, 1998; Crocetti *et al.*, 2000; Dabert *et al.*, 2001; Hesselmann *et al.*, 1999; Kampfer *et al.*, 1996; Liu *et al.*, 2000; Wagner *et al.*, 1994).

All such studies have indicated the involvement of a much wider range of organisms in the EBPR process than had been recognized by traditional methods – at least thirty different phylotypes from major phyla of the domain *Bacteria* (Liu *et al.*, 2000). Conspicuously, however, *Acinetobacter* spp. have proved to be present in very much lower numbers than would be required to account for EBPR (Bond *et al.*, 1995; Hiraishi *et al.*, 1998; Wagner *et al.*, 1994). By contrast members of the *beta*-2 subclass of the *Proteobacteria* (related to the genera *Rhodocyclus* and *Propionibacter*) and of the high GC Gram-positive *Actinobacteria* appear to dominate high performance EBPR sludges from both laboratory reactors and full-scale waste water treatment installations. These bacterial groups are currently regarded as strong candidate PAOs (Bond *et al.*, 1999; Crocetti *et al.*, 2000; Hesselmann *et al.*, 1999); indeed *Rhodocyclus*-related cells appeared to account for up to 73% of all PAOs in one full-scale EBPR process (Zilles *et al.*, 2002). It has also been suggested that competition by members of the *beta* subclass of the *Proteobacteria* other than *beta*-1 or *beta*-2 may be responsible for poor phosphate removal in instances in which the sludge population becomes dominated by cells that accumulate glycogen rather than polyP (Bond *et al.*, 1999).

The high degree of phylogenetic diversity of putative PAOs that has been revealed by the above studies may be in part related to the fact that they were carried out on a variety of experimental EBPR systems; these ranged from bench-top reactors fed with defined media of various types to full-scale plants operated under widely-differing regimes and treating waste waters of widely-differing characteristics. Nevertheless in the light of this evidence it does seem highly likely that the PAO phenotype is found in a number of different bacterial groups (Mino *et al.*, 1998). A final resolution of this issue may come about with the development of strategies that overcome the perennial difficulty of coupling the *in situ* detection of the polyP-accumulating phenotype with molecular identification of strains that possess it. This problem has meant that the majority of studies to date have relied upon indirect methods, although most recent work has involved attempts to physically enrich for polyP-containing strains and/or to combine FISH with polyP-specific chemical staining. The possibility of combining FISH with microautoradiography to identify cells that accumulate radiolabelled phosphate (Lee *et al.*, 1999) appears to offer a particularly promising route to final resolution of the question. Successful isolation of PAOs could then be followed by detailed investigation of the physiology, biochemistry, and genetics of phosphate metabolism in pure culture studies. Whilst it is likely that the maintenance of many such strains under laboratory conditions would be difficult, a number of strategies for their successful cultivation have been suggested (Mino *et al.*, 1998).

12.5 NEW APPROACHES TO BIOLOGICAL PHOSPHATE REMOVAL

Conventional EBPR technology is based upon the enrichment of PAOs through the recycling of activated sludge via alternating anaerobic and aerobic zones as discussed above in Section 12.1. In such systems phosphate removal during the

EBPR process occurs during the aerobic phase. Recent investigations have however demonstrated that phosphate uptake may not necessarily require an aerobic stage. Under anoxic conditions nitrate can provide an alternative electron acceptor for phosphate assimilation and intracellular polyP formation (Barak and van Rijn, 2000a; Barker and Dold, 1996; Egli and Zehnder, 2002; Jorgensen and Pauli, 1995; Kuba *et al.*, 1993). Nitrification–Denitrification Biological Enhanced Phosphate Removal (NDBEPR) may have a number of advantages over EBPR including: (1) elimination of the need for aeration gives considerable energy savings; (2) less biomass production may be achieved, and (3) the system maximises the amount of COD available for both nitrogen and P removal (Kuba *et al.*, 1993). Various laboratory scale studies have demonstrated that NDBEPR sequence batch reactors can perform to a similar level as the conventional EBPR process albeit at lower biomass levels with different P release and uptake rates (Kerrnjespersen and Henze, 1993). NDBEPR requires the enrichment of microorganisms capable of accumulating P in the presence of nitrate as the terminal electron acceptor. Microbial community analysis using PCR-single strand conformation polymorphism (SSCP) has revealed that those micro-organisms capable of conventional EBPR using oxygen as the terminal electron acceptor, may also be those responsible for the utilisation of nitrate in an NDBEPR system (Dabert *et al.*, 2001). Of particular significance however may be the recent identification of a strain of *Paracoccus denitrificans* and other denitrifiers capable of combined nitrate removal and phosphate uptake and accumulation as polyP without the need for alternating anaerobic/anoxic conditions (Barak and van Rijn, 2000a, 2000b). This may provide a method for combined biological P and nitrate removal in a single reactor step.

This latter observation highlights an important consideration for the development of new technologies for biological P removal (Barak and van Rijn, 2000a). Microbial polyP accumulation, by definition, must result in enhanced phosphate uptake from the external medium. Therefore any condition under which the ability to accumulate intracellular polyP is necessary for microbial survival (or at least confers a competitive advantage on cells that possess it) will result in enhanced phosphate removal. An understanding of such conditions (which are provided, for example by the EBPR regime) might then be exploited to provide alternative and possibly superior treatment options for biological phosphate removal from industrial and municipal effluents.

As an example, extensive accumulation of polyP has been detected in *Escherichia coli* in response to osmotic stress or to nutritional stress imposed by either nitrogen, amino acid or phosphate limitation (Ault-Riche *et al.*, 1998; Rao *et al.*, 1998). Similarly *Pseudomonas aeruginosa* mucoid strain 8830 accumulates intracellular polyP particularly during stationary phase and in response to phosphate and amino acid limitations (Ault-Riche *et al.*, 1998; Kim *et al.*, 1998) while both the unicellular alga *Dunaliella salina* and the yeast *Saccharomyces cerevisiae* utilise their intracellular polyP reserves to provide a pH-stat mechanism to counterbalance alkaline stress (Bental *et al.*, 1991; Castro *et al.*, 1995; Pick and Weiss, 1991;

Pick *et al.*, 1990). PolyP accumulation has also been observed upon exposure of the freshwater sponge *Ephydatia muelleri* to various organic pollutants (Imsiecke *et al.*, 1996).

It has also been postulated that intracellular polyP may play a role in the physiological adaptation of microbial cells during growth and development, and in their response to nutritional and environmental stresses (Kornberg, 1995). For example it has been demonstrated that exposure to acid pH results in the induction of polyP accumulation by a variety of environmental microorganisms. This phenomenon of low pH – stimulated polyP accumulation was first observed in the environmental yeast *Candida humicola* G-1 (McGrath and Quinn, 2000). Enhanced phosphate uptake and polyP accumulation occurred in *Candida humicola* G-1 as a consequence of growth at acid pH under fully aerobic conditions; cells grown at pH 5.5 in a medium containing glucose as carbon source removed 450% more phosphate than at pH 7.5, with a concomitant 10-fold increase in intracellular polyP accumulation (McGrath and Quinn, 2000). A further sludge isolate to be studied in detail, the bacterium *Burkholderia cepacia* AM 19, also showed maximal removal of phosphate and accumulation of polyP at pH 5.5; levels were up to 220% and 330% higher, respectively, than in cells grown at pH 7.5. During the early stationary phase of growth at pH 5.5 a maximum level of intracellular polyP that comprised 13.6% of cellular dry weight was reached. Neither phosphate starvation, nutrient limitation nor anaerobiosis was required to induce enhanced phosphate uptake and polyP formation in these isolates.

An investigation of the extent to which an acid response of this type occurred amongst the endogenous microflora of an activated sludge plant (McGrath *et al.*, 2001) revealed that some 34% of isolates were capable of enhanced phosphate uptake at pH 5.5 as opposed to pH 7.5; no significant difference in the amount of phosphate removed at either pH 5.5 was observed in 44% of the strains studied, while the remainder either grew poorly or displayed decreased phosphate uptake at the lower pH. Batch cultures using inocula from five further sewage works, of varying types and influent characteristics demonstrated that phosphate removal could be increased by 55–124% through growth at pH 5.5 (McGrath *et al.*, 2001). The pH optimum for phosphate uptake varied between individual isolates obtained from these plants but in all cases ranged from pH 5.0–6.5 (McGrath *et al.*, 2001).

These observations suggest that low pH – stimulated phosphate removal and polyP accumulation may be a widespread natural phenomenon which could potentially form the basis of a novel strategy for the 'one-step' removal of phosphate from effluents. Further research should aim to (1) assess the impact of acid pH on other key waste water treatment parameters such as nitrification, sludge settling and COD removal, (2) demonstrate the effectiveness of the process at pilot plant scale, and (3) identify those microbial groups responsible for phosphate assimilation. Furthermore, it is likely that other as-yet-unidentified environmental triggers of enhanced phosphate uptake and polyP accumulation might be similarly exploited for the removal of P from waste waters.

ACKNOWLEDGEMENTS

The work was supported by the Biotechnology and Biological Sciences Research Council, UK (grant, 81/E11490); the Queen's University Environmental Science and Technology Research Centre (QUESTOR); a Strategic Research Infrastructure (SRIF) grant for Environmental Engineering and Biotechnology at the Queen's University Belfast; and the Invest Northern Ireland RTD Centres of Excellence Programme.

REFERENCES

Akiyama, M., Crooke, E. and Kornberg, A. (1992) The polyphosphate kinase gene of *Escherichia coli* – isolation and sequence of the *ppk* gene and membrane location of the protein. *J. Biol. Chem.* **267**, 22,556–22,561.

Auling, G., Pilz, F., Busse, H.J., Karrasch, S., Streichan, M. and Schon, G. (1991) Analysis of the polyphosphate-accumulating microflora in phosphorus- eliminating, anaerobic–aerobic activated sludge systems by using diaminopropane as a biomarker for rapid estimation of *Acinetobacter* spp. *Appl. Environ. Microbiol.* **57**, 3585–3592.

Ault-Riche, D., Fraley, C.D., Tzeng, C.M. and Kornberg, A. (1998) Novel assay reveals multiple pathways regulating stress-induced accumulations of inorganic polyphosphate in *Escherichia coli*. *J. Bacteriol.* **180**, 1841–1847.

Barak, Y. and van Rijn, J. (2000a) Atypical polyphosphate accumulation by the denitrifying bacterium *Paracoccus denitrificans*. *Appl. Environ. Microbiol.* **66**, 1209–1212.

Barak, Y. and van Rijn, J. (2000b) Relationship between nitrite reduction and active phosphate uptake in the phosphate-accumulating denitrifier *Pseudomonas* sp. strain JR 12. *Appl. Environ. Microbiol.* **66**, 5236–5240.

Bark, K., Kampfer, P., Sponner, A. and Dott, W. (1993) Polyphosphate-dependent enzymes in some coryneform bacteria isolated from sewage sludge. *FEMS. Microbiol. Lett.* **107**, 133–138.

Bark, K., Sponner, A., Kampfer, P., Grund, S. and Dott, W. (1992) Differences in polyphosphate accumulation and phosphate adsorption by *Acinetobacter*-isolates from waste-water producing polyphosphate – AMP phosphotransferase. *Water Res.* **26**, 1379–1388.

Barker, P.S. and Dold, P.L. (1996) Denitrification behaviour in biological excess phosphorus removal activated sludge systems. *Water Res.* **30**, 769–780.

Beacham, A.M., Seviour, R.J. and Lindrea, K.C. (1992) Polyphosphate accumulating abilities of *Acinetobacter* isolates from a biological nutrient removal pilot-plant. *Water Res.* **26**, 121–122.

Bental, M., Pick, U., Avron, M. and Degani, H. (1991) Polyphosphate metabolism in the alga *Dunaliella salina* studied by P^{31}-NMR. *Biochim. Biophys. Acta.* **1092**, 21–28.

Bond, P.L., Erhart, R., Wagner, M., Keller, J. and Blackall, L.L. (1999) Identification of some of the major groups of bacteria in efficient and nonefficient biological phosphorus removal activated sludge systems. *Appl. Environ. Microbiol.* **65**, 4077–4084.

Bond, P.L., Keller, J., Hugenholtz, P. and Blackall, L.L. (1995) Bacterial community structures of phosphate-removing and non-phosphate-removing activated sludges from sequencing batch reactors. *Appl. Environ. Microbiol.* **61**, 1910–1916.

Bonting, C.F., Kortstee, G.J. and Zehnder, A.J. (1991) Properties of polyphosphate: AMP phosphotransferase of *Acinetobacter* strain 210A. *J. Bacteriol.* **173**, 6484–6488.

Brett, S, Guy, J, Morse, G.K. and Lester, J.N. (1997) *Phosphorus Removal and Recovery Technologies*. Selper Publications, London.

Castro, C.D., Meehan, A.J., Koretsky, A.P. and Domach, M.M. (1995) In situ ^{31}P nuclear magnetic resonance for observation of polyphosphate and catabolite responses of chemostat-cultivated *Saccharomyces cerevisiae* after alkalinization. *Appl. Environ. Microbiol.* **61**, 4448–4453.

Christensson, M., Blackall, L.L. and Welander, T. (1998) Metabolic transformations and characterisation of the sludge community in an enhanced biological phosphorus removal system. *Appl. Microbiol. Biotechnol.* **49**, 226–234.

Crocetti, G.R., Bond, P.L., Schuler, A., Keller, J., Jenkins, D. and Blackall, L.L. (2000) Identification of polyphosphate-accumulating organisms and design of 16S rRNA-directed probes for their detection and quantitation. *Appl. Environ. Microbiol.* **66**, 1175–1182.

Dabert, P., Sialve, B., Delgenes, J.P., Moletta, R. and Godon, J.J. (2001) Characterisation of the microbial 16S rDNA diversity of an aerobic phosphorus-removal ecosystem and monitoring of its transition to nitrate respiration. *Appl. Microbiol. Biotechnol.* **55**, 500–509.

Dawes, E.A. and Senior, P.J. (1973) Energy reserve polymers in microorganisms. *Adv. Microb. Physiol.* **10**, 178–203.

Deinema, M.H., Habets, L.H.A., Scholten, J., Turkstra, E. and Webers, H.A.A.M. (1980) The accumulation of polyphosphate in *Acinetobacter* spp. *FEMS. Microbiol. Lett.* **9**, 275–279.

Egli, T. and Zehnder, A.J.B. (2002) Phosphate and nitrate removal. *Curr. Opin. Biotechnol.* **5**, 275–284.

Fuhs, G.W. and Chen, M. (1975) Microbiological basis of phosphate removal in the activated sludge process for the treatment of wastewater. *Microb. Ecol.* **2**, 119–138.

Geißdörfer, W., Ratajczak, A. and Hillen, W. (1998) Transcription of *ppk* from *Acinetobacter sp.* strain ADP1, encoding a putative polyphosphate kinase, is induced by phosphate starvation. *Appl. Environ. Microbiol.* **64**, 896–901.

Greenberg, A.E., Levin, G. and Kauffman, W.J. (1955) The effect of phosphorus removal on the activated sludge process. *Sewage Ind. Wastes* **27**, 227.

Gurr, E. (1965) *The Rational Use of Dyes in Biology*. Hill, London.

Harold, F.M. (1966) Inorganic polyphosphate in biology: structure, metabolism, and function. *Bacteriol. Rev.* **30**, 772–794.

Hesselmann, R.P.X., Von Rummell, R., Resnick, S.M., Hany, R. and Zehnder, A.J.B. (2000) Anaerobic metabolism of bacteria performing enhanced biological phosphate removal. *Water Res.* **34**, 3487–3494.

Hesselmann, R.P.X., Werlen, C., Hahn, D., van der Meer, J.R. and Zehnder, A.J.B. (1999) Enrichment, phylogenetic analysis and detection of a bacterium that performs enhanced biological phosphate removal in activated sludge. *Syst. Appl. Microbiol.* **22**, 454–465.

Hiraishi, A., Ueda, Y. and Ishihara, J. (1998) Quinone profiling of bacterial communities in natural and synthetic sewage activated sludge for enhanced phosphate removal. *Appl. Environ. Microbiol.* **64**, 992–998.

Hsieh, P.C., Kowalczyk, T.H. and Phillips, N.F.B. (1996b) Kinetic mechanisms of polyphosphate glucokinase from *Mycobacterium tuberculosis*. *Biochemistry* **35**, 9772–9781.

Hsieh, P.C., Shenoy, B.C., Jentoft, J.E. and Phillips, N.F.B. (1993) Purification of polyphosphate and ATP glucose phosphotransferase from *Mycobacterium-tuberculosis* H37Ra – evidence that polyP and ATP glucokinase activities are catalyzed by the same enzyme. *Protein. Expres. Purif.* **4**, 76–84.

Hsieh, P.C., Shenoy, B.C., Samols, D. and Phillips, N.F.B. (1996a) Cloning, expression, and characterization of polyphosphate glucokinase from *Mycobacterium tuberculosis*. *J. Biol. Chem.* **271**, 4909–4915.

Imsiecke, G., Munkner, J., Lorenz, B., Bachinski, N., Muller, W.E.G. and Schroder, H.C. (1996) Inorganic polyphosphates in the developing freshwater sponge *Ephydatia muelleri*: Effect of stress by polluted waters. *Environ. Toxicol. Chem.* **15**, 1329–1334.

Ishige, K. and Noguchi, T. (2000) Inorganic polyphosphate kinase and adenylate kinase participate in the polyphosphate : AMP phosphotransferase activity of *Escherichia coli*. *Proc. Natl. Acad. Sci.-USA* **97**, 14,168–14,171.
Ishige, K. and Noguchi, T. (2001) Polyphosphate:AMP phosphotransferase and Polyphosphate:ADP phosphotransferase activities of *Pseudomonas aeruginosa*. *Biochem. Biophys. Res. Commun.* **281**, 821–826.
Ishige, K., Zhang, H. and Kornberg, A. (2002) Polyphosphate kinase (PPK2), a potent polyphosphate-driven generator of GTP. *Proc. Natl. Acad. Sci.- USA* **99**, 16,684–16,688.
Jorgensen, K.S. and Pauli, A.S.I. (1995) Polyphosphate accumulation among denitrifying bacteria in activated-sludge. *Anaerobe* **1**, 161–168.
Kampfer, P., Bark, K., Busse, H.J., Auling, G. and Dott, W. (1992) Numerical and chemotaxonomy of polyphosphate accumulating *Acinetobacter* strains with high polyphosphate – AMP phosphotransferase (PPAT) activity. *Syst. Appl. Microbiol.* **15**, 409–419.
Kampfer, P., Erhart, R., Beimfohr, C., Bohringer, J., Wagner, M. and Amann, R. (1996) Characterization of bacterial communities from activated sludge: Culture-dependent numerical identification versus *in situ* identification using group- and genus-specific rRNA- targeted oligonucleotide probes. *Microb. Ecol.* **32**, 101–121.
Kato, J., Yamamoto, T., Yamada, K. and Ohtake, H. (1993) Cloning, sequence and characterisation of the polyphosphate kinase-encoding gene (*ppk*) of *Klebsiella aerogenes*. *Gene* **137**, 242.
Kerrnjespersen, J.P. and Henze, M. (1993) Biological phosphorus uptake under anoxic and aerobic conditions. *Water Res.* **27**, 617–624.
Kim, H.Y., Schlictman, D., Shankar, S., Xie, Z., Chakrabarty, A.M. and Kornberg, A. (1998) Alginate, inorganic polyphosphate, GTP and ppGpp synthesis co-regulated in *Pseudomonas aeruginosa:* implications for stationary-phase survival and synthesis of RNA/DNA precursors. *Mol. Microbiol.* **27**, 717–725.
Kornberg, A. (1995) Inorganic polyphosphate: toward making a forgotten polymer unforgettable. *J. Bacteriol.* **177**, 491–496.
Kornberg, A. and Fraley, C.D. (2000) Inorganic polyphosphate: a molecular fossil come to life. *ASM News* **66**, 275–280.
Kornberg, A., Rao, N.N. and Ault-Riche, D. (1999) Inorganic polyphosphate: a molecule of many functions. *Annu. Rev. Biochem.* **68**, 89–121.
Kortstee, G.J.J., Appeldoorn, K.J., Bonting, C.F.C., van Niel, E.W.J. and van Veen, H.W. (2000) Recent developments in the biochemistry and ecology of enhanced biological phosphorus removal. *Biochemistry-Moscow* **65**, 332–340.
Kortstee, G.J.J., Appeldoorn, K.J., Bonting, C.F.C., Vanniel, E.W.J. and van Veen, H.W. (1994) Biology of polyphosphate-accumulating bacteria involved in enhanced biological phosphorus removal. *FEMS. Microbiol. Rev.* **15**, 137–153.
Kuba, T., Smolders, G., van Loosdrecht, M.C.M. and Heijnen, J.J. (1993) Biological phosphorus removal from waste-water by anaerobic-anoxic sequencing batch reactor. *Water Sci. Technol.* **27**, 241–252.
Kulaev, I. and Kulakovskaya, T. (2000) Polyphosphate and the phosphate pump. *Annu. Rev. Microbiol.* **54**, 709–734.
Kulaev, I.S. (1979) *The biochemistry of inorganic polyphosphates*. J. Wiley and Sons Inc., New York.
Kulaev, I.S. and Vagabov, V.M. (1983) Polyphosphate metabolism in microorganisms. *Adv. Microb. Physiol.* **15**, 731–738.
Kulaev, I.S., Vagabov, V.M. and Kulakovskaya, T.V. (1999) New aspects of inorganic polyphosphate metabolism and function. *J. Biosci. Bioeng.* **88**, 111–129.
Kulakovskaya, T.V., Andreeva, N.A., Karpov, A.V., Sidorov, I.A. and Kulaev, I.S. (1999) Hydrolysis of tripolyphosphate by purified exopolyphosphatase from *Saccharomyces cerevisiae* cytosol: Kinetic model. *Biochemistry-Moscow* **64**, 990–993.
Kumble, K.D. and Kornberg, A. (1996) Endopolyphosphatases for long chain inorganic polyphosphate in yeast and mammals. *J. Biol. Chem.* **271**, 27,146–27,151.

Lee, N., Nielsen, P.H., Andreasen, K.H., Juretschko, S., Nielsen, J.L., Schleifer, K.H. and Wagner, M. (1999) Combination of fluorescent *in situ* hybridization and microautoradiography – a new tool for structure-function analyses in microbial ecology. *Appl. Environ. Microbiol.* **65**, 1289–1297.
Levin, G.V. and Shapiro, J. (1965) Metabolic uptake of phosphorus by wastewater organisms. *J. Water Pollut. Control Fed.* **37**, 800–821.
Liu, W.T., Linning, K.D., Nakamura, K., Mino, T., Matsuo, T. and Forney, L.J. (2000) Microbial community changes in biological phosphate-removal systems on altering sludge phosphorus content. *Microbiology-UK* **146**, 1099–1107.
McGrath, J.W., Cleary, S., Mullan, A. and Quinn, J.P. (2001) Acid-stimulated enhanced phosphate uptake by activated sludge microorganisms under aerobic laboratory conditions. *Water Res.* **35**, 4317–4322.
McGrath, J.W. and Quinn, J.P. (2000) Intracellular accumulation of polyphosphate by the yeast *Candida humicola* G-1 in response to acid pH. *Appl. Environ. Microbiol.* **66**, 4068–4073.
Melasniemi, H., Hernesmaa, A., Pauli, A.S.L., Rantanen, P. and Salkinoja-Salonen, M. (1998) Comparative analysis of biological phosphate removal (BPR) and non-BPR activated sludge bacterial communities with particular reference to *Acinetobacter*. *J. Ind. Microbiol. Biotechnol.* **21**, 300–306.
Mino, T., van Loosdrecht, M.C.M. and Heijnen, J.J. (1998) Microbiology and biochemistry of the enhanced biological phosphate removal process. *Water Res.* **32**, 3193–3207.
Mullan, A., Quinn, J.P. and McGrath, J.W. (2002) A nonradioactive method for the assay of polyphosphate kinase activity and its application in the study of polyphosphate metabolism in *Burkholderia cepacia*. *Anal. Biochem.* **308**, 294–299.
Ogawa, N., DeRisi, J. and Brown, P.O. (2000) New components of a system for phosphate accumulation and polyphosphate metabolism in *Saccharomyces cerevisiae* revealed by genomic expression analysis. *Mol. Biol. Cell* **11**, 4309–4321.
Ohtake, H., Takahashi, K., Tsuzuki, Y. and Toda, K. (1985) Uptake and release of phosphate by a pure culture of *Acinetobacter calcoaceticus*. *Water Res.* **19**, 1587–1594.
Phillips, N.F.B., Horn, P.J. and Wood, H.G. (1993) The polyphosphate-dependent and ATP-dependent glucokinase from *Propionibacterium shermanii* – both activities are catalyzed by the same protein. *Arch. Biochem. Biophys.* **300**, 309–319.
Pick, U., Bental, M., Chitlaru, E. and Weiss, M. (1990) Polyphosphate-hydrolysis – a protective mechanism against alkaline stress. *FEBS* **274**, 15–18.
Pick, U. and Weiss, M. (1991) Polyphosphate hydrolysis within acidic vacuoles in response to amine-induced alkaline stress in the halotolerant alga *Dunaliella salina*. *Plant Physiol.* **97**, 1234–1240.
Rao, N.N., Liu, S. and Kornberg, A. (1998) Inorganic polyphosphate in *Escherichia coli*: the phosphate regulon and the stringent response. *J. Bacteriol.* **180**, 2186–2193.
Srinath, E.G., Sastry, C.A. and Pillai, S.C. (1959) Rapid removal of phosphorus from sewage by activated sludge. *Experientia* **15**, 339–240.
Streichan, M., Golecki, J.R. and Schon, G. (1990) Polyphosphate-accumulating bacteria from sewage plants with different processes for biological phosphorus removal. *FEMS. Microbiol. Ecol.* **73**, 113–124.
Szymona, M. and Ostrowski, W. (1964) Inorganic polyphosphate glucokinase of *Mycobacterium phlei*. *Biochim. Biophys. Acta* **85**, 283–295.
Tandoi, V., Majone, M., May, J. and Ramadori, R. (1998) The behaviour of polyphosphate accumulating *Acinetobacter* isolates in an anaerobic-aerobic chemostat. *Water Res.* **32**, 2903–2912.
Tinsley, C.R. and Gotschlich, E.C. (1995) Cloning and characterization of the meningococcal polyphosphate kinase gene – production of polyphosphate synthesis mutants. *Infect. Immun.* **63**, 1624–1630.
Trelstad, P.L., Purdhani, P., Geißdörfer, W., Hillen, W. and Keasling, J.D. (1999) Polyphosphate kinase of *Acinetobacter* sp. Strain ADP1: purification and characterization

of the enzyme and its role during changes in extracellular phosphate levels. *Appl. Environ. Microbiol.* **65**, 3780–3786.

Upton, J., Hayes, E. and Churchley, J. (1996) Biological phosphorus removal at Stratford Upon Avon, UK: The effect of influent wastewater characteristics on effluent phosphate. *Water Sci. Technol.* **33**, 73–80.

van Alebeek, G.J.W.M., Keltjens, J.T. and Vanderdrift, C. (1994) Tripolyphosphatase from *Methanobacterium-thermoautotrophicum* (Strain Delta-H). *FEMS Microbiol. Lett.* **117**, 263–268.

van Groenestijn, J.W., Bentvelsen, M.M., Deinema, M.H. and Zehnder, A.J. (1989) Polyphosphate-degrading enzymes in *Acinetobacter* spp. and activated sludge. *Appl. Environ. Microbiol.* **55**, 219–223.

van Groenestijn, J.W., Deinema, M.H. and Zehnder, A.J.B (1987) ATP production from polyphosphate in *Acinetobacter strain*-210a. *Arch. Microbiol.* **148**, 14–19.

van Loosdrecht, M.C.M., Hooijmans, C.M., Brdjanovic, D. and Heijnen, J.J. (1997) Biological phosphate removal processes. *Appl. Microbiol. Biotechnol.* **48**, 289–296.

van Niel, E.W., Appeldoorn, K.J., Zehnder, A.J. and Kortstee, G.J. (1998) Inhibition of anaerobic phosphate release by nitric oxide in activated sludge. *Appl. Environ. Microbiol.* **64**, 2925–2930.

van Veen, H.W., Abee, T., Kortstee, G.J.J., Pereira, H., Konings, W.N. and Zehnder, A.J.B. (1994) Generation of a proton motive force by the excretion of metal–phosphate in the polyphosphate-accumulating *Acinetobacter johnsonii* strain 210a. *J. Biol. Chem.* **269**, 29,509–29,514.

Vasiliadis, G., Duncan, A., Bayly, R.C. and May, J.W. (1990) Polyphosphate production by strains of *Acinetobacter*. *FEMS Microbiol. Lett.* **70**, 37–40.

Wagner, M., Erhart, R., Manz, W., Amann, R., Lemmer, H., Wedi, D. and Schleifer, K.H. (1994) Development of a ribosomal-RNA-targeted oligonucleotide probe specific for the genus *Acinetobacter* and its application for *in situ* monitoring in activated-sludge. *Appl. Environ. Microbiol.* **60**, 792–800.

Water Environment Federation (1998) *Biological and Chemical Systems for Nutrient Removal: A Special Publication.* Water Environment Federation, Alexandria, Virginia.

Weltin, D., Hoffmeister, D., Dott, W. and Kampfer, P. (1996) Studies on polyphosphate and poly-beta-hydroxyalkanoate accumulation in *Acinetobacter johnsonii* 120 and some other bacteria from activated sludge in batch and continuous culture. *Acta Biotechnol.* **16**, 91–102.

Wentzel, M.C., Loewenthal, R.E., Ekama, G.A. and Marais, G.V.R. (1988) Enhanced polyphosphate organism cultures in activated sludge systems – Part 1: Enhanced culture development. *Water SA* **14**, 81–92.

Wiame, J.M. (1947) The metachromatic reaction of hexametaphosphate. *J. Am. Chem. Soc.* **69**, 3146–3147.

Wiame, J.M. (1948) The occurrence and physiological behavior of two metaphosphate fractions in yeast. *J. Biol. Chem.* **178**, 919–929.

Wood, H.G. and Clark, J.E. (1988) Biological aspects of inorganic polyphosphates. *Annu. Rev. Biochem.* **57**, 235–260.

Yeoman, S., Stephenson, T., Lester, J.N. and Perry, R. (1988) The removal of phosphorus during wastewater treatment: A review. *Environ. Pollut.* **49**, 183–233.

Zago, A., Chugani, S. and Chakrabarty, A.M. (1999) Cloning and characterization of polyphosphate kinase and exopolyphosphatase genes from *Pseudomonas aeruginosa* 8830. *Appl. Environ. Microbiol.* **65**, 2065–2071.

Zilles, J.L., Peccia, J., Kim, M.W., Hung, C.H. and Noguera, D.R. (2002) Involvement of *Rhodocyclus*-related organisms in phosphorus removal in full-scale wastewater treatment plants. *Appl. Environ. Microbiol.* **68**, 2763–2769.

13
A review of solid phase adsorbents for the removal of phosphorus from natural and waste waters

G.B. Douglas, M.S. Robb, D.N. Coad and P.W. Ford

13.1 INTRODUCTION

The effective control of phosphorus (P) in aquatic systems is widely recognised as pivotal to short- and long-term management or restoration. Excess bioavailable phosphorus is a major factor leading to the eutrophication of natural waters and is manifested as excessive growth of aquatic plants and algae, and the degradation of water quality including the depletion of dissolved oxygen with concomitant effects on ecological health. Phosphorus is frequently the limiting nutrient in freshwater aquatic systems, and aquatic organisms have evolved efficient biochemical mechanisms for concentrating P from the surrounding waters. Reductions in the algal growth rate may only become apparent at concentrations of *ca.* 3 µg/l, which is a threshold value in some aquatic systems (Reynolds, 1999) where phosphorus is a limiting nutrient.

© 2004 IWA Publishing. *Phosphorus in Environmental Technology: Principles and Applications.*
Edited by Eugenia Valsami-Jones. ISBN: 1 84339 001 9

Phosphorus may occur in a variety of chemical forms: simple orthophosphates, condensed polyphosphates, organic forms, such as inositol phosphate and organic esters, and as colloidal and particulate forms often sorbed on or contained within Fe- and Al-hydrous oxides or as particulate phosphate minerals, usually apatite (e.g. Baldwin et al., 1995, 2001). Relative concentrations of phosphorus species may be altered by processes such as adsorption and desorption (by a buffer mechanism in response to a change in solution phosphate concentrations, e.g. Froelich, 1988), changes in physicochemical conditions such as pH or dissolved oxygen (e.g. leading to dissolution of, or desorption from, hydrous Fe–Al oxides) or biological (e.g. bacteria, algae, aquatic plant) uptake.

Large-scale methods of phosphorus removal from waste waters, such as sewage or that generated from industrial processes, include biological nutrient removal, the use of chemical coagulants and/or precipitants such as ferric iron, alum, lime, precipitation in struvite, and the application of ion exchange (Brett et al., 1997). The presence of any added precipitant may also become a site for microbial growth and subsequent uptake and transformation of dissolved P.

Solid phase adsorbents[1] are frequently utilised in the removal of phosphorus (principally as phosphate -PO_4–P) in a variety of applications as diverse as waste water treatment, maintenance of potable water supplies and constructed wetlands. Some of the most commonly used solid phase adsorbents are:

- naturally occurring minerals from soils (e.g. Fe-oxides/oxyhydroxides – Barrow, 1999, allophane – Wada, 1989),
- naturally occurring (poly-mineralic) soils or sands (e.g. Arias et al., 2000, Degens et al., 2000),
- derived from mineral deposits (e.g. wollastonite – Brooks et al., 2000) or other natural materials (e.g. shale, Drizo et al., 1999; serpentinite, Forget et al., 2002; maerl, Gray et al., 2000),
- synthetic analogues of natural minerals produced on an experimental or industrial scale (e.g. polymeric hydrogels – Kofinas and Kioussis, 2003; hydrotalcites – Shin et al., 1996),
- expanded clay aggregates (e.g. Zhu et al., 1997), or
- waste materials from industrial processes that adsorb phosphate or that may also be further modified to enhance their uptake capacity (e.g. electric arc furnace (EAF) steel slag – Drizo et al., 2002; blast furnace slag – Johansson, 1999b; Mann, 1997; red mud – Thornber and Hughes, 1987).

Byproduct materials such as fly ash, red mud and blast furnace slag (BFS), and electric arc furnace (EAF) slag, have particular potential for widespread applications as they are produced in large quantities, often present storage and/or disposal problems and have little intrinsic value. This has made industrial

---

[1] Solid phase adsorbents (or simply adsorbents): These terms are used in this chapter, and also in the P removal literature in their broader sense, to describe materials of variable quality and source, often highly heterogeneous, which may be simultaneously involved in formation of insoluble precipitates and coprecipitation as well as adsorption, via surface complexation and/or interstitial ligand exchange. In addition, the physico-chemical processes which produce P removal are often poorly characterized, and often do not display uptake and release equilibria that would be characteristic of true adsorption.

by-products the subject of extensive research in application of constructed wetland systems (CWS) to augment both the substrate phosphate uptake capacity and extend the wetland longevity (Drizo et al., 1999; Heal et al., this volume; Hylander and Siman, 2001; Pant et al., 2001).

In general, absorbents with a significant anion (phosphate) uptake capacity are enriched in Ca, Fe and/or Al (Drizo et al., 1999, 2002; Johansson, 1999). Interaction with phosphate may lead to the formation of discrete secondary minerals, such as hydroxylapatite [$Ca_5(PO_4)_3(OH)$] or vivianite [$Fe_3(PO_4)_2 \cdot 8H_2O$], ion exchange via surface adsorption (PO_4–P–Al-oxide) or internal ion-exchange (PO_4–P–hydrotalcite) or the formation of less well defined associations that may involve complex multi-component surface adsorption or precipitation (e.g. in soils) that may change with time (Barrow, 1983, 1999; Drizo et al., 1999; Johansson, 1999a). Decades-old research on soils (e.g. Barrow, 1974; Barrow and Shaw, 1975) and more recent research on slags (Drizo et al., 2002) indicates the requirement for longer-term experiments to better estimate the ultimate uptake capacity of phosphorus adsorbents.

In the following discussion aspects of phosphorus adsorbents such as their origin, mode(s) of occurrence and/or synthesis will be elucidated. In addition, their physico-chemical characteristics such as phosphate uptake capacity, stability, potential environmental impact and selectivity with respect to other (oxy)anions will be described. Many of these adsorbents have undergone a wide range of laboratory evaluation, with some being tested in applications as diverse as the removal of phosphate from influent via constructed wetlands (Kadlec and Knight, 1996; Kadlec et al., 2000; Pant et al., 2001), natural and waste waters (Baker et al., 1998), as reactive capping or barriers in aquatic systems (Douglas and Adeney, 2001; Yamada et al., 1987) to broad acre application as a soil amendment to assist in phosphate fertiliser retention (Summers et al., 1993). Unless otherwise stated, studies of individual adsorbents are laboratory-based and generally consist of either batch adsorption or column adsorption studies. In most cases uptake is measured as a function of absorbent mass, rather than surface area, which has been a standard practice.

Recycling of adsorbed phosphorus has not been considered in the assessment of materials described below. This was considered secondary to the actual performance of the material in a specific application. Methods to recover phosphorus and sometimes nitrogen from more substantially enriched phosphorus sources such as sewage effluent are now the subject of on-going research with a view to addressing the challenge of phosphorus sustainability (Morse et al., 1998; see also Jaffer and Pearce, Battistoni, Piekema, Ueno, Hobbs, all this volume). A summary of phosphorus adsorbents and their major properties is given in Table 13.1.

13.2 ADSORBENT MATERIALS

13.2.1 Hydrotalcites

Hydrotalcites are layered double hydroxides most commonly formed by the coprecipitation of divalent (e.g. Mg^{2+}, Fe^{2+}) and trivalent (Al^{3+}, Fe^{3+}) metal

Table 13.1 Summary of phosphorus adsorbents and their properties.

Adsorbent	Chemistry/mineralogy	Occurrence	Modification	Phosphate uptake	Applications	Cost
Hydrotalcites	Double-layer hydroxides, usually Mg and Al or Fe	Soils, synthetically prepared	Conversion to non-carbonate form	High but with a strong pH dependence	Waste streams, columns	Moderate to high
Allophane/imogolite	Amorphous Al and Si with Al:Si ratio $ca.$ 1:1 – 2:1.	Soils, synthetically prepared	None	High – strong pH dependence	Wetlands, filters and pasture application	Low (soils) to moderate (synthetic)
Fe-oxides	Oxides/hydrous oxides, e.g. goethite, ferrihydrite	Soils, synthetic, industrial waste	Smaller particle size/increased surface area	Low to high depending on mineralogy	Wetlands, column filters and pasture	Low as waste material
Activated alumina	Al-oxide (Al_2O_3)	Commercially produced	Pre-treatment to tailor surface properties	High	Columns, bed reactors	High – synthesis and modification
Soils, sands and gravels	Generally contain Ca-, Fe- or Al – bearing minerals.	Naturally occurring	Smaller particle size/increased surface area	Low to high – depends on chemistry/mineralogy	Constructed wetlands, pasture application	Generally low
Red mud	Complex Fe and Al oxides/oxyhydroxides, hydrotalcites if modified	Waste product of alumina refining (Bayer process)	Neutralization – e.g., seawater amendment	Low to high, but depends on form and pre-treatment	Wetlands, column filters and pasture application	Low to moderate if pre-treated
Fly ash	Mullite, aluminosilicate glasses, zeolite, hydrotalcites if modified	Waste product of coal combustion	Thermal treatment possible to produce hydrotalcites	Low to moderate depending on chemistry and modification	Wetlands, column filters and pasture application	Low to moderate if pre-treated
Expanded clay aggregates	Illitic clays and calcined carbonates	Industrially prepared by calcination	Produced as porous aggregates	Low to moderate, related to Ca, Fe concentration	Wetlands and as column filters	Moderate – industrial process
Blast/arc furnace slag	Complex mineralogy of Ca, Fe oxides	Industrial (waste) by-product	Screening/milling for particle size control	Low to high, related to Ca, Fe concentration	Wetlands and as column filters	Low as waste material
Rare earth modified clay	REE exchanged into high CEC clay	Synthetically prepared slurry	Pelletization, filters etc.	Moderate to high Depending on clay CEC	Natural and waste waters, filters, barriers	Moderate to high
Carbonates	Calcite, aragonite ($CaCO_3$), dolomite (Ca, Mg)CO_3 etc.	Naturally and as industrial by-product	May be calcined to form oxides (CaO, MgO etc.)	Low to high depends on form, surface area etc.	Natural and waste waters, filters, barriers	Low (natural) to moderate (synthetic)

cation solutions at high pH (Shin *et al.*, 1996; Taylor, 1984; Vucelic *et al.*, 1997). Hydrotalcites were first described over 60 years ago (Frondel, 1941; Feitknecht 1942). They occur naturally as accessory minerals in soils and sediments (e.g. Taylor and McKenzie, 1980), but may also be synthesised from industrial waste materials by the reaction of bauxite residue (red mud) with seawater (e.g. Thornber and Hughes, 1987, see Section 13.2.5) or by the reaction of lime with fly ash (e.g. Reardon and Della Valle, 1997, see Section 13.2.6).

Within the hydrotalcite structure there are octahedral metal hydroxide sheets that carry a net positive charge. As a consequence, it is possible to substitute a wide range of inorganic or organic anions into the hydrotalcite structure. Hydrotalcites are generally unstable below a pH of approximately 5 (Ookubu *et al.*, 1993) and may act as buffers over a wide range of solution pH (Seida and Nakano, 2002).

A number of studies have been conducted to investigate ways to exploit the anion exchange properties of hydrotalcites. These studies have focussed on the removal of phosphate and other oxyanions and humic substances from natural and wastewaters (Amin and Jayson, 1996; Misra and Perrotta, 1992; Miyata, 1980; Seida and Nakano, 2000; Shin *et al.*, 1996).

Laboratory studies of phosphate uptake using synthetically prepared Mg-Al hydrotalcites and a range of initial solution phosphate concentrations indicate an uptake capacity of from *ca*. 25–30 mg P/g (Miyata, 1983; Shin *et al.*, 1996) to *ca*. 60 mg P/g with uptake also influenced by initial phosphate concentration, pH (with maximum phosphate adsorption near pH 7), degree of crystallinity and the hydrotalcite chemistry (Ookubo *et al.*, 1993). A major obstacle to the use of hydrotalcites for phosphate removal in natural and/or waste waters is the selectivity for carbonate over phosphate, with a selectivity series in the approximate order $CO_3^{2-} > HPO_4^{2-} \gg SO_4^{2-}$, $OH^- > F^- > Cl^- > NO_3^-$ (Cavani *et al.*, 1991; Miyata 1983; Sato *et al.*, 1986; Shin *et al.*, 1996). Many hydrotalcites are also synthesised with carbonate as the predominant anion and thus require anion exchange before they are exposed to phosphate. When carbonate is also combined with sulphate, nitrate and chloride (as might commonly occur in natural or waste waters) the reduction of phosphate adsorption to the hydrotalcite is further decreased (Shin *et al.*, 1996).

13.2.2 Allophane and imogolite

Allophanes and imogolites occur in nature as secondary aluminosilicate minerals formed during weathering process in rocks and soils (e.g. Barreal *et al.*, 2001; Farmer and Russell, 1990; He *et al.*, 1998). Allophanes and imogolites can also be synthesised in the laboratory, typically from stoichiometric addition of pure Al and Si solutions (Denaix *et al.*, 1999; Farmer *et al.*, 1983; Wada, 1989). Allophanes are variously defined as a group of X-ray amorphous minerals consisting of a solid solution of Si, Al and H_2O (Ross and Kerr, 1934) or hydrous aluminosilicate minerals characterized by short-range order (van Olphen, 1971). Allophanes and imogolites are chemically similar, with allophanes generally having

Al:Si ratios 1:1 to 2:1 but up to 4:1 in Al-rich soils (Farmer and Russell, 1990), while imogolites have Al:Si ratios of 2:1 suggesting a predominance of Si—O—Al bonds for all of these minerals (Wada, 1989). Allophanes and imogolites differ structurally and morphologically occurring as nanometre-sized, high surface area spheres and hollow tubes respectively (Denaix et al., 1999; Farmer and Russell, 1990; Gustafsson, 2001).

Allophanes and imogolites are well known to sorb a range of cations (e.g. Clark and McBride, 1984), humic materials (e.g. Parfitt et al., 1977; Yuan et al., 2000) and anions (Parfitt, 1990). Phosphate adsorption by allophane and imogolite has been the subject of considerable research over at least the past four decades (e.g. Birrell, 1961; Imai et al., 1981; Jones et al., 1979; McLaughlin et al., 1981; Su and Harsh, 1993; Theng et al., 1982; Wada, 1959).

Anion (phosphate) uptake capacity in allophane and imogolite has been demonstrated in laboratory studies to vary strongly as both the function of surface charge, and hence pH (e.g. Horikawa, 1975; Su and Harsh, 1993), and ionic strength. Typically between pH 5–7 both anions and cations can be adsorbed with ion uptake increasing with ionic strength (e.g. Farmer and Russell, 1990; Wada, 1989). Above pH 8–9 anion (phosphate) uptake is minimal with cation uptake the predominant reaction (e.g. Wada, 1980).

The extent of phosphate adsorption has been demonstrated to increase with increasing Al:Si ratio in both natural and synthetic allophanes (e.g. Clark and McBride, 1984; Farmer and Russell, 1990). Phosphate is initially rapidly adsorbed to allophane with additional phosphate more weakly adsorbed with, in some cases, the eventual formation of Al-PO_4 precipitates (Parfitt, 1989). The decomposition of allophane and its partial conversion to cryolite ($NaAlF_6$) in the presence of strong phosphate solutions has also been demonstrated in other studies (e.g. Su and Harsh, 1993; Wada, 1959). The extent of phosphate adsorption has been demonstrated, by batch uptake experiments, to vary according to pH, with maxima typically occurring at or below a value of 6 (Wada, 1989), the concentration of phosphate in solution and adsorption time. Typical phosphate uptake for allophanes is between ca. 6 and 19 mg P/g, and ca. 4 mg P/g for imogolite (Parfitt, 1989 and reference therein; Rajan, 1975). Synthetic allophanes can have phosphate uptake up to 87 mg P/g (McLaughlin et al., 1981).

The ability of allophanic soils to retain phosphorus has been documented by Degens et al. (2000). This study revealed that 91% of the phosphorus in a dairy effluent applied to allophanic soils (ca. 1–12% allophane) over a period of 22 years was retained in a 75 cm depth profile with 81% retained within the upper 25 cm. Interestingly, where applied total phosphorus concentrations were ca. four-fold greater than untreated soils, available phosphorus was ca. 22-fold greater indicating a substantial exchangeable phosphorus pool. Most of the effluent (ca. 70%) pH was below 6, which provided near optimal conditions for phosphate adsorption (Wada, 1989). Studies of the plant availability of phosphate adsorbed to allophane suggest that initially adsorbed phosphate, or that bound for an extended period, might not be bioavailable, although later bound phosphate might become bioavailable (Parfitt et al., 1982, 1989).

13.2.3 Fe- and Al-oxides

The adsorption of phosphate to both Fe- and Al-oxides (including hydrous Fe- and Al-oxides) has been studied extensively using both natural (e.g. from soils, or other natural deposits, Parfitt, 1989) and synthetically prepared materials (Borggaard, 1983; McLaughlin et al., 1981). Certain Fe- and Al-oxides may also form a solid solution series such as Al in goethite (Fordham and Norrish, 1974), or amorphous oxides (Potter and Yong, 1999) or are intimately associated on an atomic scale such as in red mud (e.g. Thornber and Hughes, 1987). Because of this close association in nature, their mineralogical similarity and their common evaluation as phosphorus adsorbents, Fe- and Al oxides (and hydrous oxides) are discussed together here.

The most common Fe-oxides that occur in soils are goethite (α-FeOOH) and ferrihydrite ($Fe_5O_7OH \cdot H_2O$) and both may contain significant concentrations of other elements such as Al and Si, respectively (Schwertmann and Taylor, 1989). Many other types of oxide may also occur in nature and as part of industrial wastes such as haematite and magnetite (e.g. Gilkes et al., 1992), which have a range of phosphate uptake capacities. Gibbsite ($Al(OH)_3$) is the most widely occurring aluminium oxide in soils and bauxite deposits (Hsu, 1989).

Studies of Fe- and Al-oxides in soils suggest that phosphate is predominantly associated with Fe-oxides at a rate 4–20 times higher than that of Al-oxides (Norrish and Rosser, 1983). In contrast, comparative studies of the adsorption of phosphate on synthetic Fe- and Al-oxides consistently indicate that Al-oxides have a substantially higher phosphate uptake capacity (e.g. Borggaard, 1983; McLaughlin et al., 1981). It has also been observed that natural materials have a higher phosphate uptake capacity than synthetic materials, this property being ascribed to the larger number of crystal defects and element substitutions in natural materials (Parfitt, 1989). Phosphate uptake in natural materials has also been correlated with Fe-oxide mineralogy and specific surface area (Fontes and Weed, 1996).

An approximate order of phosphate adsorption to Fe and Al-oxide minerals is Al-gel > Fe-gel ≫ haematite, goethite, magnetite > gibbsite, with phosphate uptake capacity varying from ca. 60 mg P/g and ca. 70 mg P/g for Fe- and Al-gels respectively, to between ca. 3–6 mg P/g for haematite and goethite (Baker et al., 1998; Borggaard, 1983; McLaughlin et al., 1981). This order of phosphate adsorption may also be dependant on factors such as particle size, age, drying, degree of crystallinity of particular minerals and pH, ionic strength, the period of phosphate adsorption and initial phosphate concentration (e.g. Bolan and Barrow, 1984; Bolan et al., 1986). It has been suggested that the uptake of phosphate is primarily related to particle specific surface area and to hydroxyl buffering with the latter providing a superior measure of phosphate uptake capacity for a range of Fe- and Al-oxide/hydroxide materials and allophanes (McLaughlin et al., 1981). McLaughlin et al. (1981) also demonstrated the importance of particle specific surface area showing that kaolinite covered with an Fe-gel (approximately 2 wt%) had twice and ten times the phosphate uptake capacity of Fe-gel alone and kaolinite respectively. This increase in uptake capacity was attributed to the occurrence of more surface sites when the gel was present as a coating. In studies of the adsorption of phosphate to ferrihydrite and goethite, Parfitt (1989) demonstrated that phosphate adsorption was preceded by Si desorption with an initial rapid exchange followed by a slower exchange possibly due to diffusion of adsorbed phosphate into the mineral substrate.

Iron oxides derived from industrial processes such as the Bayer process (red mud – Section 13.2.5) and mineral sand processing have also been evaluated as phosphorus adsorbents. In a comparative study of Fe-oxide waste, neutralized acidic effluent (containing abundant Fe-oxides) derived from synthetic rutile manufacturing, and red mud, it was demonstrated that additions of these three materials to soils at between 10 and 30% substantially increased phosphate retention, with neutralized acidic effluent and red mud being the most effective (Gilkes et al., 1992).

13.2.4 Activated alumina

The potential of activated alumina as a sorbent for the removal of anions from solution has been known for over 50 years (e.g. Schwab, 1949) while the use of activated alumina to specifically remove phosphate from natural and waste waters has received considerable attention over the past three decades (e.g. Ames and Dean, 1970; Baker et al., 1998; Brattebo and Odergaard, 1986; Donnert and Salecker, 1998; Hano et al., 1997; Narkis and Meiri, 1981; Winkler and Thordos, 1971; Yee, 1966).

The interest in activated alumina is primarily due to its moderate to high absorptive capacity of ca. 20 mg P/g in optimal (pH, solution chemistry) conditions and selectivity for phosphate (i.e. $OH > PO_4-P \gg SO_4-S > NO_2-N > Cl > NO_3-N$, Fleming, 1986) and the capacity of activated alumina to simultaneously remove ortho- and polyphosphates (Cohen, 1971; Donnert, 1988) with highest adsorption of phosphate generally between pH 2–6. Column studies of a mixed activated alumina-limestone and sand system demonstrated a high estimated minimum phosphate uptake capacity of activated alumina of ca. 4.5 mg P/g at a pH between 8 and 9 (Baker et al., 1998). Mobile pilot plants were constructed in the early 1970's to assess the performance of activated alumina in the removal of phosphate from waste waters (e.g. Neufeld and Thodos, 1969), however, it appears that the high cost of this material during this period limited further research (Ames, 1970).

Factors that may affect the phosphate uptake capacity of activated alumina include the surface area of the alumina, pH, ionic strength (salinity), contact time of the phosphate solution and the presence of particulate or organic matter (Brattebo and Odergaard, 1986). It has been shown that the surface properties of the activated alumina could be modified in specific cases to minimise the effects of adsorption of non-phosphate species. Due to the amphoteric nature of activated alumina acidic, basic and neutral forms of this material may be generated by treatment with acid, alkali or a combination of these reagents (Narkis and Meiri, 1981) endowing this material with a unique range of properties. Hano et al. (1997)

investigated the combined use of alum (aluminium sulphate) with activated alumina and demonstrated that the adsorption capacity of phosphate was enhanced by up to a factor of 1.7 over alumina alone.

A major advantage of the use of activated alumina over many other materials is that it does not alter the pH in the treated effluent and that no inorganic ions are needed in the process other than those periodically required to regenerate the activated alumina substrate (Cohen, 1971). Recent research has demonstrated that phosphate may be recovered from the spent alumina by treatment with $Ca(OH)_2$ and precipitated as $Ca_3(PO_4)_2$ rather than NaOH which has been more frequently used (Donnert and Salecker, 1998).

13.2.5 Red mud (bauxite residue)

In 1992, approximately 66 million tonnes of red mud was generated worldwide as a by-product of alumina processing using the Bayer process, at an approximate rate of one tonne of red mud produced per two tonnes of alumina (Akay *et al.*, 1998). Given the high Fe and Al concentration in red mud, it is an obvious candidate to be utilised for the removal of phosphate and other oxyanions from both natural and waste waters (municipal, agricultural, industrial). An inherent characteristic of red mud is that it is strongly alkaline (pH generally >11) as a consequence of the Bayer process and thus, is usually neutralised (amended) prior to experimental use. This treatment, most frequently with gypsum, may significantly increase the phosphate adsorption capacity of red mud (e.g. Ho and Matthew, 1993). However, red mud may also contain high concentrations of dissolved salts that may have to be removed by washing for particular applications. In addition, untreated and/or amended red mud may also contain substantial concentrations of chemically leachable (bicarbonate extractable) phosphate (up to *ca.* 0.2 mg/g, Jeffery, 1996) or water extractable phosphate (maximum 0.01 mg/g, Ho and Matthew, 1993).

Red mud has frequently been demonstrated to bind phosphate from effluent or in soils. The degree of adsorption may be strongly affected by pH (e.g. see Barrow, 1982). Phosphate retention may also be dependant on the amount of gypsum added to the red mud as an amendment (e.g. Jeffery, 1996). It is likely that this high affinity for phosphate is related to the presence of high concentrations of iron-oxyhydroxides and/or aluminium species in the red mud, although little research has been undertaken to elucidate the exact mechanism of phosphate retention.

Extensive field and laboratory trials have been conducted with native red mud (or amended derivatives), mostly in south-western Western Australia, to examine potential to reduce P loss in sandy coastal plain soils (e.g. Barrow, 1982; Summers *et al.*, 1996a, b; Vlahos *et al.*, 1989). Results from the trials with red mud added to the catchment soils suggest that this material has the capacity to significantly reduce phosphate loss (Summers *et al.*, 1993) with best rates of application between 10 and 20 tonnes/ha without gypsum amendment (Summers *et al.*, 1996a, b).

Red mud has also been tested for its ability to reduce phosphate from pig farm effluents (Weaver and Ritchie, 1987), aqueous solutions (Shiao and Akashi, 1977) during cross flow filtration (Akay et al., 1998) and from drain water (Ho and Matthew, 1993). In addition, red mud was also used in commercial wastewater treatment systems where effluent is either reticulated onto amended soil or passed through an enclosed leachate drain or soaked in the system (Jeffery, 1996). Apart from the addition of gypsum, which is the most frequently applied neutralization method, red mud may also be neutralized by a variety of methods including addition of hydrochloric acid, carbon dioxide adsorption from air, ferrous sulphate and ferrous sulphate/sulphuric acid mixtures and mixing with seawater (Thornber and Hughes, 1987).

The mixing of red mud with seawater, the lowest cost and most readily available material for neutralization, results in the formation of hydrotalcites (most probably Mg-rich forms – $Mg_6Al_2(OH)_{16}CO_3 \cdot H_2O$) and hydromagnesite $(MgCO_3)_4 Mg(OH)_2 \cdot 4H_2O$, as a result of the reaction of Mg^{2+} in the seawater with free Al species under the initially strongly alkaline conditions (Gastuche et al., 1967). These secondary hydrotalcites and hydromagnesite associate with the red mud solid particles forming intimate precipitates, increasing P retention. However, the precipitates may also contribute to poor dewatering characteristics of the amended mud (Thornber and Hughes, 1987). The hydrotalcite content of the seawater-neutralized red mud is primarily related to the free Al content and typically the amount of 1–2% hydrotalcite may be formed during the neutralization process. Although no data is available on the phosphate uptake capacity of seawater-neutralized red mud, it is unlikely to be substantial due to the hydrotalcite being in a carbonate form (see Section 13.2.1). If the red mud is sufficiently neutralized and possesses a near neutral pH, the abundant Fe and Al (oxy)hydroxides may also constitute a substantial adsorptive component for phosphate.

13.2.6 Fly ash

Fly ash is a residue from fossil fuel combustion in coal-fired power stations. Approximately 90 Mt of fly ash and bottom ash was produced in 1998 by coal-fired power stations in the USA alone (Scheetz and Earle, 1998). Although some applications in building and construction do exist, at least half of the fly ash produced is not utilised, resulting in its disposal as an undesirable landfill (Amrhein et al., 1996; Higgins et al., 1976; Singer and Berkgaut, 1995; Theis and De Pinto, 1976). Given the current environmental concern regarding landfill sites, and the amounts of fly ash generated every year, solutions for its re-use are being sought.

The fly ash typically consists of silica (ca. 30–60% as SiO_2), aluminium (ca. 10–30% as Al_2O_3) and lesser amounts of calcium, iron and sulphur (Grubb et al., 2000; Singer and Berkgaut, 1995). Trace element abundances are generally high with concentrations of elements such as As, Pb, Cu, Zn, Cd, Cr and Hg typically 1–100 times above average crustal abundances (Theis, 1975). The principal mineralogical constituents of fly ash are impure aluminosilicate glasses (ca. 60–90%), with smaller amounts of quartz, mullite, residual coal and ore minerals, which frequently occur as spherical particles of complex mineralogy (Dudas and Warren, 1988; Singer and Berkgaut, 1995).

Limited research has demonstrated that fly ash (often in combination with other materials such as lime) may have the capacity to remove phosphate from waste waters (e.g. Cheung *et al.*, 1994; Drizo *et al.*, 1999; Gangoli and Thodos, 1973; Higgins *et al.*, 1976; Mann 1997; Theis and McCabe, 1979; Vinyard and Bates, 1979), natural waters (e.g. lakes – Dunst *et al.*, 1974; Tenny and Echelburger, 1970), constructed wetlands (Mann and Bavor, 1993; Wood and Hensman, 1989) and that it also has the capacity to reduce the release of phosphate from bottom sediments (Theis and McCabe, 1978). Fly ash has also been demonstrated to reduce soluble phosphate in septic tank effluent (Cheung and Venkitachalam, 2000) and reduce the concentration of water soluble P when applied as a soil amendment (Stout *et al.*, 1999). The efficiency of many of these applications, and in particular, phosphate removal from the water column, may be strongly dependant on both the type of fly ash used (e.g. Type C or Type F, Grubb *et al.*, 2000; Higgins *et al.*, 1976) and/or on solution pH. Some studies suggest enhanced removal of phosphate under moderate to strongly alkaline conditions (e.g. Fine and Jensen, 1981). Typical estimates of phosphate uptake capacity for fly ash range from 0.05 to over 3 mg P/g (Cheung *et al.*, 1994; Gangoli and Thodos, 1973; Mann, 1997; Wood and Hensman, 1989).

Related research has centred on the conversion of fly ash via a hydrothermal (or microwave) caustic/limestone/dolomite treatment process to produce a residue rich in a mixture of aluminosilicate minerals (e.g. zeolite and/or hydroxysodalite and/or hydrotalcite and/or hydrocalumite depending on the method of synthesis) with a cation exchange capacity often in excess of 300 meq/100 g (e.g. Amrhein *et al.*, 1996; Mondragon *et al.*, 1990; Reardon and Della Valle, 1997; Singer and Berkgaut, 1995) which may have a high affinity for both ammonia and trace elements. In addition, due to the net positive charge in the interlayers of some types of these materials (e.g. hydrotalcite and hydrocalumite), they have a high capacity to adsorb anions such as phosphate, borate, sulphate and arsenate.

A potentially major problem in the application of fly ash in any aquatic system is the potential for the leaching of high concentrations of trace elements and hence, toxicity, large increases in pH (for calcium-rich fly ash) and high chemical oxygen demand (Theis and De Pinto, 1976). Laboratory tests suggest that water column concentrations of 10–20 mg/l of fly ash are toxic to fish indigenous to a North American lake (Hampton, 1974).

13.2.7 Expanded clay aggregates

Expanded clay aggregates (or Light Weight Aggregates – LWA, e.g. Filtralite®, light expanded clay aggregate – LECA®) are commercially produced modified mineral substrates that have been the subject of extensive investigation in laboratory and constructed wetland systems to evaluate their potential for phosphate uptake and retention (e.g. Jenssen *et al.*, 1991; Johansson *et al.*, 1995; Johansson and Hylander, 1999). The expanded clay aggregates generally consist of an inert ceramic particle surrounded by a porous coating. These materials are produced by the high temperature (up to *ca.* 1200°C) calcination of clay minerals such as illite.

During the calcination in rotary kilns the organic matter in the clay expands, resulting in a high porosity mineral of low bulk density with higher hydraulic conductivities than similar sized sands and gravels (Maehulm, 1998).

An examination of expanded clay aggregates manufactured in the USA, Sweden and Norway suggests that their phosphate uptake capacity may vary over two orders of magnitude up to *ca.* 3.5 mg/g with the primary determinant of phosphate uptake capacity being the total metal, and in particular Ca concentration (Zhu *et al.*, 1997). Speciation analysis of adsorbed phosphate suggested that it was predominantly associated with calcium, probably as hydroxylapatite (Zhu, 1998). Field evaluation of expanded clay aggregates has principally occurred in constructed wetlands, where they have also been demonstrated to assist nitrogen transformation via adsorbed microbial biomass (Zhu, 1998), and are able to support substantial aquatic plant biomass despite the high pH (*ca.* 10) of the most Ca-rich expanded clay substrates (Jenssen *et al.*, 1996; Maelhulm *et al.*, 1995; Zhu, 1998).

13.2.8 Blast furnace slag

The use of blast furnace slag, and related slags such as basic oxygen furnace (BOF) slag and electric arc furnace slag (EAF), as a phosphate adsorbent has been investigated extensively over the past two decades (Baker *et al.*, 1998; Drizo *et al.*, 2002; Johansson, 1999a, b; Johansson and Gustafsson, 2000; Yamada *et al.*, 1986). Possible applications that have been investigated at either laboratory or field scale include the capture of phosphate released from bottom sediments (Yamada *et al.*, 1987), and as a substrate material to enhance the phosphate uptake capacity of constructed wetlands (e.g. Mann and Bavor, 1993; Sakadevan and Bavor, 1998).

A large resource of blast furnace slag is available worldwide with the United States alone producing approximately 13 Mt annually (Proctor *et al.*, 2000). Blast furnace slag may occur in both crystalline and amorphous forms and is principally composed of silica and aluminium, derived from the parent iron ore, with appreciable calcium (typically 20–45%), derived from the flux, and predominantly in a calcined form as $Ca(OH)_2$ (portlandite), CaO or a variety of calc-silicate minerals such as wollastonite – $CaSiO_3$ (Drizo *et al.*, 2002; Proctor *et al.*, 2000; Sakadevan and Bavor 1998; Yamada *et al.*, 1987).

The large proportion of CaO (and occasionally elevated MgO, generally present as periclase or Mg-silicates) results in an alkaline slag, which may impart a pH of between *ca.* 8–11 in leachates (Proctor *et al.*, 2000) when used in laboratory and field trials (e.g. Drizo *et al.*, 2002; Johansson, 1999a, b; Mann, 1997). It has been demonstrated in column experiments that the pH of constructed wetland effluents can be reduced by the application of autoclaved secondary sewage effluent with the maintenance of a lower pH expected to improve phosphate uptake capacity (Mann, 1996). An assessment of blast furnace slag as a slow release fertilizer in plant trials indicates that a substantial amount of adsorbed phosphate was available for plant growth (Hylander and Siman, 2001).

The uptake capacity of blast furnace slag may vary widely from *ca* 0.4 mg P/g (Mann and Bavor, 1993) up to 44 mg P/g (Sakadevan and Bavor, 1998). This substantial difference on phosphate uptake capacity largely reflects the initial phosphate concentrations in the batch experiments, which varied by up to 10 orders in magnitude. Phosphorus uptake capacity by slags may also reflect both the form of the slag (e.g. granulated, crystalline or amorphous, Johansson, 1999a, b), the length of the adsorption isotherm experiment, physico-chemical conditions, hydraulic conductivity and the composition of the slag with most studies identifying the formation of Ca-phosphates as the predominant form of bound phosphate (e.g. Drizo *et al.*, 2002; Johansson and Gustafsson, 2000; Yamada *et al.*, 1986). Due to their origin, blast furnace slags may have elevated metal and metalloid concentrations, however, investigation of metal and metalloid mobility using a Toxicity Characteristic Leaching Protocol (TCLP) revealed that the majority of elements are bound tightly to the slag matrix (Proctor *et al.*, 2000). Further assessment of slag leachate is necessary, however, if this material is going to be used as a slow-release fertilizer (Hylander and Siman, 2001).

13.2.9 Rare earth modified clays

Investigations conducted over three decades ago suggested that rare earth elements (REE), and in particular lanthanum (La), may be effective in reducing phosphate concentrations from lake water in pilot-scale studies (Gahler *et al.*, 1969), from waste waters (Cohen, 1971; Recht *et al.*, 1970) and for the removal of straight chain polyphosphates (McNabb *et al.*, 1968). This potential was confirmed in later studies for both artificial algal assay medium and natural pond water with optimal phosphate removal achieved at pH between 6 and 10 (Peterson *et al.*, 1974, 1976). Removal of dissolved phosphate was demonstrated to be highly efficient with La:PO_4 molar ratio of approximately 1:1 compared to a sodium aluminate ($NaAlO_2$) which was relatively inefficient, requiring a molar ratio of *ca.* 7:1 to achieve a similar level of phosphate reduction (Peterson *et al.*, 1976). Early research into the use of lanthanum to remove phosphate from secondary wastewater effluent (Melnyk *et al.*, 1974; Melnyk, 1975; Recht *et al.*, 1970) suggested it was superior to the more conventional Fe and Al salts in five ways:

1) lanthanum was effective over a wider pH range (*ca.* 4.5–8.5) than Fe (III) (*ca.* 3.5–4.5) or Al (*ca.* 5.0–6.5)
2) lanthanum precipitated polyphosphates equally as well as orthophosphates
3) the reaction of lanthanum with phosphate was stoichiometric
4) almost total removal of phosphates could be obtained
5) the solubility product of lanthanum phosphate was extremely low (K_{sp} = *ca.* −24.5 to −25.8, Firshing and Brune, 1991; Jonasson *et al.*, 1995; Liu and Byrne, 1997).

The research conducted on REE-phosphate interaction revealed that in both seawater (Byrne and Kim, 1993) and nutrient solutions for plant growth (Diatloff *et al.*, 1993) phosphate might be a significant limiting factor influencing dissolved REE concentrations. Toxicity testing of lanthanum compounds showed that this element might be toxic (e.g. Kyker and Cress, 1957; Peterson *et al.*, 1976) depending on concentration and application rate (Sax *et al.*, 1984). However, other researchers showed that this toxicity may be ameliorated in the presence of humic substances, presumably due to complexation (Stauber, 1999; NICNAS, 2001).

Potential solute toxicity was largely overcome by the incorporation of lanthanum into clay minerals with high cation exchange capacity clay such as bentonite (Douglas *et al*., 1997; Douglas, 1997). Extensive laboratory testing demonstrated that in the presence of a strong oxyanion such as phosphate, a stable mineral ($LaPO_4 \cdot nH_2O$ – rhabdophane) is formed (Douglas *et al*., 2000). This mineral is an analogue of other REE-phosphate minerals that were demonstrated to be resistant to weathering over geological time. Short duration studies suggested that the newly precipitated rhabdophane is resistant to microbial attack from endemic sediment micro-organisms (Douglas *et al*., 1997). The phosphate uptake capacity of lanthanum-substituted clays typically exceeds 10 mg P/g. Large-scale field trials of lanthanum-substituted clays suggest that they may be effective in capturing a substantial proportion of phosphate released from bottom sediments when applied as a thin (*ca*. 1 mm) reactive capping (Douglas and Adeney, 2001).

13.2.10 Carbonates

The coprecipitation of phosphate onto or with calcite (as limestone – $CaCO_3$ or other mixed or impure carbonates) in aquatic systems is a well known process (e.g. Effler and Driscoll, 1985). There is also evidence that algal biomass is often regulated in hardwater lakes due to adsorption of P during carbonate precipitation (Otsuki and Wetzel, 1972). As a consequence, carbonates or lime have been extensively tested in the removal of soluble phosphate from both natural (e.g. rivers – Jack and Platell, 1983; lakes or other impoundments – Dittrich *et al*., 1997 and references therein) and waste waters (e.g. pig farm effluents, Takeuchi and Komada, 1998; Weaver and Ritchie, 1987), stormwaters (Barbin *et al*., 1989), soils (Anderson *et al*., 1995), wetland effluents (Arias *et al*., 2000; Drizo *et al*., 1999; Forget *et al*., 2002) and in laboratory trials (e.g. Ho and Monk, 1988).

The adsorption/desorption of phosphate by limestone and its mechanism of complexation in both soils and sediments has been the subject of extensive research over a number of decades (e.g. Freeman and Rowell, 1981; Hartikainen, 1989; House and Donaldson, 1986; Jack and Platell, 1983; Millero *et al*., 2001; Moreno *et al*., 1960; Ryan *et al*., 1985; Sims and Ellis, 1983). The exact mechanism of phosphate removal from solution in the presence of carbonate minerals is thought to occur via a surface complex of calcium-carbonate-phosphate with only a small fraction of phosphate ultimately being incorporated into the primary carbonate crystal structure (e.g. Avnimelech, 1980; Freeman and Rowell, 1981), with enhanced rates of P removal achieved at both increasing pH (over the range 7–9.5) and temperature (5–35°C) (House and Donaldson, 1986). Laboratory trials of phosphate removal from solution by limestone suggest that the rate of removal (as expected) may be significantly affected by surface area, with increasing rates of adsorption with decreasing particle size (Jack and Platell, 1983) and increasing contact time between solution and solid (Ho and Monk, 1988). A recent study of phosphate adsorption/desorption on aragonite and calcite in seawater by Millero *et al*. (2001), showed that phosphate adsorption is a multi-stage process, and that the extent of adsorption increases with temperature (in agreement with House and Donaldson, 1986), but declines with salinity, primarily due to the presence of bicarbonate. In addition, it was demonstrated that short term phosphate uptake was higher on aragonite than calcite, and that carbonate minerals may act both as a sink and source of phosphate with the potential for rapid (*ca.* 1 day) release of adsorbed phosphate which was dependent on the period of pre-equilibration.

Many other forms of carbonate minerals and derivatives have been assessed for their ability to remove P from natural and wastewaters. Dolomite (Ca, Mg)CO_3 has been investigated for the removal of phosphorus from hog farm effluents (Takeuchi and Komada, 1998), and from constructed wetlands (Pant *et al.*, 2001; Roques *et al.*, 1991). In both studies the dolomite was found to have inferior performance compared to other forms of carbonate, shales and sands. Dolomite derivatives, such as partially calcined dolomite (Roques *et al.*, 1991) and granular magnesia clinker (Kaneko and Nakajima, 1998; Suzuki and Fuji, 1988), were also assessed as substrates for phosphorus removal. In the case of partially calcined dolomite, phosphate removal varied as a function of particle grain size and was typically around 3–5 mg P/g.

Rare earth element carbonates such as basic yttrium carbonate (Haron *et al.*, 1997) have also been investigated for their phosphate removal capacity. Phosphate retention was highest in the pH region 7–12 with substantial substrate dissolution at pH below 6. Phosphate adsorption occurred via carbonate exchange and was also strongly influenced by particle size with the highest phosphate uptake of *ca.* 130 mg P/g occurring in the presence of 5 μm particles. Lanthanum carbonates have also been demonstrated to remove both arsenate (Tokunaga *et al.*, 1997) and phosphate from aqueous solutions (Mills, 2000).

13.2.11 Soils, sands and gravels

Adsorption of phosphorus (as phosphate) to soils (*sensu stricto*) has been the subject of extensive research for many decades (Barrow, 1970; Bray and Kurtz, 1945; Hylander and Siman, 2001). Most studies have concentrated on the ability of soils to adsorb fertiliser phosphate and its form(s) of retention in the context of relating these properties to soil fertility. In general, although the majority of phosphorus in soils is present as phosphate (Frossard *et al.*, 1995), smaller pools of organic phosphorus are also present.

Studies of phosphorus adsorption by soils generally focus on either the whole soil *per se* or examination of specific soil components, either via sequential extraction procedures, or using synthetic mineral analogues produced in the laboratory. These latter studies reflect the mineralogical and physico-chemical heterogeneity of most soils. The major constituents that may be involved in the adsorption of phosphate by soils, sands and gravels include hydrous Fe- and Al-oxides, carbonates, clay minerals (which may constitute a substrate for the hydrous oxides, e.g. Fordham and Norrish, 1979; Norrish and Rosser, 1983) as well as adsorbing phosphate themselves (*ca.* 0.2–0.4 mg P/g) (Bar-Yossef *et al.*, 1988; Fontes and Weed, 1996) and less common soil minerals such as allophanes and imogolites. Many of these soil minerals have been discussed in detail in previous sections. Organic matter may also play a significant part in phosphorus adsorption in specific soils with some studies demonstrating a correlation between phosphate uptake and soil organic matter (e.g. Evans and Smillie, 1976), while metal-humic complexes of Fe and Al have also been observed to be influential in phosphate uptake presumably through the availability of the solubilized (complexed) metal ions (Sinha, 1971; Bloom, 1981). In contrast, while it has also been demonstrated that organic matter has little or no influence on phosphorus uptake, with hydrous Fe- and Al-oxides the main repositories of adsorbed phosphate, organic matter indirectly increased phosphate adsorption on Al-oxides by inhibiting their crystallization (Borggaard et al., 1990).

A number of studies have examined the adsorption of phosphate by soils with a view to their possible use as a specific phosphorus adsorbing material (e.g. a soil B horizon in Hylander and Siman, 2001; Johannson, 1999a; Tofflemire and Chen, 1977) as part of larger studies to compare background soil characteristics to that of selected adsorbents (e.g. Mann, 1996; Sakadevan and Bavor, 1998) or as an examination of soils treated with phosphorus-enriched waste (Degens et al., 2000; Sawhney and Hill, 1975). The phosphorus uptake capacity of natural soils varies widely as may be expected from the diversity of mineralogical and geochemical compositions with the most adsorptive soils generally enriched in one or more of Ca-, Fe- or Al- bearing mineral phases. A study of soils for the use of a possible substrates in constructed wetland systems revealed a wide phosphate uptake from ca. 0.9 mg P/g in surface soils to ca. 5.2 mg P/g in soils from an operational constructed wetland system (Sakadevan and Bavor, 1998). Short-term laboratory-based batch studies of an allophane-rich soil suggest a phosphate uptake capacity of at least ca. 5 mg P/g while field measurements of phosphorus uptake in a 75 cm deep soil profile after 22 years of application of dairy effluent indicate a current phosphorus uptake of ca. 1 mg P/g (estimated from Degens et al., 2000).

In related research, sands and gravels have also been investigated for their phosphate uptake capacity with a view to their possible incorporation as an adsorptive substrate in constructed wetlands. Arias et al. (2000) concluded that Ca was the most important parameter in determining phosphate uptake by sands, presumably due to the precipitation of calcium phosphate phases, however, under more acidic conditions, Fe and Al were also important in phosphate precipitation reactions. Extensive batch and column phosphate uptake studies suggested that traditional Langmuir isotherms, which often suggested relatively low uptake capacities (ca. 0.1 mg P/g), were not a good measurement of actual phosphate removal. This aspect of the application of Langmuir isotherms has also been recognised previously in studies of phosphate uptake by natural materials (e.g. Jenssen and Krogstad, 1988) and may reflect the heterogeneous nature of uptake sites, which may not be well described by the assumption of monolayer adsorption, and the possibility of more than one mechanisms (i.e. adsorption as well as ionic exchange and/or precipitation) may operate in tandem. In a batch study of the phosphate uptake of gravels used in constructed wetlands, Mann (1997) demonstrated a low uptake phosphorus capacity of between ca. 0.03 and 0.05 mg P/g. As in the study of Arias et al. (2000) and other research, Ca was determined to be the most important parameter in phosphate uptake.

13.3 SELECTION OF AN APPROPRIATE PHOSPHORUS ADSORBENT

As we have outlined in Section 13.2 there are at least 11 different classes of potential phosphorus adsorbents. Would-be users, principally water quality managers and/or their scientific advisers, face an initially bewildering array of options. The effectiveness of each class of adsorbent as measured by its adsorption capacity (mg P/g solid), and relative costs, cover a wide range in most instances and there are considerable overlaps between the different classes. The cost-effectiveness relationships between the different phosphate adsorbents are shown diagrammatically in Figure 13.1. In addition, there are substantial differences in the sensitivity of adsorbents to changes in pH and redox conditions. To provide some structure to a rational approach to selecting the optimum adsorbent for a particular application, it is helpful to consider several different model cases, and to look at some of the constraints which apply to different adsorbents.

13.3.1 Phosphorus load/concentration in the target water body/effluent stream

We consider three extreme cases, as summarised in Table 13.2. Firstly, an effluent stream coming from a sewage treatment plant without a specific P removal stage. Here the water volume will be high (10 to 20 ml/day) and the concentration of biologically available P is also going to be high (~5 mg/l), leading to a daily load of

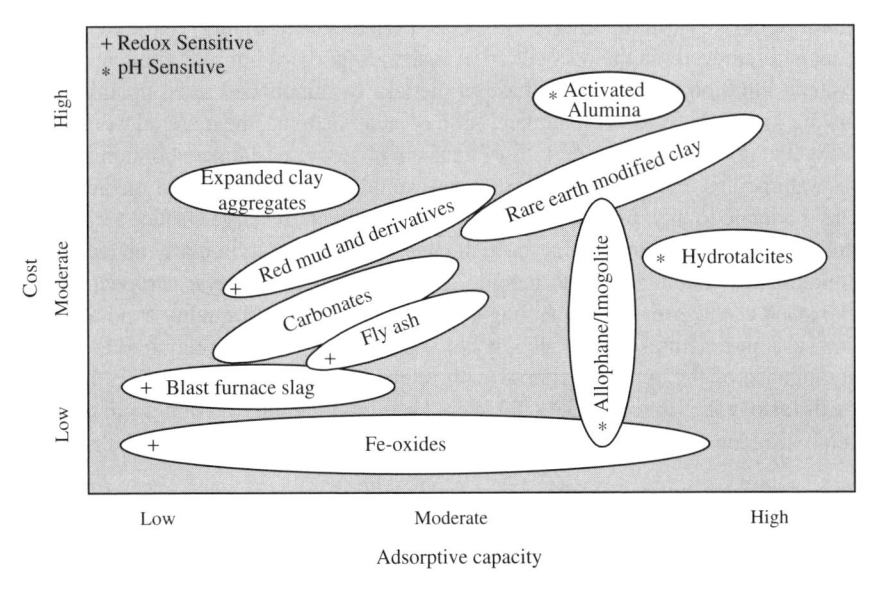

Figure 13.1 Cost-effectiveness relationship between phosphate adsorbents. Adsorbents with significant redox (+) or pH (*) sensitivity are marked.

50 to 100 kg P/day. In contrast, a domestic swimming pool may contain 0.1–1 ml, which may be replaced once per season. Most of the P comes from either atmospheric fallout of dust or human use so the load is generally low. The goal would be to avoid any signs of algal growth, which requires maintaining the P concentration below 3 µg/l. The third case is a natural waterbody such as a lake or impounded river during summer, with limited catchment input and the majority of P derived from the bottom sediments via nutrient regeneration processes. As in the domestic swimming pool, the goal in managing the natural waterbody, either for human recreational use or as a supply of drinking water, is to avoid algal growth by maintaining the bioavailable P concentration below 3 µg/l.

In the treatment of the sewage effluent stream the obvious choice is a cheap and readily available adsorbent with a capacity to produce a moderately low residual P concentration. The obvious candidate materials would be the red mud, Fe- and/or Al- oxides or carbonates. In theory, with a sufficiently large dose and adequate mixing and reaction time, these would reduce the concentration in the discharge stream to perhaps 0.1 mg P/l – a 98% reduction. In practice, however, because of logistical and practical constraints, a reduction of this magnitude is rarely achieved. A large settling pond or mechanical means of water removal such as centrifuging, hydrocycloning or other dewatering mechanism, by means of application of polymeric coagulants, would be required to retain the added material.

In contrast, the swimming pool application deals with minimal loads but requires an adsorbent with a very high P affinity to produce the very low desired residual concentration. In addition, the material chosen has to have no adverse human health effects on the water quality. In this case we would adopt an engineered material such as the hydrotalcites (subject to pH), allophane/imogolites (subject to pH), activated alumina, or rare earth modified clays, added periodically as a slurry to the swimming pool and removed via the pool filtration system or perhaps housed within an in-line column filtration system. In the case of the lake or impounded river, in addition to having no adverse effects on human health or endemic biota, the material would have to be able to retain P under periods of transient or sustained anoxia common in natural waterbodies. Furthermore, the material would ideally be of low to medium cost and moderate to high uptake capacity, given the potentially large surface area of the bottom sediments and potential for high rates of P-release, particularly under anoxic conditions. Possible materials include rare earth modified clays, carbonates, and expanded clay aggregates depending on factors such as water quality conditions. In any case, in a natural system, ecotoxicological testing would be required to ensure compliance of the applied material with relevant water quality guidelines or standards (also refer Section 13.3.4). Table 13.2 sets out the preferred options in trading-off load against residual concentration.

13.3.2 Redox sensitivity of the adsorbent

In choosing an adsorbent for a particular application, it is important to be mindful of the biogeochemical environment in which the adsorbent is to be deployed. A significant factor in the effectiveness of all the various iron containing adsorbents (Fe oxides, red mud, fly ash, blast furnace slag, and soils) is their high content

Table 13.2 Adsorbent choices on the basis of load, cost and residual concentration.

System characteristics	Criteria for choice	Most appropriate adsorbent
Sewage effluent – High through puts and loads, moderate residual P concentration (e.g. sewage treatment or industrial effluents, storm water).	Primarily low cost, and effectiveness. Large reduction on P but not necessarily to very low concentration.	Fe oxides, red mud, fly ash, expanded clay aggregates, blast furnace slag.
Swimming pool – Low load and low but transiently variable residual concentration (e.g. outdoor swimming pools).	P removal to very low residual concentration, reliability of material characteristics. No significant human health/ecotoxicological effects.	Activated alumina, rare earth modified clay, hydrotalcites.
Lake or impounded river – Generally low load in summer and low residual surface water P-concentration except when mixed (e.g. density overturn) with anoxic bottom waters.	No significant human health/ecotoxicological effects. Playoff between low-moderate cost and moderate-high adsorptive capacity due to scale of application.	Rare earth modified clay, hydrotalcites.

of Fe (III) oxyhydroxides. Under reducing conditions, which exist in anoxic sediments or the anoxic waters (hypolimnion) in seasonally stratified lakes, Fe (III) is reduced microbially to Fe (II). There is a release of adsorbed P, as well as P produced by breakdown of organic detritus, in the form of dissolved orthophosphate (Mortimer, 1941). Turbulent diffusion and density driven currents will redistribute this biologically available P to the uppermost and best illuminated water layer (epilimnion) where it will support extensive algal growth. Thus, in applications aimed primarily at limiting the algal biomass in deep (>5 m) water bodies where seasonal anoxia will occur, Fe (III) adsorbents are likely to be ineffective in the long term (weeks to months time scale) although they could produce a temporary reduction in biologically available P. Under these circumstances the appropriate adsorbents to use would be the hydrotalcites, allophane/imogolite, rare earth modified clays, or the carbonates. These all have a very low residual P concentration, have a moderate cost, and are not redox sensitive. Activated alumina has been excluded on cost grounds. However, in shallow systems, which are well mixed and well oxygenated, Fe (III) adsorbents such as Fe-oxides, blast furnace slag, and fly ash are likely to be quite effective in the long term, as long as they can achieve a sufficiently low residual P concentration. Use of red mud without pre-treatment in natural aquatic ecosystems is to be avoided due to its high alkalinity.

13.3.3 pH sensitivity of the adsorbent

Some adsorbents show an optimal pH range for P uptake, as well as the potential for desorption of phosphorus and/or dissolution of the adsorbent substrate as pH changes. As an example, both allophanes and imogolites display a wide operational pH range for adsorption of phosphorus (as phosphate) between *ca.* 4 and 8. Many natural waters, and in particular eutrophic systems with elevated phosphorus concentrations, may be subject to algal blooms, which may frequently elevate pH to above 9. Thus, a potential exists for release of exchangeable phosphate into solution where allophanes either occur naturally or have been used as an amendment to reduce bioavailable phosphorus concentrations. Alternatively, at lower pH, hydrotalcites, whether present as a discrete phase or in amended red muds or fly ash, may dissolve into their constituent phases (usually Mg and/or Al and Fe cations) with concomitant release adsorbed phosphorus. Similarly, a potential exists for Fe-based adsorbents such as red muds, Fe-oxides to dissolve and release adsorbed phosphorus at low pH.

13.3.4 Public health aspects

Many of the different classes of adsorbents are the waste products of industrial processes. Much of the impetus to explore their suitability as P adsorbents has come from a desire to find a profitable use for such materials and avoid the costs of an ecologically sustainable long-term disposal. These materials are produced in a range of industrial environments where the nature of the waste material may periodically vary and where full characterisation of all the raw materials is difficult. We have in mind especially Fe-oxides derived from pickling baths, fly ash, and blast furnace slag. In these cases, the possibility of heavy metal or other trace element contamination cannot be ruled out *a priori*. It is an elementary precaution to ensure that these materials are fully characterised, including full ecological evaluation under a range of expected operational conditions, before they are used in situations where heavy metals or other potential contaminants may be incorporated into the food chain. In any case, full ecological evaluation or reference to existing, appropriate data should be considered a benchmark prior to any application of solid phase adsorbents for phosphorus removal, particularly in applications to natural waters or where treated effluent may ultimately be discharged to the wider environment.

13.3.5 Requirement for preliminary evaluation of adsorbents at the laboratory and mesocosm scale

While the above discussion provides guidance in selecting the most effective adsorbent for a range of applications, it is impossible to specify confidently the optimum adsorbent without taking account of the particular circumstances in which it is to be applied. It is important that the target P concentration be clearly specified beforehand and be used as a prime criterion for evaluating the performance of different adsorbents. The quantitative predictive relations of Vollenweider (1976) provide a most helpful basis for defining the target P concentration based on the desired algal biomass (as measured by chlorophyll concentrations in the water column). After initial laboratory testing to ascertain that chosen adsorbent can achieve the desired P reductions, *in-situ* testing using small mesocosms (1 to 3 m diameter) which enclose the water column, and the sediments at the base of the mesocosm should be conducted before embarking on a full-scale implementation.

Mesocosm-scale trials, where appropriately replicated, and with suitable controls, also provide a convenient *in-situ* 'laboratory' to evaluate aspects of changes, if any, in benthic biota due to physical (e.g. smothering of benthic biota, gill irritation of fish), chemical (interaction of the adsorbent with the water column) or ecotoxicological (e.g. from chemical interaction or direct ingestion) effects from the application of solid phase phosphorus adsorbents.

13.4 CONCLUDING REMARKS

The goal of effective interception, containment and long-term immobilization of phosphorus from point and/or diffuse sources has gained increasing prominence over the past two decades. This is particularly so, as more efficient and environmentally-conscious methods are sought as the effects of eutrophication on natural and man-made water bodies become increasingly apparent.

Numerous phosphorus interception methods have been devised and evaluated over recent decades including constructed wetlands and reactive columns, or direct application to the water column or bottom sediment in aquatic systems. A total of 11 materials that have potential, or are currently being used as phosphorus adsorbents have been reviewed in this chapter. The application of adsorptive materials to augment the function of phosphorus interception/immobilization methods, with the aim of increasing the uptake capacity and longevity, places a number of constraints on the types of material(s) that may be utilized. These constraints generally relate to the nexus between cost, for example between low cost industrial wastes such as red muds or slags relative to more expensive adsorbents, phosphorus adsorption capacity, stability in terms of phosphorus retention, and practicality of implementation. In addition, factors such as public health requirements and the ecotoxicity of post-treatment effluents that may be generated from sources as diverse as sewage or constructed wetlands need to be carefully considered. This is particularly pertinent where downstream maintenance or improvement of ecological values is considered essential in natural systems such as wetlands or rivers. Where long-term phosphorus retention is envisaged, the long-term physical and chemical stability of the adsorbent needs to be carefully assessed being cognisant of possible re-release of adsorbed phosphate under variable redox, pH or other physico-chemical conditions, and possible site re-use or recreational use as might be envisaged for a constructed wetland. Other factors that may also be influential in the choice of an adsorbent include its ease of handling and deployment, and life-cycle factors including the ultimate disposal of spent adsorbent, if required, such as to landfill.

REFERENCES

Andrews, J.F. (1993) Modeling and simulation of wastewater treatment processes. *Water Sci. Technol.* **28**(11/12), 141–150.

Addiscott, T.M. and Thomas, D. (2000) Tillage, mineralization and leaching: phosphate. *Soil Tillage Res.* **53**, 255–273.

Ainsworth, C.C., Sumner, M.E. and Hurst, V.J. (1985) Effect of aluminium substitution in goethite on phosphorus adsorption: I. Adsorption and isotopic exchange. *Soil Sci. Soc. Am. J.* **49**, 1142–1149.

Akay, G., Keskinler, B., Cakici, A. and Danis, U. (1998) Phosphate removal from water by red mud using crossflow microfiltration. *Water Res.* **32**, 717–726.

Ames, L.L. (1970) Mobile pilot plant for the removal of phosphate from wastewaters by adsorption on alumina. *Water Pollut. Control Res. Ser. 17010 EER 05/70, US EPA*.

Ames, L.L. and Dean, R.B. (1970) Phosphorus removal from effluents in alumina columns. *J. Water Pollut. Control Fed.* **42**, R161.

Amin, S. and Jayson, G.G. (1996) Humic substance uptake by hydrotalcites and PILCS. *Water Res.* **30**, 299–306.

Amrhein, C., Haghnia, G.H., Kim, T.S., Mosher, P.A., Gagajena, R.C., Amanios, T. and de la Torre, L. (1996) Synthesis and properties of zeolites from coal fly ash. *Environ. Sci. Technol.* **30**, 735–742.

Anderson, D.L., Tuovinen, O.H., Faber, A. and Ostrokowski, I. (1995) Use of soil amendments to reduce soluble phosphorus in dairy soils. *Ecol. Eng.* **5**(2–3), 229–246.

Arias, C.A., Del Bubba, M. and Brix, H. (2000) Phosphorus removal by sands for use as a media in subsurface flow constructed reed beds. *Water Res.* **35**, 1159–1168.

Avnimelech, Y. (1980) Calcium-carbonate-phosphate surface complex in calcareous systems. *Nature* **228**, 255–257.

Babin, J., Prepas, E.F., Murphy, T.P. and Hamilton, H.R. (1989) A test of the effects of lime on algal biomass and total phosphorus concentrations in Edmonton stormwater retention lakes. *Lake Res. Manag.* **5**, 129–135.

Baker, M.J., Blowes, D.W. and Ptacek, C.J. (1998) Laboratory development of permeable reactive mixtures for the removal of phosphorus from onsite wastewater disposal systems. *Environ. Sci. Technol.* **32**, 2308–2316.

Baldwin, D.S., Beattie, J.K., Coleman, L.M. and Jones, D.R. (1995) Phosphate ester hydrolysis facilitated by mineral phases. *Environ. Sci. Technol.* **29**, 1706–1709.

Baldwin, D.S., Beattie, J.K., Coleman, L.M. and Jones, D.R. (2001) Hydrolysis of an organophosphate ester by manganese dioxide. *Environ. Sci. Technol.* **35**, 713–716.

Bar-Yosef, B., Kafkafi, U., Rosenberg, R. and Sposito, G. (1988) Phosphorus adsorption by kaolinite and montmorillinite: effect of time, ionic strength and pH. *Soil Sci. Soc. Am. J.* **52**, 1580–1585.

Barreal, M.E., Camps Arbestian, M., Macais, F. and Fertitta, A.E. (2001) Phosphate and sulphate retention by nonvolcanic soils with acidic properties. *Soil Sci.* **166**, 691–707.

Barrow, N.J. (1970) Comparison of the adsorption of molybdate, sulphate and phosphate by soils. *Soil Sci.* **109**, 282–288.

Barrow, N.J. (1974) Effect of previous additions of phosphate on phosphate adsorption by soils. *Soil Sci.* **118**, 82–89.

Barrow, N.J. (1982) Possibility of using caustic residue from bauxite for improving the chemical and physical properties of sandy soils. *Aust. J. Agr. Res.* **33**, 275–285.

Barrow, N.J. (1983) On the reversibility of phosphate sorption by soils. *J. Soil Sci.* **34**, 751–758.

Barrow, N.J. (1985) Reactions of anions and cations with variable-charge soils. *Adv. Agron.* **38**, 183–200.

Barrow, N.J. (1999) The four laws of soil chemistry: the Leeper lecture 1998. *Aust. J. Soil Res.* **37**, 787–829.

Barrow, N.J. and Shaw, T.C. (1975) The slow reactions between soil and anions. II. Effects of time and temperature on the decrease in phosphate concentration in the soil solution. *Soil Sci.* **119**, 167–177.

Birrell, K.S. (1961) Ion fixation by allophane. *NZ J. Soil Sci.* **4**, 393–414.

Bloom, P.R. (1981) Phosphorus adsorption by an aluminium-peat complex. *Soil Sci. Soc. Am. J.* **45**, 267–272.

Bolan, N.S. (1983) Phosphate adsorption by soil constituents and its effect on plant response to both phosphorus application and mycorrizal infection. *PhD thesis, Univ. W. Aust.* 197 pp.

Bolan, N.S. and Barrow, N.J. (1984) Modelling the effect of adsorption of phosphate and other anions on the surface charge of variable charge oxides. *J. Soil Sci.* **35**, 273–281.
Bolan, N.S., Syers, J.K. and Tillman, R.W. (1986) Ionic strength effects on the surface charge and adsorption of phosphate and sulfate by soils. *J. Soil Sci.* **37**, 379–388.
Borggaard, O.K. (1983) Effect of surface areas and mineralogy of iron oxides on their surface charge and anion-adsorption properties. *Clay Clay Min.* **31**, 230–232.
Borggaard, O.K., Jorgensen, S.S., Moberg, J.P and raben-Lange, B. (1990) Influence of organic matter on phosphate adsorption by aluminium and iron oxides in sandy soils. *J. Soil Sci.* **41**, 443–449.
Brattebo, H. and Odegaard, H. (1986) Phosphorus removal by granular activated alumina. *Water Res.* **20**, 977–986.
Bray, R.H. and Kurtz, L.T. (1945) Determination of total, organic and available forms of phosphorus in soils. *Soil Sci.* **59**, 39–45.
Brett, S., Guy, J., Morse, G.K. and Lester, J.N. (1997) Phosphorus removal and recovery technologies. *Centre European D'Etudes des Polyphosphates E.V.* 142 pp.
Brooks, A.S., Rozenwald, M.N., Geohring, L.D., Lion, L.W. and Steenhuis, T.S. (2000) Phosphorus removal by wollastonite: a constructed wetland substrate. *Ecol. Eng.* **15**, 121–132.
Byrne, R.H. and Kim, K.H. (1993) Rare-earth precipitation and coprecipitation behavior – the limiting role of PO_4^{3-} on dissolved rare-earth concentrations in seawater. *Geochim. Cosmochim. Acta* **57**(3), 519–526.
Cavani, F., Trifiro, F and Vaccari, A. (1991) Hydrotalcite-type anionic clays: preparation, properties and applications. *Catalysis Today* **11**, 173–301.
Cheung, K.C., Venkitachalam, T.H. and Scott, W.D. (1994) Selecting soil amendment materials for removal of phosphorus. *Water Sci. Technol.* **30**, 247–256.
Cheung K.C. and Venkitachalam, T.H. (2000) Improving phosphate removal of sand infiltration using alkaline fly ash. *Chemosphere* **41**, 243–249.
Clark, C.J. and McBride, M.B. (1984) Chemisorption of Cu(II) and Co(II) on allophane and imogolite. *Clay Clay Min.* **32**, 300–310.
Cohen, J.M. (1971) Nutrient removal from wastewater by physical-chemical processes. *U.S. EPA, Office Res. Monitor. Adv. Waste Treatment Res. Lab.* 49 pp.
Degens, B.P., Schipper, L.A., Claydon, J.J., Russell., J.M. and Yeates, G.W. (2000) Irrigation of allophanic soil with dairy effluent for 22 years: responses of nutrient storage and soil biota. *Aust. J. Soil Res.* **38**, 25–35.
Denaix, L., Lamy, I. and Bottero, J.Y. (1999) Structure and affinity towards Cd^{2+}, Cu^{2+}, Pb^{2+} of synthetic colloidal amorphous aluminosilicates and their precursors. *Colloid. Surf. A* **158**, 315–325.
Diatloff, E., Asher, C.J. and Smith, P.W. (1993) Use of GEOCHEM-PC to predict rare earth element (REE) species in nutrient solutions. *Plant Soil* **155/6**, 251–254.
Dittrich, M., Dittrich, T., Sieber, I. and Koschel, R. (1997) A balance analysis of phosphorus elimination by artificial calcite precipitation in stratified hardwater lake. *Wat. Res.* **31**(2), 237–248.
Donnert, D. (1988) Investigations on phosphorus removal from waste water. *Progress report KfK 4459, Kernforschungszezsntrum Karlsruhe.* 96 pp.
Donnert, D. and Salecker, M. (1998) Elimination of Phosphorus from wastewater. *Proc. Water Environ. Fed. Technol. Conf. Exhib.* Singapore, 201–207.
Douglas G.B. (1997) Remediation material and remediation process for sediments. International PCT phase entry PC/AU97/00892, 30 December, 1997.
Douglas, G.B. and Adeney, J.A. (2001) 2000 Canning River Phoslock™ trial. *Confidential report prepared for Water and Rivers Commission. CSIRO Land and Water* Report January, 2001, 75 pp.
Douglas, G.B., Adeney, J.A. and Zappia, L.R. (2000) Sediment remediation project: Laboratory trials 1998/9. *Confidential report prepared for Water and Rivers Commission. CSIRO Land and Water* Report 6/00, 92 pp.

Douglas, G.B., Coad, D.N. and Adeney, J.A. (1997) Sediment Remediation Project: Laboratory and field trials. *Confidential report prepared for Water and Rivers Commission. CSIRO Land and Water* Report (97-58), 86 pp.

Drizo, A., Frost, C.A., Grace, J. and Smith, K.A. (1999) Physico-chemical screening of phosphate-removing substrates for use in constructed wetland systems. *Water Res.* **33**, 3595–3602.

Drizo, A., Comeau, Y., Forget, C. and Chapuis, R.P. (2002) Phosphorus saturation potential: A parameter for estimating the longevity of constructed wetland systems. *Environ. Sci. Technol.* **36**, 4642–4648.

Dudas, M.J. and Warren, C.J. (1988) Submicroscopic structure and characteristics of intermediate-calcium fly ashes. *Mater. Res. Soc. Symp. Proc.* **113**, 309–316.

Dunst, R.C. (1982) Sediment problems and lake restoration in Wisconsin. *Environ. Int.* **7**, 87–92.

Effler, S.W. and Driscoll, C.T. (1985) Calcium chemistry and deposition in ionically enriched Onondaga Lake, New-York. *Environ. Sci. Technol.* **19**(8), 716–720.

Evans, L.J. and Smillie, G.W. (1976) Extractable iron and aluminium and their relationship to phosphorus retention in Irish soils. *Ir. J. Agr. Res.* **15**, 65–73.

Farmer, V.C., Adams, M.J., Fraser, A.R. and Palmieri, F. (1983) Synthetic imogolite: properties, synthesis and possible applications. *Clay Min.* **18**, 459–472.

Farmer, V.C. and Russell, J.D. (1990) Structure and genesis of allophanes and imogolite and their distribution in non-volcanic soils. In *Soil Colloids and their Associations in Aggregates* (eds De Boodt *et al.*), pp. 165–177.

Feitknecht, W. von (1942) Uber die Bildung von Doppelhydroxen zwischewn zwei- und dreiwertigen Metallen. *Helv. Chim. Acta* **25**, 555–569.

Firsching, F.H. and Brune, S.N. (1991) Solubility products of the trivalent rare-earth phosphates. *J. Chem. Eng. Data* **36**, 93–95.

Fine, L.O. and Jensen, W.P. (1981) Phosphate in waters: I. Reduction using northern lignite fly ash. *Water Res. Bull.* **17**, 895–897.

Fleming, H.L. (1986) Adsorption with aluminas in systems with competing waters. *Int. Conf. Fundamentals Adsorption. Santa Barbara Calif. U.S.A.* 29 pp.

Fordham, A.W. and Norrish, K. (1974) Direct measurment of the composition of soil components which retain added arsenate. *Aust. J. Soil Res.* **12**, 165–172.

Forget, C., Drizo, A., Comeau, Y. and Chapuis, C.P. (2002) Phosphorus removal by steel slag and serpentinite. *Water Res.* (submitted).

Freeman, J.S. and Rowell, D.L. (1981) The adsorption and precipitation of phosphate onto calcite. *J. Soil Sci.* **32**, 75–84.

Froelich, P.N. (1988) Kinetic control of dissolved phosphate in natural rivers: a primer on the phosphate buffer mechanism. *Limnol. Oceanogr.* **33**, 649–668.

Fontes, M.P.F. and Weed, S.B. (1996) Phosphate adsorption by clays from Brazillian Oxisols: relationships with specific surface area and mineralogy. *Geoderma* **72**, 37–51.

Frondel, C. (1941) Constitution and polymorphism of the pyroauritic and sjogrenite groups. *Am. Mineral.* **26**, 295–306.

Frossard, E., Brossard, M., Hedley, M.J. and Metherall, A. (1995) Reactions controlling the cycling of P in soils. In *Phosphorus in the Global Environment* (ed. Tiessen, H.), pp. 107–137.

Gahler, A.R., Sanville, W.D., Searcy, J.A., Powers, C.F. and Miller, W.E. (1969) *Studies on Lake Restoration by Phosphorus Inactivation*. Environ. Prot. Agency, Nat. Eutroph. Prog. Corvallis, Oregon.

Gastuche, M.C., Brown, G. and Mortland, M.M. (1967) Mixed magnesium aluminium hydroxides. I. Preparation and characterization of compounds form dialised systems. *Clay Min.* **7**, 177.

Gangoli, N. and Thodos, G. (1973) Phosphate adsorption studies. *J. Water Pollut. Control Fed.* **45**, 842–849.

Gilkes, R.J., Mosquera-Pardo, A.C., Newsome, D. and Watson, G. (1992) Iron oxide wastes from synthetic rutile manufacture: a comparison with bauxite red mud as ameliorants for P-leaching sandy soils. In *An International Bauxite Tailings Workshop, 2–6 Nov, 1992, Perth, WA*, pp. 346–354.

Goldberg, S. and Sposito, G. (1985) On the mechanism of specific phosphate adsorption by hydroxylated mineral surfaces: a review. *Commun. Soil Sci. Plant Anal.* **16**, 801–821.

Gray, S., Kinross, J., Read, P. and Marland, A. (2000) The nutrient assimilative capacity of maerl as a substrate in constructed wetland systems for waste treatment. *Water Res.* **34**, 2183–2190.

Grubb, D.G., Guimaraes, M.S. and Valencia, R. (2000) Phosphate immobilization using an acidic type F fly ash. *J. Hazardous Mater.* **76**, 217–236.

Gustafsson, J.P. (2001) The surface chemistry of imogolite. *Clay Clay Min.* **49**, 73–80.

Hampton, T.K. (1974) Evaluation of the effects of fly ash and lake restoration by fly ash on aquatic consumer organisms. *MS Thesis, Univ. Notre Dame, Indianna, USA*.

Hano, T., Takanashi, H., Hirata, M., Urano, K. and Eto, S. (1997) Removal of phosphorus from wastewater by activated alumina adsorbent. *Water Sci. Technol.* **35**, 39–46.

Haron, M.J., Wasay, S.A. and Tokunaga, S. (1997) Preparation of basic yttrium carbonate for phosphate removal. *Water Environ. Res.* **69**, 1047–1051.

Hartikainen, H. (1989) Effects of cation species on the desorption of phosphorus in soils treated with carbonate. *Z. Pflanzenernahr. Biodenk.* **152**, 435–439.

He, J.Z., Gilkes, R.J. and Dimmock, G.M. (1998) Mineralogical properties of sandy podzols on the Swan Coastal Plain, south-west Australia, and the effects of drying on their phosphate sorption characteristics. *Aust. J. Soil Res.* **36**, 395–409.

Higgins, B.P.J., Mohleji, S.C. and Irvine, R.L. (1976) Lake treatment with fly ash, lime and gypsum. *J. Water Pollut. Control Fed.* **48**, 2153–2164.

Ho, G. and Matthew, K. (1993) Phosphorus removal using bauxite refining residue (red mud). *Aust. Water and Waste water Assoc. 15th Fed. Conv.* pp. 607–613.

Ho, G. and Monk, R. (1988) Adsorption of phosphate by crushed limestone. *Draft proposal to Environmental Protection Authority of West. Aust.* 7 pp. Murdoch University.

Horikawa, Y. (1975) Electrokinetic phenomena of aqueous suspensions of allophane and imogolite. *Clay Sci.* **4**, 255–263.

House, W.A. and Donaldson, L. (1986) Adsorption and coprecipitation of phosphate on calcite. *J. Colloid Interf. Sci.* **112**, 309–324.

Hsu, P.H. (1989) Aluminum Oxides and Oxyhydroxides, In *Miner. Soil Environ.* (ed. Dixon, J.B. and Weed, S.B.), Ch. 7, pp. 331–378.

Hylander, L.D. and Siman, G. (2001) Plant availability of phosphorus sorbed to potential wastewater treatment materials. *Biol. Fertil. Soils* **34**, 42–48.

Imai, H., Goulding, K.W.T. and Talibudeen, O. (1981) Phosphate adsorption in allophanic soils. *J. Soil Sci.* **32**, 555–570.

Jack, P.N. and Platell, N. (1983) Phosphorus adsorption by limestone. Potential for Harvey River inputs to Harvey Estuary. *West. Aust. Govt. Chem. Labs.* 7 pp.

Jeffery, R.C. (1996) The chemical properties of bauxite residues with particular emphasis on their capacities to retain phosphorus and nitrogen and to act as liming materials. *Report prepared for Alcoa of Australia Limited*.

Jenssen, P.D. and Krogstad, T. (1988) Particles found in clogging layers of wastewater infiltration systems may cause reduction in infiltration rate and enhance phosphorus adsorption. *Water Sci. Technol.* **20**, 251–253.

Jenssen, P.D., Krogstad, T., Briseid, T., Norgaard, E. (1991) Testing of reactive filter media (LECA) for use in agriculture drainage systems. *International Seminar of Technical Section of CIGR on Environmental Challenges and Solutions in Agricultural Engineering. As-NLH*: pp. 160–166.

Jenssen, P.D., Maehulm, T. and Zhu, T. (1996) Construction and performance of subsurface flow constructed wetlands in Norway. *Proc. Symp. Constr. Wetlands in Cold Climates. Niagara, Ontario*.

Johansson, L. (1999a) Industrial by-products and natural substrata as phosphorus adsorbents. *Environ. Technol.* **20**, 309–316.

Johansson, L. (1999b) Blast furnace slag as phosphorus sorbents – column studies. *Sci. Total Environ.* **229**, 89–97.

Johansson, L. and Gustaffson, J.P. (2000) Phosphate removal using blast furnace slags and opoka-mechanisms. *Water Res.* **34**, 259–265.

Johansson, L. and Hylander, L. (1999) Phosphorus removal from waste water by filter media: retention and estimated plant availability of sorbed nutrients. *Pol. Acad. Sci.* **456**, 397–409.

Johansson, L., Renman, G. and Carlstrom, H. (1995) Light expanded clay aggregates (LECA) as reactive filter medium in constructed wetlands. *Proc. Ecotechnics 95 – Intl. Symp. Ecol. Eng.* pp. 214–224.

Jonasson, R.G., Bancroft, G.M. and Nesbitt, H.W. (1985) Solubilities of some hydrous REE phosphates with implications for diagenesis and seawater concentrations. *Geochim. Cosmochim. Acta* **49**, 2133–2139.

Jones, J.P., Singh, B.B., Fosberg, M.A. and Falen, A.L. (1979) Physical, chemical and mineralogical characteristics of soils from volcanic ash in northern Idaho: II. Phosphorus sorption. *Soil Sci. Soc. Am. J.* **43**, 547–552.

Kadlec, R.H. and Knight, R.L. (1996) *Treatment Wetlands*. Lewis Publishers, Boca Raton, NY.

Kadlec, R.H., Knight, R., Vymazal, J., Brix, H., Cooper, P. and Haberl, R. (2000) *Constructed Wetlands for Pollution Control*. Published by International Water Association, London, UK.

Kaneko, S. and Nakajimi, K. (1988) Phosphorus removal by crystallization using a granular activated magnesia clinker. *J. Water Pollut. Control Fed.* **60**, 1239–1244.

Kofinas, P. and Kioussis, D.R. (2003) Reactive phosphorus removal from aquaculture and poultry productions systems using polymeric hydrogels. *Environ. Sci. Technol.* **37**, 423–427.

Kyker, G.C. and Cress, E.A. (1957) Acute toxicity of yttrium, lanthanum and other rare earths. *AMA Arch. Ind. Health* **16**, 475.

Lindsay, W.L. and Moreno, E.C. (1960) Phosphate equilibria in soils. *Soil Sci. Soc. Am. Proc.* **24**, 177–182.

Liu, X. and Byrne, R.H. (1997) Rare earth and yttrium phosphate solubilities in aqueous solution. *Geochim. Cosmochim. Acta* **61**, 1625–1633.

Maehulm, T. (1998) Cold-climate constructed wetlands: Aerobic pre-treatment and horizontal subsurface flow systems for domestic sewage and landfill leachate purification. *PhD thesis, Agr. Univ. Norway. (unpubl.)*.

Maehulm, T., Jenssen, P.D. and Warner, W. (1995) Cold-climate constructed wetlands. *Water Sci. Technol.* **32**, 95–101.

Mann, R.A. (1996) Phosphorus removal by constructed wetlands: Substratum adsorption. PhD Thesis. *Faculty Sci. Technol. Univ. West. Sydney (Unpubl.)*.

Mann, R.A. (1997) Phosphorus adsorption and desorption characteristics of constructed wetland gravels and steelworks by-products. *Aust. J. Soil Res.* **35**, 375–384.

Mann, R.A. and Bavor, H.J. (1993) Phosphorus removal in constructed wetlands using gravel and industrial waste substrata. *Water Sci. Technol.* **27**, 107–113.

McLaughlin, J.R., Ryden, J.C. and Syers, J.K. (1981) Sorption of inorganic phosphate by iron and aluminium-containing compounds. *J. Soil Sci.* **32**, 365–377.

McNabb, W.M., Hazel, J.F. and Baxter, R.A. (1968) The reactions of lanthanum with the straight chain polyphosphates. *J. Inorg. Nucl. Chem.* **30**, 1585–1593.

Melnyk, P.B., Norman, J.D. and Wasserlauf, M. (1974) Lanthanum precipitation: An alternative method for removing phosphates from wastewater. *Proc. 11th Rare Earth Res. Conf.* **1**, 4–13.

Melnyk, P.B. (1975) Precipitation of phosphates in sewage with lanthanum: an experimental and modelling study. *PhD thesis*, McMaster Univ. Hamilton, Ontario.

Millero, F., Huang, F., Zhu, X., Liu, X. and Zhang, J.Z. (2001) Adsorption and desorption of phosphate on calcite and aragonite in seawater. *Aquat. Geochem.* **7**, 33–56.

Mills, D. (2000) Treatment of swimming pool water. *U.S. Patent no.* 6,146,539.

Misra, C. and Perrotta, A.J. (1992) Composition and properties of synthetic hydrotalcites. *Clay Clay Min.* **40**, 145–150.

Miyata, S. (1980) Physico-chemical properties of synthetic hydrotalcites in relation to composition. *Clay Clay Min.* **28**, 50–56.

Miyata, S. (1983) Anion-exchange properties of hydrotalcite-like compounds. *Clay Clay Min.* **31**, 305–311.

Mondragon, F., Rincon, F., Sierra, L., Escobar, J., Ramirez, J. and Fernandez, J. (1990) New perspectives for coal ash utilisation: synthesis of zeolitic materials. *Fuel* **69**, 263.

Moreno, E.C., Brown, W.E. and Osborn, G. (1960) Stability of dicalcium phosphate dihydrate in aqueous solutions and solubility of octocalcium phosphate. *Soil Sci. Soc. Am. Proc.* **24**, 99–102.

Morse, G.K., Brett, S.W., Guy, J.A. and Lester, J.N. (1998) Review: phosphorus removal and recovery technologies. *Sci. Total Environ.* **212**, 69–81.

Mortimer, C.H. (1941) The exchange of dissolved substances between water and mud in lakes (Parts I and II). *J. Ecol.* **30**, 147–201.

Narkis, N. and Meiri, M. (1981) Phosphorus removal by activated alumina. *Environ. Pollut.* **2**, 327–343.

Neufeld, R.D. and Thodos, G. (1969) Removal of orthophosphates from aqueous solutions with activated alumina. *Environ. Sci. Technol.* **3**, 661–667.

NICNAS (2001) Lanthanum Modified Clay (Phoslock™). National Industrial Chemical Notification and Assessment Scheme Full Public Report NA/899, 28 pp.

Norrish, K. and Rosser, H. (1983) Mineral Phosphates. In *Soils: An Australian viewpoint* (ed. CSIRO Div. Soils), pp. 335–361.

Oobuku, A., Ooi, K. and Hayashi, H. (1993) Preparation and phosphate ion-exchange properties of a hydrotalcite-like compound. *Langmuir* **9**, 1418–1422.

Otsuki, A. and Wetzel, R.G. (1972) Coprecipitation of phosphate with carbonates in a marl lake. *Limnol. Oceanogr.* **17**, 763–767.

Pant, H.K., Reddy, K.R. and Lemon, E. (2001) Phosphorus retention capacity of root media of sub-surface flow constructed wetlands. *Ecol. Eng.* **17**, 345–355.

Parfitt, R.L. (1989) Phosphate reactions with natural allophane, ferrihydrite and goethite. *J. Soil. Sci.* **40**, 359–369.

Parfitt, R.L. (1990) Allophane in New Zealand – a review. *Aust. J. Soil. Res.* **28**, 343–360.

Parfitt, R.L., Fraser, A.R. and Farmer, V.C. (1977) Adsorption on hydrous oxides. III. Fulvic and humic acid on goethite, gibbsite and imogolite. *J. Soil Sci.* **28**, 289–296.

Parfitt, R.L., Hart, P.B.S., Meyrick, K.F. and Russell, M. (1982) Response of ryegrass and white clover to phosphorus on an allophonic soil, Egmont black loam. *NZJ. Agr. Res.* **25**, 549–555.

Parfitt, R.L., Hume, L.J. and Sparling, G.P. (1989) Loss of availability of phosphorus in New Zealand soils. *J. Soil Sci.* **40**, 371–382.

Peterson, S.A., Sanville, W.D., Stay, F.S. and Powers, C.F. (1974) Nutrient Inactivation as a Lake Restoration Procedure. Laboratory Investigations. *U.S. E.P.A. Ecol. Res. Ser.* 660/3-74-032. 118 pp.

Peterson, S.A., Sanville, W.D., Stay, F.S. and Powers, C.F. (1976) Laboratory evaluation of nutrient inactivation compounds for lake restoration. *Lake Restor.* **48**, 817–831.

Potter, H.A.B. and Yong, R.N. (1999) Influence of iron/aluminium ratio on the retention of lead and copper by amorphous iron-aluminium oxides. *Appl. Clay Sci.* **14**(1–3), 1–26.

Proctor, D.M., Fehling, K.A., Shay, E.C., Wittenborn, J.L., Green, J.J., Avent, C., Bigham, R.D., Connolly, M., Lee, B., Shepker, T.O. and Zak, M.A. (2000) Physical and chemical characteristics of blast furnace, basic oxygen furnace, and electric arc furnace steel industry slags. *Environ. Sci. Technol.* **34**, 1576–1582.

Rajan, S.S.S. (1975) Mechanism of phosphate adsorption by phosphate clays. *NZJ. Soil Sci.* **18**, 93–101.

Reardon, E.J. and Della Valle, S. (1997) Anion sequestering by the formation of anionic clays: lime treatment of fly ash slurries. *Environ. Sci. Technol.* **31**, 1218–1223.

Recht, H.L., Ghassemi, M. and Kleber, E.V. (1970) Precipitation of phosphates from water and wastewaters using lanthanum salts. *Proc. Adv. Water Pollut. Res. 5th Intl. Conf.* **1**, 1–17.

Reynolds, C.S. (1999) Non-determinism to probability or N:P in the community ecology of phytoplankton. *Arch. Hydrobiol.* **146**, 25–35.

Roques, H., Nugroho-Juedy, L. and Lebugle, A. (1991) Phosphorus removal from wastewater by half-burned dolomite. *Water Res.* **25**, 959–965.

Ross, C.S. and Kerr, P.F. (1934) Halloysite and allophane. *U.S. Geol. Surv. Prof. Paper*, *185G*, 135–148.

Ryan, J., Curtin, D. and Cheema, M.A. (1985) Significance of iron oxides and calcium carbonate particle size in phosphate sorption by calcareous soils. *Soil Sci. Am. J.* **49**, 74–76.

Sakadevan, K. and Bavor, H.J. (1998) Phosphate adsorption characteristics of soils, slags and zeolite to be used as substrates in constructed wetland systems. *Water Res.* **32**, 393–399.

Sato, T., Wakabayashi, T. and Shimada, M. (1986) Adsorption of various anions by magnesium aluminium oxide. *Ind. Eng. Chem. Prod. Res.* **25**, 89–92.

Sawhney, B.L. and Hill, D.E. (1975) Phosphate sorption characteristics of soils treated with domestic waste water. *J. Environ. Qual.* **4**, 342–346.

Sax, N.I., Feiner, B., Fitzgerald, J.J., Haley, T.J. and Weisberger, E.K. (1984) In *Dangerous Properties of Industrial Minerals. Van Nostrand Reinhold* (ed. Sax, N.I.), 3124 pp.

Scheetz, B.E. and Earle, R. (1998) Utilization of fly ash. *Curr. Op. Sol. State Mater. Sci.* **3**, 510–520.

Schwab, G. (1949) Nature and applications of inorganic alumina chromatography. *Discuss. Faraday Soc.* **7**, 170.

Schwertmann, U. and Taylor, R.M. (1989) Iron Oxides, In *Miner. Soil Environ.* (eds. Dixon, J.B. and Weed, S.B.), Ch. 8, pp. 379–428.

Seida, Y. and Nakano, Y. (2000) Removal of humic substances by layered double hydroxide containing iron. *Water Res.* **34**, 1487–1494.

Seida, Y. and Nakano, Y. (2002) Removal of phosphate by layered double hydroxides containing iron. *Water Res.* **36**, 1306–1312.

Shiao, S.J. and Akashi, K. (1977) Phosphate removal from aqueous solutions on activated red mud. *J. Water Pollut. Control Fed.* **49**, 280–285.

Shin, H.S., Kim, M.J., Nam, S.Y. and Moon, H.C. (1996) Phosphorus removal by hydrotalcite compounds (HTLcs). *Water Sci. Technol.* **34**, 161–168.

Sims, J.T. and Ellis, B.G. (1983) Adsorption and availability of phosphorus following the application of limestone to an acid, aluminous soil. *Soil Sci. Am. J.* **47**, 888–893.

Singer, A. and Berkgaut, V. (1995) Cation exchange properties of hydrothermally treated coal fly ash. *Environ. Sci. Technol.* **29**, 1748–1753.

Sinha, M.K. (1971) Organo-metallic phosphates. I. Interaction of phosphorus compounds with humic substances. *Plant Soil* **35**, 471–484.

Stauber, J.L. (1999) Further toxicity testing of modified clay leachates using freshwater and marine organisms. *CSIRO Energy Technology Investigation Report CET/IR202R*, 22 pp.

Stout, W.L., Sharpley, A.N., Gburek, W.J. and Pionke, H.B. (1999) Reducing phosphorus export from croplands with FBC fly ash and FGD gypsum. *Fuel* **78**, 175–178.

Su, C. and Harsh, J.B. (1993) The electrophoretric mobility of imogolite and allophane in the presence of inorganic ions and citrate. *Clays Clay Min.* **41**, 461–471.

Summers, R.N., Guise, N.R. and Smirk, D.D. (1993) Bauxite residue (red mud) increases phosphorus retention in sandy soil catchments in Western Australia. *Fert. Res.* **34**, 85–94.

Summers, R.N., Guise, N.R., Smirk, D.D. and Summers, K.J. (1996b) Bauxite residue (red mud) improves pasture growth on sandy soils in Western Australia. *Aust. J. Soil Res.* **34**, 555–567.

Summers, R.N., Smirk, D.D. and Karafilis, D. (1996a) Phosphorus retention and leachates from sandy soil amended with bauxite residue (red mud). *Aust. J. Soil Res.* **34**, 569–581.

Suzuki, M. and Fuji, T. (1988) Simultaneous removal of phosphate and ammonium ions from wastewater by composite adsorbent. *Proc. Water Pollut. Control in Asia* pp. 239–245.

Taylor, R.M. and McKenzie, R.M. (1980) The influence of aluminium and iron oxides VI. The formation of Fe(II)-Al(III) hydroxy-chlorides, -sulphates, and -carbonates as new members of the pyroaurite group and their significance in soils. *Clay Clay Min.* **28**, 179–187.

Taylor, R.M. (1984) The rapid formation of crystalline double hydroxy salts and other compounds by controlled hydrolysis. *Clay Min.* **19**, 591–603.

Takeuchi, M., and Komada, M. (1998) Phosphorus removal from hoggery sewage using natural calcium carbonate. *Jap. Agr. Res. Qtly.* **32**, 23–30.

Tejedor-Tejedor, M.A. and Anderson, M.A. (1990) Protonation of phosphate on the surface of goethite as studied by CIR-FTIR and electrophoretic mobility. *Langmuir* **6**, 602–611.

Tenny, M.W. and Echelburger Jr., W.F. (1970) Fly ash utilization in the treatment of polluted waters. *Bur. of Mines Inf. Circ. 8488, Ash Util. Proc.* 237–265.

Theis, T.L. (1975) The potential trace metal contamination of water resources through the disposal of fly ash. *Proc. 2nd. Nat. Conf. Complete Water Reuse, U. S. EPA.*

Theis, T.L. and McCabe, P.J. (1978) Retardation of sediment phosphorus release by fly ash application. *J. Water Pollut. Control Fed.* 2666–2676.

Theis, T.L. and De Pinto, J.V. (1976) Studies on the reclamation of Stone Lake, Michigan. *Corvallia Env. Res. Lab., Office of Res. and Devel., US EPA. Grant No. R-801245.* 84 pp.

Theng, B.K.G., Russell, M., Churchman, G.J. and Parfitt, R.L. (1982) Surface properties of allophane, halloysite and imogolite. *Clays Clay Min.* **30**, 143–149.

Thornber, M.R. and Hughes, C.A. (1987) The mineralogical and chemical properties of red mud waste from the Western Australian alumina industry. In *Proc. Int. Conf. Bauxite Tailings*, (ed. Wagh, A.S. and Desai, P.), Kingston, Jamaica. 1–19.

Tofflemire, T.J. and Chen, M. (1977) Phosphate removal by sands and soils. *Groundwater* **15**, 377–387.

Tokunaga, S., Wasay, S.A. and Park, S.W. (1997) Removal of arsenic (V) ion from aqueous solutions by lanthanum compounds. *Water Sci. Technol.* **35**, 71–78.

Torrent, J., Barron, V. and Schwertmann, U. (1990) Phosphate adsorption and desorption by goethites differing in crystal morphology. *Soil Sci. Soc. Am. J.* **54**, 1007–1012.

Van Olphen, H. (1971) Amorphous clay minerals. *Science* **171**, 90–91.

Vinyard, D.L. and Bates, M.H. (1979) High-calcium fly ash for tertiary phosphorus removal. *Water Sewage Wks.* **126**, 62–64.

Vlahos, S., Summers, K.J., Bell, D.T. and Gilkes, R.J. (1989) Reducing phosphorus leaching from sandy soils with red mud bauxite processing residues. *Aust. J. Soil Res.* **27**, 651–662.

Vollenweider, R.A. (1976) Advances in defining critical loads for phosphorus in lake eutrophication. *Mem. Ist. Ital. Idrobiol.* **33**, 53–83.

Vucelic, M., Jones, W. and Moggridge, G.D. (1997) Cation ordering in synthetic layered double hydroxides. *Clays Clay Min.* **45**, 803–813.

Wada, K. (1959) Reaction of phosphate with allophane and halloysite. *Soil Sci.* **87**, 325–330.

Wada, K. (1980) Mineralogical characteristics of Andosols. In *Soils with variable charge. NZ Soil Sci. Soc.* (ed. Theng, B.K.G.), pp. 87–107.

Wada, K. (1989) Allophane and imogolite. In *Miner. Soil Environ.* (eds Dixon, J.B. and Weed, S.B.), pp. 1051–1087.

Weaver, D.M. and Ritchie, G.S.P. (1987) The effectiveness of lime-based amendments and bauxite residues at removing phosphorus from piggery effluent. *Environ. Pollut.* **46**, 163–175.

Winkler, B.F. and Thordos, G. (1971) Kinetics of orthophosphate removal from aqueous solutions by activated alumina. *J. Water Pollut. Control Fed.* **43**, 474 pp.

Wood, A. and Hensman, L.C. (1989) Research to develop engineering guidelines for implementation of constructed wetlands for wastewater treatment in South Africa. In *Constructed Wetlands in Wastewater Treatment* (ed. Hammer, D.), pp. 581–589.

Yamada, H., Kayama, M., Saito, K and Hara, M. (1986) A fundamental research on phosphate removal using slag. *Water Res.* **20**, 547–557.

Yamada, H., Kayama, M., Saito, K and Hara, M. (1987) Suppression of phosphate liberation from sediment by using iron slag. *Water Res.* **21**, 325–333.

Yee, W.C. (1966) Selective removal of mixed phosphates by activated alumina. *J. Am. Water Wks. Assn.* **58**, 239–247.

Yuan, G., Theng, B.K.G., Parfitt, R.L. and Percival, H.J. (2000) Interactions of allophane with humic acid and cations. *Euro. J. Soil Sci.* **51**, 35–41.

Yuan, G. and Lavkulich, L.M. (1994) Phosphate sorption in relation to extractable iron and aluminium in spodosols. *Soil Sci. Soc. Am. J.* **58**, 343–346.

Zhu, T., Jenssen, P.D., Maehlum, T. and Krogstad, T. (1997) Phosphorus sorption and chemical characteristics of lightweight aggregates (LWA) – potential filter media in treatment wetlands. *Water Sci. Technol.* **35**, 103–108.

Zhu, T. (1998) Phosphorus and nitrogen removal in light-weight aggregate (LWA) constructed wetlands and intermittent filter systems. *DSc Thesis, Agr. Univ. Norway.*

14
Removing phosphorus from sewage effluent and agricultural runoff using recovered ochre

*K.V. Heal, K.A. Smith, P.L. Younger,
H. McHaffie and L.C. Batty*

14.1 INTRODUCTION

Widespread closure of deep coal mines in industrialised countries in recent decades has resulted in increased pollution of watercourses by acidic, metal-rich mine drainage following flooding of abandoned mine workings. Schemes for remediation of such polluted mine drainage encourage the precipitation of iron, normally in settlement ponds or constructed wetlands, thereby generating large quantities of 'ochre' (i.e., $Fe(OH)_3$ and $FeO \cdot OH$) (see, for instance, Younger *et al.*, 2002). Typically, this ochre is stockpiled pending use or disposal. Although a number of possible end-uses have been considered (e.g. colouring bricks/cement and in synthesising coagulants for drinking water; Tarasova *et al.*, 2002), no single end-use has yet been identified which could consume the projected future production of ochre in the UK and Europe. In this chapter a further possible

© 2004 IWA Publishing. *Phosphorus in Environmental Technology: Principles and Applications.*
Edited by Eugenia Valsami-Jones. ISBN: 1 84339 001 9

large-scale use for ochre is described: as a low-cost reagent to remove unwanted phosphorus from sewage effluent and agricultural runoff.

Phosphorus pollution from point and diffuse sources is a serious threat to the water environment in the UK and other industrialised countries (D'Arcy et al., 2000). The release of excessive quantities of phosphorus to rivers and lakes from sewage treatment works, septic tanks and agricultural runoff causes eutrophication, frequently resulting in algal blooms, fish kills and loss of water resources (see also Burke et al.; Farmer, this volume). The costs of this form of pollution to the UK water industry are estimated at >15 £M per annum (D'Arcy et al., 2000). This cost is incurred as the direct result of expenditure in the agricultural sector, since farmers purchase fertiliser containing phosphorus to maximise crop yields from agricultural land. Clearly, substantial savings for both the agricultural and water sectors could be realised if ways were found to improve the sustainability of phosphorus use within the environment. While technologies to remove and recover phosphorus from waste waters have been developed since the 1950s (Morse et al., 1998), existing methods are prohibitively costly for many potential applications. The use of ochre recovered from mine water treatment systems offers a potentially low-cost alternative, which could both remove excess phosphorus from waste water and result in the creation of a slow-release fertiliser for subsequent agricultural use. In this chapter the potential of ochre for phosphorus removal is demonstrated and examined within two main contexts:

(1) Ochre as a substrate in constructed wetlands receiving sewage and agricultural drainage;
(2) Ochre in instream filter units in rivers with elevated phosphorus concentrations or in dosing systems in agricultural drainage ditches.

The chapter closes with comments on ongoing research on these topics.

14.2 FORMATION AND PROPERTIES OF OCHRE

14.2.1 Formation

The flooding of abandoned mines frequently results in the formation of acidic, ferruginous water due to the oxidation of pyrite in the mine workings, according to the simplified reaction:

$$FeS_{2(s)} + \frac{15}{2} O_{2(g)} + \frac{7}{2} H_2O \longrightarrow Fe(OH)_{3(s)} + 2SO_4^{2-} + 4H^+ \quad (14.1)$$

To prevent the pollution of surface watercourses with mine drainage, mine water treatment plants (MWTPs) are employed to treat the most serious discharges (Younger et al., 2002). In MWTPs, the key steps are oxidation of reduced Fe (II) to the Fe (III) form, and hydrolysis of the Fe (III) to form ferric hydroxide:

$$Fe^{3+} + 3H_2O \longrightarrow Fe(OH)_{3(s)} + 3H^+ \quad (14.2)$$

Removing phosphorus from sewage effluent and agricultural runoff 323

Figure 14.1 Ochre production by mine water treatment in the UK. (a) A high-density sludge active treatment works at Wheal Jane, Cornwall (photo courtesy of Unipure Europe Ltd). (b) An aerobic wetland treatment system at St Helen Auckland, Co Durham. (c) Ochre drying beds, for ochre recovered from sedimentation tanks at the Coal Authority's Woolley Pumping Station, West Yorkshire.

In active treatment plants, these processes are enhanced by the addition of chemicals (e.g. oxidising reagents, alkalis to raise pH and increase the rate of Fe (II) oxidation, and flocculants to assist floc formation and sedimentation). In both active and passive treatment plants, atmospheric oxidation and precipitation processes are harnessed in sedimentation tanks and/or constructed wetlands with long retention times (Figure 14.1). By these means, MWTPs accumulate large quantities of $Fe(OH)_3$ and $FeO \cdot OH$ precipitate, collectively known as 'ochre' (in the order of tens of tonnes per annum at a single site).

14.2.2 Properties

Ochre precipitated in most MWTPs has a very high water content (80–95%) (Younger *et al.*, 2002). Somewhat lower values (65–70%) can be achieved by application of high-density sludge circuits (Figure 14.1.a) and/or by air-drying of ochre for long periods (Figure 14.1.c). Since ochre which has not been air-dried is difficult to handle and transport, investigations of its phosphorus removal properties in the work reported here used the air-dried form. Selected chemical and

Table 14.1 Chemical and physical properties of air-dried ochres from two MWTPs, Scotland (Bozika, 2001).

	Polkemmet	Minto
pH (in distilled water)	7.2	6.9
% Fe[1]	65 ± 0.5	67.5 ± 3
% Al[1]	0.7 ± 0.02	0.1 ± 0.01
% Mg[1]	0.6 ± 0.01	0.8 ± 0.04
% Ca[1]	7.0 ± 0.1	11.8 ± 0.4
Dry bulk density (g cm^{-3})	1.8	0.8
Saturated hydraulic conductivity (m day^{-1})[2]	26–32	0.7–1.7

[1] Mean of triplicate samples ± standard error. Determined by atomic absorption spectrophotometry of acid digests (concentrated nitric and hydrochloric acid additions) of ashed samples.
[2] Determined in columns over 32 days using the falling head method and Darcy's law.

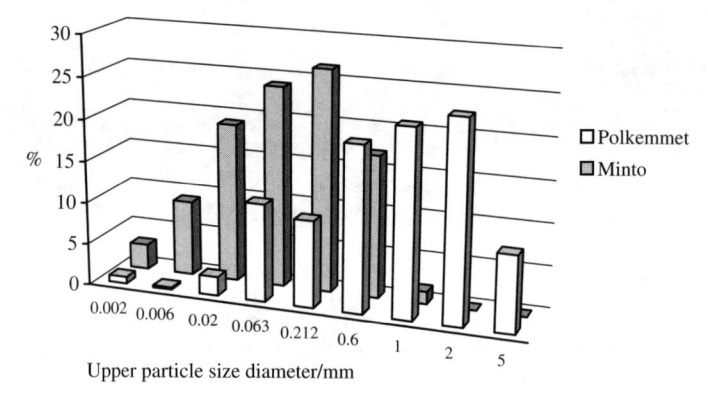

Figure 14.2 Particle size distribution of air-dried ochres from two MWTPs, Scotland (after Bozika, 2001).

physical properties of air-dried ochres from two MWTPs at Polkemmet and Minto in central Scotland, UK, are shown in Table 14.1 and Figure 14.2.

Both ochres have a similar chemical composition and mineralogy (identified by X-ray diffraction to comprise mixtures of poorly crystalline solids, ferrihydrite and goethite (α-FeO·OH)) but very different particle size distributions. Polkemmet ochre dries to a coarse, granular texture, which has a high saturated hydraulic conductivity (equivalent to coarse sand). In contrast, Minto ochre dries to a fine powder with a considerably lower saturated hydraulic conductivity. The cause of the different physical properties of the two ochres is unclear, but is thought to be related to differences in the operation of the MWTPs. At Polkemmet, hydrogen peroxide and a polymer are added to the mine water to encourage oxidation and flocculation of iron, whereas at Minto the mine water is unamended. Different physical properties of ochres influence the rate and quantity of phosphorus removal and their suitability for different phosphorus removal applications. Coarse-grained ochres, such as Polkemmet, are more suitable for phosphorus removal in filter units or in the substrate of constructed wetlands. Fine-grained ochres, such as

Minto, are difficult to contain and cause filter units to clog rapidly, but remove phosphorus rapidly from waste water because of their larger surface areas. Such ochres are more suitable for dosing applications to waters with undesirable phosphorus concentrations as long as adequate sedimentation space is provided.

14.3 CAPACITY OF OCHRE FOR PHOSPHORUS REMOVAL

14.3.1 Mechanisms of phosphorus removal

There is a good general understanding of the mechanisms of phosphorus sorption and desorption by natural soils and sediments. The predominant mechanism of phosphorus removal from water by ochre is sorption onto iron and aluminium oxides and hydroxides and calcium carbonate. Phosphorus sorption to natural goethite initially involves rapid ligand exchange with surface OH^- groups at very reactive sites and the formation of a binuclear bridging complex between a phosphate group and two surface Fe atoms (Parfitt, 1989). Weaker ligand exchange follows with less reactive sites, and finally over time there is a slower penetration of phosphate into the solid matrix via defect sites and pores (Barrow, 1983; Parfitt, 1989). The sorption of phosphate by iron oxides depends on the number of reactive surface FeOH groups and the crystallinity of the oxides (Parfitt, 1989). In acid soils phosphorus sorption is governed by the presence of amorphous hydroxides of Al and Fe(III), and by Ca and Mg in alkaline soils (Reddy et al., 1999). The sorption and desorption of phosphorus displays hysteresis, as sorbed phosphorus diffuses into the solid phase and is not re-released readily (perhaps due to *in situ* crystallisation of ferric phosphate). The long-term storage of phosphorus by ochre is thus governed by two factors: (1) the phosphorus retention capacity – the maximum number of sites available for phosphorus sorption, and (2) the phosphorus buffer intensity or strength of adsorption (Reddy et al., 1999). These factors are in turn controlled by the physicochemical properties of the sediment and the redox conditions.

Phosphorus removal from waste water by ochre may also occur by precipitation, although this is believed to be quantitatively less significant than removal by sorption. High calcium concentrations in ochre and waste water, and slightly alkaline conditions, are likely to favour the precipitation of calcium phosphate. In more acid conditions, precipitation of phosphate with Fe and Al may become more important (Arias et al., 2001).

14.3.2 Previous investigations

Previous work to examine the use of ferruginous materials for phosphorus removal from waste water has been piecemeal. In the USA, Webster and Wieder (1997) found that the addition of ochre from acid mine drainage to fertilised soils reduced phosphorus concentrations in runoff. In Northern Ireland, Wood and

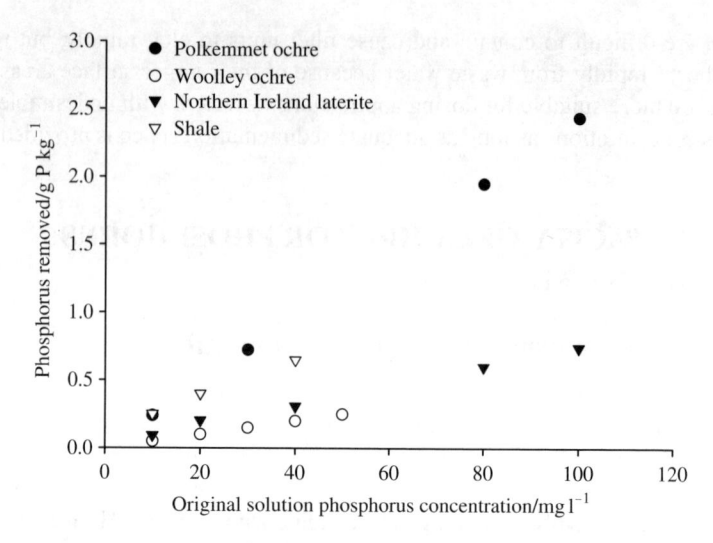

Figure 14.3 Phosphorus removal by four substrates in batch experiments (after McHaffie *et al.*, 2000; Lamont-Black *et al.*, 2001).

McAtamney (1996) showed that the use of laterite as a substrate in mini constructed wetlands removed 95% of phosphorus from landfill leachate. The treatment of dairy farm waste water has also been investigated in bucket-scale subsurface flow constructed wetlands with an iron ore substrate (Grüneberg and Kern, 2001). Other ferruginous media, such as scrap iron, peat doped with bauxite red (Roberge *et al.*, 1999) and sand and olivine coated with iron aluminium hydroxyoxides (Ayoub *et al.*, 2001), have been investigated for phosphorus removal from waste water, but few trials have been conducted at the field scale, and attempts to design novel treatment systems are limited.

14.3.3 Batch experiments

Laboratory batch experiments were used to measure the phosphorus retention characteristics of ochre. In such experiments with artificially phosphorus-enriched water, ochre has a high removal capacity for phosphorus, increasing linearly with solution concentration, compared with other wetland substrates (Figure 14.3). Parallel tests (Lamont-Black *et al.*, 2001) on ochre obtained from the Woolley MWTP ochre drying beds (the site shown in Figure 14.1.c) yielded very similar results (Figure 14.3), which demonstrate that the properties of the Polkemmet ochre are consistent with those of other mine-water derived ochres. Neither the method of ochre preparation (air-dried versus oven-dried at 105°C) nor the temperature at which the experiment was conducted (6°C versus 28°C) affected phosphorus removal (McHaffie *et al.*, 2000).

The Langmuir and Freundlich equations were fitted to the batch experiment results to characterise phosphorus adsorption by Polkemmet and Minto ochres

Table 14.2 Fitting of the Langmuir and Freundlich equations for phosphorus adsorption by Polkemmet and Minto ochres (Bozika, 2001).

Equation	Parameter	Polkemmet	Minto
Langmuir	R^2	0.98 ($p < 0.001$)	0.95 ($p < 0.001$)
	a	0.6	1.0
	b	17.8	21.5
Freundlich	R^2	0.91 ($p < 0.001$)	0.75 ($p < 0.001$)
	K	3.5	7.5
	n	2.1	1.5

(Table 14.2). Although these equations are well-known, it is worth briefly reviewing them to make clear the results reported below. The Langmuir equation was first developed to describe the adsorption of gases by solids and takes the form:

$$\frac{Ce}{x/m} = \frac{1}{ab} + \frac{Ce}{b}, \qquad (14.3)$$

where Ce is the concentration of the sorbate in solution at equilibrium (mg l^{-1}), x/m is the mass of molecules adsorbed per unit weight of material (mg g^{-1}), a is a constant related to the binding strength of molecules onto the material and b is the maximum adsorption capacity of the material (mg g^{-1}). Four assumptions underpin the Langmuir equation (Moore, 1972):

(1) the solid surface contains a fixed number of adsorption sites;
(2) each site can only bind one molecule of the adsorbing species;
(3) the energy of adsorption is the same for all sites and does not depend on the fraction occupied;
(4) there is no interaction between adsorbed molecules on adjacent sites.

The derivation of the Freundlich equation encompasses the heterogeneity of the material surface and the exponential distribution of adsorption sites and adsorption energies. It takes the form:

$$(x/m) = K \cdot Ce^{1/n} \qquad (14.4)$$

where Ce and x/m are defined as before and K and n are constants related to the strength of affinity for the adsorbed molecules by the adsorbent material. Values of n close to 1 indicate that the material has a large adsorptive capacity at high molecule equilibrium concentrations.

Table 14.2 shows that the Langmuir equation is a better fit to the batch experiment results than the Freundlich equation. Minto ochre has higher values of a and b than the Polkemmet ochre, indicating that it binds phosphorus more strongly and has a higher adsorption capacity than the Polkemmet ochre. The K and n values also suggest that the Minto ochre is more effective at removing phosphorus from solution than the Polkemmet ochre. The maximum phosphorus adsorption capacities estimated from the Langmuir equation are smaller than those determined in

Table 14.3 Maximum phosphorus adsorption capacities of different wetland substrates (after Drizo, 1998 and Mann, 1997).

Substrate	Adsorption capacity/mg P g^{-1}
Gravel	0.03–0.05
Bottom ash	0.06
Steel slag	0.38
Blast furnace slag	0.40–0.45
Fly ash	0.62
Shale	0.75
Laterite	0.75
Zeolite	1
Polkemmet ochre	26
Minto ochre	30.5

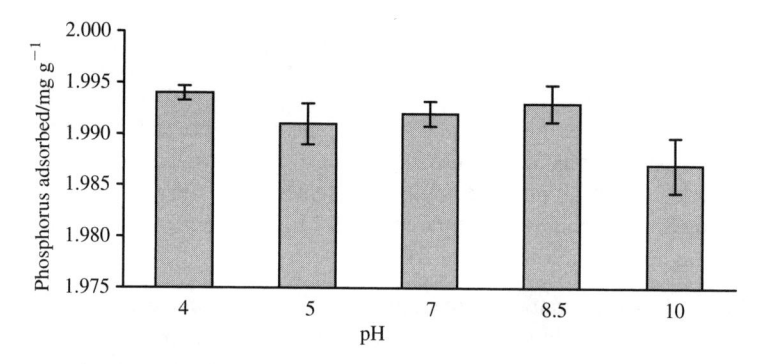

Figure 14.4 Phosphorus uptake by Polkemmet ochre from 100 mg P l^{-1} solutions of different pH in batch experiments (Bozika, 2001). Bars are the mean of triplicate samples ± standard error.

saturation experiments at 26.0 and 30.5 mg g^{-1} for Polkemmet and Minto ochres, respectively (Bozika, 2001). These values for the Polkemmet and Minto ochres are orders of magnitude higher than those measured in other wetland substrates (Table 14.3), suggesting that ochre has the potential to vastly improve the performance of constructed wetlands for phosphorus removal.

14.3.4 Effect of pH

Batch experiments with 100 mg l^{-1} phosphorus solution at pH values of 4, 5, 7, 8.5 and 10 showed that pH has no significant effect on phosphorus removal by Polkemmet and Minto ochres (see Figure 14.4 for Polkemmet ochre), probably due to the buffering capacity of the ochre, although the original study of Bozika (2001) provides no pH measurements at the end of the experiment to corroborate this.

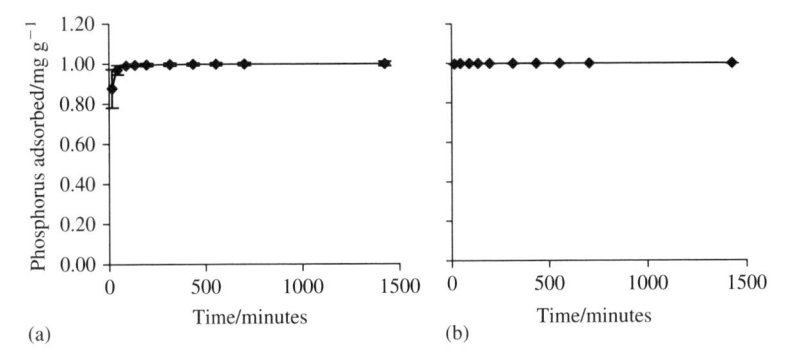

Figure 14.5 Phosphorus uptake by (a) Polkemmet and (b) Minto ochres over time from 50 mg P l^{-1} in batch experiments (Bozika, 2001).

14.3.5 Kinetics of phosphorus removal

Batch experiments (initial ratio of P solution volume to mass of ochre of 20 : 1), in which solution samples were removed at different time intervals during a 24-hour period, show that the removal of phosphorus by Polkemmet and Minto ochres is rapid (Figure 14.5). For the Polkemmet ochre, almost all phosphorus is adsorbed rapidly within the first hour, after which the solution reaches an equilibrium state. The rate of adsorption by Minto ochre is higher than by Polkemmet ochre, due to the larger available surface area for adsorption in the former that results from its fine-grained texture. Experiments in which artificial phosphorus solutions were added to ochre in beakers demonstrated that removal by Polkemmet ochre is still rapid (phosphorus concentrations decreased from 5 to <0.01 mg l^{-1} within eight minutes), even in non-agitated mixtures (McHaffie *et al.*, 2000). These results are supported by independent kinetic tests by Lamont-Black *et al.* (2001) (initial ratio of P solution volume to mass of ochre of 10 : 1), which also showed that the reaction rate is very rapid, with >98% phosphorus removal in five minutes of contact time with ochre.

14.3.6 Long-term phosphorus removal

Long-term phosphorus removal by ochre was examined by pumping artificial phosphorus solution continuously for 9 months onto a gently-angled trough packed with Polkemmet ochre to simulate a horizontal flow filtration system (McHaffie *et al.*, 2000, 2001). The effectiveness of removal remained consistently high (Figure 14.6), with the concentration of applied phosphorus dropping from 20 mg l^{-1} to <1 mg l^{-1} throughout the period. Less than one-third of the ochre in the trough was saturated after 9 months, indicating that it is a suitable medium for long-term phosphorus removal in filter systems. Most of the phosphorus removed by the ochre is unlikely to be readily remobilised since only 0.4% of that adsorbed was found to be in a plant-available form (extraction with Olsen's reagent – 0.5 M NaHCO$_3$ buffered at pH 8.5 (Allen, 1974)).

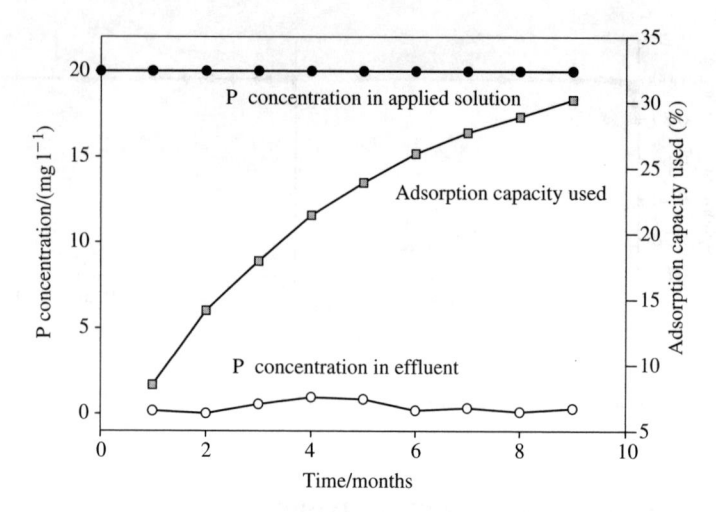

Figure 14.6 Phosphorus removal and % adsorption capacity used in Polkemmet ochre after 9 months operation of trough experiment (after McHaffie et al., 2001).

14.3.7 Desorption of adsorbed phosphorus

In developing applications for ochre-mediated phosphorus removal from waste water, it is important to quantify the magnitude of the re-release of adsorbed phosphorus to ensure that there are no adverse environmental consequences of the treatment process. Desorption experiments, in which 10 mg Polkemmet ochre containing $20\,mg\,P\,g^{-1}$ was shaken with 200 ml tap water for 24 hours, showed that only 2% of the total ochre content was released (Bozika, 2001). Re-release of adsorbed phosphorus into receiving waters does not therefore appear to be of concern, although further verification is required in field conditions.

14.4 FUTURE DEVELOPMENTS

Laboratory investigations of phosphorus removal by ochre demonstrate that it is highly effective in rapidly removing this element from waste waters. A number of potential applications exist for using ochre to treat phosphorus-rich waters, including sewage effluent, agricultural runoff and eutrophic watercourses. Scaling up laboratory findings to full-scale implementation is the next step in providing the scientific basis for using ochre for phosphorus removal from waste water. In addition, some significant issues demand further research before full-scale use of ochre for this purpose can commence.

14.4.1 Resolution of engineering issues

Recent experiences have shown that a major obstacle to the use of ochre for phosphorus-removal purposes is the difficulty of handling the material. Ochres, particularly the Minto ochre, are often very weak compared with most engineering soils and therefore are vulnerable to erosion by running water and wind disturbance, and have poor-loading capacities. If a more granular form can be prepared at reasonable cost, without compromising its phosphorus removal properties, widespread applications become possible. Granulation may be accomplished by mixing different binding agents (e.g. Portland cement and/or lime) and supporting matrices (e.g. coarse ferruginous sludge, fly ash or other waste materials) in various ratios with dried fine-grained ochre which is then wetted and oven-dried to produce granules. Similar methods have been demonstrated to be successful in creating fly ash granules (Zoumis and Calmano, 2000). The properties of granules formed by different methods will require testing (e.g. permeability, compaction characteristics, erosion potential, freeze-thaw and wet-dry susceptibility, batch experiments) in order to establish the durability and phosphorus sorption ability of these materials in the longer term.

14.4.2 Influence of phosphorus form on removal efficiency

The form in which phosphorus occurs in waste waters may affect its removal by ochre. Haygarth *et al.* (1997) showed that reactive phosphorus is associated with different fractions in river water and soil water. In soil leachate and river water in a lowland chalk catchment, the majority of reactive phosphorus occurred in the fraction that passed through a 1000 Dalton ultrafilter, but in unfiltered soil surface runoff most of the reactive phosphorus occurred in particles >0.45 µm in diameter. Most of the investigations to date have used artificial solutions, containing soluble inorganic phosphorus, but it is necessary to examine whether phosphorus removal from raw waste waters is as effective before full-scale field testing of the treatment process.

14.4.3 Treatment of sewage effluent

With stricter controls on sewage discharge into receiving waters (e.g. the EU Urban Wastewater Treatment Directive (Farmer, 2001; Farmer this volume)), the development of new methods for phosphorus removal are required. Constructed wetlands are increasingly used for tertiary treatment of effluent in sewage treatment works and also for sewage treatment in rural communities where it is uneconomic to provide full-scale chemical and biological treatment works. Although constructed wetlands are effective in removing nitrogen from sewage effluent, they are less efficient in removing phosphorus (Cooper *et al.*, 1996). The incorporation of ochre, which has been demonstrated to have a high capacity for phosphorus removal, into the substrate of constructed wetlands is therefore anticipated to improve their treatment efficiency for phosphorus.

The Polkemmet ochre is particularly suitable for such an application as it has a relatively high saturated hydraulic conductivity. The actual amount of phosphorus removed in the long-term trough experiment indicated a potential removal capacity of at least an order of magnitude greater than has been demonstrated for other

Table 14.4 Estimated operational lifetimes for constructed wetlands using 3 or 5 m² area per person with shale or Polkemmet ochre as a substrate and with three concentrations of sewage effluent, 20, 5 or 3 mg P l⁻¹ (McHaffie et al., 2001).

Substrate	Operational lifetime/years					
	3 m² per person[1]			5 m² per person[1]		
	20 mg l⁻¹	5 mg l⁻¹	3 mg l⁻¹	20 mg l⁻¹	5 mg l⁻¹	3 mg l⁻¹
Shale[2]	1.6	6.5	10.7	2.7	10.7	17.8
Ochre[3]	18.8	75	125	33.5	134	224

Constructed wetland substrate assumed to be 0.6 m deep.
[1] Each person assumed to produce 0.2 m³ sewage per day.
[2] From data in Drizo (1998).
[3] Calculated with half the measured maximum phosphorus adsorption capacity to take account of the anticipated decline in flow rates.

constructed wetland substrates (Table 14.3). Realistically, a constructed wetland might absorb only half its maximum phosphorus capacity before flow rates declined significantly, but this would still give a long operational life time compared to other substrates (Table 14.4). Field trials of ochre as a substrate in constructed wetlands treating sewage effluent are currently being conducted. In these trials water quality (including metals) and longevity of phosphorus sorption are being assessed, in addition to the phosphorus removal efficiency and hydraulic performance of the ochre.

The Minto ochre is less suitable for use as a filter material for removal of phosphorus from sewage effluent due to its lower saturated hydraulic conductivity, but it has a high capacity for phosphorus adsorption due to its fine-grained texture. A potential use of the Minto ochre for sewage effluent treatment is in a system in which the effluent is dosed with the ochre and mixed in holding tanks. For example, it is estimated that a septic tank of 7 m³ capacity serving a household of five people would require the addition of 5.1 kg of Minto ochre per week to remove the phosphorus (McHaffie et al., 2001).

14.4.4 Treatment of agricultural runoff

Approximately 40% of agricultural land in the UK (excluding rough grazing) is underlain by subsurface field drains which have been shown to be a major conduit of soluble and particulate phosphorus export to watercourses in storm events, even when best management practices are implemented (Dils and Heathwaite, 1996; Burke et al., this volume). Although it is not feasible to install waste water treatment works or constructed wetlands for every field drain, treatment with ochre may form a cheap, low maintenance means of reducing phosphorus exports from field drains. Treatment at the field scale could take the form of filter cartridges containing coarse-grained ochre. Alternatively, dosing with powdered ochre, followed by settlement, has already been demonstrated to be effective in reducing phosphorus concentrations in simulated agricultural runoff in laboratory-scale flume experiments (Sweetman, 2001). In a short-term pilot study, instream filter units, containing the coarse-grained Polkemmet ochre, reduced instream phosphorus concentrations and altered the algal community composition in a river affected by agricultural runoff (Bush, 2001). Longer-term field investigations, which include a complete assessment of the environmental benefits and impacts, especially with regard to aquatic ecology, are necessary to develop further the use of ochre for treatment of agricultural runoff.

14.4.5 Recycling of phosphorus-saturated ochre as a slow-release fertiliser

When the phosphorus removal capacity of ochre in a constructed wetland or filter unit is finally exhausted, the 'spent' ochre will require removal and disposal. A more sustainable alternative to disposal to landfill is the use of phosphorus-saturated ochre as a slow-release fertiliser. Pot experiments with other constructed wetland substrates artificially saturated with phosphorus concluded that phosphorus sorbed to crystalline slag was delivered to barley plants more efficiently than phosphorus added in conventional K_2HPO_4 fertiliser (Hylander and Simán, 2001). However, for other substrates tested in the same experiments, barley yields were lower than for the application of fertiliser phosphorus, mainly due to manganese deficiency caused by the liming effect of many substrates. A pilot pot experiment in which barley was grown in soils amended with ochre containing different amounts of sorbed phosphorus has already demonstrated that phosphorus sorbed to ochre is available for uptake to plants (Smith, 2001). Future experiments examining the use of phosphorus-saturated ochre as a slow-release fertiliser should also assess the metal composition of soil, leachate and plant material to ensure that there are no adverse environmental consequences of this application.

14.5 CONCLUSIONS

Phosphorus removal and recycling has traditionally relied heavily on high-tech solutions requiring considerable capital and maintenance costs for plant and chemicals. Laboratory experiments and small-scale field trials have demonstrated that ochre, formed as a by-product from mine water treatment, has a high capacity for phosphorus removal from waste water by sorption to the high concentration of iron oxides and hydroxides contained within the ochre. Potential applications of phosphorus removal by ochre are widespread, including sewage effluent treatment in constructed wetlands and dosing systems, and treatment of agricultural runoff by instream filter units and/or dosing and settlement systems. Furthermore, phosphorus sorbed to ochre has the potential to be recycled as a slow-release fertiliser. Full-scale implementation of phosphorus removal from waste water by ochre requires field trials to assess the performance and costs of the treatment process. The use of ochre for phosphorus removal from waste water is particularly attractive not only because treatment costs are potentially lower than more traditional solutions, but also due to the sustainability advantage of using a by-product of combating one form of aquatic water pollution (i.e. mine water pollution) to address another (eutrophication).

ACKNOWLEDGEMENTS

The authors are grateful to the Coal Authority and Northumbrian Water Limited for funding the preliminary investigations of phosphorus removal by ochre. Part of the experimental work was carried out by Eleni Bozika at the School of GeoSciences, University of Edinburgh. The support of the UK Engineering and Physical Sciences Research Council (EPSRC) for ongoing work described in Section 14.4 is gratefully acknowledged (EPSRC grant nos. GR/R73522/01 and GR/R73539/01).

REFERENCES

Allen, S.E. (1974) *Chemical Analysis of Ecological Materials*. Blackwell Scientific, Oxford, pp. 66–67.
Arias, C.A., Del Bubba, M. and Brix, H. (2001) Phosphorus removal by sands for use as media in subsurface flow constructed reed beds. *Water Res.* **35**(5), 1159–1168.
Ayoub, G.M., Koopman, B. and Pandya, N. (2001) Iron and aluminium hydroxy (oxide) coated filter media for low-concentration phosphorus removal. *Water Environ. Res.* **73**(4), 478–485.
Barrow, N.J. (1983) A mechanistic model for describing the sorption and desorption of phosphate by soil. *J. Soil Sci.* **34**, 751–758.
Bozika, E. (2001) Phosphorus removal from wastewater using sludge from mine drainage treatment settling ponds. MSc thesis, Univ. of Edinburgh.
Bush, A.M. (2001) An investigation into the use of iron ore sludge to remove phosphorus in the River Leet, in an attempt to mitigate eutrophic conditions. MSc thesis, Univ. of Edinburgh.
Cooper, P.F., Job, G.D., Green, M.B. and Shutes, R.B.E. (1996) *Reed Beds and Constructed Wetlands for Wastewater Treatment*. WRc, 184 pp.
D'Arcy, B.J., Ellis, J.B., Ferrier, R.C. and Jenkins, A. (2000) *The Environmental and Economic Impacts of Diffuse Pollution in the UK*. CIWEM/EA/SEPA joint publication, Terence Dalton, Lavenham, 181 pp.
Dils, R.M. and Heathwaite, A.L. (1996) Phosphorus fractionation in hillslope hydrological pathways contributing to agricultural runoff. In *Adv. Hillslope Process* (eds Anderson, M.G. and Brookes, S.), John Wiley & Sons Ltd., Chichester, pp. 229–252.
Drizo, A. (1998) Phosphate and ammonium removal from waste water, using constructed wetland systems. PhD thesis, Univ. of Edinburgh.
Farmer, A.M. (2001) Reducing phosphate discharges: the role of the 1991 EC urban wastewater treatment directive. *Water Sci. Technol.* **44**(1), 41–48.
Grüneberg, B. and Kern, J. (2001) Phosphorus retention capacity of iron-ore and blast furnace slag in subsurface flow constructed wetlands. *Water Sci. Technol.* **44**(11–12), 69–75.

Haygarth, P.M., Warwick, M.S. and House, W.A. (1997) Size distribution of colloidal molybdate reactive phosphorus in river waters and soil solution. *Water Res.* **31**(3), 439–448.
Hylander, L.D. and Simán, G. (2001) Plant availability of phosphorus sorbed to potential wastewater treatment materials. *Biol. Fert. Soils* **34**, 42–48.
Lamont-Black, J., Younger, P.L. and Batty, L. (2001) *Factual report of test results for use in designing phosphate removal by passive wetlands.* Department of Civil Engineering, Univ. of Newcastle, Report in Confidence.
Mann, R.A. (1997) Phosphorus adsorption and desorption characteristics of constructed wetland gravels and steelworks by-products. *Aust. J. Soil Res.* **35**, 375–384.
McHaffie, H., Heal, K.V. and Smith, K.A. (2000) *Using sludge from mine drainage treatment settling ponds for phosphorus removal from wastewaters: Report to Coal Authority.* Institute of Ecology and Resource Management, Univ. of Edinburgh. Contact: Property and Environment Manager, The Coal Authority, 200 Lichfield Lane, Mansfield, Notts. NG18 4RG, U.K.
McHaffie, H., Heal, K.V. and Smith, K.A. (2001) *Using sludge from mine drainage treatment settling ponds for phosphorus removal from wastewaters: Final report to Coal Authority.* Institute of Ecology and Resource Management, Univ. of Edinburgh. Contact: Property and Environment Manager, The Coal Authority, 200 Lichfield Lane, Mansfield, Notts. NG18 4RG, U.K.
Moore, W.J. (1972) *Physical Chemistry.* Longman, London.
Morse, G.K., Brett, S.W., Guy, J.A. and Lester, J.N. (1998) Review: Phosphorus removal and recovery technologies. *Sci. Total Environ.* **212**, 69–81.
Parfitt, R.L. (1989) Phosphate reactions with natural allophane, ferrihydrite and goethite. *J. Soil Sci.* **40**, 359–369.
Reddy, K.R., Cadillac, R.H., Flag, E. and Gale, P.M. (1999) Phosphorus retention in streams and wetlands: a review. *Crit. Rev. Environ. Sci. Technol.* **29**, 83–146.
Roberge, G., Blais, J.F. and Mercier, G. (1999) Phosphorus removal from wastewater treated with red mud-doped peat. *Can. J. Chem. Eng.* **77**(6), 1185–1194.
Smith, A. (2001) The use of sludge from mine drainage settling ponds containing different levels of phosphate as a substrate for the growth of barley. BSc thesis, Institute of Ecology and Resource Management, Univ. of Edinburgh.
Sweetman, R. (2001) The removal of phosphate from agricultural runoff with blast furnace slag and ochre. MSc thesis, Univ. of Newcastle.
Tarasova, I., Georgaki, I., Dudeney, A. and Monhemius, A.J. (2002) Re-use of iron-rich sludges. *Proceedings SWEMP 2002, 7th International Symposium on Environmental Issues and Waste Management in Energy and Mineral Production,* (ed. R. Ciccu – ISBN 88-900895-0-4), pp. 1087–1093, DIGITA, University of Cagliari, Italy.
Webster, A.K. and Wieder, R.K. (1997) Reduction of phosphate losses from fertilised soils using acid mine drainage sludge. *J. Conf. Abs.* **2**(2), Biogeomon '97, 323.
Wood, R.B. and McAtamney, C.F. (1996) Constructed wetlands for waste water treatment; the use of laterite in the bed medium in phosphorus and heavy metal removal. *Hydrobiologia* **340**, 323–331.
Younger, P.L., Banwart, S.A., and Hedin, R.S. (2002) *Mine Water: Hydrology, Pollution, Remediation,* Kluwer Academic Publishers, Dordrecht (ISBN 1-4020-0137-1), 464 pp.
Zoumis, T. and Calmano, W. (2000) Development of fly ash granules for heavy metal removal from mine waters. In *Mine Water and the Environment* (eds Rózkowski, A. and Rogoz, M.), pp. 492–504, IMWA, Katowice, Poland.

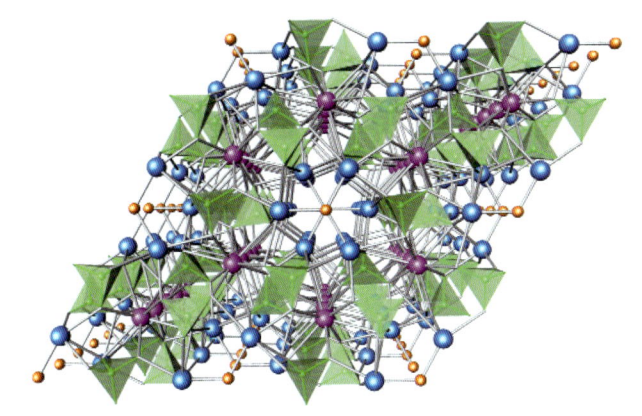

Figure 2.7 The idealised structure of apatite, viewed as a *c*-axis projection. Note the hexagonal structure, and the location of X ions (OH, F, Cl) along channels parallel to the *c*-axis.

(a)   (b)

(c)

Figure 5.5 Phosphorus deficiency symptoms in (a) barley (left), maize (right), (b) maize and (c) tomato.

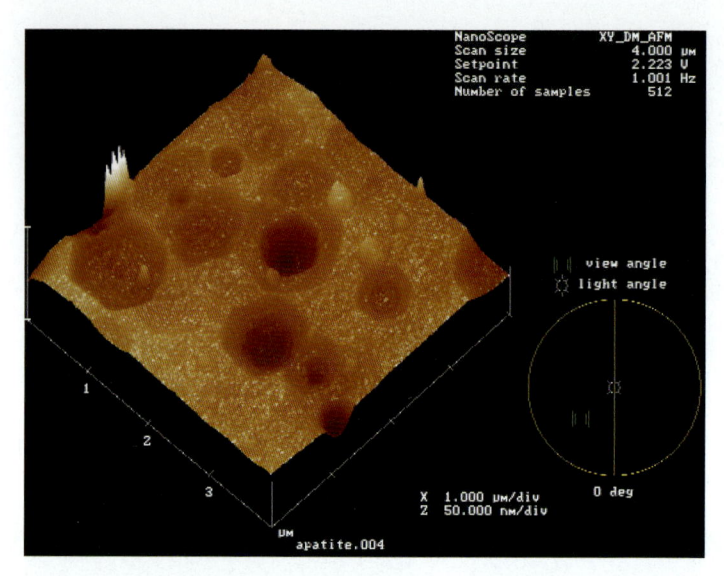

Figure 9.24 A dissolving apatite surface, showing hexagonal dissolution pits. Note that the vertical scale (Z) is 50 nanometers; X and Y scales are in μm. Data collected using Atomic Force Microscopy, courtesy of Andrew Putnis and Dirk Bosbach, Münster University.

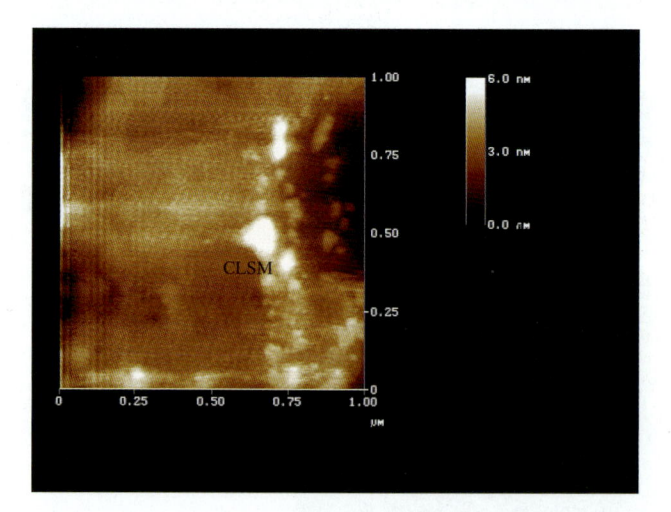

Figure 9.26c AFM image of of Pb-hydroxylapatite (light areas) precipitating on the surface of a dissolving natural apatite, showing hexagonal dissolution pits (see also Figure 9.24); note the alignment of the precipitating hexagonal crystals along crystallographic orientations of the apatitic substrate. AFM image extracted from a figure in Valsami-Jones et al. (1998).

Figure 10.3 Radial sedimentation tank treating municipal waste water.

Figure 10.5 Trickling filter treating municipal waste water.

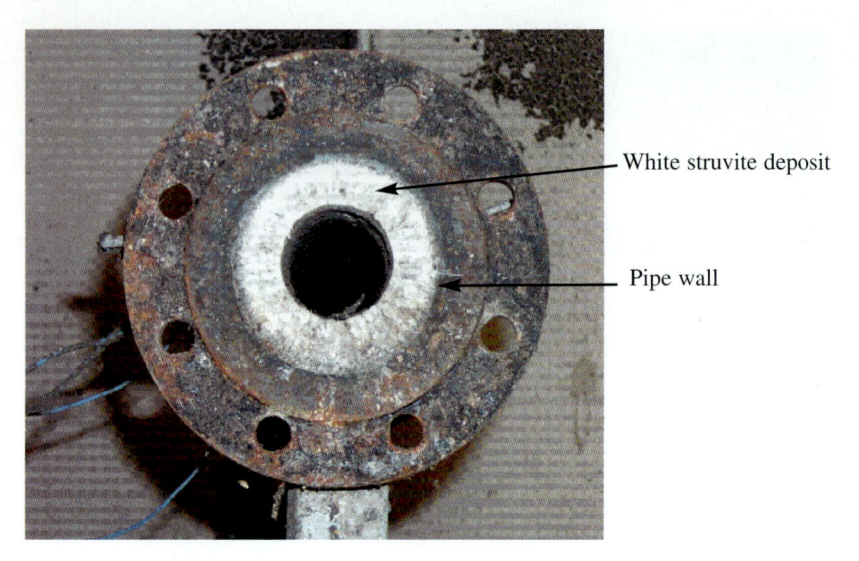

Figure 17.2 Digested sludge pipeline at Slough STW.

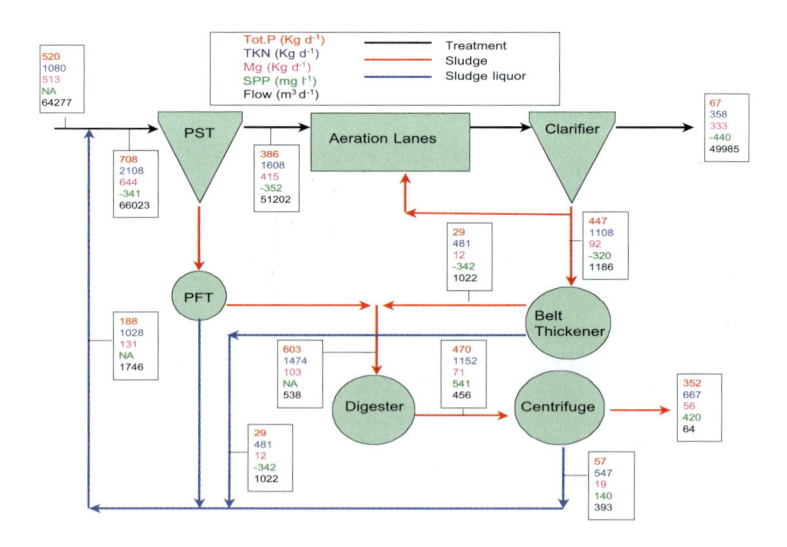

Figure 17.3 Simplified mass balance for entire site of Slough STW. For acronyms see page 427.

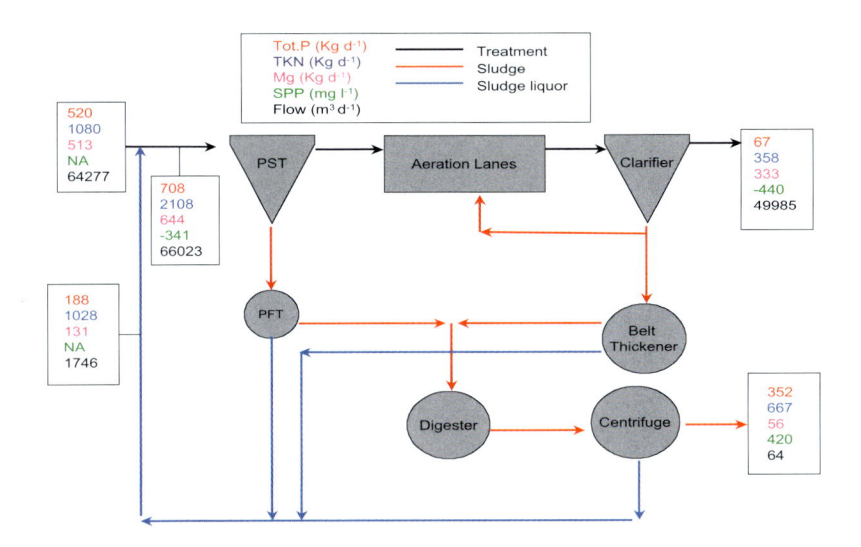

Figure 17.4 Mass balance highlighting inlet and outlet of Slough STW.
For acronyms see page 427.

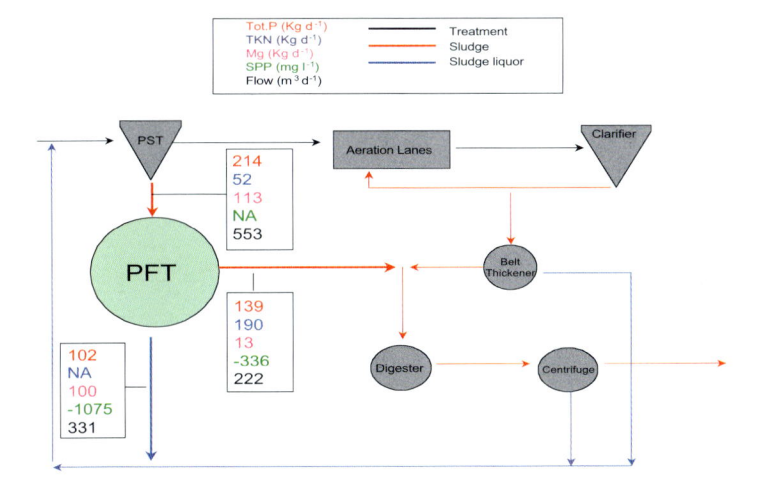

Figure 17.5 Mass balance around the primary sludge thickener, Slough STW.
For acronyms see page 427.

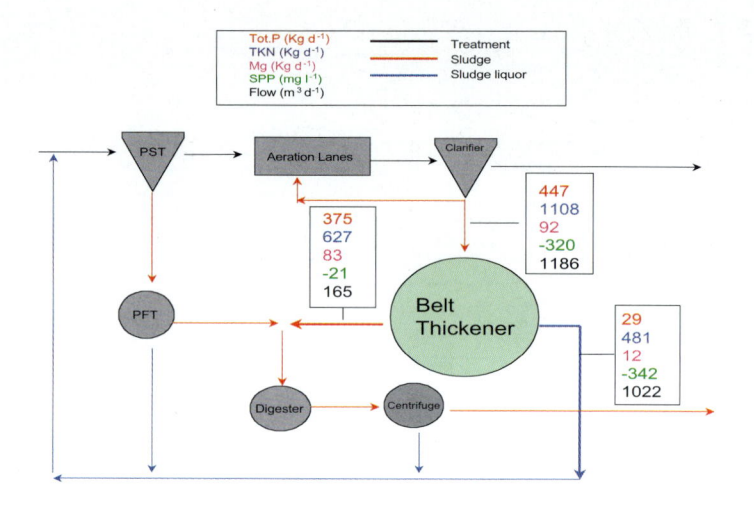

Figure 17.6 Mass Balance around the belt thickener, Slough STW. For acronyms see page 427.

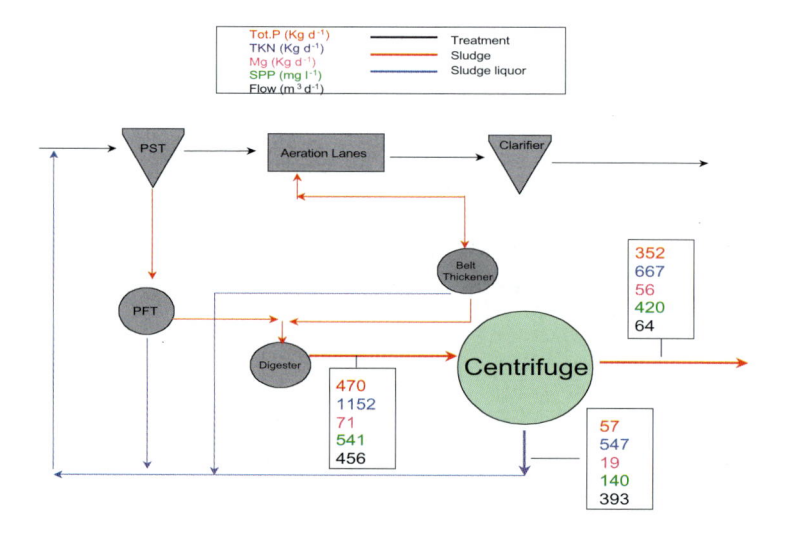

Figure 17.7 Mass balance around the centrifuge, Slough STW. For acronyms see page 427.

Figure 19.9 Calcium phosphate pellets (diameter 0.7–1.0 mm)

Figure 19.10 Carrousel 2000® at Geestmerambacht (main stream biological water treatment). 1. Pre-denitrification zone; 2. Aeration zone; 3. Division structure between pre-denitrification zone and aeration zone; 4. Surface aerator; 5. Final clarifier; 6. Return sludge pumping station.

Figure 21.4 One of four replicate SBR vessels used by Greaves et al. (2001) for the recovery of phosphorus from pig slurry. The reaction vessels are constructed from polypropylene bins with a total capacity of 65 l. Diluted pig slurry is pumped into the reaction vessels and the effluent is drained from the reaction vessels 10 cm from the base.

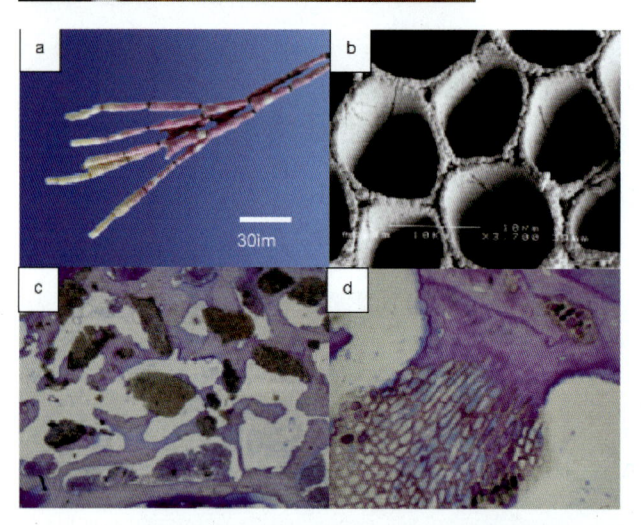

Figure 25.2. From the ocean to human bone - FRIOS® ALGIPORE®. a. A single branch of the alga Amphiroa ephedra, which grows in the sea off the coast of South Africa. b. SEM image of a granule, showing honeycomb-like structure. Pores are all interconnected and range from 10-15µm in diameter. Lateral longitudinal pores form the outer part of the plant scaffold. Size bar = 10µm. c. Bone biopsy from a patient eleven months after a sinus lift operation and augmentation with FRIOS® ALGIPORE®. Newly formed mineralised bone (violet), can be seen bridging the ALGIPORE® granules (grey). Thionine staining; original magnification x4 objective. d. Higher magnification (x40) of the same biopsy showing ingrowth of mineralised bone (purple), and non-mineralised bone (blue) within pores. Photographs courtesy of DENTSPLY Friadent, Mannheim, Germany and of Prof. Rolf Ewers, University of Vienna, Austria.

PART FOUR

Phosphorus recovery for reuse: principles, technologies, feasibility

15.	Phosphorus recovery in the context of industrial use	339
16.	Fluid dynamic concepts for a phosphate precipitation reactor design	358
17.	Phosphorus recovery via struvite production at Slough sewage treatment works, UK – a case study	402
18.	Phosphorus recovery trials in Treviso, Italy – theory, modelling and application	428
19.	The case study of a phosphorus recovery sewage treatment plant at Geestmerambacht, Holland – design and operation	470
20.	Full scale struvite recovery in Japan	496

21. Phosphorus recovery from unprocessed manure 507

22. Scenarios of phosphorus recovery from sewage for industrial recycling 521

23. Phosphorus recycling: regulation and economic analysis 529

15
Phosphorus recovery in the context of industrial use

I. Steén

15.1 INTRODUCTION

In a sustainable society renewable resources should substitute non-renewable sources wherever possible. Where this cannot be done, recovery, reuse and recycling of non-renewable resources should be implemented (European Commission, 1999). As a consequence of increasing concerns regarding over-exploitation of natural non-renewable resources, and the many different human activities that impact on the environment, legislation and action programmes to promote recycling have been adopted in many countries.

Waste and by-products from various industrial processes, sewage sludge and municipal waste are commonly disposed in landfills, have in the past often been discharged untreated to the aquatic environment, or are applied unprocessed to agricultural land. Most of these major routes for disposal of waste materials are not acceptable in a sustainable society (European Commission, 2001). Especially

© 2004 IWA Publishing. *Phosphorus in Environmental Technology: Principles and Applications.*
Edited by Eugenia Valsami-Jones. ISBN: 1 84339 001 9

when non-renewable and finite resources, used to manufacture products required by society, are concerned, then society should strive to close the product and product raw material cycles in order to achieve a sustainable system. Therefore, improved resource and waste management (European Commission, 1975), i.e. product stewardship and more efficient resource utilisation, are prerequisite in any community.

Water is an indispensable resource that moves in the continuous sun-powered water (or hydrologic) cycle and is a primary ingredient for the development and nourishment of life. However, man's activities have disturbed this cycle and a large quantity of the world's fresh water is being spoiled (European Environment Agency, 1995, 1999a, 2002). Moreover, phosphorus pollution, eutrophication of the aquatic environment (Aquatic Environment, 1999, 2000), is a critical issue, and improved phosphorus management is a key factor in resolving this (European Environment Agency, 1999b). Pathways of concern regarding the aquatic environment are discharges or releases of effluents from point sources, for instance industries and cities, and diffuse pollution, such as losses from agricultural land and forest areas (Agriculture, Environment, Rural Development: Facts and Figures – A Challenge for Agriculture, 1999; De Clerq et al., 2001; Tunney et al., 1997). Drinking water abstraction and waste water treatment are thus components of a much greater whole.

Phosphorus (P) is a non-renewable resource and an important macronutrient on which life depends and for which there is no substitute. It is a central element in many physiological and biochemical processes and, by controlling energy generation in all cells, it also controls life in plants, animals and man (see Dobrota, this volume). Phosphorus is the eleventh most abundant element in the lithosphere, but reserves are unevenly spread in the earth's crust. Almost all phosphorus used by society is mined from a comparatively small number of commercially exploitable deposits in the world. The annual total production is some 20 million tonnes (Mt) of P, derived from roughly 140 Mt of rock concentrate (IFA, 2002). Although phosphorus has many applications (Emsley, 2000), its largest use is as mineral fertiliser for agricultural production which accounts for some 80% of global consumption. Other versatile applications of phosphorus include its use as supplement to animal feeding stuffs, food additives, and in various industrial applications as diverse as detergents, water and metal surface treatment, crop protection products, pharmaceuticals, pyrotechnics, flame retardants, semi-conductors, catalysts and many more.

In the natural environment, P is supplied through the weathering and dissolution of minerals with very low solubility (see Valsami-Jones, this volume). Therefore, P can often be the critical element limiting animal and plant production in both natural and agricultural environments. Throughout the history of agricultural production, phosphate has been largely in short supply (Johnston and Steén, 2000). The importance of recycling organic waste and manures to maintain crop production has been recognized by farmers for thousands of years. In the past, the resource cycle was relatively tight, and recycling was part of daily life. Organic waste was recycled in agriculture, locally produced agricultural products were consumed locally, and waste was recycled back to local agricultural land. However, the nutrient recycling loop has been broken as agricultural produce, i.e. food, is transferred from rural to urban areas regionally, nationally and even intercontinentally. With growing populations, and thus increasing food demand and urbanization, nutrients, in the form of agricultural produce, are distributed to the cities and the resulting wastes are usually discharged to water bodies or disposed in landfills. Thus, the recycling loop was replaced by a linear throughput system and with the introduction of water as waste carrier this was further augmented.

15.2 WHY RECOVERY?

For the phosphate industry in general, improved phosphorus management is imperative (Driver *et al.*, 1999), as P is a non-renewable, non-interchangeable finite resource, although the estimated life expectancy of global reserves varies. Different estimates suggest that global P reserves might last between 100 to as much as thousands of years depending on exploitation rate, phosphate rock quality and costs for processing (Steén, 1998). Besides being a non-renewable resource, modern society has a number of other reasons and driving forces for improving the management of P. These are associated with waste management and the recently developed drive for recovery from waste flows, and waste water in particular, to alleviate both environmental and operational problems.

Among the issues related to water and waste water, improved and advanced water management is essential to ensure ample protection of water quantity and quality. The issue for the water treatment industry, in relation to protecting the aquatic environment, is mainly about reducing effluent P concentrations in discharge waters and the total P loading in rivers and lakes. Other driving forces are potential cost savings and cost recovery as well as minimization of operational problems (due to phosphate scale formation). At the same time a more integrated view would include, besides effluent standards, aspects such as energy and chemical use and sludge production and recycling/disposal (Edge, 1999; Priestley, 1990s).

In the European Union (EU) policies, such as the 6th Environment Action Programme (6EAP, 2000) and specific regulations have had, and increasingly will have, a great impact on the management of waste, sludge, water, soil and natural resources in general. The 6EAP is a progression from the 5th Environment Action Programme (5EAP, 1993) and has addressed the issue of sustainable use of natural resources. It describes the objective of its thematic resource strategy as 'to ensure that the consumption of renewable and non-renewable resources and the associated impacts do not exceed the carrying capacity of the environment and to achieve a decoupling of resource use from economic growth through significantly improved resource efficiency, dematerialisation of the economy and waste prevention'.

The purpose of the Water Framework Directive, 2000/60/EC (EC, 2000) is to establish a framework for the protection of inland surface waters, transitional waters, coastal waters and groundwater. It aims to introduce, for instance, enhanced protection and improvement of aquatic ecosystems, to prevent further deterioration and promote sustainable water use. Several measures contained in this directive will have an impact on the water industry, both related to water abstraction and discharges of effluents.

With the full implementation of the Urban Waste Water Treatment Directive, 91/271/EEC (EC, 1991b), UWWTD, in 2005, nutrient reduction targets have been established and sludge volumes will subsequently increase. Depositing of biodegradable waste must be reduced according to the Landfill of Waste Directive, 1999/31/EC (EC, 1999), so land filling of sludge will be limited.

The Sewage Sludge Directive, 86/278/EEC (EC, 1986) on sludge use in agriculture is currently under revision and standards regarding contamination of heavy metals, pathogens etc. are likely to be tightened. The P contained in sewage emanates from domestic waste water, storm water and industrial waste water that is compatible with the urban waste water system. However, the major part of the P present in sewage comes from human excreta and every adult excretes around 0.5 kg of P annually. The IPPC Directive, 96/61/EEC (EC, 1996) concerns industry production and also large animal rearing units and their emissions and effluents.

Stakeholders' views and perceptions are generally significant driving forces and the agricultural use of sludge, which would recycle much of the human dietary intake, is not without controversy. During the last 10–15 years sludge disposal routes have been the target of growing interest, and concerns have been expressed about the potential risks of agricultural use of sludge related to health and the environment. Society, in general, is much concerned about the safe use of sludge, particularly in agriculture for food production. Agricultural use of sewage sludge provides a convenient means of recycling but this use is declining and today some countries have given up the practice. If only human waste entered the sewage treatment works, and the sewage sludge was properly treated, it could be returned to land with safety (Enskog and Johansson, 1999). But, waste water treatment plants inevitably receive industrial and household waste, some of which may contain toxic and/or persistent non-biodegradable compounds, pathogens, hormones and other undesired substances. Once treated, sludge conventionally is recycled or disposed of using three main routes: recycled to agriculture (land spreading), incinerated or sent to landfill. Other, possible outlets exist, such as forestry and silviculture, land reclamation and aiding re-vegetation of, for instance, derelict land. In addition, combustion technologies may be used, including wet oxidation, pyrolysis and gasification. Each recycling or disposal route has specific inputs, outputs and impacts. And, most of these routes must be scrutinised and challenged using impact analysis.

In the EU more than 7 Mt of sewage sludge, on a dry matter basis, is produced annually and with the implementation of the UWWTD it is expected to increase to at least 9.4 Mt by 2005 (European Environment Agency 2002). Sewage sludge contains typically between 1% and 5% P (Eriksson, 2001; Otabbong, 1997). Considering a full implementation of the UWWTD, the potential total phosphorus from sludge could amount to about 300,000 tonnes of P, but 100% recovery is not feasible. Reliable technologies may allow 50 to 80% recovery of sewage phosphate. Furthermore, it is probable that recovery of P from small waste water treatment works in rural areas in particular cannot be economically justified.

Approximately 1,340,000 t P are used in the form of phosphatic mineral fertilisers in the EU annually (IFA, 2002; EFMA, 2002) and some 250,000 t are used as animal feed supplements (2001 Directory of Chemical Producers, 2002; The Chemicals Economics Handbook, 2002). These applications, together with some 110,000 (105,000–120,000) t P (2001 Directory of Chemical Producers, 2002; The Chemicals Economics Handbook, 2002) used for the manufacture of detergents, account for about 95% of total annual P consumption in West Europe (the EU, Norway and Switzerland). Of this present total of mineral P input derived from phosphate rock, there could be the potential of substituting at maximum some 15% by using recovered phosphorus from sewage. This would require full implementation of the UWWTD and 100% P recovery at every waste water treatment facility.

There are other P containing wastes that could be considered for either improved recycling or, alternatively, for recovery. One such waste that has been recycled in the past in agriculture is abattoir waste. Until recently some 20,000 t P based on chemically treated bone, was used annually in the animal feed sector in Western Europe. However, in the aftermath of the BSE (bovine spongiform encephalopathy, one of a number of transmissible spongiform encephalopathies, brain disease) crisis, this is no longer permitted in the EU, although the European Commission is currently investigating if this restriction could be terminated. Meat and bone meal are accepted and used as fertiliser for organic farming, so some of the abattoir waste is recycled in this sector of agriculture.

There is also a considerable amount of P in animal feeding stuffs imported into Europe, but the actual total quantity is difficult to estimate. However, animal excreta in Western Europe is estimated to contain around 1,600,000 t P, some of which is recycled on farm, some originates from the feed stuffs, and some from mineral feed phosphate supplements (Johnston and Steén, 2000). A fraction of the total excreta is collected when the animals are housed, whereas a large percentage is deposited directly on fields whilst animals are grazing. In areas with high animal densities, and thus local surpluses, i.e. more nutrients contained in the available manure than can be justified for application to agricultural land, this has become an issue. In regions, such as parts of the Netherlands, the Flanders region of Belgium, areas of Germany, Brittany in France, and the Po valley in Italy, there could be an option to recover P from animal manures. Even with a very conservative approximation, significant amounts of P from animal manures constitute a surplus that could be removed annually (Brouwer et al., 1995). How much this feasibly would amount to is difficult to assess, because it would depend on the quantity of manure to be processed and the economies of recovery and reuse (Greaves et al., 1999). Other aspects, such as crop P requirements, soil P status (Mengel and Kirkby, 1987), crop offtake and nutrient export of agricultural produce from farm, type of animal production and husbandry, type of manure and manure management etc. should also be considered. The impact of current and imminent agriculture and environment policies could also introduce some uncertainties regarding future estimates. A more viable alternative for excess manure may be to transport it to other areas for direct use in agriculture; however, transporting manures, especially over long distances might cost more than the nutrients are worth and could introduce other social infrastructure and environmental problems. A benefit with manure is that it is, comparatively to sewage sludge, richer in nutrients and is free from many substances occurring in waste water which may remain, after water treatment, in the sewage sludge. On average animal manures contain around 2.0–2.5% P on a dry matter basis (Eriksson, 2001).

The Nitrates Directive, 91/676/EEC (EC, 1991a), which aims at reduced nitrogen (N) load or applications of manure in vulnerable zones should, if correctly implemented, also have an indirect effect on the P load to agricultural land. But generally, animal manures have a smaller N:P ratio, contain more P relative to N, than most crops require on European soils. This has resulted in P enrichment of soils on farms with animal production and especially in areas with high animal densities. Thus, regardless of the implementation of the Nitrates Directive, P accumulation in regions with high animal densities will continue (Steén, 2001).

15.3 POSSIBLE TECHNICAL-INDUSTRIAL PATHWAYS

There is a number of removal technologies applied at municipal and industrial waste water treatment facilities, some of which could be adapted and applied to P recovery, including P from other sources such as animal manures (CEEP, 1998; Derden et al., 1998). Conventional removal methods include biological nutrient removal, with the P incorporated in the activated sludge, and chemical phosphorus precipitation, by use of ferric and ferrous salts, such as chlorides and sulphates, aluminium and other metal salts (see Parsons and Berry, this volume). For example calcium salts are common and the precipitated phosphorus is removed in the sludge. Ferric chloride precipitation is the most widely used physiochemical method for P removal in Europe. Waste water treatment can be divided into three stages, primary (mechanical), secondary (biological) and tertiary (advanced) (see Parsons and Stephenson, this volume). In the advanced stage, a combination of biological and chemical treatment for nutrient removal often takes place; this combination has to be both feasible and adequate.

Sludge is produced as a result of all water treatment processes and is the solid end product that generally contains most of the substances, including insoluble and not readily decomposable matter, contained in the influent, together with precipitation chemicals and coagulants. The major constituent of raw sludge is water and the sludge is most often further treated, via thickening, anaerobic digestion and de-watering, depending on the final disposal route. Sewage generally contains in the range of 10 to 50 mg phosphate as orthophosphate (PO_4–P) per litre. Consequently, if phosphorus is not removed, sewage effluent discharges could be a very significant source of bio-available P entering surface waters. In chemically treated sewage sludge, some 95% of the P is in inorganic form and the organically bound phosphorus rarely exceeds 10%. In anaerobically digested sludge organic phosphorus may account for 10 to 20%. In raw sludge, that has not been stabilised with salts, data from European countries suggest that the organic phosphorus content ranges between 20% and 30% of the total phosphorus (Otabbong, 1997). In a waste water treatment plant, P could be recovered directly from the waste water or from a solid phase, i.e. the sewage sludge or, possibly, from other end products of sewage processing.

Increasingly, incineration of sludge is being considered, particularly in countries where the agricultural use of sludge is small or diminishing and where environmental and financial implications of disposal at landfill are of concern. Incinerated sludge ash is a resource not only of P but also of other recoverable inorganic substances. However, the ash will inevitably contain the inorganic substances of the precipitation chemicals such as iron and aluminium, which might make recovery of P difficult. A possible use for the ash may be as fertiliser, but again any undesirable substances and other chemical as well as physical properties may make it inappropriate to use. Alternative uses of the ash may be as inorganic filler substitute in the cement and the asphalt industries, i.e. for construction purposes (Minnesota Pollution Control Authority, 1990).

Most phosphorus recovery technologies are currently in the development stage and only a few full-scale processes are in operation at present. The potential technologies for the recovery of P from waste water and/or sludge are numerous, and all have different advantages and disadvantages. There are differences in potential recovery rates, the stage in the treatment process where the recovery operation is introduced, the effect on sludge reduction rate, and the composition and quality of the recovered P compounds and their potential outlet. Furthermore, the potential for recovering other substances, the types and concentrations of undesired elements and the impact of other waste water characteristics, energy requirement or potential energy recovery/production, residual waste quality, quantity and disposal options and, not least, cost efficiency and the effect on treatment works operation, are all significant factors. In addition, a number of related issues must be considered when assessing the feasibility and economics of phosphate recovery.

Over the last few years in particular, the interest and developments in phosphorus recycling have accelerated and a number of recovery techniques have been reported in the literature (Brett *et al*., 1997; International Conference on the Recovery of Phosphates for Recycling from Sewage and Animal Wastes, 1998; Second International Conference on the Recovery of Phosphorus from Sewage and Animal Wastes, 2001). The main recovery techniques could be classified into: (1) chemical precipitation and recovery of primarily iron and aluminium phosphates from the sludge liquor, sludge or further processed sludge; (2) precipitation or crystallisation of, foremost, calcium phosphate and magnesium ammonium phosphate (struvite) from the waste water or sludge liquor; (3) other technologies applicable to waste water treatment such as membrane and ion exchange technologies; and (4) extraction of phosphorus from sludge incineration ash.

An outline of existing and potential recovery pathways shows that a number of options seem feasible; some examples are discussed in later chapters in this book. These methods have been tested on a larger scale and it appears that the costs are within acceptable limits. Also, that they do not increase environmental pressure and have additional positive impacts, such as sludge reduction and energy recovery.

15.3.1 Chemical precipitation and separation

Chemical precipitation is based on the addition of a metal salt to waste water. The metal salts are commonly iron (ferrous and ferric chloride or sulphate), aluminium (alum or hydrated aluminium sulphate, sodium aluminate etc.) and calcium (lime) compounds. In principle, these precipitation chemicals could be added either before the primary, the secondary or the tertiary treatment step or at both the primary and secondary treatment steps. The resulting iron, aluminium or calcium phosphates settle in a sludge fraction and may subsequently be separated and recovered. Current examples are the Hypro and Krepro (and Cambi/Krepro) processes. The sewage sludge is acidified and hydrolysed and, after the reaction, the solid organic fraction could be separated and used as a low cost carbon source, i.e. biofuel for energy production. In the Hypro concept, the removed phosphate is retained in the supernatant, whilst in the Krepro process (Naturvårdsverket, 1997), the phosphate fraction (ferric phosphate) is precipitated and separated by centrifugation. The remaining liquid phase contains dissolved organic matter and may be used to improve the denitrification of nitrogen.

15.3.2 Biological P removal

More than 50 years ago it was suggested that activated sludge could take up more phosphorus than its normal microbial growth requirements. In conventional sludge treatment, bacteria only use enough P to satisfy their basic metabolic requirements. This typically results in removal rates of total phosphorus between 20% and 40%. For the quantitative removal of phosphorus in the treatment plant, special bacterial consortia accumulate phosphorus in excess of their normal requirements; an improved understanding of the mechanism involved has resulted in several commercial processes. Examples are the Phostrip process, developed in the second half of the 1960s, the Modified Bardenpho process from the early 1970s, the Phoredox process and the similar A/O and Rotanox processes, the Bio-P process and others (Brett et al., 1997; McGrath and Quinn, this volume). All these processes use somewhat different technological approaches and integration of P removal in the sewage works. Generally, the P available for recovery is either contained in the remaining sludge, or in the P-rich liquor that could be separated from the sludge. The phosphorus contained in sludge could be recovered either by co-deposition within the sludge, or incineration and then recovery from the ash. The P contained in the P-rich liquor could be recovered via different techniques such as precipitation. Removal rates in full scale plants are typically around 80%, but variations between 20% and 95% have been reported.

15.3.3 Special forms of precipitation and crystallisation

Several techniques are available for precipitating the phosphate separately from the sludge bulk, which make separation and re-use very feasible. These techniques generally take place in a dedicated section of the waste water treatment plant and in the main or side streams. There is also a number of different techniques used to extract the phosphate from the waste water through precipitation in a dedicated reactor. Precipitation techniques are likely to result in a recyclable and directly useable end product. The principle of phosphate precipitation/crystallisation is not new. In the late 1930s crystalline material was identified in digested supernatant sludge and in the 1960s problems with crystalline deposits and blockage of pipes in waste water treatment plants was noticed. Current full scale operations include the DHV Crystallactor® process (see Piekema, this volume), the CSIR Fluidised Bed crystallisation, the Kurita Fixed Bed Crystallisation, the RIM-NUT ion exchange process, the Unitika Phosnix process (see Ueno, this volume) and a number of other processes (see Battistoni, this volume; Jaffer and Pearce, this volume). Most current systems are variations of the fluidised bed principle and most often use sand grains as seeding material. The precipitation is either in the form of calcium phosphates or magnesium ammonium phosphates (struvite). In the case of struvite, the formation of crystals (nucleation) often occurs spontaneously (Booker et al., 1999), but can also be aided by the presence of suitable nuclei or seed crystals, for instance sand grains. Generally, for the precipitation of calcium phosphates seed crystals, or sand are necessary (Moriyama et al., 2001). The seed crystals can be of different material depending on the process being used. These could either be added to the process, could be solid impurities in suspension, or the rough surfaces of pipe walls and equipment. The effectiveness of the precipitation is much dependent on the phosphate concentration of the solution, the pH, and the presence of compounds assisting or inhibiting crystal formation.

15.3.4 Other recovery pathways

There are numerous alternative processes of phosphorus recovery, removal, or recovery and removal in combination. A few examples could be ion exchange, biological techniques including adsorption or precipitation on bacteria surfaces, and incineration of sludge with subsequent phosphate recovery from the ashes. However, many of these are not yet applicable under real conditions.

15.3.5 Feasibility of recovery

Of concern is the feasibility of phosphorus recovery. It has been shown in pilot and full-scale operations that recovery can be technically feasible (Morse et al., 1993). The economic viability has been studied (Jeanmaire and Evans, 2001; Naturvårdsverket, 2002c) and it is suggested that P recovery is most viable with high P concentrations, low biochemical oxygen demand (BOD) in the waste water, and high sludge handling costs (Woods et al., 1999). Also the amount of chemicals required and the related costs, as well as costs for sludge or waste disposal, must not be overlooked in an economic viability calculation (Second International Conference on Recovery of Phosphorus from Sewage and Animal Wastes, 2001). The value of the recovered phosphate will be related to the outlet options, i.e. the possible reuse/application and requirement of further processing. However, it has been shown that a premium price on recovered phosphate in comparison to phosphate rock is required, if the cost/price ratio was to be acceptable. The least costly and most simple outlet suggested is to use the recovered phosphate, without further treatment, as fertiliser for agricultural purpose. Very high prices on fertilisers containing recovered phosphate have been reported in Japan, e.g. ten times higher per unit P than for phosphorus contained in mineral fertiliser produced from phosphate rock (Ueno and Fujii, 2001). This is not likely to occur in Europe, considering the farmers' financial situation and current price and quality of common and special, highly soluble, phosphatic fertilisers. Other more costly options include further processing to produce higher-value phosphate compounds. This suggests replacing rock phosphate by recovered phosphates, at current conditions, would effectively limit the value of recovered phosphates.

15.4 RECYCLING BY THE PHOSPHATE INDUSTRY

A number of routes for the re-use of recovered phosphates have been suggested in the literature (Brett *et al.*, 1997; Durrant, 1997; Durrant *et al.*, 1999), many of which appear feasible in principle. However, little evidence of practical feasibility regarding the use of recovered phosphate has been demonstrated. If broad-scale phosphate recovery and recycling is to become accepted practice, the P must be recovered in a potentially usable form. If the direct reuse as fertiliser is currently unlikely in Europe, an alternative may be further processing of the recovered phosphorus compounds. This could be done by the phosphate industry, although there are numerous other issues to be considered, such as availability, frequency in delivery and volumes, composition and quality including impurities, consistency in composition and price.

15.4.1 Technical pathways in the phosphate industry

At present, the traditional raw material for the industry is phosphate ore, either of sedimentary or igneous and metamorphic origin. Commercial phosphate ores or phosphate rocks have one property in common; their phosphate content is a calcium phosphate, i.e. a phosphate-fluorine-calcium apatitic structure. Most phosphate ores, whatever their origin, must be concentrated or beneficiated before further processing. There are two basic types of technical pathways or processes in the phosphate industry, the wet and the furnace or thermal processes. These processes are used to produce phosphoric acid, which is the primary intermediate for the production of phosphate compounds. The wet processes may be further characterized by the acid used to dissolve the phosphate rock. Sulphuric, nitric or hydrochloric acid may be used, of which sulphuric acid is by far the most common means of producing phosphoric acid (Becker, 1988; UNIDO and IFDC, 1998). The wet processes accounts for more than 90% of the current phosphoric acid production globally. The furnace process, i.e. the electric furnace reduction process, currently in use, produces elemental phosphorus which is used as raw material in a wide range of organic chemistry. Alternatively, it may be oxidised to form high purity phosphoric acid for further high-grade phosphate applications. Because of the cost of electricity, the electric furnace process is generally only competitive for producing premium phosphate products and not for the production of technical or fertiliser phosphates.

The phosphate industry has continuously restructured by both concentration and rationalisation since the 1970s. Much production has been discontinued, for example the number of phosphoric acid plants in Europe halved during the 1980s and today phosphoric acid production is some 30% of what it was at its peak in the mid 1970s (IFA, 2002). In 1977 there were more than ten phosphoric acid plants in the UK, now there is none in the country that was the cradle of the phosphate industry. The reasons for this development are not surprising. The rock producing countries have changed increasingly from producing a low cost bulky commodity that is expensive and energy-intensive to transport, to producing an added value, premium product. Though this change has been capital intensive, it has brought economies of scale.

In Western Europe today, there is only one electric furnace plant in operation and a handful of wet process phosphoric acid operations. Consequently, several European countries are lacking a phosphoric acid operation. Should recovered phosphate be treated by traditional means, the present industry structure will certainly have a major impact on the cost of using recovered phosphates particularly regarding their long distance transport. Relevant to this calculation will be the concentration of P in the recovered phosphate, the volumes transported, energy cost of transportation per nutrient unit, and the related impact on the environment. Another important issue is the nature of the recovered phosphates, and the concentration of any impurities, and consequently the suitability of the recovered phosphate as an alternative raw material in the traditional phosphate treatment processes, including the production of different phosphate compounds. Generally, it would be possible to substitute phosphate rock by recovered phosphates but if the substitution of sources is to be successful, a number of criteria have to be met and a number of questions answered. Some of the topics to be considered by the phosphate industry are liability and dependability, cost-effectiveness, consistency in raw material composition and quality, volume and availability and impact of product quality. Finally, the public perception of 'waste' being used as fertilisers or even transformed into food supplements and additives is likely to have a major impact on any decisions taken about recycling phosphorus by the industry.

15.4.1.1 *The electric furnace process*

There is only one West European producer of elemental phosphorus using an electric furnace, Thermphos International in the Netherlands. Currently, the company has decided to replace 17,500 tonnes of the phosphorus intake by recovered materials. By experience there are some limitations regarding alternative raw materials and their suitability for use in this process (Schipper *et al.*, 2001). The P concentration must be adequate and a limit on the amount of certain impurities, such as iron, is important. The material must be dry and while small amounts of organic substances and ammonium can be tolerated, they must not interfere with the technical processes. Several sources of recycled phosphates have been tested and one that contains relatively little impurities and could meet the above requirement is manure incineration ash. However, its zinc and copper content currently limits its use by Thermphos. Certain recovered phosphates, such as calcium and aluminium phosphates from waste water treatment plants, have proved usable in the thermal process. Currently, calcium phosphate pellets produced in the Crystallactor® process at the waste water treatment plant of Geestmerambacht (see Piekema, this volume) in the Netherlands are being recycled by Thermphos.

15.4.1.2 The wet process

Most of the principles that apply to the use of alternative raw materials in the thermal process also apply to the wet process of producing phosphoric acid. Generally, phosphate rock is a complex raw material that affects the operation of the processing plant in numerous ways, some of which may be unpredictable. As a result, a thorough evaluation of quality factors is vital before selection of a raw material for the wet process is made, or change from one source to another can take place. Often, the impurities and non-desired elements contained in the phosphate rock will remain to some extent in the phosphoric acid produced. Therefore, the use for which the phosphoric acid is required is important for the choice of rock or a substitute. Currently, the Kemira GrowHow Oy's apatite mine in Finland, the only phosphate mine in Western Europe, extracts phosphate rock pure enough to produce phosphoric acid of food grade quality without further purification. For most phosphoric acid producers, further purification could be considered, if less desirable phosphorus sources are used, but not many cost-effective processes are available or used by the industry. If, however, the main objective is to produce phosphate fertiliser, then the requirements for purity are not as high as for food and feed grade phosphoric acid; a lower standard phosphoric acid can also be used in some technical purposes. Current requirements for phosphate rock to use in the wet process are that the iron and aluminium concentrations, in particular, should not be too high and that other metals, such as copper and magnesium, should not exceed certain critical concentrations. The raw material must be dry, because the water balance in phosphoric acid production is critical, and organic matter must be limited, as this might create significant problems especially in the filtration phase. Incineration ashes could be a possible source of phosphate and, depending on origin and solubility, it might not be necessary to solubilise with strong acids. However, some waste and sewage sludge incineration ashes might contain too much iron and other metals for this process.

15.4.2 Options and constraints

There are several options as well as concerns related to the use of recovered phosphate by the phosphate industry. Some of the technical and economic constraints described above might be overcome in the manufacturing process but a major constraint is the structure of the industry. Many of the issues that need to be resolved have been identified and appropriate development work considered and, in the best cases, this is already underway. Additional critical topics for the industry are the traceability and safety of raw materials used and liability and dependability.

Some of the issues, however, are not of a technical nature but have more to do with perception and some have also become political. The public view on recycling phosphate and other materials is often positive, but then concern is expressed over the use of recovered materials for the production of food and feed, for products that are used for skin application, and for use in the treatment of clothing and textiles. It could be suggested that this is irrational, but only if science can guarantee zero risk or provide enough evidence for no risk of pathogens and contaminants entering the food chain acceptability may be viable. In the past, attitudes towards recycling sludge and residues from food production were positive, due to the general sensitivity of the public to environmental issues. However, in the aftermath of recent food scares, public opinion has changed to one that considers it better to avoid such uses. Other critical issues are those of traceability and safety of inputs used in the production of food and other consumer goods. Consequently, the public perception of the raw materials used, production methodologies, ethics etc. can have a major impact on the future success of a whole industry.

15.4.3 Fertiliser and soil

The literature abounds with statements on the feasibility for the fertiliser industry to use or incorporate recovered phosphate in its products and for the agricultural sector to use recovered phosphate directly as a fertiliser (Gaterell *et al.*, 2000). It is also often suggested that agriculture would benefit by a slow release phosphatic fertiliser and that some of the recovered phosphates would have a premium value for this purpose. However, due to the nature of soil phosphates and the reactions of phosphate applied to soil, there is no need or desire for a slow release phosphatic fertiliser. Slow release phosphate from recovered phosphorus would not qualify therefore for a premium price to cover the extra cost of recovery and distribution. Depending on the composition of the recovered phosphate, its solubility and thus availability to plants varies. The majority of applied phosphate, independent of its origin, would be subject to normal soil chemical and biological reactions. The chemistry of phosphorus in the soil is complex because it may be associated with many different compounds to which it is bound with a range of bonding energies or strengths (Johnston and Steén, 2000; see also Haygarth and Condron, this volume). When phosphatic fertilisers, whether organic or inorganic, are added to soil, only a fraction of the phosphorus is taken up immediately by the crop. The remainder becomes adsorbed and possibly after further reactions absorbed to soil particles. The speed at which this sorption and other reactions occur very much depends on the type and size of the soil particles and the presence of other elements such as aluminium, iron and calcium, soil acidity and organic matter. The phosphorus in many of the recovered phosphates is not as soluble as in most mineral fertilisers and farmers might be reluctant to directly use recovered phosphates as fertiliser until more is known about their value as source of phosphorus for crop production. Some comparative studies of fertiliser value of recovered phosphatic compounds versus mineral fertiliser phosphates have been done. Generally, the results from these studies show that recovered phosphates directly used as fertiliser can have similar or inferior effectiveness to mineral fertiliser phosphate (Enskog and Johansson, 1999; Naturvårdsverket, 1997; Richards and Johnston, 2001). For the direct application of recovered phosphates to become successful, more studies under field conditions are required; furthermore the physical and chemical characteristics of the recovered phosphates must be acceptable to the farmer. This means that the declared P content, plant availability, particle size, strength and shape, and conditioning would need to be similar to those of a mineral fertiliser. Another option suggested is to produce or incorporate recovered phosphates into mineral fertilisers. Kemira GrowHow Oy has quite successfully tested the incorporation of chicken manure incineration ash in its production of mineral fertilisers. But, much more work is required before the use of recovered phosphates is practised widely. Besides the issues related to the direct use of recovered phosphates, and/or their incorporation into other fertiliser products, it must be recognised that the phosphorus status of European soils has improved. Today some 15 to 70%, depending on country, of soils test high or very high in readily available phosphorus (Johnston and Steén, 2000). Thus the need for phosphate fertilisers is likely to continue to decline, as it has done for the last 30 years, and farmers will look for the most cost effective materials to use. Although, there is a need to further tighten the phosphorus cycle in agriculture, and the use of recovered phosphates offers such an opportunity, other environmental and ecological effects must be taken into account.

Specific to the fertiliser industry, as for other sectors of the phosphate industry, is that the requirements on traditional raw materials for mineral fertiliser manufacture would apply also to substitutes (Steén, 1997, 1998a). And, regarding the use of recovered phosphates, these must be free from organic matter and the concentration of certain metals must be very low. Moreover, consistency in composition and availability are also very important. Fertiliser use of phosphate recovered from animal manures has been suggested as a plausible solution to remove surplus manure from regions with high animal densities to other regions. However, considerations regarding this solution would include the appropriateness of this removal strategy to resolve a political issue of animal production.

15.4.4 Inorganic feed phosphate

The inorganic feed phosphate industry in Europe is at present very reluctant to use recovered phosphate as a substitute for its current rock sources to produce phosphoric acid. Even if all the technical constraints and requirements on the raw material could be met, and chemically pure phosphatic feed supplements meeting all requirements on the finished product produced, this sector of the industry would still be very hesitant to use such raw materials. Not least for the public perception reasons discussed previously. Currently however, there is a ban in the EU on using meat and bone meal as raw material for the manufacture of calcium phosphate for feeding purposes (EC, 2001). This was imposed after the BSE outbreak, as it is perceived that it cannot be guaranteed that the prions (an abnormal form of a protein) causing BSE/TSE, and found in certain ruminant tissues, such as bone marrow, would be inactivated by the acid treatment. The risk assessments made included the risk for cross-contamination at feed mills, during transportation, and on the farm but a full quantitative risk assessment has not yet been made. This is mainly an issue of credibility in the public view, ethics in animal production and safe feed and food. Consequently, this industry sector can only afford to use raw material that can be traced back to its source of origin, that is guaranteed free from contaminants and undesired substances and that meets all relevant regulations and perhaps most important of all, the stakeholders' concerns.

15.4.5 Detergents and other high grade applications

The detergent industry sector has suggested that by 2010 it could be possible to achieve up to 25% substitution of current phosphate input by recovered phosphates in detergents (International Conference on the Recovery of Phosphates for Recycling from Sewage and Animal Wastes, 1998). This could possibly also apply to other high-grade applications. For the manufacture of detergents, this would be equal to about 25,000 to 30,000 t P, which is only a small part of the amount of phosphorus that can be available as recovered phosphate. Consequently, other sectors must consider using recovered phosphorus when this becomes available, provided that the quality meets appropriate standards. These standards may be achieved by further treatment and purification of recovered phosphates. Recovered phosphate from sewage generally may contain little impurities and contaminants such as heavy metals (see, for example, Ueno, this volume) and could therefore, be useful for high-grade applications. Moreover, it is believed that this type of uses would be more acceptable to the wider public than uses close to or integrated in the food and feed chain.

15.4.6 Other applications

For technical applications such as phosphoric acid for metal surface treatment, phosphorus in crop protection agents etc. appropriately treated recovered phosphates might be optional phosphorus sources. Suppliers of garden soil and composts could possibly enrich their products with recovered phosphates. However, a small initial survey among garden soil suppliers revealed some reluctance to incorporate recovered phosphate into their products. The main concerns were the risk of inconsistency of quality and composition, impurities, contaminants and undesired substances, and the solubility and plant availability of the phosphate.

15.5 FUTURE PROSPECTS

To conclude, there are many driving forces for P-recovery, but the progress regarding implementation and turning into reality is slow. Does this mean the driving forces are not strong enough? Research has been going on for decades and there will always be more to study and investigate, even though today there is much knowledge and suitable technologies available. Costs are considerable and this is always the case with emerging technologies (see Köhler, this volume).

There are numerous policies, both at national level as well as international commitments, relevant to this discussion, but there is no specific policy or political route for P-recovery. In spring 2001, the Swedish government commissioned the Environment Protection Agency to propose an action plan to achieve a practical and feasible degree of P-recycling, whilst at the same time retain protection of human health and the environment. Initially, a P recycling rate of 75% by 2010 was proposed. This high recycling rate cannot be implemented by reliance on land spreading of sewage sludge only. The Swedish EPA presented the investigations and the report (Naturvårdsverket 2002a, 2002b, 2002c) in 2002 and finally suggested that by 2015 a minimum 60% of the P in sewerage should be used on productive land, at least half of it on agricultural land. With this proposal, the current development and construction plans for full scale P-recovery from sewage and potentially from incineration ash were discontinued, and, presently, research in P-recovery in Sweden has stalled.

In Germany the Federal Environment Agency have recently, March 2003, initiated a discussion on recovery and recycling of phosphate. This seems promising and has already encouraged funding of research and development work and possible construction of technical installations.

To conclude, more than anything else, it is evident that there is a need for a clear and targeted political strategy regarding improved management of P (as well as other resources) otherwise the development will continue at very slow pace.

REFERENCES

2001 Directory of Chemical Producers (2002) Europe Volume 2, SRI International, California.
Agriculture, Environment, Rural Development: Facts and Figures – A Challenge for Agriculture (1999) Europe, European Communities, Luxembourg.
Aquatic Environment *1999* (2000) European Environment Agency, Copenhagen.
Brett, S., Guy, J., Morse, G.K. and Lester, J.N. (1997) *Phosphorus Removal and Recovery Technologies*. Selper Publications, London.
Becker, P. (1988) *Phosphate and Phosphoric Acid: Raw Materials, Technology and Economics of the Wet Process*, 2nd edn. Marcel Dekker, NY.
Booker, N.A., Priestley, A.J. and Fraser, I.H. (1999) Struvite formation in wastewater treatment plants: opportunities for nutrient recovery. *Environ. Technol.* **20**(7), 777–782.
Brouwer, F.M., Godeschalk, F.E., Hellegers, P.J.G.J. and Kelholt, H.J. (1995) *Mineral Balances at Farm Level in the European Union*, onderzoeksverslag **135**, Agricultural Economics Research Institute, The Hague.
CEEP (1998) *Phosphate recovery from animal manure, the possibilities in the Netherlands*, prepared by van Voorneburg, F., van Ruiten, L.H.A.M. and ten Have, P.J.W., Report CEEP, Brussels.
De Clercq, P., Gertsis, A.C., Hofman, G., Jarvis, S.C., Neeteson, J.J. and Sinabell, F. (eds), (2001) *Nutrient Management Legislation in European Countries*, Department of Soil Management and Soil Care, Gent.

Derden, A., Vaesen, A., Konings, F., ten Haeve, P. and Dijkmans, R. (1998) *Beste Beschikbare Technieken voor het be-en verwerken van dierlijke mest*, Vlaams BBT-Kenniscentrum, Academia Press, Gent.
Driver, J., Lijmbach, D. and Steén, I. (1999) Why recover phosphorus for recycling and how. *Environ. Technol.* **20**(7), 651–662.
Durrant, A.E. (1997) *The Feasibility of Recovering Phosphate from Waste Water in a Form Suitable for Use as a Raw Material by the Phosphate Industry*, Imperial College of Science, Technology and Medicine, Centre for Environmental Technology, University of London, London.
Durrant, A.E., Scrimshaw, M.D., Stratful, I. and Lester, J.N. (1999) Review of the feasibility of recovering phosphate from wastewater for use as a raw material by the phosphate industry. *Environ. Technol.* **20**(7), 749–758.
EC (1975) Council Directive (75/442/EEC) on waste, the European Commission, Brussels.
EC (1986) Council Directive (86/278/EEC) on the protection of the environment, and in particular of the soil, when sewage sludge is used in agriculture, the European Commission, Brussels.
EC (1991a) Council Directive (91/676/EEC) concerning the protection of waters against pollution by nitrates from agricultural sources, the European Commission, Brussels.
EC (1991b) Council Directive (91/271/EEC) concerning urban waste water treatment, the European Commission, Brussels.
EC (1996) Council Directive (96/61/EEC) concerning integrated pollution, prevention and control, the European Commission, Brussels.
EC (1999) Council Directive (99/31/EC) on the landfill of waste, The European Commission, Brussels.
EC (2000) Council Directive (2000/60/EC) establishing a framework for Community action in the field of water policy, the European Commission, Brussels.
EC (2001) Regulation No. 999/2001 of the European Parliament and of the Council of May 22, 2001, laying down the rules for the prevention, control and eradication of certain transmissible spongiform encephalopathies, the European Commission, Brussels.
Edge, D. (1999) Perspectives for nutrient removal from sewage and implications for sludge strategy. *Environ. Technol.* **20**(7), 759–763.
EFMA (2002) *Forecast of Food, Farming and Fertilizer Use in the European Union 2002 to 2012*. The European Fertilizer Manufacturers' Association, Brussels.
Emsley, J. (2000) *The Shocking History of Phosphorus*. Macmillan Publishers Ltd, London.
Enskog, L. and Johansson, L. (1999) *Bra fosforprodukter från avloppsslam*, Rapport 29, Stockholm Vatten, Stockholm.
Eriksson, J. (2001) *Concentrations of 61 trace elements in sewage sludge, farmyard manure, mineral fertiliser, precipitation and in oil and crops*, Report 5159, Swedish Environmental Protection Agency, Stockholm.
European Commission (1999) *EU focus on waste management*. Office for Official Publications of the European Communities, Luxembourg.
European Commission (2001) *Disposal and recycling routes for sewage sludge*. Office for Official Publications of the European Communities, Luxembourg.
European Environment Agency (1995) *Europe's Environment: The Dobris Assessment*. European Environment Agency, Copenhagen.
European Environment Agency (1999a) *Environment in the European Union at the turn of the century*. European Environment Agency, Copenhagen.
European Environment Agency (1999b) *Nutrients in European Ecosystems*. Environmental assessment report 9, European Environment Agency, Copenhagen.
European Environment Agency (2002) *Environmental Signals 2002, Benchmarking the Millennium*. Office for Official Publications of the European Communities, Luxembourg.
Gaterell, M.R., Gay, R., Wilson, R. and Lester, J.N. (2000) An economic and environmental evaluation of the opportunities for substituting phosphorus recovered from wastewater treatment works in existing UK fertiliser markets. *Environ. Technol.* **21**, 1067–1084.

Greaves, J., Hobbs, P., Chadwick, D. and Haygarth, P. (1999) Prospects for the recovery of phosphorus from animal manures: a review. *Environ. Technol.* **20**(7), 697–708.

IFA (2002) *Fertilizer Consumption Statistics*. International Fertilizer Industry Association, Paris.

International Conference on the Recovery of Phosphates for Recycling from Sewage and Animal Wastes (1998) *Summary of Conclusions and Discussions*, Warwick.

Jeanmaire, N. and Evans, T. (2001) Technico-economic feasibility of P-recovery from municipal wastewaters. *Environ. Technol.* **22**(11), 1355–1361.

Johnston, A.E. and Steén, I. (2000) *Understanding Phosphorus and its Use in Agriculture*. EFMA, Brussels.

Mengel, K. and Kirkby, E.A. (1987) *Principles of Plant Nutrition*, 4th edn. International Potash Institute, Bern.

Minnesota Pollution Control Agency (1990) *Sewage Sludge Ash Use in Bituminous Paving*. Report to Legislative Commission on Waste Management, Minnesota DOT, Minnesota Pollution Control Agency, Metropolitan Waste Control Commission, Minnesota.

Moriyama, K., Kojima, T., Minava, Y., Matsumoto, S. and Nakamachi, K. (2001) Development of artificial seed crystal for crystallization of calcium phosphate. *Environ. Technol.* **22**(11), 1245–1252.

Morse, G.K., Lester, J.N. and Perry, R. (1993) *The Economic and Environmental Impact of Phosphorus Removal from Wastewater in the European Community*. Selper Publications, London.

Naturvårdsverket (1997) *Fosfor och energi ur avloppsslam*, Rapport 4822, Naturvårdsverket, Stockholm.

Naturvårdsverket (2002a) *Aktionsplan för återföring av fosfor ur avlopp*, Rapport 5214, Naturvårdsverket, Stockholm.

Naturvårdsverket (2002b) *System för återanvändning av fosfor ur avlopp*, Rapport 5221, Naturvårdsverket, Stockholm.

Naturvårdsverket (2002c) *Samhällsekonomisk analys av system för återanvändning av fosfor ur avlopp*, Rapport 5222, Naturvårdsverket, Stockholm.

Otabbong, E. (1997) Agronomic value and behaviour of sewage sludge in incubation and pot experiments, report 30, Department of Soil Sciences, The Swedish University of Agricultural Sciences, Uppsala.

Priestley, A.J. (no date, but internal evidence suggests 1990s) *Report on Sewage Sludge Treatment and Disposal – Environmental problems and research needs from an Australian perspective*, CSIRO, Division of Chemical and Polymers, http://www.eidn.com.au/ukcsirosewagesludge.htm, Victoria.

Richards, I.R. and Johnston, A.E. (2001) *The effectiveness of different precipitated phosphates as sources of phosphorus for plants*. http://www.nhm.ac.uk/mineralogy/phos/Richardsjohnston.pdf, London.

Schipper, W.J., Klapwijk, A., Potjer, B., Rulkens, W.H., Temmink, B.G., Kiestra, F.D.G. and Lijmbach, A.C.M. (2001) Phosphate recycling in the phosphorus industry. *Environ. Technol.* **22**(11), 1337–1345.

Second International Conference on the Recovery of Phosphorus from Sewage and Animal Wastes (2001) *Proceedings*, Nordwijkerhout.

Steén, I. (1997) *Recycled Plant Nutrients from Industrial Processes as Resources for Manufactured Fertilizers*, Proceedings of 10th International Symposium of CIEC 1996 – Recycling of Plant Nutrients from Industrial Processes (eds Schnug, E. and Szlabolcs, I.), Braunschweig-Völkenrode.

Steén, I. (1998a) *New Development in the Fertilizer Industry*, Proceedings Volume III 11th International World Fertilizer Congress CIEC 1997 – Fertilization for Sustainable Plant Production and Soil Fertility (eds Van Cleemput, O., Haneklaus, S., Hofman, G., Schnug, E. and Vermoesen, A.), Gent.

Steén, I. (1998) *Phosphorus availability in the 21st century: Management of a Non-Renewable Resource*. Phosphorus and Potassium No 217, CRU Publishing, London.

Steén, I. (2001) *Summary of general standards in the North European Countries*. NJF Seminar 322 optimal nitrogen fertilization – tools for recommendation, DIAS Report plant production no. 84, The Danish institute of agricultural sciences, Foulum.

The 5th Environment Action Programme (1993) *Towards Sustainability*. Official Journal of the European Communities, No C 138/7, 17.5.1993, the European Commission, Brussels.

The 6th Environment Action Programme (2001) *Environment 2010 – Our future, Our Choice*, Com (2001)31, the European Commission, Brussels.

The Chemical Economics Handbook (2002) SRI International, California.

Tunney, H., Carton, O.T., Brookes, P.C. and Johnston, A.E. (eds) (1997) *Phosphorus Loss from Soil to Water*. CAB International, Oxon.

Ueno, Y. and Fujii, M. (2001) Three years experience of operating and selling struvite from full-scale plant. *Environ. Technol.* **22**(11), 1373–1381.

United Nations Industrial Development Organization (UNIDO) and International Fertilizer Development Center (IFDC) (1998) *The Fertilizer Manual*, 3rd edn. Kluwer Academic Publishers, Dordrecht, The Netherlands.

Woods, N.C., Sock, S.M. and Daigger, G.T. (1999) Phosphorus recovery technology modeling and feasibility evaluation for municipal wastewater treatment plant. *Environ. Technol.* **20**(7), 663–679.

16
Fluid dynamic concepts for a phosphate precipitation reactor design

D. Mangin and J.P. Klein

16.1 INTRODUCTION

Precipitation is a complex process in which a chemical reaction proceeds to initiate the formation of a solid. The solid particles generated can be amorphous or crystalline. However, as far as fundamental concepts are concerned, crystallinity has no influence in reactor design. The two stages of precipitation, chemical reaction and particle formation, each have their own kinetics. Reactor design and characterization will then need to consider the respective kinetics of these two processes. The hydrodynamic conditions in the precipitation reactor can also affect strongly the size and quantity of the produced particles. The effect of the hydrodynamic conditions is particularly important, if the precipitation kinetics are faster than, or of the same order of magnitude as, the mixing kinetics. Indeed, in such case, the mixing kinetics will govern to a large extent the global process.

© 2004 IWA Publishing. *Phosphorus in Environmental Technology: Principles and Applications.* Edited by Eugenia Valsami-Jones. ISBN: 1 84339 001 9

This can induce, for instance, high supersaturation levels and consequently high nucleation rates at the reactant feed points, if the reaction occurs significantly before complete mixing is achieved. This, in turn, will result in the formation of smaller than expected particles. Mixing effects have to be considered at two different scales: (a) the global homogeneity at reactor scale (i.e. macromixing), and (b) the intimate mixing of the reactants at microscopic scale (i.e. micromixing). These two steps of macromixing and micromixing, and the methodologies developed to evaluate their respective intensities, are described in the next section. Correlation equations, allowing estimates of their duration, will also be given.

In a precipitation reaction, primary nucleation is probably the mechanism most affected by hydrodynamics. However, other mechanisms, such as secondary nucleation, growth, agglomeration or breakage are also dependent upon hydrodynamics. The impact of the hydrodynamic conditions on the precipitation mechanisms is analysed in a later section.

The final three sections deal with the precipitation reactor design and modelling. Two main types of reactors are being presented: mechanically stirred reactors (which are commonly used in the field of precipitation) and fluidised bed reactors (which have recently been used for phosphate recovery by precipitation).

Although the concepts considered here are general, this chapter serves to support later chapters presenting reactor designs for phosphate recovery from waste water. It may also serve as a guide for the development of future reactor designs.

16.2   FLUID DYNAMICS AND MIXING

16.2.1   Fluid segregation

Let us consider a solution of a reactant B, which is poured into a reactor containing a solution of a reactant A. The mixing of the two solutions will go through several steps.

The first step corresponds to 'macromixing.' Large 'fluid aggregates' of size L_S form, stretch and split in the reactor without significantly exchanging matter with each other. These aggregates each have almost uniform composition, albeit different from one another. At this stage, the fluid is considered to be totally segregated and is called a 'macrofluid.'

One can generally consider that the 'micromixing' process relates to the decrease in segregation, described in the following steps.

In the second step, the macrofluid aggregates progressively diminish in size, essentially under the action of turbulent shear. At the same time, the contact surface, and consequently the viscous friction between the aggregates, increases with time. This second step will proceed until the viscous friction forces become so high that the action of the turbulence becomes negligible. The fluid aggregates have then reached the so-called Kolmogorof microscale λ_K (minimum size of the turbulent eddies – in m) given by:

$$\lambda_K = \left(\frac{v^3}{\varepsilon}\right)^{1/4} \qquad (16.1)$$

where v (m^2 s^{-1}) is the fluid kinematic viscosity and ε (W kg^{-1}) is the local dissipated specific power.

Then, in the third step, the mixing is performed by laminar shear and molecular diffusion until a final size is reached, below which only diffusion occurs. This size corresponds to the Batchelor microscale (minimum size of the fluid aggregates – in m):

$$\lambda_B = \frac{\lambda_K}{\sqrt{Sc}}, \qquad Sc = \frac{v}{D_m} \qquad (16.2)$$

where D_m is the molecular diffusivity (m^2 s^{-1}).

Finally, the 'microfluid' stage corresponds to a final state in which there is perfect mixing at molecular level and no concentration fluctuations. Table 16.1 gives an order of magnitude of the different segregation scales in water.

A real fluid presents a segregation state intermediate between microfluid and macrofluid. For the reaction between A and B to occur, it is necessary that the micromixing has sufficiently progressed to allow significant contact between the reactants. Each step has its own duration, and micromixing can locally be significant before macromixing is achieved throughout the entire reactor.

A local inhomogeneity can often appear near the reactant feed point, if the reactant B is fed into the reactor through a pipe. This phenomenon, of great importance in precipitation, can be studied by considering an intermediate process between macro and micromixing called 'mesomixing.'

16.2.2 Macromixing – residence time distribution

Macromixing can be defined as the process which results in uniformity of the local average value of the concentrations of all the species present in the vessel.

16.2.2.1 Macromixing time in a batch stirred vessel

In a batch stirred vessel, the macromixing kinetics can be evaluated by 'mixing time' measurement. A tracer is injected at a given point into the vessel and its

Table 16.1 The different scales of turbulence in water calculated with $\varepsilon = 1$ W/kg.

v	D_m	L_S	λ_K	λ_B
10^{-6} m^2 s^{-1}	10^{-9} m^2 s^{-1}	$5 \cdot 10^{-3}$ m	32 µm	1 µm

concentration is monitored at another point. The measurement method can be either conductimetry (with KCl, NaCl or $CaCl_2$ in aqueous solution as tracer), colorimetry or discoloration of an indicator, involving an acid–base or an oxidation–reduction reaction. The response curve, expressed in terms of time normalized concentration C/C_∞ is shown in Figure 16.1, for a well agitated vessel. It is possible to identify the circulation time t_c which corresponds to the time separating two consecutive peaks. This time is related to the pumping capacity of the stirrer (see Section 16.5.2). It illustrates the movement of large fluid aggregates of size L_S which are progressively destroyed. The mixing time t_m is the time required for the tracer to be homogenised in the entire reactor volume. Its determination is subjective, since an infinite time is in fact required to reach perfect homogenisation. Commonly, t_m is defined for a given residual fluctuation of concentration (fluctuations of $\pm 5\%$, $\pm 1\%$ or $\pm 0.5\%$ around the equilibrium concentration, for instance). However, it is important to state that this mixing time measures only macromixing since the probe used (whatever its nature) is not capable of recording what happens at the molecular scale.

An evaluation of the macromixing time t_m (in s) can also be obtained from the following correlation for a turbulent flow regime (Nienow, 1997):

$$t_m = 5.9 \left(\frac{d_V^2}{\bar{\varepsilon}} \right)^{1/3} \left(\frac{d_V}{d_a} \right)^{1/3} \tag{16.3}$$

where $\bar{\varepsilon}$ (W kg^{-1}) is the average specific power input, d_V (m) is the vessel diameter and d_a (m) is the agitator diameter.

Equation 16.3 assumes that the stirred vessel is correctly designed (see Section 16.5.6). It can be used for different types and sizes of agitators.

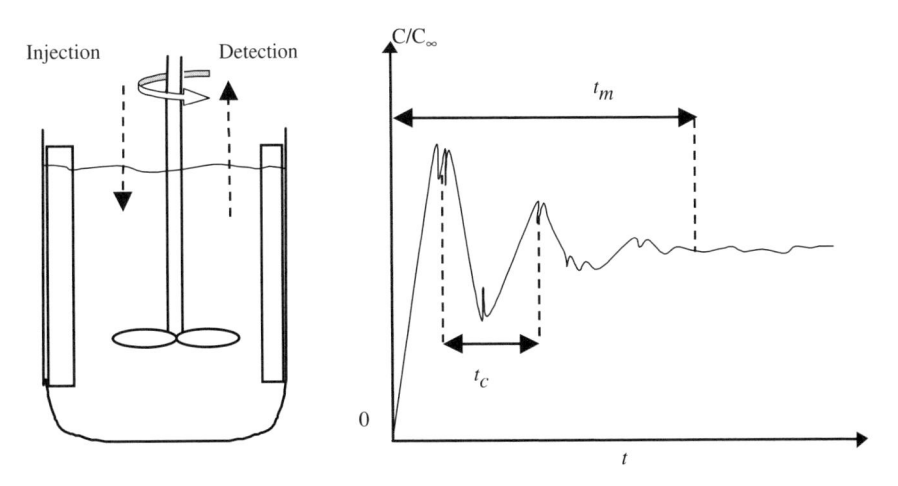

Figure 16.1 Measurement of (macro)mixing time t_m and of circulation time t_c.

16.2.2.2 Attainment of a macroscopic flow model in a continuous reactor

In the particular case of a continuous stirred reactor, the macromixing kinetics can still be evaluated through the macromixing time t_m estimated from batch experiment, provided this mixing time is far shorter than the mean residence time $\bar{t}_s$, or the mean holding time τ (see below Equations 16.6 and 16.7).

More generally, the hydrodynamic behaviour of continuous reactors is characterised at macroscopic scale by Residence Time Distribution (RTD) measurements (Froment and Bischoff, 1990; Levenspiel, 1972; Schweich and Falk, 2001; Villermaux, 1993). A tracer is injected in the inflow and its concentration is monitored in the outflow. Several injection methods are possible, but the most common is to inject an impulse of tracer, i.e. to inject a known amount of tracer during a very short time. The tracer must be such that it follows the flow inside the reactor in order to be representative of the movement of the fluid aggregates. This method allows attainment of a statistical analysis of the time spent by the fluid aggregates in the reactor, i.e. their residence time. Indeed, the residence time is a distributed parameter, since its value varies from one fluid aggregate to another. In the case of an impulse injection, the concentration vs. time curve obtained at the reactor outflow yields directly the RTD function $E(t_s)$. $E(t_s)$ is such, that the product $E(t_s)dt_s$ represents the fraction of fluid leaving the reactor after a residence time between t_s and $t_s + dt_s$. The fraction of fluid that has remained in the reactor for a period less than t_1 can then be derived from:

$$\int_0^{t_1} E(t)dt \qquad (16.4)$$

When time $t = \infty$, all tracer will exit the system, and this can be expressed by:

$$\int_0^{\infty} E(t)dt = 1 \qquad (16.5)$$

Finally, the mean residence time $\bar{t}_s$ can be calculated from:

$$\bar{t}_s = \int_0^{\infty} t E(t)dt \qquad (16.6)$$

Theoretically, this mean residence time should be equal to the mean holding time τ given by:

$$\tau = \frac{V}{Q} \qquad (16.7)$$

where V is the reactor volume (m^3) and Q is the volumetric flow rate (m^{3}s^{-1}). The key requirements in order to perform a correct RTD measurement are:

(1) The tracer injection must be such that it induces no perturbation. This can be difficult if a large quantity of tracer is required for its concentration to be measurable at the outflow.

(2) It is preferable to continuously monitor the outflow concentration, so as to detect possible bypasses.
(3) The tracer must be: (a) very soluble in the medium studied; (b) chemically inert with respect to the reactants, so as to avoid chemical degradation or pollution of the production; and (c) easy to measure at parts per million concentrations, to avoid the need of injecting too large quantities.

Radioactive tracers satisfy the above criteria, but their use in the industrial field may be restricted. To avoid such difficulties, it is possible to work with a model reactor fed with water and to use a tracer detected by conductivity.

The significance of measuring RTD is that firstly it allows assessment of a potential reactor malfunction. Such malfunction can be due to the existence of bypasses or dead zones. The shape of the RTD curve, a mass balance on the tracer, or a comparison between the calculated mean residence time (Equation 16.6) and the expected holding time (Equation 16.7) allow the detection of bypasses or dead zones (Schweich and Falk, 2001; Villermaux, 1993).

Furthermore, RTD allows modelling of the macroscopic flow in the reactor. A real reactor is generally described as an association of ideal zones, i.e. plug flows with or without axial dispersion, and perfectly mixed zones, in series or in parallel, potentially including bypasses, dead zones (with or without exchange with other zones) and recycling.

In the perfectly (macro)mixed reactor, the macromixing is maximum, since fully uniform concentration throughout the reactor is assumed. Conversely, the perfect plug flow reactor can be represented by a tube in turbulent flow conditions, whereby all the fluid elements move at a uniform velocity through the reactor. The concentrations are uniformly distributed in a cross section, but no intermixing occurs between two successive fluid elements moving along the flow path.

Let us consider, for example, the reactor shown in Figure 16.2a, which consists of a settling zone above a mixing zone (see also Section 16.5.7). This reactor is operated in continuous mode with respect to the liquid phase, which enters in the mixing zone and flows out at the top of the settler. The macroscopic flow model may then be composed of a perfectly mixed reactor followed by a plug flow reactor (Figure 16.2b). The theoretical RTD corresponding to this model is given in Figure 16.2c. This should have to be confirmed by tracer injection.

In general, a complete macroscopic flow model requires several parameters. The agreement between model and experiment can be improved by increasing the number of parameters. However, the number of parameters must remain reasonable with respect to the information contained in the RTD curve, and a maximum number of 3 parameters is generally acceptable.

16.2.3 Micromixing

Chemical reaction and nucleation are molecular-scale processes, and thus micromixing has to be considered in addition to macromixing. Even if the reactor is perfectly macromixed, mixing at the molecular scale can directly influence the course of chemical reaction and of nucleation.

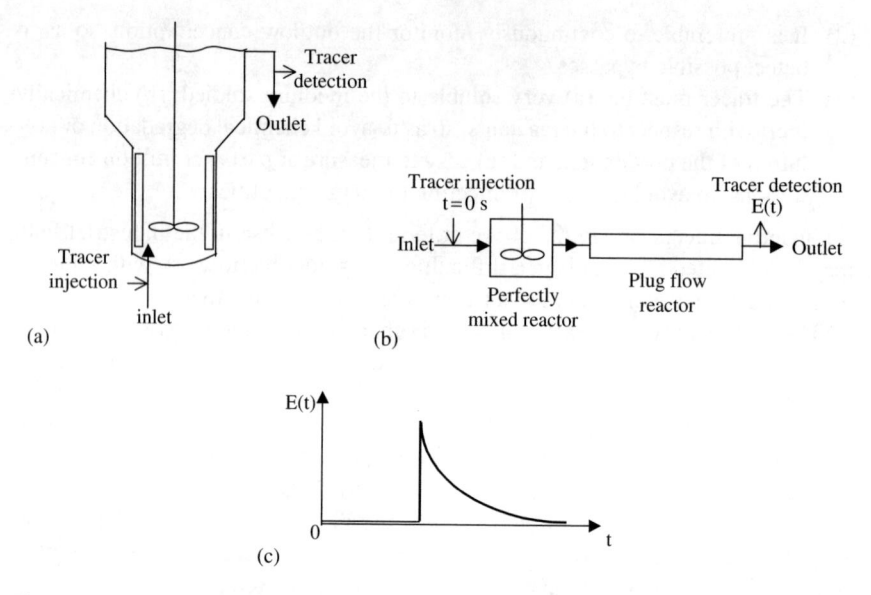

Figure 16.2 a) Schematic representation of reactor setup; b) hydrodynamic model; c) residence time distribution.

Micromixing was earlier (Section 16.2.1) described as the process of segregation decrease. This leads to the definition of two micromixing time constants:

(1) The first relates to the incorporation mechanism, also called mixing by engulfment, which concerns the scale change of fluid aggregates from $12\lambda_K$ to λ_K. This time constant is given by (Schweich and Falk, 2001):

$$t_{\mu i} = 17.2 \left(\frac{v}{\varepsilon}\right)^{1/2} \qquad (16.8)$$

(2) In the range between Kolmogorov scale λ_K and Batchelor scale λ_B, micromixing obeys a molecular diffusion process accelerated by deformation. The characteristic time constant for this second mechanism is of the form (Schweich and Falk, 2001):

$$t_{\mu d} = \frac{\text{Sc}}{(985 + 0.0175\,\text{Sc})} \left(\frac{v}{\varepsilon}\right)^{1/2} \qquad (16.9)$$

It should be emphasised that in both the above equations, ε is the *local* dissipated specific power, which can vary significantly from one point of the reactor to another.

The mechanism described by the highest time constant governs the micromixing process. For example, Table 16.2 gives order of magnitude values of these two

Table 16.2 Micromixing times in water calculated for $\varepsilon = 1\,\text{W}\,\text{kg}^{-1}$.

$t_{\mu i}$	$t_{\mu d}$
17×10^{-3} s	1×10^{-3} s

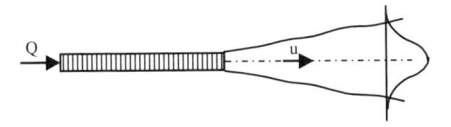

Figure 16.3 Radial dispersion of the concentration downstream from a feed pipe.

micromixing time constants in water, calculated for a specific input power $\varepsilon = 1\,\text{W}\,\text{kg}^{-1}$. It appears that $t_{\mu i} > t_{\mu d}$, which implies that engulfment controls micromixing in these conditions.

16.2.4 Mesomixing

In situations of semi-batch or continuous precipitation, a local inhomogeneity of scale larger than the molecular scale but smaller than the whole reactor scale, can appear near the reactant feed point. This phenomenon, intermediate between micromixing and macromixing, corresponds to the so-called 'mesomixing', introduced by Baldyga and Bourne (1992). Mesomixing describes the interaction of a plume of fresh feed with its surroundings. Two mechanisms have been identified:

(1) The feed stream is dispersed radially, transverse to its own flow direction, under the action of local turbulent dispersion (Figure 16.3). The characteristic time constant for this mechanism, t_{mesoD}, is given by (Baldyga and Bourne, 1992):

$$t_{mesoD} = \frac{Q}{u\,D_t} \qquad (16.10)$$

where Q (m³ s⁻¹) is the volumetric feed rate, u (m s⁻¹) is the local average velocity and D_t (m² s⁻¹) is the turbulent diffusivity.

(2) The large eddies of fresh solution in the course of dispersion are then desintegrated by inertial convective motions of fluid. During this process, the segregation scale is reduced from the integral scale of turbulence L_S towards the Kolmogorov scale λ_K (see Section 16.2.1). The characteristic time constant for this process is of the form (Baldyga et al., 1995; Corrsin, 1964):

$$t_{mesoC} \approx 2\left(\frac{L_S^2}{\varepsilon}\right)^{1/3} \qquad (16.11)$$

where ε (W kg⁻¹) is the local dissipated specific power.

The dominant mechanism is identified by comparing the two time constants. Thus, the slowest mechanism, i.e. the one described by the highest characteristic time, controls the mesomixing:

$$\begin{cases} t_{meso} = t_{mesoD} & \text{if } t_{mesoD} > t_{mesoC} \\ t_{meso} = t_{mesoC} & \text{if } t_{mesoC} > t_{mesoD} \end{cases} \quad (16.12)$$

The micromixing process occurs within the large eddies undergoing mesomixing. As a result, mesomixing determines the environment for micromixing. Here again, comparison of mesomixing and micromixing time constants allows a controlling mechanism to be identified. Thus, if $t_\mu \leqslant t_{meso}$, the local mesomixing effects may be significant. Inversely, if $t_\mu \gg t_{meso}$, micromixing alone controls the overall process.

16.3 INTERACTION BETWEEN MIXING AND PRIMARY NUCLEATION

Primary nucleation can be of homogeneous or heterogeneous type. Homogeneous primary nucleation corresponds to nuclei formation directly in the supersaturated solution. Heterogeneous primary nucleation occurs on foreign surfaces, which can be, for example, dust in suspension in the supersaturated solution or parts of the reactor itself. In both cases of primary nucleation, the nucleation rate B_{prim} is given by an equation of the form:

$$B_{prim} = A_{prim} \exp\left[\frac{-K_{prim}}{(\ln S)^2}\right] \quad (16.13)$$

where A_{prim} (m^{-3}s^{-1}) and K_{prim} (−) are kinetic constants depending on the mechanism under consideration.

S is the local supersaturation ratio. Its definition can vary according to two alternative precipitation scenarios (Klein and David, 2001):

(1) The chemical reaction leads to a rather soluble molecule P, which then crystallizes. An example of this is salicylic acid precipitation from sodium salicylate and sulfuric acid. S is then given by:

$$S = \frac{a_P}{a_{Pe}} \quad (16.14)$$

where a_{Pe} is the activity of the solute P in solution at equilibrium and a_P is the activity of P in the supersaturated solution (a_P can be deduced from the equilibrium constant of the chemical reaction).

(2) The chemical reaction directly leads to the solid. This is the case of an ionic reaction leading to a sparingly soluble salt. For the reaction,

$$xA^{z+} + yB^{z'-} \Leftrightarrow A_xB_y \downarrow \qquad (16.15)$$

S can be defined as:

$$S = \frac{a_A^x a_B^y}{K_S} \qquad (16.16)$$

where a_A is the activity of the cation A^{z+} in the supersaturated solution, a_B is the activity of the anion $B^{z'-}$ in the supersaturated solution and K_S is the solubility product.

The two primary nucleation processes are highly non-linear. In each case, the nucleation kinetics are very slow for small values of S (i.e. in the metastable zone of the nucleation mechanism under consideration) and become very high if S exceeds a critical value, S_{crit} (i.e. the metastable zone limit). This is due to the fact that these nucleation processes involve an activation energy. Nucleation is facilitated by the presence of foreign surfaces, with the result that heterogeneous primary nucleation requires a lower supersaturation to spontaneously occur: S_{crit} (heterogeneous) $< S_{crit}$ (homogeneous) (see Figure 16.5 in Section 16.4.1). However, if S exceeds S_{crit} (homogeneous), the homogeneous primary nucleation dominates (Blandin et al., 2001).

In reaction precipitation, high supersaturation levels can be easily reached, and it is common (especially with sparingly soluble components) for supersaturation to exceed, at least locally in the reactor, the metastable zone limit of the homogeneous primary nucleation. Consequently, very high primary nucleation rates, essentially of homogeneous type, can be obtained. The primary nucleation can take place so rapidly that it may occur before complete mixing can be achieved. This is described by the term 'mixing effects', which signifies that the mixing mechanisms influence the course of the primary nucleation and consequently the quantity of crystals produced and the final crystal size distribution (nucleation and growth are discussed in more detail by Koutsoukos and Valsami-Jones, this volume). The impact of these mixing effects depends on the characteristic time constants of the different processes involved.

At the beginning of the precipitation, three processes are competing:

(1) Micromixing, which brings the reactants into contact.
(2) Chemical reaction, which leads to supersaturation.
(3) Primary nucleation.

The appropriate micromixing time, t_μ, which will be a function of the local specific power input, can be estimated by the Equations 16.8 or 16.9.

The chemical reaction time t_R is inversely proportional to the reaction rate constant k_R:

$$t_R \propto \frac{1}{k_R} \qquad (16.17)$$

The chemical reaction can occur very rapidly, especially for ionic reactions, which can practically always be considered to be at equilibrium ($t_R \to 0$).

Finally, the characteristic time constant for primary nucleation t_{prim}, can be taken to be equal to the time required to form a characteristic number of nuclei. The variable t_{prim} is then inversely proportional to the nucleation rate B_{prim}:

$$t_{prim} \approx \frac{\bar{N}}{B_{prim}} \qquad (16.18)$$

where $\bar{N}$ can be the average concentration of particles present in the suspension (in number of particles per m³ of suspension).

If $t_{prim} \gg t_\mu$ (chemical reaction is assumed instantaneous), then the fluid can be considered as being at microfluid state, and no micromixing effects occur. Inversely, if, $t_{prim} \leq t_\mu$, micromixing affects or controls the overall process. The fluid is still partially segregated at molecular scale during the nucleation step, and micromixing proceeds between these fluid aggregates and their environment. In addition, macro and mesomixing, which both determine environment concentrations for micromixing have to be taken into account.

Let us now examine the mixing effects that can occur under different reactor configurations, and how they can be avoided or at least controlled. In each case, chemical reactions will be considered as instantaneous, forming via mixing of two ionic solutions containing the reactants A and B, and leading to the product P with or without a soluble intermediary.

16.3.1 Batch precipitation and single jet semi-batch precipitation in a stirred vessel

In batch precipitation, a volume V_{0B} of solution containing reactant B is rapidly poured into a volume V_{0A} of solution of reactant A. Mixing effects can be avoided if the two volumes are equal ($V_{0B} \approx V_{0A}$) (Marcant and David, 1991). In these conditions, molecular mixing takes place when the added fluid has been dispersed through the entire vessel allowing all A to be in contact with B. In fact, the macromixing time is generally higher than the nucleation time but, during the macromixing stage, few molecules of either reactant are in contact and primary nucleation is then negligible. The overall mixing process is thus controlled by micromixing, which can be faster than the nucleation, if the concentrations are sufficiently small.

If a small volume of a concentrated solution of reactant B is injected ($V_{0B} \ll V_{0A}$), three main scenarios are possible:

(1) Macromixing and/or mesomixing cannot be achieved before micromixing permits the contact between significant quantities of A and B. Then, micromixing will take place within an environment more concentrated in reactant B, than would have been if complete macro and meso mixing occurred. In addition to these limitations by macro and/or mesomixing, micromixing is likely to be also limiting, if nucleation is faster.

(2) Macro and mesomixing are achieved before micromixing permits significant contact between A and B. Only micromixing can be limiting here, if nucleation is fast.
(3) Macro, meso and micromixing can be achieved before significant nucleation occurs; the solution then acquires a microfluid state, as defined in Section 16.2.1.

In case 1 and potentially in case 2, the process is limited by mixing kinetics. In fact, in both these cases, the process resembles a situation where B was mixed with only part of A at the early stages of the process. Marcant and David (1991) have shown that, in these conditions, the local supersaturation can be higher than what would have been obtained with complete macro, meso and micromixing. Consequently, mixing effects can induce local supersaturations and subsequent local nucleation rates much higher than the mean supersaturation and the mean nucleation rate expected after complete mixing.

A situation similar to the above occurs in semi-batch single jet precipitation, where a solution of reactant B is progressively fed to a reactor previously filled in one step with solution of reactant A. As previously, very high local supersaturations and primary nucleations may be reached due to mixing effects. This phenomenon is influenced by the reactant concentrations, the reactant feed rate, the feed tube diameter, the stirring speed and the position of the feed point. Decreasing the concentrations, the feed rate and the feed tube diameter can reduce the mixing effects. With regard to the influence of the stirring speed and the feed point position, many different views have been reported in the literature, and these have been reviewed in detail by Tavare (2000).

An increase of the stirrer speed will accelerate all the mixing steps, but will not necessarily lead to the total suppression of mixing effects. If mixing effects subsist, the time constants of the different mixing mechanisms indicate that an increase of the stirring speed might cause a shift from micro to mesomixing as the dominant mechanism. It is therefore not necessary that an increase of the stirring speed will systematically reduce the mixing effects and the primary nucleation induced. This may explain why different trends are observed in the literature. Moreover, the impact of the stirring speed is difficult to study, since the stirring speed affects not only primary nucleation, but also almost all the subsequent stages of crystallisation (i.e. secondary nucleation, diffusion growth, agglomeration and breakage) with, sometimes, contradictory trends (Section 16.4).

The feed point position is probably the most sensitive operating parameter and a change to the location of the feed point can lead to drastically different results, in the presence of mixing effects, although no systematic rules can emerge from the literature. Changing the feed point position, while keeping the stirring speed constant influences the primary nucleation only. This is because the rates of the other crystallization mechanisms are not sensitive to the mixing effects (Section 16.4). Therefore, changing the feed point whilst retaining a constant stirring speed can be a test of the state of mixing in the reactor. If a change of the feed point location causes no modifications to the conversion curve or the crystal size distribution, then there is no influence of mixing on the course of precipitation.

Batch or semi-batch operation modes are often used to determine the precipitation kinetics in laboratory reactors. Kinetic parameters obtained in such way are only valid if the mixing effects are avoided.

Mixing effects are difficult to avoid in industrial reactors. In practice, it is recommended to position the feed point near the stirrer, preferably in its region of discharge, to prevent the stirrer from fouling. The objective of such choice is so as to minimize primary nucleation. Other configurations may induce lower primary nucleation and consequently lead to larger particles. But this particular configuration of the feed point, near the stirrer, allows better control of the mixing phenomena and facilitates subsequent scaling up. It is also recommended to work with moderate concentrations. Finally, it may be interesting to use several small feeding tubes instead of one single large tube.

16.3.2 Double jet semi-batch precipitation and continuous precipitation in a stirred vessel

In double jet semi-batch precipitation (i.e. a set-up that involves no suspension removal) or in double jet continuous precipitation (i.e. a set-up where suspension is continuously removed), the two reactants are fed in simultaneously, by means of two distinct tubes. Two main configurations of the tubes are then possible. The reactants can be fed close together (Figure 16.4a), or far apart from each other (Figure 16.4b). These two configurations lead to completely different behaviours.

In the first case, where the two feed pipes are close to each other and located in the zone of discharge of the stirrer, the two fresh and concentrated feed solutions are directly mixed together. High local supersaturation may then occur leading to elevated rates of nucleation and a small mean crystal size. Inversely, if the two feed points are far away from each other, the two plumes of fresh feed are mixed

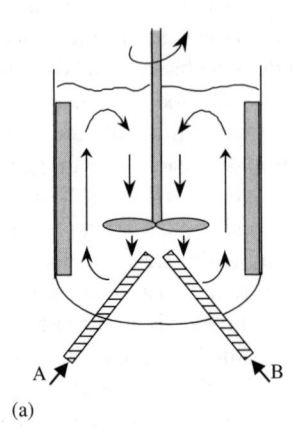

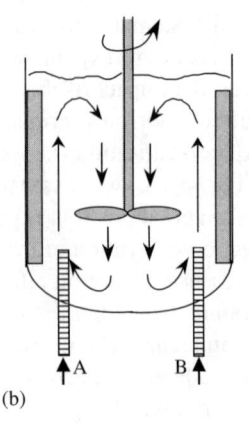

Figure 16.4 Double jet precipitation in a vessel equipped with an axial flow propeller (a) the two feed pipes are close together, and near the discharge of the stirrer; (b) the two feed pipes are far apart from each other.

with the bulk without directly exchanging mass between each other. In this way, a large dilution of both reactants occurs. Supersaturation levels much lower than in the previous case are obtained, leading to a larger mean crystal size.

An informative literature review on the influence of the operating parameters can be found in the papers of Tavare (2000) and David and Marcant (1994).

When the two feed points are close together, the effect of stirring speed can be difficult to interpret. This is due to the fact that an increase of the stirring speed can enhance the mixing of the two streams inducing higher local supersaturations, but can also favour a decrease of the local environment concentrations, by dilution with the bulk, and consequently a decrease of the local supersaturation. Since this operating mode, with the feed points close by, is usually chosen to produce small particles, a turbine-type stirrer, producing high local specific power in the discharge region (i.e. rapid local mixing) but low circulation rates (slow dispersion in the entire vessel), should be more appropriate. Note that such stirrer is also likely to globally favour the production of fine particles during the subsequent crystallization steps, notably because of the high secondary nucleation rates it can induce.

When the two feed points are far apart, the objective is to mix each plume of fresh feed with the entire bulk, and to avoid, as much as possible, direct mixing between the plumes. Furthermore, potential mixing effects between the bulk and the plumes pose fewer problems in double jet mode than in single jet mode (Section 16.3.1), because the reactant concentrations in the bulk are generally smaller. In these conditions, an increase of the stirring speed, which induces higher circulation rates and favours the macromixing of the fresh streams in the entire vessel, should result in lower supersaturation levels, lower nucleation rates and therefore, larger crystals. To favour macromixing, a profiled propeller of large diameter, producing large circulation rates, will be preferable to a Rushton turbine (see Section 16.5.5 for a description). In general, the use of a profiled propeller is always recommended for the production of large particles.

Whether the feed points are close or distant, it can be expected that the median crystal size is likely to increase with decreasing concentrations of the feeds. Low reactant concentrations are obtained in the bulk, if growth can consume the supersaturation brought by the two feed flows. It is therefore advisable to operate continuous precipitation with a long residence time, a high solid concentration (which is an independent parameter) and low reactant concentrations in the feed solutions. Regarding semi-batch operation, it is recommended that precipitation be seeded, fed with low flow rates and with solutions of low concentration. Under ideal conditions, no primary nucleation is expected if the solubility of the precipitate is sufficient. Sparingly soluble precipitates require attention because they easily reach high supersaturation levels; it is then difficult to avoid local nucleation induced by mixing effects.

To obtain large particles, it is necessary to reduce mean supersaturation and especially localized supersaturations. For this, it is recommended:

(1) to work with two feed points far from each other;
(2) to use a propeller favouring macromixing;

(3) to feed the precipitator with low concentration solutions;
(4) to work with long residence times and high solid concentrations (to act as seeds) in continuous mode;
(5) to work with low reactant flow rates and large amounts of seed in semi-batch mode;
(6) to increase solubility (by changing solvent, for instance).

Primary nucleation should then be avoided. Nevertheless, mixing effects and a local nucleation are likely to subsist with sparingly soluble components.

16.3.3 Fluidised bed reactors

A fluidised bed reactor (FBR) consists in general of a vertical cylinder, filled with solid particles which are brought to fluidised state by the reactant solutions fed at the bottom of the cylinder. All the crystallization processes can, a priori, occur along the column, the ideal case being in having growth only on the fluidised particles.

Nucleation, particularly primary nucleation, is undesirable because it creates fine particles, which do not have enough time to grow and are driven in the effluent by the upward flow. This is likely to reduce the global efficiency of the reactor. Agglomeration of the fines on the particles of the bed, and recycling of the effluent can reduce the efficiency drop. However, it is preferable, when possible, to act directly on primary nucleation.

FBRs can be operated with or without effluent recycling. In either case, the liquid phase behaves approximately like a plug flow reactor. Consequently, in the absence of mixing effects, the different inlet flows are rapidly mixed within a cross-section at the bottom of the column. From then on the system behaves as if the inlet flows were premixed. However, mixing effects are likely to occur. Local supersaturation levels, higher than the mean value expected if premixing was perfect, may then be reached.

To avoid, or at least reduce, primary nucleation at the bottom of the column:

(1) The reactant concentrations must be very low. It is important to note that in a FBR, the reactants are not mixed over the entire reactor volume but only within a cross-section. This should impose the use of lower inflow reactant concentrations than in continuous stirred reactor.
(2) It is recommended to operate the FBR with a recycling loop. With very high recycling flow rate, the FBR tends to become a perfectly mixed reactor. It is then possible to work with concentrations close to those of continuous stirred reactors.
(3) It may be appropriate to distribute the supersaturation over the entire reactor volume by injecting the reagents at different heights in the column (Seckler et al., 1996).

As mentioned above, whatever the chosen operating mode, sparingly soluble components require attention, because the metastable zone limit may easily be exceeded even with very low concentrations and consequently, avoiding primary nucleation becomes difficult, especially in the presence of mixing effects.

16.4 INTERACTION BETWEEN MIXING AND OTHER PRECIPITATION MECHANISMS

Up to this point, we have discussed only the initial stages of precipitation, and the meso/micromixing process, which brings the reactants in contact. This fast primary nucleation can then be affected by mixing effects, which generate locally and temporally high supersaturation. On the other hand, the stages that follow, i.e. growth, agglomeration, breakage and even secondary nucleation, are, normally, slow enough not to be influenced (or at least not significantly) by temporary local supersaturations induced by the mixing effects. Their rates are therefore simply a function of the average supersaturation induced by the macroscopic flow.

Consequently, interactions between mixing and other processes are independent of mixing effects. Their dependence on mixing is of a different nature. Indeed, it is the mechanisms themselves, which are affected by the hydrodynamic conditions, i.e. their rates are a direct function of the energy dissipation (which is not the case for primary nucleation). Note that, since the energy dissipation in the vicinity of the stirrer can be much higher than in the bulk of the vessel, especially if a turbine-type agitator is used, it is important to consider the zone near the stirrer, in addition to the average volume, when these processes are studied.

16.4.1 Secondary nucleation

Secondary nucleation corresponds to the formation of nuclei by detachment from the growing particles. Three mechanisms have been identified.

16.4.1.1 Surface secondary nucleation

Surface secondary nucleation can be described by considering a two-dimensional nucleation mechanism on the surface of the particles already present (Mersmann, 1996). The new surface nuclei are then detached by the shear exerted by the fluid or by impacts undergone by the particles. The rate of this nucleation mechanism, B_{surf}, is given by:

$$B_{surf} = A_{surf} \, S_{cryst} \, \exp\left[\frac{-K_{surf}}{\ln S}\right] \quad (16.19)$$

where A_{surf} ($m^{-5} s^{-1}$) and K_{surf} (−) are kinetic constants, S (−) is the supersaturation ratio and S_{cryst} (m^2) is the surface of the particles already present.

The influence of hydrodynamics does not appear explicitly in Equation 16.19. Further research is needed to quantify this influence, though little effect is expected, since minimal energy dissipation is required to bring the particles in suspension.

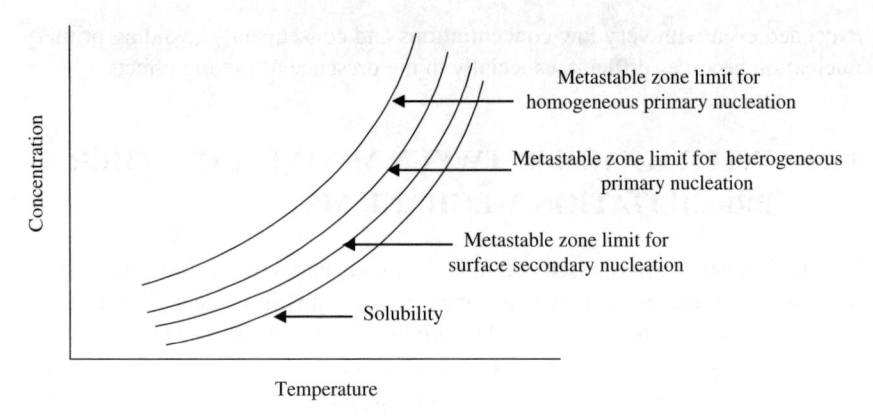

Figure 16.5   Metastable zone limits for all the different nucleation mechanisms.

Therefore, although hydrodynamics may play a role, most likely supersaturation is by far the principal parameter controlling surface secondary nucleation.

In the same way as homogeneous and heterogeneous primary nucleation, surface secondary nucleation is an activated process and has its own metastable zone limit (see Section 16.3). Amongst the different nucleation mechanisms, surface secondary nucleation requires the lowest supersaturation to spontaneously occur (Figure 16.5). A comparison of the respective intensities of the various nucleation mechanisms shows that homogeneous primary nucleation is by far predominant, when its own metastable zone limit is exceeded (Blandin et al., 2001). At lower supersaturation, close to the metastable zone limit for heterogeneous primary nucleation, heterogeneous primary nucleation and surface secondary nucleation can be of the same order of magnitude (Nallet et al., 1998).

Knowing the metastable zone limit for the surface secondary nucleation is particularly important in continuous precipitation. Indeed, continuous precipitation is generally performed at sufficiently low supersaturation so that primary nucleation is avoided (or strongly reduced if it cannot be avoided because of mixing effects). However, surface secondary nucleation can still be an important source of fines if its critical supersaturation is exceeded. This probably explains the high rate of fines obtained by Hirasawa et al. (2002) during magnesium ammonium phosphate (MAP) precipitation.

16.4.1.2   Secondary nucleation by contact; attrition

Secondary nucleation by contact is the result of macro-attrition of the particles. It is essentially caused by impacts sustained by the particles (due to collision with other particles or more likely impact with the stirrer or other parts of the reactor). Several authors, such as Gahn and Mersmann (1997, 1999) and Marrot (1996, 1997) have studied the impact of a crystal with a flat hard object and have proposed a mechanistic modelling of the phenomenon. In practice, the secondary nucleation by contact rate in a stirred vessel, B_{att} is simply evaluated through an empirical equation of the form (Ottens and de Jong, 1973):

$$B_{att} = K_{att}\, \varepsilon^h (C - C^*)^f (M_T)^g \qquad (16.20)$$

where K_{att} is a kinetic parameter, function of temperature, ε (W kg^{-1}) is the specific power input, C (mol m^{-3}) is the working concentration, C^* (mol m^{-3}) is the concentration at saturation (($C - C^*$) being the absolute supersaturation), and M_T (in kg of solid per m^3 of solution) is the suspended particle concentration. Exponents f and g have values between 0.5 and 3, and 0.5 and 2, respectively. Exponent h can range between 0 and 1.

The empirical parameters in Equation 16.20 strongly depend on the nature of the suspended particles and on the hydrodynamics, i.e. the reactor and the stirrer. The significance of the stirrer is evident when considering that a turbine type agitator will cause much more attrition than a marine or a profiled propeller.

Unlike other types of nucleation, contact nucleation is not an activated process and does not require a critical supersaturation to occur. It can happen at any supersaturation, and its rate increases moderately with increasing supersaturation. Attrition affects primarily particle edges, which re-grow, if supersaturation persists (Gahn and Mersmann, 1997, 1999). The higher the supersaturation, the smaller the size of the attrition fragments that can survive, according to Gibbs law.

At high supersaturation, secondary nucleation by contact is generally negligible with respect to the other nucleation mechanisms. When it is the only form of nucleation, i.e. at low supersaturation, secondary nucleation by contact can be moderate in a fluidised bed reactor or in a stirred vessel equipped with a profiled propeller, but can become significant if a turbine type agitator is used.

16.4.1.3 Apparent secondary nucleation

Apparent secondary nucleation occurs in seeded precipitation, when fines, produced by attrition during seed preparation, are attached to the seed particles by static electricity. These fines will detach and grow when the seed is introduced in solution. The final crystal size distribution will then be affected. Mixing has no influence on this phenomenon.

16.4.2 Crystal growth

Crystal growth can be described by the film model, with the mechanism of mass transfer across the diffusion layer surrounding the crystal, in competition with the surface integration mechanism of the solute molecule into the crystal (see also Koutsoukos and Valsami-Jones, this volume). The linear growth rate G (m s^{-1}) of a crystal of characteristic size L, is then given by:

$$G = \frac{dL}{dt} = \frac{\Phi_s M_s k_c}{3 \rho_s \Phi_v} \eta_r (C - C^*)^j \qquad (16.21)$$

where $\Phi_s(-)$ and $\Phi_v(-)$ are the surface and volumetric shape factors respectively, M_s (kg mol^{-1}) is the molecular weight of the solid, ρ_s (kg m^{-3}) is the solid density, K_c (mol$^{(1-j)}$m$^{(3j-2)}$s^{-1}) is the kinetic factor of the integration growth rate, j is the integration growth rate exponent, usually ranging between 1 and 2 (Klein et al., 1989), and $\eta_r(-)$ is the effectiveness factor introduced by Garside (1985) and defined by:

$$\eta_r = \frac{G_{experimental}}{G_{integration}} \quad (16.22)$$

where $G_{experimental}$ is the actual growth rate and $G_{integration}$ is the growth rate in the absence of transfer limitation.

Garside (1985) showed that η_r was solution of:

$$\left[\frac{k_c}{k_d}(C-C^*)^{j-1}\right]\eta_r + \eta_r^{1/j} - 1 = 0 \quad (16.23)$$

where k_d (m s^{-1}) is the mass transfer coefficient.

The mass transfer coefficient k_d depends on the degree of agitation, the physical characteristics of the liquid, the reactor dimensions, and the crystal size. Several equations to calculate k_d have been reported in literature.

For a stirred vessel, the following correlation of Herndl and Mersmann (1981) is appropriate:

$$Sh = 2.0 + 0.6\, Re_p^{1/3}\, Sc^{1/3} \quad (16.24)$$

where the Sherwood number is: $Sh = \dfrac{k_d L}{D_m}$

the Schmidt number is: $Sc = \dfrac{\nu}{D_m}$

and the Reynolds number for the particles is:

$$Re_p = 0.139 \left(\frac{(\bar{\varepsilon}L^4)}{\nu^3}\right)^{4/9} \left(\frac{N_a d_a^2}{\nu}\right)^{0.133} \quad (16.25)$$

where D_m (m^2 s^{-1}) is the molecular diffusion coefficient of the solute in the solution, L (m) is the particle size, $\bar{\varepsilon}$ (W kg^{-1}) is the average specific power input, V (m^2 s^{-1}) is the kinematic viscosity, N_a (s^{-1}) is the agitator rotation speed and d_a (m) is the agitator diameter.

The correlation proposed by Levins and Glastonbury (1972) is also appropriate for a stirred vessel.

With a fluidised bed, one can use the Froessling equation:

$$Sh = 2.0 + 0.6 \ Re_p^{1/2} \ Sc^{1/3} \qquad (16.26)$$

with the Reynold number for the particles being: $Re_p = \dfrac{u_r L}{v}$

where the dimensionless numbers Sh and Sc are defined as above and u_r (m s^{-1}) is the relative particle rate, with respect to the liquid.

When the mass transfer kinetics are slow with respect to the integration kinetics, the global crystal growth process is controlled by mass transfer, i.e. is 'diffusion controlled'. The growth rate is then totally dependent on the hydrodynamic conditions.

Given that k_d decreases as crystal size increases, growth is integration controlled, when particles are small (from nuclei size until around few μm – depending on the supersaturation – in an aqueous solution and a classical stirred vessel) and becomes mass transfer controlled when the particles become larger. Similarly, a decrease in supersaturation can cause a shift from an integration-controlled mechanism, to a diffusion-controlled mechanism. For these reasons, it is essential to take into account both mechanisms when a crystallization process is being studied from the beginning to the end.

For growth controlled by diffusion, only k_d is needed to be known, and this can be evaluated by means of the above equations (with $k_c \ \eta_r = k_d$ and $j = 1$ in Equation 16.21). Conversely, when growth is controlled by the integration mechanism, or governed by both the integration and mass transfer processes, k_c and j have to be known.

16.4.3 Agglomeration

Agglomeration is the aggregation of primary particles, followed by development of crystalline bridges between the particles (David et al., 1991, 1995; Seyssiecq et al., 1998). Indeed, during precipitation, the supersaturation can allow strong attachment between the aggregating crystals. Agglomeration differs from aggregation and flocculation, which can occur in saturated or undersaturated solutions and which involve only weak bonding forces such as van der Waals or hydrophobic forces. Note that a distinction can also be made between two types of agglomerates obtained under supersaturation: (a) the agglomerates produced from crystal–crystal collisions, and (b) polycrystals, dendrites and twins which result from crystal growth under non-ideal conditions. It is often difficult to distinguish between these two different origins, simply by observing the particles under the microscope. However, here, only the first process based on crystal–crystal collision will be considered and referred to as agglomeration.

The rate of agglomeration of two particles of size L_i and L_j may be given by:

$$r_{ij} = K_{ij} \ N_i \ N_j \qquad (16.27)$$

where N_i and N_j are the concentrations (number of particles per unit of volume) of particles of size L_i and L_j, respectively, and K_{ij} is the agglomeration kernel, i.e. the agglomeration kinetic 'constant'.

According to David *et al.* (1991, 1995) and Seyssiecq *et al.* (1998), the agglomeration process is divided into three steps:

(1) collision (mainly of binary type at ordinary solid concentrations),
(2) association, i.e. probability that the collision succeeds in keeping the particles together for a sufficient time,
(3) consolidation by crystal growth.

Steps (1 + 2) correspond to aggregation in the absence of supersaturation. An extensive literature review of aggregation kernels is given in Seyssiecq *et al.* (2000).

16.4.3.1 Collision frequency

In precipitation reactors, flow is generally turbulent. This leads to two different collision mechanisms, depending on whether the particles are smaller or larger than the turbulence microscale (David *et al.*, 1995).

For particles smaller than the turbulence microscale (also called Kolmogorov microscale – see Section 16.2.1) the collisions between particles occur within the small eddies created by the turbulence. The flow inside these eddies is laminar. Aggregation is then induced by the shear field and concerns only particles embedded in the same eddy. The simple kernel developed by Saffman and Turner (1956) corresponds to this case:

$$K_{ij} = \left(\frac{8\pi}{15}\right)^{1/2} \left(\frac{\bar{\varepsilon}}{\nu}\right)^{1/2} (r_i + r_j)^3 \qquad (16.28)$$

where r_i and r_j are the radius of the two particles under consideration.

When the particles under consideration are too large to be contained in small eddies of Kolmogorov microscale size, their mechanism of encounter is different. Particles are transported by larger eddies of various sizes, which interpenetrate and project particles against one another. In order to represent this turbulent mechanism of encounter, some authors, such as Abrahamson (1975), proceeded by analogy to the kinetic theory of gases:

$$K_{ij} = 2^{3/2} \pi^{1/2} \left(\bar{u}_i^2 + \bar{u}_j^2\right)^{1/2} (r_i + r_j)^2 \qquad (16.29)$$

where $\bar{u}_i^2$ and $\bar{u}_j^2$ are the mean-square velocities of the two particles under consideration.

Note that whatever the case, the probability of two particles colliding increases with increasing energy dissipation.

16.4.3.2 *Association*

To describe association, the collision rate can be corrected by introducing an efficiency coefficient accounting for hydrodynamic interactions (e.g. drag forces) and physico-chemical interactions between particles (e.g. attractive van der Walls forces) (David *et al.*, 1995; Higashitani *et al.*, 1983; Kuboi *et al.*, 1984). The efficiency coefficient can be a simple non-dissociation probability, linked to the fragmentation mechanism described in Section 16.4.4.

16.4.3.3 *Consolidation*

If two colliding particles stay together long enough, a crystalline bridge may be formed between the particles (step (3), above). The above aggregation equations can then be corrected by a sticking frequency function of supersaturation (David *et al.*, 1991, 1995; Seckler *et al.*, 1996).

16.4.4 Fragmentation

Fragmentation is the breakage of the particles into large fragments. It differs from attrition by the size of the fragments produced. The fragmentation mechanism applies essentially to particles undergoing agglomeration. Indeed, primary crystals are more likely to undergo macro-attrition leading to numerous micro-fragments.

In terms of fundamental models, fragmentation associated to agglomeration can be taken into account either through the non-dissociation probability of the global agglomeration process (Section 16.4.3), or by considering an independent fragmentation kernel (Seyssiecq *et al.*, 2000). A priori, the first option appears more suitable, since the bonds formed last, are more likely to be broken first.

Fragmentation is also essential to explain an important observation: it is often reported that particles agglomerating in a turbulent medium cannot exceed a certain maximum size (Seyssiecq *et al.*, 2000). Thus, the probability of non-dissociation appears to become null, when two particles collide and form an agglomerate of size larger than a certain limiting size (Brakalov, 1987; David *et al.*, 1991, 1995).

Fragmentation is induced by shocks and/or shear stress and is thus directly governed by the hydrodynamic conditions prevailing in the reactor. Fragmentation alone increases with increasing energy dissipation and leads to a decrease of the final size. Conversely, if we consider the global process of agglomeration + fragmentation, it is difficult to predict the effect of mixing on the final size since we have competition between two antagonistic processes. In practice, a turbine type impeller will induce high fragmentation rates, especially in the discharge region of the turbine, where the shear rate is very high. A profiled propeller is preferable, in order to reduce fragmentation and produce larger agglomerates.

16.5 STIRRED VESSELS

Mechanically stirred reactors are the most frequently used type of precipitation reactors (also called precipitators). Because of the complexity of the phenomena in competition in the precipitator, the stirrer is one of the key components for the control of the final particle size. The stirrer has essentially three missions:

- to homogenize the reactants at macro and micro-scale,
- to maintain the solid particles in homogeneous suspension,
- to control the crystallization mechanisms, which to a large extent depends upon hydrodynamic conditions.

The stirrers are classified according to their properties:

- power input,
- induced shear stress and pumping capacity,
- aptitude to satisfy the missions cited above.

16.5.1 Power required for agitation

The power P transmitted to the fluid by the stirrer is characterised by a dimensionless number referred to as the power number, N_P, such that:

$$P = N_P \rho N_a^3 d_a^5 \qquad (16.30)$$

where ρ (kg m^{-3}) is the liquid density, N_a (s^{-1}) is the stirrer rotation speed and d_a (m) is the stirrer diameter.

The flow regime in the vessel is also characterized by a dimensionless number referred to as the Reynolds number and given by:

$$Re = \rho \frac{N_a d_a^2}{\mu} \qquad (16.31)$$

where μ (Pa s) is the liquid viscosity.

A relationship exists between the power number and the Reynolds number. Figure 16.6 shows typical N_P vs. Re curves, for baffle reactors. The curve corresponds to the three classical flow regimes of fluid mechanics:

- the turbulent regime for Re $> 10^4$. In general, the precipitators will be operated in that regime, since it is, by definition, a random flow regime leading to maximum mixing. In that case, there is a constant value for the power number at all Reynolds numbers: N_P = constant.
- the laminar regime, for Re < 10. In that case: $N_P = \dfrac{A}{Re}$, where A is a constant, function of the type of stirrer.
- the intermediary regime, where behaviour is more complex.

It is worth noting that the relationship between dimensionless numbers N_P and Re is conserved if the impeller size is increased by maintaining geometric similitude, i.e. by keeping equal ratios of all feature dimensions of the impeller. This is due to the use of dimensionless numbers; the use of dimensionless numbers is particularly interesting for scaling-up.

Thus, A in laminar regime, and especially N_P in turbulent regime, are approximately constant for a given stirrer type, regardless of size. N_P in turbulent regime

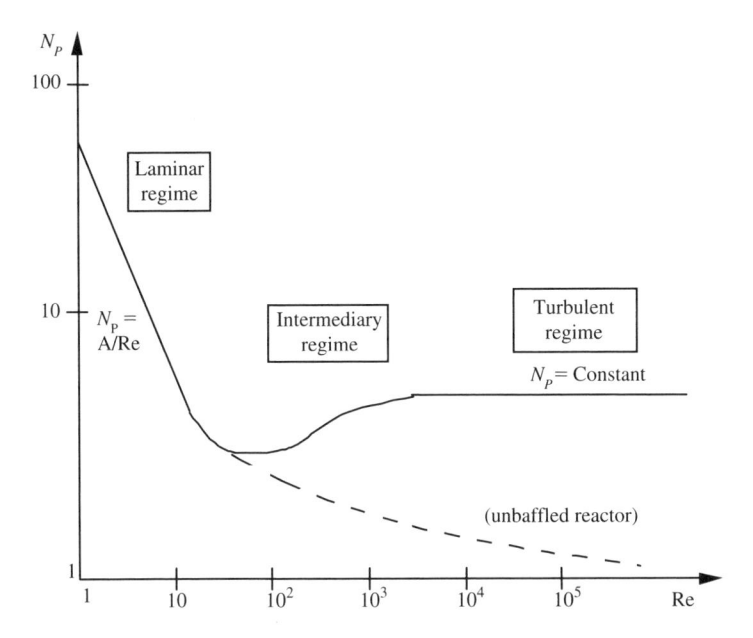

Figure 16.6 Relation between N_P and Re in a typical stirred baffle reactor.

is specific to the stirrer type under consideration. The N_P values, relative to standard impellers, are known; some examples will be given in Section 16.5.5.

Note also that N_P is specific to the impeller and not the combination of impeller + vessel. A condition is, however, that the impeller should not be too close to the walls of the vessel. A recommended upper limit is: $d_a/d_V = 0.6$ (d_V being the vessel diameter).

The average input power per unit volume is given by:

$$\bar{\varepsilon} = \frac{P}{V} \tag{16.32}$$

where V (m^3) is the tank volume.

In practice, typical values for input power per unit volume range from 200 to 1000 W m^{-3}, a value between 200 and 500 W m^{-3} being generally chosen with a profiled propeller.

16.5.2 Pumping capacity and circulation flow rate

The impeller induces a fluid circulation in the vessel. Again, the pumping capacity or discharge rate of the impeller, i.e. the volumetric flow rate crossing the impeller section, Q, is characterised by a dimensionless number such that:

$$Q = N_Q N_a d_a^3 \tag{16.33}$$

where N_Q (−) is the pumping capacity number or discharge coefficient.

Similar to the power number N_P, the number N_Q is constant in the turbulent regime (for Re $> 10^4$) and independent of the impeller size, provided geometric similitude is observed. Values for standard impellers are known, and some will be given in Section 16.5.5. N_Q is also not affected by the vessel walls when $d_a/d_V \leq 0.6$. Conversely, above this limit, the variation of N_Q with d_a/d_V can become very important. In general, N_Q is more sensitive to the set-up geometry than N_P and is particularly sensitive to the position of the stirrer with respect to the bottom of the tank.

One should not confuse the impeller discharge rate with the total circulating flow rate in the vessel. Indeed, the total flow rate is the sum of the discharge rate and of an induced flow rate, carried by the discharge rate outside of the impeller. The total flow rate can be estimated as being about 1.5 times higher than the discharge flow rate.

16.5.3 Flow and shear stress

Shear stresses are developed in a fluid when a layer of fluid moves faster or slower than a nearby layer of fluid or a solid surface. Shear stresses therefore depend on the fluid velocity field in the vessel. In the turbulent regime, the fluid velocity at a given point fluctuates around its mean value because of fluid segregation into transient random eddies. Shear stress also results from the behaviour of these eddies.

The energy transmitted to the fluid by the impeller permits fluid circulation and induces a decrease in size of the turbulent eddies, before being finally converted into heat through viscous shear at microscale level. The local shear rate is then linked to the local energy dissipation.

The shear rate and the energy dissipation are higher near the stirrer than in the rest of the vessel. The heterogeneity of the energy dissipation, for a given power input, depends on the impeller type (Section 16.5.5).

16.5.4 Fundamental scale-up criteria

Laboratory or pilot stirred vessels are scaled-up (or scaled-down from industrial geometry) by applying a geometric similitude. Two systems are geometrically similar when the ratios of all corresponding feature dimensions are equal. However, the geometric similarity is not sufficient to control mixing conditions. The correct rotational speed for the impeller still needs to be selected. What speed should be chosen, at similar geometry, to obtain the same performances at industrial scale as at laboratory or pilot scale? The choice is difficult, since it is impossible to maintain all parameters constant. The two scale up criteria most frequently used are:

- Scale up can be based on the principle of power per unit volume being equal. The relation between the laboratory scale (subscripts Lab) and the industrial scale (subscripts Ind) is then:

$$\frac{N_{a,Lab}^3 \, d_{V,Lab}^3}{H_{L,Lab}} = \frac{N_{a,Ind}^3 \, d_{V,Ind}^3}{H_{L,Ind}} \qquad (16.34)$$

Table 16.3 Stirring speed as a function of the reactor size, maintaining either the power input or the stirrer peripheral velocity constant.

Reactor volume V (m^3)	5×10^{-3}	30
Agitator diameter d_a (m)	9.2×10^{-2}	1.68
Stirring speed N_a (in rpm) for $\bar{\varepsilon} = 1\,\text{kW m}^{-3}$	317	46
Stirring speed N_a (in rpm) for $\pi\, N_a\, d_a = 4\,\text{m s}^{-1}$	830	46

where H_L is the liquid height, or:

$$N_{a,Lab}^3\, d_{V,Lab}^2 = N_{a,Ind}^3\, d_{V,Ind}^2 \tag{16.35}$$

with $H_L = d_V$ (see Section 16.5.6).

When the above equation holds and no mixing effects occur, the performances of the mixing and its impact on the precipitation mechanisms should be similar, whether at lab or industrial scale, especially with axial flow propellers, since the mechanisms involved are then essentially dependent on the average power input per volume unit.

- Scale up can also be done by maintaining the peripheral velocity of the impeller. In these conditions, the criterion becomes:

$$N_{a,Lab}\, d_{V,Lab} = N_{a,Ind}\, d_{V,Ind} \tag{16.36}$$

The pumping performances and the local shear rates in the discharge region of the impeller should then be similar and independent of scale.

In practice, with axial flow propeller, the first criterion is generally preferred. Note that in both cases, the rotational speed will have to be much smaller in the industrial vessel compared to the lab vessel (Table 16.3).

16.5.5 Stirrer types

Impellers can be divided into two main categories, depending on the flow they generate inside the vessel: radial flow impellers and axial flow impellers.

16.5.5.1 Radial flow impellers

Radial flow impellers have blades, which are parallel to the axis of the drive shaft. They are essentially turbines, exemplified by the Rushton turbine, which is a flat-blade turbine equipped with 6 blades (Figure 16.7a). A typical flow pattern, illustrated in Figure 16.8a, shows that the vessel is separated into two zones.

Turbines generate low circulating flow rates with very intensive turbulence in the discharge zone and quite low turbulence in the rest of the vessel. The shear rates and the energy dissipation are thus very heterogeneously distributed in the vessel. The main characteristics of a Rushton turbine are given in Table 16.4.

(a) Rushton turbine

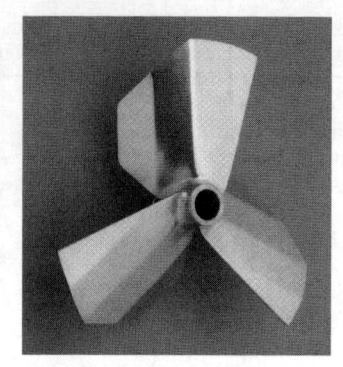

(b) Profiled propeller of type TT from Mixel® (France)

(c) Pitched-blade turbine – 4 blades angle = 45°

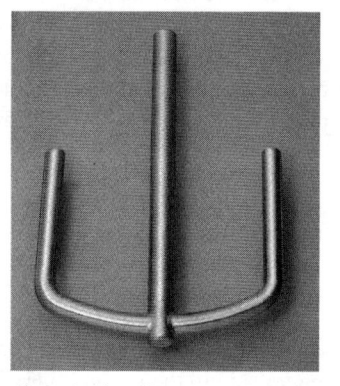

(d) Anchor impeller

Figure 16.7 Different types of impellers.

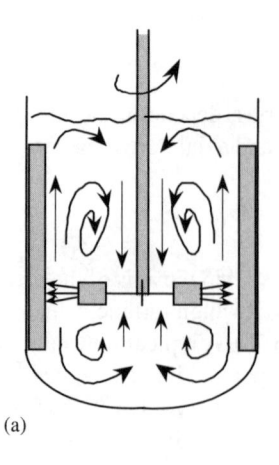

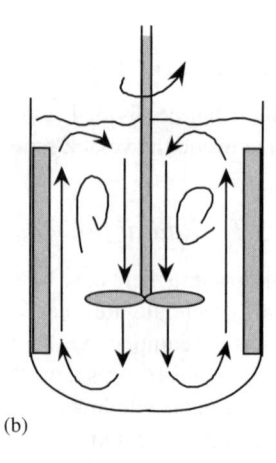

Figure 16.8 Typical fluid circulation pattern for: (a) a radial flow impeller; and (b) an axial flow impeller.

Table 16.4 Recommended propeller diameter, power number, N_P, and pumping capacity number, N_Q, of typical impellers.

Impeller type	Recommended ratio d_a/d_V	N_P in turbulent regime	N_Q in turbulent regime
Rushton turbine (blade height = $0.2d_a$; blade length = $0.25d_a$)	0.5	5–6.3	0.68–0.75
Marine propeller (pitch = d_a)	0.33	0.32–0.37	0.5–0.55
Profiled propeller Mixel® TT	0.55–0.6	0.8	0.96
Pitched blade turbine (4 blades; angle = 45°; blade width = $0.2d_a$)	0.5–0.6	1.25–1.5	0.75

The high N_p values and low N_Q values illustrate that turbines use high power input to generate small circulating flow rates, most of the energy input being degraded by turbulent shear around the turbine.

Such impellers are only used in a precipitation processes where fine particles are wanted, since the very high local shear rate in the outflow zone promotes secondary nucleation and fragmentation, which both play against the formation of large particles.

16.5.5.2 Axial flow impellers

Axial flow impellers are essentially propellers. They are always installed so that the outflow is directed towards the bottom of the vessel (except in some cases of evaporative crystallization). A typical flow pattern for axial flow impellers is shown in Figure 16.8b.

Propellers generate high circulating flow rates. The shear rate is low, even in the vicinity of the impeller, and rather homogeneously distributed in the entire vessel. Indeed, little of the energy input is degraded in the zone near the impeller and most of it is used to maintain fluid circulation. The energy dissipation occurs by turbulence all along the flow circulation.

The 'old' marine-type propeller is losing popularity. This is because it generates linear fluid velocities, which increase when moving away from the axis. The reaction exerted on the blade extremity is important and strengthening of the blade fixation is needed, which makes the propeller heavier. The shear is also not negligible at the blade extremity.

Conversely, profiled propellers (Figure 16.7b) are designed to produce linear fluid velocities approximately constant all along the blades, which avoids the problems of the marine-type propeller.

Table 16.4 gives the characteristics of the marine propeller and of the Mixel® TT propeller shown in Figure 16.7b. The recommended propeller diameters are also given, with respect to the tank diameter. The propeller diameter has to be such that the sections of the downward and upward flows are approximately equal.

Axial flow is correct in the vessel, provided the liquid height H_L is approximately equal to the tank diameter d_V. If $H_L/d_V \gg 1$, it is recommended to equip the axis with several propellers.

16.5.5.3 Pitched-blade turbines and anchor impellers

Pitched-blade turbines (Figure 16.7c; characteristics given in Table 16.4) produce axial flow, but also significant shear. Their behaviour, with respect to the precipitation mechanisms, is then rather similar to that of the flat-blade turbines. The fluid flow induced by an anchor impeller (Figure 16.7d) is principally circular, in the direction of rotation of the anchor. The mass exchange between the bottom and the top of the vessel is very poor. Despite their simplicity and their relatively common use, anchor impellers are not appropriate for precipitation.

16.5.6 Stirred vessel design

A typical design of a reactor equipped with a profiled propeller is given in Figure 16.9. This design is appropriate for batch or continuous operating mode. Similar design can be used with turbine type impeller, the recommended turbine diameter being in that case such that $d_a/d_V = 0.5$.

The reactor is equipped with four baffles to avoid the formation of a vortex. The space between the baffles and the tank wall prevents the accumulation of solid particles on the baffles. Note that the number of baffles has to differ from the number of impeller blades to reduce the risk of entering in resonance. The small

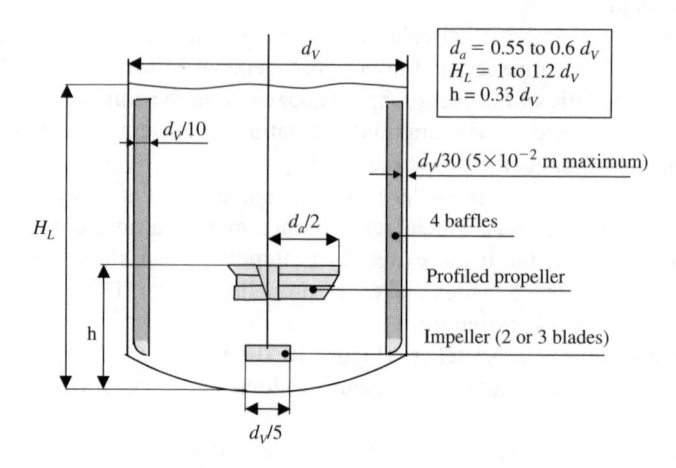

Figure 16.9 Typical design of a reactor stirred by a profiled propeller.

impeller added at the axis extremity aims to avoid deposition at the bottom flow gate. A correct flow circulation is obtained when the liquid height is approximately equal to the tank diameter ($H_L = d_V$), and when the propeller is positioned at $h = (1/3)H_L$ from the bottom of the tank. Finally, the recommended specific input power ranges from 200 to 1000 W m^{-3} (Section 16.5.1).

16.5.7 Application to phosphate removal

Owing to the low level of the phosphate concentration in waste water, the precipitator used for phosphate removal has to be operated in continuous mode, with respect to the liquid phase, and batchwise, with respect to the solid phase. Consequently, a simple stirred vessel cannot suit, and a settling zone has to be added in order to keep the solid particles in the reactor. The reactor design shown in Figure 16.10 has been tested for continuous magnesium ammonium phosphate (MAP) precipitation in the authors' laboratory, with promising results (Regy et al., 2002). The lower part of the reactor is the reaction zone, designed according to the concepts described above. The reactant feed points are positioned far apart from each other in the discharge region of the propeller, to minimize mixing effects and to prevent the propeller from fouling. The upper part is the settling zone. Its section is sized so as to equalize the upward liquid flow velocity with the settling terminal velocity, so that the smallest particles are kept in suspension.

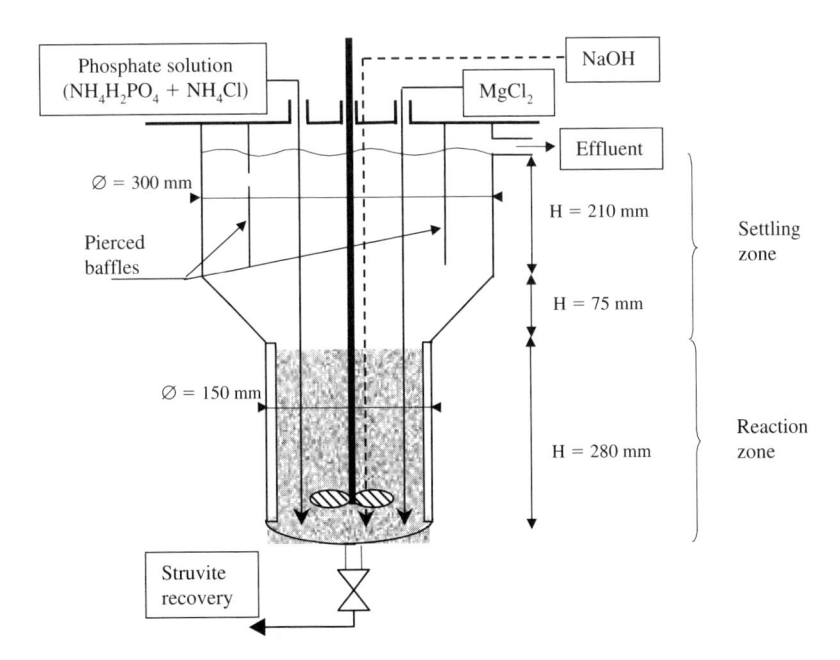

Figure 16.10 Pilot stirred reactor designed for struvite recovery.

For example, if the settling zone diameter is $d_{sz} = 300 \times 10^{-3}$ m and the effluent flow rate is $Q = 2.6L\,h^{-1}$, the upward flow velocity is:

$$U = \frac{4Q}{\pi d_{sz}^2} = 10.4 \times 10^{-6}\ m\,s^{-1} \qquad (16.37)$$

Using Stoke's law (introduced later: Equation 16.43 in laminar regime) to express the particle terminal settling velocity U_{max}, it can be shown that struvite particles of size over 5 μm should remain in the reactor ($\rho_S = 1710\,kg\,m^{-3}$; $\rho_L = 1000\,kg\,m^{-3}$; $\mu = 10^{-3}\,Pa\,s$; the laminar regime can be confirmed by calculating Re_p *a posteriori*).

The apparatus presented here is simple to operate, and can accept fluctuations in the operating conditions, provided the settler section is sufficiently large. Other designs, combining a mixed reaction zone with a settler can also be envisaged. In particular, a DTB Swenson® type geometry, i.e. an annular settling zone with a central agitated zone equipped with a propeller 'blowing' towards the top, might be appropriate to investigate.

16.6 FLUIDISED BED REACTORS

Fluidised bed reactors (FBRs) are usually made of a simple vertical cylinder. For industrial crystallization purposes, there is also the Oslo-Krystal type-crystallizer, which is structurally more complex but obeys the same fluid dynamic principles.

Let us consider a fluid, which passes upwards through a bed of particles. At low flow rate, the particles are packed and immobile. If the flow rate is increased, the pressure drop across the bed increases. This pressure drop is due to the frictional forces between the fluid and the particles. The increase of the pressure drop will continue, until the frictional drag on the particles becomes equal to their apparent weight (actual weight, minus buoyancy). At that stage, the particles can undergo a rearrangement so that they offer less resistance to the flow of the fluid. If the flow rate is increased further, the individual particles are separated from one another and become freely supported in the fluid. The bed is said to be fluidised. Further increase of the flow rate causes an expansion of the bed, while the pressure drop remains approximately constant. Finally, there is a maximum flow rate above which the particles are driven upwards with the fluid. This phenomenon, referred to as entrainment of particles, happens when the fluid velocity reaches the terminal settling velocity of a single particle.

The design of a FBR and the choice of the operating conditions lie in several criteria:

- the minimum fluidising velocity, i.e. the minimum superficial velocity of flow at which fluidisation takes place,
- the terminal settling velocity of the particles, i.e. the maximum superficial velocity of flow at which entrainment of particles takes place,

Phosphate precipitation reactor design

- the bed void space for a given fluid superficial velocity, which leads to the particle bed height,
- the pressure drop across the particle bed which leads to the energy dissipation, which is itself an important parameter of the precipitation laws.

The superficial velocity of flow (U) is simply the velocity the fluid would have through the column in the absence of particles. It is linked to the total inlet fluid flow rate (Q) and to the fluidiser cross-sectional area (Ω) by:

$$U = \frac{Q}{\Omega} \qquad (16.38)$$

The above design criteria can be estimated by means of correlations. Although theses correlations are only approximate, they enable us to evaluate the operating limits of the apparatus. The main operating parameters to consider are:

- the inlet flow rate Q
- the mass of solid particles m_S
- the particle size d_p
- the solid density ρ_S
- the fluid density ρ_F

Generally, the FBR is operated by controlling the flow rate, with all the other parameters being fixed. The range of inflow-rates, to generate correct fluidisation, is then quite narrow.

16.6.1 Minimum fluidising velocity

The following formula, deduced from Ergun equation (1952), is recommended to estimate the minimum fluidising velocity:

$$U_{mf} = \frac{h_k}{2h_b}(1 - e_{mf})\frac{\mu a_p}{\rho_F}\left[\sqrt{1 + \left(\frac{4h_b}{h_k^2}\frac{e_{mf}^3}{(1-e_{mf})^2}\frac{\rho_F(\rho_S - \rho_F)g}{\mu^2 a_p^3}\right)} - 1\right] \qquad (16.39)$$

where h_k: Kozeny constant, $h_k = 4.5$ for spherical particles
h_b: Burke-Plummer constant, $h_b = 0.3$
e_{mf}: void fraction of the bed at the incipient fluidisation point
μ: fluid viscosity (kg m^{-1} s^{-1})
ρ_F: fluid density (kg m^{-3})
ρ_S: solid density
g: gravitational acceleration (m s^{-2})
a_p: specific area of the particles (m^{-1}).
a_p is the ratio between the actual surface and the actual volume of the particles.
For spherical particles of diameter d_p, we have:

$$a_P = \frac{6}{d_p} \qquad (16.40)$$

To obtain U_{mf}, we need to know e_{mf}. A common approximation (leading to a crude U_{mf}) involves taking e_{mf} as being equal to the void space of the packed bed; e_{mf} can then be evaluated by:

$$e_{mf} = 1 - \frac{\rho_S}{\rho_{pb}} \qquad (16.41)$$

where ρ_{pb} is the apparent density of the packed bed, i.e. the mass of particles in the bed (m_S) divided by the apparent volume of the bed ($h_{pb} \times \Omega$):

$$\rho_{pb} = \frac{m_S}{h_{pb}\Omega} \qquad (16.42)$$

h_{pb} being the height of the packed bed.

A typical value of e_{mf} for spherical particles is around $e_{mf} = 0.4$; e_{mf} can, however, be much higher with coarse or needle-like particles.

Finally, it is important to note that U_{mf} is independent of the fluidiser size. As a consequence, U_{mf} can be directly measured in a lab-scale fluidiser if more precision is required.

16.6.2 Terminal settling velocity of a single particle; entrainment of particles

The fluid flow rate through a fluidised bed is limited firstly by U_{mf} and secondly by entrainment of solids by the liquid. This upper limit to superficial fluid velocity, U_{max}, is approximated by the terminal settling velocity of the particles, which, for a spherical particle, can be estimated by:

- if $Re_p < 1$ (laminar regime):

$$U_{max} = \frac{g d_p^2 (\rho_S - \rho_F)}{18\,\mu} \qquad (16.43)$$

- if $1 < Re_p < 1000$ (intermediary regime):

$$U_{max}^{1.4} = \frac{0.072\, g\, d_p^{1.6} (\rho_S - \rho_F)}{\rho_F^{0.4}\, \mu^{0.6}} \qquad (16.44)$$

- if $1000 < Re_p < 4 \times 10^5$ (turbulent regime):

$$U_{max}^2 = \frac{3\, g\, d_p (\rho_S - \rho_F)}{\rho_F} \qquad (16.45)$$

where Re_p is the particle Reynolds number, for $U = U_{max}$, defined by:

$$Re_p = \frac{\rho_F\, U_{max}\, d_p}{\mu} \qquad (16.46)$$

In practice, U_{max} is calculated first, and the condition on Re_p is verified, *a posteriori*, to allow validation of the calculated U_{max}.

16.6.3 Bed expansion

In a fluidised bed beyond U_{mf}, the particle separation increases with increasing fluid superficial velocity. Richardson and Zaki (1954) showed that the relationship between the fluid superficial velocity (U) and the bed void space (e) is of the form:

$$\frac{U}{U_{max}} = e^n \qquad (16.47)$$

where exponent n ranges from 4.6 in laminar regime to 2.4 in turbulent regime.

It is worth noting that exponent n can also be determined experimentally, by plotting log(U) vs. log(e). Extrapolating the plot to $e = 1$ also permits to evaluate U_{max}.

Knowledge of the bed void space is of great importance for the fluidiser design, since it allows calculation of the fluidised bed height:

$$h = \frac{m_S}{(1-e)\rho_S \Omega} \qquad (16.48)$$

16.6.4 Pressure drop and energy dissipation

The force balance across the fluidised bed dictates that the fluid pressure loss across the bed of particles is equal to the apparent weight of the particles per unit area of the bed. Thus, across a bed of height h and of void space e, the pressure drop ΔP is given by:

$$\Delta P = h(1-e)(\rho_S - \rho_F)g \qquad (16.49)$$

This relation applies from the incipient fluidisation point, until entrainment of solids takes place.

The pressure drop is due to the frictional forces between the fluid and the particles. The energy dissipation linked to this phenomenon can be calculated from (Seckler et al., 1996):

$$\bar{\varepsilon} = \frac{U \Delta P}{\rho_F \, h \, e} \qquad (16.50)$$

where $\bar{\varepsilon}$ (W kg^{-1}) is the average energy dissipation.

Note that the energy dissipation can significantly affect the different precipitation mechanisms.

16.6.5 Particle segregation

In precipitation, the nucleation, growth, agglomeration and fragmentation processes can induce a spread of the particle size distribution. The most common approach to determine U_{mf} involves using an average particle diameter in equation 16.39. For a binary population this average diameter can be given by:

$$\frac{1}{\overline{d_p}} = \frac{X_1}{d_{p1}} + \frac{X_2}{d_{p2}} \qquad (16.51)$$

where X_1 and X_2 are the volume fractions of component 1 and 2, respectively.

However, such approach is only consistent if the size distribution remains sufficiently narrow to have a correct mixing of the particles within the bed. If the size distribution is wide, the bed becomes completely segregated, i.e. consisting of mono-size particle layers. In that case, the minimum fluidising velocity will be equal to the velocity calculated for the larger particles. A problem will then arise, if this velocity is higher than the terminal velocity of the smallest particles, since the smallest particles will then be entrained by the flow.

16.6.6 Application to phosphate removal

Fluidised bed type precipitators are often chosen to study phosphate removal at lab and pilot scale or to process phosphate removal at industrial scale. The Crystallactor®, for instance, is a fluidised bed reactor (Piekema and Giesen, 2001; Piekema, this volume). The choice of this technology lies in the fact that it permits different residence times for the liquid and solid phases.

The FBRs used for phosphate removal are mainly made of a vertical cylinder. The reactants are fed at the bottom of the column through several nozzles. The seed (i.e. the initial particle bed) can be either sand, or pellets of the material to be precipitated. Growth of the fluidised particles occurs and the effluent, impoverished in the precipitating components, flows out at the top of the column. In principle, the effluent should be free of any particles, but it may drive some fines. At regular intervals, a quantity of the largest particles, or the totality of the particles, is discharged from the reactor bottom and fresh seed material is added. Thus, the FBR is operated in continuous mode for the liquid phase and batchwise for the solid phase. In order to allow flexibility with respect to the waste water flow rate that has to be treated, it is recommended to operate the reactor with a circulation loop. This also permits reduction of the operating supersaturation and consequently the risk of having intensive primary nucleation (Section 16.3.3) and secondary nucleation (Section 16.4.1). The main inconvenience is then that the reactor has to be larger.

16.7 AIR MIXING

In the processes of phosphate removal by precipitation, air injection permits CO_2 stripping and induces an increase of the pH, which favours phosphate precipitation. The stripping of CO_2 can be done prior to entering the precipitation reactor, but may also be done directly within the precipitation reactor itself. In that case, let us examine the impact of the air injection on the reactor hydrodynamics.

16.7.1 Air injection in a mechanically stirred reactor

Air is injected through a distributor placed at the bottom of the vessel. In comparison to a stirred reactor without gas injection, several additional points have to be considered.

The stirring speed has to be sufficient to assure circulation of the gas bubbles within the reactor and a relatively homogeneous distribution of the air in the entire vessel. The mechanical agitation should then permit enhanced contact between the gas and liquid phases.

The power dissipation in the vessel changes and becomes the sum of two terms:

- The first term corresponds to the power transmitted by the agitator. It has been shown that, in the presence of gas, this energy is lower than that transmitted by the agitator in the absence of gas. However, in practice, equation 16.30, which relates to agitation in the absence of gas and which overestimates the power consumption, is generally sufficient to allow choice of agitator motor.
- The second term corresponds to the power transmitted by the injected gas. This is equal to the power dissipated by the gas discharge. Assuming an isothermal expansion, we have:

$$\varepsilon_G \approx \frac{\rho_G Q_G}{V_L} \frac{RT}{M_G} \ln\left(\frac{P_d}{P}\right) \quad (16.52)$$

where ε_G (W m^{-3}) is the power dissipated per unit volume of liquid, ρ_G (kg m^{-3}) is the gas density, Q_G (m^3 s^{-1}) is the volumetric gas flow rate, V_L (m^3) is the volume of liquid in the vessel, R is the ideal gas constant, T (K) is the absolute temperature, M_G (kg mol^{-1}) is the molecular weight of the gas, P_d (Pa) is the gas pressure in the distributor and P (Pa) is the pressure over the vessel (generally $\cong$ 1 atm).

Lastly, although knowledge of the gas retention can be important, formulas proposed in the literature are too specific and not very reliable. Experimental measurement are therefore preferable.

Most of the studies of gas–liquid reactors reported in literature deal with turbine-type agitators. Although such agitators are well adapted to favour gas–liquid reactions, they are generally not recommended for precipitation of large particles. Besides, the gas–liquid mass transfer kinetics, which are crucial in gas–liquid reaction, should not play a dominant role in our case of interest, since they are likely to be more rapid than crystallization kinetics.

16.7.2 Air lift

If no mechanical agitation is used, an air lift technique may be envisaged. The principle of air lift is shown in Figure 16.11. Air is injected in the central tube of the reactor. A flow circulation is then generated, due to the density difference which exists between the flux of the central zone and the flux of the peripheral zone. Thus, air lift provides flow circulation of the type induced by an axial-flow

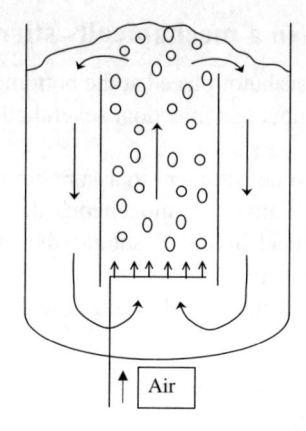

Figure 16.11 The principle of an air lift reactor.

propeller. Mixing is then better than that provided by air injection in a simple vessel not equipped with a central tube. However, it remains much less efficient than mechanical mixing. In particular, the shear rate near the reactor walls is smaller and the risk of encrustation is more significant.

16.7.3 Air injection in a FBR

This technology has been applied to phosphate removal by Unikita ltd (Ueno and Fujii, 2001) in Japan, and by Münch and Barr (2000) in Australia for struvite crystallization.

When air is introduced in a bed of particles initially fluidised by liquid alone, one observes either an expansion (if d_p is over about 2.5 mm) or a contraction of the bed (if d_p is under about 2.5 mm). Several equations, such as the correlation proposed by Han *et al.* (1990), permit evaluation of the bed expansion in the presence of gas as well as the gas retention.

More details on the hydrodynamics of three-phase fluidised bed can also be found in Wild and Poncin (1996).

16.8 MODELLING OF PRECIPITATION REACTORS

Modelling of the precipitation processes is based on the population balance equation, which links the particle size distribution to the precipitation kinetics. In this equation, the population of particles is described by the number density function Ψ. The product $\Psi(L,t)dL$ is the number of particles of size between L and $L + dL$ per unit volume, at time t. The population balance consists of an instantaneous balance of the particles entering and leaving a differential size range $[L, L + dL]$. Its formulation is well described by Randolph and Larson (1988). The general

Phosphate precipitation reactor design

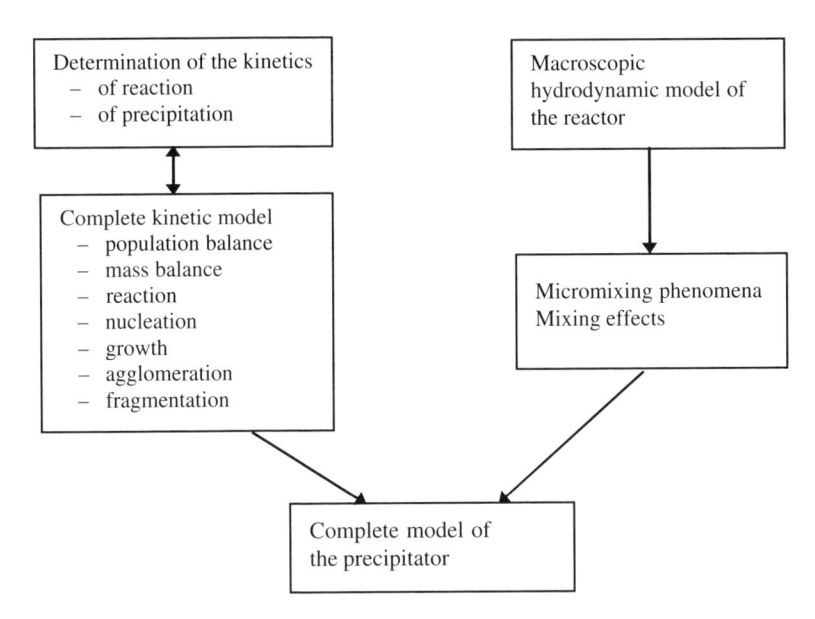

Figure 16.12 Methodology for the modelling of a precipitation reactor.

equation, for a well mixed (continuous or batch) precipitator of volume V, can be written as follows:

$$\frac{1}{V(t)}\frac{\partial[\Psi(L,t)V(t)]}{\partial t} + \frac{\partial[\Psi(L,t)G(L,t)]}{\partial L} + \frac{Q_{out}\Psi_{out}(L,t) - Q_{in}\Psi_{in}(L,t)}{V(t)}$$

$$= B(L,t) - D(L,t) \qquad (16.53)$$

where $G(L,t)$ is the linear growth rate, Q_{in} and Q_{out} are the inlet and outlet volumetric flow rates, $B(L, t)$ is the birth rate (particles appearing in a specific size interval by primary and/or secondary nucleation, by agglomeration of smaller particles and/or by fragmentation of larger particles) and $D(L,t)$ is the death rate (particles disappearing from the size interval of interest by agglomeration with other particles and/or by fragmentation into smaller particles).

A literature review of the methods used to solve Equation 16.53 is given in the paper of Blandin *et al.* (2001). For problems associated to agglomeration, a Monte Carlo approach can also be used.

The complete model of a precipitator combines the hydrodynamic model of the reactor with the population balance, reaction kinetics and the mass balance (Figure 16.12).

The reaction and precipitation kinetics can be determined from laboratory batch experiments. Kinetic constants can be obtained by fitting theoretical models on experimental measurements of supersaturation and particle size distribution, which are then both a function of time. The experimental conditions can be varied (temperature, concentrations, etc), to isolate the influence of each mechanism (reaction only, nucleation, growth, agglomeration) as much as possible. A complete model of precipitation is too complicated to be fitted on experiments influenced simultaneously by all variables. It is also due to the complexity of the kinetic model, that it is essential to work in laboratory reactors that are *highly efficient in terms of mixing*. Laboratory experiments performed under imperfect mixing conditions cannot lead to reliable kinetic models. In particular, mixing effects have to be avoided. A solution to the mixing effect problems occurring in precipitation can be to use a T-mixer (Blandin et al., 2001).

The macroscopic hydrodynamic model of the *industrial reactor* is determined by Residence Time Distribution (RTD) measurements. A real reactor is then generally described as the association of several ideal reactors (Section 16.2.2).

To account for (micro and meso) mixing effects, a great number of mixing models have been developed in literature. The simple IEM (interaction by exchange with the mean) model and the engulfment model are the most used micromixing models (see Baldyga et al., 1995, for a literature review). The IEM model, for instance, assumes that the fluid consists of aggregates of negligible volume and uniform concentration in species i, C_i, exchanging mass with their environment of average concentration $\overline{C}_i$ during a time scale equal to t_μ. The concentration evolution of an aggregate is thus given by:

$$\frac{dC_i}{dt} = \frac{(\overline{C}_i - C_i)}{t_\mu} + r_i(C_i) \tag{16.54}$$

where t represents the time spent by the aggregate (i.e. its age) in the reactor, and $r_i(C_i)$ is the reaction rate (no precipitation is considered here). The concentration in each aggregate can then be followed as a function of its age.

However, it is rare that micromixing governs alone the mixing conditions and local inhomogeneities of larger scale are often observed. A solution is then brought by the segregated feed model developed by Villermaux (1989). Let us consider a double feed continuous or semi-batch reactor. The model assumes that the reactor is divided in three zones: two segregated zones, located in the vicinity of the feed points, and the bulk. The feed zones exchange mass with each other (if they are sufficiently close), and with the bulk by micromixing (characteristic time t_μ) and mesomixing (characteristic time t_{meso}). Note that these time constants have to be adjusted. Such modelling has been applied by Marcant (1996) to barium sulfate precipitation, and by Zauner and Jones (2000) to calcium oxalate precipitation. It should be noted that the zones of the segregated model complete the macroscopic flow model obtained by RTD measurement. Thus, even if the reactor is found to be perfectly macromixed with respect to RTD measurements, these zones have to be considered to account for micromixing and local inhomogeneities.

Another model has been developed by Baldyga et al. (1995) to study the competition between different mixing mechanisms in a double feed semi-batch precipitator.

This model, based on the turbulence theory, involves circulation loops to account for macromixing, together with inertial convective mesomixing and engulfment micromixing.

Compartmental modelling, which consists of splitting the reactor into several ideal zones, is the classic methodology used in chemical engineering to model reactor hydrodynamics. The segregated feed model described above follows the same approach. The main advantage of compartmental modelling is that it permits a relatively easy implementation of the complex population balance in the model. It can also permit consideration of an additional zone near the stirrer. The separate consideration of such a zone is interesting, especially if a turbine-type agitator is used. Indeed, in that case, the energy dissipation near the stirrer is much higher than in the average volume of the vessel, and this can affect precipitation mechanisms. However, the approach by compartmental modelling might loose popularity in the future to the benefit of computational fluid dynamic (CFD) techniques. Recently, progress has been made in the modelling of complex flow scenarios by CFD and the application of CFD techniques to model precipitation is promising (Wei *et al.*, 2001).

16.9 CONCLUSION AND PERSPECTIVES

Phosphate removal by precipitation of either calcium phosphate or magnesium ammonium phosphate is a very promising process. However, its industrial implementation may require some additional studies. In particular, it seems necessary to know better the different mechanisms actually involved during precipitation and their interactions with the hydrodynamics. Improvement of knowledge in that field should be a precious help for the optimization of the reactor.

The first problem potentially encountered in such a process may come from the excessive production of fine particles. Indeed, the efficiency of the process may drop significantly, if the fines are driven out with the outflow. These fines are due to excessive homogeneous primary and/or surface secondary nucleation rates. As a consequence, to control the production of fines, it is essential to improve our knowledge of the kinetics of these two mechanisms or at least to determine which mechanism, between primary and secondary nucleation, is at the origin of the generated fines. If primary nucleation cannot be avoided, particular attention will be required because of the probable existence of mixing effects. A solution to the problem of fines can also be brought by agglomeration, if it encourages fine particles to stick together, or allows coating of the larger particles by the fines. However, if most of the agglomerates are found to be formed by growth under non ideal conditions, leading to twins or polycrystals, then agglomeration will not induce any improvement. Further research in agglomeration would be interesting.

Another problem may be the encrustation or fouling of the reactor wall and of the different parts present inside the reactor, i.e. the deposit of a hard layer of solid on different reactor surfaces. Fouling may be initiated by heterogeneous primary nucleation. A correct mixing and an optimized location of the reactant feed point, both of which should avoid high localized supersaturation near the reactor wall, should prevent the reactor from fouling. This has to be tested.

Thus, laboratory/pilot experiments seem still necessary. It is important to note that most of the precipitation mechanisms may also be strongly influenced by the presence of soluble and insoluble impurities. Therefore, experiments with real effluents are essential.

The industrial reactor will have to be operated in continuous mode, with respect to the liquid phase and batchwise with respect to the solid phase. This specific operating mode requires equipping the reactor with a settling zone. In that case, a stirred reactor seems preferable to a fluidized bed, since it allows a separation between the mixing and settling tasks and provides an additional degree of freedom. A simple design is proposed in section 16.5.7. Other geometries, with an annular settling zone for example, may be appropriate to investigate. Pilot experiments will be necessary to validate the design. The pilot reactor will be designed by homothetic scale down of the industrial reactor to be investigated. The study of the precipitation mechanisms may require working with reactant concentrations different from those of the real effluents in order to minimize the mixing effects. The optimization of the reactor and its hydrodynamics should be done with real effluents. Finally, the impact of a potential CO_2 stripping by direct injection of air at the bottom of the reactor will also have to be tested.

NOTATION

B_{att} contact secondary nucleation rate ($m^3\,s^{-1}$)
B_{prim} primary nucleation rate ($m^3\,s^{-1}$)
B_{surf} surface secondary nucleation rate ($m^3\,s^{-1}$)
C working concentration ($mol\,m^{-3}$)
C^* solubility concentration ($mol\,m^{-3}$)
d_a agitator diameter (m)
D_m molecular diffusivity ($m^2\,s^{-1}$)
D_t turbulent diffusivity ($m^2\,s^{-1}$)
d_V vessel diameter (m)
e bed void space (−)
e_{mf} void fraction of the bed at the incipient fluidisation point (−)
$E(t)$ residence time distribution (s^{-1})
g gravitational acceleration ($m\,s^{-2}$)
G linear crystal growth rate ($m\,s^{-1}$)
h agitator position (measured from the bottom) (m)
h bed height (m)
H_L liquid height in the vessel (m)
k_d mass transfer coefficient ($m\,s^{-1}$)
K_{ij} agglomeration kernel ($m^3\,s^{-1}$)
k_R reaction rate constant (var.)

K_S	solubility product (var.)
L	crystal size (m)
L_S	Large 'fluid aggregate' size (m)
m_S	mass of solid particles in the fluidised bed (kg)
N_a	agitator rotation speed (s^{-1})
N_P	power number (−)
N_Q	pumping capacity number (−)
P	power transmitted to the fluid by the stirrer (W)
Q	volumetric flow rate (m^3 s^{-1})
Q	volumetric flow rate crossing the impeller section (m^3 s^{-1})
Re	agitation Reynolds number (−)
Re_p	Reynolds number of the particles (−)
r_{ij}	agglomeration rate (m^{-3} s^{-1})
r_i, r_j	particle radius (m)
S	supersaturation ratio (−)
Sc	Schmidt number (−)
Sh	Sherwood number (−)
t	time (s)
t_m	macromixing time (s)
t_μ	micromixing time (s)
t_{meso}	mesomixing time (s)
t_{prim}	characteristic time constant for primary nucleation (s)
t_R	characteristic time constant for chemical reaction (s)
$\bar{t}_S$	mean residence time (s)
u	local average velocity (m s^{-1})
U	fluid superficial velocity ($= Q/\Omega$) (m s^{-1})
U_{mf}	minimum fluidising superficial velocity (m s^{-1})
U_{max}	maximum fluidising superficial velocity (m s^{-1})
V	reactor volume (m^3)
ΔP	pressure drop across the bed (Pa)

greek letters

ε	local dissipated specific power (W kg^{-1})
$\bar{\varepsilon}$	average specific power input (W kg^{-1}) or (W m^{-3})
λ_B	Batchelor microscale (m)
λ_K	Kolmogorof microscale (m)
μ	liquid viscosity (Pa s)
ν	fluid kinematic viscosity (m^2 s^{-1})
ρ	liquid density (kg m^{-3})
ρ_F	fluid density (kg m^{-3})
ρ_S	solid density (kg m^{-3})
τ	mean holding time (s)
Ψ	number density function (m^{-1} m^{-3})
Ω	fluidiser cross-sectional area (m^2)

REFERENCES

Abrahamson, J. (1975) Collision rates of small particles in a vigorously turbulent fluid. *Chem. Eng. Sci.* **30**, 1371–1379.

Baldyga, J. and Bourne, J.R. (1992) Interactions between mixing on various scales in stirred tank reactors. *Chem. Eng. Sci.* **47**(8), 1839–1848.

Baldyga, J., Podgorska, W. and Pohorecki, R. (1995) Mixing precipitation model with application to double feed semibatch precipitation. *Chem. Eng. Sci.* **50**(8), 1281–1300.

Blandin, A. F., Mangin, D., Nallet, V., Klein, J.P. and Bossoutrot, J.M. (2001) Kinetics identification of salicylic acid precipitation through experiments in a batch stirred vessel and a T-mixer. *Chem. Eng. J.* **81**, 91–100.

Brakalov, L.B. (1987) A connection between the orthokinetic coagulation capture efficiency of aggregates and their maximum size. *Chem. Eng. Sci.* **42**(10), 2373–2383.

Corrsin, S. (1964) The isotopic turbulent mixer. Part II. Arbitrary Schmidt number. *A. I. Ch. E. J.* **10**, 870–877.

David, R. and Marcant, B. (1994) Prediction of micromixing effects in precipitation: case of double-jet precipitators. *A. I. Ch. E. J.* **40**(3), 424–432.

David, R., Marchal, P., Klein, J.P. and Villermaux, J. (1991) Crystallization and precipitation engineering–III. A discrete formulation of the agglomeration rate of crystals in a crystallization process. *Chem. Eng. Sci.* **46**(1), 205–213.

David, R., Marchal, P. and Marcant, B. (1995) Modelling of agglomeration in industrial crystallization from solution. *Chem. Eng. Technol.* **18**, 1–8.

Ergun, S. (1952) Fluid flow through packed columns. *Chem. Eng. Progr.* **48**, 89–94.

Froment, G.F. and Bischoff, K.B. (1990) *Chemical Reactor Analysis and design*, 2nd edn. Wiley, New York.

Gahn, C. and Mersmann, A. (1997) Theoretical prediction and experimental determination of attrition rates. *Chem. Eng. Res. Des.* **75**(A2), 125–131.

Gahn, C. and Mersmann, A. (1999) Brittle fracture in crystallization processes. Part A. Attrition and abrasion of brittle solids. *Chem. Eng. Sci.* **54**, 1273–1282.

Garside, J. (1985) Industrial crystallization from solution. *Chem. Eng. Sci.* **40**(1), 3–26.

Han, J.H., Wild, G. and Kim, S.D. (1990) Phase holdup characteristics in three phase fluidized beds. *Chem. Eng. J.* **43**, 67–73.

Herndl, G. and Mersmann, A. (1981) Fluid dynamics and mass transfer in stirred suspensions. *Chem. Eng. Comm.* **13**, 23–37.

Higashitani, K., Yamauchi, K., Massuno, Y. and Osokawa, G.J. (1983) Turbulent coagulation of particles dispersed in a viscous fluid. *Chem. Eng. Jpn.* **16**, 299–304.

Hirasawa, I., Kaneko, S., Kanai, Y., Hosoya, S., Okuyama, K. and Kamahara, T. (2002) Crystallization phenomena of magnesium ammonium phosphate (MAP) in a fluidized-bed-type crystallizer. *J. Cryst. Growth* **237–239**, 2183–2187.

Klein, J.P., Boistelle, R. and Dugua, J. (1989) Cristallisation: aspects théoriques. *Techniques de l'Ingénieur* **J 1500**, 1–21.

Klein, J.P. and David, R. (2001) Reaction crystallization. In *Crystallization Technology Handbook*, 2nd edn. (eds Mersmann, A.), pp. 513–561, Marcel Dekker Inc., New York.

Kuboi, R., Nienow, A.W. and Conti, R. (1984) Mechanical attrition of crystals in stirred vessels. In *Industrial Crystallization 84* (eds Jancic, S.J. and de Jong, E.J.), pp. 211–216, Elsevier, Amsterdam.

Levenspiel, O. (1972) *Chemical Reaction Engineering*, 2nd edn. Wiley, New York.

Levins, D.M. and Glastonbury, J.R. (1972) Application of Kolmogoroff's theory to particle-liquid mass transfer in agitated vessels. *Chem. Eng. Sci.* **27**, 537–543.

Marcant, B. (1996) Prediction of mixing effects in precipitation from laser sheet visualisation. *13th Symposium on Industrial Crystallization* vol. 2, pp. 531–538, Toulouse, France.

Marcant, B. and David, R. (1991) Experimental evidence for and prediction of micromixing effects in precipitation. *A. I. Ch. E. J.* **37**(11), 1698–1710.

Marrot, B. and Biscans, B. (1996) Impact attrition of sodium chloride crystal in saturated solution. Single particle impact experiment. Comparison with indentation measurements. *13th Symposium on Industrial Crystallization* Vol. 2, pp. 491–496, Toulouse, France.

Marrot, B. (1997) Attrition dans les cristallisoirs. Etude expérimental de l'impact d'un cristal en suspension contre une cible. Détermination de lois phénomènologiques. PhD thesis, INP. Toulouse, France.

Mersmann, A. (1996) Supersaturation and Nucleation. *Chem. Eng. Res. Des.* **74**(A7), 812–820.

Münch, E.V. and Barr, K. Controlled struvite crystallization for removing phosphorus from anaerobic digester sidestreams. *Water Res.* **35**(1), 151–159.

Nallet, V., Mangin, D. and Klein, J.P. (1998) Model identification of batch precipitations: application to salicylic acid. *Computers chem. Engng.* 22 (Suppl.), S649–S652.

Nienow, A.W. (1997) On impeller circulation and mixing effectiveness in the turbulent flow regime. *Chem. Eng. Sci.* **52**(15), 2557–2565.

Ottens, E.P.K. and de Jong, E.J. (1973) A model for secondary nucleation in a stirred vessel cooling crystallizer. *Ind. Engng. Chem. Fundam.* **12**, 179–184.

Piekema, P. and Giesen, A. (2001) Phosphate recovery by the crystallisation process: experience and developments, *Second International Conference on Recovery of Phosphates from Sewage and Animal Wastes*, Noordwijkerhout, Holland.

Randolph, A.D. and Larson, M.A. (1988) *Theory of Particulate Processes*, 2nd edn. Academic Press, San Diego, USA.

Regy, S., Mangin, D., Klein, J.P., Lieto, J. and Thornton, C. (2002) Phosphate recovery in the waste water by crystallisation: state of the art and stirred reactor technology, *LAGEP internal report*, www.nhm.ac.uk/mineralogy/P-recovery/

Richardson, J.F. and Zaki, W.N. (1954) Sedimentation and fluidisation: Part I, *Trans. IchemE.* **32**, 35–53.

Saffman, P.G. and Turner, J.S. (1956) On the collision of drops in turbulent clouds. *J. Fluid Mech.* **1**, 16–30.

Schweich, D. and Falk, L. (2001) Réacteurs Réels et Hydrodynamique. In *Génie de la Réaction Chimique*, 1st edn. (ed Schweich, D.), pp.137–177, Tec et Doc, Lavoisier, Paris.

Seckler, M.M., Van Leeuwen, M.L.J., Bruinsma, O.S.L. and Van Rosmalen, G.M. (1996) Phosphate removal in a fluidized bed. Part II. Process optimization. *Water Res.* **30**(7), 1589–1596.

Seyssiecq, I., Veesler, S., Boistelle, R. and Lamérant, J.M. (1998) Agglomeration of gibbsite Al(OH)$_3$ crystals in Bayer liquors. Influence of the process parameters. *Chem. Eng. Sci.* **53**(12), 2177–2185.

Seyssiecq, I., Vessler, S., Mangin, D., Klein, J.P. and Boistelle, R. (2000) Modelling gibbsite agglomeration in a constant supersaturation crystallizer. *Chem. Eng. Sci.* **55**, 5565–5578.

Tavare, N.S. (2000) Mixing, reaction and precipitation: The Barium Sulphate Precipitation System. In *Mixing and Crystallization* (eds Sen Gupta, B. and Ibrahim, S.), Kluwer Academic Publishers, Dordrecht.

Ueno, Y. and Fujii, M. (2001) 3 years operating experience selling recovered struvite from full-scale plant. *Second International Conference on Recovery of Phosphates from Sewage and Animal Wastes*, Noordwijkerhout, Holland.

Villermaux, J. (1989) A simple model for partial segregation in a semi-batch reactor. *A. I. Ch. E. Annual Meeting*, San Francisco, Paper 114a.

Villermaux, J. (1993) *Génie de la Réaction Chimique, Conception et Fonctionnement des Réacteurs*, 2nd edn. Tec et Doc, Lavoisier, Paris.

Wei, H., Wei, Z. and Garside, J. (2001) Computational fluid dynamics modeling of the precipitation process in a semibatch crystallizer. *Ind. Eng. Chem. Res.* **40**, 5255–5261.

Wild, G. and Poncin, S. (1996) Hydrodynamics in three-phase sparged reactors. In *Three-Phase Sparged Reactors* (eds Nigam, K.D.P. and Schumpe, A.), pp. 84–85, Gordon & Breach, Amsterdam.

Zauner, R. and Jones, A.G. (2000) Scale-up of continuous and semi-batch precipitation processes. *Ind. Eng. Chem. Res.* **39**, 2392–2403.

17

Phosphorus recovery via struvite production at Slough sewage treatment works, UK – a case study

Y. Jaffer and P. Pearce

17.1 INTRODUCTION

Slough sewage treatment works (STW) is located 30 miles west of London and is operated by Thames Water Utilities Ltd. It serves a population equivalent of 260,000 of which approximately 115,000 is from industrial sources, most of which originates from various packaged food and confectionary processors.

In 1996 a new effluent treatment stream was built incorporating primary settlement tanks, raw sludge thickeners and a biological nutrient removal (BNR) plant. This stream replaced two ageing process streams and was designed to treat 60% of the total flow. A conventional carbonaceous activated sludge plant followed by nitrifying trickling filters treats the remainder. Since commissioning, the performance of the BNR plant has allowed its throughput to be increased to 75% of the total flow, which equates to a maximum flow of $863 \, l \, s^{-1}$.

© 2004 IWA Publishing. *Phosphorus in Environmental Technology: Principles and Applications.*
Edited by Eugenia Valsami-Jones. ISBN: 1 84339 001 9

The BNR design adopted was a three stage biological nutrient removal configuration (Randall *et al.*, 1992), comprising of an anaerobic zone, through which settled sewage and return activated sludge flow, an anoxic zone and an aerobic zone, with a mixed liquor recycle returning a portion of nitrified mixed liquor to the anoxic zone. The entire site will be required to remove phosphorus by 2004 when a discharge consent of $1\,mg\,l^{-1}$ total phosphorus will be enforced.

The new plant encountered problems with phosphate scale formation, which was a major driver for considering removal of phosphate in the form of a potentially recoverable product.

17.2 SITE DESCRIPTION

The new sludge treatment stream at Slough comprises of separate primary and secondary sludge thickening streams, which are blended with imported sludges prior to anaerobic digestion. The primary sludge is thickened in gravity thickeners, producing a primary sludge of 5–6% dry solids. The sludge liquors are returned to the head of the works. Surplus activated sludge, from the return activated sludge lines of both activated sludge plants, is pumped to a holding tank prior to mechanical thickening to 5% dry solids on belt thickeners. The thickener liquors are returned to the head of the works.

Both thickened sludge streams are pumped to a mechanically mixed blending tank, where they are mixed with primary and secondary sludges from numerous smaller outlying STWs. The overall blended sludge is fed to anaerobic digesters where approximately 40% of the volatile solids are destroyed during the 15-day retention time at a temperature of 35°C. Biogas is used on site for combined heat and power production.

Sludge discharged from the digesters flows to a lagoon, which acts as buffer storage prior to dewatering by centrifuges. Dewatered sludge cake is stored on site prior to disposal to agricultural land. The centrifuge liquors (centrate), comprising of sludge liquors and a portion of the polymer carrier water, discharge to a collection sump from where they are pumped back to the head of the works.

The site area at Slough STW is relatively long and thin, bounded to the North by a motorway (M4). This has resulted in the need for long pipelines for sludge transfer between units and return of liquors to the head of the works. A site plan is shown in Figure 17.1.

Pumping between the various batch tanks and process units is not continuous which results in sludge remaining static for periods of several hours. This may result in cooling of sludge, which exits the anaerobic digesters at 35°C, but which may remain in contact with steel pipework at ambient ground temperatures for prolonged periods.

Within 6 months of commissioning of the BNR plant problems were encountered with struvite precipitation in the centrate return pipeline and on the impellers of the liquor return pumps. After eight months, the digested sludge main, between

Slough STW Site Layout and Schematic Sludge/Liquor Pumped Mains

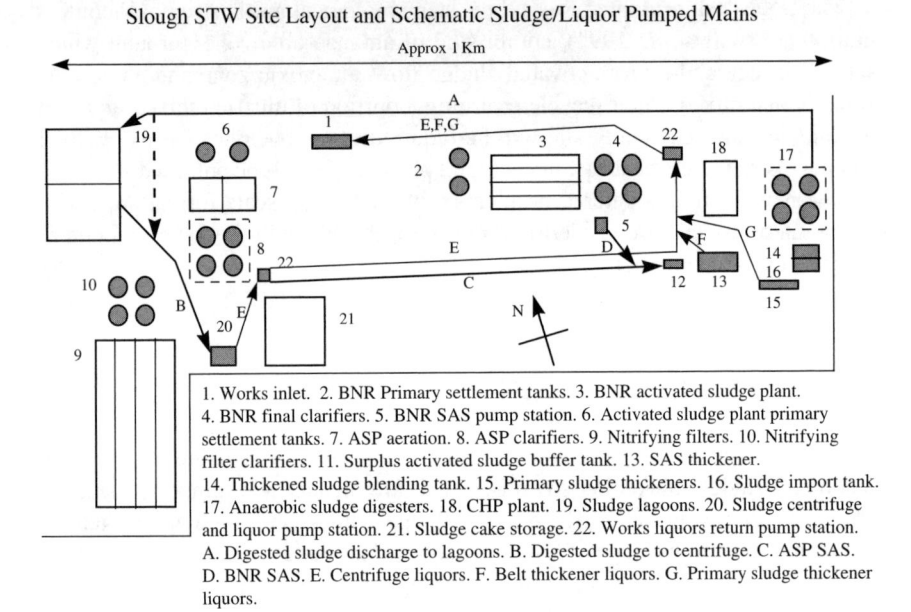

1. Works inlet. 2. BNR Primary settlement tanks. 3. BNR activated sludge plant. 4. BNR final clarifiers. 5. BNR SAS pump station. 6. Activated sludge plant primary settlement tanks. 7. ASP aeration. 8. ASP clarifiers. 9. Nitrifying filters. 10. Nitrifying filter clarifiers. 11. Surplus activated sludge buffer tank. 13. SAS thickener. 14. Thickened sludge blending tank. 15. Primary sludge thickeners. 16. Sludge import tank. 17. Anaerobic sludge digesters. 18. CHP plant. 19. Sludge lagoons. 20. Sludge centrifuge and liquor pump station. 21. Sludge cake storage. 22. Works liquors return pump station. A. Digested sludge discharge to lagoons. B. Digested sludge to centrifuge. C. ASP SAS. D. BNR SAS. E. Centrifuge liquors. F. Belt thickener liquors. G. Primary sludge thickener liquors.

Figure 17.1 Slough STW site layout and schematic (some process units omitted for clarity). For acronyms see Appendix.

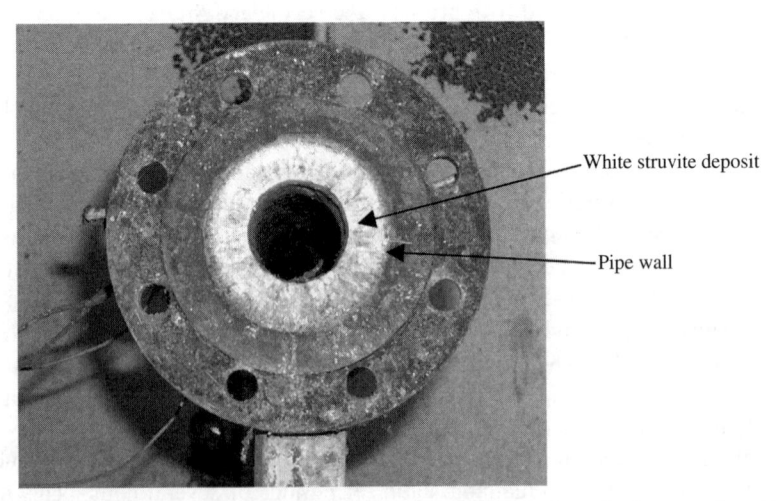

Figure 17.2 Digested sludge pipeline at Slough STW. [This figure is also reproduced in colour in the plate section after page 336.]

the digesters and the digested sludge holding tank became so severely constricted, that the existing pumps could not develop sufficient pressure to move the sludge.

Several of the lines were acid cleaned with dilute sulphuric acid and several lines have been bypassed with the use of temporary over ground piping. These problems plus a monthly requirement to remove struvite from the liquor return pump impellers, result in additional operational costs. These costs are associated with plant downtime and maintenance as well as increased pumping costs associated with restrictions to flow in the sludge and liquor lines. Figure 17.2 illustrates how severe struvite deposition within pipes can be. The white areas are struvite deposits. In this case the struvite has reduced the diameter of the pipe to such an extent, that 70% of the pipe capacity has been lost.

17.3 PHOSPHORUS FORMS THROUGH THE SEWAGE TREATMENT PROCESS

Several authors have noted struvite precipitation downstream of anaerobic digestion, some as long ago as 1939 (Rawn et al., 1939). The phenomenon is not restricted to plants with a BNR process followed by anaerobic digestion; industrial experience suggests that it can also occur on conventional non-nutrient removal sites. The BNR process, however, will exacerbate any problem by increasing the amount of soluble phosphorus and magnesium in the digested sludge and liquors (Carliell-Marquet and Wheatley, 2002; Williams, 1999).

The mechanism by which BNR increases the soluble phosphorus in anaerobic digesters is one example of many transformations that phosphorus may undergo through the sewage treatment process. Sewage from domestic sources enters the sewerage network as a mixture of orthophosphates, polyphosphates and organic phosphates associated with the particulate fraction. Under anaerobic conditions in the sewer, the polyphosphates and some of the organic phosphates are hydrolysed to the degree that on arrival at the STW typically 70% of the total phosphorus is present as orthophosphate (Pearce, 2002). An average of 2.5 g phosphorus per head of contributing population has been established, from STWs data in the Thames region (unpublished data). This gives typical crude sewage concentrations of $11\,\text{mg}\,\text{l}^{-1}$ total phosphorus and $8\,\text{mg}\,\text{l}^{-1}$ PO_4–P.

During primary sedimentation, inlet orthophosphate will pass through the settlement tanks. In the highly anaerobic conditions of the primary sludge layer, some hydrolysis of settled organic particulate phosphorus will occur. This may result in an increase in orthophosphate concentrations across the PSTs.

All biological sewage treatment processes produce excess biological sludge, and so all will remove an amount of PO_4–P from the settled sewage in proportion to the biomass production. At the same time, organic phosphorus associated with the fine and colloidal solids fraction that is not removed in the PST, will be hydrolysed as it passes through the process. The net result is a small reduction of 10–20% of total phosphorus across the process but often a net increase in PO_4–P. The PO_4–P proportion of the total phosphorus in the effluent from the biological process has been observed as typically greater than 90% (unpublished data from internal reports).

McGrath and Quinn (this volume) describe how phosphorus is accumulated in the activated sludge flocs, and the role of polyphosphate accumulating organisms

(PAOs). The fate of the intracellular phosphorus, once surplus activated sludge is wasted from the BNR process, is of key importance for the recovery of phosphorus.

The stored polyphosphate in the surplus activated sludge will be rapidly hydrolysed when the thickened surplus sludge is mixed with the thickened raw sludge in the blending tank prior to digestion. By this time, both sludges will have become highly anaerobic and the volatile fatty acids formed in the fermenting raw sludge will trigger polyphosphate breakdown in the PAOs increasing the bulk PO_4–P concentration. This solublisation of phosphorus will continue in the highly reducing environment of the anaerobic digester. The breakdown of polyphosphates also releases magnesium and potassium, which are present as counter ions on the polyphosphate chains. Jardin and Popel (1994) have estimated the amount of polyphosphate release during sludge digestion by measuring the increase in potassium concentration and concluded that polyphosphate hydrolysis was essentially complete within 2–3 days retention in the anaerobic digester.

For a typical BNR plant, a theoretical mass balance suggests that the expected amount of phosphorus hydrolysis would be in the region of 700–800 $mg l^{-1}$ PO_4–P in the digested sludge. Reported and observed values are much less than this at 100–200 $mg l^{-1}$ PO_4–P. A high degree of inorganic precipitation of phosphorus is therefore thought to be occurring.

However, even these lower values found in practice, are sufficient to produce significant struvite scaling problems. Combined with the high ammoniacal nitrogen concentrations and elevated soluble magnesium concentrations present in the digester, and the buffered pH of 7.3–7.5, the potential for struvite precipitation is large. If left unchecked this can cause severe operational difficulties and return high loads of phosphorus to the waste water treatment plant inlet.

The soluble phosphorus remaining in all of the return liquor streams may represent a 25–45% recycle of the phosphorus load arriving at the works on a daily basis. This recycle load therefore requires a higher degree of phosphorus capture via the BNR plant to maintain effluent quality.

The balance of the received phosphorus load leaves the site in the dewatered sludge cake. Some of this will still be in organic form, predominantly intercellular, arising from non degraded cells from the activated sludge, and some will still be in soluble form in the liquid phase of the sludge cake, but most is likely to be inorganically precipitated as proposed by Jardin and Popel (1994) and Carliell and Wheatley (1997). There are a large number of potential inorganic precipitants in sludge; for example, any iron added to the waste water treatment process for odour removal will end up in the sludge, similarly iron associated with the incoming waste water will also most likely be associated with the sludge stream. Most sites, even where no iron is added, will have iron present in the raw sludge at a concentration of 0.5–1.5%, on a dry weight basis. In hard water areas, such as the Thames valley, calcium and magnesium are concentrated in the sludge to 5–6% and 0.5–1% respectively (see Table 17.1). Aluminium present in industrial discharges may also contribute to the precipitation of phosphorus though levels in UK sludges are relatively low compared to countries, such as Germany, where zeolites have been used to replace phosphates as detergent builders. Digested sludge also contains high concentrations of sulphur as insoluble metal sulphides

Table 17.1 Slough digested sludge.

Dry solids % wet weight	2.81
Volatile solids % wet weight	1.97
Total P mg kg^{-1} DS	35,300
Total S mg kg^{-1} DS	12,500
Total Fe mg kg^{-1} DS	12,000
Total Ca mg kg^{-1} DS	50,000
Total Mg mg kg^{-1} DS	5,500
pH	7.29
Alkalinity (at pH 4.5) mg l^{-1} as CaCO$_3$	3717

(typically 1–1.5% sulphur by dry weight). Therefore some of the potential precipitants discussed above, especially iron, are likely to be unavailable for precipitation of phosphorus as they will already be strongly bound as insoluble sulphides.

There is also the potential to precipitate phosphorus as struvite. At the typical operating pH range of the digesters some struvite formation must occur. It is likely that the suspended solids in the sludge will enmesh the small crystals formed if they do not nucleate on the solids themselves.

17.4 PHOSPHORUS MASS BALANCE

During the summer of 2000, a comprehensive mass balance for phosphorus, nitrogen and magnesium was conducted at Slough STW. The aim was to establish streams, which were best suited to phosphorus recovery.

An accurate mass balance is difficult to achieve for a number of reasons. Flow metering is not installed on all of the streams. Direct flow measurement was available for incoming sewage flow, settled sewage flow to the BNR, effluent flow from both streams, thickened SAS flow, thickened primary sludge flow and centrifuge feed flow. All other flows had to be either deduced or calculated from pump run hours and estimates of pump outputs.

Similarly with sampling, it is possible to obtain time weighted composite 24-hour samples at a limited number of locations. At other sampling points grab samples only are possible, and these can give highly variable results if there is a lot of variation in the characteristics of the stream and/or if they are not homogenous. For example, sludge thickener liquors may contain abnormally high concentrations of solids and this will impact on the results of all solids related parameters. Also grab samples are taken at approximately the same time, therefore the outlet samples of a process unit may reflect the feed conditions of several hours ago and not necessarily the conditions pertaining to the inlet sample. Composite samples were obtained where possible and seven sets of grab samples were taken on seven different days within the period of composite sampling of the major streams. The following points were sampled:

Composite samples: Crude sewage inlet to works (prior to return liquors), settled sewage to BNR plant and both final effluent flows.

Grab samples: Primary sludge thickener feed, thickened sludge and liquors, trickling filter sludge, BNR and non BNR return activated sludge, blended surplus activated sludge, thickened surplus activated sludge, surplus activated sludge thickener liquors, imported sludge, anaerobic digester feed and product sludge, centrifuge liquors and centrifuge cake.

Given the above methodology there is a large potential for error, some of which may be compounded by use in subsequent flow calculations. Flow meter accuracy is also assumed though this was not checked. Figures 17.3–17.7 show the derived mass balances.

The soluble phosphorus level in the SAS is relatively low, at 65 mg l^{-1}. A previous study, by Williams (1999), showed 190 mg l^{-1} of soluble phosphorus, at Slough STW. The reduction in soluble phosphorus could be due to a number of factors. The recirculation in the BNR lanes operated at only 50% capacity during this work. The reduction in recirculation would increase the concentration of nitrates in the SAS. The nitrates would provide a source of oxygen, thereby reducing the tendency for the SAS to go anaerobic. The flow rate of the SAS has also increased since the previous work (1186 m^3 compared to 700 m^3) so the hydraulic retention time in the SAS buffer tank has decreased. Jardin and Popel (1996) and Pitman (1999) have both shown that hydraulic retention times in storage facilities affect the amount of phosphorus released from the sludge. When the SAS is subjected to anaerobic conditions, e.g. due to long retention times, the phosphorus stored as polyphosphate chains is hydrolysed and released (Levin and Shapiro, 1965). It should be noted that the soluble phosphorus content of the SAS in this study was based on the results of one grab sample. However, the soluble phosphorus content of the belt thickener liquors (15.7 mg l^{-1} compared to 190 mg l^{-1}) also confirms the attenuated release in phosphorus.

The total phosphorus in the anaerobic digester feed averages 1066 mg l^{-1} and the soluble phosphorus concentration averages 95 mg l^{-1} (based on centrifuge liquor data). This gives a phosphorus release of 9%. The degree of phosphorus solubilisation at Slough is lower than figures reported by Murakami *et al.* (1987) and Knocke *et al.* (1988). Both authors found a phosphorus release of 60%. However, Wedi and Wilderer (1994) also found a relatively low phosphorus release of 4–10% at full-scale. During laboratory trials on the same sludge, Wedi and Wilderer (1994) found a release of 20%. They explained the difference by concluding the phosphorus was released during digestion, but became chemically bound in the sludge. This refixation of phosphorus could explain the relatively low concentrations of phosphorus in the centrifuge liquor (167.0 mg l^{-1}) compared to the phosphorus levels in the centrifuge cake (6531 mg l^{-1}). The phosphorus released in the anaerobic digesters at Slough STW could have been refixed as struvite and other phosphates, which were held in the sludge and removed with the centrifuge cake.

The digested sludge, centrifuge liquor and centrifuge cake are identified by the model of Lowenthal *et al.* (1994) as streams which have the potential to form struvite. These streams all have high concentrations of soluble phosphorus. The centrifuge cake has the highest struvite precipitation potential (SPP), as it has the highest magnesium concentration of any of the streams. The digested sludge also has a high SPP, which could be due to the high concentration of phosphorus. The

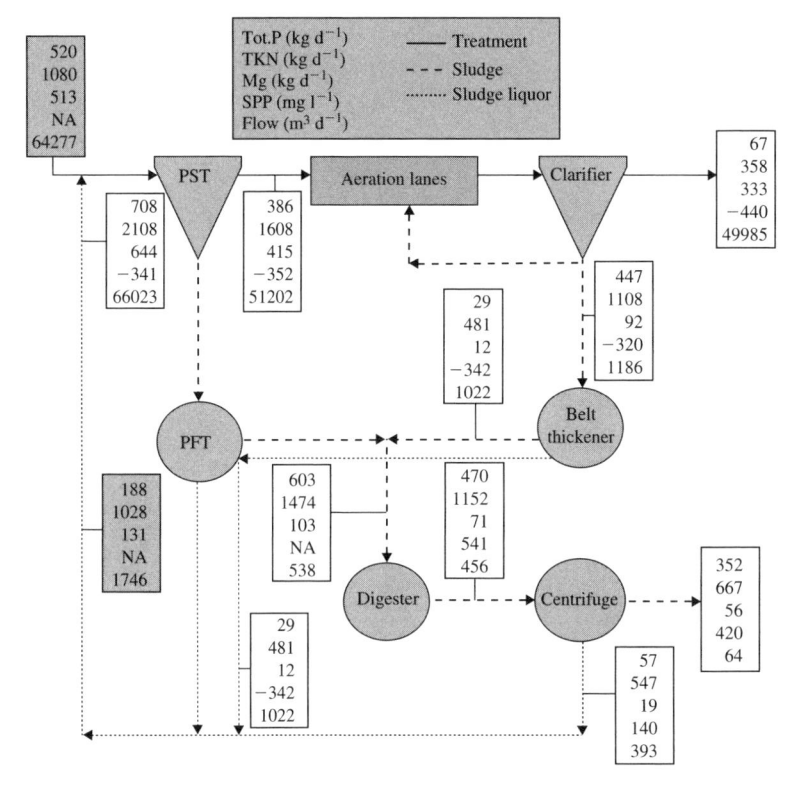

Figure 17.3 Simplified mass balance for entire site of Slough STW. For acronyms see Appendix. [This figure is also reproduced in colour in the plate section after page 336.]

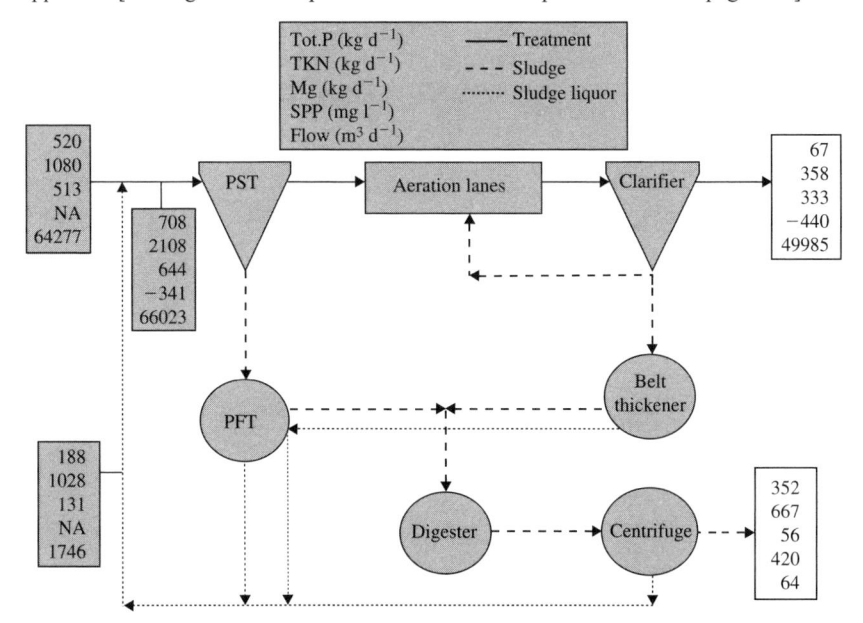

Figure 17.4 Mass balance highlighting inlet and outlet of Slough STW. For acronyms see Appendix. [This figure is also reproduced in colour in the plate section after page 336.]

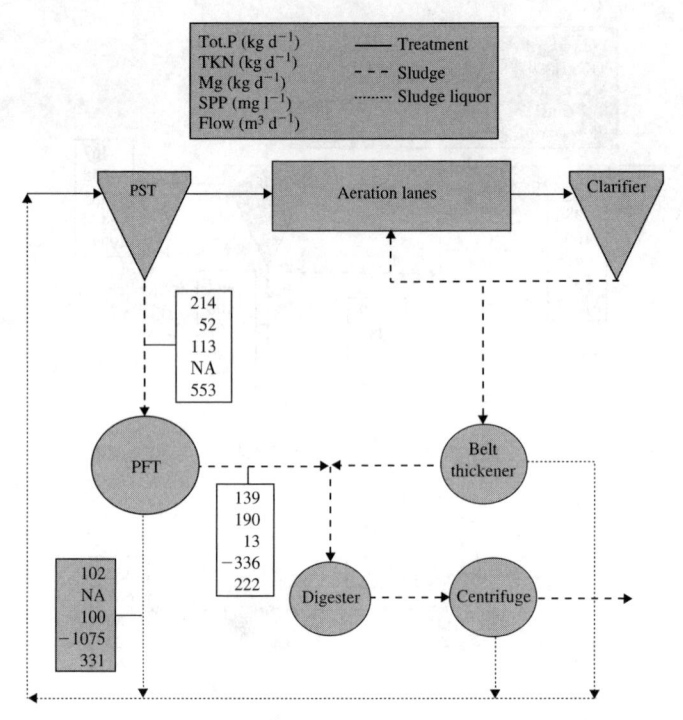

Figure 17.5 Mass balance around the primary sludge thickener, Slough STW. For acronyms see Appendix. [This figure is also reproduced in colour in the plate section after page 336.]

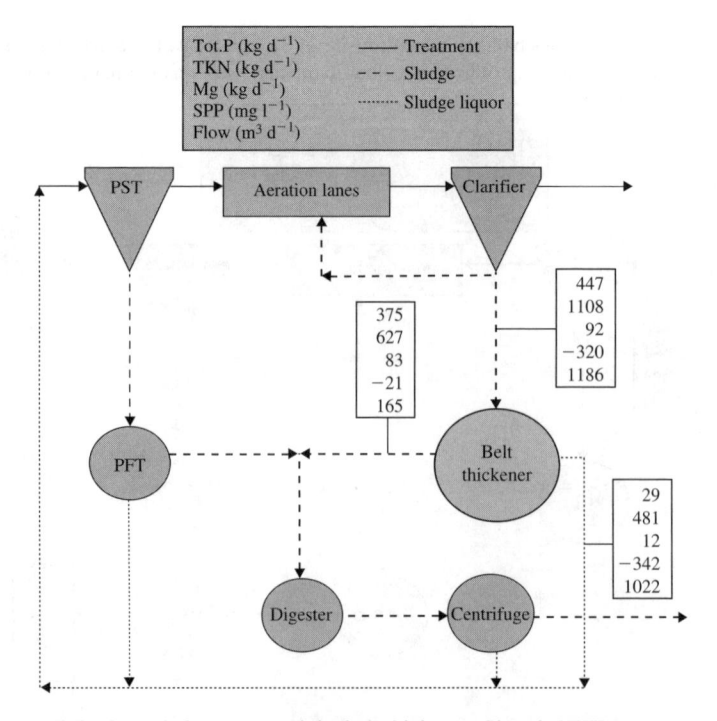

Figure 17.6 Mass balance around the belt thickener, Slough STW. For acronyms see Appendix. [This figure is also reproduced in colour in the plate section after page 336.]

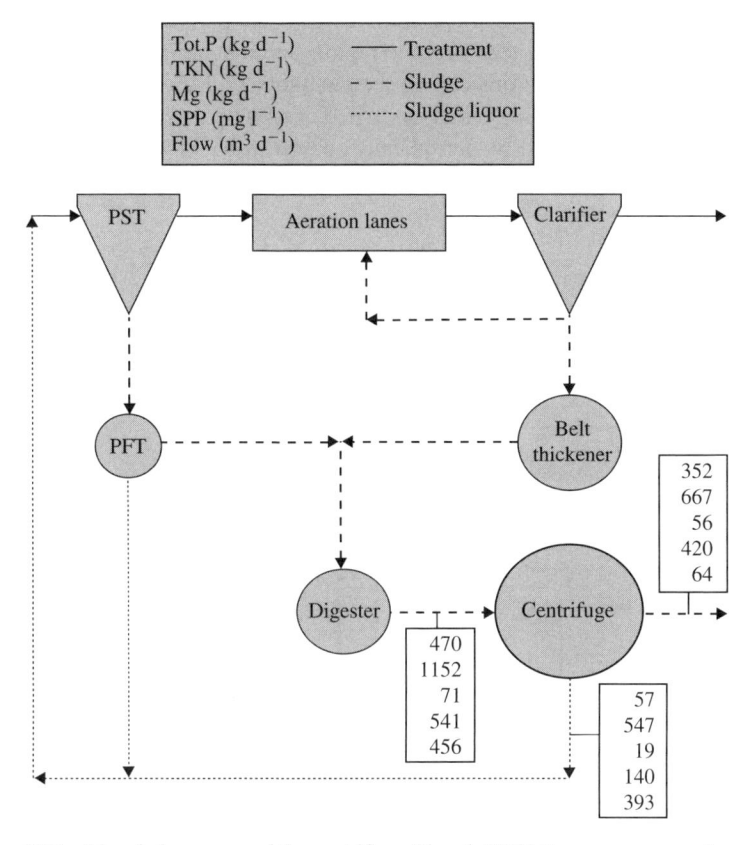

Figure 17.7 Mass balance around the centrifuge, Slough STW. For acronyms see Appendix. [This figure is also reproduced in colour in the plate section after page 336.]

centrifuge liquor has a positive SPP as well, though at 140 mg l^{-1}, it is not as high as the other two streams. The magnesium concentration in the centrifuge liquors is at least 50% less than the magnesium concentration in the other two SPP positive streams.

It was decided to use the centrifuge liquors in a series of bench-scale experiments, to determine how easy it would be to precipitate struvite from them.

17.5 BENCH-SCALE WORK

17.5.1 Pilot plant

A bench-scale pilot plant was operated to see if struvite could be precipitated from the centrate liquors. The bench-scale reactor was constructed using a Water Research Council (WRC) porous pot apparatus. The apparatus consisted of a porous vessel (BTSK346) inside an impermeable vessel (BTSK347). A diagram of the bench scale rig can be seen in Figure 17.8.

Bird and Tole Ltd, UK supplied both vessels. Two peristaltic pumps (Watson and Marlow) were used to feed centrate liquors and magnesium chloride into the reactor. The reactor conditions can be seen in Table 17.2. Aeration was supplied by an aquarium aeration unit, which consisted of a pump and two aeration stones. The aeration stones had to be periodically soaked in 0.5 M hydrochloric acid to remove struvite deposits.

17.5.2 Pilot plant feed

The reactor was fed with centrate liquors obtained from Slough STW. The liquors were collected in 25 l plastic containers and stored at room temperature. The pH

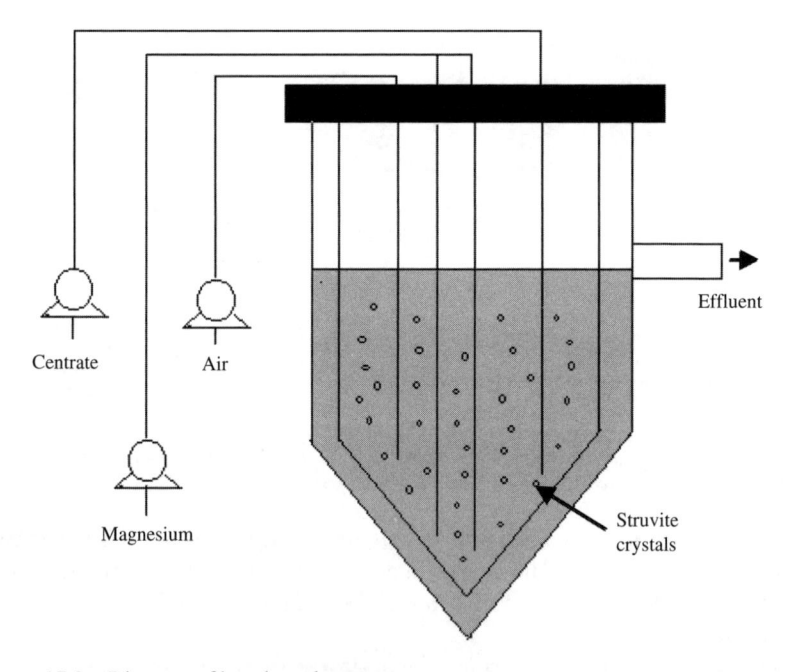

Figure 17.8 Diagram of bench-scale reactor.

Table 17.2 Reactor conditions.

Parameter	Value	Units
Volume	3.5	litres
Centrate feed	7	ml min^{-1}
Chemical feed	12	ml min^{-1}
Total feed	19	ml min^{-1}
HRT of centrate only	8.31	hours
HRT of centrate + MgCl$_2$	3.07	hours
Aeration	440	ml min^{-1}

of the liquors were recorded using a Jenway 100 pH probe and then adjusted to 9.0, using 20% sodium hydroxide solution. The date the samples were obtained and the day the pH of the sample was adjusted were recorded. Samples of the liquors, prior to and after pH adjustment, were taken and sent to the Thames Water Centre Laboratories for analysis. Samples were taken before and after pH adjustment to determine if raising the pH affected the chemical composition of the liquors. The analyses performed by the laboratories are listed in Table 17.3.

The reactor was also fed with magnesium chloride, which was supplied in a liquid form, of purified quality, at a concentration of 400 kg m^{-3}. Solutions of the required concentration of magnesium were prepared using deionised water and tested (Dr Lange kit, LCK 326) to verify the amount of magnesium present. Deionised water was used to prevent the addition of extra magnesium and calcium ions, due to the hardness of the water in the Thames region.

17.5.2.1 Analytical results

The effluent from the reactor was sampled on a daily basis (apart from weekends), after at least four HRTs. A sample of effluent was sent to Thames Water Centre Laboratories for analysis.

On the spot analysis was carried out on the effluent samples for pH, ammoniacal nitrogen (NH_4–N), soluble phosphorus (P–PO_4), total phosphorus (T-P) and total magnesium (T-Mg). Spot analyses were carried out using Dr Lange test kits.

Table 17.4 shows the average composition of centrifuge liquors before and after pH adjustment. Levels of soluble phosphorus, magnesium and calcium dropped

Table 17.3 Laboratory analysis of centrate liquors.

Determinant	Units
TKN	(mg l^{-1})
NH_4–N	(mg l^{-1})
T-P	(mg l^{-1})
P–PO_4	(mg l^{-1})
T-Mg	(mg l^{-1})
Mg–$MgCl_2$	(mg l^{-1})
Ca	(mg l^{-1})
Alkalinity	(mg l^{-1})
SS	(mg l^{-1})

Table 17.4 Analysis of centrate liquor before and after pH adjustment.

	T-P (mg l^{-1})	P–PO_4 (mg l^{-1})	T-Mg (mg l^{-1})	Mg–$MgCl_2$ (mg l^{-1})	N–NH_4 (mg l^{-1})	TKN (mg l^{-1})	Ca_s (mg l^{-1})	Alkalinity as $CaCO_3$ (mg l^{-1})	pH
Before	99.6	92.1	13.15	11.5	530.5	547.0	26.0	4888.0	7.5
After	100.8	85.6	5.9	1.7	574.8	654.0	6.0	2322.5	9.0

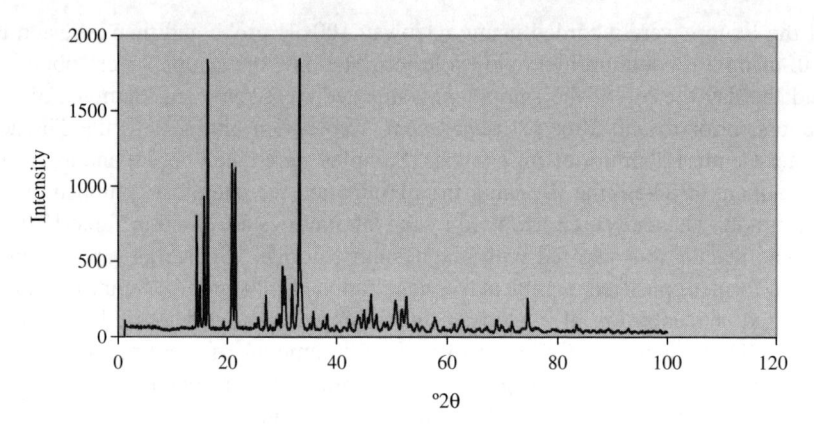

Figure 17.9 XRD pattern of crystals produced in pilot plant at Slough STW. The pattern is characteristic of struvite.

after raising the pH to 9.0. Raising the pH must cause the magnesium and calcium in the sample to react with phosphorus, forming precipitates. The ratio of calcium: magnesium in the centrifuge liquor is roughly 2 : 1. This decreases to a 1 : 1 ratio after the pH has been raised. This reduction in the ratio indicates that calcium is preferentially reacting with the phosphorus compared to magnesium. The concentration of total phosphorus remained the same.

The magnesium levels in the centrifuge liquor samples were too low to remove the remaining phosphorus as struvite. The pilot plant was therefore dosed with magnesium chloride ($MgCl_2$) to provide a source of magnesium ions and maximise struvite production. The pilot plant was started using centrifuge liquors and a dose of 252 mg l^{-1} of $MgCl_2$. Crystals were seen on the surface of the porous pot within 24 hrs. Samples of crystals were taken and analysed by X-ray diffraction (XRD) at the Natural History Museum's laboratories, UK. The results of the XRD can be seen in Figure 17.9. The crystals produced were validated to be struvite and were of a high purity. Once it was established that struvite could be formed from the centrifuge liquors at Slough STW, the magnesium dose to the pilot plant was altered to determine an optimum dosing regime. Figure 17.10 shows that phosphorus removal increases with magnesium addition. The level of phosphorus removal reaches approximately 97% at magnesium molar addition of 5.7 mM, but does not go higher than this, even though the magnesium dose is doubled.

At high dosages of magnesium, the phosphorus is probably removed as struvite. This can be seen by the parallel phosphorus and magnesium removal at higher doses of magnesium. At magnesium doses below 3.4 mM, phosphorus is still removed, but perhaps not solely as struvite. Figure 17.10 also shows that the molar removal of ammonia exceeds the molar removals of phosphorus and the molar usage of magnesium, i.e. there is excessive loss of ammonia

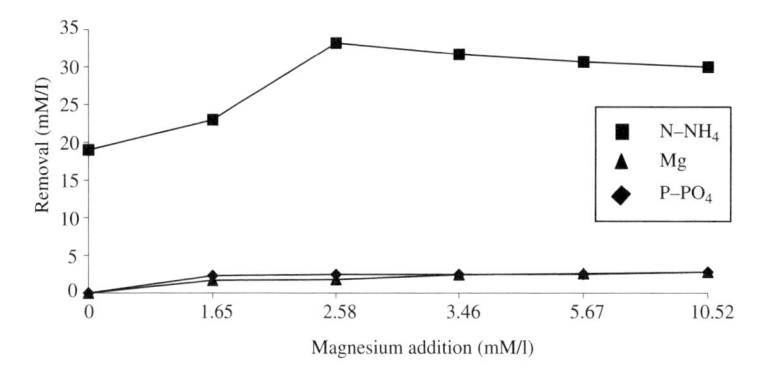

Figure 17.10 Molar removal of phosphorus and ammonium and molar usage of magnesium against molar addition of magnesium.

compared to phosphorous and magnesium. The excess is probably due to loss of ammonium as gas via air stripping of the reactor contents at an elevated pH. On a large scale plant the loss of ammonia gas could present problems in terms of odour production.

The experiments conducted during this work showed that struvite could be rapidly formed from the centrifuge liquors at Slough. Struvite crystals were produced within 24 hours of starting the pilot plant. Raising the pH of the liquors to 9.0 was an effective method of forming struvite. Magnesium in the storage containers was removed before even entering the reactor. A pH of 9.0 was found to be optimum by Siegrist et al. (1992). It was found that 95% of the total phosphorus could be removed from the centrifuge supernatant as struvite, by the addition of at least a 1.05:1 molar ratio of magnesium to phosphorus i.e. a magnesium dose of about 83 mg l^{-1}. This ratio of 1.05:1 was also found by Fujimoto et al. (1991). Other authors, Siegrist et al. (1992) and Salutsky et al. (1972), found a higher ratio of 1.3:1 was required to guarantee phosphorus removal as struvite. The lower ratio necessary for struvite precipitation during this series of experiments was probably due to the lack of any competing reactions. When the pH of the centrifuge liquor was raised to 9.0 in the storage containers, 77% of the calcium was removed, before entering the reactor. On average 6 mg l^{-1} calcium entered the reactor. At a dosing regime of 83 mg l^{-1} this equates to a 0.04:1 molar ratio of calcium : magnesium. Hwang and Choi (1998) found that for effective struvite formation, the ratio of calcium to magnesium should be less than 1.

17.6 THE REACTOR

The reactor used at Slough STW was a modified lamella clarifier. The lamella plates were removed from the reactor and the central baffles were extended to create a reaction zone and two settling zones. The reactor was then sandblasted to smooth the rough surface and painted in a two-part epoxy resin. The surface of the reactor was relatively smooth, but pitted. All the associated pipe work of the clarifier was removed and replaced. A new inlet and outlet to the reactor was created. The ancillary equipment was then added. Centrate is pumped into the reactor, where the conditions are manipulated by adding sodium hydroxide to raise the pH and magnesium to increase the ratio of magnesium : phosphorus (see Figure 17.11). The reactor is separated into a reaction and settling zones by baffles. The overflow weirs in the settling zone can be raised or lowered, to increase or decrease the settling area. The contents of the reactor are mixed by forcing a stream of air through the fluid, from the base of the reaction zone. The contents of the reaction zone eventually move into the settling zone where the organic solids in the centrate theoretically rise to the surface and are eliminated with the effluent. The crystals of struvite have a density higher than water (1.7 compared to 1.0), so sink to the bottom of the settling zone, where they are collected.

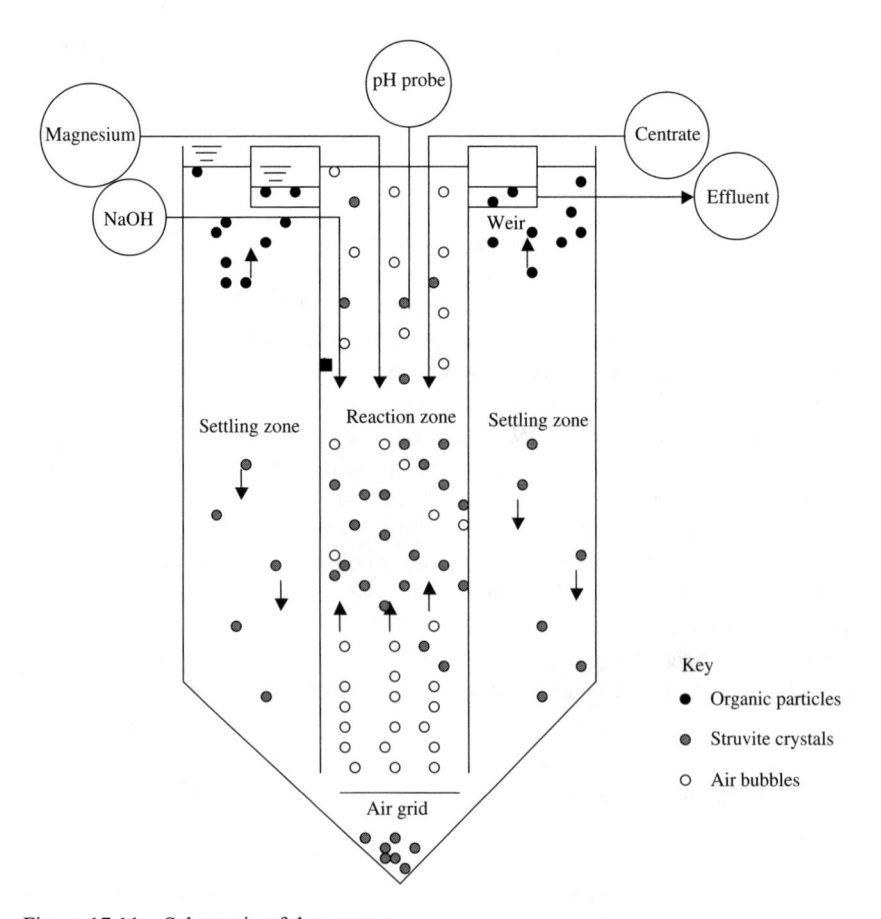

Figure 17.11 Schematic of the reactor.

17.6.1 Influent

The influent to the reactor was taken from the centrate liquor sump. It was pumped directly from the sump via a chopper pump into the top of the reaction zone. The maximum flow to the reactor was 7.2 m^3 hr^{-1}, but typically the influent flow was 5.4 m^3 hr^{-1}. The maximum flow that can be treated is around 40% of the total flow of centrate at Slough STW. The characteristics of the centrate are detailed below and are based on an average of 30 samples:

Total phosphorus	111 mg l^{-1}	Soluble phosphorus	86 mg l^{-1}
Total magnesium	22 mg l^{-1}	Soluble magnesium	13 mg l^{-1}
Soluble calcium	57 mg l^{-1}	Suspended solids	408 mg l^{-1}
Ammoniacal nitrogen	590 mg l^{-1}	pH	7.7
Alkalinity (as CaCO$_3$)	2375 mg l^{-1}		

17.6.2 Aeration

The purpose of the aeration was to provide suitable mixing of the contents in the reaction zone. Also it was hoped the aeration would help increase the pH of the reactor contents, by stripping out carbon dioxide.

Initially the air was supplied via three grids above the base of the reactor. The grids were made from 1.5 inch PVC pipe, with 1 mm diameter holes drilled into them at 50 mm intervals. Air supply to each grid could be controlled and adjusted separately. The grids provided coarse, evenly distributed aeration. A compressor was also attached to the base of the reactor, to periodically 'air-lift' the crystals back into the aerated reaction zone.

However within a month, the grids had become encrusted with struvite and aeration became limited. The air grids were removed and aeration was provided via two injection points at the base of the reactor, which were previously used for the air-lift. The initial blower was also replaced with a rotary vane compressor (Becker UK, model DT 4.25K). To avoid problems with the struvite settling at the base of the reactor and the air merely blowing a hole through the middle of the settled struvite, other air injection points were fitted around the bottom section of the reactor. The inclusion of more injection points allows a greater degree of mixing. The aeration was set at 140 l min^{-1} to provide an upflow velocity of 0.01 m min^{-1} (Munch and Barr, 2001).

17.6.3 Correction of pH

The pH was monitored using a S410 pH probe (LTH Electronics Ltd.) connected to a MPD53 monitor (LTH Electronics Ltd.). The system allows the pH to be set to a desired range and is connected to a supply of sodium hydroxide, so that the pH can be maintained. Most of the trials were conducted without the use of pH correction, as the incoming centrate had an average pH of 7.6. This pH seemed to be sufficiently high to give reasonable struvite formation.

17.6.4 Magnesium source

To date, the magnesium used in the pilot plant has been obtained from magnesium chloride. Magnesium chloride is easy to handle and has good solubility, therefore most of the magnesium is readily available. The magnesium chloride is provided in a 31% solution, with a density of $1300\,kg\,m^{-3}$. The magnesium chloride solution is slightly acidic, so it can lower the pH of the reactor contents. The magnesium was dosed at different rates, but typically at a 1.3 : 1 ratio of magnesium : phosphorus. Typical flow rate for magnesium chloride was $100\,ml\,min^{-1}$.

17.6.5 Struvite recirculation

Initially the reactor was operated without any crystal recirculation. The crystals produced were about 0.1–0.3 mm in length (see Figure 17.12). In an attempt to increase the size of the crystals, a recycle system was set up. Crystals from the base of the reactor are pumped, via a peristaltic pump, back to the top of the reaction zone. The peristaltic pump was set to operate for 6 minutes every 30 minutes.

Figure 17.12   SEM of struvite crystals from Slough STW.

17.6.6 Struvite collection

The flow of crystals from the base of the reactor to the top of the reaction zone can also be diverted so the crystals can be collected into a conical silo container. The silo has a shut off valve at the base, so the struvite slurry can be retained in the silo. Retention of the struvite allows the crystals to settle and any associated water and organic matter can be decanted from the top. Once the water has been removed, the valve is opened and the struvite is placed into a skip, for storage.

17.7 PRELIMINARY RESULTS

Due to continuous development work, extensive runs have not yet been possible on the reactor. The results achieved so far have been on relatively short run times, i.e. a few days continuous use. Table 17.5 summarises the conditions the reactor has been operated in.

Most of the work so far has been conducted with no pH correction, as the use of hydroxide to alter pH is expensive. In general, phosphorous removal was best when the pH of the reactor was above 8.5. At this elevated pH, soluble phosphorus removals of over 90% could be achieved. Total phosphorus removal rates at this pH averaged above 80%. Both these results were obtained by dosing magnesium chloride at a molar ratio of at least 1.3 : 1 Mg : P. If the dose rate of magnesium was reduced, so that the molar ratio was below 1.3 : 1 Mg : P, but all other conditions were the same, then soluble phosphorus removal rates were about 60%.

If the reactor was operated at an ambient pH of 7.7, then a magnesium dose of 1.3 : 1 Mg : P achieved 68% soluble phosphorus removal and 69% total phosphorus removal. If the magnesium dose was increased to 2 : 1 Mg : P or above, then above 80% soluble phosphorus removal could be achieved. Table 17.6 highlights some of these results.

17.8 QUALITY OF STRUVITE

The struvite collected has the consistency of fine sand. The struvite settled quickly but the size of the crystals did not seem to increase beyond 0.3 mm in length. Some

Table 17.5 Summary of pilot reactor conditions.

	Range	Typical
Reaction zone volume	$5.8\,m^3$	–
Settling zone volume	6.8 or $13.6\,m^3$	$6.8\,m^3$
Airflow rate	$60-140\,l\,s^{-1}$	$100\,l\,s^{-1}$
Upflow velocity	$0.4-0.1\,m\,min^{-1}$	$0.07\,m\,min^{-1}$
Influent flow rate	$7.2-5.4\,m^3\,hr^{-1}$	$5.4\,m^3\,hr^{-1}$
pH	7.5–9.0	7.8
Mg dose	$65-200\,ml\,min^{-1}$	–

of the crystals have a pitted surface (see Figure 17.13), which may be due to abrasion or dehydration of the crystals.

17.8.1 Metal content

The struvite was analysed for heavy metal content. If the product is to be used as a fertiliser, then it must meet certain standards. At present, the EU Directive (86/278/EEC) sets limits for heavy metals in sludge. The limits in the current directive are shown in Table 17.7, along with proposed revisions. The proposed revisions are due to be implemented in 2006. The heavy metal content of the struvite produced at Slough STW is shown in Table 17.8.

Table 17.6 Summary of pilot reactor operation results.

Mg:P	pH	HRT (min)	Influent flow ($m^3 hr^{-1}$)	Upflow velocity ($m\,min^{-1}$)	Soluble P removed (%)	Total P removed (%)
0.8:1	7.3–7.5	48	7.2	0.04	59	NA
1.2:1	7.7	64	5.4	0.07	68	69
2.1:1	7.3–7.5	64	5.4	0.08	81	83
2.3:1	7.3–7.5	64	5.4	0.08	81	80
2.4:1	7.3–7.5	48	7.2	0.08	75	NA
1.1:1	8.8–8.9	64	5.4	0.04	60	54
1.2:1	8.8–8.9	64	5.4	0.07	76	72
1.3:1	8.8–8.9	64	5.4	0.04	91.5	88.5
1.7:1	8.8–8.9	64	5.4	0.08	94	87.5

Figure 17.13 Struvite crystal surface, Slough STW.

As can be seen from the two tables, the heavy metal content of the struvite produced at Slough STW, is well below the current limits. The metal content of the struvite is also below the proposed revisions to Directive 86/278/EEC.

17.8.2 Microbial content

The struvite was also tested for *Escherichia coli* (*E. coli*). At present if sludge is to be used on land it must contain less than 500 CFU (colony forming units) *E. coli* per gram of product. The struvite produced at Slough contains <1 CFU *E. coli* per wet gram and 1.03 log 10 *E. coli* per dry gram. These values for *E. coli* content are below the required limits. However, it should be noted that the method used during this analysis has only been validated using sewage sludges and biosolids and it may be an unsuitable method for recovering *E. coli* from the struvite matrix.

17.8.3 Organic content

The organic content of the struvite appears to be insignificant. Most of the organics are removed in the settling zone of the reactor. The organic matter in the centrate has a density less than water, so rises in the settling zone and flows over the weirs. The struvite has a density greater than water so sinks in the settling zone. This separation is effective. Additionally, the struvite slurry removed from the base of the reactor can be washed while it is being held in the silo. Any organic matter clinging to the struvite crystals can be washed and decanted. This additional washing would remove most of the last organic traces from the product.

Table 17.7 Heavy metal limits for sludge used in agriculture.

Metal	Current limit (mg kg^{-1})	Revised limit (mg kg^{-1})
Chromium	No limit	1000
Copper	1000–1750	1000
Mercury	16–25	10
Nickel	300–400	300
Lead	750–1200	750
Cadmium	20–40	10

Table 17.8 Heavy metal content of struvite produced at Slough STW.

Metal	mg kg^{-1}	Metal	mg kg^{-1}
Calcium	5592	Boron	29.0
Magnesium	>100000	Potassium	843.0
Copper	37.8	Chromium	<3.0
Zinc	<10	Manganese	74.0
Cadmium	<0.3	Iron	558
Mercury	<0.1	Nickel	<3.0
Lead	<7.0	Molybdenum	<3.0
Selenium	<430 µg kg^{-1}		

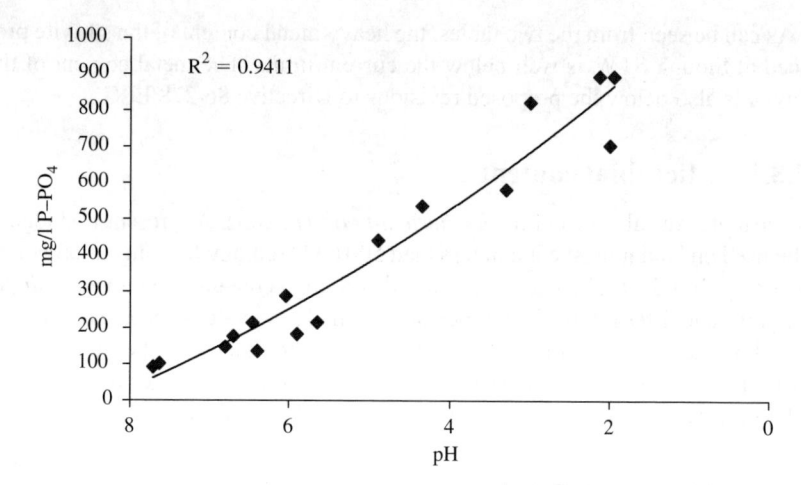

Figure 17.14 The relationship between phosphorus release and pH of sludge at Slough STW.

17.8.4 Production rates

The daily rate of struvite production at Slough is currently 53 kg a day, based on a flow rate of 5.4 m^3 hr^{-1} and a magnesium dose of 1.3 : 1 Mg : P. The phosphorus content of the digested sludge liquor averages at 86 mg l^{-1} and removal rate averages 60%. If the dose of magnesium is increased to 2 : 1, then the average removal rate is 80%, resulting in a daily production rate of 70 kg.

Another way to increase the amount of struvite produced could be to increase the amount of phosphorus in the digested sludge liquor. According to the mass balance conducted at Slough, 75% of the phosphorus in the digested sludge is bound in the sludge cake. There is, therefore, six times more phosphorus in the centrate cake than there is in the centrifuge liquor. This phosphorus could be released from the digested sludge prior to dewatering, by dropping the pH of the digested sludge. One way of decreasing the pH is to add acid to the sludge.

At a pH of 7.6 the amount of soluble phosphorus in the centrate is 90 mg l^{-1}. If the pH is dropped to 6.0 then the phosphorus content of the digested sludge liquor could increase to almost 300 mg l^{-1}. Figure 17.14 shows the release of phosphorus from digested sludge by the addition of 5M hydrochloric acid. However, this is an expensive option and would only be viable if there was a lucrative market for struvite.

If the pH of the digested sludge at Slough STW was dropped to 6.0, then the struvite production could be 185 kg, assuming no pH correction and a magnesium dose of 1.3 : 1. Table 17.9 summarises the potential production rates of struvite, under different scenarios.

17.8.5 Economics

The main operational costs of running the reactor at Slough STW are due to the costs of chemical addition and the cost of aeration. The cost of sodium hydroxide

Table 17.9 Potential production rates at Slough STW.

Mg:P ratio	1.3:1	2:1	1.3:1	2:1
pH	7.6	7.6	7.6	7.6
Flowrate	5.4	5.4	5.4	5.4
Incoming P	100	100	300	300
% P Removal	60	80	90	95
Kg struvite/day	62	82	185	247
Tonnes/year	23	30	68	90
Revenue (€/year)	3170	4140	9380	34070

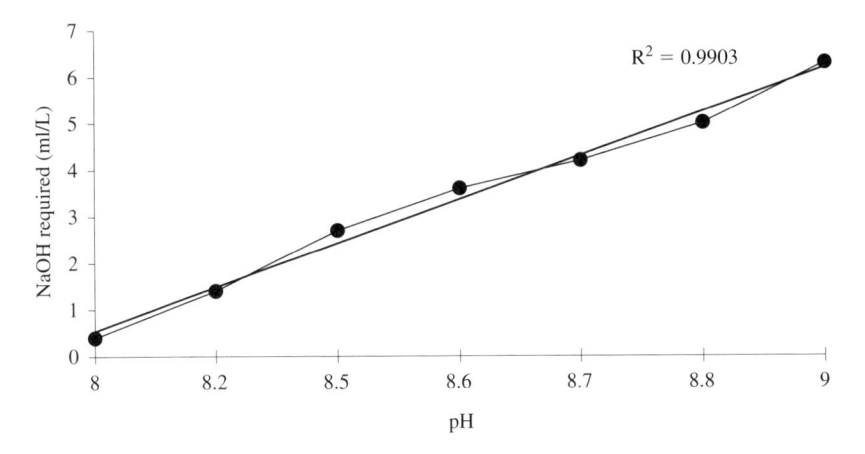

Figure 17.15 Sodium hydroxide required to alter pH of centrate at Slough STW.

addition, to raise the pH of the centrate liquors to a value of 9.0, is the most expensive chemical addition. Figure 17.15 shows how much sodium hydroxide is required to raise the pH of the centrate liquor. The amount of sodium hydroxide required to raise the pH to 8.5 is relatively small. However, to raise the pH from 8.0 to 9.0, requires more than twice the amount required to raise the pH from 8.0 to 8.5. If we assume the daily flow of centrate is $350\,m^3$, then annually the amount of sodium hydroxide required to raise the pH to 8.5 is $350\,m^3$. To raise the pH of the same amount of centrate to 9.0 would require $5500\,m^3$ of sodium hydroxide. At €62 m^{-3} for the sodium hydroxide this equates to a cost of €21,400 and €341,400 per year. This cost is clearly prohibitive.

The cost of adding magnesium will depend on the Mg:P ratio required, the flowrate of centrate being treated and also the incoming concentration of phosphate. Table 17.10 summarises the cost of magnesium addition under different scenarios.

From Table 17.10, it can be seen that the precipitation of struvite from centrate liquors at Slough STW is not commercially viable. The cost of the chemicals exceeds the revenue that might be obtained from the sale of the struvite. However, the main reason for having a sidestream process which recovers phosphorus from

Table 17.10 Summary of potential magnesium chloride costs at Slough STW.

Incoming P–PO$_4$ (mg l^{-1})	86	86	300	300
Magnesium dose (Mg : P)	1.3 : 1	2 : 1	1.3 : 1	2 : 1
Cost of magnesium (€/m^3 of treated centrate)[a]	0.36	1.27	0.56	1.95
Revenue from struvite (€/m^3 of treated centrate)[b]	0.12	0.41	0.16	–

Notes: [a] Assuming magnesium chloride costs £300 m^{-3}.
[b] Assuming struvite could be sold for £200 tonne.

the centrate liquor is not necessarily commercial. The main objective is to reduce the amount of phosphorus recycled back to the inlet of the works. In the Thames region, it can be difficult to operate BNR plants, as the carbon to phosphorus ratios are not always favourable. There is usually not enough carbon to drive the removal of phosphorus during BNR. At Slough STW 26% of the phosphorus load to the works originates from return liquor streams. Therefore, anything, which reduces this recycle of phosphorus and subsequently the load of phosphorus to the works, is beneficial.

17.9 FURTHER WORK

17.9.1 Magnesium source

One way of reducing the cost of chemical addition is to use alternative sources of magnesium, for example magnesium hydroxide. If magnesium hydroxide is used, it will function as a source of magnesium ions and a source of hydroxide ions, to raise the pH. Magnesium hydroxide is less expensive than magnesium chloride, in terms of percentage magnesium, by weight. Also, by using magnesium hydroxide the amount of struvite produced might increase, as the pH is more favourable. However, using magnesium hydroxide to serve both functions means that the magnesium dose or the pH cannot be optimised independently of each other.

Using magnesium hydroxide may also have potential disadvantages. Magnesium hydroxide is difficult to handle, it is relatively insoluble, and comes as a suspension. This suspension has to be continuously stirred to prevent the magnesium hydroxide settling. Also magnesium hydroxide will not disassociate as rapidly as magnesium chloride. Longer disassociation times may require longer hydraulic retention times for the reaction to occur.

17.9.2 Aeration rate

At present the aeration rate is set at 140 l min^{-1}, to give the upflow velocity of 0.1 m min^{-1}. This rate has not yet been optimised. It may be possible to get better separation of the organic matter and the struvite produced by increasing the airflow rate. Also, by increasing the aeration rate the pH of the centrate liquors might increase to a more favourable value. This has been demonstrated well by Battsitoni et al. (1997). At present, the aeration rate used makes little difference in the pH of the centrate liquors. The aeration does raise the pH level, but only by 0.2 units.

17.9.3 Solids content

The solids content of the reactor also needs optimising. At present the reactor is de-sludged i.e. struvite removed form the reactor, when the solids content is about 10%. We do not know if this is an ideal indicator, as it is difficult to tell how much struvite has settled to the bottom of the reactor and is no longer being lifted by the air into main body of the reaction zone.

17.9.4 Effluent recycle

At present, the reactor at Slough STW has no effluent recycle, but one may be beneficial. Any unused magnesium could be fed back into the reactor and have a chance to react again. This may be of use, especially at high magnesium to phosphorous dosing rates, to prevent a high concentration of magnesium being recycled back to the head of the works.

At the moment, effluent from the reactor is combined with other sludge liquor streams and returned back to the sewage treatment works inlet. The flow from the reactor is only 30% of the total flow of centrate at Slough. The centrate flow, in turn, is only about 0.6% of the total inlet flow at Slough. The crude sewage is being monitored and at present there is no significant increase in the magnesium levels coming into the works, so there is no likelihood of an increase in the struvite precipitation potential, further downstream. If however, larger volumes were being used to recover phosphorus, e.g. if a combination stream of surplus activated sludge liquor was combined with digested sludge liquor to give a stream with a high phosphorus and ammonia concentration, problems may occur. If a combined stream was used, the effluent from the reactor would equate to 2% of the total flow into the works. An increase in magnesium concentration in a flow of this size might make a difference to the overall concentration in the works.

17.9.5 Hydraulic retention time

Currently, the hydraulic retention time is 64 minutes. Due to the size of the reactor and the capacity of the inlet pump, it is not possible to reduce this time any further. However it is necessary to see if the same phosphorus removal efficiencies can be achieved at reduced retention times.

17.10 PROPOSED USE OF PRODUCT

The struvite recovered from Slough STW will ideally be used as an agricultural fertiliser. A recent study by Richards and Johnston (2001) has examined the effectiveness of recovered struvite as a fertiliser. Recovered fertilisers were compared to monocalcium phosphate (MCP), a source of water soluble phosphorus that is generally considered to be fully plant available. The report concludes that the recovered struvites are as good a source of phosphorus as the MCP.

Thames Water has started to investigate potential markets for the struvite produced at Slough STW. Thames has looked at using the struvite as a product enhancer, by adding it as a supplement to existing product lines, and also looked at using struvite as a stand-alone product. Product development is in the initial stages, but it is envisaged that the recovered struvite will be manufactured into a struvite based fertiliser.

REFERENCES

Battistoni, P., Fava, G., Pavan, P., Musacco, A. and Cecchi, F. (1997) Phosphate removal in anaerobic liquors by struvite crystallisation without addition of chemicals: preliminary results. *Water Res.* **31**(11), 2925–2929.

Carliell, C.M. and Wheatley, A.D. (1997) Metal and phosphate speciation during anaerobic digestion of phosphorus rich sludge. *Water Sci. Technol.* **36**(6–7), 191–200.

Carliell-Marquet, C.M. and Wheatley, A.D. (2002) Measuring metal and phosphorus speciation in P-rich anaerobic digesters. *Water Sci. Technol.* **45**(10), 305–312

Council of European Communities (2000) Working Document on Sludge, 3rd draft (86/278/EEC). ENV.E.3/LM.

Fujimoto, N., Mizuochi, T. and Togami, Y. (1991) Phosphorus fixation in the sludge treatment system of a biological phosphorus removal process. *Water Sci. Technol.* **23**, 635–640.

Hwang, H.J., Choi, E. (1998) Nutrient control with other sludges in anaerobic digestion of BPR sludge. *Water Sci. Technol.* **38**(1), 295–302.

Jardin, N. and Popel, J. (1994) Phosphate release of sludges from enhanced biological P-removal during digestion. *Water Sci. Technol.* **30**(6), 281–292.

Jardin, N. and Popel, J. (1996) Behaviour of waste activated sludge from enhanced biological phosphorus removal during sludge treatment. *Water Environ. Res.* **68**, 965–973.

Knocke, W.R., Sen, D., Nash, J. and Randall, C.W. (1988) Phosphorus control issues in the digestion and dewatering of biological phosphorus removal sludges. *Proceedings of the 1988 CSCE-ASCE National Conference*, Vancouver.

Levin, G.V. and Shapiro, J. (1965) Metabolic uptake of phosphorus by wastewater organisms. *J. Water Pollut. Con. Fed.* **37**, 800–821.

Loewenthal, R.E., Kornmuller, U.R.C. and van Heerden, E.P. (1994) Modelling struvite precipitation in anaerobic treatment systems. *Water Sci. Technol.* **30**(12), 107–116.

Munch, E.V. and Barr, K. (2000) Controlled struvite crystallisation for removing phosphorus froma anaerobic digester sidestreams. *Water Res.* **34**, 1868–80.

Murakami, T., Koike, N., Taniguchi, N. and Esumi, H. (1987) Influence of return phosphorus load on performance of the biological phosphorus removal process. *Proceedings of IAW-PRC Specialised Conference, Rome, September/October 1987*. Pergamon Press, London.

Pearce, P. (2002) Control options for the efficient pre-precipitation of phosphorus from wastewater. Conference Proceedings, IWA Congress, Melbourne.

Pitman, A.R. (1999) Management of biological nutrient removal plant sludges – change the paradigms? *Water Res.* **33**(5), 1141–1146.

Randall, C.W., Barnard, J.L. and Stensel, H.D. (1992) *Design and Retrofit of Wastewater Treatment Plants for Biological Nutrient Removal*. Technomic Publishing Co, ISBN 87762-922-6.

Rawn, A.M., Banta A.P. and Pomeroy, R. (1939) Multiple stage sewage digestion. *Am. Soc. Civ. Eng. Trans.* **105**, 93–132.

Richard, I.R. and Johnston, A.E. (2001) The effectiveness of different precipitated phosphates as sources of phosphorus for plants. *IACR, Rothamstead, UK and Ecopt, Suffolk, UK*.

Salutsky, M.L., Dunseth, M.G., Ries, K.M. and Shapiro, J.J. (1972) Ultimate disposal of phosphate from wastewater by recovery as fertiliser. *Effl. Water Treat. J.* **22**, 509–519.

Siegrist, H., Gajcy, D., Sulzer, S., Roeleveld, P., Oschwald, R., Frischknecht, H., Pfund, D., Morgeli, B. and Hugerbuhler, E. (1992) Nitrogen elimination from digester supernatant with magnesium-ammonium-phosphate. *Proceedings of the 5th* Gothenburg Symposium, September 28–30. Springer-Verlag, Berlin.

Wedi, D. and Wilderer, P.A. (1994) Full scale investigations on enhanced biological phosphorus removal – P-release in the anaerobic reactor. *Water Sci. Technol.* **29**(7), 153–156.

Williams, S. (1999) Struvite precipitation in the sludge stream at Slough wastewater treatment plant and opportunities for phosphorus recovery. *Environ. Technol.* **20**, 743–747.

APPENDIX

Acronyms used

ASP	Activated Sludge Plant
BNR	Biological Nutrient Removal
CHP	Combined Heat and Power
DS	Dry Solids
HRT	Hydraulic Retention Time
PAO	Polyphosphate accumulating organisms
PFT	Picket Fence Thickener
PST	Primary Settling Tank
RAS	Return Activated Sludge
SAS	Surplus Activated Sludge
SPP	Struvite Precipitation Potential
SS	Suspended solids
STW	Sewage Treatment Works
TKN	Total Kjeldahl nitrogen (ammonia and organic N)
XRD	X-ray Diffraction

18
Phosphorus recovery trials in Treviso, Italy – theory, modelling and application

P. Battistoni

18.1 INTRODUCTION

Phosphorus (P) recovery in municipal waste water treatment plants is mainly carried out on supernatants from the anaerobic sludge digestion section (see Parsons and Stephenson, this volume). To reach a high content of orthophosphate in this stream, biological nutrient removal must be applied in the waste water treatment line. This has to be planned taking into consideration that the P release has to take place in a controlled way in one single point of the process. This approach allows for the formation of P-rich side streams for treatment. This scenario is typical in parts of Europe, as well as in North America and South Africa, but not in Italy, mainly due to the low P content in sewer networks as well as the low content of readily biodegradable carbon content (RBCOD[1]) in the waste water.

---

[1] RBCOD = readily biodegradable chemical oxygen demand.

© 2004 IWA Publishing. *Phosphorus in Environmental Technology: Principles and Applications.*
Edited by Eugenia Valsami-Jones. ISBN: 1 84339 001 9

For these reasons, optimisation of a P recovery process from municipal waste water is linked to the evaluation of problems concerning the management of the biological nutrient removal (BNR) process. To better understand the problem, this chapter considers different aspects of the process in the following order:

- Italian sewer waste water and WWTPs[2];
- improvement of BNR technology and performance;
- P recovery tests;
- fluidised bed reactor (FBR) using air stripping;
- planning of the Treviso full scale P-recovery process (PRP) plant, and its interaction with the BNR process.

The aim of this chapter is to explain the principles employed for sizing and planning the FBR plants devoted to P recovery via precipitation, and their use in standard and waste treatment integrated BNR plants.

18.2 ITALIAN SEWER WASTE WATER AND WWTPs

18.2.1 Process adopted in large WWTPs

The situation in Italy, concerning existing waste water treatment plants operating nutrient removal, can be evaluated from an analysis done in 1996, considering 13 of the 20 regions and plants with a capacity > 50.000 PE (people equivalent).

More specifically, of the 157 plants considered, 77% use the primary settling option, while the other 23% do not use this approach. Of all the plants considered, 52% reach an effective N removal, regardless of whether the process schemes consider only nitrification (N), which applies to 27% of the plants, pre denitrification–nitrification (DN), relevant to 66% of the plants, or BNR (7%). In other words,

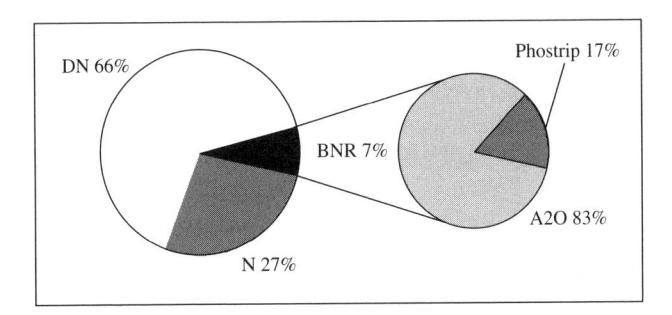

Figure 18.1 Distribution of plants performing nitrogen and/or P removal in Italy. DN: denitrification–nitrification, N: nitrification only, BNR: biological nutrient removal, Phostrip, A2O: P removal processes (see Chapter 12).

---

[2] WWTP(s) = waste water treatment plant(s).

only 6 plants in the whole country are working as BNR plants, although this number is continuously increasing due the need to reach the limits imposed by the EC directive 271/91. The process predominance between the existing BNR plants is A2O (anaerobic anoxic oxic process) (Deakyne et al., 1984).

18.2.2 Chemical and physical characteristics of waste water

The nutrient content of the incoming waste water in Italian plants has to be linked to the network situation, in other words to surface water infiltration, and thus to the dilution of the stream. The scenario can be described using a dimensionless dilution coefficient (f), defined as follows:

$$f = \frac{Q}{\alpha\, DI\, P} \qquad (18.1)$$

where:
Q = flow rate (m³/d)
P = plant capacity (evaluated on an effective mass load basis (PE))
α = collection coefficient (0.8 m³ waste water/m³ taken from water distribution)
DI = per capita water consumption (0.25 m³/PEd)
f = global dilution index

The global dilution index can be linked to the COD content as follows (see Figure 18.2):

$$COD = 513 f^{-0.9} \qquad (18.2)$$

No meaningful relationship can be seen between the global dilution index and the P content. This evidence leads to the conclusion that the P content in waste water is linked to other factors.

Data for the main pollutants (Table 18.1) demonstrate a wide variation, which is caused not only by dilution, but also by the industrial discharges in the sewer. It is evident, however, that the phosphorous concentration is always low (on average 5.7 mg/l in 57 plants at >50.000 PE capacity), mainly due to detergent reformulation imposed in Italy.

The COD/N_{TOT} ratio is not affected by the dilution caused by surface water or groundwater in the sewer, and can be used to evaluate the feasibility of a BNR process in a given stream: a value of 9 can be assumed as minimum (Table 18.1). As an example, taken from the literature, a ratio COD/N_{TOT} of 7.1–8.3 is needed to use UCT process[3], 9.1 for the modified UCT, 12.5–14.3 for the Bardenpho-5. From an analysis of the Italian scenario, it is clear that in an important proportion of situations (30–40%), the COD/N_{TOT} ratio assumes values lower than 9, and

---

[3] UCT process = University of Cape Town process.

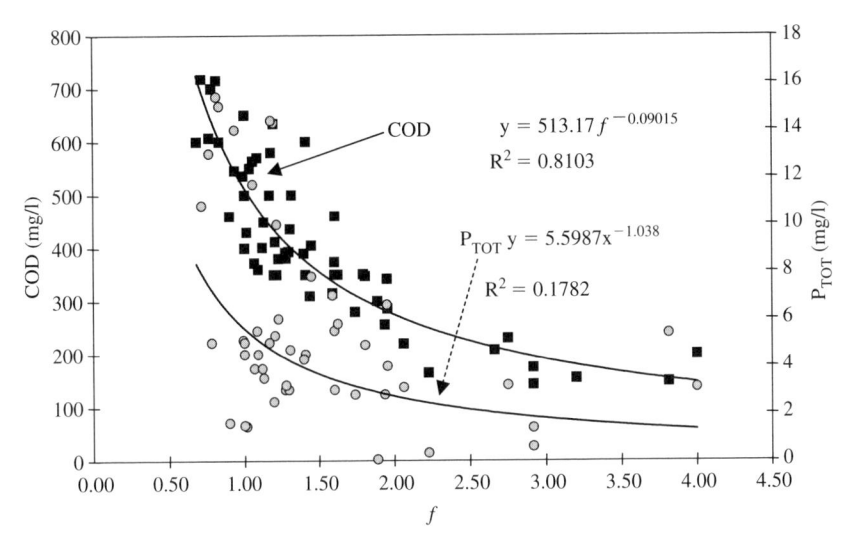

Figure 18.2 Chemical Oxygen Demand (COD) and phosphorus content (P_{TOT}) vs. global dilution factor (f).

Table 18.1 Statistical data concerning the main pollutants in sewer networks.

	COD (mg/l)	N_{TOT} (mg/l)	P_{TOT} (mg/l)	COD/N_{TOT}
Number of analyses	85	85	71	85
Average	482	42	5.7	13.7
Minimum	143	7	0.1	5.2
Maximum	1000	108	22	62.6
Standard deviation	208	21	4.6	10.0

therefore an increase of biodegradable carbon in the stream is needed by using an external carbon source to allow the use of a BNR approach.

Additional information about P concentration and its speciation can be found in data from periodic measurements taken in the inflow of municipal WWTPs (Battistoni et al., 2001a). In particular, the relationship between P concentration and other sewer characteristics, including global dilution factor can be seen in Table 18.2. Moreover, a link between TSS[4] and particulate P (defined as the difference between P_{TOT} and PO_4–P concentrations) is shown in Figure 18.3. The slope of the trend can be used to evaluate the P percentage present in total solids. This percentage is characteristic of each plant and does not show annual variation.

---

[4] TSS = total suspended solids.

Table 18.2 Sewer characteristics in four municipal WWTPs.

Plant	Dimension PE	Process	f	RBCOD/COD	COD/N_{TOT} Avg.	SD.	P_{TOTin} (mg/l) Avg.	SD.	PO_{4in}–P(mg/l) Avg.	SD.	P/TS[5] %	R^2
a	60,000	BNR	3.0	0.07	8.1	3.4	2.3	0.7	0.8	0.4	0.80	0.49
b	100,000	BNR	1.5	0.20	11.3	2.1	3.4	0.9	1.4	0.5	0.55	0.31
c	85,000	D-N	1.7	0.27	8.2	2.6	4.0	1.5	1.6	0.5	1.48	0.60
d	70,000	BNR	4.6	0.20			1.7	0.7	1.0	0.4	1.20	0.52

[5] TS = total solids

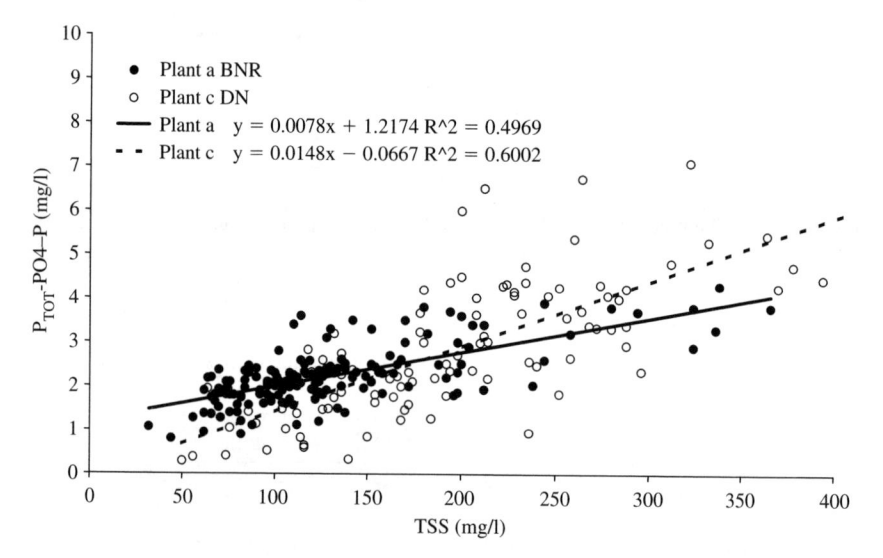

Figure 18.3 Particulate P [P_{TOT}-(PO_4–P)] correlation with total suspended solids (TSS).

18.2.3 Phosphorus mass balance in real WWTPs

A comparison of P removal performance in WWTPs b and c in Table 18.2, was made during the year 2000 (Battistoni et al., 2001a). A good P removal is achieved in both plants, even though the P content in dewatered sludge is quite low (2.1% of TS for both plants, see Table 18.3). Periodic tests of P concentration were carried out to verify the presence of phosphorus accumulating organisms (PAO) and denitrifying organisms, and to determine the fraction of poly-P with respect to total P. From these tests, the BNR activated sludge (where P is 2.3% of TS, and 3.2% of TVS[6]) contains approximately 35% of the total P in the sludge, which can be linked to BNR activity. This fraction changes during the year, from 10% (in periods in which the plant had some malfunctions) to 52%, when the

[6] TVS = total volatile solids.

Table 18.3 Phosphorus statistical parameters in two WWTPs.

WWTP	BNR (plant b)		DN (plant c)	
	Avg.	SD.	Avg.	SD.
P_{TOT} influent (mg/l)	3.4	0.9	4.0	1.5
PO_4–P influent (mg/l)	1.4	0.5	1.6	0.5
P_{TOT} effluent (mg/l)	1.4	0.5	1.3	0.7
PO_4–P effluent (mg/l)	0.9	0.5	0.9	0.5
$P_{\%TS}$ in waste activated sludge	2.1	1.0	2.1	0.6
$P_{\%TS}$ in dewatered sludge	1.6	0.3	1.5	0.1

Table 18.4 Sludge production comparison between two WWTPs.

	WWTP type	Unit	BNR-b	DN-c	Δ(*)(%)
Biol. process	$Y_{observed}$[7]	A	0.57	0.54	5
Waste water line	Specific sludge production	B	17.7	23.3	−24
		A	0.56	0.74	−24
Sludge line	Specific sludge production	B	9.5	12.5	−24
		A	0.31	0.40	−23

(*) as difference between BNR and DN process; Unit A = Kg TS kg COD_r^{-1}; r: removed. Unit B = Kg TS $PE^{-1} y^{-1}$.

[7] $Y_{observed}$ = thermodynamic coefficient representing the biomass produced for unit of substrate transformed in operative conditions of a process.

plant was operating in stable steady state conditions. No P release was found in sludge from the denitrification plant, as would be expected. The low P content in the sludge can be related to the low P concentration in the incoming waste water, as reported in the literature (Popel and Jardin, 1993). As a conclusion, in the case of small to medium capacity plants (≤100,000 p.e.), with low influent P content, P removal can be achieved within the EC law limits (2 mg/l), when the COD/N_{TOT} ratio is appropriate, without any particular concern about the P feedback coming from anaerobic or thickening supernatants.

In this scenario, sludge production on annual basis (year 2000) presents important differences between the two plants. In particular, the BNR plant shows lower specific production than the denitrification plant. This can be seen by comparing their calculated production of mixed and disposed sludge, as Kg TS/PEy (Table 18.4).

Even though these parameters can be affected by the sewer characteristics and by the specific yields of each process (i.e. the TVS removal in anaerobic digestion), they confirm the hypothesis that chemical P precipitation combined with the biological process leads to a sludge overproduction. This can be explained only in part by the added mass of chemicals. In fact, the addition of aluminium ions corresponds to an average concentration in the influent of 2 mg Al/l, which, in the case of a global precipitation of $AlPO_4$, leads to an 8% sludge overproduction on a weight dry basis. Although the Y_{obs} (Kg TS/Kg COD_r) values show that an apparent slightly higher sludge production is present in the BNR process compared to the

Table 18.5 Sludge production from water line in different WWTP (Woods et al., 1999).

Flow rate	m^3/d	15,000	50,000	50,000	50,000	50,000
PE	–	50,100	83,500	167,000	83,500	167,000
P$_{TOT}$	mg/l	5	5	5	8	8
Fe/P	mol/mol	1.5	1.5	1.5	1.5	1.5
BOD/P	w.w	40	20	40	12.5	25
Water line sludge production						
DN and P chem. prec.	Kg TS PE^{-1}y^{-1}	19.0	21.3	19.0	25.7	21.2
BNR	Kg TS PE^{-1}y^{-1}	17.3	18.0	17.3	20.0	18.3
BNR + P recovery	Kg TS PE^{-1}y^{-1}	16.0	15.4	16.0	15.4	16.0

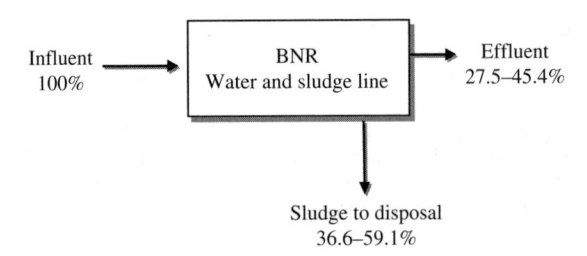

Influent 100% → BNR Water and sludge line → Effluent 27.5–45.4%

Sludge to disposal 36.6–59.1%

Figure 18.4 Black box P mass balance in b-plant (Table 18.4).

denitrification process (about 5%), due to differences in SRT[8] (9 d in BNR, 18.6 d in DN). The results obtained can be compared to other data reported in literature for full scale BNR and denitrification plants (e.g. Woods et al., 1999), with chemical P precipitation and P recovery from secondary streams (Table 18.5).

The data are from 16 operational plants (European and North American). Specific sludge production of the water line is a function of the process performed. In particular, specific sludge production for denitrification processes with chemical P precipitation is much lower (19–21.3 Kg TS/PE y) than the equivalent in the Italian case (23.3 Kg TS/PE y), while the sludge production for the BNR is the same (17.3–18.0 Kg TS/PE y) as the one discussed above (17.7 Kg TS/PE y). It can therefore be concluded that, if P recovery is present in BNR plants, sludge production can be additionally reduced by 8–14%.

A further analysis can be also conducted for plant b, on the basis of a black box phosphorus mass balance, over a period of four years of management (1996–1999). In this case the influent P mass loading is 100 Kg P/d (SD 13.8 Kg/d) but the residual P in the effluent ranges from 27.5 to 44.7%, revealing that poor performance is always obtained (Figure 18.4). This situation can be represented by a yearly average total P content consistent with EC limits for plants with a capacity ≤ 100,000 PE (where the yearly average P in effluent ranges from 0.6 to 1.6 mg P/l). However, for the enhancement of P removal in a plant

---

[8] SRT = sludge retention time.

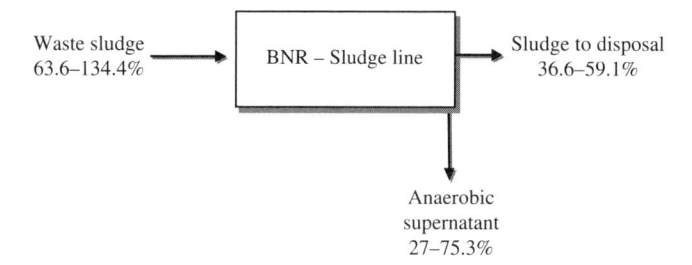

```
Waste sludge          ┌─────────────────┐      Sludge to disposal
63.6–134.4%     ──▶   │ BNR – Sludge line│ ──▶  36.6–59.1%
                      └─────────────────┘
                              │
                              ▼
                          Anaerobic
                         supernatant
                          27–75.3%
```

Figure 18.5 Black box P mass balance in sludge line of b-plant (Table 18.4).

of >100,000 PE chemical addition is required. The main reason for the low performance can be found in the high P feedback to water line, that can reach up to 75.3% of sewer P mass loading (Figure 18.5).

18.2.4 Phosphorus release in the sludge line

In BNR plants, P is released in the anaerobic step when adequate concentration of VFA[9] or, more generally, RBCOD, is present. The amount of the release is linked to:

- the treatment scheme;
- the operative conditions of the single unit operations;
- the physical–chemical characteristics of the inflow.

The resulting variation is always significant, and it can be equal to the P removed in the water line. Phosphorus is present mainly in sludge line, thus an adequate treatment of the sludge or, at least, of the supernatants, before their recycling to the headworks, has to be carried out.

The planning strategy of the sludge line must follow the double aim of containing the release and concentrating it in a single point of the line for optimum removal. This can be achieved using a process scheme in which the waste activated sludge (WAS) is thickened separately from the primary sludge (PS). Then, it can be treated to avoid the supernatant release (composting, landfill, incineration), or processed further (anaerobically), so as to concentrate the phosphorous in the supernatants and thus produce a lower flow to treat (Randal et al., 1992).

This approach uses primary sludge as C source, which can then be used for the P release step. Thus, both the settling and the gravitational thickening of the mixed sludge have to be avoided in a BNR process approach. Evidence in this field comes from the monitoring of full scale plants (Table 18.6) in which the gravitational thickening of WAS alone, in low times (HRT[10] = 10–25 h), leads to low P release (2–4%, example A and D in Table 18.6), while the settling and the thickening of

---

[9] VFA = volatile fatty acids.
[10] HRT = hydraulic retention time.

Table 18.6 Phosphorus release in primary sedimentation and thickening.

WWTP	Process	Unit operation	P released (%)	P (mg/l)	References
A	BNR	WAS thickening	4	20	Randal et al., 1992
A_1	BNR	WAS thickening	22	12	Randal et al., 1992
B	BNR	Mixed primary sedimentation	43		Randal et al., 1992
C	BNR	WAS + PS thickening	42	60–100	Murakami et al., 1987
D	BNR	WAS thickening	1.9	9.8	Tanaka et al., 1987
D_1	BNR	WAS + PS thickening (1/1)	37	64–131	Tanaka et al., 1987
D_2	BNR	WAS + PS thickening (2/1)	34	64–131	Tanaka et al., 1987

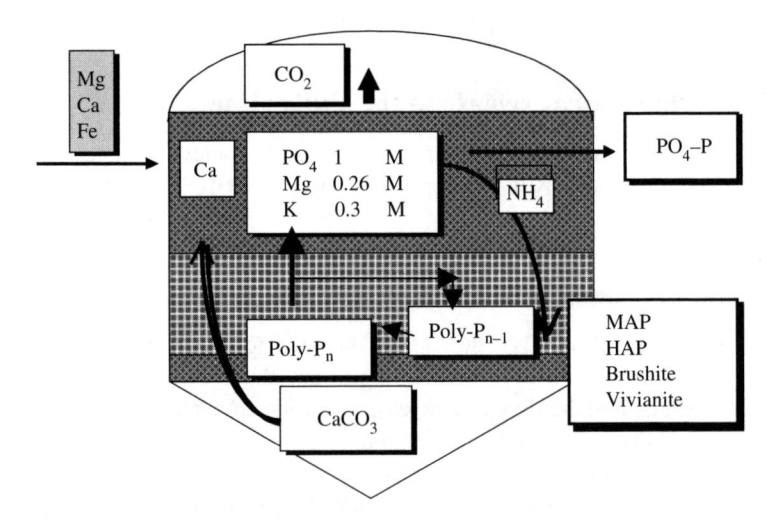

Figure 18.6 Phosphorous release and re-precipitation mechanisms in an anaerobic digester. MAP: magnesium ammonium phosphate, HAP: hydroxylapatite.

mixed sludge shows release in the order of 40% (examples A_1, B, C, D_1 and D_2 – Table 18.6).

During anaerobic digestion P release is mainly due to the high concentration of VFA, which are generally present in this process. Moreover, biologically active ions, such as potassium (K) and magnesium (Mg), which play a role in the stabilisation of the negative charge of intracellular polyphosphates, are also widely released (Battistoni, 1999). This release follows the molar ratios, which ranges from 0.3 M K/M P to 0.26 M Mg/M P (Popel and Jardin, 1993; Wentzel et al., 1992).

The P total released in the digester can differ from the actual amount recycled into the headworks, because re-precipitation processes may occur (Figure 18.6). Thus, the maximum release observed in conventional plants is up to 20%, while it

Table 18.7 Phosphorus release in anaerobic digestion.

WWTP		P (%)	P (mg/l)	References
A	BNR	<10	<100	Popel and Jardin, 1993
B	BNR	~60	300	Murakami et al., 1987
C	BNR*	27	250	Randal et al., 1992
D	BNR	63	–	Popel and Jardin, 1993
E	BNR	–	290	Webb et al., 1995
F	–	–	32	Vandaele et al., 2000

BNR* WAS in dissolved air flotation and anaerobic digestion.

reaches 43–63% in BNR plants at medium-high P content in the sludge (Table 18.7). It can rise up to 95% in plants where the P content of the sludge is 7% of TS.

The actual quantity released, as observed in different plants, can vary widely from one to another. This can be due to:

- the hardness of the influent, which controls the re-precipitation processes of phosphorous inside the digester; and
- the amount of supernatant compared to influent volume, which can control whether or not prethickening of the sludge before digestion is used.

The re-precipitation can occur (Figure 18.6) mainly via formation of struvite (MAP: $MgNH_4PO_4$) or hydroxylapatite (HAP: $Ca_5(PO_4)_3OH$). Some other mineral phases are hypothesised, such as vivianite ($Fe_2(PO_4)_3 \cdot 8H_2O$) and brushite ($CaHPO_4 \cdot 2H_2O$) (Nancollas, 1984). The struvite precipitation, which is thermodynamically unfavourable compared to hydroxylapatite ($\log K_{sp} = -12.6$ vs -57.8 respectively), is hypothesised to be kinetically promoted due to the high pH of the supernatant (7.4–7.9) as well as due to alkalinity inhibiting hydroxylapatite crystallisation (Jenkins et al., 1971).

Recent studies (Wild et al., 1996) have demonstrated that zeolites, derived from detergents, may also play an important role in the final removal of P in digesters, particularly since they may represent 7% of the solids fed to the digester. Zeolites can work both as cation sources, due to their ion exchange capacity, which allows them to provide Ca and Al ions, and also as nucleation seeds for P salts. A wide range of literature can be found concerning actual plant observations (Table 18.7).

18.3 IMPROVEMENT OF BNR TECHNOLOGIES AND PERFORMANCE

In the Italian scenario, the improvement of BNR process application in WWTPs requires two main approaches: the conservation of carbon substrates in sewers, to enhance nutrient removal, and the use of new technologies to supply volatile fatty acids or readily biodegradable COD from an external renewable carbon source.

18.3.1 The use of internal carbon

The use of internal carbon sources is preferred for positive impacts on management costs and waste sludge production (Isaacs and Henze, 1995). To this end, elimination of primary sedimentation or elutriation and fermentation of primary sludge involve widespread solutions in actual biological nutrient removal plants (Lotter & Pitman, 1992). However, attention must be paid to the quality and quantity of carbon recovered to enhance performance. The products of primary sludge fermentation are mainly short chain fatty acids (SCFAs) (Pitman et al., 1992). The best operative conditions are 2–3 d for elutriation or 1–10 d for fermentation (Pavan et al., 1994; Rabinowitz & Oldham, 1985; Rozzi et al., 1995).

18.3.2 The use of external carbon

The use of an external carbon source is more difficult due to two types of problems: the availability of sufficient cheap carbon sources, and the need for a constant composition of carbon substrates. To satisfy these requirements, addition of industrial and agricultural waste (brewery waste, molasses or corn-silage) can be considered. When industrial waste is not available, methanol, or even better, acetic acid may be employed; this however represents a very expensive solution and could be a good alternative only if investment costs for waste water treatment plant upgrading are avoided. A novel external carbon source is represented by the anaerobically fermented organic fraction of municipal solid waste (OFMSW). The integrated treatment of waste water and solid waste has the multiple benefits of: a good final disposal for OFMSW, enhancement of P removal of BNR processes and negative operational costs with respect to alternative external carbon sources. A comprehensive study of this methodology tested in large scale pilot plants, the use of hydrolysis products and guidelines for the design of the fermenter and the system of anaerobic fermentation have recently been published (Cecchi et al., 2002). The main novel aspects of the OFMSW fermentation are:

- OFMSW hydrolysis products in mesophilic conditions (HRT 3–6 d) mainly consist of VFA (15,000 mg/l), where acetic acid is predominant (80–85%), and the lactic acid content can reach up to 17,000 mg/l (Llabres et al., 1999; Pavan et al., 2000);
- the hydrolysis products of OFMSW allow high phosphate release and high denitrification rates; both these can be attributed to the high presence of VFA (Pavan et al., 1998);
- an alternative to OFMSW is the mesophilic fermentation of source separated mixtures of vegetables and fruits wasted by supermarkets (Traverso et al., 2000); this produces a hydrolysis product containing a soluble COD equal to 43% of the total influent COD, where VFA represent 93%;
- the remaining solid COD from the above processes is used for anaerobic co-digestion, with waste activated sludge contributing to improve the thermal balance of the digester (Cecchi et al., 2002).

Table 18.8 Characteristics of internal and external carbon sources.

Characteristics of the products

Substrate	Process	$COD_{TOT}/NH_3-N\%$	$SCOD/NH_3-N\%$	$COD_{TOT}/TKN\%$	$VFA/SCOD\%$*	$COD_{conv.}\%$**	Ref.
PC	HF	–	11.3	–	67	11	Aesoy et al., 1994
PC	HF	–	18	–	60–70	–	Isaac & Henze, 1995
PC	HF	–	18	35	60–70	10–13	Kristensen et al., 1992
P	FE	16–26	–	–	–	–	Lotter & Pitman, 1992
OFMSW	F	250	167	–	30	66	Pavan et al., 1994
Blend	F	–	12–32	–	55–73	12.5	Ghosh et al., 1975
P	F	–	–	–	–	3.8–3.9	Chu et al., 1994
Blend	F	–	–	–	27–40	2.5–10.5	Bhattacharya et al., 1996
Vegetable and fruit mixtures	F	313	147	94	28	40	Traverso et al., 2000

PC = primary and chemical sludge; P = primary sludge; OFMSW = organic fraction of municipal solid waste. HF = hydrolysis and fermentation; FE = fermentation and elutriation; F = fermentation; SCOD = soluble COD.
* VFA measured in terms of COD; ** ratio between soluble COD in the effluent and total COD in the feed.

The comparison between the use of internal and external carbon sources requires the consideration of the source availability, to assure a continuous enhancement of nitrogen removal and the COD/N_{TOT} ratio, to obtain an exact evaluation of the carbon fraction available for the process (Table 18.8). On the basis of these aspects, the primary sludge fermentation process produces nearly 4 g C/PE d as a substrate to enhance the nutrient removal, while the fermentation of OFMSW produces up to 16 g C/PE d (Battistoni et al., 1998a). This indicates that an internal carbon source can be used when enough carbon exists in sewer and only a fast, cheap transformation is necessary, while the fermentation of OFMSW is the right approach when, for different reasons, not enough carbon exists in sewers to enhance nutrient removal.

18.4 PHOSPHORUS RECOVERY TEST

The optimisation of phosphorus recovery is strongly linked to the exact characterisation of the stream to be treated, and to the availability of instrumentation, which can allow understanding of the process in detail. On the basis of the first aspect, by using the main stream (liquid effluent of the WWTP), a good performance is difficult to be assured, as is saving of additional reagent costs; however, the anaerobic supernatant, presenting high phosphate, ammonium, calcium and magnesium concentrations, is the most favourable candidate for treatment. To predict the type of phosphorus salt formation or the mechanism of recovery (precipitation, homogeneous nucleation, etc.), a mass balance is necessary as all information is obtainable through thermodynamic evaluation. In this way a methodological protocol to evaluate experimental results is derived. The most comprehensive method is to elaborate phosphorus analysis for every type of process: precipitation, nucleation, fluid bed reactor, packed bed reactor, etc., according to Figure 18.7.

$$\eta\ \% = \frac{P_{TOT\ in} - P_{TOT out}}{P_{TOT\ in}} \cdot 100$$

$$L\ \% = \frac{P_{TOT out} - P_{SOL out}}{P_{TOT\ in}} \cdot 100$$

$$X\ \% = L\ \% + \eta\ \%$$

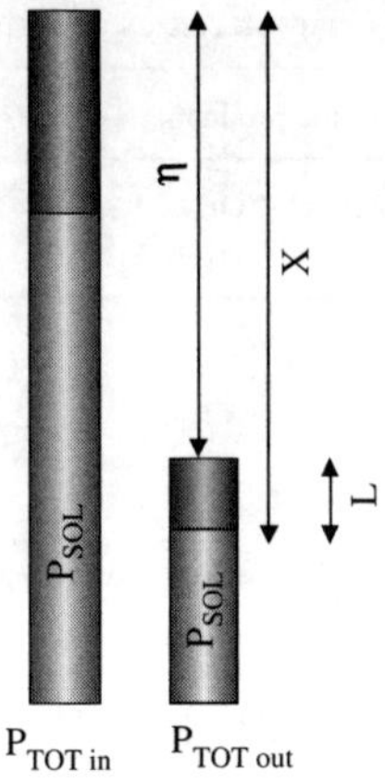

Figure 18.7 Schematic representation of phosphate mass balance in a waste water system.

In Figure 18.7, P_{TOT} is the phosphate measured in unfiltered acidified samples, while P_{SOL} is the phosphate measured in 0.45 μm filtrate, both in the influent and the effluent of the test.

In this way, it is possible to distinguish crystallised or nucleated phosphate (η), phosphate present as small particles (L) called *fines*, and the global phosphate conversion (X). This methodology, which is widely used (Battistoni et al., 1998b; Seckler et al., 1996a,b,c), allows for the comparison of the results of different researchers.

18.4.1 Natural time evolution

Natural time evolution of an anaerobic supernatant was noted during the cooling of digested sludge in open lagoons, with natural degassing of CO_2 occurring (Salutsky et al., 1972). The same phenomenon can be systematically investigated and used to forecast phosphorus removal (Battistoni et al., 1997, 1998b).

A typical time evolution called natural *ageing* of the supernatant can be observed in Figure 18.8. During a time varying from four to eight days, phosphate decreases without any addition of reagents. This behaviour is linked to a pH increase. Both behaviours can be interpreted as consequence of a loss of alkalinity (200–300 mg $CaCO_3$/l) due to the passage from an overpressure situation in the digester to the ambient value. On the basis of the typical composition of anaerobic supernatant (Table 18.9, Battistoni et al., 1998b) conversions up to 80% can be observed (Table 18.10, run 1b). By monitoring calcium and magnesium concentrations, it is possible to assess that phosphate is removed as struvite (MAP), hydroxylapatite (HAP), or a mixture of the two (Table 18.10, runs 1–1b).

18.4.2 Air stripping

Air stripping is a technique that was used in the past to lower the buffer effect in the supernatant and to save reagents costs in BNR plants where anaerobic supernatants

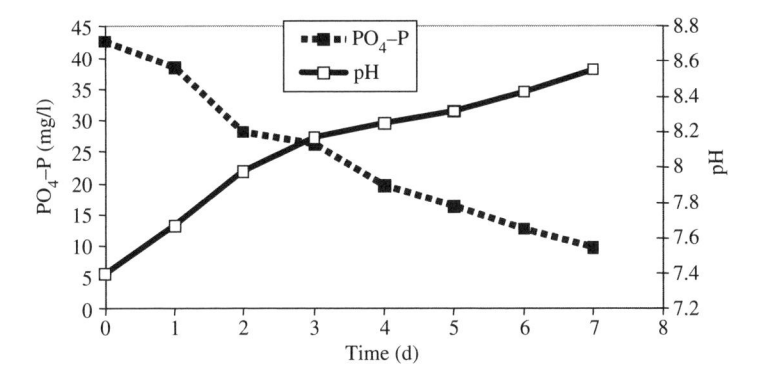

Figure 18.8 Natural ageing of an actual supernatant (Battistoni et al., 1998b).

Table 18.9 Chemical–physical characteristics of supernatants.

Parameters	Average	SD.
pH	7.7	0.1
SCOD* mg/l	1320	223
PO_4 mg/l	139	16
NH_3 mg/l	914	127
Ca mg/l	153	15
Mg mg/l	24	3
HCO_3 mg $CaCO_3$/l	3550	325
CO_3 mg $CaCO_3$/l	0	–

* SCOD = soluble COD.

Table 18.10 Natural ageing performance.

RUN	Ca/Mg M M^{-1}	Ca/PO_4 M M^{-1}	Mg/PO_4 M M^{-1}	X %	HAP %M	MAP %M	Δ Alkalinity mg $CaCO_3$/l
1	4.5	1.4	0.3	77	73	17	−265
1a	1.2	1.4	1.2	79	6.5	93	−305
1b	4.6	2.3	0.5	80	100	–	−275

are treated (Pitman et al., 1991). Air stripping allows for the acceleration of phosphorus precipitation by reducing the process time from eight days to no more than one-hundred minutes (Battistoni et al., 1998b). In this case no reagent addition is required to precipitate phosphorus, but the strong action of the stripping allows for higher conversions (X = 85–91%) and the same blend of HAP and/or MAP, as in natural aging. The use of quartz sand (5% w.w.) during air stripping tests showed a nucleation performance of 78% in comparison to a conversion of 89%, and a reduction of the loss of fines.

These results suggest that the air stripping technique is an efficient way to induce phosphorus removal from supernatants in a fluidised bed reactor. The preliminary results obtained were very encouraging, reaching a nucleation yield of

80%, even if the FBR reactor (bench scale) was managed in batch mode, using external gradual or continuous aeration (Battistoni et al., 1996).

18.4.3 Supersaturation curves

Supersaturation curves were first introduced by Joko (1984). They represent a good method of evaluating the actual events taking place in the different processes of phosphorus removal. An example of adopted methodology is shown in Figure 18.9, where two supersaturation curves (S_1 using anaerobic supernatant of a real WWTP, S_2 adding Ca ions to the supernatant) are related to experimental points obtained in FBR fed continuously, during tests of natural ageing of supernatants (supplied from real WWTP or synthetic) and air stripping tests (Battistoni et al., 1998b). Looking at S_1 curve, it can be concluded that FBR and natural ageing work in metastable conditions, while air stripping in supersaturation conditions. With a systematic approach in the comparison of experimental results employing synthetic supernatants and supersaturation curves (Battistoni et al., 1998b) the following conclusions can be drawn:

- alkalinity and magnesium exert a significant inhibitory effect on HAP formation, requiring a higher pH to reach supersaturation at concentrations above 500 mg $CaCO_3$/l and 10 mg/l Mg; the inhibition of HAP precipitation by bicarbonate and magnesium ions has been described by Jenkins et al. (1971).
- alkalinity was not found to have an effect on MAP formation, with precipitation operating at an alkalinity of 3000 mg $CaCO_3$/l;
- the optimal performance of FBR, related to the minimum loss of particulate phosphate, can be interpreted considering that all FBR points are located farthest from the reference supersaturation (Figure 18.9, system S_1 and S_2);

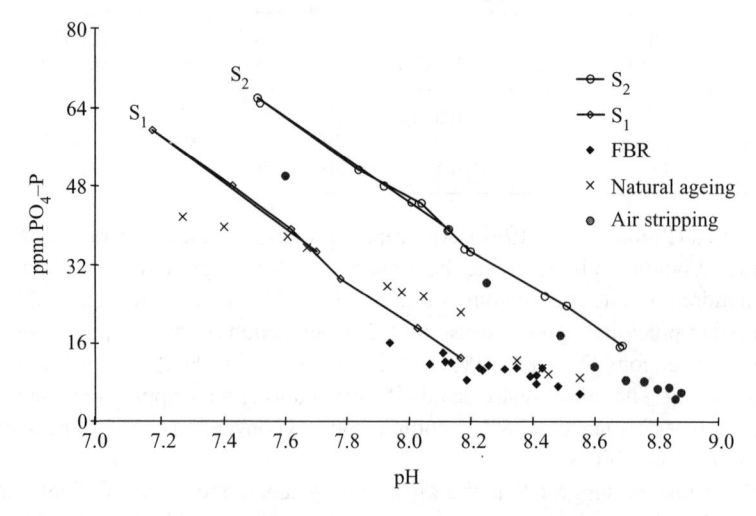

Figure 18.9 Supersaturation curves for two systems (S1 and S2), shown together with a range of operative conditions (Battistoni et al., 1998b). See text for details.

- air stripping results, with or without quartz sand, are proximate to the S_1 and S_2 curves, hence, a strong precipitation associated with crystallisation is equally possible in batch test notwithstanding the use of quartz sand as seed material;
- finally, natural ageing points are in an intermediate position, consequently, spontaneous precipitation or self nucleation is possible.

The conclusions of these studies on practical application of the process in phosphorus removal from anaerobic supernatants can be summarised as follows:

- the anaerobic supernatant, supplied from a full scale A2O process, has generally enough Ca and Mg content to satisfy phosphate removal without the addition of chemical reagents;
- a continuous FBR to crystallise MAP and/or HAP can employ air stripping to produce the operative pH required as a feasible and reliable process;
- no pre-treatment is necessary to remove residual suspended solids in a supernatant supplied directly from sludge centrifugation; high performance in phosphate removal is obtained with minimum loss in particulate phosphate, consequently no filtration step is necessary on the effluent.

18.5 FBR AND AIR STRIPPING

Fluidised bed reactors (FBR) are commonly used to crystallise (precipitate) salt out of a solution. An FBR plant can be the optimal way for P removal and recovery since a reusable P salt can be obtained. The use of air stripping and FBR for the treatment of the anaerobic supernatant was studied by Battistoni et al. (2002) using bench and demonstrative plants, synthetic and real supernatants, with the aim to optimise the design of FBR and its utilities. The recycling flow rate and hydrodynamics are important aspects to better understand the results, and will be discussed next.

18.5.1 Flow scheme of FBR with air stripping

An FBR plant using air stripping can be divided in two sections: a stripping tank and a fluidised bed reactor. In the pilot study described here, the anaerobic supernatant, as supplied from the actual plant, was initially stored in a tank (3.0 m^3 volume, Figure 18.10) to allow a daily continuous supply. From the storage tank, the supernatant was then fed to the stripper with the recycle flow rate and an air flow rate needed for CO_2 stripping. The reactor is a Perspex column (ϕ_{in}[11] = 0.09 m, height 2 m, volume 12.8 l) filled with 9.5 kg of virgin silica sand (0.21–0.35 mm, ϕ_m[12] = 0.265 mm) to obtain a 1 m tall compressed bed. At the bottom of the column a Perspex cylinder (ϕ_{in} = 0.09 m, height 0.5 m) filled with gravel with decreasing size distribution, acts as a filter, blocking the sand that comes back towards the pumps, and allows a homogeneous distribution of the

---
[11] ϕ_{in} = inner diameter of reactor.
[12] ϕ_m = average diameter of sand particles.

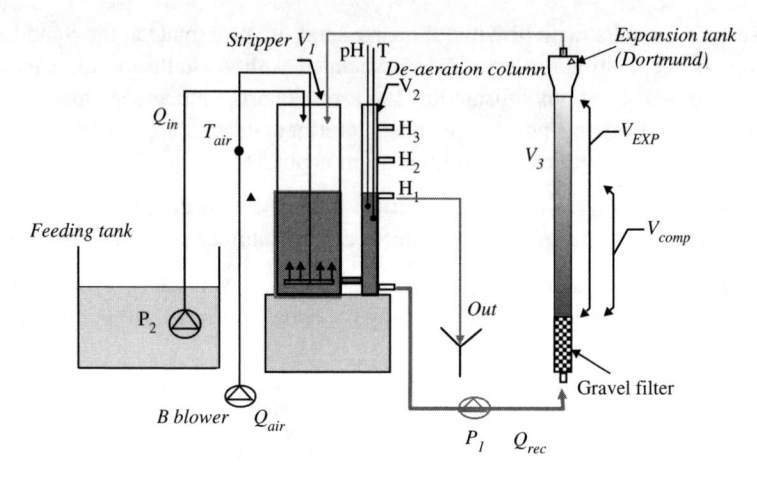

Figure 18.10 Layout of pilot plant (Battistoni et al., 1998b).

Table 18.11 Main volumetric characteristics of FBR.

Apparatus	Volume, geometry and flow rate
Feeding tanks	Variable volume up to 3.01 m³
FBR reactor	Compressed bed height: 1.5 m free volume 3.2 l
	1.0 m – 6.4 l; 0.5 m – 9.5 l
Expansion tank	Volume 18 l
Deaeration column	Inner diameter 0.08 m; Height 1.6 m
Stripper	h.h. 0.5 m – volume 40.0 l, 0.8 m – volume 62.6 l, 1.0 m, volume 77.7 l
B – blower	Power 0.25 kW
P_1 – Signal to PC	Flow rate 0.5–2.5 m³/h
	T_{air}, $T_{stripper}$, $pH_{stripper}$

h.h. = hydraulic head.

stream to the reactor. At the top of the column an expansion tank (Dortmund zone) is provided in order to prevent the sand from coming out of the reactor.

The stripping section consists of a stripper and a connected de-aeration column. The effluent from the de-aeration column is combined with the recycle flow. Three levels of liquid in the stripper can be chosen, through three outlet ports located at different levels, so as to provide different hydraulic heads (h.h., H = 0.5, 0.8, 1.0 m) and different volumes of the stripping section and the de-aeration column. A temperature probe and a pH probe were positioned in the de-aeration column, while a second temperature probe was positioned in the air pipeline. The recycle pump had a flow rate ranging from 50 to 750 l/h. In Table 18.11 the main characteristics of FBR system are reported.

A possible location of the pilot plant may be the dewatering station of a BNR plant. To optimise the reactor design, several experimental runs in continuous mode must be performed. Each set is characterized by a constant feeding flow rate of the anaerobic supernatant, while the airflow rate can be changed to allow the system to operate at different pH. Furthermore, the recycle flow rate adopted allows operation with a constant volume of the expanded bed in all experiments (up to the Dortmund zone).

18.5.2 FBR hydrodynamic principles

In an FBR with air stripping system, four hydraulic retention times can be identified (see Equation 18.3–18.6) where Q_i is the feed flow rate, Q_{RIC} the recycle flow rate, V_1 the stripping column volume, V_2 the volume of the de-aeration column, V_3 the column free volume (including tube, Dortmund zone and filter volumes), V_{EXP} the volume of the fluidised bed and ε the bed porosity.

$$HRT_T = \frac{(V_1 + V_2 + V_3)}{Q_i} \quad (18.3)$$

$$HRT_{stripp} = \frac{(V_1 + V_2)}{Q_i} \quad (18.4)$$

$$HRT_{FBR} = \frac{V_3}{Q_{RIC}} \quad (18.5)$$

$$HRT_{EXP} = \frac{V_{EXP}}{Q_{RIC}} \varepsilon \quad (18.6)$$

$$n = \frac{HRT_T}{HRT_{FBR}} \quad (18.7)$$

$$t_c = n \cdot HRT_{EXP} \quad (18.8)$$

In the upward movement of the effluent inside the FBR, the number of passages (n) is defined as the ratio between the mean hydraulic retention time in the plant and the time spent in the expanded bed. The contact time (t_c) on sand grains depends on the number of cycles and the time needed for one single passage (Battistoni et al., 2001b).

Bed porosity (ε) is defined as the free space to the expanded bed volume ratio. Since an FBR pilot plant operates at a fixed volume of the expanded bed, the porosity decreases while the pilot operating time increases, due to the growth of sand grains. To calculate the bed porosity, a weighted average sample of bed (bottom, top and medium) is collected after every run. The sand is air dried for 48 h and 20 g are loaded in a glass containing a known volume of distilled water. The increment of water volume can be related to the dried sand and used to calculate its specific volume (V_{SP} = vol/g sand, Battistoni et al., 2002). The porosity of the expanded bed can be calculated according to Equation 18.9:

$$\varepsilon = \frac{V_{EXP} - (V_{SP} \cdot M)}{V_{EXP}} \quad (18.9)$$

where M is the mass of sand grains (virgin sand plus precipitates).

18.5.3 FBR performance

A good performance in P removal is obtained when a high precipitation efficiency (η%, see Figure 18.7) is coupled with a low difference between conversion (X%) and particle growth. This means that only a small amount of P is lost (as fines, L) from the effluent, and filtration of the effluent is avoided, simplifying the system. A further consideration is the presence of calcite, or the potential lack of growth on the seed material or of the recovery product, if homogeneous nucleation is performed. This will depend strongly on the supernatant composition.

A lot of experimental work, both at bench and demonstrative scale, using FBR with air stripping and without reagent addition, was done by Battistoni *et al.*, (1998a, 2001b, 2002). The main results obtained can be summarised considering three variables: the supernatant composition, the nucleation and conversion performance and the P salt formation.

18.5.3.1 Supernatant composition

When anaerobic supernatants are directly used, any treatment to remove the suspended solids, lost in the dewatering station, is avoided. It is worth noting the variations occurring into the system for some of the measured parameters (PO_4–P, Ca, Mg and HCO_3). In particular, a good homogeneity exists in each experiment, but a comparison among all tests shows a considerable variability in the characteristics of the effluent of anaerobic digester. This is due to changes in the sludge handling during the treatment plant management. Moreover, significant differences can be observed by comparing the average values reported in Table 18.12 of supernatants from different experiments. This introduces a further complication: the stoichiometry required to precipitate HAP or MAP on the basis of phosphate concentration is satisfied in all three situations with regards to Ca ions, but is not satisfied with regards to Mg ions. This consideration allows the conclusion that changes in supernatant characteristics can be common, and they may produce a noticeable effect in phosphorus recovery process.

18.5.3.2 Nucleation and conversion performance

The process performance was evaluated using anaerobic supernatants taken from a full-scale BNR plant (Battistoni *et al.*, 1998c). The Ca/PO_4 molar ratio was 2.6 ± 0.4, exceeding the stoichiometric request (1.7) for HAP formation, while

Table 18.12 Comparison of different characteristics of experimental supernatants.

Reference	P_{TOTin} mg l^{-1}	NH_3 mg l^{-1}	HCO_3 mg$_{CaCO_3}$ l^{-1}	Ca mg l^{-1}	Mg mg l^{-1}
Battistoni *et al.*, 1998a	45	914	3550	153	24
Battistoni *et al.*, 2001	22	1055	3720	200	54
Battistoni *et al.*, 2002	28	713	2900	178	36

the 0.7 ± 0.1 Mg/PO$_4$ molar ratio was insufficient to guarantee a complete MAP formation (a ratio of 1.0 is required for stoichiometry), despite the high concentration of ammonia. The process behaviour was studied mainly considering changes in influent and air flow rates. Run time ranged from a minimum of 1 d for an enriched effluent, to a maximum of 4 d for real supernatant. The mass balance demonstrates high performance in phosphate removal with a maximum loss of fines L at 3.5% and an average value of 1.7 ± 0.9% (Table 18.13).

Using a supernatant with a lower P concentration (Battistoni et al., 2001b) the precipitation yields ranged from 53% to 80%, depending on the operative pH. The precipitation efficiency (L, defined in Figure 18.7) was normally very low: in 2 of the 15 runs performed, it was less than 5%. This was the cut-off value chosen to integrate the plant with a post filtration operation. In 3 of the 15 runs performed, precipitation loss ranged from 7 to 8%, while in 3 other cases it was higher than 10%.

Using the same supernatant, with a P content of 28 mg/l (Battistoni et al., 2002) precipitation efficiency was always a function of the operative pH, and a loss of fines up to 20% was obtained (Table 18.13).

18.5.3.3 P salt and co-precipitation

The evaluation of P salt distribution on seed material as MAP or HAP can be calculated either from Equation 18.10, as precipitation percentage of P–MAP on total removed P, or from Equation 18.11 as precipitation percentage of P–HAP on total removed P:

$$P\text{–MAP\%} = 100 \frac{(Mg_{in} - Mg_{out})_{mol}}{(P_{TOTin} - P_{TOTout})_{mol}} \quad (18.10)$$

$$P\text{–HAP\%} = 100 \frac{(P_{TOTin} - P_{TOTout})_{mol} - (Mg_{in} - Mg_{out})_{mol}}{(P_{TOTin} - P_{TOTout})_{mol}} \quad (18.11)$$

The excess amount of calcium removed is computed to be due to CaCO$_3$ formation. The kind of P salt removed changes dramatically with supernatant composition. In particular, when high P content is used, a mixture of MAP and HAP is observed: 35% MAP, 65% HAP, when operating with a Ca/Mg molar ratio of about 1.8 (Battistoni et al., 1998c). Increasing the Ca/PO$_4$ molar ratio from 2.6 up to 5.8 or the Mg/PO$_4$ ratio from 0.7 up to about 1.6, resulted in a complete formation

Table 18.13 FBR operative conditions and performance.

Reference	P_{TOTin} (mg l^{-1})	pH	Q_{air} m^3/h	Flowrate l/h	η (%)	X (%)
Battistoni et al., 1998c	45	8.1–8.5	0.9–3.0	1–19	70–83	73–85
Battistoni et al., 2001b	22	8.04–8.78	1.5–12.3	18–190	53–80	60–91
Battistoni et al., 2002	28	8.04–8.44	7.5–13.0	148–327	46–65	67–77

of HAP or MAP respectively. When P_{TOT} concentration in the supernatant decreases from 45 to 28–22 mg/l, the formation mainly of MAP is observed (60–100%), but $CaCO_3$ co-precipitates. In that case, the seed material was found to contain 64 to 92% w.w. of its mass as $CaCO_3$ (Battistoni et al., 2001–2002).

Different analyses of the spent sand grain were carried out:

- chemical analysis of the grain;
- comparisons of the total amount of phosphorus removed calculated by a mass balance carried out on experimental results and spent matrix;
- particle size distribution on virgin and spent silica sand;
- thermal differential analysis on grains;
- scanning electron microscopy analysis (Battistoni et al., 2001b).

The data obtained confirm the following predictions:

- the value of the process, after all phosphorus was removed from the liquid phase, would depend on the precipitated product;
- the precipitated salt could contain one or more of: MAP, HAP and calcite;
- the morphological aspect of spent sand varied with different zones of precipitation being present, classified as sheet type, mainly formed by MAP, and sphere type mainly formed by calcite.

However, direct evidence (e.g. by X-ray diffraction) of crystallisation on sand, of HAP, or other calcium phosphate, was not achieved. Further studies, based on X-ray determinations, may lead to a better understanding of the kind of salts formed on the seeds.

18.5.4 FBR mathematical models

On the basis of the results obtained, some model approaches can be proposed, notwithstanding the uncertainty on salt type formed.

Several models are available in the literature to describe the phosphorus precipitation process; they can be distinguished by:

- models based on primary nucleation mechanism (e.g. nucleation is caused by pure supersaturation);
- models based on secondary nucleation mechanisms (e.g. nucleation and growth take place on introduced seeds, in metastable supersaturation conditions).

The work of Seckler et al. (1996 a, b, c) belongs to the first group, due to the high operative pH in their studies. Nevertheless, they propose a theoretical model for fine particle aggregation on sand grains. The following equation, derived from a particle number balance, and expressing the decrease in the fines concentration by aggregation with grains in a fluidised bed:

$$\frac{dN_i}{dt} = -BJ_{ii}\beta \qquad (18.12)$$

where B is the collision efficiency, N is the initial particle concentration, J_{ii} is the collision frequency, β is the supersaturation and t is the reaction time, which is expressed by:

$$t = \frac{\varepsilon x}{v_{sup}} \qquad (18.13)$$

where ε is the bed porosity, x is the axial position and v_{sup} is the superficial velocity. The supersaturation is defined as:

$$\beta = \frac{1}{5} \ln \frac{\left(Ca^{2+}\right)^3 \cdot \left(PO_4^{3-}\right)^2}{K_{SP}} \qquad (18.14)$$

where K_{SP} is the solubility product of amorphous calcium phosphate.

The collision efficiency B is derived from the following expression, which describes the influence of the energy dissipation rate (E):

$$B = B_0 \left(\frac{E}{E_0}\right)^\alpha \qquad (18.15)$$

where E_0 is a reference value and B_0 and α are characteristic parameters of the precipitating system.

By integrating Equation 18.12 from $t = 0$ to $t = t_{out}$, the phosphate removal efficiency η_{ag} (by aggregation only) can be calculated, by assuming the particle size of the fines to be constant in time:

$$\eta_{ag} = \frac{N_{i,in} - N_{i,out}}{N_{i,in}} \qquad (18.16)$$

It has been shown (Seckler et al., 1996 a, b, c), both theoretically and experimentally, that the aggregation can be increased by distributing the supersaturation more evenly throughout the reactor, while the breakage can be reduced by choosing fluidisation conditions where the energy dissipation rate in the bed is minimised (see also Mangin and Klein, this volume).

A model based on a secondary nucleation mechanism developed by DeRooij et al. (1984), described the formation of different calcium phosphates on a seed material, under well defined experimental conditions, such as a fixed temperature (T = 37°C), ionic strength (I = 0.10 mol/l), Ca/P ratio (Ca/P = 1.333) and for various pH values (5 < pH < 8). The model obtained is described by the following equation:

$$\frac{dm}{dt} = ks \left(IP^{1/v} - K_{SP}^{1/v} \right)^p \qquad (18.17)$$

where m is the quantity of the precipitated phase in moles, k is the rate constant for precipitation, s a factor proportional to the number of sites available for growth, p the effective order of reaction, v the number of ions in the formula unit,

IP the ionic product of supersaturated solution and K_{SP} the thermodynamic solubility product. The Gibbs free energy related to the transfer from supersaturated bulk solution to an assumed saturated solution at the surface of the developing solid phase, depends on the ratio between IP and K_{SP} (De Rooij et al., 1984).

Kaneko and Nakajima (1988) described the crystal growth of HAP, under metastable conditions in synthetic water solutions, using an approximation of Equation 18.17, expressed as:

$$\frac{dC}{dt} = ksC^2 \qquad (18.18)$$

where C is the molar concentration and t is the retention time.

The precipitation of MAP and HAP from an anaerobic supernatant in fixed FBR works has been shown to occur in conditions of metastability (Battistoni et al., 1998b), and thus the approximation made by Kaneko and Nakajima (1988) is suitable. This kinetic equation can be integrated from 0 to t, leading to the empirical saturation model in the contact time (t_c) (Equation 18.19) of precipitation efficiency η (Battistoni et al., 2000):

$$\eta = E_m \frac{t_c}{t_{1/2} + t_c} \qquad (18.19)$$

Moreover, a maximum precipitation efficiency (E_m) and a half time ($t_{1/2}$) equations can be introduced:

$$E_m = \frac{C_0 - C_t}{C_0} \qquad (18.20)$$

$$t_{1/2} = \frac{1}{(C_0 - C_t)k} \qquad (18.21)$$

where C_t is the final concentration and C_0 the initial concentration.

The empirical double saturation model thus follows from the theoretical model of precipitation in conditions of limiting supersaturation for what concerns the saturation in t_c. The behaviour as a function of pH has been reported in literature but never described in a theoretical model. In a previous experiment (Battistoni et al., 2001b), the precipitation efficiency was found to be variable, although no explanation for this observation was reported. In more recent work (Battistoni et al., 2002), results obtained from a large-scale pilot plant were extended to improve the validity of the model proposed, even at lower contact time. Therefore a double saturation model, taking into account both pH and t_c, was devised, which exhibits an excellent agreement with all experimental results.

$$\eta\% = 100 \frac{(\text{pH} - 7.325)}{(\text{pH} - 7.325) + 0.371} \cdot \frac{t_c}{t_c + 0.0196} \qquad (18.22)$$

The value of R^2 is 0.99 while the standard error is 4.9 considering a data base of 44 runs (Battistoni et al., 2002). The introduction of the double saturation model (Equation 18.22) in a pH range which can be easily maintained in waste waters (from 8.1 to 9.1) allows the use of the most suitable contact time to be reached in order to achieve a predetermined precipitation efficiency. It can be seen that for a pH equal to 8.5, in order to get a precipitation efficiency η of 70%, a t_c equal to 0.2 h is needed, while an efficiency of 80% is very difficult to achieve at a pH below 9. Loss, L, is given by the difference between X and η. High loss is undesirable, since it implies loss of particulate phosphorus in the form of fines. In such situation, a complex plant would be necessary, that would provide a filtration step, thus increasing the value of η. Using all experimental runs (44 runs, in Battistoni et al., 2002), the following equation is obtained, which describes the behaviour of X as a function of pH, by using a saturation model:

$$X\% = 100 \frac{(pH - 7.21)}{(pH - 7.21) + 0.38} \qquad (18.23)$$

for which $R^2 = 0.99$, S.E. $= 5.2$ and n $= 44$.

The physical meaning of the semi-saturation constant (0.38, Equation 18.23) is that at a pH of 7.59 (pH−7.21 = 0.38), 50% of the conversion can be obtained, while a pH value of 8.73 is needed for a conversion of 80%.

Furthermore, conversion and nucleation must be considered as concomitant phenomena on the basis of a double saturation model (Equation 18.22). Thus, phosphorus removal cannot be optimised by considering a nucleation process only. In fact, if the conversion is much higher than the nucleation efficiency, a meaningful loss of fines is obtained. It can be observed that to minimise the loss of fines, nucleation must be performed for a contact time higher than 0.4 h and a pH above 8.0. (Figure 18.11). The supersaturation model can be verified by applying it in a predictive mode using the experimental results of Seckler and co-workers (Seckler et al., 1996b, 1996c). The experimental values fall near or within the area defined by the two curves calculated with the double saturation model by using a highest and lowest contact time (Figure 18.12).

All of these analyses reveal how phosphorus removal through nucleation on sand can be related to contact time. In order to achieve a minimum contact time of 0.4 h in a small size reactor, a number n of passages into fluidised bed have to be performed. If only one passage is desired, the alternative could be a reactor with an expanded bed volume (V_{EXP}) equal to the flow rate (in m³/h, with $\varepsilon = 0.4$).

The results described above also demonstrate that optimisation of a single pass process by chemical addition (to produce pH increase) is not a feasible process, as it will result in high loss of fines (20–40%). To avoid such loss a filtration step is required, thus introducing complications and costs into the process.

These considerations demonstrate that phosphorus nucleation process in FBR would be best applied in the secondary streams of the waste water treatment plants. Therefore the water treatment line has to be planned in order to carry out biological nutrient removal (BNR process) and to concentrate the phosphorus release in a minimum flow rate of supernatant avoiding mixing between primary and activated sludge before the anaerobic digestion. A FBR will fix phosphorus treating dewatering centrate. A CO_2 stripping stage must be included, to allow control of the operative pH.

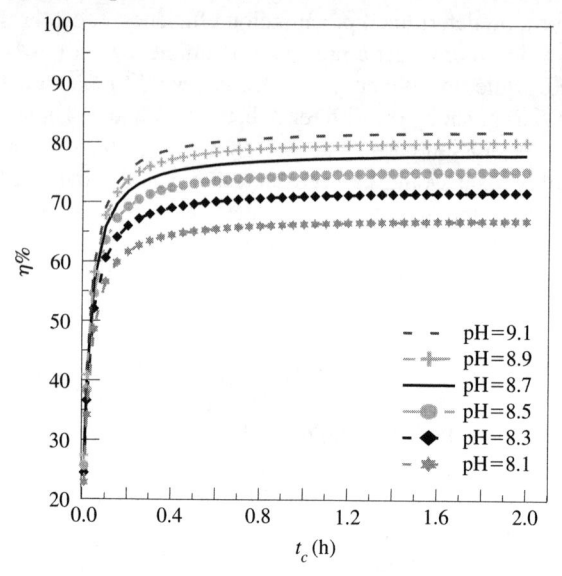

Figure 18.11   Conversion and nucleation vs contact time (Battistoni *et al.*, 2002).

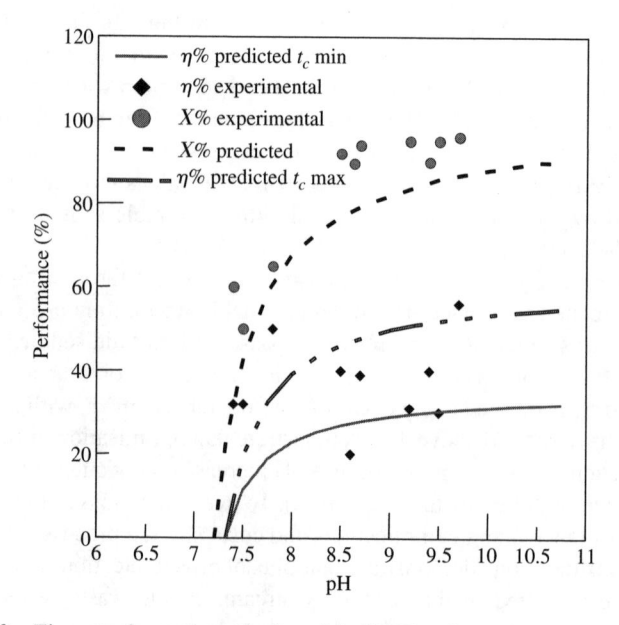

Figure 18.12   The use of a mathematical model of FBR to interpret reference data (Battistoni *et al.*, 2002).

18.5.5 Mass and economic balances of FBR

Specific managing costs for PRP (P-recovery process, assuming struvite is the precipitating phosphate in the reactor and using the model described above) are expressed in this section in Euros per m³ of supernatant treated; they can be estimated on the basis of a large pilot scale experience (Battistoni et al., 2001b). The costs considered are: energy, maintenance, sand consumed, recovery value of struvite (sales as fertiliser at half the price of a low release P fertiliser) and process management; energy costs are calculated using the high cost of energy in Italy (0.103 €/kWh).

It should be noted that the global specific cost (Table 18.14) has to be a function of supernatant P content. For this reason, three cases are considered: A) equivalent to 150 mg P_{TOTin}/l; B) to 100 mg P_{TOTin}/l; and C) to 50 mg P_{TOTin}/l, where P removal performance is 80% in all three cases. Their conversion in specific cost in relation to P removed (€/Kg P_r) determines three different values with an increasing trend as a function of decreasing P content. The main expense is due to energy consumption (feeding, recycle and air supply).

Experience with the management of a full scale PRP plant will help to determine the global cost with more precision.

A first comparison of specific costs of PRP with those from other processes found in the literature can be done, if we take a cost of 4.5–5.5 €/Kg (Ir. Gaastra, pers. comm.) of P removed as hydroxylapatite in the plant of Geestmerambacht sewage works. This figure is given without any detailed analysis (i.e. cost for reagents), while a cost of 1.95 €/m³ has been calculated in detail in the demonstrative plant of DHV Crystallactor (Van Dijk and Braakensiek, 1985) treating 30 m³/d of supernatants. However, the importance of investment costs to build the reactor, which were not reported, should be considered when comparing with the low cost of a simple process such as the PRP (where there is no requirement for filtration or need for addition of chemicals).

Costs can be further reduced if costs of managing (31–38%) can be minimised. To conclude, the costs of the PRP process appear modest even without considering the possible use of spent sand.

Table 18.14 Managing cost of PRP process.

	Case A (€/m³)	Case B (€/m³)	Case C (€/m³)
Energy consumption (feeding, recycle, air)	0.13	0.13	0.13
Sand consumption			
P_{TOTin} 150 mg/l	0.37		
P_{TOTin} 100 mg/l		0.24	
P_{TOTin} 50 mg/l			0.12
Disposal of spent sand			
P_{TOTin} 150 mg/l	−0.23		
P_{TOTin} 100 mg/l		−0.15	
P_{TOTin} 50 mg/l			−0.08
Analytical control	0.09	0.09	0.09
Managing	0.16	0.16	0.16
Global costs (€/m³)	0.52	0.47	0.42
Global costs (€/Kg P_r)	4.33	5.88	10.50

18.5.6 Some guidelines for FBR design

The most important aspects to be considered in the planning of a P-recovery process (PRP) facility are the following:

- the PRP has to be linked to a civil waste water treatment plant which operates biological phosphorus removal;
- the recovery unit has to be located as close as possible to the sludge facilities in order to reduce precipitation problems in the pipes;
- the hydraulic loading must be predicted or calculated on the basis of the supernatant from sludge facilities produced in the plant;
- the phosphorus mass loading must be calculated on the basis of the phosphorus maximum release during sludge treatment;
- the plant must be provided with a storage tank to avoid the fluctuations in the flow rate and P mass loadings coming from the discontinuous operation mode of the dewatering section;
- a pH control system must be provided to check the condition of the system;
- on-line analysis of phosphate is necessary to evaluate the efficiency of the process;
- proper devices for air distribution and supply in the stripping section must be selected; it is better to use a system independent from the blower of the water line;
- selected pumps with adjustable flow rates capable of fluidising the sand bed are essential for the precipitation process;
- the FBR influent has to be clarified, so it is necessary to provide a section for solids removal (decanter). This is particularly recommended if the dewatering station works with a belt press instead of a centrifuge, because this undergoes a lower loss of solid matter;
- a by-pass line must be provided to avoid the passage of untreated supernatants into the FBR in case the plant does not work.

In particular, when designing an FBR, the following must be taken into consideration:

- the aim of the facility is the precipitation of phosphorus salts and the reuse of the salt in agriculture as low cost fertiliser;
- the sand used as seed has to be of specific (certified) grain size distribution so that the flow rate required to fluidise the bed can be calculated;
- the upper section of the reactor must be designed to avoid loss of fine particles (Dortmund zone);
- the hydraulic retention time must be able to assure a contact time (t_c) equal to or higher than 0.4 hours;
- the reactor must be provided with ports to allow for the substitution of the spent sand with virgin sand and for sampling.

Additionally in the design of the stripping section, the following devices must be considered:

- the stripper section must be designed with the possibility to operate at various heights, or with variable volumes, which will guarantee the required contact time (0.4 hours) at the porosity of the expanded bed;

- it is necessary to include in the design a free zone of at least 1 m from the maximum hydraulic head to prevent foam from coming out;
- a proper device to collect and remove foam must be provided;
- the stripper section produces gaseous emissions so the plant must be indoors and contain a biofilter for the treatment and the removal of odours.

18.6 CASE STUDY: TREVISO CITY WASTE WATER TREATMENT PLANT

The actual plant of the city of Treviso (northern Italy) is the result of an upgrading that brought the potentiality of the plant from 20,000 PE to 70,000 PE. This is the result of recent legislation, which imposes stringent limits on nutrient concentration in waste water effluents. The pre-existing section of the plant (employing a total oxidation scheme) is used to treat 20,000 PE, while the main load fraction (50,000 PE) is treated by the new BNR section. The plant was planned using modular characteristics, in order to apply various process configurations. The primary sedimentation is avoided, and the old primary settler is used as the sand removal section and flow rate partition. This is to preserve more COD for nitrogen and phosphorus removal steps. The process scheme is reported in Figure 18.13. In order to promote these steps, and considering that the waste water to be treated is very poor in terms of RBCOD, integration with the OFMSW fermentation process was considered. This was done using a demonstration area which operates the source separated OFMSW fermentation at mesophilic conditions, producing a VFA-rich stream to add into the water line, and a solid residue to be sent to sludge anaerobic digestion to improve biogas production. The supernatants coming from the dewatering section are treated in a PRP unit, using an FBR as a precipitation reactor.

BNR section description The main plant characteristics and operative conditions of the new line in the Treviso waste water treatment plant are reported in Table 18.15. The BNR section is capable of biologically removing carbon, nitrogen and phosphorus operating with the configuration of the Phoredox (modified Johannesburg). First, the influent is degritted and the sand is wasted while the organic fraction is recycled at the headworks. Then, the influent is directly sent to biological treatment. The single tanks are designed in a modular way so that it is possible to change the tank volume and to obtain a high operating flexibility. In this way the biological process can support changes in influent load and in its characteristics. The first section is a pre anoxic zone where only recycle activated sludge is sent for denitrification. The following sections are anaerobic and anoxic; the fermented OFMSW is added here. More precisely, the addition is planned in the anaerobic zone but can be carried out also in the anoxic and pre-anoxic ones. The oxidation/nitrification tank has also been designed to operate at low loading conditions.

The sludge treatment section operates essentially with an anaerobic mesophilic single-stage digestion. The sludge is sent to digestion after blending and thickening with the solid fraction coming out of the fermentative OFMSW area. The sludge

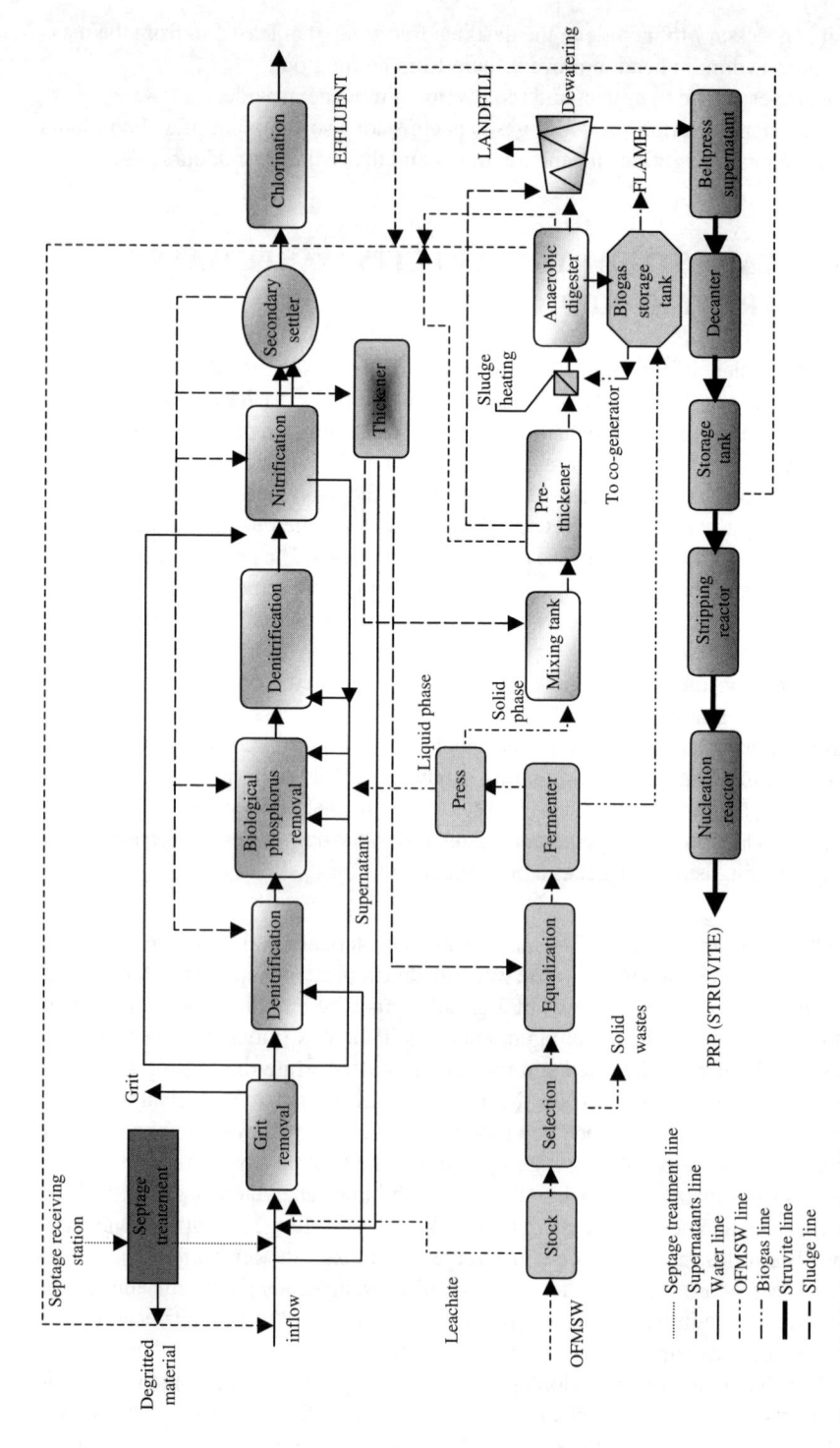

Figure 18.13 The design of the new waste water treatment plant of Treviso city (Veneto Region, Italy).

Table 18.15 The WWTP of Treviso. Design data and new sections volume.

Waste water treatment design data	
Influent average flow rate (Q_a), m³/d	14,000
Maximum flow rate (Q_m), m³/d	21,600
Peak flow rate (Q_p), m³/d	21,000
Section volumes	
Degritting, m³	181
Pre-treatments, m³	628
Pre-anoxic zone, m³	400–1,200
Anaerobic zone, m³	700–1,200
Anoxic zone volume, m³	1,600–2,200
Oxidation/nitrification zone, m³	5,500
Total volume biological reactor, m³	9,000
Chlorination, m³	250
Secondary settlers, m³	2480
Sludge treatment	
Thickener, m³	210
Anaerobic digester, m³	2 200

coming out from anaerobic digestion is added to a polyelectrolyte and sent to mechanical dewatering. The supernatant from the thickening and dewatering sections can be treated in the struvite section. Phosphorus is removed by seeded precipitation and the effluent of the PRP is recycled as supernatant in the water stream.

OFMSW area description The demonstrative area for OFMSW fermentation has been described by Cecchi *et al.* (2002). Its main characteristics are: a refining step for the OFMSW, using a simplified sorting line (to eliminate bulky fraction, ferrous and packaging), an anaerobic fermenter that works with an HRT of 3–6 days, a screw-press, which splits the OFMSW into two fractions – the liquid streams and the solid streams that are sent respectively to waste water treatment and to the anaerobic mesophilic co-digestion section.

18.6.1 PRP demonstrative area

18.6.1.1 Design elements

The PRP demonstrative area was planned on the basis of the tests carried out with small and medium scale pilot plants. It can be used as a full scale device, but at the same time it is equipped with all the probes and facilities typically used in a bench scale reactor. This allows for the rapid change of operative conditions as well as for the acquisition of all the data needed in a scientific experiment. The plant is mainly composed of three sections: pre-treatment, stripping and FBR (Figure 18.14).

The pre-treatment section (Figure 18.15) has the dual role of removing the suspended solids, which leave the dewatering section of the anaerobically digested sludge, and of stocking the supernatants (48 m³) to assure continuous feeding of the FBR, even if the dewatering section works only a few days per week.

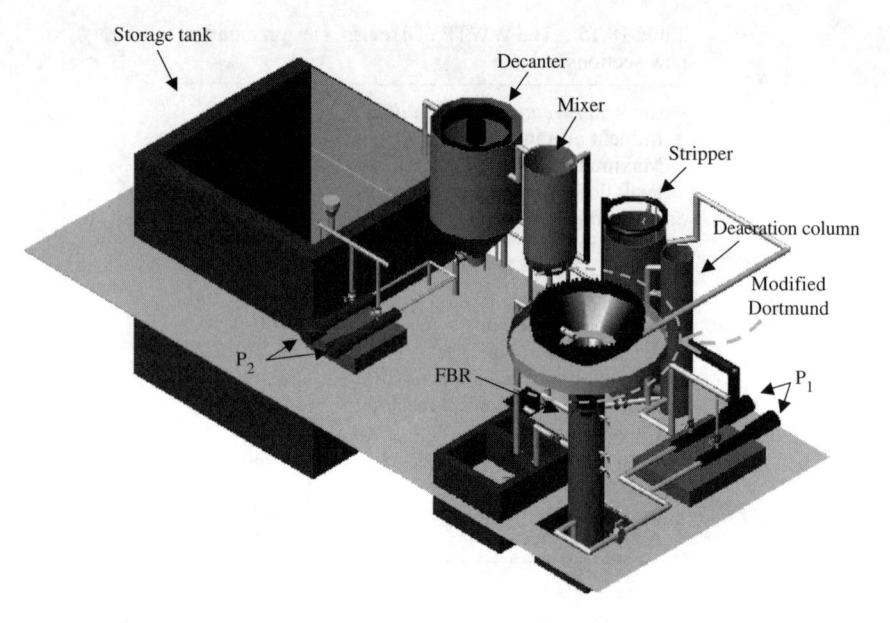

Figure 18.14 P-recovery process (PRP) demonstrative area, Treviso city plant.

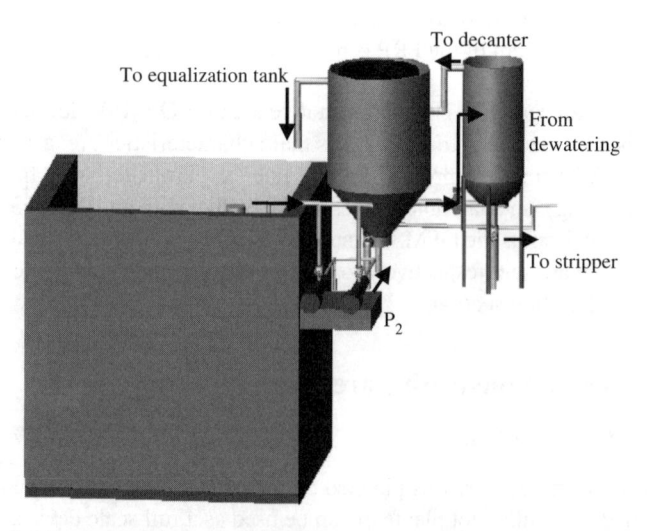

Figure 18.15 Pre-treatment section, PRP, Treviso city plant.

The pumping section (Figure 18.16) assures several possibilities in the plant management, mainly linked to the contact time variation (changing the hydraulic level in the stripper), the operative pH, change in the air flowrate, the bed expansion (normally $0.9\,m^3$) and change in the recycle flowrate (Figure 18.17).

Main information about plant sizing is given in Table 18.16. The volumes used allow a change in the contact time from 0.4 to 1 h, in connection with bed porosity

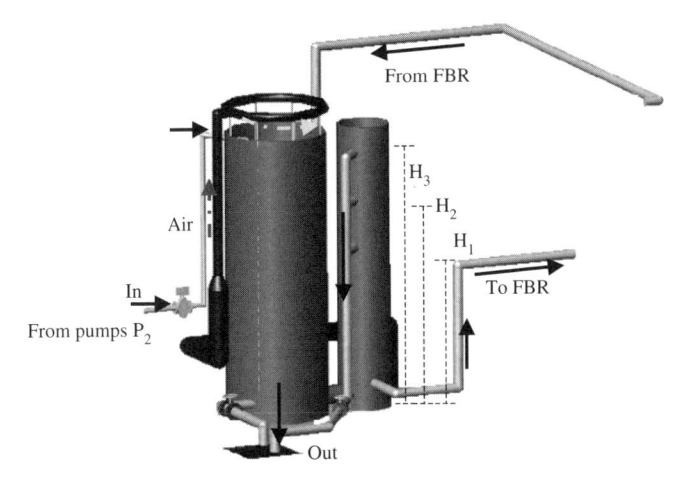

Figure 18.16 Stripping section, PRP, Treviso city plant.

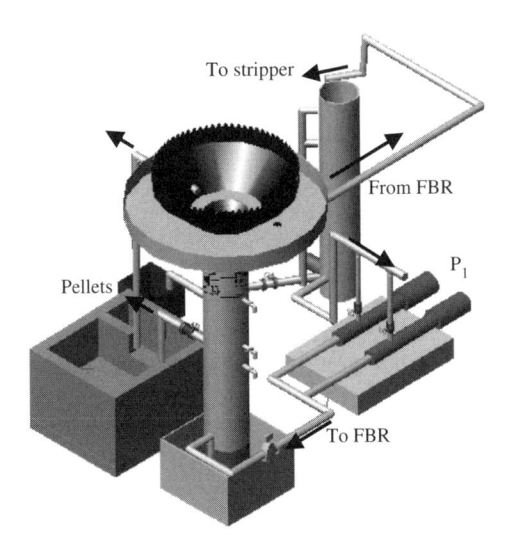

Figure 18.17 FBR section, PRP, Treviso city plant.

and inlet flowrate (Figure 18.18). In this way, the plant can treat a supernatant flowrate from 1 to 2 m^3/h, assuring, at the same time a contact time higher than the critical value of 0.4 h. The process control is done also using several on-line probes, such as flowrate measurements, pH, Oxidation Reduction Potential (ORP) (Table 18.17). In this way it is possible to achieve:

- a correct monitoring of the operative parameters;
- the possibility to perform and follow changes of parameters in real time;
- the correct mass balance management of phosphorous, as a result of the knowledge of the exact working time of the plant.

Table 18.16 Main section volumes, PRP Treviso city plant.

Section	Volume (m³)	Section	Volume (m³)
Pre-treatment		Deaeration column	
Mixer	4.7	H_1	0.3
Decanter	4.1	H_2	0.4
Equalisation tank	48	H_3	0.5
Stripping column		FBR	
H_1	1.3	Dortmund zone	1.3
H_2	1.7	Expanded bed	0.9
H_3	2.1	Gravel filter	0.3

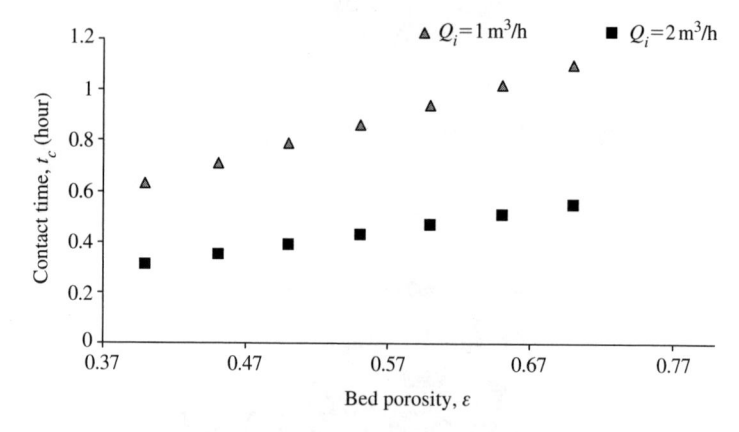

Figure 18.18 Contact time variation with bed porosity of the PRP.

Table 18.17 On line monitoring apparatus, PRP Treviso city plant.

Parameter	Point of measure			
Q (m³/h)	From dewatering (Q_{dw})	Influent to stripping column (Q_i)	Influent to FBR (Q_r)	Air supplied (Q_{air})	–
T (°C)	–	Stripping column (T_s)	–	Air supplied (T_a)	–
pH	–	–	–	–	Deaeration column (pH)
ORP (mV)	–	–	–	–	Deaeration column (ORP)

18.6.1.2 Interaction between OFMSW and PRP areas

The use of fermentate enhances phosphorus removal from main stream, assuring a high phosphorus content in waste activated sludge. The blending of residual solids of OFMSW with waste activated sludge can cause a phosphorus release in gravitational thickening in addition to what is normally expected in a dewatering section.

Table 18.18 Chemical and physical characteristics of thickening supernatant, Treviso city plant.

Period	Q (m³/d)	Q_{OFMSWrs} (m³/d)	pH	Alkalinity (mg CaCO$_3$/l)	PO$_4$–P (mg/l)	NH$_3$–N (mg/l)
A						
Average	48	–	6.8	684	48	92
Min	0	–	6.4	497	35	75
Max	90	–	6.9	990	50	120
B						
Average	62	1.9	6.6	1320	63	406
Min	36	0.0	6.4	1065	45	210
Max	76	9.8	6.9	1884	90	450

Table 18.19 Chemical and physical characteristics of the digester supernatant, Treviso city plant.

Period	NH$_3$–N (mg/l)	PO$_4$–P (mg/l)	Alkalinity (mg/l)
A			
Average	392	33	1877
Min	248	14	1470
Max	687	47	2268
B			
Average	685	45	2243
Min	653	21	1896
Max	706	56	2724

This means that the two types of supernatants must be verified and that it is necessary to ascertain if both must be treated in the PRP plant. As a result, the WWTP of Treviso is completely different from other European plants using a BNR process.

In particular, the supernatant from the gravitational thickener during period A (thickening of waste activated sludge alone) shows an alkalinity ranging from 500 to 1000 mg CaCO$_3$/l, while after the addition of the residual solids of OFMSW this increases to values up to 1900 mg CaCO$_3$/l (Table 18.18). This, in turn, means that fermentation takes place in the sludge and that phosphorus and ammonia releases occur.

Also the physicochemical characteristics of digested sludge, and hence of the supernatant produced in the dewatering section, show a meaningful increase in alkalinity, ammonia and phosphate content compared with the effluent prior to the OFMSW addition (Table 18.19). In conclusion, in traditional BNR WWTP avoiding the mixture of primary and waste activated sludge, phosphorus is released in all the supernatants produced after anaerobic digestion (i.e. post gravitational thickening, centrate of dewatering station etc); while in the Treviso BNR WWTP using external carbon source as OFMSW, the treatment of supernatant produced by mixing waste sludge with residual solid of fermentate OFMSW becomes important.

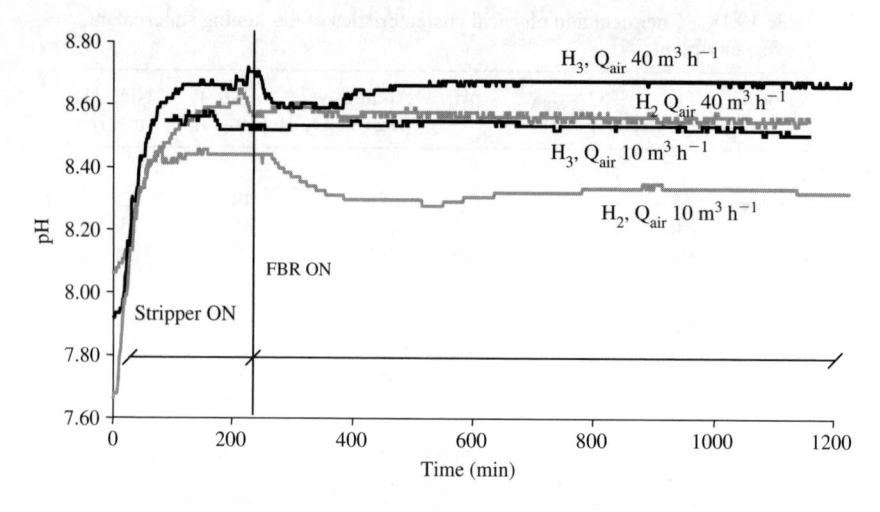

Figure 18.19 Operative pH at different hydraulic head values ($H_1 = 1.7$ m, $H_2 = 2.2$ m, $H_3 = 2.7$ m) and air flow conditions PRP, Treviso city plant.

18.6.1.3 Operative feasibility of PRP

The reliability of PRP is mainly controlled by three interconnected variables during normal plant management: the operative pH, which is a function of the air flow rate and the hydraulic head in the stripping section; the exact knowledge of the operative pH is therefore important when trying to establish that the phosphorus removal process is similar to that observed in previous pilot trials.

The pH can be modified at a constant hydraulic head if air flow rates are varied, and vice versa as shown in Figure 18.19. The pH signal is also normally constant, although a significant drift may be observed in 24 hours, mainly due to foam formed as a result of air stripping.

A test of the mathematical model previously described in a large scale PRP pilot is initially tested performing short runs at different operative conditions (Table 18.20). Results at low (run 1), medium (run 2) and high (run 3) phosphorus, ammonia and alkalinity concentrations, appear substantially different. In the first and second runs a conversion (X) and nucleation (η), lower than expected, are observed. In the third run a good agreement between model and observation is obtained for conversion efficiency (X). However the loss of fines is still not satisfactory. These results can be interpreted as implying that the environmental conditions (ion type and concentrations) exert a significant influence and a further test of the model is necessary. It is therefore concluded that, prior to its employment at industrial scale, the FBR system must work as a pilot plant with the following aims:

- to verify the model of the process;
- to determine the best operative conditions;
- to confirm the managing cost (particularly if using two types of supernatants).

Table 18.20 Short runs performance of PRP, Treviso city plant.

			Run 1	Run 2	Run 3
Influent	P_{TOT}	mg l^{-1}	15.7	24.8	59.0
	PO_4–P	mg l^{-1}	15.0	22.6	42.4
	NH_3–N	mg l^{-1}	219	106	323
	HCO_3	mg l^{-1}	1265	720	1500
Operative parameters	Q_r	m^3 h^{-1}	4.0	6.2	6.0
	Q_{in}	m^3 h^{-1}	0.6	1.4	1.0
	Q_{air}	m^3 h^{-1}	20	45	10
	$T_{stripper}$	°C	26	28	12
	pH	–	8.5	8.4	8.3
	h (head)	m	2.2	2.2	2.2
	HRT_t	h	7.6	3.3	4.6
Effluent	P_{TOT}	mg l^{-1}	10.0	11.60	26.9
	PO_4–P	mg l^{-1}	6.9	9.75	14.2
Performance of experimental results	X	%	56.1	60.6	75.9
	η	%	36.2	53.2	54.4
	L	%	19.9	7.5	21.5
Forecasted (model) performance	X	%	76.5	75.8	74.3
	η	%	74.3	72.4	71.3
	L	%	2.2	3.4	3.1

18.7 LONG TERM PERFORMANCE OF THE REACTOR IN TREVISO

The results of a continuous four-month P-recovery test, in steady state conditions, are discussed here. To start the recovery test, spent sand was replaced with 475 kg of new sand ($\Phi_m = 0.09$ mm). Table 18.21 shows average values and standard deviation of influent and effluent P concentration, operative parameters, and the actual performances obtained, as well as predicted performances based on the model described earlier. The obtained percentage of conversion (X%) is on average 61%; the difference with the expected value can be attributed to scaling up problems and to variations in pH from 8.0 to 8.7. The precipitation efficiency (L%) is quite low (5.3% on average), even though higher values were observed in pilot tests (Battistoni et al., 2001b).

Chemical and sieve analysis of sand particles were carried out on spent sand taken from FBR (Table 18.22). The sand particles increased in size, with increasing volume of anaerobic supernatant (Φ_m ranges from 0.09 mm for the virgin sand to 0.14 mm after the treatment of about 2,200 m^3). The proportion of secondary precipitates reached 55% of the total mass of the spent sand. The nucleated material is composed of Ca phosphate salts (perhaps a mixture of mainly $Ca_3(PO_4)_2$ and minor hydroxylapatite) ranging from 52 to 65% on weight basis, and of a variable percentage of struvite (from 12% to 24%). Calcite was not identified by X-ray diffraction of the precipitates. Mass balance of the newly formed precipitates shows

Table 18.21 Long term period performances of PRP, Treviso city plant.

			Avg.	SD.
Influent	P_{TOT}	mg l^{-1}	110	66
	PO_4–P	mg l^{-1}	94	55
Operative parameters	Q_{in}	m^3 h^{-1}	1.3	0.5
	pH		8.5	0.2
Effluent	P_{TOT}	mg l^{-1}	48	31
	PO_4–P	mg l^{-1}	38	27
Forecasted (model) performance	X	%	75	2.7
	η	%	72	3.0
	L	%	3.0	0.7
Performance of experimental results	X	%	61	14
	η	%	56	15
	L	%	5.3	4.9

Table 18.22 Chemical analysis on spent sand.

Treated volumes m^3	Φ_m mm	Precipitate proportion (PP) %w.w.	MAPPP %w.w.	CPPP %w.w.	CaCO$_3$PP %w.w.	Other PP %w.w.
634	0.14	21	12	52	5	31
1900	0.16	51	24	60	0	16
2194	0.18	55	13	65	1	21

MAP = struvite, CP = calcium phosphate.

Table 18.23 Comparison of the spent sand with commercial fertilisers.

Treated volumes m^3	P %w.w.	N %w.w.	P–N fertiliser (L.748/84)	P fertiliser (L.748/84)
634	2.5	0.1	2.2% P	4.4% P
1900	7.6	0.7	3% N	–
2194	8.1	0.4	–	–

also presence of other, perhaps organic substances, probably introduced in the FBR reactor, together with the solids lost by the belt press. The percentage of P and N of the product (sand plus precipitated material) collected at the end of the long-term experimentation, permits to classify the spent sand as a basic P fertiliser according to the Italian law (L.748/84). To classify the spent sand as a P–N composite fertiliser, a nitrogen content of at least 3% is necessary: this condition would require a 55% of the precipitate to be struvite (Table 18.23). Such a result can only be obtained with external addition of magnesium salts.

The X-ray analysis (Figure 18.20) shows a partially crystalline deposit on the spent sand from Treviso PRP, similar to that on pellets obtained from Geestmerambacht DHV Crystallactor (see also Piekema, this volume). Both are

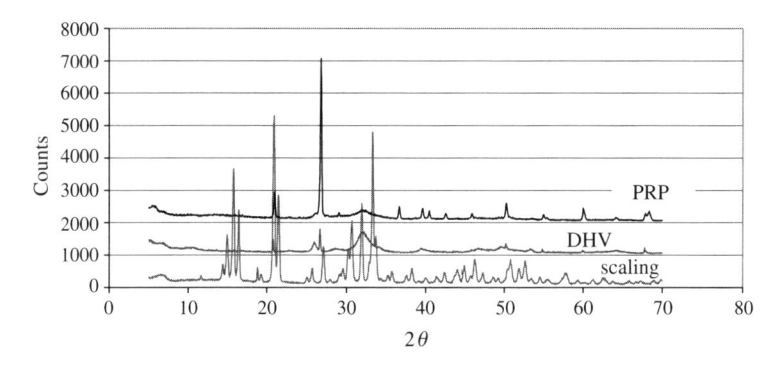

Figure 18.20 X-ray diffraction pattern of precipitates from Treviso (PRP = precipitate on sand, scaling = precipitate on the diffuser), compared with precipitates from the Geestmerambacht DHV Crystallactor.

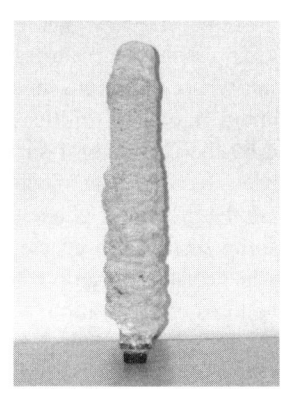

Figure 18.21 Scaling on the air diffuser of the Treviso reactor.

consistent with the presence of poorly crystalline P salts, however hydroxylapatite and tricalcium phosphate cannot be resolved. On the basis of ionic mass balance obtained from chemical analysis, $Ca_3(PO_4)_2$ formation is more probable. Scaling formed on the diffuser in the stripping column of the Treviso PRP (Figure 18.21), had a clear crystalline structure mainly consisting of struvite and hydroxylapatite (Boccadoro, 2003). This reveals the necessity of a periodic maintenance of the diffusers, to guarantee the efficiency of the stripping section.

18.8 PERSPECTIVE AND CONCLUSIONS

In this paragraph, the advantages of the air stripping FBR process in terms of its applicability and future developments are summarised. A BNR process for nutrient removal and phosphorous recovery (and potentially reuse) is considered an indispensable tool for sewage sludge minimisation and for avoiding sludge overproduction.

A FBR with air stripping constitutes an easy and low cost approach to achieve phosphorus removal, without sludge overproduction and reagent additions. A next logical step would be to recover the phosphorus for external use.

In a scenario where a low content of P in sewer exists (like in Italy), the use of a PRP plant is necessary to provide an effluent, which satisfies the stringent P law limit (1 mg/l). Such scheme is preferable in smaller scale WWTPs (<100,000 PE), because it can guarantee safe management and avoidance of major P_{TOT} fluctuations in the main stream effluent.

The employment of the PRP process does not require any modification of the existing flow scheme present in the sludge line, however, some attention must be paid to the planning in order to minimise the supernatant flow rate to treat.

The particular process used, operates in metastable conditions and requires a large number of passages in FBR to allow P salt growth on seed material without significant losses of fines. Alternatively, supersaturation conditions, at higher pH, but with increased managing costs, requires a more complex system where filtration of fines is essential.

The simplicity of the system presented here allows for both the production of a cheaper and more robust plant, where minimum investment is required, the flow scheme is simple, and a minimum number of utilities are involved. The operations of the plant may be affected by the presence of suspended solids in the supernatant, lost by thickening or dewatering. Suspended solids determine a pollution of the seed material, hindering the growth of phosphate crystals and resulting in scum production in the stripping section. An efficient unit for solids separation must therefore be included if the dewatering station. To limit supernatant flow rate and to avoid dilution of phosphorus concentration, a good separation of washing waters from centrate must be realised at the belt press station. Summarising, the substantial benefits of the PRP process can be outlined as:

- the recovery of phosphate released in the sludge line;
- the good performance of the BNR process;
- no requirement of chemicals to obtain the operative conditions;
- recovery of phosphorus and perhaps ammonia for agricultural purposes;
- reduction of waste sludge production by the abatement of nutrients otherwise incorporated in the sludge;
- operational benefits, due to avoidance of clogging of piping serving the dewatering supernatant elimination.

At present, not all the possible aspects of PRP process employment are fully understood, those remaining to be investigated include:

- the competition between calcite, MAP and HAP precipitation;
- optimisation of growth on the quartz (or other) seed, bed porosity and contact time required by the process;
- optimisation of performance in homogeneous nucleation (i.e. without a seed) in order to save managing costs and to produce a more recyclable product.

ACKNOWLEDGEMENTS

The author wishes to thank Dr. Ing. Raffaella Boccadoro and Dr. Gaia D'Alessi for their important collaboration in the preparation of the manuscript.

REFERENCES

Aesoy, A. and Odegaard H. (1994) Nitrogen removal efficiency and capacity in biofilms with biologically hydrolysed sludge as a carbon source. *Water Sci. Technol.* **30**(6), 63–71.

Battistoni, P. (1999) Cap. 6 La progettazione di impianti di depurazione a cicli integrati acque-reflue rifiuti solidi. Analisi tecnico economica. In '*Una gestione Integrata del ciclo dell'acqua e dei rifiuti. Fondamenti, stato dell'arte, ingegneria di processo*' (ed F. Angeli) – Proaqua- Milano, pp. 130–159.

Battistoni, P., De Angelis A, Prisciandaro, M., Boccadoro, R., and Bolzonella, D. (2002) P removal from anaerobic supernatants by struvite crystallisation: long term validation and process modelling. *Water Res.* **36**(8), 1927–1938.

Battistoni, P., Fava G., Pavan, P., Musacco, A. and Cecchi, F. (1997) Phosphate removal in anaerobic liquors by struvite crystallisation without addition of chemicals. Preliminary results. *Water Res.* **31**, 2925–2929.

Battistoni, P., Pavan, P., Cecchi, F., Mata-Alvarez J. and Majon M. (1998a) Integration of civil wastewater and municipal solid waste treatments. The effect on biological nutrient removal processes. In *European Conference on New Advances in Biological Nitrogen and Phosphorus Removal for Municipal or Industrial Wastewaters*, 12–14 October, Narbonne, France, pp. 129–137.

Battistoni, P., Pavan, P., Cecchi, F. and Mata-Alvarez, J. (1998b) Effect of composition of anaerobic supernatants from an anaerobic, anoxic and oxic (A_2O) process on struvite and hydroxyapatite formation. *Ann. Chim.* **88**, 761–772.

Battistoni, P., Pavan, P., Cecchi, F. and Mata-Alvarez J. (1998c) Phosphate removal in real anaerobic supernatants: modelling and performance of a fluidised bed reactor. *Water Sci. Technol.* **38**(1) 275–283.

Battistoni, P., Pavan, P., Prisciandaro, M. and Cecchi, F. (2000) Struvite crystallisation: a feasible and reliable way to fix phosphorus in anaerobic supernatants. *Water Res.* **34**, 3033–3041.

Battistoni, P., Pavan, P., Prisciandaro, M. and Cecchi, F. (2001b) Phosphorus removal from a real anaerobic supernatant by struvite crystallisation. *Water Res.* **35**, 2167–2178.

Battistoni, P., Pezzoli S., Bolzonella, D. and Pavan, P. (2001a) The AF-BNR-SCP process as a way to reduce global sludge production comparison with classical approaches on a full scale basis. In *Proceeding of: Spec. Conf. On Sludge Management*: regulation, treatment, utilisation and disposal. October 25–27, 2001, Acapulco, Mexico.

Bhattacharya, S.K., Madura R.I., Walling, D.A. and Farrell, J.B. (1996) Volatile solids reduction in two phase and conventional anaerobic sludge digestion. *Water Res.* **30**(5), 1041–1048.

Boccadoro, R. (2003) Ottimizzazione del processo di cristallizzazione in letto fluido degli ortofosfati presenti in surnatanti anaerobici. PhD Thesis. University of Ancona, Italy.

Cecchi, F. and Battistoni, P. (2003) "Use of hydrolysis products of the OFMSW for biological nutrient removal in waste water treatment plants" In *Biomethanization of the organic fraction of municipal solid wastes*. Ed. "IWA Publishing", Cap 8, pp. 229–261, Portland Press.

Choi, E., Lee, H.S., Lee, J.W. and O, S.W. (1996) Another carbon source for BNR systems. *Water Sci. Technol.* **34**(1–2), 363–369.

Chu, A., Mavinic, D.D., Kelly, H.G. and Ramey, W.D. (1994) Volatile fatty acid production in thermophilic aerobic digestion of sludge. *Water Res.* **28**(7), 513–1522

Deakyne, C.W., Patel, M.A. and Krichten, D.J. (1984) Pilot plant demonstration of biological phosphorus removal. *J. Water Pollut. Con. F.* **56**, 867–873.

DeRooij, J.F., Herghebaert, J.C. and Nancollas, G.H. (1984) A pH study of calcium–phosphate seeded precipitation. *J Colloid Interf. Sci.* **100**(2), 350–358.

Ghosh, A., Conrad, J.R. and Klass, L. (1975) Anaerobic acidogenesis of wastewater sludge. *J. Water Pollut. Con. F.* **47**(1), 30–44.

Kaneko, S. and Nakajima, K. (1988) Phosphorus removal by crystallization using a granular activated magnesia clinker. *J. Water Pollut. Con. F.* **60**(7), 1239–1244.

Kristensen, G.H., Jorgensen, P.E., Strube, R. and Henze M. (1992) Combined pre-precipitation, biological sludge hydrolysis and nitrogen reduction – a pilot demonstration of integrated nutrient removal. *Water Sci. Technol.* **25** (5–6), 1057–1066.

Isaacs, S.H. and Henze M. (1995) Controlled carbon source addition to an alternating nitrification-denitrification wastewater treatment process including biological P removal. *Water Res.* **29**(1), 77–89.

Jenkins, D., Ferguson, J.F. and Menar, A.B. (1971). Chemical process for phosphate removal. Review paper. *Water Res.* **5**, 369–389.

Joko, I. (1984) Phosphorus removal from wastewater by the crystallisation method. *Water Sci. Technol.* **17**, 121–132.

Llabres, P., Pavan, P., Battistoni, P., Cecchi, F. and Mata-Alvarez, J. (1999) The use of organic fraction of municipal solid waste hydrolysis products for biological nutrient removal in wastewater treatment plants. *Water Res.* **33**(1), 214–222.

Lotter, L.H. and Pitman, A.R. (1992) Improved biological phosphorus removal resulting from the enrichment of reactor feed with fermentation products. *Water Sci. Technol.* **25**(5–6), 943–953.

Murakami, T., Koike S., Taniguchi, N. and Esumi, H. (1987) Influence of return flow phosphorus load on performance of the biological phosphorus removal process. In *Biological phosphate removal from wastewaters* (ed. Ramadori, R.), pp. 237–247, Pergamon Press (Oxford).

Nancollas, G.H. (1984) The nucleation and growth of phosphate minerals. In *Phosphate Minerals* (eds Nriagu, J.O. and Moore, P.B.), pp. 137–154. Springer-Verlag, London.

Pavan, P., Battistoni, P., Bolzonella, D., Innocenti, L., Traverso, P. and Cecchi F. (2000) Integration of wastewater and OFMSW treatment cycles: from pilot scale to industrial realization. The new full scale plant of Treviso (Italy). In *Proc. 4th Int. Symp. On Environmental Biotechnology*, 10–12 April, Noordwijkerhout, The Netherlands.

Pavan, P., Battistoni, P., Musacco, A. and Cecchi, F. (1994) Mesophilic anaerobic fermentation of SC-OFMSW: a feasible way to produce RBCOD for BNR processes. In *Proc. Int. Symp. On Pollution of the Mediterranean Sea*, 2–4 November, Nicosia, Cyprus, pp. 561–570.

Pavan, P., Battistoni, P., Traverso, P., Musacco, A. and Cecchi F. (1998) Effect of addition of anaerobic fermented OFMSW (organic fraction of municipal solid waste) on biological nutrient removal (BNR) process: preliminary results. *Water Sci. Technol.* **38**(1), 327–334.

Pitman, A.R., Deacon, S.L. and Alexander, W.V. (1991) The thickening and treatment of sewage sludge to minimize phosphorus release. *Water Res.* **25**(12), 1285–1294.

Pitman, A.R., Lötter, L.H., Alexander, W.V. and Deacon S.L. (1992) Fermentation of raw sludge and elutriation of resultant fatty acids to promote excess biological phosphorus removal. *Water Sci. Technol.* **25**(4–5), 185–194.

Pöpel, H.J. and Jardin, N. (1993) Influence of enhanced biological phosphorus removal on sludge treatment, *Water Sci. Technol.* **28**(1), 263–271.

Rabinowitz, B. and Oldham, W.K. (1985). Excess biological P removal in the activated sludge process using primary sludge fermentation. In *Proc. Annual Conf. Of the Canadian Society for Civil Engineering*, Saskatoon, pp. 387–397.

Randal, C.W., Barnard J.L. and Stensel, H.D. (1992) Water Quality Management library Vol. 5. Design and retrofit of wastewater treatment plants for biological nutrient removal. Technomics ed. Lancaster USA.

Rozzi, A., Bortone G., Canziani, R., Andreottola G., Ragazzi M., Pugliese M. and Tinche A. (1995) Mesophilic and psychrophilic fermentation of primary sludge for RBCOD production in a Phostrip® plant. In *Mediterraneanchem – Int. Conf. On Chemistry and the Mediterranean Sea*, May 23–27, Taranto, Italy.

Salutsky, M.L., Dunseth, M.G., Ries, K.M. and Shapiro, J.J. (1972) Ultimate disposal of phosphate from wastewater by recovery as fertiliser. *Effl. Water Treat. J.* October, pp. 509–5019.

Seckler, M.M., Bruinsma, O.S.L. and Van Rosmalen, G.M. (1996a) Phosphate removal in a fluidised bed – I. Identification of physical processes. *Water Res.* **30**, 1585–1588.

Seckler, M.M., Bruinsma, O.S.L., and Van Rosmalen, G.M. (1996b) Calcium phosphate precipitation in a fluidised bed in relation to process conditions: a black box approach. *Water Res.* **30**, 1677–1685.

Seckler, M.M., Leeuwen, M.L.J., Bruinsma, O.S.L. and Van Rosmalen, G.M. (1996c) Phosphate removal in a fluidised bed – II. Process optimization. *Water Res.* **30**, 1589–1596.

Tanaka, T., Kawakami A., Yoneyama Y. and Kobayashi, S. (1987) *Study of the reduction of returned phosphorus from a sludge treatment process.* In *Biological phosphate removal from wastewaters* (ed. Ramadori R.), pp. 201–211, Pergamon Press (Oxford).

Traverso P., Pavan P., Innocenti L., Bolzonella D., Mata-Alvarez J., and Cecchi F. (2000) Anaerobic fermentation of source separated mixtures of vegetables and fruits wasted by supermarkets. In *Proc. 4th Int. Symp. On Environmental Biotechnology*, 10–12 April, Noordwijkerhout, The Netherlands.

Vandaele, S., Bollen F., Thoeye, C., Novembre E., Verachtert H. and Van Impe J.F. (2000) *Ammonia removal from centrate of anaerobically digested sludge: state of the art of biological methods* 1st World Water congress of the international water association, July 3–7 Paris.

Van Dijk, J.C. and Braakensiek, H. (1985) Phosphate removal by crystallization in a fluidised bed. *Water Sci. Technol.* **17**,133–142.

Webb, K.M., Bhargava S.K., Priestley, A.J., Booker N.A. and Cooney E., (1995) Struvite ($MgNH_4PO_4 \cdot 6H_2O$) precipitation: potential for nutrient removal and re-use from wastewaters, *Chemistry in Australia* **62**(10), 42–44.

Wentzel, M.C., Ekama, G.A. and Marais, G.V.R. (1992) Processes and modelling of nitrification denitrification biological excess phosphorus removal systems – a review. *Water Sci. Technol.* **25**, 59–82.

Wild D., Kisliakova, A. and Siegrist, H. (1996) P-Fixation by Mg, Co and zeolyte A during stabilization of excess sludge from enhanced sludge from enhanced biological P-removal. *Water Sci. Technol.* **34**(1–2), 391–398.

Woods, N.C., Sock, S.M. & Daigger, G.T. (1999). Phosphorus recovery technology modeling and feasibility evaluation for municipal waste water treatment plants. *Environ. Technol.* **20**(7), 663–679.

19

The case study of a phosphorus recovery sewage treatment plant at Geestmerambacht, Holland – design and operation

P.G. Piekema

19.1 INTRODUCTION

At the beginning of the 1990s, the Dutch government announced new discharge limits for municipal waste water treatment plants (MWWTPs), to comply with European agreements on reduction of nutrients in the river Rhine and the North Sea. In 1994 the following discharge limits for total phosphorus became effective: for MWWTP of <20,000 p.e. (people equivalent) there was no limit, whereas for MWWTPs of >20,000 p.e. the limit was set at 2 mg/l (average of 10 samples) and for MWWTPs of >100,000 p.e. the limit was 1 mg/l (average of 10 samples).

For total nitrogen new limits became effective in 1998: MWWTPs of <20,000 p.e. had a limit of 15 mg/l y.a. (yearly average), whereas MWWTPs of >20,000 p.e. had a limit of 10 mg/l y.a.

© 2004 IWA Publishing. *Phosphorus in Environmental Technology: Principles and Applications.*
Edited by Eugenia Valsami-Jones. ISBN: 1 84339 001 9

Due to the new discharge limits, many MWWTPs in the Netherlands had to be upgraded in the 1990s. A few of these were included in a governmental subsidy program aimed to stimulate innovative nutrient removal technologies. The program supported the development of phosphate recovery (P-recovery) technologies, and the Geestmerambacht MWWTP was selected to include a full scale demonstration plant for a combined process of biological phosphate removal and crystallization[1] of re-usable calcium phosphate.

This MWWTP is located in the province of North-Holland and is operated by the Waterboard 'Uitwaterende Sluizen in Hollands Noorderkwartier'. The design and construction of the upgraded plant was the responsibility of the consultancy firm DHV Water in Amersfoort. Apart from the P-recovery demonstration plant, the upgrading also included:

- expansion of the biological capacity from 160,000 p.e. to 230,000 p.e.;
- expansion of hydraulic capacity from 3500 m^3/h to 5000 m^3/h;
- improvement of total nitrogen removal;
- implementation of mechanical sludge dewatering.

This chapter describes the design and full scale operating results of the Geestmerambacht MWWTP, focusing on the P-recovery process.

19.2 OVERVIEW OF THE PLANT BEFORE UPGRADING

The old plant was designed as an extended aeration activated sludge system with a very simple and robust treatment scheme (Figure 19.1): screening, activated sludge treatment and final clarification. Since the food/micro-organisms ratio (F/M ratio) was

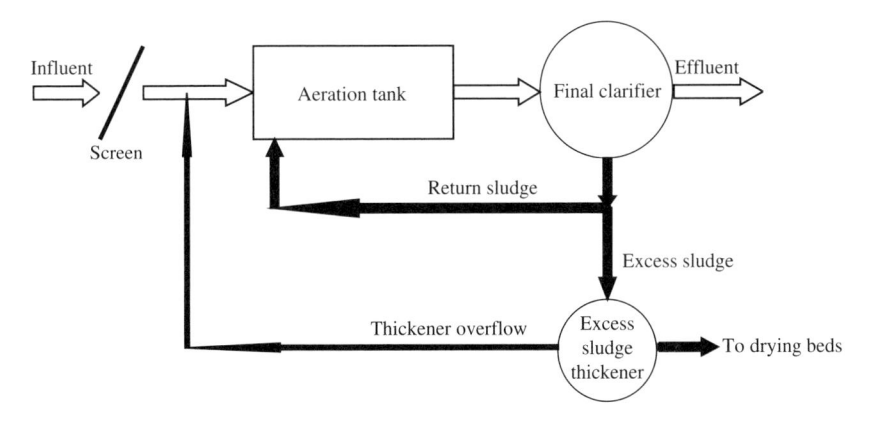

Figure 19.1 Flow scheme of the Geestmerambacht MWWTP before upgrading.

---

[1] The term crystallization is used in this chapter to indicate precipitation of a (potentially crystalline) inorganic phase.

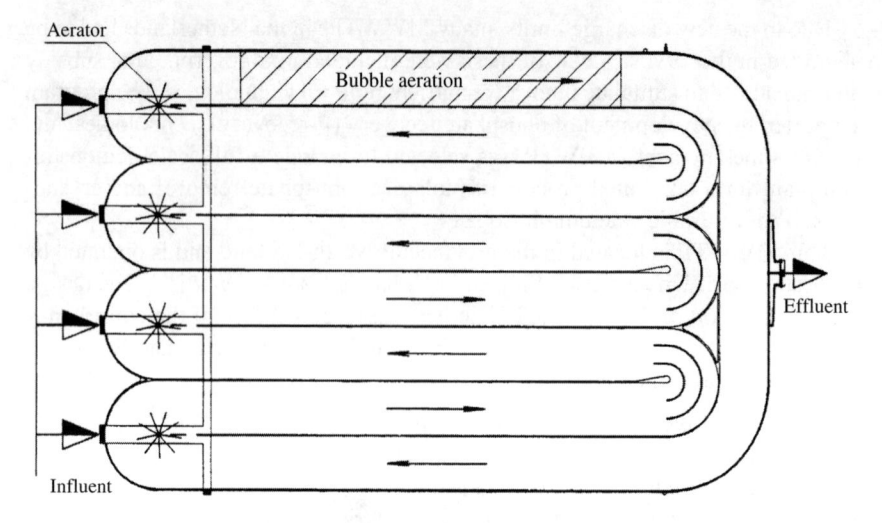

Figure 19.2   Carrousel® of the Geestmerambacht MWWTP before upgrading.

Table 19.1   Yearly average effluent quality of the Geestmerambacht MWWTP before upgrading in 1989.

Parameter	Unit	Value
COD	mg/l	58
BOD	mg/l	5
Total N	mg/l	12
Total P	mg/l	6
Suspended solids	mg/l	19

very low, the sludge was stabilized simultaneously in the extended aeration process. Excess sludge was thickened by gravity and dewatered on sludge drying beds.

The extended aeration process consisted of one 25,000 m³ oxidation ditch of the Carrousel® type, shown in Figure 19.2. The activated sludge was settled in four final clarifiers with a total settling surface of 5400 m².

The effluent quality (yearly average) in the year 1989, when the plant operated under the old design, is summarized in Table 19.1. Due to insufficient oxygenation capacity, the nitrogen content of the effluent mainly consisted of Kjeldahl-N (Kjeldahl-N is the sum of ammonia and organic N).

19.3   KEY FEATURES OF THE PLANT AFTER UPGRADING

At the start of the design process a feasibility study was executed, comparing different options for upgrading. On the basis of the results from the feasibility study,

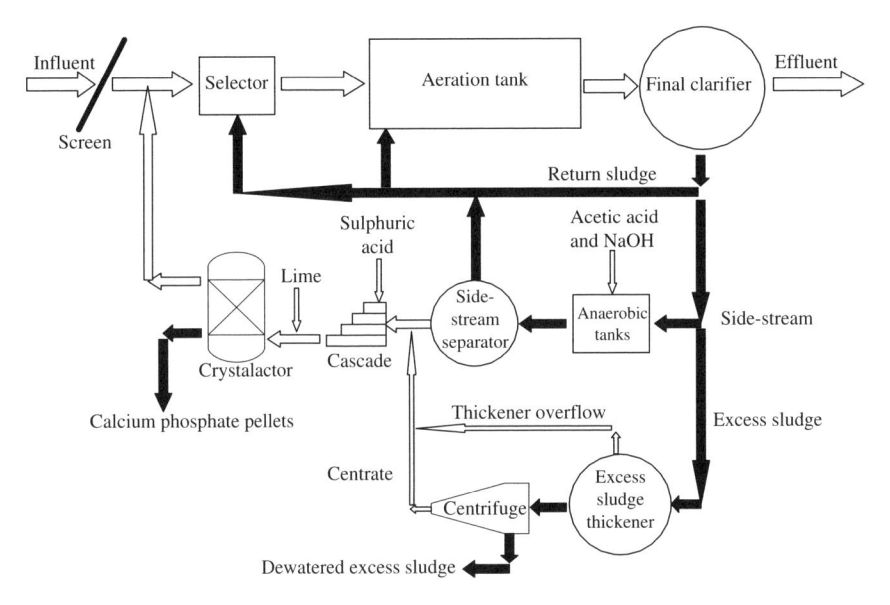

Figure 19.3 Flow scheme of the Geestmerambacht MWWTP after upgrading.

Table 19.2 Design load of the Geestmerambacht MWWTP after upgrading.

Parameter	Unit	Value
Yearly average flow	m^3/d	35,000
Daily flow during dry weather	m^3/d	25,000
Peak flow during dry weather	m^3/h	2,500
Peak flow during rain	m^3/h	5,000
Biological capacity	p.e.	230,000
COD load	kg/d	22,150
BOD load	kg/d	8,520
Total N load	kg/d	2,050
Total P load	kg/d	340
Suspended solids load	kg/d	10,500

it was decided to maintain the main features of the simple and robust extended aeration type process, with some modifications to improve the sludge settling characteristics and the nitrogen removal. The phosphate removal and recovery process would be realized in a side stream (sometimes also referred to as 'sludge line') and therefore, in principle, be independent from the other processes. Excess sludge thickening by gravity would be maintained and followed by mechanical sludge dewatering by centrifuge. Figure 19.3 shows the flow scheme of the upgraded plant. The design load of the upgraded plant is summarized in Table 19.2 and the main dimensions in Tables 19.3, 19.4 and 19.5.

Table 19.3 Main dimensions of the upgraded Geestmerambacht MWWTP: water treatment.

Process installations	Unit	Capacity (totals)
Screens	m^3/h	5,000
Selector	m^3	750
Aeration tanks	m^3	50,000
Integrated pre-denitrification zone (see Section 19.8 and Photo 19.3)	m^3	2,500
Aeration capacity	$kg\,O_2/h$	1,500
Final clarifiers	m^2	7,930
Return sludge pumping	m^3/h	5,840

Table 19.4 Main dimensions of the upgraded Geestmerambacht MWWTP: excess sludge treatment.

Process installations	Unit	Capacity (totals)
Excess sludge pumps	m^3/h	200
Gravity sludge thickeners	m^2	550
Thickened sludge pumps	m^3/h	50
Thickened sludge buffer tank	m^3	330
Dewatering centrifuges	m^3/h	50
Dewatered sludge pumps	m^3/h	10
Dewatered sludge silos	m^3	300

Table 19.5 Main dimensions of the upgraded Geestmerambacht MWWTP: P-recovery process.

Process installations	Unit	Capacity (totals)
Return sludge feed pumps	m^3/h	300
Anaerobic tanks	m^3	1,000
Sludge separator/thickener	m^2	570
Thickened return sludge pumps	m^3/h	200
Cascade feed pumps (phosphate rich water)	m^3/h	250
Cascade (CO_2 stripping)	m^3/h	250
Crystalactor® feed pumps	m^3/h	700
Crystalactors®	m^2	14

19.3.1 Description of the side-stream P-recovery process

Figure 19.3 shows a simplified flow scheme of the side-stream phosphate recovery process, as designed for the Geestmerambacht WWTP (see also Figure 19.4).

Part of the return sludge is pumped to an anaerobic tank where acetic acid is dosed to induce a quick and complete release of phosphate, via microbial activity enhancement. Subsequently, the mixed liquor is separated in a gravity

Figure 19.4 Side-stream P-recovery installation at Geestmerambacht. 1. Anaerobic tanks; 2. Sludge separator/thickener; 3. Cascade for carbonate stripping; 4. Crystalactors®, 5. Storage of sand; 6. Storage of lime; 7. Selector (is part of main water treatment).

separator/thickener into two fractions, a thickened sludge and a supernatant. The thickened sludge is returned to the aeration tank where it takes up phosphate. The supernatant is treated in an acidic stripping cascade to remove carbonates, in order to prevent them from disturbing the crystallization process. Subsequently the supernatant, containing a high concentration of phosphate, is treated with lime, and calcium phosphate is crystallized on sand pellets in a Crystalactor®. The treated supernatant is returned to the aeration tank.

19.4 DESIGN PHILOSOPHY OF THE P-RECOVERY PROCESS

19.4.1 Choice of recovered product

At the time the design was made (1991), most phosphate removal processes in MWWTPs were of the chemical type, which involved dosing of iron and aluminium salts. The reuse potential of the iron or aluminium phosphate precipitates was considered low, due to their perceived low solubility. It was therefore concluded that in order to produce a product with high recovery potential calcium phosphate should be produced. The possibilities of magnesium phosphate or MAP (magnesium ammonium phosphate or struvite) were relatively unknown at that time.

The precipitation of calcium phosphate is a process primarily governed by pH. Unfortunately, at higher pH, the process always results in the additional formation of large amounts of calcium carbonate, as in normal municipal waste water the total concentration of inorganic carbon (mostly bi-carbonate) is much higher than the concentration of phosphate.

To enhance the recovery potential, it was decided that crystals rather than sludge should be produced, and that impurities should be avoided. Therefore the Crystalactor® was selected as the chemical P-recovery process, preceded by removal of carbonate to prevent dilution of the product and disturbance of the crystallization by the precipitation of calcium carbonate.

19.4.2 Biological pre-concentration of phosphate

At the time the new Geestmerambacht MWWTP was being designed, one other full scale Crystalactor® and cascade were in operation on the effluent of a small MWWTP in the Netherlands. There was, thus, experience with the removal of carbonate, and crystallization of calcium phosphate, although the main disadvantage of the process was that for the treatment of all effluent (with low phosphate concentration 4–6 mg P/l) large dimensions of the treatment units and high chemical usage were needed. Also, to produce low final effluent P-concentrations, a filtration stage followed the Crystalactor® to remove the 'carry-over', which represents an extra cost. Carry-over is a term used to describe the fine particles that form spontaneously in the water and do not attach to the seed material. These fine particles typically represent 5–15% of the removed phosphate, and tend to be washed out of the Crystalactor®, although they can easily be removed from the effluent by sand filtration.

For the Geestmerambacht MWWTP the following new philosophy was used: The biological phosphate process allows complete phosphate uptake during the aerobic phase in the main water treatment line, without the use of chemicals. The final effluent quality is therefore determined by the biological phosphate uptake. The principle of phosphate release during the anaerobic phase can be used to produce a concentrated phosphate flow. Chemical treatment of this side stream allows smaller treatment units and lower chemical usage. Also, a high phosphate removal efficiency and removal of carry-over are not necessary, as the treated side-stream can be returned to the main water line.

In order to reach the highest phosphate concentrations possible, anaerobic treatment of return sludge (higher dry solids concentration than in the aeration tank) was considered as the best option. Commercially available acetic acid was selected as feed in the anaerobic phase, as it induces a fast and complete phosphate release and does not dilute the side stream. In principle other fatty acids can also be applied, as long as these are in a concentrated form. Some form of dosing fatty acids in the side stream anaerobic phase is essential, as it is also the most important step in the process where the growth of PAOs (Phosphate Accumulating Organisms) is stimulated. To reduce investment, it was decided not to incorporate any large anaerobic tank in the main water treatment.

19.4.3 Phosphate balance

The balance of phosphate loads in the WWTP is an important tool in the design of the side-stream process. All process units influence the P-balance and flows that have to be treated by the side-stream. Phosphorus levels in the effluent will depend on the efficiency of the Crystalactor®, efficiency of sludge separator/thickener, level of phosphate 'luxury' uptake by the biomass, phosphate removal by the excess sludge and phosphate release in the excess sludge treatment.

The side-stream process has two 'sinks' for phosphate: the excess sludge and the Crystalactor® (or other chemical treatment). It is possible to balance and to manipulate these two sinks. In the case of the Geestmerambacht MWWTP the objective was to remove phosphate with the Crystalactor® as much as possible, in view of the reuse prospects of the calcium phosphate pellets.

The new excess sludge treatment included gravity-thickening and mechanical dewatering by means of decanters. Phosphate is released due to the anaerobic conditions in the excess sludge-thickeners (no lime or metal salts are dosed in the sludge treatment). This released phosphate causes high phosphate concentrations in the overflow of the excess sludge-thickeners and particularly in the centrate of the decanters. These flows are suitable for treatment in the Crystalactor® economically, as there was no dosing of acetic acid needed to induce these releases. The design therefore included pumps and piping to transport the thickener overflow and centrate to the side-stream P-recovery process.

19.4.4 Tests executed during design

Since this was the first full scale plant consisting of combined biological P-concentration and Crystalactor® P-recovery, a number of tests were needed to support the design:

1. Biological pilot test at Wageningen University to determine the effectiveness of biological phosphate removal in the main water treatment, anaerobic residence times, acetic acid dosing requirements, and behaviour under varying hydraulic and phosphate loading (Rensink *et al.*, 1991).
2. Thickener tests, full scale (at other MWWTP) and bench scale (application of solid flux theory), to determine the relation between solid loading rate on the thickener and suspended solids (SS) concentration of the thickened sludge (not published).
3. Pilot test with acid stripping cascade to determine the necessary dimensions of the cascade for the required carbonate removal (not published).
4. Pilot test with 2 cm Crystalactor® at the Geestmerambacht MWWTP, with phosphate enriched effluent from the old biological treatment. This was necessary to check any disturbance that may be caused to the Crystalactor® by the content of a biologically treated effluent, and to determine crystallization efficiency, loading rate, effect of circulation and production of carry-over (STOWA, 1992).

19.5 DESIGN OF THE BIOLOGICAL PART OF THE SIDE-STREAM

19.5.1 Design of phosphate mass balance

Figure 19.5 shows the design phosphate mass balance for the Geestmerambacht MWWTP. The balance is expressed in loads (kg P/d) and includes both dissolved phosphate loads and phosphate loads bound to the sludge. The design process starts with an overall balance, to determine the P-load that has to be removed by the Crystalactor® in the side-stream. From that point onwards the side-stream P-balance is calculated 'back to front'.

Step 1: Overall balance

Recovered P-load = Influent load − effluent load − P load in dewatered sludge

The overall balance should be made on the basis of seasonal average loads. Peak loads do not have to be included as the biological phosphate uptake has substantial buffering capacity. Slower variations as differences between winter and summer, (i.e. in excess sludge production) or seasonal fluctuations in influent load (due to industrial seasonal loads or tourist influx) have to be taken into account. The

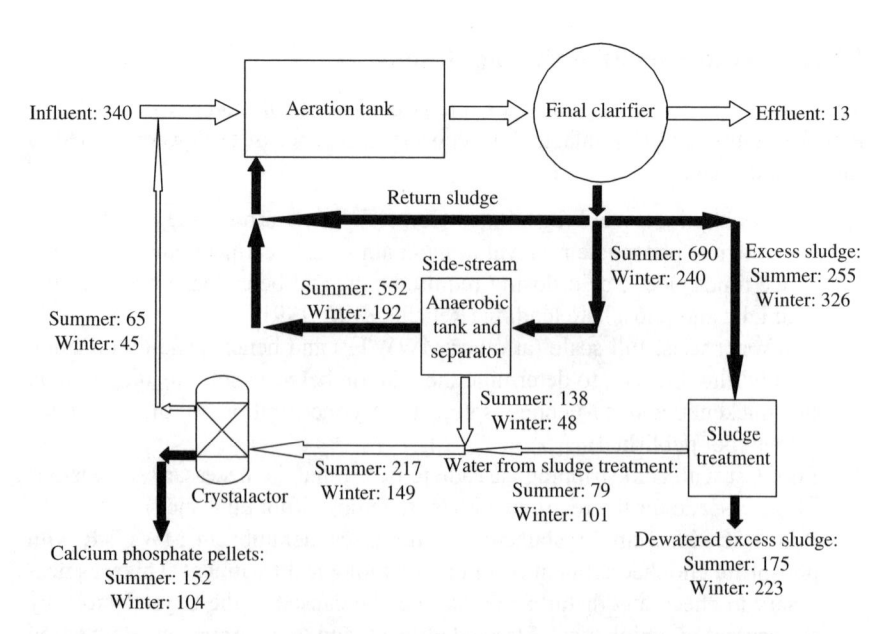

Figure 19.5 Simplified phosphate balance for the Geestmerambacht MWWTP (expressed in loads kg P/d and including both dissolved phosphate loads and phosphate loads bound to the sludge).

design influent load for the Geestmerambacht MWWTP was found to be stable at 340 kg P/d and not show any seasonal variations.

The effluent load depends on the discharge limits. In the Netherlands these are defined as average of 10 samples. For large MWWTPs as the one at Geestmerambacht, effluent is analyzed every day. From full scale experience at other MWWTPs in the Netherlands (applying chemical phosphate removal) it was known that in order to meet a 10 day moving average of 1 mg P/l at all times, the yearly average has to be around 0.5 mg P/l. Using a average dry weather flow of 25,000 m^3/d, the effluent P-load is calculated at 13 kg P/d.

The last component in the mass balance is the P-load removed from the plant by the dewatered sludge. The production of excess sludge depends on temperature and therefore can vary considerably between winter and summer. Many activated sludge models in use in different countries can calculate the excess sludge production. In the Netherlands the calculation method originally described by Chudoba et al. (1985) is popular. For Geestmerambacht, the design excess sludge production was calculated as max. 10,850 kg SS/d (in winter) and min. 8,500 kg SS/d (in summer).

Sludge retention in the gravity thickeners is around two days. From measurements at other MWWTPs with biological phosphate removal, and also from the biological pilot test, it was clear that during such long retention time, virtually all 'luxury uptake' is released. The P-content of the dewatered sludge is therefore only the content related to normal (non PAO) biological growth. An average content of 23 g P/kg SS was deduced from the operational data of the old Geestmerambacht MWWTP. For calculation purposes, a content of 20 mg P/kg SS was used, giving the results for the two phosphate exits, dewatered sludge and pellets, as shown in Figure 19.5.

Step 2: Efficiency of Crystalactor®

High efficiency of the Crystalactor® in the side-stream has a positive effect on the dimensions of the process units up-stream and on the chemical usage there. Nevertheless, high efficiencies cause a higher cost of the Crystalactor® itself. There is a minimum SS load to be treated in the side-stream to assure sufficient selective pressure for the PAOs (see Step 7). For the side-stream Crystalactor® design the balance between economics and technology lies at an efficiency of 70–80% (see Section 19.6.2). The remaining phosphate is returned to the main water treatment. For the design, an efficiency of 70% was used. For example, in order to reach a removal in the Crystalactor® of 152 kg P/d in summer, the P-load to the Crystalactor® has to be: 152/0.7 = 217 kg P/d.

Step 3: Phosphate release in the excess sludge treatment

The overflow (supernatant) from the separator/thickener in the side-stream does not have to supply the full P-load for the Crystalactor®, if the load from the sludge treatment is used beneficially. The released P-load depends on the processes used in the sludge treatment (e.g. digestion, iron/lime dosage, aeration). Measurements on other full-scale installations with gravity excess sludge thickeners and mechanical dewatering showed virtually complete release of 'luxury uptake'. A large part of the released phosphate remains in the thickened sludge layer causing higher phosphate concentrations in the centrate than in the thickener overflow. Variations frequently occur, but as an order of magnitude estimate, it may be assumed that half of the released phosphate is present in the thickener overflow and the other half in the centrate. To cope with the variations and uncertainties, facilities were installed to transport both flows (thickener overflow and centrate) to the side-stream in a controlled manner (variable speed pumps).

The P content of the sludge in the aeration tank is another important factor determining the amount of P release in the sludge treatment, as well as the P release in the side-stream, and to a certain extent, the effluent quality (see step 7). For Geestmerambacht the design value for the P content in the aeration tank sludge was set at 30 g P/kg SS (see Step 6). An amount of 10 g P/kg SS represents the released P-loads as mentioned in Figure 19.5. One can calculate, using a SS concentration in the thickener from 0.6% to 2.5%, and in the centrifuge from 2.5% to 20%, that the expected P concentration in the thickener overflow is around 40 mg P/l and in the centrate around 120 mg P/l. These values are in agreement with measurements from full-scale installations with a functional biological phosphate removal.

Step 4: P-load from the side-stream separator/thickener

The difference between the P-load needed in the Crystalactor® and the P-load delivered from the sludge treatment is the load that has to be produced in the overflow/supernatant of the side-stream separator/thickener. One can see that the production in the sludge treatment is an important part of the total load, even in the most critical design situation (summer): around 35% of the needed P-load to the Crystalactor® is produced by the excess sludge treatment.

Step 5: P-load from the side-stream anaerobic tanks

The dissolved P-load to be produced in the side-stream anaerobic tanks depends on the hydraulic and thickening efficiency of the side-stream separator/thickener. In the case of Geestmerambacht the separator was designed on a 50 kg SS/(m^2 · d) basis, producing a thickening result of 1.5% SS in the thickened sludge (see also Section 19.5.3). The average SS concentration in the feed to the separator/thickener is 0.6% SS. Therefore, the flow of thickened sludge is 0.6/1.5 = 40% of the feed flow, resulting in an overflow/supernatant equal to 60% of the feed flow. This is called the hydraulic efficiency of the separator/thickener, which is equal to the efficiency of phosphate transported to the Crystalactor®. For example, the dissolved P-load to be produced in the anaerobic tanks in summer has to be: 138/0.6 = 230 kg P/d.

The hydraulic efficiency can be improved by choosing a low dry solids load on the separator/thickener producing a higher SS concentration in the thickened sludge. However, too low loads cause long sludge retention times in the separator/thickener, which can cause problems with odor or floating sludge (anaerobic sludge enriched by acetic acid). A low loading rate also presents an extra investment cost in a large separator/thickener. Therefore normally the separator functions as a high loaded thickener, or an intermediate unit between a final clarifier and an excess sludge thickener. The separation efficiency can also be improved by choosing mechanical thickening instead of a gravity separator. However, the effect of suspended solids in the overflow and the effect of polyelectrolyte in both overflow and return sludge has to be considered.

Step 6: P-load, SS-load and hydraulic flow to the side-stream anaerobic tanks

The biological pilot tests led to the conclusion that P-release from the sludge into the side-stream, with dosing of acetic acid, varied between 5 and 15 g P/kg SS. A higher release would lead to a higher P-content in the sludge in the aeration tank. A high P-content in the aeration tank signifies that the PAOs are working hard (fully loaded with P). This may cause higher phosphate concentrations in the effluent by overload of the PAOs, which also can cause a higher level of P-release in the final clarifier. With this in mind, a P-release of 10 g P/kg SS was chosen, giving a P-content of 30 g P/kg SS in the aeration tank and 20 g P/kg SS after complete release (P-content due to normal biological growth). On the basis of these assumptions, the amount of SS to be treated in the side-stream can be calculated. In summer, the dissolved P-load from the anaerobic tank averages at 230 kg P/d. The total P-load, bound to the sludge, in the feed to the anaerobic tank is therefore 230 * 30/10 = 690 kg P/d. The sludge to be treated follows from 690 * 1000/30 = 23,000 kg SS/d.

Finally, the average SS concentration in the return sludge is needed to define the hydraulic flow to the side-stream. This depends on the flow and its variations of the influent in the main water treatment and the return sludge and should be evaluated in the design of the activated sludge system. An evaluation of the flow variations for Geestmerambacht resulted in a SS concentration in the return sludge varying over the day from 5 to 8 g/l, with 24 hours average 6 g/l (sludge content in the aeration tank is 4–5 g/l). An optimal automatic control of the return sludge flow should be used to increase the SS content and therefore also the dissolved phosphate concentrations in the side-stream as much as possible. However, the formation of a sludge blanket in the final clarifiers to increase the SS content of the return sludge should not be applied (see also Section 19.7). For design purposes, the safe value of 6 g/l was used. The flow to be treated in the side-stream during summer is calculated as 23,000/6 = 3830 m^3/d (=160 m^3/h). The average concentration of dissolved phosphate after the anaerobic tanks is calculated as 6 g SS/l * 10 mg P/kg SS = 60 mg P/l.

Step 7: Calculation of all SS loads and hydraulic flows and selective pressure on PAOs

From the described P-balance, P-release in the anaerobic tanks and SS concentration, the SS loads and hydraulic flows in all process units of the side-stream can be calculated. Since Geestmerambacht was the first full-scale installation, an extra safety margin was incorporated: design summer feed flow was set at 200 m^3/h and the maximum available feed pumping capacity to the side-stream at 300 m^3/h. These additional facilities are useful in any WWTP where deviations or variations from the design influent conditions may be expected.

Finally, the lower limit of feed flow to the side-stream is not only defined by the P-balance in winter, but also by the criterion that a minimum selective pressure has to be present to secure sufficient PAOs in the sludge. This however, is more difficult to calculate than the phosphate balance. On the basis of early results from the Pho-strip™ process in the USA, it was concluded that the lower flow limit should be 10% of influent flow (= 0.1*25,000 = 2500 m^3/d or approximately 100 m^3/h). The biomass has to pass a minimum number of times through the side-stream anaerobic tanks before being wasted as excess sludge. This can be expressed as a ratio between the SS load (in kg SS/d) treated in the side-stream anaerobic tanks, and the excess sludge production (in kg SS/d). A clear design value for this ratio is not available from research. However, as a rule of thumb, the ratio should be at least one (i.e. the biomass passes once through the side-stream before being wasted), so around 11,000 kg SS/d should be treated in the side-stream during winter (= 76 m^3/h).

For the design presented here, it was decided to set the flow to the side-stream as variable, between 100 and 300 m^3/h, and average design value at 200 m^3/h. On this basis, the P-load balance had to be recalculated for winter time, because the values for winter in the side-stream, as mentioned in Figure 19.5, are too low. For the minimum flow of 100 m^3/h during winter, one can recalculate by trial and error that the P-content in the sludge of the aeration tank will be around 28 g P/kg SS. The P-load in winter from the excess sludge treatment is reduced to around 80 kg P/d, while the P-load from the separator/thickener increases to around 70 kg P/d.

19.5.2 Design of side-stream anaerobic tanks

The biological pilot tests led to the conclusion that P-release with dosing of acetic acid is complete within 2.0–2.5 hours. Some extra time may be needed to remove nitrate if present in the return sludge. To ensure sufficient time in all situations, a design residence time of 5 hours at average design flow was set, for a total volume of 1,000 m^3. The residence time at maximum flow was set at 3.3 hours. The total volume was divided in two round tanks of 500 m^3, which was a cost effective choice, but also gave the possibility of using the tanks in parallel as well serial (for semi plug flow). After start-up during full-scale operation, it was concluded that there is no measurable difference in operation between parallel or serial operation of the two tanks.

The pilot tests also indicated that with an acetic acid dosing ratio of 20–30 g COD/kg SS a P-release of 10 g P/kg SS could be reached in a limited time. For design purposes, the average dosing ratio was set at 25 g COD/kg SS and the maximum dosing ratio at 50 g COD/kg SS. The maximum value may be needed if high nitrate levels were present in the return sludge. Normally however, this will not be the case, as the main water treatment should be designed for complete and reliable nitrate removal (see Section 19.8).

In the Netherlands, commercially available sodium acetate has a similar cost to separately purchased acetic acid and caustic soda. Since strict pH control is absolutely necessary in the anaerobic tanks, it was decided to apply acetic acid as P-release nutrient, and caustic soda separately, for pH-control. The strict control of pH is necessary because at pH lower than 6.5 the rate of P-release is reduced, while at pH higher than 7.5 the presence of calcium (commonly present in waste water) and high concentration of phosphate may cause precipitation of calcium phosphate in the anaerobic tanks. This is undesirable, as the phosphate has to remain dissolved until it reaches the Crystalactor®. The risk of precipitation in the anaerobic tanks also prohibits the use of lime at that point, for pH control, as this would increase the concentration of calcium.

In the detail engineering of the chemical dosing, the basic requirements of chemical handling and dosing have to be carefully considered (e.g. safety, freezing risk, heat generation, mixing requirements, control equipment).

19.5.3 Design of separator/thickener

Table 19.6 shows the results from full scale and bench scale tests for the separation and thickening of the anaerobic sludge. At the time of design, tests were needed since gravity thickeners were being used in the Netherlands only at low loading rates, but no data were available on loading rates in the magnitude of 40–100 kg SS/(m²·d).

From these data, and from preliminary investment estimates, it was concluded that a design solids loading rate of 50 kg SS/(m²·d) would be suitable at the average design flow, resulting in stable operation and not increasing the chemical usage to high values if maximum flow conditions occur (solids loading rate at maximum flow = 75 kg SS/(m²·d)).

From pilot tests of the Crystalactor® at Geestmerambacht it was known that suspended solids in the water higher than 250 mg/l disturb the crystallization

Table 19.6 Relation between SS loading and thickening result from tests.

SS loading rate [kg SS/(m²·d)]	Thickening result [g/l]
25	20–25
50	15
75	12
100	10

process, leading to precipitation of calcium phosphate on the suspended solids rather than on the seed material. Therefore, several extra measures were incorporated in the design to safeguard the Crystalactor® from high levels of suspended solids:

- The inlet construction is adjustable in height. This provides the possibility of filtration of the incoming flow in the sludge layer.
- A scum baffle and skimming device is present, to prevent floating matter from escaping with the overflow.
- The height of the sludge layer is controlled automatically by means of a sludge blanket detection unit. This instrument adjusts the thickened sludge pump capacity.
- The turbidity of the supernatant is measured continuously. The turbidity monitor controls a valve which can bypass the Crystalactor® if necessary.

The presence of a gravity separator/thickener proved to be a useful feature, to act as sludge separator before chemical treatment by the Crystalactor®. At another MWWTP, a centrifuge was designed to allow separation of the sludge before the Crystalactor®, and this proved to be very difficult to operate in a stable way. The centrifuge was also used to thicken the sludge as much as possible, as part of the sludge was transported as excess sludge to a digester with capacity problems. The two different objectives (clean, solids-free centrate and high degree of thickening) were very difficult to combine without the addition of polyelectrolytes. Focusing on a clean centrate produced a low thickening degree, whereas achieving a high thickening degree resulted in high solids in the centrate. Dosing of polyelectrolyte improved the performance of the centrifuge, but the presence of polyelectrolyte proved to disturb the crystallization process (in the form of a high proportion of carry-over).

19.6 DESIGN OF CRYSTALACTOR® IN THE SIDE-STREAM

19.6.1 Carbonate stripping cascade

Carbonate removal in a cascade by stripping of carbon dioxide at low pH is a well-known and robust process, used mainly in drinking water production. At Geestmerambacht, where it is applied in treated waste water with high phosphate concentrations, it was decided to execute a pilot test to determine the optimal pH, the number of steps required and falling height in each step. The flow to be treated and the concentration of total inorganic carbon can vary within a wide range:

- Flow between 40 and 250 m^3/h (average design value 150 m^3/h);
- Total inorganic carbon between 3 and 10 mmol/l (=300–1000 mg/l as $CaCO_3$).

The pilot tests with the Crystalactor® indicated that the total inorganic carbon should be kept under 1.5 mmol/l to avoid disturbance during crystallization. For the design of the cascade, an effluent requirement of 1.0 mmol/l of total inorganic carbon was used, to be met also under maximum hydraulic loading and maximum influent carbonate concentration.

Pilot tests of the cascade led to the conclusion that the most economical situation would be:

- pH equal to or lower than 5;
- Falling height in each step around 0.2 m;
- Depth of each basin around 0.15 m;
- Maximum overflow rate 150 m^3/h per meter weir;
- Carbon dioxide removal efficiency around 10% in each step.

Figure 19.6 shows the relation between the number of steps and the carbon dioxide concentration. From the figure, one can conclude that around 20 steps are sufficient to reduce the inorganic carbon content from 10 mmol/l to 1 mmol/l. For the design, a total number of 25 steps was chosen.

From the separator/thickener, the overflow is collected in a pump well, pumped through an in-line mixing unit, where acid is dosed, to the top of the cascade. The cascade is covered to prevent odour problems that may arise from the formation and stripping of hydrogen sulphide, and also to prevent the growth of algae. To design the ventilator, a ratio of 10 to 1 air to water was used. The exhaust air was discharged in the open, but could be connected, if needed, to one of the existing biological air filters. This never proved necessary during full-scale operation.

For acid dosing, sulphuric acid (96%) was selected, as concentrated and cost effective acid source. The amount needed can be calculated through normal acid-base chemical balance calculations, taking into account that the inorganic carbon and the phosphate ions constitute the main acid–base buffer.

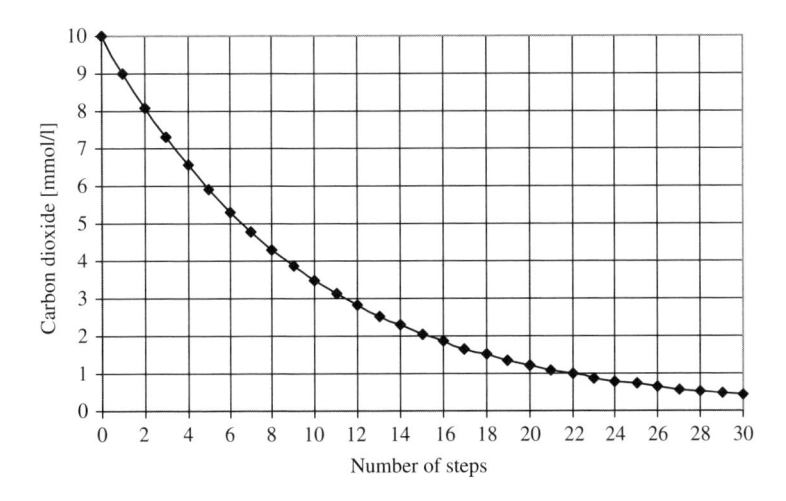

Figure 19.6 Relation between number of cascade steps and effluent CO_2 concentration.

19.6.2 CRYSTALACTOR®

19.6.2.1 Principle

Figure 19.7 shows the design principle of the Crystalactor®. The reactor consists of a cylindrical vessel, partially filled with a suitable seed material, e.g. filter sand. The water is pumped through the reactor in an upward direction, and at such a high velocity (30–50 m/h) the pellet bed is kept in a fluidized state. A sharp phase separation exists between the fluidized bed and the supernatant. At the bottom of the reactor, influent and chemicals are injected through separate nozzles. By introducing sufficient resistance over these nozzles, the required even distribution of water and chemicals over the base is obtained. The high level of turbulence caused by the horizontal outflow of water from the nozzles is used for the mixing of water and chemicals.

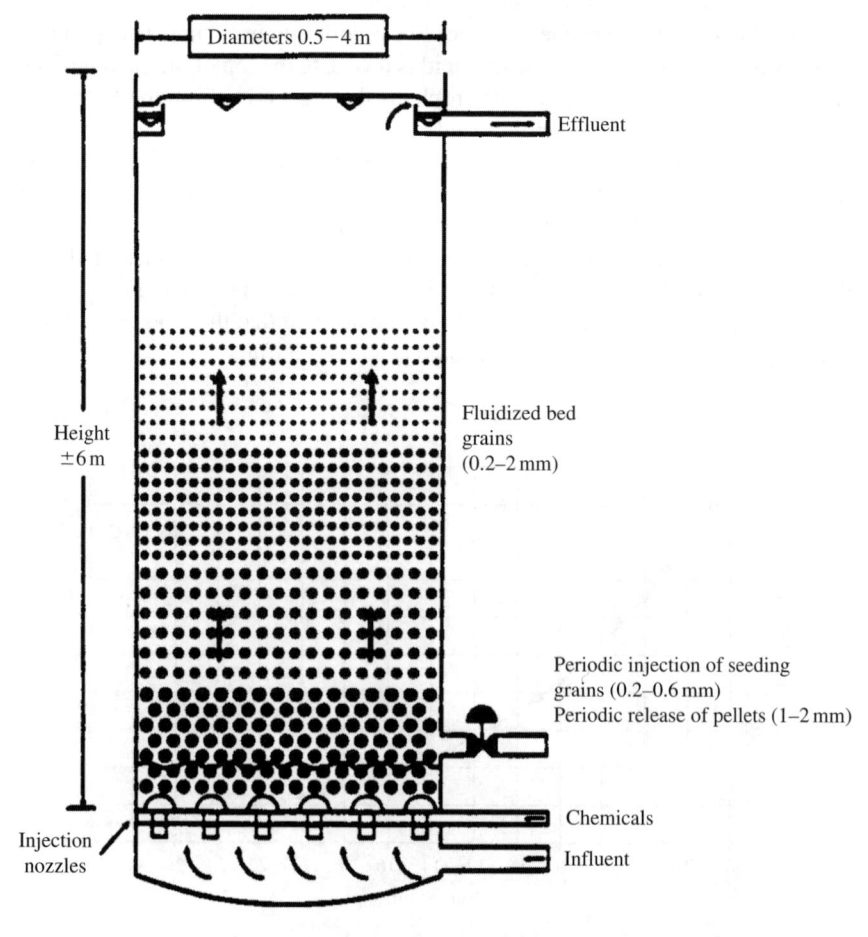

Figure 19.7 The design principle of the Crystalactor®.

The fluidized bed provides a large crystallization surface, and in a fast reaction calcium phosphate crystallizes on the seed material. The pellets grow and the heavier pellets accumulate at the base of the reactor where they can be removed periodically and replaced by smaller-diameter seed material. As only a fraction of the pellets is removed, this procedure can take place during full operation. More information about the fundamentals and theory of the Crystalactor® can be found in van Dijk and Wilms (1991) and van Dijk and Braakensiek (1984).

The two main advantages of crystallization in a fluidized bed compared with precipitation are:

- The fluidized bed provides a very large crystallization surface enabling the process to operate at a high rate, while at the same time cementation of the pellets is prevented. Consequently, the system is compact.
- The process conditions are chosen in such a way that the ions crystallize directly from the water onto the crystalline substrate. In this way, pure, almost water free pellets are produced (dewatering of the pellets is not necessary). Impurities like organics, suspended matter and other ions are hardly incorporated in the pellets, making industrial reuse attractive (P-content of the pellets: around 10%).

19.6.2.2 Results from pilot testing

Pilot trials are always necessary to test a crystallization process using the actual effluent to be treated, since many different components in the effluent may have an unexpected influence on the process. The driving force for the crystallization of calcium phosphate is pH, and the saturation with respect to the two components, phosphate and calcium ions. Since the phosphate ion is the one to be removed, calcium ions need to be present in excess. The needed excess of calcium ions and optimal pH can vary with the influent phosphate concentration and other characteristics of the water to be treated and is not completely predictable from theoretical models. The tests on Geestmerambacht showed:

- relatively low optimal pH: 8.0 to 8.5;
- Ca/P ratio: 2 : 1 to 3 : 1 mol/mol.

These values have to be met in the effluent of the Crystalactor® (a pH monitor for pH control is installed in the effluent line or in the reactor above the fluidized bed).

Apart from these essential effluent conditions (after crystallization), the conditions at the bottom of the Crystalactor® are important, since the initial crystallization occurs there. A high (local) value of supersaturation will promote spontaneous precipitation and produce a large amount of carry-over. Therefore, a high degree of initial mixing of inlet water and chemical (lime suspension) has to be attained at the bottom. The mixing cannot be realized by turbulence alone, as a very high degree of turbulence causes shearing of the pellet crystals (see Mangin and Kline, this volume). Several specially shaped nozzles to inject either the influent or chemicals are distributed over the surface of the flat bottom to ensure sufficient mixing.

Even with sufficient mixing, the crystallization of calcium phosphate is a sensitive process, and the pilot tests revealed that the supersaturation at the bottom of the reactor should be kept low to avoid a high production of carry-over. A low level of supersaturation at the bottom, in combination with a high influent phosphate concentration, leads to a low phosphate crystallization efficiency in a once-through situation (without re-circulation of effluent). Circulation of effluent over the reactor reduces the phosphate influent concentration and allows several passes for the water to be treated through the fluidized bed, which improves the removal efficiency (see Figure 19.8). With an effluent circulation ratio of 2 to 3 a crystallization efficiency of 70–80% could be reached. Circulation ratio is defined as circulated flow divided by influent flow. An effluent circulation ratio of 2 means that the flow through the reactor is three times the influent flow (influent flow + circulation flow = 1 * influent flow + 2 * influent flow = 3 * influent flow).

Despite the beneficial effect of recirculation of effluent on the crystallization efficiency, circulation poses a risk in carry-over to be passed through the reactor several times, presenting extra surface for new precipitation and build-up of amorphous material. This can be prevented by filtration in the circulation line, but this is costly. The pilot test showed that the circulation flow could be mixed with the acidic influent flow before being fed to the reactor. Due to the acid present in the influent, the carry-over in the circulation flow is dissolved rapidly, within 30 seconds, as long as the pH of the mixture is lower than 7.5.

19.6.2.3 Design

Considering the conditions at Geestmerambacht, a crystallization efficiency of 70–80% was considered appropriate. As a precaution, a circulation factor of 3 was set, giving an average design influent flow of 150 m^3/h, and a total flow to be

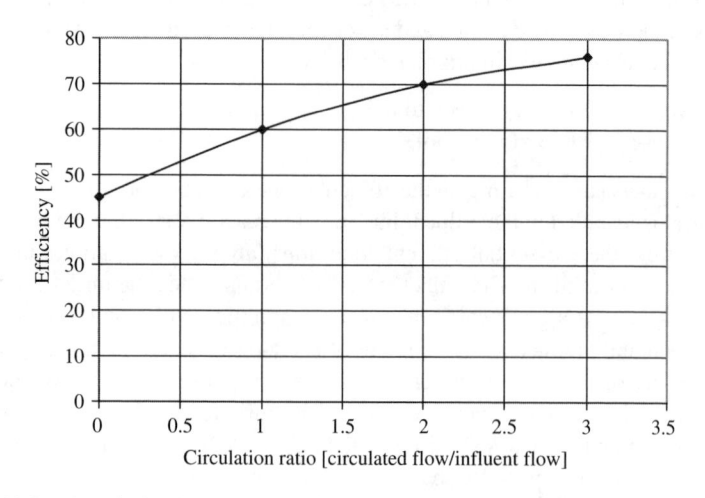

Figure 19.8 Correlation between circulation and crystallization efficiency.

treated of 600 m³/h. At a maximum influent flow of 250 m³/h, a circulation factor of almost 1.5 could still be maintained.

Best test results with respect to fluidization, shear of pellets and turbulence, were reached at a hydraulic surface load of around 40 m³/(m²h). This defines the total seed material surface area required as $600/40 = 15\,m^2$.

The available diameters designed into the Crystalactor® increase from 1.0 meter with steps of 0.25 m up to a maximum of 3.5 m. For practical reasons, a two-reactor set-up was chosen, each with a diameter of 3.0 m, giving a total surface of 14 m². The height of the fluidized sand bed was designed to be approximately 4 m (or approximately 2 m in fixed state). The total height of the reactor was around 8 m.

19.6.2.4 Dosing of lime

Lime is commercially available as stable suspensions, as lime powder and as quicklime powder. For small quantities, the suspension is the most cost effective choice, as no large storage and suspension production facilities are needed. For larger requirements (as in Geestmerambacht) lime powder is the most cost effective choice. Quicklime would be even more cost-effective, but is more difficult to handle and produces particles in suspension, which are not fine enough for the fast crystallization reaction. The lime powder should therefore be of a fine and high quality type (purity higher or equal to 93%). Lime powder that is suitable for drinking water treatment is a reliable choice.

Production of lime suspension, dilution and dosing is always a process that needs attention, as it frequently causes clogging problems in the dosing lines and dosing equipment. To reduce these problems, the lime dosing nozzles in the reactor can be removed from the reactor and cleaned during normal operation of the reactor, and the connections between the dosing nozzles and dosing lines are made from flexible material. Using exclusively filtered and decarbonated effluent from the MWWTP as suspension, and adding dilution water, prevents precipitation of calcium carbonate in the dosing lines and equipment. The decarbonation of the dilution water is realized in the same manner as for the influent of the Crystalactor®: dosing of sulphuric acid and stripping of carbon dioxide in a small scale packed column with intensive air–liquid contact. By chemical calculations the risk on precipitation of other components in the dosing equipment (e.g. calcium sulphate) has to be examined.

During full scale operation, the use of stripped water for the lime suspension and dilution has proved to be a successful way of making lime dosing reliable and operator friendly.

19.6.2.5 Seed material and phosphate pellets

During full-scale operation of the reactor, the larger pellets can be removed close to the bottom of the reactor, and new seed material introduced at the top of the reactor.

Figure 19.9 Calcium phosphate pellets (diameter 0.7–1.0 mm). [This figure is also reproduced in colour in the plate section after page 336.]

For seed material different options are available. Mostly normal filter sand is used as the most cost effective option. The diameter of filter sand should range between 0.2 and 0.45 mm. The dosing equipment includes a vibrating conveyor and a washing unit, which removes the small fraction of light material always present in the filter sand before it enters the reactor.

The phosphate pellets can grow up to a diameter of around 0.8 mm. The pellets flow under gravity into a filtering container, where they are drained under gravity (see Figure 19.9). After draining the typical content is as follows (in percentage of weight):

- moisture: 10%;
- seed material (filter sand): 30%;
- calcium phosphate: 50% (of which 20% calcium, 20% oxygen and 10% P);
- other (mostly Mg and carbonate): 10%.

19.7 CONTROL OF SLUDGE SETTLING

In order to reach low effluent phosphate concentrations, while the phosphate content in the sludge is high, it is of importance to:

- achieve a low content of suspended solids in the effluent;
- prevent P-release in the final clarifier;
- maintain good sludge settling properties.

The first two demands have led to a safe design with respect to the choice of final settling surface (low loaded: maximum $0.65\,m^3/(m^2 h)$; able to handle high sludge contents and high sludge volume indices) and return sludge capacity (flexible flow from 2000–5800 m³/h). The last demand has led to the incorporation of a selector in the main water treatment. In the selector, a high loading of organics from influent on the sludge is maintained (100–200 g COD/kg SS) for a short period of time (10–15 minutes). This is a reliable way to stimulate the floc forming bacteria to outcompete most types of filamentous bacteria present in the biological treatment of municipal waste water. The selector is of the anaerobic type for two reasons:

- An aerobic selector would remove easily degradable COD that is needed for denitrification;
- A pure anaerobic selector (also virtually no nitrate present) stimulates phosphate release and therefore the presence of PAOs.

Even though the residence time in the selector is much shorter than in 'normal' anaerobic tanks in the main stream of a biological phosphate removal process, it is believed that the presence of fatty acids from the influent under anaerobic conditions still plays an important role in the selective pressure in favour of the PAOs. The anaerobic selector is therefore in an essential part of the side-stream P-recovery process, both for stimulating PAOs, and maintaining favourable sludge settling properties.

19.8 CONTROL OF NITROGEN REMOVAL

In order to comply with total nitrogen discharge limits, the MWWTP of Geestmerambacht was designed for nitrification and denitrification. In the resulting low-loaded activated sludge system, nitrification is normally complete, as long as sufficient oxygen supply is maintained. Complete nitrate removal is more difficult to maintain under all circumstances, as it is affected by oxygen input (too much oxygen causes high nitrate levels) and sufficient availability of COD under denitrifying conditions. This is especially true for classic Carrousel® systems, such as the old design (Figure 19.2), that have the advantage of a high recirculation ratio enabling denitrification up to very low nitrate concentrations, but have the disadvantage that the oxygen input in the Carrousel® affects the size of the denitrifying (anoxic) zones and availability of COD for denitrification.

Moreover, the presence of nitrate is 'felt' by PAOs as is the presence of oxygen, and nitrate is a major risk with respect to disturbance of the anaerobic conditions needed for the P-release in the selector and side-stream anaerobic tanks. With this in mind, it is recommendable to incorporate extra measures in the design to prevent and/or limit the occurrences of high nitrate concentrations. In the case of Geestmerambacht the following features were installed:

- To stabilize nitrate removal, the new aeration tank was constructed as a Carrousel 2000® system, which incorporates an integrated pre-denitrification zone (see Figure 19.10 and Figure 19.11). The pre-denitrification zone is always anoxic and ensures the availability of easily degradable influent COD for denitrifica-

Figure 19.10 Carrousel 2000® at Geestmerambacht (main stream biological water treatment) 1. Pre-denitrification zone; 2. Aeration zone; 3. Division structure between pre-denitrification zone and aeration zone; 4. Surface aerator; 5. Final clarifier; 6. Return sludge pumping station. [This figure is also reproduced in colour in the plate section after page 336.]

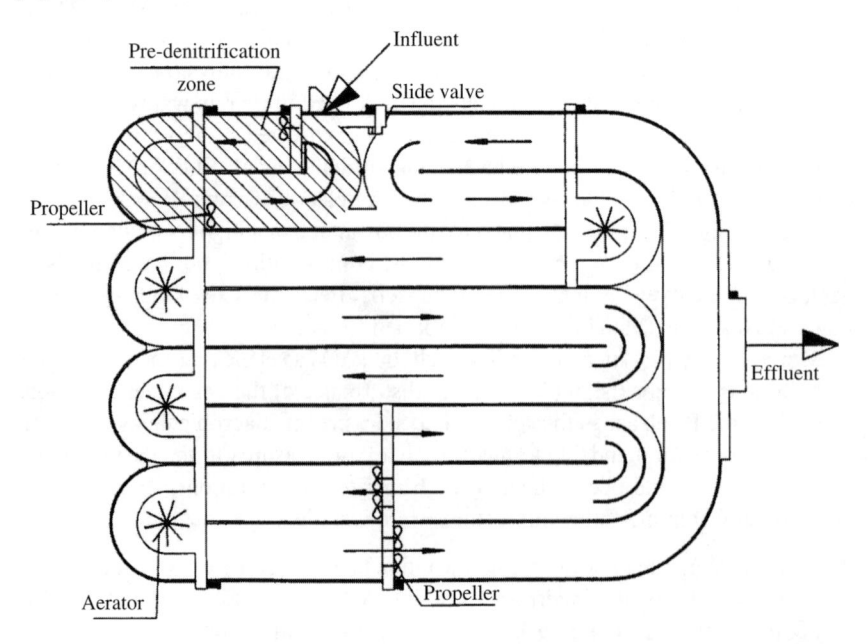

Figure 19.11 Carrousel 2000® with internal pre-denitrification reactor (after upgrading).

tion under all circumstances. The last part of nitrate removal up to very low concentrations is still accomplished in the aerated channels of the Carrousel 2000® system.
- To achieve complete nitrate removal, effective control of the oxygen supply is essential. The oxygen input has to be accurate and available to respond quickly to changes in oxygen demand. Therefore the speed of two of the four vertical shaft aerators was made adjustable, and controlled by two oxygen meters. Also propellers were installed in the aeration zone to maintain a minimum liquid velocity during low oxygen demand. To obtain detailed information about (variations in) effluent ammonium and nitrate concentrations, on-line monitoring of ammonium and nitrate was installed.

19.9 FULL SCALE RESULTS

The first aim of the demonstration plant was to prove the full-scale realization of recovery and re-use of phosphate. During the first three years of plant operation, the phosphate industry was hesitant about reuse of phosphate pellets produced from waste water, and the pellets were landfilled. In 1997 and 1998 the pellets were recycled in the chicken feed industry. From 1999 onwards the pellets are being used by the company Thermphos (The Netherlands), as additional P-source to raw phosphate ore in the production of elemental phosphorous. Thermphos uses the thermal process, to which no disturbances were noticed from incorporation of the Crystalactor® pellets.

The side-stream process and Crystalactor® proved to be a technically feasible and reliable process for the full-scale removal and recovery of phosphate. A number of start-up problems were encountered, which showed that specific knowledge and experience are needed to operate the recovery process successfully. Table 19.7 shows a summary of the loads and results in 2000 (yearly averages).

As the table shows, the results with respect to the removal of suspended solids, organics, nitrogen and phosphate are very good. Influent phosphate load is relatively low compared to the concentrations assumed in the design phase, but so is the amount removed via the dewatered sludge.

Average flow to the side-stream was at its minimum value: $100 \, m^3/h$ with average sludge content of 6.5 g/l. Therefore, the loading rate of the side-stream separator/thickener was low (around $25 \, kg \, SS/(m^2 d)$) and the thickening result high at 25 g/l.

Average P-concentration in the overflow of the side-stream anaerobic tanks was very close to the design value of 60 mg P/l. Dosing of acetic acid was around 15 g COD/kg SS and therefore more economical than expected in the design phase. Average P-concentration of water to the cascade (mixture of side-stream water and water from sludge treatment) was around 55 mg P/l. Average removal efficiency of the Crystalactor® was around 65% at a circulation ratio between 1.5 and 2.0. Only one Crystalactor® was in operation in 2000 and was found to be sufficient to handle the actual loads. Average P content in the pellets was 10–12%.

Table 19.7 Full scale yearly average results of MWWTP Geestmerambacht in 2000 (personal communication from Simon Gaastra of Water Board Uitwaterende Sluizen, The Netherlands)

Influent		Actual (2000)	Design
Average flow	m^3/d	34,348	35,000
COD	kg/d	18,023	22,150
BOD	kg/d	6,620	8,250
Total N	kg/d	1,501	2,045
Total P	**kg/d**	**230**	**340**
Suspended solids	kg/d	9,726	10,500
pe (136 g TOD)	–	182,963	231,600
Effluent			
COD	mg/l	31	60
BOD	mg/l	4	5
Total N	mg/l	3	10
Total P	**mg/l**	**0.3**	**0.5**
Suspended solids	mg/l	7	12
Sludge production			
Sludge production	kg/d	7,180	9,700
P in sludge	g P/kg DS	17	20
P removal via sludge	**kg/d**	**118**	**194**
P-balance			
Influent	kg/d	230	340
Effluent	kg/d	10	13
Dewatered sludge	kg/d	118	194
P recovery via pellets	**kg/d**	**101**	**133**

19.10 COST

Total cost, including all consumables, operation and yearly capital cost, is around 6 € per kg P removed/recovered, which is close to the value expected during design. Chemical P-removal and biological mainstream P-removal at other MWWTPs has a cost of around 3.5 € per kg P removed. The difference in cost between P-recovery in the side-stream and mainstream biological P-removal, is approximately equal to the cost of all chemicals required in the side-stream P-recovery process. Other consumables, as well as investment, are more or less comparable. The operating Waterboard considers a cost level of around 5 € per kg P removed feasible, by optimizing the dosing of chemicals and making better use of the capacity available. The remaining excess between more cost-effective ways of P-removal and the process at Geestmerambacht should be assigned to realizing P-recovery rather than P-removal at full scale. In addition to recovery of phosphate, the extra cost involved produces the advantage that very low and stable effluent P-concentrations can be achieved, of better quality than by mainstream chemical or biological phosphate removal.

Thermphos (the phosphate industry) bears the cost of transport of the pellets but does not contribute to other cost components. Unfortunately, the value of phosphate in raw material (phosphate ore) is still many times lower than the cost involved in producing the phosphate rich pellets from municipal waste water.

19.11 NEW DEVELOPMENTS

The most promising new development with respect to the P-recovery by crystallization is the production of magnesium phosphate pellets. Pilot tests on an effluent comparable to the side-stream effluent at Geestmerambacht, showed that crystallization driven by magnesium is technically feasible without the prior removal of carbonates from the water. Tests where carbonate concentration was up to 10 mmol/l, showed no disturbance of the crystallization by the formation of magnesium carbonate. Further testing is still needed, especially to investigate the possible competitive formation of calcium carbonate or calcium phosphate, if magnesium is dosed in calcium-rich effluent. The reuse potential of such material should also be considered.

If crystallization is possible without the need of prior removal of carbonates, flows with lower phosphate concentration can be treated by crystallization in a more cost-effective way. For example, the supernatant flow from an anaerobic tank in the mainstream biological P-removal line could be treated, thus substantially reducing the addition of chemicals in the process.

REFERENCES

Chudoba, J., Cech, J.S. and Chudoba, P. (1985) The effect of aeration tank configuration on nitrification kinetics. *J. Water Pollut. Con. F.* **57**, 1078–1083.
Rensink, J.H., Eggers, E. and Donker, H.J.G.W. (1991) High biological nutrient removal from domestic wastewater in combination with phosphorus recycling. *Water Sci. Technol.* **23**, 651–657.
STOWA, Dutch Association for applied Research in Water Management (1992). Phosphate removal from municipal wastewater by crystallization: Results of pilot scale research, STOWA publication 92–11 (in Dutch).
Van Dijk, J.C. and Braakensiek, H. (1984) Phosphate removal by crystallization in a fluidised-bed. *Water Sci. Technol.* **17**, 133–142.
Van Dijk, J.C. and Wilms, D.A. (1991) Water treatment without waste material – fundamentals and state of the art of pellet softening. *J. Water SRT-Aqua* **40**, 263–280.

20
Full scale struvite recovery in Japan

Y. Ueno

20.1 INTRODUCTION

There are no economic reserves of phosphorus ore in Japan. As a result, phosphorus can only be acquired by import; the quantities imported reach about one million tons per year (JPCFMA, 2001). The imported ore is primarily used to make chemical fertilisers. Inevitably, some of the phosphorus in the fertilisers may ultimately drain into surface waters. Similarly, phosphorus containing domestic and industrial waste waters drain into surface waters and the released phosphorus may cause eutrophication in enclosed water bodies, such as lakes and bays.

Effluent quality standards, in terms of phosphorus content acceptable for discharge into enclosed water bodies, have been upgraded in Japan recently (fifth total pollutant load regulation, a law concerning special measures for conservation of lake water quality). As a result, advanced waste water treatment facilities, for phosphorus removal, have been introduced in sewage treatment plants when their effluent discharges into enclosed water bodies.

The majority of sewage treatment plants, operate two alternative treatment processes: either a process involving addition of a coagulant to the activated sludge, or an anaerobic/aerobic activated sludge process. Alternatively, an anaerobic

© 2004 IWA Publishing. *Phosphorus in Environmental Technology: Principles and Applications.* Edited by Eugenia Valsami-Jones. ISBN: 1 84339 001 9

digestion process is used for stabilisation and volume reduction of sludge in many sewage treatment plants.

However, phosphorus absorbed in the sludge is released when organic matter is being decomposed in the digestion process. The effluent produced, following sludge treatment, is returned to the main sewage stream, thereby increasing its phosphorus load, thus resulting in higher concentration of discharged phosphorus.

In this chapter examples are presented of the introduction of an additional treatment unit, in which phosphorus in the supernatant from the sludge treatment process is removed via struvite precipitation. Struvite is recovered in the form of nearly pure crystals. This process aims primarily to reduce eutrophication caused by sewage treatment plants discharging directly into enclosed water bodies in Japan.

20.2 FULL SCALE STRUVITE RECOVERY IN JAPAN

At present, two full scales struvite recovery plants exist within Japanese sewage treatment works: one at the Fukuoka City West Waste Water Treatment Centre, and the other at Shimane Prefecture Lake Shinji East Clean Centre. Their details are described below.

20.2.1 The full scale struvite recovery plant at the Fukuoka City West Waste Water Treatment Centre (Tomoda, 1999)

20.2.1.1 Circumstances of introduction

The treated waters discharging from the five Waste Water Treatment Centres of Fukuoka City drain out into the Hakata Bay, which is an enclosed water body; as a result eutrophication was observed in the Bay.

In order to control the growth of phytoplankton that causes eutrophication, it was considered that reduction of phosphorus, which is a limiting nutrient in Hakata Bay, would be the most effective option.

When selection of the most appropriate additional waste water treatment for Fukuoka City, was being made, the following conditions had to be considered:

- The new treatment method should not require additional treatment space.
- The amount of sludge generated should be equal to a conventional activated sludge process.
- The new method should provide effective resource and energy savings.
- The new process should feature easy operation and maintenance.

Considering the above, it was decided to adopt the anaerobic and aerobic activated sludge process for sewage treatment. It was also decided to introduce a

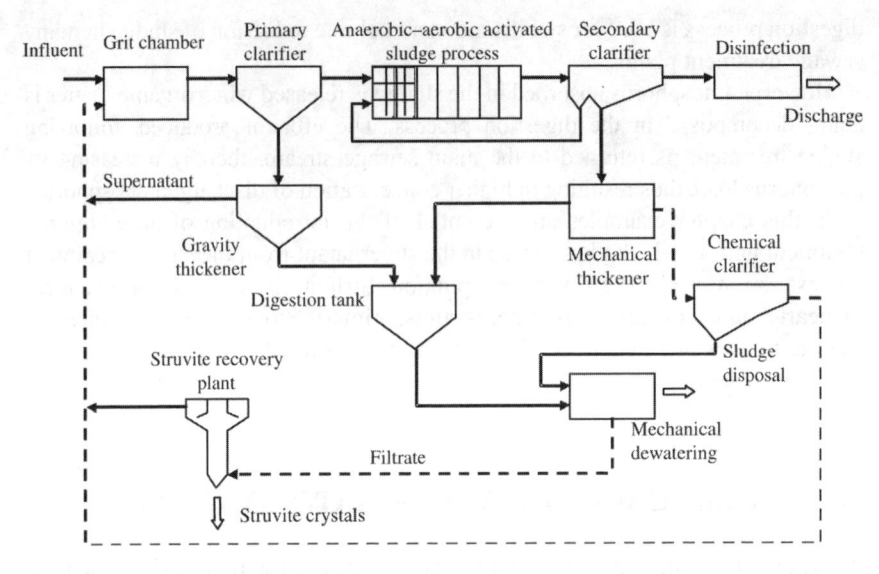

Figure 20.1 Schematic diagram of WWTC.

struvite recovery plant as a means of removing phosphorus from the supernatant of the sludge treatment process.

20.2.1.2 *Summary of the Fukuoka City West Waste Water Treatment Centre*

The struvite recovery plant of the Fukuoka City West Waste Water Treatment Centre (hereafter WWTC) was implemented in July 1997.

A schematic diagram of the WWTC is shown in Figure 20.1 and a schematic diagram of the struvite recovery plant is shown in Figure 20.2. The main specifications of the struvite recovery plant are listed in Table 20.1 and the main characteristics of the supernatant treated are shown in Table 20.2.

A summary of the process flow of the struvite recovery plant is described below:

- The filtrate is continuously and quantitatively fed to the nucleation zone by the feed pump.
- The precipitation reaction in the struvite recovery plant is accelerated by mixing of the effluent situated in the nucleation zone by the circulation pump.
- Aeration is carried out in the nucleation zone by the blower, in order to accelerate the precipitation process.
- Magnesium chloride is dosed to induce struvite precipitation by reacting with phosphorus.
- The pH in the nucleation zone is adjusted within the alkaline range by dosing of sodium hydroxide.
- Wet struvite is dried by a fluidised bed dryer.
- Struvite is stored in a storage hopper.

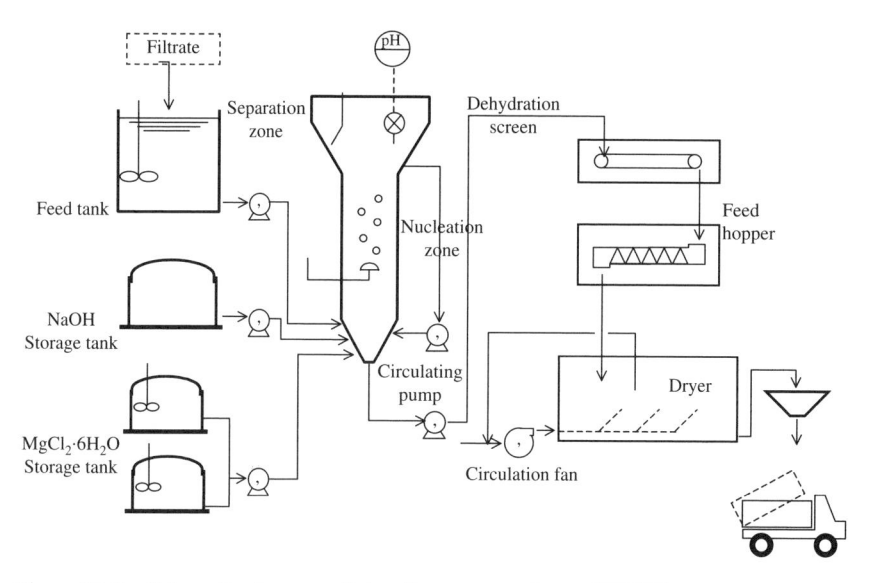

Figure 20.2 Schematic diagram of struvite recovery plant at WWTC.

Table 20.1 Specifications of the struvite recovery plant at WWTC.

Operation	Quantity	Specification
Struvite recovery plant ($170\,m^3\,d$)	6	Type: Gas–liquid mixture circulation-based Vertical cylinder type Effluent treated: Filtrate Dimensions: Diameter of nucleation zone 200 mm Diameter of seperation zone 600 mm Total height 7000 mm
Dehydration screen	1	Type: Screen belt conveyer type Treatment rate: $1.2\,m^3\,hr$
Drier	1	Type: Fluidised bed drier
Hopper	1	Type: Square damper discharge type Effective volume: $13\,m^3$

20.2.1.3 *Analysis and use of struvite pellets*

The composition of the struvite produced at the WWTC is shown in Table 20.3. Magnesium, ammonium, and phosphorus contents were similar to the theoretical values for struvite, and the concentration of hazardous metals in the struvite were below the levels permitted for commercial fertilisers.

Although the struvite recovered does not contain potassium, it has components such as nitrogen and phosphorus. It was therefore appropriate to be registered as a fertiliser in the category of 'High Performance Complex Fertilisers; Fuku-MAP21' in July 1994.

Table 20.2 The main characteristics of the anaerobic supernatant at WWTC (Ookuma et al., 1999).

	pH	PO_4–P(mg l^{-1})	NH_4–N(mg l^{-1})	Mg (mg l^{-1})
Average	7.9	120	710	11.2
Maximum	8.1	160	840	18.4
Minimum	7.8	92	600	6.5

Table 20.3 Composition of struvite generated at WWTC.

Component	Struvite generated at WWTC1 (%)	Struvite generated at WWTC2 (%)
Magnesium (Mg)	9.6	9.5
Nitrogen (N)	4.9	5.4
Phosphorus (P)	12.7	12.7
Calcium (Ca)	0.117	0.136
Silicon oxide (SiO_2)	0.290	0.320
Iron (Fe)	0.039	0.013
Mercury (Hg)	N.D.(less than 0.00005)	N.D.(less than 0.00005)
Zinc (Zn)	N.D.(less than 0.002)	N.D.(less than 0.002)

At present, all of the struvite recovered is sold to fertiliser companies as raw material for chemical fertilisers.

20.2.2 The full scale struvite recovery plant at the Shimane Prefecture Lake Shinji East Clean Centre (Ueno and Fujii, 2001)

20.2.2.1 Circumstances of introduction

In 1988, Lakes Shinji and Nakaumi were designated as being protected by legislation concerning conservation of lake water quality. The following year local effluent standards, more stringent than the national uniform standards of water pollution control law, were reinforced.

The Shimane Prefecture Lake Shinji East Clean Centre (hereinafter SECC) started an advanced waste water treatment operation involving coagulant-addition activated sludge recycling in a modified process, with a present-day treatment capacity of 45,000 m^3 d^{-1} from April 1994, as shown in Figure 20.3.

The plant performs nitrogen removal, as part of a two-stage process which also includes recirculation. An anaerobic and aerobic activated sludge process was also carried out to remove phosphorus, and at the same time poly-aluminium chloride (PACl) was occasionally added at a rate of about 5 mg l^{-1} of Al in an aeration tank, when treatment became unstable.

A schematic diagram of the sludge treatment process of SECC is shown in Figure 20.4.

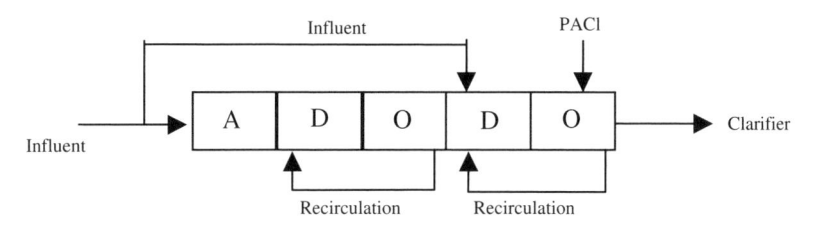

Full scale struvite recovery in Japan 501

Figure 20.3 Schematic of the coagulant-addition activated sludge recycling process operating at SECC.

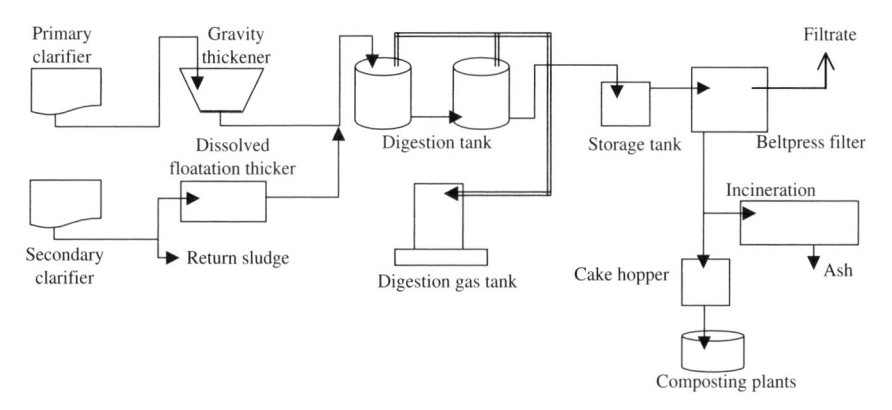

Figure 20.4 Schematic diagram of the sludge treatment process of SECC.

The amount of sludge generated by sewage treatment plants increases, following the increase of sewered population. Accordingly, while the reduction of sludge (anaerobic digestion) is designed for the primary purpose of waste disposal, the potential of a beneficial to the environment process development, if possible, is considered. Excess phosphorus in sludge is released in anaerobic digestion, and then returned to the sewage effluent treatment process as filtrate of dehydration of sludge.

At SECC, the phosphorus load originated from the supernatant represents about 70% of the inflow sewage. As it was important to reduce phosphorus concentration in the filtrate in order to stabilise phosphorus removal at the sewage water treatment process, dosing of polyferric sulfate had been carried out, followed by dehydration of the sludge, in order to fix phosphorus in the sludge. Further, a large amount of PACl was required to be dosed in the aeration tank in order to attain the target discharge value for total phosphorus in the final effluent.

The sludge treatment process produces compost for agricultural use. A reduction of the amount of PACl used was therefore intended, because high levels of Al in the sludge result in a deterioration of the quality of the compost.

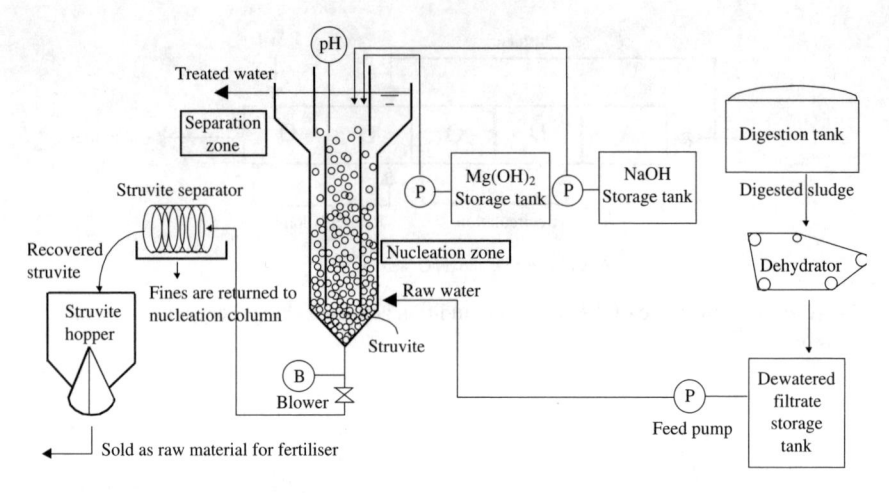

Figure 20.5 Schematic diagram of struvite recovery plant at the SECC.

Introduction of the struvite recovery plant was decided for the following overall reasons:

- Stabilisation of the total phosphorus concentration in the final effluent by reduction of the phosphorus load in the supernatant of the sludge treatment process.
- Reduction of chemical consumption such as PACl and polyferric sulfate, as well as reduction of the amount of sludge generated by the use of such chemicals.
- Production of a phosphorus rich fertiliser (struvite) and utilisation of this 'waste' as a resource.

20.2.2.2 Summary of the struvite recovery plant at the SECC

The original struvite recovery plant (two units of 500 m^3 d^{-1} and 150 m^3 d^{-1}) has been in operation since 1998 at the SECC. Another struvite recovery plant with a capacity of 500 m^3 d^{-1} was newly constructed in September 2000 and thus, the present plant capacity is 1150 m^3 d^{-1}.

A schematic diagram of the struvite recovery plant is shown in Figure 20.5, whereas its main specifications are listed in Table 20.4 and the main characteristics of the supernatant treated are shown in Table 20.5.

Summary of an operational flow chart for the struvite recovery plant is described below.

- Filtrate from the sludge treatment process is continuously fed to the nucleation zone.
- Magnesium hydroxide is added at magnesium to phosphorus ratio of 1 : 1, and pH is adjusted to a value between 8.2–8.8 with addition of sodium hydroxide.

Table 20.4 Specification of the struvite recovery plant at the SECC.

Operation	Quantity	Specification
Struvite recovery plant (500 m^3 d^{-1})	2	Type: Vertical type double cylinder reaction tower Effluent treated: Filtrate Dimensions: Diameter of nucleation zone 1430 mm Diameter of separation zone 3600 mm Total height 9000 mm
Struvite recovery plant (150 m^3 d^{-1})	1	Type: Vertical type double cylinder reaction tower Effluent treated: Filtrate Dimensions: Diameter of nucleation zone 960 mm Diameter of separation zone 2600 mm Total height 5500 mm
Struvite separator	1	Type: Rotational separator Effluent: Mixed size struvite particles Treatment rate: 6 m^3 hr^{-1}
Hopper	1	Type: Automated square cut gate hopper Effective volume 10 m^3

Table 20.5 The main characteristics of the anaerobic supernatant at SECC.

	pH	PO$_4$–P(mg l^{-1})	T-P(mg l^{-1})	NH$_4$–N(mg l^{-1})
Average	7.7	117	120	261
Maximum	8.0	149	147	329
Minimum	7.5	89	91	194

- Fine crystals of struvite are grown in fluidised bed by mixing with air.
- Struvite particles are periodically (about ten days) discharged from the bottom of the reactor column; the larger struvite pellets are recovered separately, whereas the finer struvite particles are returned to the reaction column to provide new seed material for continuation of the process.
- Recovered struvite pellets are stored in a hopper where the water content is reduced to less than 10% prior to transportation to fertiliser companies.

20.2.2.3 Operational situation of the struvite recovery plant at the SECC

The PO$_4$–P concentration in influent and effluent of the struvite recovery plant is shown in Figure 20.6. The influent PO$_4$–P concentration was 100–140 mg l^{-1}, whereas the effluent PO$_4$–P concentration was 10 mg l^{-1} or less, which equates to minimum removal rate of 90%. This result indicates that the struvite formation reaction proceeds without any problems.

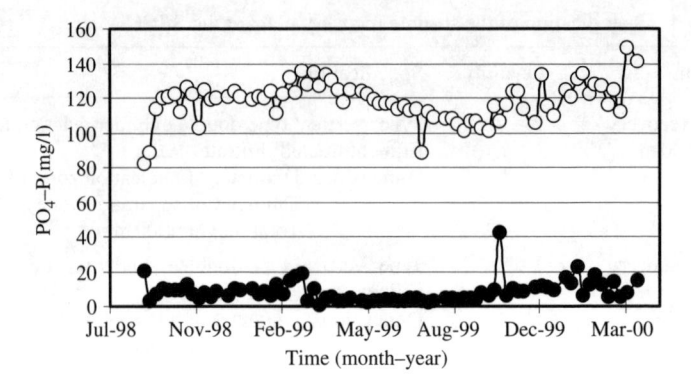

Figure 20.6 The PO$_4$–P concentration in influent and effluent of the struvite recovery plant at the SEEC. (Influent ○; Effluent ●)

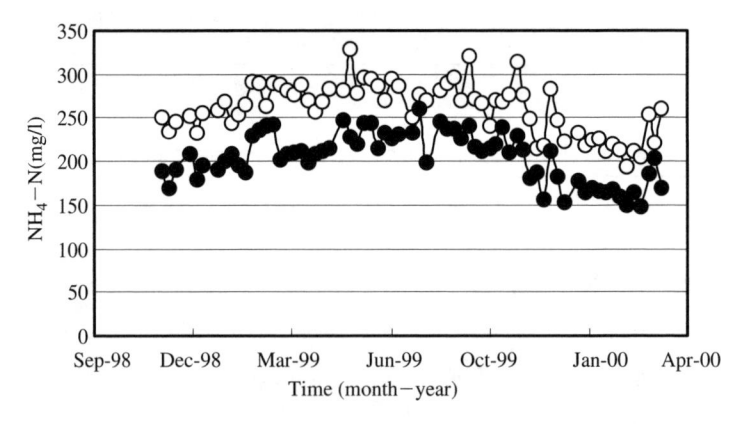

Figure 20.7 The NH$_4$–N concentration in influent and effluent of the struvite recovery plant at the SEEC. (Influent ○; Effluent ●)

The NH$_4$–N concentration in the influent and effluent of the struvite recovery plant is shown in Figure 20.7. The influent NH$_4$–N concentration was 200–300 mg l^{-1}, whereas the effluent NH$_4$–N concentration was 150–250 mg l^{-1} or less, and therefore the removal rate of NH$_4$–N is about 20%. It was predicted, based on the stoichiometry of struvite, that the mole amount of NH$_4$–N removal would be approximately the same to that of PO$_4$–P in the struvite recovery process. However, the operational results indicate that NH$_4$–N was removed at the same or even higher concentration to PO$_4$–P. Since struvite formation reaction is taking place under weakly alkaline conditions, a small excess of NH$_4$–N is considered to be stripped by the struvite.

Total phosphorus (T-P) concentration results for the struvite recovery plant are presented in Figure 20.8. The influent T-P concentration was 120–140 mg l^{-1}, and the effluent T-P concentration was 10–60 mg l^{-1} or less, and thus the removal rate of T-P is about 70%.

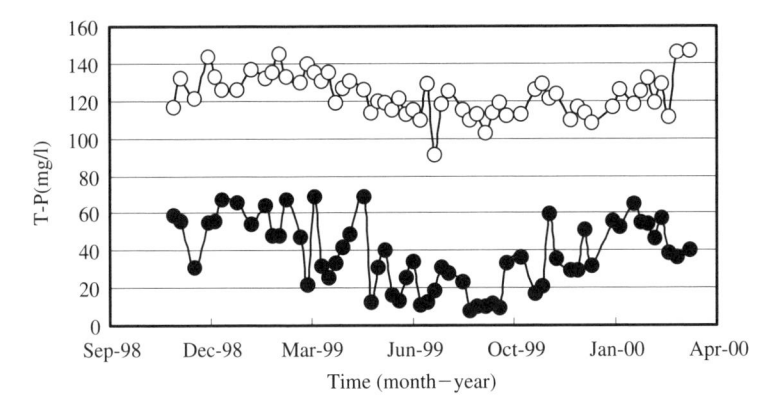

Figure 20.8 The T-P concentration in influent and effluent of the struvite recovery plant at the SEEC. (Influent ○; Effluent ●)

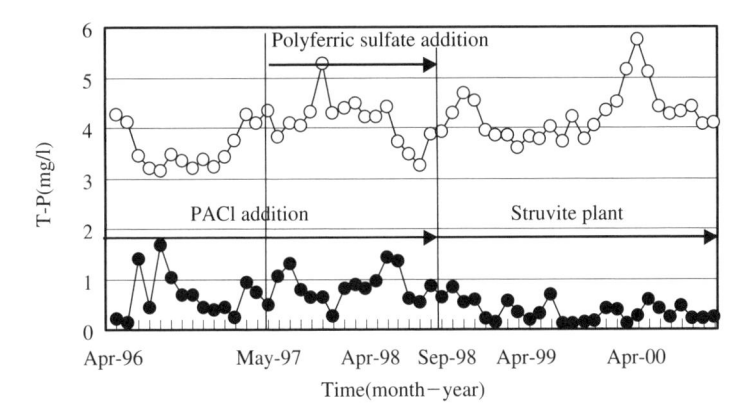

Figure 20.9 The T-P concentration in influent and effluent of the SECC before and after the introduction of the struvite recovery plant. (Influent ○; Effluent ●)

The T-P concentration in influent and effluent of the SECC before and after the introduction of the struvite recovery plant is shown in Figure 20.9. The influent T-P concentration of the SECC was about 4 mg l^{-1}. There was less fluctuation in the effluent T-P concentration during the operational period of the struvite recovery plant, compared to when polyferric sulfate and PACl were used. The T-P concentration ranged between 0.3–0.6 mg l^{-1} under the struvite recovery process. However during polyferric sulfate and PACl addition term the concentration peaked at 1.8 mg l^{-1}.

20.2.2.4 *Analysis and use of struvite pellets*

Composition of the struvite produced at the SECC is shown in Table 20.6.

Table 20.6 Composition of struvite generated at SECC.

Component	Struvite generated at SECC (%)	Theoretical value (%)
Magnesium (Mg)	9.7	9.8
Nitrogen (N)	5.7	5.7
Phosphorus (P)	12.9	12.7
Arsenic (As)	0.000048	–
Cadmium (Cd)	0.000006	–
Chromium (Cr)	N.D. (less than 0.02)	–
Lead (Pb)	N.D. (less than 0.92)	–
Mercury (Hg)	N.D. (less than 0.00005)	–
Nickel (Ni)	N.D. (less than 0.01)	–

The magnesium, ammonium, and phosphorus content were similar to the theoretical values and the concentration of hazardous metals were less than those permitted in commercial fertilisers. The present struvite recovery plant can produce 500–550 kg d^{-1} of struvite, which is then sold to fertiliser companies.

The fertiliser companies buying the produced struvite do not use it on its own, but mix it with other inorganic and organic materials and adjust the proportion of nitrogen, phosphorus and potassium.

The fertilisers are widely used for paddy rice, vegetables and flowers; in particular it is claimed to significantly improve the taste of paddy rice.

20.3 SUMMARY

Two full scale struvite recovery plants currently in operation in Japan were introduced in this chapter. At present, there are several experimental and pilot treatment units in various other localities, aiming to remove and recover phosphorus. These are not, as yet, committed to introduce a full scale plant. It is expected that future operations will improve and expand on the production of recovered products.

Phosphorus discharge load contained in domestic waste water in Japan is estimated to be at about 60,000 tons year^{-1}. If all phosphorus could be recovered, the amount would be equivalent to about 20% of Japan's current phosphorus ore import. It can thus be concluded that sewage may eventually become a promising phosphorus resource.

REFERENCES

Japan Phosphatic & Compound Fertilisers Manufacturers Association (JPCFMA) (2001) Research and study of demand trend of compound fertiliser and phosphoric fertiliser data **31**, 14 pp.

Tomoda, M. (1999) The use of struvite as a fertiliser in Fukuoka City. *Jap. Sewage Works Assoc. Conf.* **36**, 42–46.

Ookuma, T, Fujii, T and Higuchi, K. (1999) The operation management of the struvite plant. *Jap. Sewage Works Assoc. Conf.* **36**, 800–802.

Ueno, Y. and Fujii, M. (2001) Three years experience of operating and selling recovered struvite from full-scale plant. *Environ. Technol.* **22**, 1373–1381.

21
Phosphorus recovery from unprocessed manure

P. Hobbs

21.1 INTRODUCTION

The first question that springs to mind when presented with this title is, 'why recover phosphorus in all its forms from wastes, surely it is good for the land and the potential expensive extraction procedures make its use prohibitive?' First, in principle, phosphorus is good for the land until present in excess, which mostly results from the application of fertiliser and animal manure in intensive agriculture. Watercourses then become susceptible to eutrophication and more so during storms by soil particle transfer. Hence, climate change associated with extremes of weather, i.e. storms, may become a significant factor for eutrophication of waterways. Second, the world's resources of high grade phosphorus are declining and lower grade, with a higher heavy metal content, will be used. Estimates currently show that within 80 years resources may expire (Johnston and Steén, 2000). Currently, 127 million tonnes of the 150 million tonnes extracted annually from the earth are used by agriculture (CEEP, 1998). Phosphorus is often present in the soil at higher levels than those required for plant growth especially in intensive

© 2004 IWA Publishing. *Phosphorus in Environmental Technology: Principles and Applications.*
Edited by Eugenia Valsami-Jones. ISBN: 1 84339 001 9

agricultural systems. The chemical nature of phosphorus is such, that adhesion to particles is often the means of transfer to watercourses (see for example Haygarth and Condron, this volume).

Environmentally favourable approaches that use minimal energy are necessary to meet society's phosphorus demands and those of many farmers who currently operate with low financial margins. Furthermore, from a chemical engineering perspective it is better to treat wastes at the point of highest concentration, i.e. at the source. For intensive pig and poultry 'factory farms' this may present a possible option, because of the expense and logistical difficulties of spreading large quantities of manure to land.

Within the bounds of current agricultural practice, there are means by which phosphorus release can be minimised from manure containing the highest concentrations which originate from poultry and pig production. Pig and poultry are monogastric species that cannot digest some forms of organic phosphorus whereas ruminant species produce phytase in their digestive system that can release six molecules of phosphate from each molecule of phytol. An efficient extraction requires that all forms of phosphorus in manure are readily changed into extractable species and that extraction is relatively easy; the resolution of these two requirements is complex.

21.2 PROFILE OF PHOSPHORUS SPECIES IN DIFFERENT LIVESTOCK MANURE

There are approximately 90 million tonnes of livestock manure produced annually from housed livestock in the UK which contains about 226,392 tonnes of phosphorus as P_2O_5 (98,847 tonnes of phosphorus) (Table 21.1) according to the MAFF Fertiliser Recommendations for Agriculture and Horticultural Crops (MAFF, 2000). Approximately 70% of the dietary phosphorus is excreted by livestock and will be

Table 21.1 Total phosphorus as P_2O_5 present in manure from housed livestock in the UK for year 2000.

Animal	UK housed livestock numbers (year 2000*)	Averaged phosphorus (as P_2O_5) per animal kg	Total phosphorus as P_2O_5 tonnes yr^{-1}
Dairy	2,318,00	19	44,042
Beef	5,178,000	5.8	30,032
Pigs	6,483,000	11.25	72,934
Poultry		per 1000 birds kg	
Broiler	106,050,000	435	46,132
Layer	28,686,000	545	15,634
Pullet	9,461,000	435	4116
other flocks	24,776,000	545	13,503
Total all livestock			226,392

* MAFF 2000, Fertiliser recommendations for agriculture and horticultural crops.

present as different chemical species. As with many degrading organic wastes, the phosphorus species present are dependent on the chemical and microbial environment. Manure varies, from that with high water content known as slurry, to more solid farmyard manure (FYM) resulting from bedding livestock with straw. Each will have a different effect on the concentrations of phosphorus species during storage, which will be dependent on the degree of oxygen diffusion into the manure which is in turn dependent on water content. Generally, an anaerobic environment will increase the soluble phosphate concentration.

There has been a range of analytical methods used for identifying the phosphorus profile of different manure. Each method tends to have its own means of identification that does not clearly recognise differences in organic or inorganic as soluble or insoluble species (Barnett, 1994; Hedley and Stewart, 1982; McAuliffe and Peech, 1959). Chemical identification of phosphorus forms in different types of manure has proven elusive, primarily because of the highly reactive nature of the trivalent PO_4^{3-} group, and its propensity towards hydrolysis from its organic species during extraction and identification procedures. Therefore, current understanding of phosphorus speciation in excreted material mostly relies upon analytical definitions according to the fractionation procedures of Hedley and Stewart (1982) and McAuliffe and Peech (1959), and to the molybdate-reactive dependent detection methodology of Murphy and Riley (1962). Increasing amounts of organic phosphorus will be added to the inorganic phosphorus detected in the analytical methods from resin extract, Olsen bicarbonate, sodium hydroxide and sulphuric acid extractions (see Haygarth and Condron, Evans and Johnston, this volume, for discussions on phosphorus extraction methods). Because of the labile nature of degrading organic wastes, it is better to visualise each form in dynamic equilibrium with some more stable inorganic insoluble species. Gerritse and Zugec

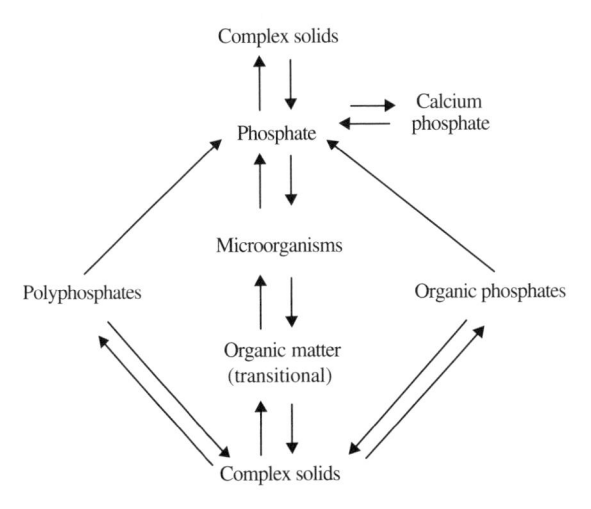

Figure 21.1 Model for the phosphorus cycle in pig slurry, based on the distribution of radioactive labelled phosphorus from Gerritse and Zugec (1977).

(1977) recognised these forms in a model (Figure 21.1) based on ^{32}P isotope distribution. In their approach, $H_3{}^{32}PO_4$ was added to pig slurry and the distribution of the isotope measured between the inorganic, organic and phosphorus contained in the micro-organisms. It was concluded that the changes in pig slurry are driven by microbial activity. About 10 to 20% of organic phosphorus is in microbial cellular material and a similar amount in solution. The remaining fraction is present in the solid material. Further, inorganic phosphorus increased with greater content in the feed. The composition of manure from livestock varies between monogastric animals and ruminants and is dependent on their diets. Generally monogastric livestock are unable to utilise some organic forms of phosphate, but the diet can be supplemented with phytase to access this source. Phosphorus transfer to the animal body and from the body as excretion is dependent on the concentration of dietary phosphate, the presence of vitamin D and the age and health of the animal. The phosphorus content in the excreta will depend on a range of factors, which can be managed for maximum uptake and minimum excreta concentration. The forms of phosphorus in excreta are shown in Table 21.2. It was recognised that all phosphorus forms were in dynamic equilibrium and that mostly originate from viable or discarded microbial nuclear and lipid-associated material that derives from cell walls of microbes involved in digestion (Gerritse and Zugec, 1977). During storage of slurry the environment is mostly anaerobic, however the oxygen availability in the hindgut of pigs can be regulated within the animal by oxygen transfer from the digestive artery system.

Gerritse and Zugec (1977) used $^{32}PO_4^{3-}$ also to determine the concentration of phosphorus species and the kinetic rate of transfer from inorganic, microbial and organic species in pig slurry. They concluded that there was a turnover of about 10–20 weeks for the biological uptake and release of dissolved inorganic and organic phosphates that does not include the exchange with mineral phosphates, for both aerobically and anaerobically stored pig slurry. The amount of soluble inorganic phosphorus was dependent on the relative concentration of calcium in the diet. Higher ratios of calcium reduced the soluble forms of inorganic phosphorus. Of the 10–20% of soluble organic phosphorus, about half was associated with large molecular weight species and recognised as DNA complexes with

Table 21.2 Forms of phosphorus in fresh faeces.

Waste source	Inorganic (%)	Residual (%)[b]	Acid soluble organic (%)	Lipid P (%)
Broiler litter	34.8	11	53.4	0.9
Layer	49.3	17.3	33.2	0.6
Pig faeces	54.7	15.2	29.7	0.4
Cow faeces[a]	63.2	27.7	7.8	1.4

[a] based on that from dry matter.
[b] mostly from nucleic acids.
Data from Barnett (1994).

polyphosphates, calcium and copper. Such complexes proved stable but were soluble after heating to 100°C (Gerritse and Eksteen, 1978).

21.3 EXTRACTION OF PHOSPHORUS FROM WASTES AND MANURE

The principles of inorganic phosphate precipitation are reasonably well understood, but problems are still experienced for phosphate present in waste water. For example, pipes carrying waste water in the sewage industry in the alkaline regions of South East England precipitate struvite from the sewage liquor to form a hardened crystal layer that eventually blocks the flow and takes either concentrated acid or considerable labour to remove (see for example Jaffer and Pearce, this volume). Development of controlled precipitation has led to large-scale precipitation reactors being installed at sewage works in The Netherlands (see Piekema, this volume). Such starting liquors contain fewer solids compared with livestock slurry, which require other approaches. There are three main techniques of phosphorus removal: (1) Conventional approaches, which use biological phosphorus removal. This involves both microbial accumulation in the sludge and, when financially or practically possible, the addition of a chemical reagent, to assist the process that is essentially aerobic. (2) Sequencing aerobic and anaerobic reactors where microbial communities digest the waste to produce insoluble forms of phosphorus. (3) Incineration or burning of some manure types. This has proved successful for turkey and chicken manure in the UK. Here the ash contains phosphorus, and the energy produced, is used to drive turbines to produce electricity. Given the success of the latter process, further research into reducing phosphorus in these manure types by other means may be restricted.

During storage of slurry, separation of solids occurs either to the bottom of the store or to the top in the case of cattle slurry. This results in a zone where few solids are present and it is from this liquid fraction that phosphorus can be extracted. Only in recent years has the extraction of phosphorus been attempted from livestock wastes other than by using aerobic treatments. About 60% of phosphorus is removed when the solids are filtered from pig slurry and extraction or co-precipitation is increased by oxidation. Phosphorus is mostly concentrated in the solid fraction by micro-organisms and precipitation. Anaerobic treatment can be easily achieved by storing manure, and the high oxygen demand of the waste ensures that a reducing chemical atmosphere will prevail that will solubilise the organic forms of phosphorus. However, separation from the organic waste is necessary before some purification process such as crystallisation can occur.

21.3.1 Precipitation/aerobic techniques

These methods include co-precipitation with sludge that can also be induced or enhanced by the addition of chemical agents. The waste water industry in particular uses calcium, iron and aluminium salts to precipitate phosphorus. These salts have not been investigated for use with livestock manure, possibly because of their costs and potential effects on crops once spread on the land. One of the most insoluble phosphorus compounds is struvite that not only extracts phosphate but also nitrogen as ammonium magnesium phosphate ($NH_4MgPO_4 \cdot 6H_2O$) from solution during its formation. Struvite is precipitated by aeration, which can be increased by the addition of magnesium chloride. Results show that 62–95% of phosphate can be precipitated. In addition, aeration increases the pH and may achieve an optimum precipitation at pH 9. Enhanced aeration achieved a pH of 8.5 with pig waste water and both magnesium and calcium were precipitated with phosphate with an efficiency of 65%. Sedimentation rates were high at 3 m hr^{-1} (Suzuki et al., 2002). The efficiency of phosphorus extraction can be increased with the addition of chemical reagents. A 60–70% efficiency was achieved for magnesium oxide and 95% when magnesium chloride was used on an anaerobic digester effluent (Celen and Turker, 2001). In raw pig slurry, 76% efficiencies of extraction were achieved with magnesium chloride that rose to 91% with pH adjustment to nine (Burns et al., 2001). However, the economics are questionable although the phosphate extracted is of value, the economy of the process is in doubt because of the high cost of alkali treatment.

Phosphorus present in calf manure has been successfully extracted in a facility in the Netherlands where potassium, which has similar chemical characteristics as ammonium, replaces ammonium to produce potassium magnesium phosphate ($KMgPO_4 \cdot 6H_2O$), an analogue of struvite (Schuiling and Andrade, 1999). The process utilises waste water from the calf manure that has been denitrified to remove ammonia. Effluent is supplemented with magnesium oxide and is then passed into three continuously stirred and connected tanks and a clarifier to remove the precipitate. However, in the presence of suspended solids greater than $1 \text{ g} \cdot l^{-1}$ the process becomes inefficient. Unfortunately, no reason was given for this interference.

21.3.2 Improved aerobic/anaerobic phosphorus removal

While extraction of phosphorus from wastes reduces potential releases to watercourses, the ability to recycle or return phosphorus for use via industrial electrothermal processing to other uses requires a higher grade of material. Phosphorus extracted should have a reduced metal content, but in the case of pig manure produced from a diet that has additions of copper and zinc as nutrient supplements, this can become problematic. Economics change rapidly in the pig and poultry industries and costs need to be kept to a minimum to encourage waste treatment. The use of flocculating agents or aluminium or ferric salts to improve precipitation of phosphorus would increase cost and waste volume.

With diluted manure or sewage, biological phosphate removal is based on the capacity of some microorganisms to store phosphate as polyphosphate. This occurs in an aerobic environment, and phosphorus can be extracted in the sludge from the waste. When an anaerobic stage is included, then phosphorus is released and the micro-organisms take up poly-hydroxy alkanoates as internal energy storing compounds (Stante et al., 1997). In sewage sludge it has been shown that phosphate-releasing bacteria can act as seed for crystallisation (Dick et al., 2001).

21.3.3 Sequencing Batch Reactors

The sequencing batch reactor (SBR) process primarily involves a fill-and-draw waste treatment system that enables microbes to immobilise the phosphorus as intercellular material that is then allowed to form sediment enriched in phosphorus. The upper more liquid phase is depleted in phosphorus and can be removed. Normally aeration and sedimentation clarification processes are carried out sequentially in the same SBR. The cycle processes, which are fill, react, settle and draw, are controlled in time to achieve the objectives of the operation. Each process is associated with particular reactor conditions (turbulent/quiescent, aerobic/anoxic) that promote selected changes in the chemical and physical nature of the waste water. A complete cycle begins with the fill process, when slurry is added to the system, and ends with the draw process, when a treated effluent is removed from the system. A cycle of operation may last between 12 to 24 hours for the treatment of diluted pig manure. The SBR system can be controlled by automatic switches and valves that sequence different operations.

21.3.3.1 Processes in the SBR

Labile carbon plays an indirect but significant role in phosphorus removal in the SBR. Labile carbon can be either soluble or suspended and can be related to chemical oxygen demand (COD). Labile carbon is oxidised to CO_2 while nitrogen compounds are reduced by a combination of nitrification and denitrification to nitrogen gas. However, phosphorus is contained within the sludge by a combination of biological removal by phosphorus accumulating microbes (PAMs) and chemical precipitation. The SBR has to be seeded with PAMs as most bacteria only accumulate about 20–40% of the total phosphorus present, sufficient for their metabolic requirements (Brett *et al.*, 1997). Many of the early SBR studies on liquid pig manure were not concerned with phosphorus removal, but with process control, and nitrogen and COD reduction in the supernatant or liquor (Fernandes *et al.*, 1991; Fernandes and Mckyes, 1991). Operation at 5°C reduced the SBR performance for the removal of available carbon, identified as chemical oxygen demand (COD), and ammonium, with over 95% removal of both from the supernatant at 10°C and only a 2% increase to 97% at 21°C (Fernandes, 1994), but there was no reference to phosphorus removal. Further studies demonstrated that over 95% of phosphorus can be removed by a SBR after centrifugation of piggery waste water (Bortone *et al.*, 1992). Problems such as foaming within the SBR can be controlled by the addition of vegetable oil (Edgerton *et al.*, 2000).

To improve the SBR's phosphorus removal efficiency above 90%, control of both pH (Cheng *et al.*, 1999) and oxidation–reduction potential (ORP) (Ra *et al.*, 1998) was demonstrated as important. Biological nutrient removal was also realised in a two-stage process using real-time control (Ra *et al.*, 1998). The effluent from the first tank that cycled through oxic and anoxic phases was transferred to a second tank after settling of the PAMs and the nitrate rich liquor was aerobically treated to remove nitrogen as a gas. Here the oxidation–reduction potential was used as a self-adjusting mechanism to control the batch treatment time producing higher and consistent nitrogen and phosphorus removal efficiencies of over 95% for variable waste compositions. Higher extraction or removal efficiencies were achieved for nitrogen when fractions of sludge were added to the reactor to act as a carbon source (Ra *et al.*, 2000). No change in the removal efficiency of 90% was achieved for phosphorus by adding acetate or fermented pig waste using a time controlled SBR process (Lee *et al.*, 1997). A removal efficiency of over 90% was achieved for total organic carbon (TOC), ammonium and phosphorus by controlling the process by pH change for a two stage SBR with the addition of 'Ringlace' media to immobilise bacteria (Cheng *et al.*, 1999). A full-scale SBR yielded better extraction efficiencies of over 98% of nitrogen, phosphorus and COD (Tilche *et al.*, 1999). With such high extraction efficiencies that are presumably achieved with the larger volume, it does bring into question the validity of quoting the hydraulic retention times (HRTs) of laboratory based systems that have different HRTs even for the same process.

Modifications of the SBR to include a separate biofilm for nitrification have been successful for piggery waste water with a low COD/N ratio (Bortone *et al.*, 1994). A settling phase was allowed to separate a supernatant to be nitrified in an external biofilm reactor. The nitrified effluent was returned to the SBR, where nitrates were removed and used as electron acceptors for luxury P-uptake and organic carbon oxidation. However, phosphorus uptake was lower than when oxygen was the main electron acceptor (Bortone *et al.*, 1994). The conventional anaerobic–aerobic SBR system shows a similar potential for phosphorus removal by denitrifying organisms (Kuba *et al.*, 1993). A hybrid upflow anaerobic filter integrated with a nutrient removal system, removed over 90% of the nitrogen, phosphorus and COD from piggery wastewater (Tilche *et al.*, 1994).

Cationic flocculating agents have been used to increase the concentration of phosphorus in pig slurry (Zhang and Lei, 1998). Milk of lime (macerated quicklime in water) was used in an experiment with a SBR for nitrification and an up-flow sludge blanket (USB) reactor for de-nitrification. Again best results were achieved at high pHs with an optimal phosphorus removal at pH 10.9. Calcium aluminates and lime have been used to precipitate phosphorus and heavy metals in treated pig manure, however, only small reductions in phosphorus removal were achieved.

21.3.4 Enhanced biological phosphorus removal

Modifications to the standard SBR approach have shown that an increase in the removal of nitrogen, phosphorus and COD content of the liquor is possible, especially for the two-stage SBR developed by Bortone *et al.* (1992) and Ra *et al.* (1998). Bortone *et al.* (1992) introduced a second anoxic/oxic cycle to the treatment of each influent batch in order to increase the oxidation of organic material and the conversion of nitrate and ammonium to nitrogen gas. This extended hydraulic retention time (HRT) to 10 days. This operating sequence was very effective, removing up to 93% of COD, nitrogen and phosphorus. Ra *et al.* (1998) reduced the precipitation of metal phosphates in the two stage SBR (TSSBR). The

TSSBR separates the nitrification and denitrification processes into distinct reaction vessels. This reduced the metal content in the phosphate.

Koch and Oldham (1985) demonstrated that it was not the redox value which was important, but the rate of change of the redox curve that controlled nitrogen and phosphorus removal. Two detectable features in the redox curve (dashed line Figure 21.2) are important. They are termed the 'nitrogen break point' and the 'nitrate knee'. These correspond to nitrification and denitrification during the aerobic and anaerobic stages respectively. Other important features in Figure 21.2 are labelled a, b and c. Point (a) represents the rapid response of the redox potential to the start of aeration. Little or no measurable response is observed in the dissolved oxygen (DO) curve (dotted line), until after nitrification is complete at the nitrogen break point. Point (b) is a plateaux at which time the reactor contents have become fully oxidised. Point (c) is the rapid decrease in redox potential following the cessation of aeration. The DO curve reaches zero as the redox curve levels prior to the nitrate knee. However, oxidative processes still occur during this phase, but result from nitrate being used as a terminal electron acceptor. Wareham *et al.* (1993) demonstrated that the nitrate knee also correlated with an increase in the release of phosphorus (solid line) to a soluble form in the supernatant (Figure 21.2). Phosphorus release was subject to the uptake of short chain fatty acids (SCFAs) produced under anoxic conditions by fermentative bacteria after oxidative terminal electron acceptors, such as nitrate were diminished (Megank and Faup, 1988). Wareham *et al.* (1993) used this feature to define the optimum time for the addition of acetate to enhance phosphorus release from the PAMs without competition for substrate from the denitrifying community (Toerien *et al.*, 1990). These features have been used in the development and control of the TSSBR for the treatment of pig slurry (Ra *et al.*, 1998). In the first TSSBR reactor vessel the influent input is determined by point (c) on the redox curve (Figure 21.2). When the redox potential reaches a predetermined low value the influent flow is stopped and the reactor is not aerated for 3 h. Aeration commences and continues

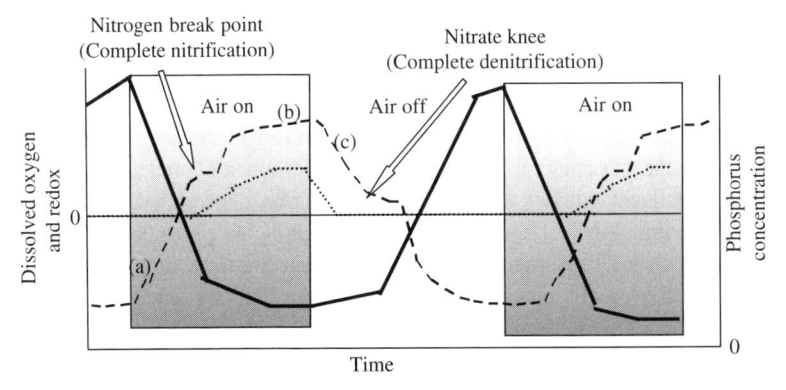

Figure 21.2 Schematic of the redox, dissolved oxygen (dotted line) and phosphorus changes (solid line) in single SBR showing significant points in the redox curve (broken line).

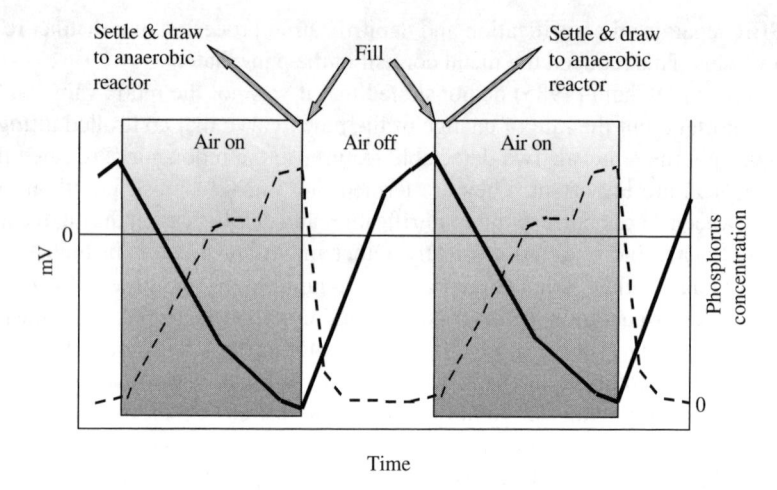

Figure 21.3 Schematic of the redox (broken line) and soluble phosphorus concentration (solid line) changes in the TSSBR.

until after the nitrogen break point where the redox curve becomes level (b). At point (b) aeration and stirring are stopped and the mixed liquor is allowed to settle for a short period. The nitrate-enriched, phosphate-depleted supernatant, which also contains soluble metal ions, is drawn off and fed into the second reactor that is anoxic (Figure 21.3). Here the nitrates are reduced to nitrogen gas by the denitrifying community using the remaining soluble carbon in the absence of competition from the PAMs. The cycle can then be repeated by feeding a fresh batch of pig slurry into the first reactor.

21.3.5 Biological phosphorus pump

In a recent development by Greaves *et al.* (2001), the TSSBR approach has been modified using a single reactor to concentrate the phosphorus in the biomass of the SBR over several cycles (Figure 21.4). The final product is phosphorus rich liquid that has a considerably reduced carbon and metal content. The major difference between that of the TSSBR (Ra *et al.*, 1998) approach and that in the Greaves *et al.* (2001) process is that the PAMs retain phosphorus from the previous influent batch, while the supernatant that contains the carbon and metals is drawn off. Subsequent influent batches introduce additional phosphorus that is incorporated into the growing PAMs. This cycle is repeated until the phosphorus capacity of the PAMs reaches a maximum (Figure 21.5). When the maximum capacity of the PAMs was achieved, the air was turned off without drawing supernatant from the SBR and without feeding further influent. The PAMs then release phosphorus to produce an enriched supernatant. Sufficient carbon should be present for phosphorus release as aeration was stopped at the beginning of the plateaux (point b in Figure 21.2) in the redox curve. Once the phosphorus enriched supernatant has been drawn off, the cycle can continue and the process of phosphorus enrichment can continue.

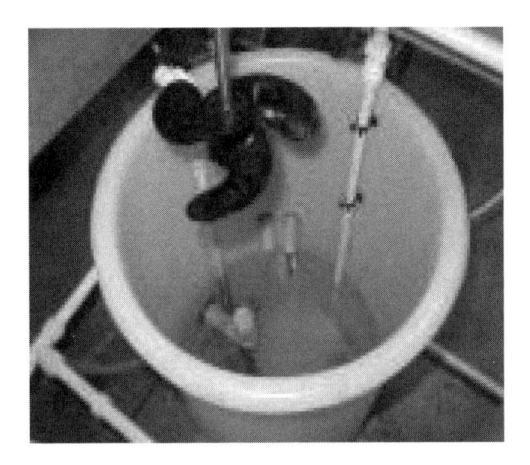

Phosphorus recovery from unprocessed manure 517

Figure 21.4 One of four replicate SBR vessels used by Greaves *et al.* (2001) for the recovery of phosphorus from pig slurry. The reaction vessels are constructed from polypropylene bins with a total capacity of 65 l. Diluted pig slurry is pumped into the reaction vessels and the effluent is drained from the reaction vessels 10 cm from the base. [This figure is also reproduced in colour in the plate section after page 336.]

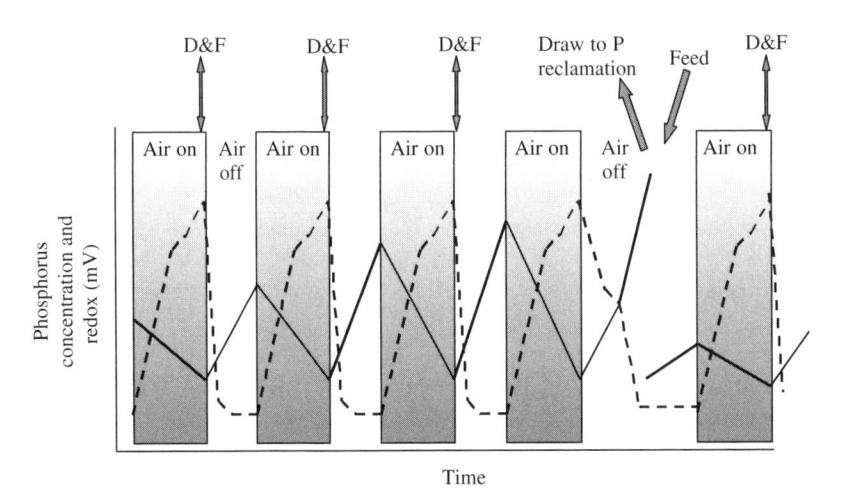

Figure 21.5 Schematic of the redox (broken line) and soluble phosphorus concentration (solid line) changes in the phosphorus pump approach illustrating the aerobic and anaerobic processes and how they are sychronised with the draw (D) of supernatant when the phosphorus is located in the PAMs followed by immediate filling (F) with the new waste to be treated. The point at which the concentrated phosphorus is drawn off is also shown by the rapid drop in concentration.

The research by Greaves *et al.* (2001) has demonstrated that from a starting liquor containing 30 mg P l^{-1}, a threefold enrichment was achieved for phosphorus in the supernatant, and reductions of carbon (87%), nitrogen (88%) and heavy metal content were observed. Final concentrations of over 100 mg l^{-1} phosphorus were achieved in the enriched supernatant and concentrations in the effluent reduced to less than $3\,mg\,l^{-1}$.

Phosphorus in the final enriched supernatant can be crystallised by precipitation to produce Mg or Ca phosphates. A major advantage of the process of extracting the supernatant of the phosphorus pump is that heavy metals are reduced. Therefore the crystal product can be sufficiently low in iron, copper and zinc, which is necessary for the recycling of phosphorus to be feasible (e.g. Schipper *et al.*, 2001).

21.4 CONCLUDING REMARKS

Phosphorus extraction methods for different forms of organic waste and manure types have been developed. Some are able to meet UK water legislative requirements of less than $3\,mg\,P\,l^{-1}$ for discharge to natural waters. Enhanced biological phosphorus removal has the potential to succeed, especially with the use of redox and pH profiles for process control. Of the sources of phosphorus in different livestock manure in the UK, diluted pig and dairy slurries and dirty waters are the most suitable. They contain a total of ca. $26\,kt\,yr^{-1}$ of phosphorus of which 5 kt is recoverable. Currently, phosphorus recovery from these effluents is mostly by struvite formation, where purity is not an issue and it can be used as fertiliser in agriculture or horticulture. To recycle pure phosphorus economically will be more difficult and depend upon several factors. The enriched phosphorus from any extraction process should be in a useable and relevant form. For example, phosphorus should be able to react with calcium for electrothermal purification to produce a high grade material (e.g. Schipper *et al.*, 2001).

To produce phosphorus with an economic and environmental advantage has been the ultimate goal of several research approaches. Increasingly research has been directed towards modified microbial communities to produce purer and more concentrated forms of phosphorus. Some considerable effort has been put into identifying those microbial species that are effective at accumulating intercellular phosphorus, but its still unclear as to the relevance of acinetobacter species (Stephenson, 1987) (see also McGrath and Quinn, this volume).

Currently, chemical extraction is effective and is used by the water industry to remove phosphorus from waste waters. Researchers have successfully demonstrated phosphorus extraction from pig ($30\,g\,P\,kg^{-1}$) and laying chicken manure ($24\,g\,P\,kg^{-1}$) that contain the highest amounts of phosphorus (Barnett, 1994). Extraction of phosphorus from liquid manure sources is only possible for liquid forms in the approach by Greaves *et al.* (2001) where pig slurry is diluted because of the high organic content which reduces the effectiveness of PAMs. This approach of biologically concentrating phosphorus from pig manure shows the potential of recycling phosphorus from animal manure in an appropriate form to e.g. be purified using the electrothermal process.

Current and future concerns: When developing any biological phosphorus removal plant, the major requirements are to improve the performance of PAMs, specifically making their function more reliable and effective. This will involve determining the maximum capacity of the PAMs to store phosphorus. Such information can be integrated into developing the phosphorus pump along with other physical characteristics such as size of, pH or organic content. In the context of scaling up of a SBR, there will be a requirement to optimise a range of processes. The rates of change of the species of phosphorus should be determined to help optimise extraction.

REFERENCES

Barnett, G.M. (1994) Phosphorus forms in animal manure. *Bioresource Technol.* **49**, 139–147.

Bortone, G., Gemelli, S., Rambaldi, A. and Tilche, A. (1992) Nitrification, denitrification and biological phosphate removal in sequencing batch reactors treating piggery wastewater. *Water Sci. Technol.* **26**, 977–985.

Bortone, G., Malaspina, F., Stante, L. and Tilche, A. (1994) Biological nitrogen and phosphorus removal in an anaerobic/anoxic sequencing batch reactor with separated biofilm nitrification. *Water Sci. Technol.* **30**, 303–313.

Brett, S., Guy, J., Morse, G.K. and Leicester, J.N. (1997) *Phosphorus Removal and Recovery Technologies*, Selper Publications, London.

Burns, R.T., Moody, L.B., Walker, F.R. and Raman, D.R. (2001) Laboratory and *in-situ* reductions of soluble phosphorus in swine waste slurries. *Environ. Technol.* **22**, 1273–1278.

CEEP, Centre Europeen d'Etudes des Polyphosphates (1998) *International Conference on Phosphorus Recovery for Recycling* CEFIC (European Phosphate Industry Council), Warwick University UK.

Celen, I. and Turker, M. (2001) Recovery of ammonia as struvite from anaerobic digester effluents. *Environ. Technol.* **22**, 1263–1272.

Cheng, N., Lo, K.V. and Yip, K.H. (1999) pH as a real-time control parameter in swine wastewater treatment process. *Adv. Environ. Res.* **3**(2), 166–178.

Dick, R.E., Devine, P.G., Quinn, J.P. and Allen, S.J. (2001) Biologically driven phosphate precipitation in biosludges. *Proceedings of the 2nd International Conference on the Recovery of Phosphorus from Sewage and Animal Wastes*, Noordwijkerhout, The Netherlands, 12th and 13th March.

Edgerton, B.D., McNevin, D., Wong, C.H., Menoud, P., Barford, J.P. and Mitchell, C.A. (2000) Strategies for dealing with piggery effluent in Australia: the sequencing batch reactor as a solution. *Water Sci. Technol.* **41**, pp. 123–126.

Fernandes, L. (1994) Effect of temperature on the performance of an SBR treating liquid swine-manure. *Bioresource Technol.* **47**, 219–227.

Fernandes, L. and Mckyes, E. (1991) Theoretical and experimental-study of a sequential batch reactor treatment of liquid swine manure. *Trans. ASAE* **34**, 597–602.

Fernandes, L., Mckyes, E., Warith, M. and Barrington, S. (1991) Treatment of liquid swine manure in the sequencing batch reactor under aerobic and anoxic conditions. *Can. Agr. Eng.* **33**, 373–379.

Gerritse, R.G. and Eksteen, R. (1978) Dissolved organic and inorganic phosphorus compounds in pig slurry:effect of drying. *J. Agr. Sci.* **90**, 39–45.

Gerritse, R.G. and Zugec, I. (1977) The phosphorus cycle in pig slurry measured from $^{32}PO_4$ distribution rates. *J. Agr. Sci.* **88**, 101–109.

Greaves, J., Haygarth, P.M. and Hobbs P.J. (2001) A novel biological phosphorus pump for livestock wastes. *Proceedings of the 2nd International Conference on the Recovery of Phosphorus from Sewage and Animal Wastes*, Noordwijkerhout, The Netherlands, 12th and 13th March.

Hedley, M.J. and Stewart, J.W.B. (1982) Method to measure microbial phosphate in soils. *Soil Biol. Biochem.* **14**, 377–385.

Johnston, A.E. and Steén, I. (2000) *Understanding Phosphorus and its Use in Agriculture*. European Fertilizer Manufactures Association, Brussels, 36pp. http://www.efma.org/publications/phosphorus/understanding%20phosphorus/contents.asp

Koch, F.A. and Oldham, W.K. (1985) Oxidation-reduction potential – A tool for monitoring, control and optimisation of biological nutrient removal systems. *Water Sci. Technol.* **17**, 259–281.

Kuba, T., Smolders, G., Vanloosdrecht, M.M. and Heijnen, J.J. (1993) Biological phosphorus removal from waste-water by anaerobic-anoxic sequencing batch reactor. *Water Sci. Technol.* **27**, 241–252.

Lee, S.I., Park, J.H., Ko, K.B. and Koopman, K. (1997) Effect of fermented swine wastes on biological nutrient removal in sequencing batch reactors. *Water Res.* **31**, 1807–1812.

MAFF (2000) *Fertiliser Recommendations for Agriculture and Horticultural Crops* (RB209) Norwich, UK.

McAuliffe, C. and Peech, M. (1959) Utilisation by plants of phosphorus in farm manure. 1. Labelling phosphorus in sheep manure with P32. *Soil Sci.* **68**, 179–184.

Megank, M.J.T. and Faup, G.M. (1988) Enhanced biological phosphorus removal from waters, In *Enhanced Biological Phosphorus removal from waste waters* (ed. Wise, D.L.) CRC Press, Boca Raton, FL, USA.

Murphy, J. and Riley, J.P. (1962) A modified single solution method for the determination of phosphate in natural water. *Anal. Chim. Acta* **27**, 31–36.

Ra, C.S., Lo, K.V. and Mavinic, D.S. (1998) Real-time control of two-stage sequencing batch reactor system for the treatment of animal wastewater. *Environ. Technol.* **19**, 343–356.

Ra, C.S., Lo, K.V., Shin, J.S., Oh, J.S. and Hong, B.J. (2000) Biological nutrient removal with an internal organic carbon source in piggery wastewater treatment. *Water Res.* **34**, 965–973.

Schipper, W.J., Klapwijk, A., Potjer, B., Rulkens, W.H., Temmink, B.G., Kiestra, F.D.G. and Lijmbach, A.C.M. (2001) Phosphate recycling in the phosphorus industry. *Environ. Technol.* **22**, 1337–1345.

Schuiling, R.D. and Andrade, A. (1999) Recovery of struvite from calf manure. *Environ. Technol.* **20**, 765–768.

Stante, L., Cellamare, C.M., Malaspina, F., Bortone, G. and Tilche, A. (1997) Biological phosphorus removal by pure culture of Lampropedia spp. *Water Res.* **31**, 1317–1324.

Stephenson, T. (1987) Acinetobacter: Its role in biological phosphate removal. In *Biological phosphorus removal from wastewaters* (ed. Ramadori, R.), Pergamon Press, Oxford, England, pp. 313–316.

Suzuki, K., Tanaka, Y., Osada, T. and Waki, M. (2002) Removal of phosphate, magnesium and calcium from swine wastewater through crystallization enhanced by aeration. *Water Res.* **36**, 2991–2998.

Tilche, A., Bortone, G., Forner, G., Indulti, M., Stante, L. and Tesini, O. (1994) Combination of anaerobic-digestion and denitrification in a hybrid upflow anaerobic filter integrated in a nutrient removal treatment-plant. *Water Sci. Technol.* **30**, 405–414.

Tilche, A., Bacilieri, E., Bortone, G., Malaspina, F., Piccinini, S. and Stante, L. (1999) Biological phosphorus and nitrogen removal in a full scale sequencing batch reactor treating piggery wastewater. *Water Sci. Technol.* **40**, 199–206.

Toerien, D.F., Gerber, A., Lotter, L.H. and Cloete, T.E. (1990) Enhanced biological phosphorus removal in activated-sludge systems. *Adv. Microb. Ecol.* **11**, 173–230.

Wareham, D.G., Hall, K.J. and Mavinic, D.S. (1993) Real-time control of aerobic-anoxic sludge-digestion using ORP. *J. Environ. Eng.-ASCE* **119**, 120–136.

Zhang, R.H., and Lei, F. (1998) Chemical treatment of animal manure for solid–liquid separation. *Trans. ASAE* **41**, 1103–1108.

22
Scenarios of phosphorus recovery from sewage for industrial recycling

A. Klapwijk and H. Temmink

22.1 INTRODUCTION

This chapter focuses on recent industry-instigated research in the Netherlands aiming to enable phosphorus (P) recycling. Thermphos International (Vlissingen, the Netherlands), one of the biggest international producers of elemental phosphorus, has the objective to replace 17.5 kt P/a of their total intake (about 20%) by recovered phosphate (Schipper *et al.*, 2001). One of the possible sources considered is phosphate from sewage. In the Netherlands, about 70% of all the waste water phosphate (14 kt P/a) was in 1998 removed in sewage treatment plants, of which about 15% by biological P-removal and 55% by chemical P-removal (CBS, 2000). The maximum amount therefore that may be recycled from sewage is approximately 12 kt P/a; this represents a substantial part of the quota of recovered phosphorus that Thermphos has set.

However, at present, the cost of recovered phosphate is much higher than that of phosphate rock. From an economic perspective, P-recovery is therefore not attractive to the water authorities. However, there may be other overriding advantages

© 2004 IWA Publishing. *Phosphorus in Environmental Technology: Principles and Applications.*
Edited by Eugenia Valsami-Jones. ISBN: 1 84339 001 9

Table 22.1 Requirements set by Thermphos for recovered phosphate materials (STOWA, 2001).

Parameter	Limit
Dry solids	>75%
P_2O_5	>18% based on dry solids
Fe	≤0.5% if P_2O_5 content is 20%
Zn	<20 ton/year^{-1}
Cu/Cr/Ni/Co/V	<2 ton/year^{-1}
SO_4	<0.5% based on dry solids

that would allow implementation of phosphate recovery from sewage. For example, an important technological advantage is that less phosphate will be returned to the sludge treatment facilities such as thickeners and digesters, resulting in a better P-removal efficiency (see for example Wild et al., 1997). At the same time, the problem of undesirable P precipitates, such as struvite ($MgNH_4PO_4 \cdot 6H_2O$) can be diminished (Jaffer and Pearce, this volume). P-recovery will also allow a decrease of the costs for treatment and disposal of sludge, due to the reduction by 4–7% of sludge volume on dry solid base, or 12–48% based on incineration ash (Etienne et al., 2001). Finally, sludge with a lower P content may find better use as a secondary material for other industries, for instance, the cement industry (Van Wersch, 2001).

In this chapter the possibilities to recover phosphate from sewage for the elemental P industry, with a focus on the Netherlands, are reviewed. Several scenarios are evaluated and the end products of these scenarios are compared to the quality that is required by Thermphos (Table 22.1).

In general, two possibilities to recover P from sewage exist. The first is to use the end product of the sludge treatment. The second is to extract and precipitate phosphate from the sludge before it is treated, and to use the precipitate as a secondary source of phosphate.

22.2 WASTE WATER AND SLUDGE TREATMENT

In municipal waste water treatment plants (Figure 22.1) phosphate is chemically or biologically removed from sewage and incorporated into the primary and secondary sludge (see also Parsons and Stephenson, this volume). For chemical precipitation of phosphorus, Fe-salts are typically used, and in some cases Al-salts (see also Parsons and Berry, this volume). Precipitation takes place in a pre-settling tank or in the biological reactor.

In biological phosphate removal (bio-P) plants (Figure 22.2) an environment is created for the proliferation of bacteria that accumulate phosphate in excess of normal metabolic requirements (see also McGrath and Quinn, this volume). For this purpose, firstly an anaerobic compartment is installed, where phosphate is released into soluble form; the effluent is then directed to anoxic/aerobic

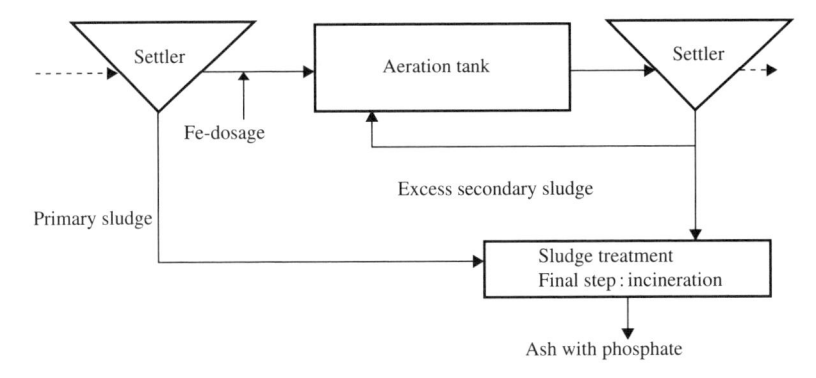

Figure 22.1 Schematic of a waste water treatment plant.

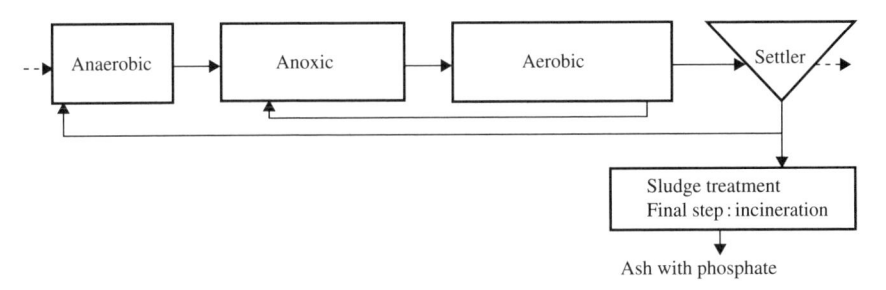

Figure 22.2 Layout of a plant with biological P-removal.

Table 22.2 Total mass of the sludge treated in the Netherlands and dry mass (%) of the end product (CBS, 2000).

Sludge disposal and utilization	Sewage sludge (Dry mass) (k ton)	Dry mass (%)
Incineration	220	100
Thermal drying	82	90
Composting	31	>65
Wet oxidation	10	50
Other	4	–
Total	347	

compartments, where the bacteria are engaged in luxury uptake of phosphate. The result is incorporation of the bulk of phosphorus into the sludge. This end result is similar to chemical precipitation; however the excess sludge that is produced has a high P-content, but a much lower Fe content.

The primary and secondary sludge is treated by different methods such as thickening, digestion and drying. The final treatment step is usually incineration, thermal drying, composting or wet oxidation. Table 22.2 presents the total dry mass of sludge being processed in the Netherlands together with the percentage of dry mass that can be generated by each treatment method.

22.3 P-RECOVERY FROM THE END PRODUCT OF SLUDGE TREATMENT

As the recovered material should have a dry mass percentage of at least 75% (Table 22.1), only the end product of incineration and thermal drying are interesting sources of phosphate (Table 22.2). However, the percentage of P in the end product of thermal drying is only 7%, which is much lower than the minimum P-content the P industry is looking for. The percentage of P in the end product of incineration is also relatively low (15–17%), but may be worthy of further consideration. Firstly, it should be realized that the low P-content in the incineration ash is the result of chemical precipitation. It can therefore be expected that ash from bio-P removal sludge may have a much higher P-content. Furthermore, bio-P sludge will have a lower Fe-content. Therefore the quality of ash, derived exclusively from bio-P treatment plants is also evaluated with respect to the requirements set by the P industry.

In Table 22.3 the quality of secondary sludge ash derived from mixed chemical and biological P-removal plants, and sludge ash from bio-P plants only, are compared with the requirements set by the elemental P-industry. The Fe, Cu and Zn content of ash of the sludge mixture are much higher than these requirements and this ash cannot be used by the P industry.

The ash from bio-P plants has a higher P-content, which is of the same order as phosphate rock (30–40%). However, the Fe, Cu and Zn contents are higher than the requirements set by the P industry. On the basis of maintaining an appropriate bulk Cu content, no more than 0.5 tons of bio-P sludge ash can be reused by Thermphos (Table 22.1). However, that is only 3% of their reuse objective.

In some cases phosphate is precipitated by Al-salts instead of Fe-salts. Although Al is not a problem for Thermphos, it is assumed that high levels of Zn and Cu can be expected, in the same concentration range as ash from bio-P plants. Therefore it is most likely that ash from plants with Al-precipitation is not a good secondary material for the P industry.

In general, we can conclude that ash from incinerated sludge does not provide a good source of secondary phosphates for recycling by the P industry.

Table 22.3 Quality of ash compared with requirements by Thermphos.

	P_2O_5 (g/kg ash)	Zn (mg/kg ash)	Fe (g/kg ash)
Sewage sludge ash (primary + secondary sludge)[1]	190	3500	100
Bio-P sludge ash (secondary sludge)[2]	360	3100	16
Requirements for reuse by the P industry	>250	<100	<10

[1] based on STOWA (2001).
[2] based on analysis of sludge from one bio-P plant.

22.4 P-EXTRACTION BEFORE SLUDGE TREATMENT

Another route to recover phosphate from sewage is extraction before the sludge enters the sludge treatment facilities. This discussion will further focus on the feasibility of P-extraction from sludge in bio-P plants specifically. At a bio-P plant (Figure 22.2), phosphate can be extracted at two locations. The first (Figure 22.3) is the sludge originating from the first, anaerobic compartment. The second (Figure 22.4) is from the activated sludge mixture derived from the last, aerobic compartment.

The first option has the advantage that phosphate is already available in soluble form. In this approach (Figure 22.3), the excess sludge is not withdrawn from the settling tank, but from the anaerobic compartment. In the anaerobic compartment

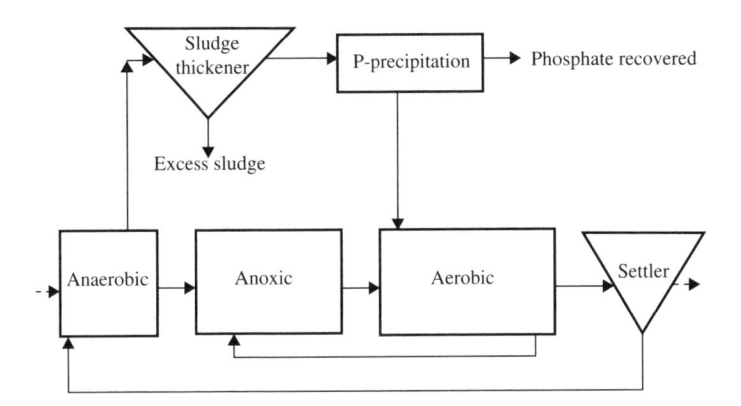

Figure 22.3 A bio-P plant with phosphate extraction from sludge of the anaerobic compartment.

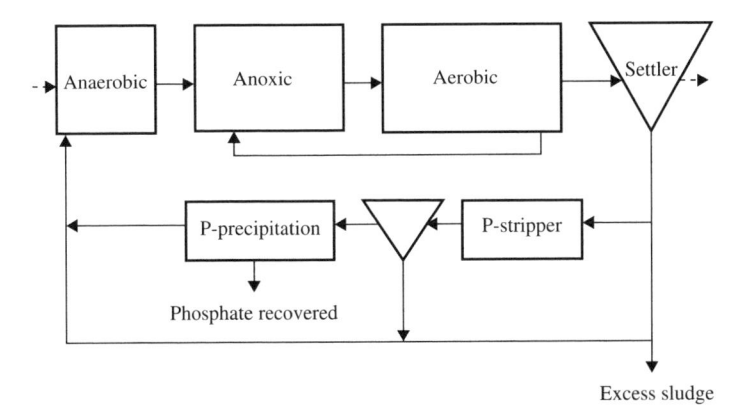

Figure 22.4 A bio-P plant with phosphate extraction from sludge of the aerobic compartment.

the phosphate has been released from the sludge. This phosphate containing liquor can be separated from the sludge in a thickener, and then directed to a precipitation reactor, prior to being returned to the activated sludge plant.

In a process developed recently (BCFS, Van Loosdrecht et al., 1998) this approach has been used in cases where biological P-removal is not sufficient to achieve the effluent requirements. However in this approach, the sludge is also returned to the activated sludge plant and sludge is discharged from the settler. For phosphate recovery, it would be more efficient to discharge the excess sludge from the anaerobic compartment, otherwise too much P leaves the system in the excess sludge and the percentage of P that is recovered is lower.

In the second option (Figure 22.4) (Levin and Shapiro, 1965), phosphate has to be released from the return sludge in an additional anaerobic compartment where extra carbon source (for example acetate) has to be added. This results in a higher P-concentration in the effluent than with the first option.

After phosphate has been extracted from the sludge, and a phosphate-rich liquor is available, the phosphate has to be precipitated. A method that has been developed in the Netherlands is the Crystalactor®, where the phosphate is converted into P-pellets (Gaastra et al., 1998; Piekema, this volume). A disadvantage of this technique is that it is rather costly, about 7300€ per ton removed P, which is more expensive than phosphate rock. The high costs are, among other reasons, caused by the need to acidify the P-rich liquor to remove carbonates, to prevent contamination of the precipitate by calcium carbonate. However, following acidification, the liquor requires a pH increase, again by chemical addition, to allow P-precipitation. The Crystalactor® uses sand as a seed for phosphate nucleation, which also has to be removed at the phosphate plant.

As alternative precipitation agents, lime or Al-salts can be used. Lime is inexpensive, but yields very small precipitation particles that are difficult to separate and dry. Precipitation as struvite is attractive to the water industry (see Jaffer and Pearce, this volume), but the ammonia in the precipitate will be converted into NO_x of N_2O during the production process used by Thermphos, in which case a costly gas treatment system is needed. Precipitations with aluminium salts or polyaluminium chloride (PAC) are interesting options. The costs are likely to be lower than the costs of the Crystalactor®, but still more expensive than phosphate rock.

In Table 22.4 the quality of the precipitate of the Crystalactor® and of a potential Al-precipitate are compared with the requirements set by the elemental P-industry. The data for the Crystalactor® are based on STOWA (2001). The data for the Al-precipitate are produced by a preliminary experiment in which P from bio-P sludge was released by adding acetate under anaerobic conditions. After separation of the sludge, phosphate was precipitated using $Al_2(SO_4)_3$. The Cu, Zn and Fe content were calculated from the ratio of their concentrations to the P concentration. Based on these metal contents, the quality meets the requirements of Thermphos.

By recovering phosphate from bio-P sludge it is not possible to retrieve all the removed phosphorus. Part of the phosphate is used for normal microbial metabolic needs and this phosphate will not be released in an anaerobic phase. Based on the COD/P ratio in Dutch sewage, it can be estimated that approximately

Table 22.4 Quality precipitate compared with the requirements of Thermphos.

	P_2O_5 (g/kg ash)	Cu (mg/kg ash)	Zn (mg/kg ash)	Fe (g/kg ash)
Phostrip phosphate precipitate (Crystalactor®)[1]	260	1.6	35	–
Aluminum phosphate precipitate[2]	474	<30	<300	2
Requirements Thermphos	>250	<500	<1,000	<10

[1] STOWA (2001).
[2] based on experiments with bio-P sludge from the sewage treatment plant of Bennekom.

65–75% of the phosphate that is removed from the sewage, and is accumulated in the sludge can be recovered, and that 25–35% will leave the waste water treatment plant with the excess sludge.

22.5 DISCUSSION

We evaluated several scenarios for reuse of phosphate in sewage by the P industry. From the perspective of the industry, sludge incineration ash is not appropriate as secondary P-material, even though from the perspective of the waste water authorities this would be an option easy to implement. The scenario in which phosphate is extracted from bio-P sludge, before it enters the sludge treatment, is attractive for the P industry but is difficult to implement by the water authorities. This is firstly because the phosphate needs to be removed biologically, not chemically, and secondly because the phosphate should be released from the sludge before this enters the sludge treatment facilities. Until now, we have assumed that phosphate recovery should be carried out locally at the plant itself, although, perhaps, it is also feasible for this to be centralized in works treating sludge collected from several treatment facilities.

There are, however, several problems to be addressed before phosphate from sewage may be produced at full-scale, for reuse by the P industry. It is proposed that the way forward may be a pilot scale facility, using the approach of Figure 22.4, and operated to test Al-precipitation. If promising, this should then be followed by a full-scale demonstration. This will aid a better assessment of the future of P-recovery.

22.6 CONCLUSIONS

- Ash from sludge incineration (a mixture of primary and secondary sludge) does not meet the requirements of the P-industry in the Netherlands because of its high iron, copper and zinc content.
- The copper and zinc content of ash from bio-P sludge is also too high; this ash is therefore unlikely to be useful as secondary material for the P industry.

- Recovery of phosphate from bio-P activated sludge followed by precipitation in a Crystalactor®, or with Al-salts, results in an end product that can be reused by the P industry.
- Pilot plant research focused on recovery by release from bio-P followed by Al-phosphate precipitation is proposed as the next step towards reuse of P from sewage by the P industry.

REFERENCES

CBS (2000) Waterkwaliteitsbeheer. Zuivering van afvalwater, 1998, *CBS*, Voorburg/Heerlen, 59 pp.

Donnert, D. and Salecker, M. (1999) Elimination of phosphorus from municipal and industrial waste water. *Water Sci. Technol.* **49**(4–5), 195–202.

Etienne, P., Marie-Line, L. and Mathieu, S. (2001) Excess sludge production and costs due to phosphorus removal, second international conference on the recovery of phosphorus from sewage and animal wastes, Noordwijkerhout, The Netherlands, 12–13 March.

Gaastra, S., Schemen, R., Pakker, P. and Bannink, M. (1998) Full scale phosphate recovery at sewage treatment plant Geesterambacht, Holland, international conference on phosphorus recovery from sewage and animal wastes, 6–7 May, Warwick, UK.

Jaffer, Y., Clark, T-A., Pearce, P.A. and Parsons, S.A. (2001) Pilot struvite precipitation reactor at slough sewage works, UK., second international conference on the recovery of phosphorus from sewage and animal wastes, Noordwijkerhout, The Netherlands, 12–13 March.

Levin, G.V. and Shapiro, J. (1965) Metabolic uptake of phosphorus by waste water organisms. *J. Water Pollut. Control Fed.* **37**, 800–821.

Schipper, W.J., Klapwijk, A., Potjer, B., Rulkens, W.H., Temmink, B.G., Kiestra, F.D.G. and Lijmbach, A.C.M. (2001) Phosphate recycling in the phosphorus industry. *Environ. Technol.* **22**, 1337–1345.

STOWA (2001) Hergebruik van fosfaatreststoffen uit rwzi's, STOWA rapport, 2001–08.

Van Loosdrecht, M.C.M., Brandse, F.A. and Vries, A.C. de (1998) Upgrading of waste water treatment processes for integrated nutrient removal – The BCFS® process. *Water Sci. Technol.* **37**, 209–217.

Van Wersch (2001) personal communication.

Wild, D., Kisliakova, A. and Siegrist, H. (1997) Prediction of recycle phosphorus loads from anaerobic digestion. *Water Res.* **31**(9), 2300–2308.

23
Phosphorus recycling: regulation and economic analysis

J. Köhler

23.1 INTRODUCTION

This chapter takes a rather different viewpoint from the others in this book: that of economic, as opposed to scientific or engineering approaches. This analysis does not examine the question of whether the recycling of phosphorus is feasible or practicable. Rather, it asks the questions: what are the pressures that have made the recycling of phosphorus an important practical issue? What are the commercial possibilities for different technologies and their different products?

One result should be made clear at the beginning. Recent surveys on the behalf of the phosphate industry (Jeanmaire, 2001; Jeanmaire and Evans, 2001) and the European Commission on sludge recycling (CEC 2002) examined the costs associated with recycling of phosphorus and found that in the few cases where full scale phosphorus recycling plants were operating in waste water treatment plants (WWTPs), the costs of investment and operation significantly outweighed the

© 2004 IWA Publishing. *Phosphorus in Environmental Technology: Principles and Applications.*
Edited by Eugenia Valsami-Jones. ISBN: 1 84339 001 9

revenues from selling products of the recycling operation. Therefore, this technology has not developed because of expectations of large profits. It has also been pointed out that WWTPs are often not run as a commercial operation, but as a service provided by public authorities with a large subsidy. The evidence in the relatively sparse literature on the economics of waste water treatment will be examined below. Recycling of phosphorus cannot be examined just as a commercial investment decision. While the commercial calculation plays a part in the analysis, it is necessary to take a much broader view.

This chapter will survey the regulation of WWTPs and waste water streams, together with the literature that has used economic methodologies to examine WWTP operation and phosphorus recycling. Section 23.2 will consider the various reasons why recycling phosphorus may be desirable. This also includes an overview of the potential uses to which phosphates, either separated or as a component of sludge output from water treatment works or WWTPs, may be put. Section 23.3 briefly reviews the relevant regulations and surveys the literature that has applied economic analysis to these issues. Section 23.4 proposes a system view which enables the many factors in the economics of phosphorus recycling to be considered in a comprehensive and consistent way. Section 23.5 concludes with an assessment of the economics of phosphorus recycling.

23.2 WHY RECYCLE PHOSPHORUS?

It is well known that phosphorus in waste water are an issue because of eutrophication. Köhler (2001) and Dils *et al*. (2001) summarise the environmental impacts and discuss the policy issues. Köhler (2001) looks at detergent phosphates and considers policies to reduce the use of phosphates in detergents. Eutrophication is the nutrient enrichment of water, which can lead to the growth of algae and cyanobacterial blooms in surface waters. These are unsightly, often have an unpleasant odour and can be toxic. Eutrophication is caused by inputs of nutrients, phosphorus and/or nitrogen, into surface water ecosystems that are far higher than the natural level. The main sources of phosphorus in Western Europe are animal manure and fertilisers used in intensive livestock agriculture and human waste in urban waste water. Dils *et al*. (2001) quote figures of 24% from household wastes, 34% from livestock 10% from detergents and 7% from industry for the EU. Phosphates used in domestic laundry detergents may make a significant contribution to the phosphorus content of urban waste water in some areas. Phosphates, and STPP (sodium tripolyphosphate) in particular, perform a vital function in modern synthetic detergents, although there are substitutes, of which the most successful is zeolite combined with polycarboxylate.

Dils *et al*. (2001) report the findings of the EEA report *Nutrients in European Ecosystems* (EEA, 1999): 'European lakes and reservoirs are heavily affected by anthropogenic nutrient pollution, the (ecological) condition of many lakes and reservoirs being far from satisfactory, with widespread impacts on recreational and other water uses. Concentrations of phosphorus, which is generally the limiting nutrient in freshwater situations, were found to be elevated in many areas. As regards rivers, while phosphorus concentrations were found in general to have fallen significantly in recent years, they were still excessive at practically all monitoring stations, generating unwanted plant growth and giving rise to blue-green algal blooms in slow-flowing waters. For transitional (estuaries) and marine waters, where phosphorus is one of the key factors controlling plant growth, whilst recognising the difficulties in understanding changes due to natural causes, eutrophication was considered a widespread, transboundary phenomenon, affecting marine biodiversity, fish and shellfish stocks, human health and recreational uses of coastal zones.'

In order to control eutrophication, many countries have acted to control the use of STPP in laundry detergents, sometimes by voluntary industry action rather than legislation. Laundry detergent formulations using STPP are no longer sold in Germany, Italy, Switzerland, Austria and Norway in Europe, as well as the US and Japan. In terms of policies for detergent phosphates, any further controls by regulation or taxation would be very unlikely to influence the extent of cyanobacterial blooms and algae. Since each local problem has to be resolved by regional action, policies such as a general tax on detergents are not relevant as environmental policies.

The environmental problems have also led to legislation in many countries imposing controls on the level of phosphate in outflows from WWTPs. In the EU, the decisive step was the adoption of the directive on waste water treatment (CEC, 1991) which requires all WWTPs serving significant centres of population to install nutrient removal plant, where phosphorus and/or nitrogen is removed, but generally both. This results in the production of either phosphate rich sludges or phosphates as a byproduct of the removal process. There is then a question of what to do with these phosphate rich products. Drinking water treatment works are faced with similar issues (UKWIR, 1999).

There are further reasons for extracting phosphate from waste water flows. Driver *et al.* (1999) describe the imbalance between phosphorus concentration in urban areas through the sewage system and the need for phosphorus in agriculture. Dils *et al.* (2001) describe the current state of phosphate reserves and look at the uses of phosphates. In agriculture, the use of inorganic chemical phosphate fertilisers is declining in the UK. The use of animal manures and sludges has to be carefully controlled to maintain a sustainable nutrient balance. Günther (1997) considers the flows associated with phosphorus and the resulting imbalances in more detail. Driver *et al.* (1999) and Ueno and Fujii (2001) also make the point that phosphate rock is a non-renewable resource, with a considerable heavy metal content which requires treatment. Dils *et al.* (2001) argue that phosphorus recycling is necessary to achieve a sustainable system of phosphorus use. The quality of phosphate rock as a raw material is deteriorating as the natural resources are steadily depleted so that recycled phosphorus or phosphorus rich sewage sludges generally have a far lower heavy metal content. Phosphate extracted in WWTPs may therefore be used as an input to the phosphate industry (Schipper *et al.*, 2001). Durrant *et al.* (1999) review the potential for recovering struvite from WWTPs for use as a raw material or as a slow release fertiliser. Pervious chapters in this book also address aspects of phosphorus recovery.

Parsons *et al.* (2001) and Williams (1999) describe the problems caused by struvite deposits in WWTPs, which result in considerable expenditures. These problems can also be ameliorated by phosphorus removal, to reduce the uncontrolled formation of struvite.

A further reason for phosphate extraction is that there may be a need for reduction in phosphorus content of sludge from WWTPs. If the sludge is to be recycled by spreading on agricultural land, the permissible rate of sludge application may be limited by the phosphorus content (Edge, 1999). Another potential application is the use of sludges in cement manufacture, where the sludge is incinerated as a fuel source and the ashes used in the cement. Jeanmaire and Evans (2001) point out that the short term resistance of the cement is reduced by the presence of phosphate and they quote an OFEFP technical document (OFEFP, 1991) recommending a limit value of 0.5% P_2O_5 by weight in cement. This could also require phosphorus extraction from sludges.

Phosphates extracted from WWTPs can be useful products. Gaterell *et al.* (2000) find considerable potential for the use of struvite directly as a fertiliser. Ueno and Fujii (2001) report that phosphate from WWTPs can be sold as an input to horticultural/domestic fertiliser (see also Ueno, this volume).

23.3 REGULATION AND ECONOMICS OF WWTPs AND PHOSPHATES

23.3.1 Regulation of WWTPs

Sewage treatment, when viewed as an economic activity, is a 'natural monopoly'. This is because there are large cost savings (so-called 'economies of scale') from having a single plant and sewage infrastructure serving any geographical area. As in water, gas or electricity supply, the provision of competing sewage piping systems leading to competing sewage plants for each household would involve much larger expenditures than having a single system. This has the implication that economic forces result in a single operator in a particular geographical area. If this operator is a private company, they enjoy the advantages of being a monopolist i.e. they can potentially charge high prices for their service. Therefore, WWTPs that are privately operated are subject to government regulation of their services and prices. Rees (1998) examines the competitive characteristics of the water sector. She considers that sewage treatment is characterised by numerous local monopolies, although sewage capacity construction is competitive. She concludes: '...the social, developmental and environmental importance of the water sector means that continued public regulation will be inevitable.' Rees (1998 p.104).

A second important point is that sewage treatment is a public health issue. WWTPs were introduced as a means of reducing the danger to the population from the large quantities of sewage and industrial waste water produced by the growth of urban centres during industrial development (Driver *et al.*, 1999). They were a part of government public health provision, owned and operated by the municipal government in the US and Japan as well as the EU.

The recent trend towards privatisation of public services has had a major impact on the operation of WWTPs, most fundamentally in the UK, where the complete system has been taken over by the new water companies. Neto (1998) compares the UK and French approaches towards privatisation. The French system has seen decentralised privatisation policies, often with contracting out of WWTP management. He concludes that the UK system requires an effective regulatory machinery to prevent excessive profits for the water companies and to ensure adequate infrastructure investment. This investment was one of the major problems of public sewage system provision, with underfunding by governments leading to decaying infrastructures. Also, the new requirements (discussed in the next section) for more extensive treatment of WWTP discharges are requiring considerable new investment. The French approach permits a weaker regulatory system. Spulber and Sabbaghi (1998) perform a system optimisation analysis of water provision and treatment and discuss regulation compatible with privatisation. They argue for franchise competition as a more efficient form of water service provision than public provision of water services.

In recent years, the situation in the EU has been driven by the EU Directive of 1991 concerning urban waste water treatment (CEC 1991). This requires the installation of plant to dramatically reduce the phosphorus content of WWTP discharges. Similar legislation applies in Japan (Ueno and Fujii, 2001). In the US, the so-called Clean Water Act (EPA 2002) controls discharges of pollutants into US waters and has provided the basis for funding of WWTPs in the US. This legislation, especially in the EU and Japan, implies the removal of a high proportion of phosphorus from WWTP discharge streams. These requirements have generated the development of many new treatment processes in the last 10 years. These processes may generate phosphorus-enriched sludge or phosphate products. The problem then is what to do with these byproducts, which may contain other potentially harmful constituents such as heavy metals, pathogens and dioxins. The relevant EU policies and legislation on sludge disposal are considered later in this chapter.

23.3.2 Economic analyses

A consequence of the public provision of sewage services has been that commercial economic considerations did not apply to WWTP operation or sewage infrastructure. Although there are few economic analyses in the literature, those that have been undertaken are unanimous in arguing that this has led to economically inefficient operation and pricing of sewage services. There is a more extensive literature on the economics of water resources in general; Spulber and Sabbaghi (1998) and Renzetti (1999) are suitable starting points. There are a few papers that employ modern microeconomic theory, with its emphasis on contracts, bargaining and optimisation under conditions of limited information and a few that use econometric techniques for data analysis of the economics of WWTP operations.

Feinermann et al. (2001) consider the case of a coastal city generating effluent, which can be treated and recycled as an input to agricultural production. They consider three alternative regimes: 1) a social planner maximising the combined payoff (welfare function) of the city and agriculture 2) regulation of the price paid by the city to discharge/treat the effluent and 3) bargaining over the price and quality of effluent. Because the effluent can be used in agriculture, it has value to farmers, so the 'polluter pays principle', whereby the city households would bear all the cost of treatment or effluent discharge, is economically inefficient.

Spulber and Sabbaghi (1998) is an extensive treatment of the theoretical issues of water resource management and policy. Their emphasis is on the reuse of water rather than the treatment of sewage. They include a theoretical application to water pollution of the policy instruments developed in environmental economics: Pigovian taxes, regulated standards, tradeable effluent permits, subsidies etc. These tools of analysis could also be applied to policies specifically for WWTPs. Thomas (1995) uses optimal taxation and regulation theory to consider incentives for industry to invest in waste water treatment. The optimal policy regime is that most (less efficient) firms should simply be taxed for discharging polluting waste water. The most efficient firms should be regulated and given investment subsidies to generate the optimal quantity of waste water.

Strudler and Strand (1983) consider a charge per unit volume of sewage instead of a fixed charge. Using data from Maryland, USA, they conclude that this policy would reduce sewage production and therefore sewage treatment costs. Bhansali et al. (1992) find that the funding provisions of the US Clean Water Act have led to considerable subsidies in the provision of sewage treatment. They find that the consequence of this has been that firms have substituted discharges to WWTPs for other inputs to production, creating an (economically) inefficient extra demand for WWTPs. Since firms' direct discharges are regulated and require capital intensive treatment plant, they argue that the act is biased against small firms.

Renzetti (1999) looks at municipal water supply and sewage treatment in Ontario. He estimates cost functions and water demand functions and evaluates the differences between estimated (actual) prices and marginal costs. He finds that prices charged by municipalities are much lower than the economically efficient costs. This encourages excessive water consumption, overexpansion of water supply and sewage treatment and discourages technological innovation in water conservation and sewage treatment. Ashton (2000) used a panel data model to estimate cost functions and calculate operating cost efficiencies of UK privatised WWTPs. He reports dispersion in cost efficiencies across the different companies, partly due to variations in the operating environment. This confirms the results of Stuart (1993, 1994), who performed an estimation of water and sewage plant efficiencies using econometric efficiency frontier techniques.

This relatively limited literature finds that pricing of sewage treatment services is usually inefficient, especially in government owned and operated WWTPs. Many authors argue for greater privatisation of waste water treatment, but all of these analyses suggest that continued regulation of prices and standards will be required to control water services prices.

23.3.2.1 Costs of tertiary treatment and phosphorus recovery in WWTPs

Since the processes for phosphorus recovery have only recently been developed, there are few full scale installations for phosphorus recovery in WWTPs and relatively little experience of their operation. Reports of costs of the different processes should be treated as preliminary. Consequently, it is not possible at the current state of the technologies to give a full picture of the relative advantages and disadvantages (operational or economic) of the various processes that are now being developed. Costs of processes are quite often mentioned in the reports of phosphorus recovery technologies, but these usually form incomplete economic analyses.

Jeanmaire (2001) surveys the available cost information on phosphorus recycling. He argues that struvite sale prices may vary widely, depending on such factors as: cost of the sludge disposal processes, cost of reagents, cost of fertilisers etc. Struvite sale prices are often close to chemical fertiliser prices, which suggests that fertilisers set an upper limit to struvite sales prices. He also finds that recovered phosphate is often much more expensive than phosphate rock. This cost information is summarised in Table 23.1, taken from Jeanmaire (2001).

From this information and the survey that was undertaken of WWTP operations it is argued that the market value of the recovered phosphate is insufficient to

Table 23.1 Costs of recycled phosphate compared to traditional sources.

Comparison for P recycling in the industry

Optimised costs recovered P (S.Gaastra)- average value	P rock costs (Northern Europe)
10.2 €/kg P that is to say ~10,200 €/ton of P	30£/t P rock or ~45US $/t P rock[1] that is to say ~200£/ton of P ≈ 320 €/ton of P

Comparison for P recycling as fertiliser

Crook[1] (2001)	Jaffer et al. (2001)	Kurita Ltd	Treviso Battistoni	Fertiliser average price K.Smith-M.Marks[6] Web site ADEME[7]	Fertiliser average price DAP
86 £/t of recovered product[2]	200£/t str[3]	150–200 US$/t str[4]	0.34 €/kg str[5]	300£/t P_2O_5[6] 3300 F/t P_2O_5[7]	233£/t P_2O_5
?	2667€/t P estimated costs	1235–1647 €/t P real costs	2833€/t P estimated costs	1104€/t P 1153€/t P	858 €/t P

[1] P rock contains ~35% of P_2O_5 that is to say ~15% of P.
[2] Purity of the recovered product not transmitted.
[3] 1 kg of struvites contains ~12% P (this number only comprises 1/3 of the cost of reagents.
[4] 1kg of struvites contains ~13,5%P.
[5] 1 kg of struvites contains ~12%P.
[6] *K.Smith-M.Marks:* 300 £/t P_2O_5.
[7] *Web Site ADEME:* 3300 F/t P_2O_5.
Source: Jeanmaire (2001, p. 22).

motivate investment in phosphorus recovery. Different countries have different motivations for undertaking such investment, e.g. struvite deposition in WWTPs is a particular problem in some UK WWTPs. Jeanmaire (2001) concludes that a general statement about the economics of phosphorus recovery is not possible; it depends on the national context.

This book provides updates of some of these numbers. Jaffer and Pearce in Chapter 17 above have a more recent analysis of the profitability of struvite precipitation from a WWTP. They still find that the process is not commercially viable, as the cost of input magnesium exceeds the projected revenues from sale of struvite. Battistoni in Chapter 18 above reports a range of costs of the struvite precipitation process at Treviso of 4330 to 10,500€/ton phosphorus removed. Piekema in Chapter 19 reports on the upgrading of the Geestmerambacht WWTP using Crystalactor® technology. Total cost, including all consumables, operation and yearly capital cost, is around 6€ per kg P recovered. While this is higher than the figures in Table 23.1, it includes capital as well as operating costs and he compares this to a figure of 3.5€ per kg P for other WWTPs. Part of the difference is accounted for by the production of recovered phosphorus as opposed to just phosphorus removal.

Paul et al. (2001) compare operating costs of WWTPs in France, for phosphorus removal by Enhanced Biological Phosphorus Removal (EBPR), chemical precipitation with ferric chloride and combined EBPR and chemical precipitation plants. The total costs are shown in Table 23.2.

Liberti et al. (2001) report an economic comparison of three methods for phosphorus recovery: chemical precipitation, ion exchange combined with chemical precipitation and their REM NUT process. Assuming prices for struvite of 500€/t and 33€/t figure for Fe-phosphate, revenues would cover >40% of operating and maintenance costs for the REM NUT process compared to as low as <3% for chemical schemes.

Nawamura et al. (2001) report a cost of 56 Yen (approximately 0.5€) /kg struvite, using sea water to replace magnesium input to reduce the cost. Subramanian and Arnot (2001) describe a two-stage process for recovery of phosphorus. The first stage is an anaerobic membrane bioreactor, followed by an adsorption column to recover the phosphorus. Overall phosphorus removal was 80% at a cost of $155 m^{-3} of feed sludge. Çelen and Türker (2001) calculate from laboratory tests

Table 23.2 Total operating costs for different removal processes; for typical French waste water with BOD/P 25; Readily Biodegradable COD/COD 8% and 80% phosphorus removal.

Process	Total cost ($€·kg^{-1}$)$P_{influent}$
EBPR	0.26
Chemical precipitation	1.76
Combined plant	0.80

Source: Paul et al. (2001, p.1370).

the cost of chemicals to produce struvite. They report costs of 6.4–8.07 $/kg NH_4^+–N, which compares with costs of 4.55–14.9$/kg NH_4^+–N from their survey of other calculations. Crook (2001) reports that phosphorus removal by Enhanced Biological Nutrient Removal (EBNR), producing calcium phosphate, would require a price of 134 €/t for the phosphate product for the process to be economically viable. Evans (2001) examines the financial consequences for biosolids recycling to land of recovering a proportion of the phosphorus from the biosolids. The conclusion is that the savings for the recycling operation would be around £1/t dry solids of original biosolids produced. In terms of the phosphorus recovered, it is of a similar order to the UK dockside price of imported phosphate rock. Where biosolids are recycled to land, most of the financial justification for recovering phosphorus would have to come from the value of the recovered phosphates and/or from the cost avoidance of preventing struvite scaling.

Karlsson (2001) analyses a process for recovering Fe phosphate. Based on a cost of €4/kg phosphate, the process is more expensive than spreading sludge on agricultural land, but cheaper if the sludge is disposed of in landfill or incinerated. Von Munch *et al.* (2001) propose construction of Unitika's Phosnix process in Australia. Assuming a sales price of €0.3/kg struvite, they calculate a return on equity of 44% over 5 years, with profits returned in the second year of operation. Jeanmaire and Evans (2001) calculate a saving in WWTP operation of £101 per tonne phosphorus recovered, half the market price of phosphate rock (£207/tP) at port of entry to the UK quoted by Gaterell *et al.* (2000). Ueno and Fuji (2001) report that struvite is sold as domestic/horticultural fertiliser at €250/tonne, with transport costs covered by customers.

23.3.3 Sludge spreading and other uses for recycled phosphates

Tertiary treatment plants produce phosphorus rich sludge, which has to be dealt with somehow. Similar considerations apply to sludges produced by drinking water treatment works (UKWIR 1999). There are a range of alternative uses or possibilities for the sludge and/or the products of phosphorus removal: struvite, Fe phosphate, Ca phosphate etc. The phosphorus rich sludge can be disposed of by use as:

- an alternative to mineral fertiliser – spread on agricultural land, forestry, brownfield sites, parks or gardens etc.;
- it can be incinerated, either just as waste or to provide power and in the case of the cement industry, to provide power and to use the ashes in the cement production process;
- it can be put into landfill or used in construction such as pavements;
- it has been dumped at sea, but this practice has been abandoned.

The phosphate products can be used as:

- a feedstock to the phosphate industry, to replace phosphate rock;
- a mineral fertiliser e.g. struvite;
- an input into mineral fertiliser manufacture or mixing with compost etc.

The most cost effective method of disposal is usually recycling as fertiliser, either by sludge spreading or by application of struvite on agricultural land. A similar use in the US is for reclaiming brownfield sites such as disused mine spoil heaps. There is a large literature on the disposal of sludge; it is also closely regulated because of the potential environmental problems: too high nutrient (N and/or P) content (Netherlands, Flanders and potentially in the UK), heavy metal content, parthogens and dioxins. Smith (1996) presents an extensive treatment of the environmental impacts of spreading sludge.

The EU has regulated sludge processes and is continuing to develop legislation. CEC (2002) is an extensive review of the subject, surveying acceptance of spreading by the different member states, the relevant regulations and an assessment of economic impacts of EU legislation. Spinosa (2001) summarises current EU regulation. Sewage sludge output from WWTPs or other sources is treated as a waste under the provisions of European Directive 91/156, the Waste Basis Directive. Also for the EU, the Urban waste water Directive 91/271, Directives 89/369 (prevention of the atmospheric pollution deriving from new incineration plants of urban wastes), 94 /67 (incineration of hazardous wastes), 00/76 (incineration of wastes) also apply. Directive 86/278 on the protection of the environment when sewage sludge is used in agriculture will soon be replaced by a new directive. This new directive will differentiate between advanced and conventional treatments for sludge hygenisation and odour reduction. After advanced treatment, there are no restrictions on spreading on agricultural land, urban areas, trees, land reclamation etc. After conventional treatment, sludge can be used only by deep injection and with time restrictions on grazing time, harvesting and public access. Limit values for heavy metals are reduced, in particular from 20–40 to 10 mg Cd/kg dry matter and values for organic micropollutants and dioxins are proposed.

Directive 00/76 requires that incineration takes place at 850°C (1100°C for wastes with more than 1% halogenated organic substances) and automated cutoffs must be provided. These requirements including measurement and control imply capital intensive incineration plants, although incineration may still be cost effective in large urban areas with high transportation costs for spreading or landfill. There are also strict regulations for the disposal of sludge in landfill. Given the limited availability of landfill sites, Directive 99/31 introduced targets for the reduction of biodegradable municipal waste to be landfilled. Spinosa (2001) considers that while European environmental politics is pushing towards recycling of sludge, the severe restrictions make this very difficult. Given the stringent requirements on the other disposal methods, the costs associated with sludge use/disposal are rapidly increasing in the EU.

Dils *et al.* (2001) also mention the following policy measures: the Nitrate Directive (91/676/EEC), which addresses nitrogen inputs to surface and ground waters from agricultural sources and the Habitats Directive (92/43/EEC), which requires controls on activities affecting waters of high nature conservation value, including control of nutrients in discharges where appropriate. They consider that the more recent Water Framework Directive (2000/60/EC) is likely to be a future driver for both point and diffuse source nutrient control in catchments where eutrophication is a concern. In 1998, under the OSPAR Convention, a strategy to combat eutrophication in marine waters was adopted and is likely to drive measures to control point and diffuse nutrient inputs to waters identified as 'problem areas'. In addition, the 1992 Convention on Biological Diversity, signed at Rio, has led to national habitat and species action plans that include measures to control eutrophication (Dils *et al.*, 2001, Section 4).

CEC (2002) reports that national regulations are often stricter than Directive 86/278. In Flanders and the Netherlands, much stricter limitations on heavy metals are in force, such as to effectively prevent the spreading of sewage sludge. Switzerland has recently announced a complete halt to sludge spreading. Sweden and the UK have voluntary agreements and sludge spreading is widely practiced in the south east of the UK in arable farming. Spreading is quite widely practiced in Germany, while there is limited debate in Spain, Italy or Greece.

Towers (1994) looks at policy in Scotland, where the limiting factor is the heavy metal content. In the Borders, Scottish sludge production would require disposal of sludge on 0.2% of land in the Borders region, but on 3.8% of land in the more urban Strathclyde region. This target will be difficult to achieve in the latter case. Sawkins and Dickie (1999) report on the regulation of the Scottish water industry.

Sweden has been at the forefront of dealing with phosphorus recycling. Wallgren (2001) summarises Swedish policy, which was an early example of the installation of phosphorus extraction to reduce eutrophication. This then generated the phosphorus rich sludge and there is now a significant problem about what to do with this sludge. Some of it is spread on land, but there is continuing resistance to this (CEC, 2002; Kvarnström and Nilsson, 1999). Eckerberg (1997) analyses the policy response to eutrophication in Scandinavia; he finds that the targets for reducing nitrogen and phosphorus loading from agriculture have not been met. Voluntary measures have dominated the economic impact of policy changes. Reduction of production subsidies and subsidies on manure and set-aside may help, while taxes on fertilisers have been too small to have an impact.

In their cost analysis, CEC (2002) finds that landspreading is more cost effective (110–160€/ton of dry matter) than landfilling or incineration (260–350€/ton of dry matter) on average. This includes the external costs of the environmental impacts. The agronomic value of the sludge can be 10–30% of the cost of landspreading. Sludge management costs are low compared to overall water management costs, although this may no longer be the case in 2003. Linster (1991, p. 333) compared heavy metals regulation in different OECD (Organisation for Economic Cooperation and Development) countries. The UK and US were exceptional in not having limits on heavy metals in sludge, but only in soils. The UK is, of course, subject to the EU directives and now has a considerably more strict regulation.

There is also an issue of the acceptance by farmers, food companies and the general public of sludge spreading in agriculture (CEC, 2002). In the Netherlands and Flanders, the excessive supply of nutrients from agricultural wastes as well as sewage has resulted in sludge spreading being stopped. In Sweden, the perception of health risks by food retailers and the public has led to an intensive debate. Kvarnström and Nilsson (1999) surveyed operators of sewage treatment works.

They report a price of between 11 and minus 20 SEK/kg P for recycled phosphates, compared with a fertiliser price of 10 SEK/kg P. Farmers are hesitant to use sludge, due to pressures from retailers and there is no significant incentive to do so in many cases where the costs of sludge spreading are the same as chemical fertilisers.

Hall (1992) reported costs of £250 million/year for UK sludge treatment, around 50% of sewage treatment costs. Fertilizer Society (1995) argues that treatment to remove heavy metals would make sewage sludge more expensive than mineral fertilisers.

23.3.4 Conclusions from the literature

The evidence presented in section 23.3 has some clear lessons. Whether private or public, WWTPs and their output are closely regulated and the standards are continually tightening. This implies more and more treatment of the sewage and of the products, which will be more and more expensive. There are some ways of using the products which have a commercial value, but this is generally a fraction of the treatment costs. Therefore, the economic problems of recycling phosphorus are:

- given that regulation is driven by public health concerns and availability of locally varying disposal options, what are the capital and operating costs of the possible compliant treatment technologies;
- are there markets for the sludge or recycling products and are there possibilities to expand markets by e.g. reducing consumer/retailer/farmer resistance;
- given various disposal and recycling options, what is the most cost effective way of meeting the regulatory requirements;
- what is the best economic way to own and operate and regulate WWTPs;
- who should pay for waste water treatment and how should charges be decided;
- what are the related policy measures, in particular for use of fertiliser, that will contribute to the levels of phosphorus in the environment and should therefore be considered when developing policy for recycling phosphorus.

This may seem a long list of issues which are far removed from phosphorus recycling chemistry and technology, but the above survey illustrates many of the interactions between them. It is a typical feature of environmental issues that they involve considerations of many different scientific and economic aspects, such that a proper assessment of the impacts and costs requires a very broad approach. The next section suggests a way of thinking about this problem, by treating phosphorus flows using a systems analysis approach.

23.4 A SYSTEM APPROACH

Günther (1997) has a system view of the flow of phosphorus through the economy and ecosystem. While his conclusions are not particularly relevant to the

Phosphorus recycling: regulation and economic analysis 541

current analysis (a change of the socio-political system is proposed), considering phosphorus flows using a system approach is a very good way of taking a broader view of the economics (or science) of phosphorus recycling. Spulber and Sabbaghi (1998), Chapter 7, also use a system approach to perform their analysis of water reuse and recycling.

Figure 23.1 applies the idea of identifying phosphorus flows, with an emphasis on flows relevant to WWTPs and phosphorus recycling. This diagram enables the different impacts associated with the various scientific possibilities for recycling phosphorus to be identified. The following elements of flows of phosphorus that impact directly on WWTPs and phosphorus recycling operations can be seen:

- the flow of waste water into the WWTP;
- extraction of phosphate during WWTP processing, leading to reduced struvite deposition in the WWTP in some circumstances and/or improved return streams to the head (Jeanmaire and Evans, 2001);
- production of recycled phosphates either during the waste water treatment or from sludge treatment;

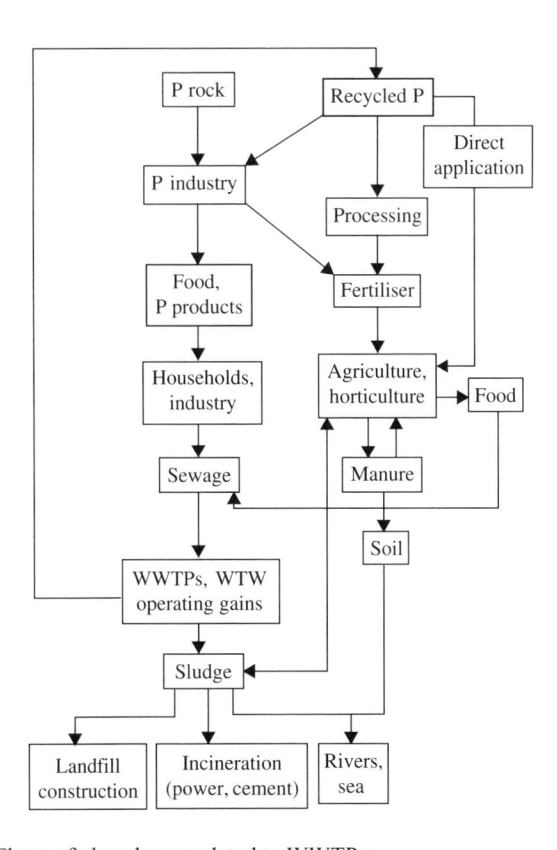

Figure 23.1 Flows of phosphorus related to WWTPs.

- production of sludge, with different possible phosphorus content depending on whether phosphates are extracted.

As listed in a previous section, the recycled phosphates can be used as feed to the phosphate industry, combined with mineral fertiliser or used directly in agriculture or horticulture. Sludge can be spread on land, dried and used in landfill or construction or incinerated.

This then provides a basis for analysing the economics of recycling phosphorus. Given regulatory requirements that limit phosphorus in WWTP discharges, phosphorus must be extracted. This will require extra treatment, involving increased capital and operating costs. The problem is to find the most cost efficient method of extraction. Therefore, the extra costs must be estimated and also the extra potential benefits or revenues from the various recycling possibilities. This requires an estimate of the size of the market and potential sales price. In the case of e.g. sludge, considerable effort and expense may be involved in developing a market for the sludge and transporting the sludge. If the sludge or phosphate require further treatment to be used, e.g. as input into the phosphate industry, the costs of this further treatment must also be included. Jeanmaire (2001) and Jeanmaire and Evans (2001) have a table that considers the benefits of phosphorus recovery, reproduced here as Table 23.3, which lists some of these expenses and benefits. However, Figure 23.1 shows that a broader view may be required to allow for all the possible consequences. The different possibilities can be compared using the usual investment appraisal calculations: Net Present Value or Rate of Return on capital etc.

The regulation of prices that may be charged for waste water treatment also should take into account the full range of operating possibilities. Thomas (1995) and Feinermann *et al.* (2001) are examples of how this can be done. It is important to emphasise here that local conditions vary considerably, e.g. in the incidence of struvite deposition in the WWTPs, in the availability of land for sludge spreading, in the national regulations and policy on incineration. This also applies to the design of charging regimes, which influence the demand for waste water services.

Table 23.3 The expenses and receipts of recovering phosphates in WWTPs.

Expenses	Financial benefits
- Investment and depreciation costs - Consumption of reagents and of energy - Additional labour costs/time - Specific training of the operating staff	- Sale of the recovered phosphates - Savings due to the suppression of problems of struvite deposits - Savings due to the optimisation of returns to head - Possible reduction of transportation distances for agricultural reuse - Savings on the landfill of incineration ashes

Sources: Jeanmaire (2001), Jeanmaire and Evans (2001).

23.5 CONCLUSIONS: ECONOMICS AND POLICIES

Phosphorus recycling is a relatively new topic of analysis. Consequently, there are as yet few analytical economic studies. There are some studies of water treatment provision, which do come to some definite conclusions. The main conclusion has been that the pricing of waste water treatment has been determined by public health and regulatory considerations. This has resulted in economically inefficient pricing regimes.

The studies of the costs of operation of phosphorus recycling in WWTPs give various different numbers for the different technologies and recycling possibilities. Because the technologies are new, much of the available information is based on laboratory analysis or pilot plants, which are not a complete picture of the costs of full scale investment and operation. However, the estimates mostly show that phosphorus recycling will increase the net costs of waste water treatment, except where there are specific local economic drivers (such as high sludge disposal costs or phosphorus-related operating problems and associated costs). One situation which might reverse this would the exhaustion of the phosphate rock raw material, although this is still a longer term prospect.

The economic problem of phosphorus recycling is then how to minimise the cost of treatment, allowing for the costs and revenues of the recycled products, and how to design regulations that provide incentives for optimising costs and revenues.

There is a long history of regulation of waste water treatment, which has already expanded to include the various recycling possibilities. EU regulations now demand operation of WWTPs such that it is necessary to extract phosphates. In the EU and in Japan, the regulatory regime is becoming ever stricter, with the implication that costs will increase. This means that recycling opportunities should be identified allowing for local conditions and where economically and environmentally beneficial, should be exploited in order to keep the increase in costs down.

These various factors make a wide analysis of the problem necessary. It is not sufficient just to consider the costs of phosphorus extraction in WWTPs. The system approach outlined in the previous section shows how this may be undertaken.

REFERENCES

Ashton, J.K. (2000) Cost efficiency in the UK water and sewerage industry. *Appl. Econ. Letters* **7**, 455–458.

Bhansali, A., Diamond, C. and Yandle, B. (1992) Sewage treatment as an industry subsidy. *Econ. Geogr.* **68**(2), 174–87.

CEC (1991) Council of the European Communities (1991) Council Directive of 21 May 1991 concerning urban waste water treatment (91/271/EEC). *Official J. Eur. Comm.* No. L135/40–52.

CEC (2000) Working document on sludge 3rd Draft European Commission, Brussels, available at http://europa.eu.int/comm/environment/sludge/

CEC (2002) Disposal and recycling routes for sewage sludge. Synthesis report for DG Environment B/2, available at http://europa.eu.int/comm/environment/sludge/

Çelen, I. and Türker, M. (2001) Recovery of ammonia as struvite from anaerobic digester effluents. *Environ. Technol.* **21**,1067–1084.
Crook, A. (2001) Experimental investigation of effective and economic methods to recover phosphorus from returned activated sludge. *Proceedings cd, 2nd International Conference on Recovery of Phosphates from Sewage and Animal Wastes*, Holland, CEEP.
Dils, R., Leaf, S., Robinson, R. and Sweet, N. (2001) Phosphorus in the environment – why should recovery be a policy issue? *Proceedings cd, 2nd International Conference on Recovery of Phosphates from Sewage and Animal Wastes*, Holland, CEEP.
Driver, J., Lijmbach, D. and Steen, I. (1999) Why recover phosphorus for recycling, and how? *Environ. Technol.* **20**(7), 651–662.
Durrant, A.E., Scrimshaw, M.D., Stratful, I. and Lester, J.N. (1999) Review of the feasibility of recovering phosphate from wastewater for use as a raw material by the phosphate industry. *Environ. Technol.* **20**(7), 749–758.
Eckerberg, K. (1997) Comparing the local use of environmental policy instruments in Nordic and Baltic countries – the issue of diffuse water pollution. *Environ. Politics.* **6**(2), 24–47.
Edge, D. (1999) Perspectives for nutrient removal from sewage and implications for sludge strategy. *Environ. Technol.* **20**(7), 759–763.
EEA (1999) *Nutrients in European Ecosystems.* European environment agency, Copenhagen, Denmark.
EPA (2002) *Clean Water Act.* http://www.epa.gov/region5/water/cwa.htm
Evans, T. (2001) Implications of Within-WwTP. P-Recovery for biosolids management: biosolids volumes, N:P ratio & recycling (Agronomic, LCA and Economic Implications) – A European Perspective. *Proceedings cd, 2nd International Conference on Recovery of Phosphates from Sewage and Animal Wastes*, Holland, CEEP.
Feinerman, E., Plessner, Y. and Eshel, D.M.D. (2001) Recycled effluent: should the polluter pay? *Am. J. Agricultural Econ.* **83**(4), 958–971.
Fertilizer Society (1995) *Opportunities and constraints in the recycling of nutrients.* Proceedings Fertilizer Society; No. 372, Peterborough: Fertilizer Society.
Gaterell, M.R., Gay, R., Wilson, R. and Lester, J.N. (2000) An economic and environmental evaluation of the opportunities for substituting phosphorus recovered from wastewater treatment works in existing UK fertilizer markets. *Environ. Technol.* **21**, 1067–1084.
Günther, F. (1997) Hampered effluent accumulation process: phosphorus management and societal structure. *Ecol. Econ.* **21**(2), 159–174.
Hall, J.E. (1992) Treatment and use of sewage sludge. In *The Treatment and Handling of Wastes* (ed. Bradshaw, A.D., Southwood, R. and Warner, F.), London: Chapman & Hall for The Royal Society.
Jaffer, Y., Clark, T.-A., Pearce, P.A. and Parsons, S.A. (2001) Assessing the potential of full scale phosphorus recovery by struvite formation. *Proceedings cd, 2nd International Conference on Recovery of Phosphates from Sewage and Animal Wastes*, Holland, CEEP.
Jeanmaire, N. (2001) *Recycling of removed phosphorus: Analysis of the potential interest in wastewater treatment plants.* Report prepared for the Centre Européen d'Etudes des Polyphosphates – a European Chemical Industry Council (CEFIC) sector group.
Jeanmaire, N. and Evans, T. (2001) Technico-economic feasibility of P-recovery from municipal wastewaters. *Environ. Technol.* **22**(11), 1355–1362.
Karlsson, I. (2001) Full scale plant recovering iron phosphate from sewage at Helsingborg Sweden. *Proceedings cd, 2nd International Conference on Recovery of Phosphates from Sewage and Animal Wastes*, Holland, CEEP.
Köhler, J. (2001) *Detergent phosphates and detergent ecotaxes: a policy assessment.* Report prepared for the Centre Européen d'Etudes des Polyphosphates – a European Chemical Industry Council (CEFIC) sector group.

Kvarnström, E. and Nilsson, M. (1999) Reusing phosphorus: engineering possibilities and economic realities. *J. Econ. issues* **33**(2), 393–402.

Liberti, L., Petruzelli, D. and De Florio, L. (2001) REM NUT Ion exchange plus struvite precipitation process. *Environ. Technol.* **22**(11), 1313–1324.

Linster, M. (1991) The impact of sewage sludge on agriculture. In *Towards sustainable agricultural development* (ed. Young, M.D.), pp. 320–336, Belhaven Press, London.

Nawamura, Y., Kumashiro, K., and Ishiwatari, H. (2001) A pilot plant study on using seawater as a magnesium source for struvite precipitation. *Proceedings cd, 2nd International Conference on Recovery of Phosphates from Sewage and Animal Wastes*, Holland, CEEP.

Neto, F. (1998) Water privatization and regulation in England and France: a tale of two models. *Nat. Resour. forum* **22**(2), 107–117.

OFEFP (1991) *Incinération des boues d'épuration*. Office Fédéral de l'Environnement des Forêts et du Paysage (Suisse) cahier de l'environnement n°156, August 1991.

Parsons, S.A., Wall, F., Doyle, J., Oldring, K. and Churchley, J. (2001) Assesing the potential for struvite recovery at sewage treatment works. *Environ. Technol.* **22**(11), 1279–1286.

Paul, E., Laval Marie-Line and Spérandio, M. (2001) Excess sludge production and costs due to phosphorus removal. *Environ. Technol.* **22**(11), 1363–1372.

Rees, J.A. (1998) Regulation and private participation in the water and sanitation sector. *Nat. Resour. forum* **22**(2), 95–105.

Renzetti, S. (1999) Municipal water supply and sewage treatment: costs, prices, and distortions. *Can. J. Econ.* **32**(3), 688–704.

Sawkins, J.W. and Dickie, V.A. (1999) Regulating Scottish water (regulatory reform in the Scottish water industry: recent progress and future prospects). *Utilities Policy* **8**(4), 233–246.

Schipper, W., Klapwijk, B., Potjer, B., Rulkens, W., Temmink, H., Kiestra, F. and Lijmbach, D. (2001) Phosphate recycling in the phosphorus industry. *Environ. Technol.* **22**(11), 1337–1345.

Smith, S.R. (1996) *Agricultural recycling of sewage sludge and the environment*. Wallingford: CAB International.

Spinosa, L. (2001) Evolution of sewage sludge regulations in Europe. *Water Sci. Technol.* **44**(10), 1–8.

Spulber, N. and Sabbaghi, A. (1998) Economics of water resources: from regulation to privatization, 2nd edn. Natural Resource Management and Policy series. Boston; London and Dordrecht: Kluwer Academic.

Stuart, M. (1993) Modelling water costs 1992–1993. OFWAT Research paper no. 2, London.

Stuart, M. (1994) Modelling sewage treatment costs 1992–1993. OFWAT Research paper no. 4, London.

Strudler, M and Strand, I.E. Jr. (1983) Pricing as a policy to reduce sewage costs. *Water Resour. Res.* **19**(1), 53–56.

Subramanian, S. and Arnot, T.C. (2001) An integrated bioreactor/adsorption process for phosphorus recovery from wastewater. *Proceedings cd, 2nd International Conference on Recovery of Phosphates from Sewage and Animal Wastes*, Holland, CEEP.

Thomas, A. (1995) Regulating pollution under asymmetric information: the case of industrial wastewater treatment. *J. Environ. Econ. Manage.* **28**(3), 357–373.

Towers, W. (1994) Towards a strategic approach to sewage sludge utilization on agricultural land in Scotland. *J. Environ. Plan. Manage.* **37**(4), 447–460.

Ueno, Y. and Fujii, M. (2001) 3 years operating experience selling recovered struvite from full-scale plant. *Environ. Technol.* **22**(11), 1373–1381.

UKWIR (1999) *Recycling of water treatment works sludges*. UK Water Industry Research Report no. 99/SL/09/1, London.

von Münch, E., Benesovsky-Scott, A., Josey, J. and Barr, K. (2001) Making a business from struvite crystallisation for wastewater treatment: turning waste into gold

Proceedings cd, 2nd International Conference on Recovery of Phosphates from Sewage and Animal Wastes, Holland, CEEP.

Wallgren, B. (2001) Swedish policy on phosphorus recovery. *Proceedings cd, 2nd International Conference on Recovery of Phosphates from Sewage and Animal Wastes*, Holland, CEEP.

Williams, S. (1999) Struvite precipitation in the sludge stream at slough wastewater treatment plant and opportunities for phosphorus recovery. *Environ. Technol.* **20**(7), 743–747.

PART FIVE

Novel biotechnologies

24.	Bacterial precipitation of metal phosphates	549
25.	Developments in the use of calcium phosphates as biomaterials	582
26.	Agronomic-based technologies towards more ecological use of phosphorus in agriculture	610
27.	Biodegradation of organophosphate nerve agents	629

24
Bacterial precipitation of metal phosphates

L.E. Macaskie, P. Yong and M. Paterson-Beedle

24.1 INTRODUCTION

Over the past twenty years increased awareness of the potential environmental impact of liquid wastes, and their safe disposal, has been paralleled by consideration of novel technologies to take over at the point where traditional physico-chemical treatments fall short.

Other chapters in this book have described approaches to biological methods of phosphate waste water remediation. These methods can provide an alternative approach to the 'classical' method of phosphate precipitation using a metal precipitatant, such as iron or calcium, which requires bulk addition of metal to the water body. However, although metal phosphates should be highly insoluble (e.g. Valsami-Jones, this volume), based on their solubility products alone, in practice the solution inorganic ionic matrix, the pH and the presence of complexing agents can reduce the available metal concentration, making precipitation more difficult. Complexing agents need not be man-made; in practice the presence of natural (bi)carbonates and organic polymers such as humates and fulvates in natural waters can provide an effective means of metal sequestration.

© 2004 IWA Publishing. *Phosphorus in Environmental Technology: Principles and Applications.*
Edited by Eugenia Valsami-Jones. ISBN: 1 84339 001 9

Chemical precipitation of phosphate with metals is convenient and economic but is not applicable to very low phosphate (or metal) concentrations and the resulting wet sludge is difficult to handle and transport and requires compaction for final disposal. Methods such as reverse osmosis are effective for metal removal, but costly. This chapter focuses on the removal of, and also the use of, phosphorus and some of its compounds in environmental biotechnology. With respect to precipitation of metal phosphates, it is clear that a phosphate waste could be used to co-treat a heavy metal waste given the right microorganism and conditions, and this chapter will focus on developments in this concept.

In the case of heavy metals, various separation and ion exchange technologies are already well established, but the cost of solvents and ion exchange resins may be substantial. Natural zeolites are a cost-effective ion-exchange option, but availability of the zeolite must be assured for the projected lifetime of the plant since reproducible operation is an absolute requirement. Metal extraction using natural zeolites, although cheap and effective, may be subject to the same constraints as chemical ion exchange processes: a generally low metal selectivity, and sensitivity to low pH and the occurrence of co-contaminating components (other metal cations or anionic counterion species) which are often present to excess in the solution. Biologically-based processes are attractive if they can overcome one or more of these constraints and if they can be operated economically, effectively and reproducibly (Eccles, 1995, 1999). Many potentially useful biomaterials (biomasses) have been identified with respect to metal uptake, but very few have been evaluated under conditions approximating to real processes with respect to the above criteria. In the case of radioactive wastes, the radiotoxicity and longevity of many radionuclide species requires a highly effective means of their removal. Most potential bioprocesses have addressed only uranium, with scant attention paid to the removal of the transuranic elements, whose chemistry and relatively low concentrations make them more challenging, and recalcitrant to removal (Lloyd and Macaskie, 2000, 2002; Macaskie and Lloyd, 2002).

In addition to phosphate discharges, the release of metals into the environment is controlled by law. For effective metal phosphate biodeposition, the biomass needs to be able to 'channel' phosphate in such a way as to overcome the constraints to its bioprecipitation as the corresponding metal phosphate, which is not so likely to occur in bulk solution where the concentrations of both metal and phosphate may be low. The principles of phosphate precipitation and mineral formation are discussed by Koutsoukos and Valsami-Jones (this volume). In bulk solution, such as natural water or waste water, the concentrations of metal and phosphate may never be enough to exceed the solubility product and, even if this requirement is fulfilled, precipitation must first overcome an energy barrier (Mann, 1994, 1997). Microorganisms can promote metal phosphate deposition in two ways. Firstly, they can provide suitable nucleation sites to initiate metal phosphate precipitation, and secondly, they can provide a concentrated localised supply of phosphate through their metabolic processes. Such processes have the effect of increasing the local concentration of phosphate in the cellular microenvironment, which then allows the phosphate concentration to exceed the solubility product in the presence of a low concentration of metal. Effectively, the microbial contribution is one of accelerating phosphate biodeposition. This chapter will highlight some examples where such approaches have been used in the removal of metal and phosphates from contaminated waters. Where inorganic phosphate is generated by bacteria using substrates supplied into the aqueous metal feed, this would ideally require the phosphate substrate to be a natural product which is widely and cheaply available.

In some cases the phosphate supply could even be inorganic phosphate itself, with the microbial strain acting as a facilitator of bioprecipitation, while in other cases the formation of one metal phosphate could promote the deposition of another metal via co-precipitation processes on bacterial surfaces. This chapter will expand on these various mechanisms, showing case histories as real examples where possible.

24.2 HEAVY METAL BIOREMEDIATION: WHY SELECT A PHOSPHATE-BASED PRECIPITATION PROCESS?

For metal cation removal dead biomass has potential as a metal sorbent. Biosorption occurs *via* interactions with various ligand groups associated with the surfaces of microbial cells. These essentially chemical reactions have been the subject of many investigations, summarised in various reviews (e.g. Gadd, 1996, 1997, 2000; Kratochvil and Volesky, 2000; Pümpel and Schinner, 1997; Schiewer and Volesky, 1996; Volesky, 1994; Volesky and Holan, 1995). A primary consideration is economic biosorbent production, ideally as a waste from another process, e.g. spent brewer's yeast was used for biosorption of uranyl ion (Omar *et al.*, 1996; Riordan *et al.*, 1997), while metal accumulating fungal biomass such as *Aspergillus* is available in large quantities from commercial fermentations (Yakubu and Dudeney, 1986). Some algal biomasses (e.g. *Sargassum natans*: Kuyucak and Volesky, 1989) have commercial potential but require harvesting from the ocean. The biosorption capacity of biomass is often rather low, although examples exist where a particular waste has been treated by biosorption-based processes (e.g. McCready and Lakshmanan, 1986). In some cases the cellular ligands responsible for metal binding have been identified; in some examples these can comprise phosphate components of the microbial cell wall, giving an essentially phosphate-based biosorption process (Andres *et al.*, 1993, 1994, 1995) but, in general, interest in commercialisation of this type of technology is not strong (Gadd, 2000). Usually the metal complexing ligand is a component of the microbial cell wall and, in this respect, metal uptake is a 'passive' process. One exceptional example has been reported recently, where a bacterial surface protein (the bacterial 'S layer') functions in metal uptake. The organism, identified as *Bacillus sphaericus* (strain JG-A12) was originally isolated from a uranium contaminated waste site (Hennig *et al.*, 2001, Panak *et al.*, 2000; Selenska-Pobell, 2002; Selenska-Pobell *et al.*, 1999, 2002). This organism produces a large quantity of S-layer protein at the cell surface and it seems likely that this comprises a metal resistance mechanism: phosphorus residues were implicated in metal binding (Selenska-Pobell, 2002) and this protein, although similar to the S-layer proteins of related organisms, but apparently unique among them, is phosphorylated (Raff, 2002; Selenska-Pobell *et al.*, unpublished work). In this case, selection for metal resistance has effectively resulted in the 'placement' of metal sequestering phosphate groups in an array on the cell surface. The maximum capacity of the cells for metals has not yet been established but the organism has been proposed as a good candidate for *in situ* remediation of uranium contaminated waters (Selenska-Pobell, 2002; Selenska-Pobell *et al.*, 2002). Since the protein is exported from the cell, the capability would exist for engineering of its metal binding capacity, and overproduction using molecular techniques, as well as for its immobilisation in ceramic columns. Immobilised whole cells have been used as an effective biosorbent for Cu, Pb, Al and Cd, as well as for uranium, in columns within a bacteria-based ceramic matrix (Raff, 2002). The use of whole cells would overcome the need for protein purification and, since only the pre-formed outermost S-layer protein component of the cell is used, cell vitality is probably not required and there would be no requirement to present the water under physiologically compatible conditions.

Living microbial cells can accumulate substantial quantities of toxic heavy metals. The target metals or, indeed, co-contaminants present in the waste can be toxic to the biomass (although many metal resistance mechanisms are described, e.g. Collard *et al.*, 1994; Hobman *et al.*, 2000; Taghavi *et al.*, 1997). Living biomass requires physiologically-compatible carrier solutions. Also, biomass growth may add to the organic loading of the final solid waste and to the cost of its transportation to the site of reprocessing (if not desorbed of surface-bound metal *in situ*), incineration or, for radioactive waste, its final burial. Hence, the use of non-growing biomass is preferable, ideally harnessing pre-formed proteins or enzymes, which essentially use the cell body or the cell surface as an immobilisation matrix.

Potentially useful hybrid technologies that could overcome many of the above problems are based on the generation of precipitant ligands, either biomass-bound ligands or cell-associated enzymes, that promote formation of insoluble metal deposits which remain attached to the microbial cells. This metal deposition over and above simple biosorption, can be termed biomineralisation and, by this approach, it is common for the deposited metal precipitate to exceed the biomass dry weight by several-fold. Importantly, the enzymatic step(s) leading to metal precipitation can often be decoupled from microbial growth, allowing metal removal from solutions too toxic to permit biomass growth; the process constraint is then that of the metal accumulating step *per se* and 'resting' (growth-decoupled) or even non-viable biomass can be used. One example of biomineralisation, occurs *via* a metal detoxification reaction in *Alcaligenes eutrophus* CH34 (now *Ralstonia metallidurans* CH34). Here, a plasmid-encoded metal efflux from the cells occurs, along with proton uptake (antiport); following metal uptake, the resulting localised alkalinization causes the precipitation of metal hydroxides and carbonates exocellularly (Diels *et al.*, 1993; Taghavi *et al.*, 1997). When immobilised in a mixed-species biofilm in a moving bed sandfilter, *R. metallidurans* was detected in the reactor after several months of continuous operation in metals removal from an on-line reactor at an industrial plant (Diels *et al.*, 2001). Other, parallel studies using a long-term reactor with a similar inoculum and challenged with another,

Ni-containing industrial waste water gave an interesting result. After 8 months *R. metallidurans* was a major component of the biofilm (Pümpel *et al.*, 2001) and a metal phosphate (nickel phosphate, arupite, $Ni_3(PO_4)_2 \cdot 8H_2O$) was clearly identified as a major insoluble species (Pümpel *et al.*, 2003). None of the inoculated bacteria had a known mechanism of phosphate-mediated bioprecipitation and it seems likely that in this case an unknown microbial component of the consortium, introduced *via* the waste water, was able to promote metal phosphate deposition using inorganic phosphate present in the water (Pümpel *et al.*, 2003), with nickel phosphate precipitation assisted by local alkalinization promoted by the activity of *R. metallidurans* (Pümpel *et al.*, 2003). The bioreactor was as described by Pümpel *et al.* (2001), and using this system 1 mg/l of Ni^{2+} was removed within a few minutes retention time (Pümpel *et al.*, 2003).

An alternative method of biomineralization uses the sulphate dissimilatory pathway of sulphate-reducing bacteria (SRB), and precipitation of metal cations as the corresponding insoluble metal sulphides. This has been used commercially (Barnes *et al.*, 1991) and more recently has been considered within an integrated process for the treatment of solid/bound metals for *in situ* soil remediation or biodecontamination processes, where metal sulphide bioprecipitation is applied subsequently to the leaching of bound metals via the activity of oxidising/leaching thiobacilli (Eccles, 1998, 1999; White and Gadd, 1996; White *et al.*, 1998).

As an alternative to sulphide, biogenic phosphate (i.e. phosphate produced by bacterial cells) has also been used as a precipitant ligand, which has seen real application to remediation of mine waste water contaminated with uranium (Macaskie *et al.*, 1997). In this case the low pH (pH 3.5–4) prevented bacterial growth. This demonstrates the use of resting cells, in which only the enzyme mediating the metal precipitation is required to be active; indeed, excess inorganic phosphate was found in the solution (Macaskie *et al.*, 1997), reflecting the broad operational pH range of the acid phosphatase (Tolley *et al.*, 1995). Such growth-decoupled, ligand producing systems are rarely described (c.f. the sulphate-reducing process is growth-associated) and lend themselves to specific applications where the waste stream is presented under non physiologically-permissive conditions.

24.3 CASE HISTORY: METAL PHOSPHATE BIOMINERALIZATION BY *SERRATIA* SP.

The approach to phosphate-based biomineralization processes relies upon the liberation of inorganic phosphate at a high concentration at the cell surface *via* the enzymatic cleavage of phosphate donor molecule, e.g. glycerol 2-phosphate. The high local concentration of phosphate allows the solubility product of the metal phosphate to be exceeded locally and thus metals can be scavenged from very dilute concentrations in the bulk solution. The most-studied system is the acid phosphatase (PhoN)-mediated metal uptake by a *Citrobacter* sp. that was originally isolated from a lead-polluted site in the UK (Macaskie and Dean, 1982). The organism was originally identified commercially, on the basis of standard biochemical tests,

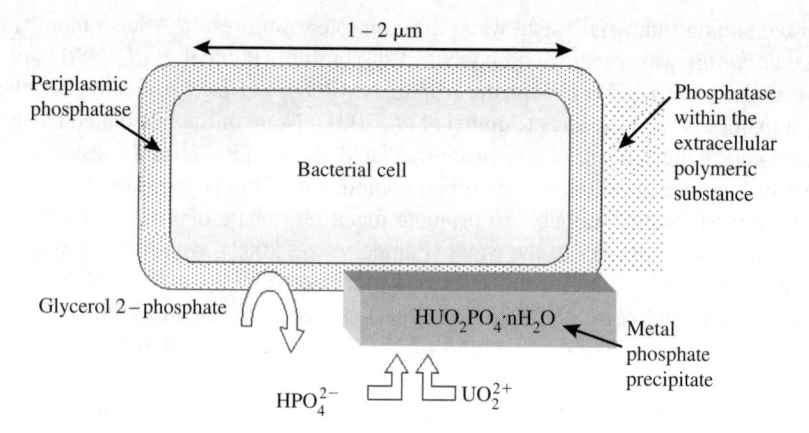

Figure 24.1 Schematic representation of uranium accumulation on the cell surface of *Serratia* sp. Phosphatase enzyme is localised within the periplasmic surface 'compartment' and also within the extracellular polymeric materials (EPM) (see later). When substrate (glycerol 2-phosphate) is cleaved to give inorganic phosphate (HPO_4^{2-}), the phosphate precipitates with uranyl ion (UO_2^{2+}) to give a cell bound deposit of HUO_2PO_4.

but application of modern molecular and biochemical methods reassigned the organism to the genus *Serratia* (Pattanapipitpaisal *et al.*, 2002). Early studies showed that metal and phosphate were deposited stoichiometrically, as e.g. HUO_2PO_4 (Macaskie *et al.*, 1992a), and therefore metal deposition (Figure 24.1) is a measure of phosphate liberation and hence phosphatase activity. Figure 24.1 shows a schematic representation of the activity of phosphatase in the release of inorganic phosphate, which then precipitates with uranyl ion, UO_2^{2+}, as cell-bound HUO_2PO_4 (Macaskie *et al.*, 1992a). The process of uranium uptake is shown in Figure 24.2. The role of phosphatase was confirmed by immunogold labelling studies (Figure 24.3), which showed that a phosphatase deficient mutant could not accumulate UO_2^{2+} (Jeong *et al.*, 1997). The parent cells (shown in Figure 24.3), which contained high levels of the enzyme, accumulated a large amount of uranium, e.g. their own weight as uranium, in only a few hours (Macaskie, 1990; Figure 24.2) The role of the enzyme in metal accumulation was conclusively proved: *Escherichia coli* does not normally contain the phosphatase gene *phoN*, or accumulate heavy metals, but introduction of the corresponding *Salmonella phoN* into an *E. coli* host allowed uranyl phosphate accumulation comparably to the naturally-occurring *Serratia* sp. (Basnakova *et al.*, 1998a). However the use of genetically modified strains may be unattractive industrially (since containment is necessary) and, since the metal accumulating *Serratia* strain can express very high activity when grown appropriately, e.g. as a biofilm on solid support matrices (Finlay *et al.*, 1999; Macaskie *et al.*, 1995, 1997; Nott *et al.*, 2001), genetic improvement is probably unnecessary. The major drawback of this technique is the requirement for glycerol 2-phosphate to be added to the metal solution, which would be uneconomic in practice.

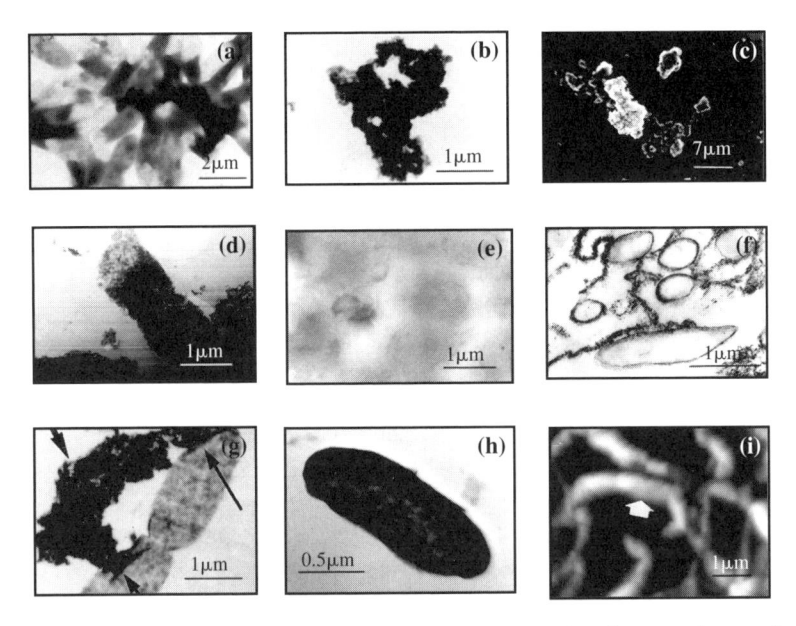

Figure 24.2 Uranyl phosphate bioaccumulation on the cells. (a) Electron micrograph of cells before uranyl ion addition. The cells have not been fixed or stained and are very indistinct. Each cell is 1–2 μm long. (b) Cells following exposure to UO_2^{2+} in the presence of glycerol 2-phosphate. The cells have become densely stained due to the accumulation of the electron-opaque material, hydrogen uranyl phosphate. (c) Scanning electron micrograph showing a mass of biomineralized cells. (d) A single cell that has been split during preparation. This shows clearly that the biomineral deposit is confined to the cell surface. The internal part of the cell contains little electron opaque material. (e) Sections of cells before exposure to uranyl ion are, like whole cells, pale and indistinct. (f) Following challenge with uranyl ion and glycerol 2-phosphate the biomineral is seen as a distinct layer on the surface of the cell sections, which has become detached as a ribbon-like material. (g) In whole cells the biomineralized wall can be seen to be shed as an intact coat. (h) A cell 'intercepted' during the wall detachment process. In this example the uranium was loaded to approximately twice the biomass dry weight. It is not clear from electron micrographs whether the wall detachment is an artefact since electron microscopy uses dried material examined in a vacuum. (i) Atomic force microscopy (AFM) can visualise native specimens. The sample shown was dried briefly in air and examined in air using AFM. Note similarity of the cell arrowed to the cell shown in h. The lighter the colour the higher the elevation of the specimen area above the baseline layer; images are processed by computer and presented as a photographic image.

A study targeting the application of this method to uranium mine water remediation concluded that the requirement for a phosphate substrate would be the single factor limiting industrial application (Roig *et al.*, 1995) and despite attempts to persuade the metal accumulating strain to extensively hydrolyse cheap substrates, such as tributyl phosphate (Jeong *et al.*, 1994) or phytic acid (unpublished), the organism proved conservative in its substrate portfolio. However this provides a good model system on which to base alternative processes.

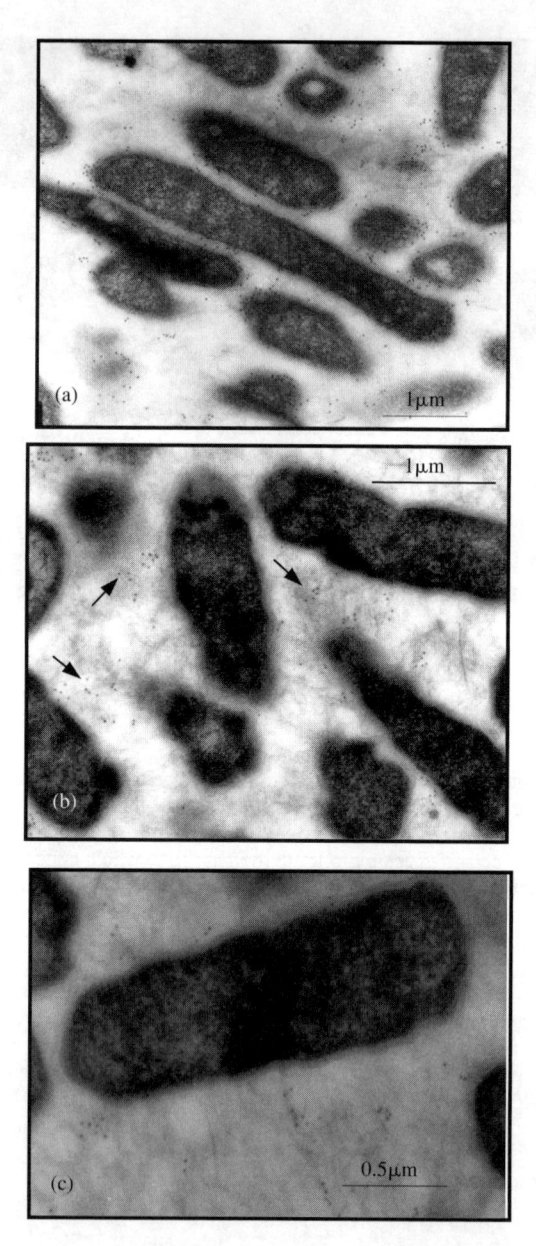

Figure 24.3 Immunogold labelling of the cells, to show the location of the phosphatase. (a) This cell (centre) shows phosphatase clearly localised within the periplasmic space (the region between the outer and cytoplasmic membranes), shown most clearly at the ends of the cell. (b) These cells show phosphatase apparently localized outside the cell, within the extracellular polymeric substance (EPS). (c) In this cell the phosphatase is clearly visible as small black dots within a fibril of EPS extruding from the long axis of the cell.

A simplistic model was developed (Figure 24.1) whereby periplasmically-localised enzyme (identified by immunogold labelling: Figure 24.3) liberated inorganic phosphate to promote metal deposition at the cell surface with cell membranes (presumably phospholipids) identified as nucleating surfaces (Jeong et al., 1997). However several inconsistencies suggested that this simple model alone could not explain metal accumulation. The optimum pH of the enzyme is from 5–7 with lower activity below pH 4.5 and the K_m (the substrate concentration giving half-maximal activity, which is a measure of the substrate affinity, Jeong and Macaskie, 1995) increased several-fold as the pH fell, attributable to protonation of the glycerol 2-phosphate substrate (Jeong et al., 1998; Tolley et al., 1995). Accordingly, preliminary attempts using immobilized cells to remediate acid mine drainage waters (pH 3.5–4) containing uranium were unsuccessful. However phosphate liberation was detected in the column outflow, increasing with time, and in parallel with a decrease in the pH of the unbuffered mine water input solution from 4.0 to 3.5 (Macaskie et al., 1997). Since phosphate liberation continued to increase with time even at the low bulk solution pH, this suggested that the liberated phosphate, or carboxyl groups within the extracellular polymeric matrix, acted as localised buffers for pH-stasis. Simple precipitation experiments had shown that uranyl phosphate formation was retarded at low pH (Tolley et al., 1995) but after 2 days in a continuous flow system (Figure 24.4) using immobilized cells, uranyl ion was removed at an efficiency of 70% (input concentration was 0.2 mM; flow rate was one column volume h^{-1}) and continued at steady-state (Macaskie et al., 1997). This could be interpreted in terms of a period of time being required for initiation of nucleation foci prior to steady-state crystal growth of $HUO_2PO_4 \cdot 4H_2O$ (identified by X-ray powder diffraction analysis: Yong and Macaskie, 1995a). The need for nucleation had been previously suggested by the observation that metal removal was greatly enhanced if the columns were allowed to stand overnight in the test solution before initiation of flow-through conditions (Macaskie et al., 1992b). These tests showed that a simplistic enzymatic model could not explain metal biomineralization, despite the finding that the phosphatase enzyme was crucial to this process.

24.4 A CONCEPTUAL MODEL FOR METAL PHOSPHATE BIOMINERALIZATION

The processes by which metal phosphates precipitate has been described elsewhere in this book (Koutsoukos and Valsami-Jones, this volume), while Mann (1994, 1997) gives excellent overviews in the literature. However, in confirmation that a simplistic model cannot adequately explain metal *bio*mineralization, it was found that low-phosphatase-activity columns could remove metal as effectively as normal columns if a quiescent period was incorporated prior to starting the flow (Macaskie et al., 1992b). The possibility of nucleation sites other than the membrane phospholipids was sought. Experiments using extracted extracellular

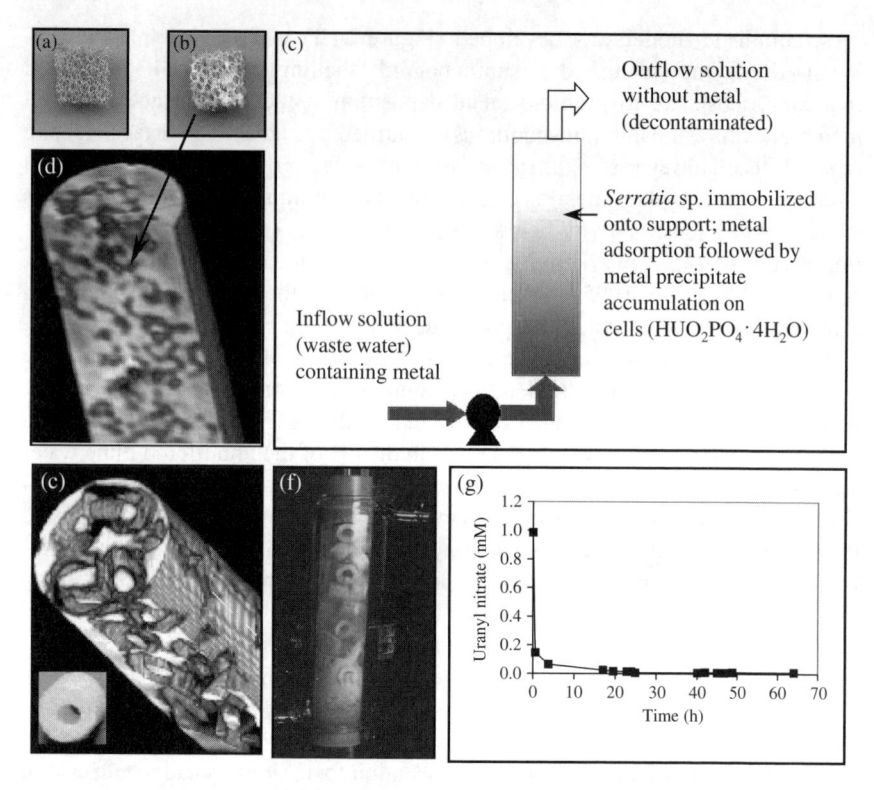

Figure 24.4 Photograph of a single polyurethane reticulated foam (1 cm³). (a) Before and (b) after biofilm colonisation. (c) Diagram of uptake of metal by a packed-bed reactor containing *Serratia* sp. biofilm immobilized onto a support. (d) A 3D MR image of the distribution of fluid occupying the pore space of a packed-bed column (diameter 1.5 cm and length 9 cm), containing *Serratia* sp. biofilm immobilized onto cubes of polyurethane foam, challenged with a sodium glycerol 2-phosphate solution (5 mM) in sodium citrate buffer (2 mM), pH 6.0. (e) A 3D MR image of the distribution of Cu^{2+} (10 mM) doped water occupying the pore space of a packed-bed column (diameter 1.5 cm and length 9 cm) containing ceramic raschig rings; insert is a photograph of a single ceramic raschig ring (length 7 mm, external diameter 6.6 mm and internal diameter 2 mm). (f) Photograph of a packed bed reactor containing ceramic raschig rings challenged with uranyl nitrate (1 mM), glycerol 2-phosphate (5 mM) in sodium citrate buffer (2 mM), pH 6.0. (g) Accumulation of uranyl phosphate in reactor containing *Serratia* sp. biofilm immobilized onto polyurethane foam and challenged with uranyl nitrate (1 mM), glycerol 2-phosphate (5 mM) in sodium citrate buffer (2 mM), pH 6.0.

polymeric substances (EPS) from the organism showed uranyl ion uptake to 25% of the polymer dry weight (Bonthrone et al., 2000). Analysis of the polymer for phosphate showed a substantial amount, and examination of the material using ^{31}P nuclear magnetic resonance gave chemical shifts in accordance with phosphate groups in the lipid A component of the cell surface lipopolysaccharide (LPS) (Bonthrone et al., 2000; Macaskie et al., 2000) as described in the literature (Strain et al., 1983a, b). Addition of Cd^{2+} to the preparation promoted a chemical shift downfield and the production of a single peak from two original major peaks, suggesting Cd-mediated crosslinking of the EPS strands that was confirmed by a change in the morphology of the EPS (observed by electron microscopy) from amorphous and diffuse to structured and electron opaque (Bonthrone et al., 2000). This suggested phosphate groups of the lipid A component as the nucleation sites. Attempts to confirm this using uranium were constrained by the interference of the paramagnetic uranium species with proton NMR, but upon addition of UO_2^{2+} to the EPS in the NMR tube the solution NMR signal disappeared and a yellow precipitate was observed which was confirmed as uranyl phosphate by X-ray powder diffraction analysis (Macaskie et al., 2000). The experiment was also done using whole cells examined under the atomic force microscope (AFM), where the formation of material 'standing proud' around the cells was clearly seen (Macaskie et al., 2000, Figure 24.2i). The solubility of $NH_4UO_2PO_4$ is much lower than that of HUO_2PO_4 (Yong and Macaskie, 1995a) and an accelerated biomineralization was seen in the presence of NH_4^+ under the AFM in parallel to batch cell uptake tests and electron microscopy (Macaskie et al., 2000).

Coordination of metal to phosphate groups within the EPS provides the rationale for the nucleation process, while the observation of enzyme held 'tethered' around the cells (Figure 24.3c) suggests that phosphate would be fed-in continuously via enzymatic activity within the EPS. A developed model would suggest that metal deposits are held within the EPS while additional phosphate is fed in via phosphatase activity, leading to consolidation and growth of the metal phosphate crystals. Since cell loadings of 9 g of uranium per g of dry biomass have been observed (Macaskie, 1990), it is surprising that access to the cell surface is not blocked by the metal phosphate surrounding the cells. Examination by electron microscopy showed a substantial phosphatase pool entrapped within the extracellular 'fuzz' seen by electron microscopy in conjunction with immunogold labelling (Macaskie et al., 2000) (Figure 24.3). In accordance with the two enzyme localizations seen by immunogold labelling (Figure 24.3), two very similar but distinct phosphatase isoenzymes were seen by enzymological (Jeong et al., 1998) and physiological (Jeong and Macaskie, 1995, 1999) studies, but their immunological cross-reactivity (Jeong et al., 1998) makes assignment to the periplasmic and exopolymeric pools difficult. The native enzyme was recovered as a high molecular mass complex (Jeong et al., 1998), in accordance with its probable association with the exocellular material.

A conceptual model can now be developed, in which incoming metal coordinates with the phosphate groups of the lipid A component of the lipopolysaccharide in juxtaposition to the matrix-bound enzyme. Metal incorporation into the nascent crystal creates a localized depletion, facilitating entry of more metal down the concentration gradient generated. The incoming metal encounters a high local concentration of phosphate liberated outward via phosphatase activity and the kinetics of metal phosphate precipitation are accelerated by the presence of a pre-formed nucleation site on the LPS. Such a model could permit the accumulation of metals from very low concentration bulk solutions provided that the initial nucleation events and steady-state are established initially. This conceptual model provides the 'starting point' for rational development of further methods and models for phosphate-based metal removal.

24.5 A QUANTITATIVE MODEL FOR METAL PHOSPHATE BIOMINERALIZATION

Once the barrier to nucleation is overcome, metal uptake proceeds at steady-state, provided that substrate is continually fed-in to a column containing immobilized cells (Figure 24.4). Examination of the metal-loaded biofilm by confocal laser scanning microscopy (using the fluorescence of uranyl ion) and environmental scanning electron microscopy suggested that the metal layer is restricted to the top of the biofilm (Paterson-Beedle and Macaskie, unpublished). In the example shown, the cells were immobilized as a biofilm on polyurethane reticulated foam and development of metal phosphate precipitate within the foam matrix was visible *in situ* using magnetic resonance imaging (Nott *et al.*, 2001). Phosphate release and metal removal are stoichiometric, although a slight excess of phosphate is required for maximum metal removal. This intrinsic inefficiency, related to the kinetics of metal phosphate crystallization within a flow-through reactor, can be given a dimensionless numerical value, the exact value of which depends on the metal species and the components of the ionic matrix of the carrier solution (e.g. the presence of anionic counterions and the pH). For any given set of conditions this value will be constant assuming that the column remains at steady-state.

Modelling trials were carried out for entrapped, chemically-coupled and biofilm-immobilized cells (Finlay *et al.*, 1999; Macaskie *et al.*, 1995; Yong and Macaskie, 1997, 1999). The activity of flow-through columns (in terms of both product liberation and metal removal) can be expressed in terms of the available biomass and phosphatase specific activity, the substrate concentration and the flow rate (1/t, where t is the residence time) by an integrated form of the Michaelis-Menten equation with the additional incorporation of the 'inefficiency value' to allow for the metal crystallization characteristics (Macaskie *et al.*, 1997). This approach enables predictions to be made of the activity of the column when parameters are varied and the model also allows extrapolation of benchscale models to actual process situations. The predictive model was tested initially to determine the interfering effect of nitrate (which behaves as competitive-type inhibitor: Yong and Macaskie, 1997), and is often present to excess in waste solutions (Macaskie, 1991). The inhibitory effect could be overcome by increasing the substrate concentration or the flow residence time in accordance with the model (Yong and Macaskie, 1997). A similar approach was used to predict the effect of sulphate (against which the system was more robust: Yong and Macaskie, 1999) and the model was used to define the bioreactor activity against real acid mine drainage water which contained approx. 35 mM SO_4^{2-} (Yong and Macaskie, 1999) in addition to 0.2 mM uranyl ion and a variety of other metallic co-contaminants (Macaskie *et al.*, 1997). These two examples demonstrate the feasibility of a mathematical description, which is essential for scale-up predictions.

24.6 PHOSPHATE BIOMINERALIZATION FOR THE REMOVAL OF TRANSURANIC ELEMENTS: THE NEED FOR NUCLEATION PROCESSES

The model system was shown to apply to the removal of many common metal cations from solution, but in some specialized cases, such as transuranic elements, the concentration of metals is very low and it is difficult to achieve effective removal by precipitation processes alone. The deposition process can be promoted by first laying down a 'priming' deposit of another metal phosphate. The elements considered were ^{241}Am, ^{239}Pu and ^{237}Np, with the most common valences of (III), (IV) and (V), respectively. It was found that Am(III) was desolubilized easily by phosphatase-containing cells (but not a phosphatase deficient mutant) (Macaskie et al., 1994), since it behaves chemically like La(III) which was removed effectively as LaPO$_4$ by the biomineralization system (Tolley et al., 1995). However problems were foreseen in the removal of Pu(IV) because in vitro precipitation tests had shown that the analogous Th(IV) did not precipitate readily as its phosphate (Tolley et al., 1995). In order to suppress hydroxide formation, the Pu(IV) can be held in solution as its citrate complex but the high strength of the actinide (IV)-citrate complex (several orders of magnitude greater than for the hexavalent actinide-citrate complexes: Macaskie, 1991) resulted in a very low free metal concentration and correspondingly slow phosphate precipitation. X-ray powder diffraction analysis of accumulated thorium phosphate showed an amorphous deposit even when the solubility of thorium phosphate was reduced by the incorporation of ammonium ion (Yong and Macaskie, 1995b), but accumulation of thorium phosphate was very slow, possibly attributable to the requirement for nucleation (above). Both HUO$_2$PO$_4$ and LaPO$_4$ formed highly crystalline material on the cells but only LaPO$_4$ could substitute as a nucleating substrate for Th(HPO$_4$)$_2$. The XRD pattern of LaPO$_4$ acquired some additional peaks, attributable to a new phase, perhaps a Th(HPO$_4$)$_2$ (Yong and Macaskie, 1998). This principle was applied successfully to the removal of Pu(IV) (Macaskie et al., 1994) but the removal obtained (50%) was not sufficiently promising for industrial application. Preliminary studies using Np(V) showed negligible removal.

It was concluded that a simple phosphate-based biomineralization system, although useful for uranium and americium removal, was insufficient for the effective removal of Pu(IV) and Np(V) per se. The tests with LaPO$_4$-nucleation followed by enhanced deposition of thorium phosphate (Yong and Macaskie, 1998) suggested that a non-homologous crystal 'priming layer' could facilitate deposition of the actinide species, with formation of a hybrid crystal (Yong and Macaskie, 1998). Indeed, 'biogenic' crystals of 'priming' metal deposit can be as good as, or superior to chemically-prepared material. For example, in the case of microbially-generated FeS the 'bio-crystal' was reported to be a superior sorbent for nuclide species as compared to 'geochemical' FeS (Watson and Ellwood, 1994). Extracellular fibrils of HUO$_2$PO$_4$ were observed under the electron microscope (Jeong et al., 1997) and in the case of FeS discrete intertwined fibrils (Ellwood *et al.*, 1992) gave a very high surface area, which produced a metal sorbent 10–100 times better than the 'geochemical' FeS counterpart (Watson and Ellwood, 1994).

These concepts were applied to the removal of Pu(IV) and Np(V) from solution. Using cells of *Serratia* sp. pre-coated with LaPO$_4$ effective removal of both nuclides was observed whereas none was removed in the absence of the 'priming' deposit (Macaskie and Basnakova 1998), the role of which was confirmed as follows (Figure 24.5). It is assumed that the nascent metal phosphate is not easily crystallized from solution, unless a nucleating surface is present. During the 'priming' step at slow flow rate (100% removal of, in this case, La^{3+} from solution as LaPO$_4$) it is likely that the substrate (glycerol 2-phosphate) is depleted at the front-end of the column, and metal desorption is confined to that region; effectively a large amount of LaPO$_4$ may be present but the surface area to volume ratio of the highly localized deposit may be small and, with this, the ability to capture nascent actinide phosphate. In contrast, a column that is pre-challenged with La^{3+} and substrate at a rapid flow rate will have nucleating surfaces distributed throughout the length of the column and thus be potentially more effective at actinide phosphate removal (since the nucleating surface is spread along the length of the column: Figure 24.5). This was confirmed using both Pu and Np, where the extent of actinide removal increased relative to the flow rate of the La^{3+} through the column during the priming step (Macaskie and Basnakova, 1998) according to the schematic model shown in Figure 24.5.

24.7 CO-PRECIPITATIVE METAL REMOVAL

Although most, if not all, heavy metals have insoluble phosphates, phosphatase-mediated bioprecipitation is not universal. Indeed, in the case of Ni^{2+} (Bonthrone *et al.*, 1996), Co^{2+} (Paterson-Beedle and Macaskie, 2003) and Cr^{3+} (Pattanapipitpaisal *et al.*, 2002) negligible metal was removed from solution even in the presence of excess liberated inorganic phosphate. However supplementation of the solution with uranyl ion resulted in the removal of UO$_2^{2+}$ into HUO$_2$PO$_4$ and co-precipitation of nickel (Bonthrone *et al.*, 1996) but not cobalt (Paterson-Beedle and Macaskie, 2003) phosphates. Similarly, in the case of Cr^{3+} removal of the cation was not promoted even in the co-presence of La^{3+} (Pattanapipitpaisal *et al.*, 2002). Removal of Ni^{2+} was also observed in a sandfilter which had been inoculated with metal biosorbing and bioprecipitating bacteria (Pümpel *et al.*, 2003). In this case nickel phosphate was clearly identified on bacterial cell surfaces. It was not possible to identify the metal accumulating bacteria conclusively but the inoculated strains were mostly Gram positive; the Gram positive bacterial cell surface contains different types of phosphate groups (e.g. teichoic acids) which could serve a coordination and nucleation function. There is clearly a case for the examination of Gram positive species for phosphatase activities which could fulfil similar roles in metal accumulation, although in this case no organic phosphate was provided and any phosphate present was introduced as a component

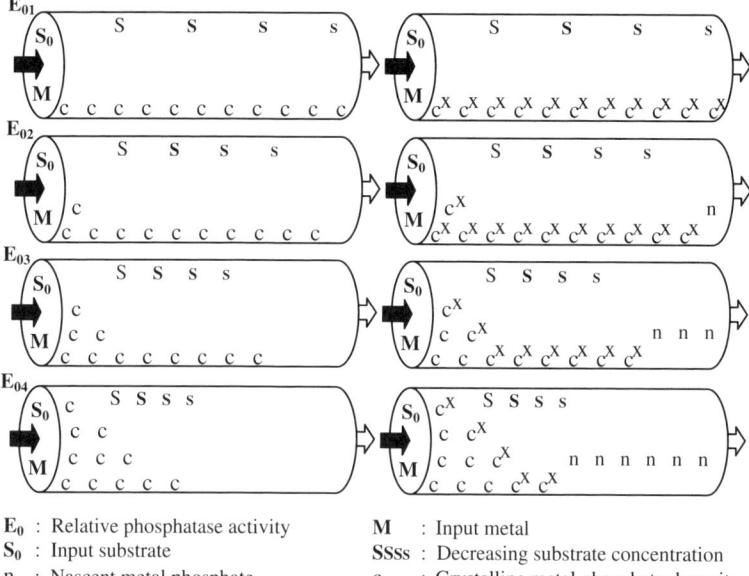

Bacterial precipitation of metal phosphates 563

Figure 24.5 Schematic representation of biomineralization of metals having phosphates which are not readily insoluble. Each cylinder represents a column of immobilized cells (see Figure 24.4c) challenged in two stages. The left side represents columns challenged with 'priming' metal (stage 1) and the right side represents columns challenged with the 'target' metal (stage 2). E_0 is the phosphatase activity of the cells in the column. E_{01} is cells of low phosphatase activity and E_{02} to E_{04} are increasingly high phosphatase activities. In the bottom example, cells of E_{04} are used. Since the phosphatase activity is high, it would be expected that the target metal would be removed very effectively but in fact the reverse happens. The model makes the assumption that the phosphate of the 'target' metal can only precipitate if there is a nucleating surface laid down previously ('priming deposit': c). For a fixed concentration of substrate (at constant flow rate) with a low enzyme activity (E_{01}) the substrate is not consumed immediately at the point of entry into the column but residual substrate is able to pass to distal column areas for cleavage and metal phosphate deposition, and the nucleating metal phosphate of the 'priming' metal is distributed along the length of the column. When the target metal is then provided, its precipitate is able to 'find' nucleation sites along the length of the column and therefore is removed effectively (top line). In the case of high enzyme activity columns (E_{04}) most of the substrate is consumed at the point of entry into the column and hence the priming metal phosphate occurs as a 'slug' near the point of flow input. The surface area of this 'slug' is less than the same amount of metal phosphate spread throughout the length of the column. Also, the column section containing the nucleating phosphate is short, and therefore the 'chance' of nascent target metal phosphate encountering a nucleating surface for the same flow residence time is less. Hence, more of the phosphate of the target metal cannot find a nucleating surface on which to reside, is left free (n) and is not retained by the column (bottom line). Thus, the removal of target metal is *inversely* proportional to the activity of the column when a priming step is used. A similar argument describes the use of columns of one common phosphatase activity challenged during the priming step at variable flow rates. At a slow flow rate (corresponding to the bottom line) all of the priming metal is confined to the

of the waste water. It is possible that cellular teichoic acids could serve as the phosphate donor, e.g. polyglycerol phosphate in *Bacillus subtilis*.

In a solution supplemented with glycerol 2-phosphate (sodium salt) as the substrate and phosphate donor, it is likely that the precipitated salt of uranium would be $NaUO_2PO_4 \cdot 4H_2O$, since this is several orders of magnitude less soluble than the protonated form, $HUO_2PO_4 \cdot 4H_2O$. This was suggested by measurement of the sodium content of the uranyl phosphate crystal using proton induced X-ray emission analysis (Nott et al., 2001) but it is not possible to discern between the two species using X-ray powder diffraction analysis since the spectra are very similar. Yong and Macaskie (1995a) suggested that in the presence of NH_4^+ ions, the ammonium form of the crystals is made. This is, similarly, highly insoluble and this was reflected in a shorter delay before the onset of precipitation, observed directly using AFM and by parallel uranyl ion uptake (Macaskie et al., 2000). Recent studies have shown that it is possible to achieve some selectivity into the crystal. For example, Na^+ and Cs^+ should be incorporated similarly since they are both members of the group I of the periodic table. Preliminary co-challenge experiments using a 5:1 excess of Na^+ to Cs^+ in the solution (5 mM and 1 mM respectively) presented to immobilized cells of *Serratia* sp. in the presence of 1 mM UO_2^{2+} showed 50%, not 20%, removal of the Cs^+ which suggests that some enrichment occurred (Paterson-Beedle and Macaskie, 2003). Full analyses of the biocrystals are not yet available, but Figure 24.6 shows that inclusion of Cs^+ in the 5 mM glycerol 2-phosphate/1 mM uranyl nitrate feed solution resulted in a disruption of the regular morphology of the crystals, presumably due to the different morphology of the Cs^+ containing crystals as compared to the ones containing the Na^+ ion.

24.8 USE OF METAL PHOSPHATE AS AN ION EXCHANGER

Hydrogen uranyl phosphates (HUP) and the phosphates of some other metals such as zirconium can behave as ion exchange materials (Clearfield, 1988; Clearfield and Tindwa, 1979). In the example of $HUO_2PO_4 \cdot 4H_2O$ the incoming metal cation displaces protons from the interlamellar spaces within the 'host' crystal (Clearfield, 1988) and it was shown that in this way Ni^{2+} was removed effectively into hydrogen uranyl phosphate previously laid down on to the biomass (Basnakova and Macaskie, 1997; Basnakova et al., 1998). Immobilized cell columns accumulated Ni^{2+} in a molar ratio of 1:2:2, consistent with the formation of $Ni(UO_2PO_4)_2$, which was

---

column input region and the removal of target metal is poor. Conversely, at a high flow rate (corresponding to the top line), the substrate is not all converted within the flow residence time at the column input end, and the nucleating sites are distributed along the length of the column, giving more effective removal of the target metal. This was shown experimentally for the removal of neptunium and plutonium phosphates by Macasakie and Basnakova (1998). Column blockage by a 'slug' of metal phosphate at the input end is shown in Figure 24.4f.

Figure 24.6 Cells of *Serratia* sp. and their heavy metal biominerals. (a) A scanning electron micrograph of a biofilm of *Serratia* on reticulated foam (see Figure 24.4b,d) shows the bacteria held within a meshwork of extracellular polymeric substances. For SEM the sample is dried and examined in a vacuum. This causes the heavily hydrated extracellular polymers to dehydrate and collapse, becoming very difficult to see by SEM. b-f. Using environmental scanning electron microscopy (ESEM) the sample can be visualised in the native, hydrated state. (b) This shows that the cells are surrounded by heavy gelatinous layers of EPS. (c) On deposition of hydrogen uranyl phosphate the gelatinous appearance of the cells transforms to a crystalline appearance, with the cells forming interweaving fibrils composed of many cells joined together. (d) A higher magnification shows the presence of regular striations of $NaUO_2PO_4 \cdot 4H_2O$. The sodium ion originates from the sodium salt of the glycerol 2-phosphate phosphate donor (see text). (e) When Cs^+ is incorporated into the flow, a mixed Cs and Na hydrogen phosphate is formed (see text), which changes the morphology of the crystals (viewed under ESEM; c.f. b and d) such that individual cells are visible and the EPS has disappeared. (f) When strontium is incorporated into the flow (see text) the regular structures are lost and the cells become hidden under the layers of precipitate.

confirmed by complementary analytical techniques (Basnakova *et al.*, 1998). Although Co^{2+} was not removed by co-precipitation (see above) the ion exchange method promoted removal of the Co^{2+} ion. For example, reactors were 'primed' with HUP to a loading of ca. 190 mg uranium deposited *per* reactor (ca. 100% of bacterial dry weight) and then challenged with Co^{2+}. The theoretical capacity of a reactor containing 190 mg uranium (in the form of HUP) would be 55.3 mg of Co^{2+} at a molar ratio of Co:U of 1:2, i.e. for the formation of $Co(UO_2PO_4)_2$. A loss of Co-removing capacity after 226 ml was observed, which corresponded to ca. 50.4 mg of deposited Co^{2+} (Paterson-Beedle and Macaskie, 2003). These results suggest a possible application of biogenic metal phosphates in the essentially chemical removal of problematic metal species. However, this approach is still in the early stages of development. For example, biogenic zirconium phosphate did not remove metals, even though according to the literature this should be possible. Zirconium phosphate exists in more than one crystalline form (see Basnakova and Macaskie, 1999) and it is likely that a less reactive form was laid down on the cells. Since use of zirconium is a non-hazardous alternative to the use of uranyl phosphate as a 'host' crystal, more studies are warranted into the possibility of chemical processing of the Zr-biomineral to obtain the useful 'host' species for coprecipitation.

24.9 USE OF ALTERNATIVE PHOSPHATE DONORS

Most of the studies on directed metal phosphate biomineralization have utilised glycerol 2-phosphate as a useful test compound. This is relatively inexpensive but an economic assessment on the use of this compound to promote metal remediation from mine drainage waters concluded that the cost of the organic phosphate 'donor molecule' was the single factor which limited practical application of this approach (Roig et al., 1995). Many organophosphate molecules are toxic (see earlier) but the phosphonate group of compounds, which has C—P (phosphonate) bonds instead of C—O—P (phosphate ester bonds) comprises pesticides, herbicides, detergent additives, antibiotics and flame retardants (see Rodriguez and Fraga, 1999; Nowack, this volume). These compounds are generally resistant to chemical hydrolysis. There have been reports of their microbially mediated degradation and phosphate release, but the degradation is slow (McGrath et al., 1995, 1998; Ohtake et al., 1996) and does not necessarily involve initial breakage of the phosphonate bonds. Use of these agents to support metal bioprecipitation does not appear to have been evaluated, but practical applications would be limited. In contrast, the phosphotriester compound tributyl phosphate (TBP), a solvent and plasticizer, which is also used industrially as a metal extractant, was shown to support the removal of uranyl ion from solution by immobilized cells of a population of *Pseudomonas* and *Comamonas* species (Thomas and Macaskie, 1996). The mixed population was immobilized as a biofilm on a microcarrier support, with TBP degradation, phosphate release, and uranyl ion removal observed in a continuous-flow system. A yellow precipitate was observed, which was confirmed by X-ray powder diffraction analysis, to be uranyl phosphate (Thomas and Macaskie, 1996). TBP biodegradation was shown to be sensitive to Cu^{2+} and Cd^{2+}, and inorganic phosphate ($>10\,mM$); sulphate ions (10–$100\,mM$) were also inhibitory but, in contrast to the *Serratia* system (Yong and Macaskie, 1997), the TBP hydrolase activity was insensitive to nitrate (Thomas and Macaskie, 1998): a potential advantage in the treatment of nitrate-rich waters. The microcarrier-immobilized biomass decontaminated uranium-bearing mine waste water at the expense of TBP hydrolysis in the presence of 35 mM sulphate in the water: at pH 4.5 79% of the UO_2^{2+} was removed, at a flow rate of 1.4 ml/h using a 7 ml test column (Thomas and Macaskie, 1998). However the microbiology of the system proved intractable. When single colony isolates were obtained the cultures lost the ability to hydrolyse TBP irretrievably, after eight serial subcultures (Thomas et al., 1997a). The parent mixed culture was more stable, continuing to utilize TBP and liberate phosphate after subculture, but the growth rate, and TBP consumption, occurred either rapidly (2 mM removed after 72 h) or slowly (0.4 mM removed after 72 h), with phosphate production paralleling loss of TBP. However intermediate rates were never observed in a total of 98 experiments. The rate of growth always paralleled TBP utilization, and the rapid TBP utilization was marked by the presence of a 24 Kb DNA fragment which was never present in the slow-growing cultures (Thomas et al., 1997b). In subsequent tests single colony isolates (12 Pseudomonas strains) could be stabilized in the presence of ampicillin and examination of the isolates in pure culture showed the presence of a similar DNA band (22–24 Kb) throughout (Thomas et al., 1997b). However attempts to produce a restriction map of the DNA fragment were unsuccessful. The strains were stored at −70°C and later attempts to repeat the studies using pure cultures were unsuccessful; no growth was observed at the expense of TBP (Berne and Macaskie, unpublished). The cultures were pooled into a chemostat provided with very low concentrations of carbon source and inorganic phosphate and with TBP as the sole additional carbon and phosphorus source. Over a period of several weeks a culture evolved which gave growth over and above that attributable to the 'background' level of nutrients alone, and the growth of which corresponded to the appearance of a 22–24 Kb DNA fragment (Berne and Macaskie, unpublished). However subsequent attempts to repeat this selection using the same frozen stocks ~1 year later were unsuccessful. A similar instability was reported in the case of a yeast strain reported to utilize TBP (McGrath, personal communication). Attempts to obtain new enrichment cultures from natural environments contaminated with organophosphorus pesticides were unsuccessful from samples obtained from a site in the Pennines, England but were successful from sites in Northern Ireland. However a new strain from an Irish site was, similarly, unstable (McGrath, personal communication). It must be concluded that although the use of TBP as a phosphate donor has potential, this will not be practically possible until a stable system can be developed.

One alternative potential phosphate donor, phytic acid (inositol phosphate: Figure 24.7), a major plant phosphate component (up to 80% of the phosphate content of legumes and cereals (Reddy et al., 1989)) has received little attention with respect to removal of metals, despite this compound comprising up to 50% of the organic phosphate in soils (Anderson, 1980; Dalal, 1977; Harley and Smith, 1983). Its breakdown is slow, probably attributable to the high insolubility of its (e.g.) calcium salt. The role of phytic acid as a possible phosphate donor in soils and subsurface has been postulated as a way of immobilizing toxic metals (Banaszak et al., 1999). It was reported that 30–48% (Greaves and Webley, 1965) and up to 63% (Richardson and Hadobas, 1997) of culturable soils and rhizosphere microorganisms utilize phytate. Only a low proportion of organisms could use phytate as a carbon source but 39–44% could use it as a phosphate source (Richardson and Hadobas, 1997), probably reflecting the high molar phosphate content of the molecule (Figure 24.7).

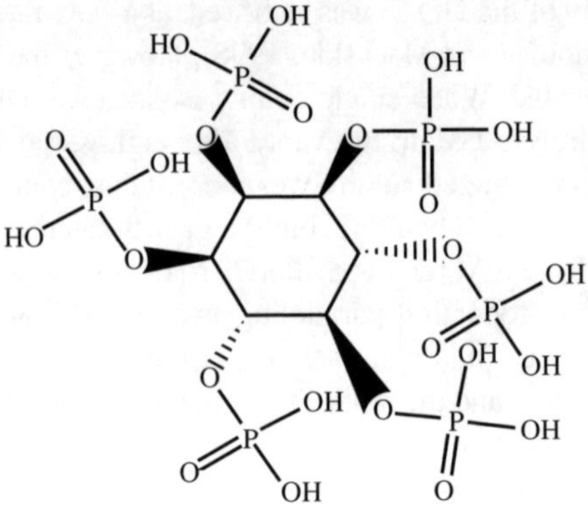

Figure 24.7 The structure of phytic acid. Note that there are 6 mols of available phosphate per mol of phytic acid, hence this is potentially a very useful phosphate donor, especially since phytic acid is a natural plant product (see text).

A more recent study of strains isolated from petrol-contaminated soil gave two strains showing close homology to *Pseudomonas* spp., using 16S rRNA gene homology. These strains grew, albeit slowly, using phytate as the sole source of carbon, and were resistant against up to 1 mM Cd^{2+}. Studies to evaluate the role of phytate hydrolysis, and the use of the liberated phosphate as a metal precipitant, are in progress (Badar, Qureshi and Ahmed, personal communication) and if an association is confirmed, the ability to hydrolyse phytate could be added to the portfolio of microbial metal resistance mechanisms. Nash *et al.* (1998) have suggested that there may be a role for phytic acid added to soils, to facilitate *in situ* metal bioprecipitation; the half life of phytate in synthetic groundwater is 100–150 years *per se* (Jensen *et al.*, 1996) but micoorganisms can accelerate the rate of hydrolysis by several orders of magnitude (Suzuki and Kamatani, 1995). Although phosphatases are very abundant in nature, these enzymes are virtually unable to hydrolyse phosphomonoesters in phytic acid (Kerovuo *et al.*, 2000). Phytases, a specific class of enzymes hydrolysing phytic acid, have been reported in a number of soil bacteria including *Bacillus subtilis, Pseudomonas putida* and *P. mendocina* (Richardson and Hadobas, 1997), and this property is probably widespread (Cosgrove *et al.*, 1970; Dvorakova, 1998). The phytases hydrolyse phytic acid to less phosphorylated *myo*-inositol phosphates, and in some case to free *myo*-inositols, releasing phosphate. In microorganisms, phytase is induced in response to phosphate starvation and, like many other phosphatases, probably acquires phosphate for growth, in this case at the expense of plant-derived phytic acids. Several phytases have been cloned and characterized, including an enzyme from *Escherichia coli* (Greiner *et al.*, 1993), which shares a highly conserved sequence motif found at the active sites of acid phosphatases (Ullah *et al.*, 1991). More recently, highly thermostable phytases have been isolated and cloned from *Bacillus* species (Ha *et al.*, 2000; Kerovuo *et al.*, 1998; Kim *et al.*, 1998a, b) and elucidation of the crystal structure has shed light on the reaction mechanism (Ha *et al.*, 2000; Kerovuo *et al.*, 2000). Although the degradation of phytic acid in soils may be slow (above), the newly-identified enzyme released phosphate from phytic acid to completion after 60 min, but with only 3 mol/mol of phosphate released out of possible 6 mol/mol (see Figure 24.7) (Kerovuo *et al.*, 2000), which could have implications for the use of phytic acid as an effective phosphate feed for metal bioprecipitation. A detailed analysis of the reaction pathway is outside the scope of this chapter but is described by Kerovuo *et al.* (2000). However several authors point out that the new enzyme, PhyC, is significantly different from other phytases (Kerovuo *et al.*, 1998; Kim *et al.*, 1998a, b) and it would be hoped that a survey of the possible use of this enzyme in bioremediation would balance the reaction rate versus the degree of dephosphorylation obtained. The loss of six phosphate groups is involved; phytases are generally divided into 3- and 6-phytases (the position of the phosphate cleaved first), while cleavage of an axial phosphate at the 2-position is very uncommon (Kerovuo *et al.*, 2000). This suggests that the reaction kinetics, important with respect to development of a flow-through system and its application to real wastes (Macaskie *et al.*, 1997), may be multi-stage and difficult to define. Presumably a second round of phosphate cleavage could be initiated before all molecules had yielded their first phosphate group, leading to a complex, multi-stage and possibly overlapping set of kinetics which would be very difficult to model. Nevertheless, the use of phytic acid should be explored with respect to metal waste remediation, particularly since some organisms were reported to grow in 1 mM Cd^{2+} in the presence of phytate. However some precipitation of calcium phytate occurred and the extent of this, and hydrolysis of the organophosphate molecule, has not yet been determined (Badar, Qureshi and Ahmed, unpublished).

24.10 USE OF INORGANIC PHOSPHATE AS THE PHOSPHATE DONOR

The above discussions have centred around the use of organic phosphates as the phosphate donor for metal precipitation. However many waste waters contain phosphate and the harnessing of these phosphate wastes to support heavy metal removal could be an environmentally-friendly solution to two waste problems at the same time. The concentration of phosphate in influent sewage in the UK is about 10 mg/l (0.32 mM) (Upton, 1998). Waste water treatment companies serving a population of more than 100,000 are required to reach a target of 2 mg/l (64 μM) phosphate, while for those serving populations of 10,000–100,000 the limit is 1 mg/l (32 μM) (Upton, 1998). Various chemical methods for phosphate removal are possible, but have disadvantages (Cooper *et al.*, 1994). Conventional activated sludge treatment without chemical precipitation can remove 20–40% of influent phosphate (Streichan *et al.*, 1990) but this falls well short of the target level and additional process steps are required. This may involve incorporation of an additional bioprocess step, which can increase the level of phosphate removal to 80–90% (Streichan *et al.*, 1990). These processes are called Enhanced Biological Phosphate Removal

(EBPR) and involve the activities of microorganisms found in activated sludge which take up phosphate in excess of their requirements and store this in the form of polyphosphate (Beacham, 1992; Fuchs and Chen, 1975; McGrath and Quinn, this volume; Mino et al., 1998), which can be later mobilized, with inorganic phosphate release. Various microorganisms are implicated in EBPR activities (Bond et al., 1995; Christensson et al., 1998; Wagner et al., 1994a). Strains of one major genus, *Acinetobacter*, are readily isolated from plants operating EBPR (Boswell et al., 1999; Mc Grath and Quinn, this volume; van Groenestijn et al., 1985, 1989) and the mixed cultures give a positive result using *Acinetobacter*-specific oligonuceotide probes in activated sludge (Wagner et al., 1994b). The microbial ecology of EBPR determined by various culture-independent methods is described elsewhere (Auling et al., 1991; Bond et al., 1995, 1999; Dabert et al., 2001; Wagner et al., 1994a, b). In fact, *Acinetobacter* spp. are relatively low numerically in EBPR sludges (Bond et al., 1995; Kampfer et al., 1996) suggesting the potential of a wide range of organisms in processes which rely on phosphate storage and its subsequent mobilisation to yield phosphate which could be used as a metal precipitant.

In order to apply polyphosphate (polyP) cycling to heavy metal removal it is necessary to consider the metabolism of polyP metabolising species, as typified by *Acinetobacter* spp. (Figure 24.8). This organism does not grow anaerobically, but is able to take up acetate anaerobically and synthesize the carbon storage polymer polyhydroxybutyrate (PHB), at the expense of energy supplied by the hydrolysis of cellular polyP, which in this context acts as a 'high energy' phosphate store. Inorganic phosphate is released as a waste product. On transfer to aerobic conditions the cells have a competitive advantage due to the presence of the

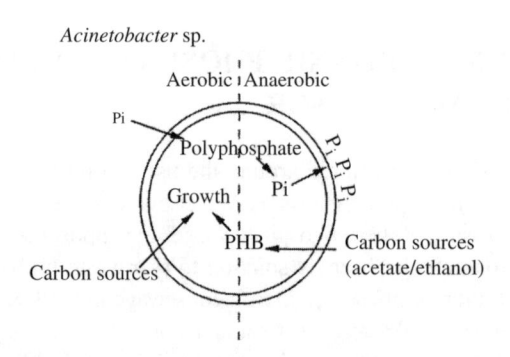

Figure 24.8 The metabolism of *Acinetobacter* sp. The cell is able to exploit a cycling of aerobic and anaerobic conditions. Aerobically, phosphate is taken up from waste water and polyphosphate is synthesised at the expense of ATP generated *via* the aerobic metabolism of carbon sources taken up from the water, and also cellular reserves of poly ß-hydroxybutyrate (PHB), which are mobilized aerobically. Upon switching to anaerobic conditions, the organism cannot grow anaerobically but can still take up carbon sources. These are metabolized and stored as PHB intracellularly. The energy to drive PHB synthesis is derived from that stored within the polyP molecule and released upon hydrolysis of polyP. Inorganic phosphate is liberated as a concentrated 'pulse' anaerobically and is made available to support heavy metal bioprecipitation.

stored PHB which can be broken down to provide energy for growth and for the synthesis of more polyP. In this way, the cells are maximally poised to exploit both aerobic and anaerobic conditions, although growth only occurs oxically. With repeated cycles cells can adapt such that the magnitude of the PHB and polyP fluxes increase over time (Boswell et al., 2001).

In general, EBPR processes use a configuration where the influent mixed liquor is exposed to an anaerobic stage before the standard aerobic treatment in which enhanced phosphate uptake occurs. In the 'Phostrip' process (Barnard, 1974, 1976; Levin, 1964) phosphate stored in sludge microorganisms is 'stripped' (released) anaerobically, producing a low-volume concentrated solution of inorganic phosphate that can be precipitated chemically. There is no reason why the phosphate concentrate should not be applied to processes of heavy metal removal. Indeed, it was shown using ^{31}P nuclear magnetic resonance that the presence of heavy metals can enhance polyP mobilization by *Acinetobacter* cells (Boswell et al., 1999; Suresh et al., 1986). Successful phosphate release is critically dependent on a regime of aerobic/anaerobic cycling in order to exploit the biochemistry of the organism (see Figure 24.8). This can be done in two ways. One approach utilized a bioreactor subjected to temporally separated aerobic/anaerobic cycles (Dick et al., 1995) while a continuous process was shown to be possible using spatially separated sequential aerobic and anaerobic vessels (Boswell et al., 2001). In the first example (Dick et al., 1995), *Acinetobacter* sp., isolated from a mixed liquor activated sludge process, were immobilized in agar or agarose beads and held within a reactor vessel operating in fill- and draw- mode. The 8 hour cycle consisted of 2.5 h of anaerobic incubation with a N_2 gas sparge followed by 5.5 h of aerobic incubation. The draw and refill period at the end of each cycle (during which the medium was changed from aerobic growth medium/phosphate feed to anaerobic metal-containing feed) was approximately 15 minutes. Phosphate was consumed in the aerobic pulse and liberated in the anaerobic pulse, concomitant with the stoichiometric removal of metals from the anaerobic feed solution (Table 24.1).

In experiments where *Acinetobacter* sp. W9 was exposed to anaerobic periods in the presence of uranyl ion for up to 16 days, no loss of activity was observed by assay for phosphate release or uranyl ion removal, giving confidence in this approach as a feasible method for long-term metal removal. At the end of this period, samples of gel removed from the reactor were yellow with a yellow-green fluorescence under ultra violet light, typical of uranyl phosphate, while X-ray

Table 24.1 Phosphate release and metal removal by immobilized cells of *Acinetobacter* sp. strain W9.

Phosphate release (mM)	UO_2^{2+} removal (mM)	Cd^{2+} removal (mM)
(mean ± SD)	(mean ± SD)	(mean ± SD)
0.056 ± 0.039	0.062 ± 0.009	0.076 ± 0.022

Data from Dick et al. (1995). SD: standard deviation. The concentration of metal supplied in the anaerobic feed was 0.2 mM. Phosphate release was determined during metal-unsupplemented anaerobic periods.

powder diffraction analysis confirmed that the deposited material was hydrogen uranyl phosphate, indistinguishable from that laid down by the *Serratia* sp. described earlier (Dick *et al.*, 1995). Since metal removal was reproducible and stable over extended periods a continuous reactor system was developed (Boswell *et al.*, 2001). Here, the effluent from an aerobic chemostat was passed into an anoxic vessel and dosed with La^{3+} as the test metal. The effluent from this passed into a settler in which the heavy, La-loaded cells (i.e. those which had liberated the most phosphate) were removed under gravity. The lighter, lesser-loaded cells passed back into the aerobic vessel for the next cycle. Comparison of the La^{3+} concentration entering the anaerobic vessel and leaving the settler confirmed the continuous removal of 95% of the La^{3+} from an input concentration of 0.3 mM (Boswell *et al.*, 2001). Examination of the cells under the electron microscope showed cell surface metal deposition in the same way as described earlier for the *Serratia* sp. (Dick *et al.*, 1995) but the extent to which cell surface phosphate groups are involved in metal phosphate nucleation processes was not determined. Hence the use of an *Acinetobacter*-based process has promise in the development of a biotechnological process. However it was found that the magnitude of the phosphate flux, and hence the degree of metal removal per cycle, decreased during long-term storage and maintenance of the *Acinetobacter* strains used (Boswell, Dick and Macaskie, unpublished). Since other organisms have been suggested to be involved in EBPR a search for other metal removing strains would be worthwhile.

Recent studies have indicated that inorganic phosphate could be used to enhance metal precipitation onto biomass directly, once nucleation and crystal growth have been initiated. This was illustrated using the *Serratia* system to manufacture hydroxylapatite (see Sammons *et al.*, this volume) (Table 24.2).

The column reactor (4.5 cm i.d. × 20 cm; fluid volume 250 ml) was packed with 100 biofilm-coated foam cubes (1 cm³ cube with dry biomass ~3.3 mg/cube; phosphatase specific activity 4,500 nmol product/min/mg protein). The column was challenged (up flow at 16 ml/h) with solutions comprising 10 mM $CaCl_2$, 20 mM citrate and 25 mM TAPSO buffer (pH 8) in the presence of 25 mM glycerol 2-phosphate (solution 1) or 10 mM glycerol 2-phosphate and 15 mM NaH_2PO_4 (solution 2). After three days of pre-nucleation, each sample was taken after 500 ml had passed and was assayed for Ca^{2+} and PO_4^{3-} as described previously (Yong *et al.*, 2003).

Table 24.2 indicates that there is scope for using inorganic phosphate to part-replace the requirement for glycerol 2-phosphate, thus decreasing the process cost. The extent to which this can be achieved clearly warrants further study. Calcium phosphate precipitation is not the best model system, since calcium phosphate is

Table 24.2 Calcium removal by biofilm-immobilized cells of *Serratia* sp. in a flow-through reactor.

	Phosphate in outflow solution (mM)	Calcium precipitated (mM)
Solution 1	17.8 ± 0.4	4.5 ± 0.4
Solution 2	21.8 ± 0.4	5.8 ± 0.4

more soluble (i.e. less readily removed by the cells) than the phosphates of other heavy metals such as Cd^{2+}, and this mixed-phosphate approach should be pursued with a view to heavy metal bioremediation processes. In this context, an earlier study showed that although the metal precipitating *Serratia* sp. could not release substantial phosphate from tributyl phosphate *per se* (Jeong *et al.*, 1994), incorporation of TBP into the flow relieved part of the requirement for glycerol 2-phosphate in a continuous, steady-state process (Michel *et al.*, 1986). Whether the phosphate came from the TBP itself or whether the phosphate triester facilitated metal bioprecipitation in some way, increasing the efficiency of utilization of the phosphate released from the glycerol phosphate was not determined, but would also warrant further study.

24.11 CONCLUSIONS

As described above, microbial metal phosphate precipitation processes are biocatalytic and inorganic precipitation combined reaction processes, in which the component reactions are:

Phosphatase + phosphate donor $\longrightarrow$
Phosphatase-donor molecule complex $\longrightarrow$ Phosphate

Phosphate + Metal ion $\longrightarrow$ Metal phosphate precipitation,

where the phosphate donor molecule can be, for example, a simple phosphomonoester (glycerol 2-phosphate), more complex phosphomonoester (phytic acid), phosphotriester (parathion, tributyl phosphate) or polyphosphate (polyP).

Metal precipitation is mediated by the activity of cellular phosphoesterases which continue to function in resting cells to release inorganic phosphate from the organic phosphate substrates. These are metal biomineralization processes, in which the microbial cells function not only to 'generate' the precipitant but also promote the nucleation reactions by provision of cellular ligands to initially chelate the metals and provide 'seeds' for further metal deposition processes. The 'seed' can be a similar metal phosphate or can be a precipitate of a different metal phosphate, laid down in a prior nucleation and priming step. In this way, deposition of 'difficult' metals can be 'driven' and, indeed, other microbial components of a mixed population, while not necessarily accumulating metals *per se*, can promote the biocrystallization of metal phosphates by promoting an upshift of the pH into a region where metal phosphates are more likely to precipitate-out on the primary phosphate-depositing organisms (Pümpel *et al.*, 2003). 'Helper' functions can be seen in other ways. Extant metal phosphate biomineral can be used as the host crystal for nucleation of other, more recalcitrant metals. This is especially useful in the case of metals whose concentration is usually too low to precipitate readily from the bulk solution (e.g. Macaskie and Basnakova, 1998).

The concept of a 'structured data space' to define the parameters for biosorption was put forward by Pümpel and Schinner in 1995 in an attempt to define 'strategic' questions that need to be asked in order to predict whether biosorption of a metal will occur. Metal biosorption is a prerequisite to subsequent biocrystallization reactions. The components of the cell wall and the biological space on and around the cells are utilised to overcome the activation energy barrier and also to promote spatial localization for the formation of the solid phase from a supersaturated solution. The metal mineralization process could be accelerated by increasing the degree of supersaturation or by lowering the interfacial energy. This would be achieved in biomineralization by speeding up the release rate of the precipitant ligand, e.g. by using cells with high phosphoesterase/polyphosphatase activity and/or increasing the substrate concentration, and also by mediating the nucleation with extant organic polymeric substrates or the pre-engineering of nucleation sites in the exocellular space of the cells.

Although bacterial cells can be made to express very high enzyme activity by the manipulation of their growth conditions, or by genetic recombination, in practice the biomineral deposition rate is ultimately limited by the kinetics of the chemical deposition process, which has been defined in terms of an 'inefficiency factor' (Macaskie et al., 1995, 1997) which is dependent on the metal, the insolubility of its phosphate, the presence of organic or inorganic ligands in the solution matrix, and the pH. For example the presence of nitrate or sulphate can interfere with enzymatic phosphate liberation (Yong and Macaskie, 1997, 1999). The reaction rate can be increased by introducing more substrate into the challenge solution, or the efficiency can be increased by decreasing the flow rate through flow-through columns. This was shown to overcome the inhibition in accordance with mathematical predictions (Yong and Macaskie, 1997, 1999) which were shown to be applicable to the treatment of real mine waste water containing uranium in the presence of an excess of sulphate ion and also other metal cations (Yong and Macaskie, 1999). However this would increase the cost of a process, which may already be financially unattractive. Also, the persistence of residual phosphate in the process outflow would be problematic, requiring treatment of the phosphate solution before its final disposal to prevent adverse environmental consequences. In addition, the biomass itself comprises an additional organic waste, which requires disposal. In cases of biosorption the biomass becomes saturated and the metal can be eluted from the biomass using mineral acids or chelating agents such as citrate. In the case of biomineralization, this is unnecessary because the biomass does not become saturated and the metal is available for recycle as a solid concentrate, for example, the sludge removed from a metal-bioaccumulating sandfilter system was thickened into filter cake for pyrometallurgical reprocessing in a shaft furnace (Diels et al., 2001). For radioactive solid wastes long-term burial is one option but the presence of the biomass organic component is highly undesirable since bio-degradation of this by natural microbial populations could re-solubilize the radionuclides (Eccles, pers. comm.) In this case desorption could be a more useful option but this may yield a liquid effluent which, in many cases, gives a greater problem since chelating agents loaded with bound metals are difficult to bio-degrade (see Thomas et al., 2000). Since the proportion of radionuclide phosphate to biomass can be up to 9–10 times the weight of the biomass itself (Macaskie, 1990), destruction of the organic components by heat treatment may be the most feasible option.

REFERENCES

Anderson, G. (1980) Assessing inorganic phosphorus in soils. In *The Role of Phosphorus in Agriculture* (eds Khasawneh, F.E., Sample, E.C. and Kamprath, E.J.), Madison, WIS: Am. Soc. Agron. pp. 411–432.

Andres, Y., MacCordick, H.J., and Hubert, J.C. (1993) Adsorption of several actinide (Th, U) and lanthanide (La, Eu, Yb) ions to *Mycobacterium smegmatis. Appl. Microbiol. Biotechnol.* **39**, 413–417.

Andres, Y., MacCordick, H.J. and Hubert, J.C. (1994) Binding sites of sorbed uranyl ion in the cell wall of *Mycobacterium smegmatis FEMS Microbiol. Lett.* **115**, 27–32.

Andres, Y., MacCordick, H.J. and Hubert, J.C. (1995) Selective biosorption of thorium ions by an immobilized mycobacterial biomass. *Appl. Microbiol. Biotechnol.* **44**, 271–276.

Auling, G., Pilz, F., Busse, H.J., Karrasch, S., Streichan, M. and Schon, G. (1991) Analysis of the polyphosphate-accumulating microflora in phosphorus-eliminating, anaerobic-aerobic activated sludge systems by using diaminopropane as a biomarker for rapid estimation of *Acinetobacter* spp. *Appl. Environ. Microbiol.* **57**, 3585–3592.

Banaszak, L.J., Rittman, B.E. and Reed, D.T. (1999) Subsurface interactions of actinide species and microorganisms: implications for the bioremediation of actinide organic mixtures. *J. Radioanal. Nucl. Chem.* **241**, 385–435.

Barnard, J.L. (1974) Cut P and N without chemicals. *Water Wastes Eng.* **11**, 22–44.

Barnard, J.L. (1976) A review of biological phosphorus removal in the activated sludge process. *Water SA* **2**, 136–144.

Barnes, L.J., Janssen, F.J., Sherren, J., Versteegh, R.O., Koch, R.O. and Scheeren, P.J.H. (1991) A new process for the microbial removal of sulphate and heavy metals from contaminated waste extracted by a geohydrological control system. *Chem. Eng. Res. Design.* **69A**, 184–186.

Basnakova, G. and Macaskie, L.E. (1997) Microbially enhanced chemisorption of nickel into biologically synthesized hydrogen uranyl phosphate: A novel system for the removal and recovery of metals from aqueous solutions *Biotechnol. Bioeng.* **54**(4), 319–328.

Basnakova, G. and Macaskie, L.E (1999) Accumulation of zirconium and nickel by *Citrobacter* sp. *J. Chem. Technol. Biotechnol.* **74**(6), 509–514.

Basnakova, G., Spencer, A.J., Palsgard, E., Grime, G.W. and Macaskie, L.E. (1998a) Identification of the nickel uranyl phosphate deposits on *Citrobacter* sp. cells by electron microscopy with electron probe X-ray microanalysis and by proton induced X-ray emission analysis. *Environ. Sci. Technol.* **32**(6), 760–765.

Basnakova, G., Stephens, E.R., Thaller, M.C., Rossolini, G.M. and Macaskie, L.E. (1998b) The use of *Escherichia coli* bearing a *phoN* gene for the removal of uranium and nickel from aqueous flows. *Appl. Microbiol. Biotechnol.* **50**(2), 266–272.

Beacham, A.M., Seviour, R.J. and Lindrea, K.C. (1992) Polyphosphate-accumulating abilities of *Acinetobacter* isolates from activated sludge. *Water Res.* **26**, 121–122.

Bond, P., Hugenholtz, P., Keller, J. and Blackall, L. (1995) Bacterial community structures of phosphate-removing and non phosphate-removing activated sludges from sequencing batch reactors. *Appl. Environ. Microbiol.* **61**, 1910–1916.

Bond, P.L., Erhart, R., Wagner, M., Keller, J. and Blackall, L.L. (1999) Identification of some of the major groups of bacteria in efficient and nonefficient biological phosphorus removal activated sludge systems. *Appl. Environ. Microbiol.* **65**, 4077–4074.

Bonthrone, K.M., Basnakova, G., Lin, F. and Macaskie, L.E. (1996) Bioaccumulation of nickel by intercalation into polycrystalline hydrogen uranyl phosphate deposited via an enzymatic mechanism. *Nature Biotechnol.* **14**(5), 635–638.

Bonthrone, K.M., Quarmby, J., Hewitt, C.J., Allan, V.J.M., Paterson-Beedle, M., Kennedy, J.F. and Macaskie, L.E. (2000) The effect of the growth medium on the composition and metal binding behaviour of extracellular polymeric material of a metal-accumulating *Citrobacter* sp. *Environ. Technol.* **21**, 123–134.

Boswell, C.D., Dick, R.E. and Macaskie, L.E. (1999) The effect of heavy metals and other environmental conditions on the aerobic phosphate metabolism of *Acinetobacter johnsonii Microbiol.* **145**, 1711–1720.

Boswell, C.D., Dick, R.E., Eccles, H. and Macaskie, L.E. (2001) Phosphate uptake and release by *Acinetobacter johnsonii* in continuous culture and coupling of phosphate release to heavy metal accumulation. *J. Ind. Microbiol. Biotechnol.* **26**, 333–340.

Christensson, M., Blackall, L.L. and Welander, T. (1998) Metabolic transformations and characterisation of the sludge community in an enhanced biological phosphorus removal system. *Appl. Microbiol. Biotechnol.* **49**, 226–234.

Clearfield, A. (1988) Role of ion exchange in solid state chemistry *Chem. Rev.* **88**, 125–148.

Clearfield, A. and Tindwa, R.M. (1979) Exchange of large cations and charged complexes with amine intercalates of zirconium phosphates. *Inorg. Nucl. Chem. Lett.* **15**, 251–254.

Collard, J.M., Corbisier, P., Diels, L., Dong, Q., Jeanthon, C., Mergeay, M., Taghavi, S., Van der Lelie, D., Wilmotte, A. and Wuertz, S. (1994) Plasmids for heavy metal resistance in *Alcaligenes eutrophus* CH34: mechanisms and applications *FEMS Microbiol. Rev.* **14**, 405–414.

Cooper, P., Day, M. and Thomas, V. (1994) Process options for phosphorus and nitrogen removal from wastewater. *J. Inst. Water Environ. Manage.* **8**, 84–92.

Cosgrove, D.J., Irving, G.C.J. and Bromfield, S.M. (1970) Inositol phosphate phosphatases of microbial origin. The isolation of soil bacteria having inositol phosphate phosphatase activity *Aust. J. Biol. Sci.* **23**, 339–343.

Dabert, P., Sialve, B., Delgenes, J.P., Moletta, R. and Godon, J.J. (2001) Characterisation of the microbial 16S rDNA diversity of an aerobic phosphorus-removal ecosystem and monitoring of its transition to nitrate respiration. *Appl. Microbiol. Biotechnol.* **55**, 500–509.

Dalal, R.C. (1977) Soil organic phosphorus. *Adv. Agron.* **29**, 83–87.

Dick, R.E., Boswell, C.D. and Macaskie, L.E. (1995) Uranyl phosphate accumulation by *Acinetobacter* spp. In *Biohydrometallurgical Processing* (eds Jerez, C.A., Vargas, T., Toledo, H. and Wiertz, J.V.) The University of Chile, pp. 177–186.

Diels, L., Van Roy, S., Taghavi, S., Doyen, W., Leysen, R. and Mergeay, M. (1993) The use of *Alcaligenes eutrophus* immobilized in a tubular membrane reactor for heavy metal recuperation. In *Biohydrometallurgical Technologies The Minerals, Metal and Material Society* (eds Torma, A.E., Apel, M.L. and Brierley C.L.), pp.133–144.

Diels, L., Spaans, P.H., Van Roy, S., Hooyberghs, L., Wouters, H., Walter, E., Winters, J., Macaskie, L.E., Finlay, J., Pernfuss, B., Woebking, H., Pümpel, T. and Tsezos, M. (2001) Heavy metals removal by sand filters inoculated with metal sorbing and precipitating bacteria. In *Biohydrometallurgy: Fundamentals, Technology and Sustainable Development* (eds Ciminelli, V.S.T. and Garcia, Jr. O.), Elsevier, Amsterdam. pp. 317–326.

Dvorakova, J. (1998) Phytase: sources, preparation and exploitation. *Folia Microbiol.* **43**, 323–338.

Eccles, H. (1995) Removal of heavy metals from effluent streams: why select a biological process? *Int. Biodeterior. Biodegr.* **35**, 5–16.

Eccles, H. (1998) Metal contaminated soil – is natural attenuation acceptable? *Biochem. Soc. Trans.* **26**(4), 657–661.

Eccles, H. (1999) Treatment of metal-contaminated wastes: why select a biological process? *Trends Biotechnol.* **17**(12), 462–465.

Ellwood, D.C., Hill, M.J., and Watson, J.H.P. (1992) Pollution control using microorganisms and magnetic separation. (eds Fry, J.C., Gadd, G.M., Herbert, R.A., Jones, C.W.

and Watson-Craik, I.), Cambridge University Press, Cambridge UK *Soc. Gen. Microbiol. Symp.* No. **48**, pp. 89–112.
Finlay, J.A., Allan, V.J.M., Conner, A., Callow, M.E., Basnakova, G. and Macaskie L.E. (1999) Phosphate release and heavy metal accumulation by biofilm-immobilized and chemically-coupled cells of a *Citrobacter* sp. pre-grown in continuous culture. *Biotechnol. Bioeng.* **63**, 87–97.
Fuchs, G.W. and Chen, M. (1975) Microbiological basis of phosphate removal in the activated sludge process for the treatment of wastewater. *Microb. Ecol.* **2**, 119–138.
Gadd, G.M. (1996) Influence of microorganisms on the environmental fate of radionuclides. *Endeavour* **20**(4), 150–156.
Gadd, G.M. (1997) Roles of micro-organisms in the environmental fate of radionuclides. Health Impacts of Large Releases of Radionuclides *Ciba Found. Symp.* **203**, 94–104.
Gadd, G.M. (2000) Bioremedial potential of microbial mechanisms of metal mobilization and immobilization. *Curr. Opin. Biotechnol.* **11**(3), 271–279.
Greaves, M.P. and Webley, D.M. (1965) A study of the breakdown of organic phosphates by microorganisms from the root region of certain pasture grasses. *J. Appl. Bacteriol.* **28**, 454–465.
Greiner, R., Konietzny, U. and Jany, K.D. (1993) Purification and characterisation of two phytases from *Escherichia coli*. *Arch. Biochem. Biophys.* **303**, 107–113.
Ha, N.C., Oh, B.C., Shin, S., Kim, H.J., Oh, T.K., Kim, Y.O., Choi, K.Y. and Oh, B.H. (2000) Crystal structures of a novel, thermostable phytase in partially and fully calcium-loaded states. *Nat. Struct. Biol.* **7**, 147–153.
Harley, J.L. and Smith, S.E. (1983) *Mycorrhizal Symbiosis*. Academic Press, London.
Hennig, C., Panak, P.J, Reich, T., Rossberg, A., Raff, J., Selenska-Pobell, S., Matz W., Bucher, J.J., Bernhard, G. and Nitsche, H. (2001) EXAFS investigation of uranium(VI) complexes formed at *Bacillus cereus* and *Bacillus sphaericus* surfaces. *Radiochim. Acta.* **89**(10), 625–631.
Hobman, J.L., Wilson, J.R., and Brown, N.L. (2000) Microbial mercury reduction. In *Environmental Microbe-Metal Interactions* (ed. Lovley, D.R.), ASM Press Washington DC pp. 177–197.
Jensen, M.P., Nash, K.L., Morss, L.R., Appelman, E.H. and Schmidt, M.A. (1996) Immobilisation of actinides in geomedia by phosphate precipitation. In *Humic and Fulvic Acids: Isolation, Structure and Environmental Role* (eds Gaffney, J.S. Marley, N.A. and Clark, S.B.), Washington DC: American Chemical Society, *Am. Chem. Soc. Symp. Ser.* **651**, pp. 272–285 .
Jeong, B.C and Macaskie L.E. (1999) Production of two phosphatases by a *Citrobacter* strain in batch and continuous culture *Enz. Microb. Technol.* **24**, 218–224.
Jeong, B.C. and Macaskie, L.E. (1995) *PhoN*-type acid phosphatases of a *Citrobacter* spp.-resistance to heavy metals and affinity towards phosphomonoester substrates. *FEMS Microbiol. Lett.* **130** (2–3), 211–214.
Jeong, B.C., Kim, H.W., Owen, S.J., Dick, R.E. and Macaskie, L.E. (1994) Phosphoesterase activity and phosphate release by a *Citrobacter* sp. *Appl. Biochem. Biotechnol.* **47**, 21–32.
Jeong, B.C, Hawes, C, Bonthrone, K.M. and Macaskie, L.E. (1997) Localization of enzymically enhanced heavy metal accumulation by *Citrobacter* sp. and metal accumulation in vitro by liposomes containing entrapped enzyme. *Microbiol.* **143**(7), 2497–2507.
Jeong, B.C., Poole, P.S., Willis, A.C. and Macaskie, L.E. (1998) Purification and chacterization of acid-type phosphatases from a heavy-metal-accumulating *Citrobacter* sp. *Arch. Microbiol.* **169**(2), 166–173.
Kampfer, P., Erhart, R., Beimfohr, C., Bohringer, J., Wagner, M. and Amann, R. (1996) Characterization of bacterial communities from activated sludge: culture-dependent numerical identification versus in situ identification using group-specific and genus-specific rRNA-targeted oligonucleotide probes *Microbiol. Ecol.* **32**, 101–121.

Kerovuo, J., Lauraeus, M., Nurminen, P., Kalkkinen, N. and Apajalahti, J. (1998) Isolation, characterization, molecular gene cloning and sequencing of a novel phytase from *Bacillus subtilis. Appl. Environ. Microbiol.* **64**, 2079–2085.

Kerovuo, J., Rouvinen, J. and Hatzack, F. (2000) Analysis of *myo*-inositol hexakisphosphate hydrolysis by *Bacillus* phytase: indication of a novel reaction mechanism. *Biochem. J.* **352**, 623–628.

Kim, Y.O., Kim, H.K., Bae, K.S., Yu, J.H. and Oh, T.K. (1998a). Purification and properties of a thermostable phytase from *Bacillus* sp. DSS 11. *Enz. Mic. Technol.* **22**, 2–7.

Kim, Y.O., Lee, J.K., Kim, H.K., Yu, J.H. and Oh, T.K. (1998b) Cloning of the thermostable phytase gene (*phy*) from *Bacillus* sp. DS11 and its overexpression in *Escherichia coli. FEMS Microbiol. Lett.* **162**, 185–191.

Kratochvil, D. and Volesky, B. (2000) Multicomponent biosorption in fixed beds. *Water Res.* **34** (12), 3186–3196

Kuyucak, N. and Volesky, B. (1989) Accumulation of cobalt by marine alga. *Biotechnol. Bioeng.* **33**, 809–814.

Levin, G.V. (1964). Sewage treatment processes. U.S. Patent application No. 3236766.

Lloyd, J.R. and Macaskie, L.E. (2000) Bioremediation of radionuclide-containing wastewaters. In *Environmental Microbe-Metal Interactions* (ed. Lovely, D.R.), ASM Press, Washington DC. pp. 277–327.

Lloyd, J.R. and Macaskie, L.E. (2002) Biochemical basis of microbe-radionuclide interactions. In *Interactions of Microorganisms with Radionuclides* (eds Keith-Roach, M. and Livens, F.R.), Elsevier, Amsterdam pp. 313–342.

Macaskie, L.E. (1990) An immobilized cell bioprocess for the removal of heavy metals from aqueous flows. *J. Chem. Technol. Biotechnol.* **49**, 357–379.

Macaskie, L.E. (1991) The application of biotechnology to the treatment of wastes produced from the nuclear fuel cycle: biodegradation and bioaccumulation as a means of treating radionuclide-containing streams. *Crit. Rev. Biotechnol.* **11**, 41–112.

Macaskie, L.E. and Dean, A.C.R. (1982) Cadmium accumulation by micro-organisms. *Environ. Technol. Lett.* **3**, 49–56.

Macaskie, L.E. and Basnakova, G. (1998) Microbially enhanced chemisorption of heavy metals: A method for the bioremediation of solution containing long lived isotopes of neptunium and plutonium. *Environ. Sci. Technol.* **32**(1), 184–187.

Macaskie, L.E. and Lloyd, J.R. (2002) Microbial interactions with radioactive wastes and potential applications. In *Interactions of Microorganisms with Radionuclides* (eds Keith-Roach, M. and Livens, F.R.), Elsevier, Amsterdam pp. 3343–381.

Macaskie, L.E., Empson, R.M., Cheetham, A.K., Grey, C.P. and Skarnulis, A. J. (1992a) Uranium bioaccumulation by a *Citrobacter* spp., as a result of enzymically mediated growth of polycrystalline HUO_2PO_4. *Sci.* **257**, 782–782.

Macaskie, L.E., Clark, P.J., Gilbert, J.D. and Tolley, M.R. (1992b) The effect of aging on the accumulation of uranium by a biofilm bioreactor, and promotion of uranium deposition in stored biofilms. *Biotechnol. Lett.* **14**(6), 525–530.

Macaskie, L.E., Jeong, B.C. and Tolley, M.R. (1994) Enzymatically accelerated biomineralization of heavy-metals – application to the removal of americium and plutonium from aqueous flows. *FEMS Microbiol. Rev.* **14**(4): 351–367.

Macaskie, L.E. Empson, R.M., Lin, F. and Tolley, M.R. (1995) Enzymically-mediated uranium accumulation and uranium recovery using a *Citrobacter* sp. immobilized as a biofilm within a plug-flow reactor. *J. Chem. Technol. Biotechnol.* **63**, 1–16.

Macaskie, L.E., Yong, P., Doyle, T.C., Roig, D.M. and Manzano, T. (1997) Bioremediation of uranium-bearing wastewater: biochemical and chemical factors influencing bioprocess application. *Biotechnol. Bioeng.* **53**, 100–109.

Macaskie, L.E., Bonthrone, K.M., Yong, P. and Goddard, D.T. (2000) Enzymatically mediated bioprecipitation of uranium by *Citrobacter* spp. a concerted role for exocellular lipopolysaccharide and associated phosphatase in biomineral formation. *Microbiol.* **146**, 1855–1876.

Mann, S. (1994) Molecular recognition in biomineralization. *Nature* **332**, 119–124.
Mann, S. (1997) Biomineralization: the form(id)able part of bioinorganic chemistry. *J. Chem. Soc. Dalton Trans.* **21**, 3953–3961.
McCready, R.G.L. and Lakshmanan, V.I. (1986) Review of biosorption research to recover uranium from leach solutions in Canada. In *Immobilization of Ions by Biosorption* (eds Eccles, H. and Hunt, S.) Ellis Horwood, Chichester, Sussex, UK. pp. 219–226.
McGrath, J.W., Wisdom, G.B., McMullan, G., Larkin, K.J. and Quinn, J.P. (1995) The purification and properties of phosphonoacetate hydrolase, a novel carbon-phosphorus bond cleaving enzyme from *Pseudomonas fluorescens* 23F. *Eur J. Biochem.* **234**, 225–230.
McGrath, J.W., Hammerschmidt, F. and Quinn, J.P. (1998) Biodegradation of phosphonomycin by *Rhizobium huakuii* PMY 1. *Appl. Environ. Microbiol.* **64**, 356–358.
Michel, L.J., Macaskie, L.E. and Dean, A.C.R. (1986) Cadmium accumulation by immobilized cells of a *Citrobacter* sp. using various phosphate donors *Biotechnol. Bioeng.* **28**, 1358–1365.
Mino, T., van Loosdrecht, M.C.M. and Heijnen, J.J. (1998) Microbiology and biochemistry of the enhanced biological phosphate removal process. *Water Res.* **32**, 3193–3207.
Nash, K.L., Jensen, M.P. and Smith, M.A. (1998) Actinide immobilisation in the subsurface environment by *in situ* treatment with a hydrolytically unstable organophosphorus complexant: uranyl uptake by calcium phytate. *J. All. Comp.* **271–273**, 257–261.
Nott, K.P., Paterson-Beedle, M., Macaskie, L.E. and Hall, L. D. (2001) Visualisation of metal deposition in biofilm reactors by three-dimensional magnetic resonance imaging (MRI*). Biotechnol. Lett.* **23**(21), 1749–1757.
Ohtake, H., Wu, H., Imazu, K., Ambe, Y., Kato, J. and Kuroda, A. (1996) Bacterial phosphonate degradation, phosphite oxidation and polyphosphate accumulation. *A. Res. Conserv. Recyc.* **18**, 125–134.
Omar, N.B., Merroun, M.L., Gonzalez-Munoz, M.T. and Arias, J.M. (1996) Brewery yeast as a biosorbent for uranium. *J. Appl. Bacteriol.* **81**(3), 283–287.
Panak, P.J., Raff, J., Selenska-Pobell, S., Geipel, G., Bernhard, G. and Nitsche, H. (2000) Complex formation *of* U(VI) with *Bacillus*-isolates from a uranium mining waste pile. *Radiochim. Acta* **88**(2), 71–76.
Paterson-Beedle, M. and Macaskie, L.E. (2003) Removal of cobalt, strontium and caesium from aqueous solutions using microbially-driven mixed crystal growth. In *Proc. International Biohydrometallurgy Symposium*, Athens, Greece, September 2003, 127–135.
Pattanapipitpaisal, P., Mabbett, A.N., Finlay, J.A., Beswick, A.J., Paterson-Beedle, M., Essa, A., Wright, J., Tolley, M. R., Badar, U., Ahmed, N.A., Hobman, J. L., Brown, N.L. and Macaskie, L.E. (2002) Reduction of Cr(VI) and bioaccumulation of chromium by Gram positive and Gram negative microorganisms not previously exposed to Cr-stress. *Environ. Technol.* **23**(7), 731–745.
Pümpel, T. and Schinner, F. (1997) Metal biosorption: a structured data space? *Res. Microbiol.* **148** (6) 514–515.
Pümpel, T., Ebner, C., Pernfuss, B., Schinner, F., Diels, L., Keszthelyi, A., Stankovic, A., Macaskie, L.E., Tsezos, M. and Wouters, H. (2001) Treatment of rinsing water from electroless nickel plating with a biologically active moving bed sandfilter. *Hydrometallurgy* **59**, 383–393.
Pümpel, T., Macaskie, L.E., Finlay, J.A., Diels, L. and Tsezos, M. (2003) Nickel removal from nickel plating wastes using a biologically active moving-bed sandfilter. *BioMetals* **16**(4), 567–581.
Raff, J. (2002) Wechselwirkungen der Hüllproteine von Bakterien aus Uranabfallhalden mit Schwermetallen. PhD Thesis FZR-Report-458 ISSN 1437-322X.
Reddy, N.R., Pierson, M.D., Sathe, S.K. and Salunkhe, D.K. (1989) *Phytates in Cereals and Legumes*. CRC Press, Boca Raton Fl.
Richardson, A.E. and Hadobas, P.A. (1997) Soil isolates of *Pseudomonas* spp. that utilize inositol phosphates. *Can. J. Microbiol.* **43**, 509–516.

Riordan, C., Bustard, M., Putt, R. and McHale, A.P. (1997) Removal of uranium from solution using residual brewery yeast: Combined biosorption and precipitation. *Biotechnol. Lett.* **19**(4), 385–387.

Rodriguez, H. and Fraga, R. (1999) Phosphate solubilizing bacteria and their role in plant growth promotion. *Biotechnol. Adv.* **17**, 319–339.

Roig, M.G., Macaskie, L.E. and Kennedy, J.F. (1995) Biological rehabilitation of metal-bearing wastewaters. Final Report EU Environment Programme 1991–1994; Contract EV5V-CT93-0251. The European Commission, Brussels.

Schiewer, S. and Volesky, B. (1996) Modeling multi-metal ion exchange in biosorption. *Environ. Sci. Technol.* **30**(10), 2921–2927.

Selenska-Pobell, S. (2002) Diversity and activity of bacteria in uranium waste piles. In *Interactions of Microorganisms with Radionuclides* (eds Keith-Roach, M. and Livens, F.R.), Elsevier, Amsterdam pp. 225–253.

Selenska-Pobell, S., Flemming, K., Tzvetkova, T., Raff, J., Schnorpfeil, M.N. and Geissler, A. (2002) Bacterial communities in uranium mining waste piles and their interaction with heavy metals. In *Uranium in the Aquatic Environment* (eds Merkel, B.J., Planar-Friedrich, B. and Wolkersdorfer, C.), Springer-Verlag, Berlin pp. 455–464.

Selenska-Pobell, S., Miteva, V., Boudakov, I., Panak, P., Bernhard, G. and Nitsche, H. (1999) Selective accumulation of heavy metals by three indigenous *Bacillus* isolates, *B. cereus*, *B. megaterium* and *B. sphaericus* in drain waters from a uranium waste pile. *FEMS Microbiol. Ecol.* **29**, 59–67.

Strain, S.M., Fesik, S.W. and Armitage, I.M. (1983a) Characterization of lipopolysaccharide from a heptoseless mutant of *Escherichia-coli* by c-13 nuclear magnetic-resonance. *J. Biol. Chem.* **258**(5), 2906–2910.

Strain, S.M., Fesik, S.W. and Armitage, I.M. (1983b) Structure and metal binding properties of lipopolysaccharide from heptoseless mutants of *Escherichia coli* studied by ^{13}C and ^{31}P nuclear magnetic resonance *J. Biol. Chem.* **258**, 13466–13477.

Streichan, M., Golecki, J.R. and Schon, G. (1990) Polyphosphate-accumulating bacteria from sewage plants with different processes for biological phosphorus removal. *FEMS Microbiol. Ecol.* **73**, 113–124.

Suresh, N., Roberts, M.F., Coccia, M., Chikarmane, H.M. and Halvorson, H.O. (1986) Cadmium-induced loss of surface polyphosphate in *Acinetobacter lwoffii*. *FEMS Microbiol. Lett.* **36**, 91–94.

Suzuki, M. and Kamatani, A. (1995) Mineralization of inositol hexaphosphate in aerobic and anaerobic marine sediments: implications for the phosphorus cycle. *Geochim. Cosmochim. Acta.* **59**, 1021–1026.

Taghavi, S.M., Mergaeay, M., Niels, D. and Van der Lelie, D. (1997) *Alcaligenes eutrophus* as a model system for bacterial interaction with heavy metals in the environment. *Res. Microbiol.* **148** (6), 536–551.

Thomas, R.A.P. and Macaskie, L.E. (1996) Biodegradation of tributyl phosphate by naturally-occurring microbial isolates and coupling to the removal of uranium from aqueous solution. *Environ. Sci. Technol.* **30**, 2371–2375.

Thomas, R.A.P. and Macaskie, L.E. (1998) The effect of growth conditions on the biodegradation of tributyl phosphate and potential for the remediation of acid mine drainage waters by a naturally-occurring mixed microbial culture. *Appl. Microbiol. Biotechnol.* **49**, 202–209.

Thomas, R.A.P., Morby, A.P. and Macaskie, L.E. (1997a) The biodegradation of tributyl phosphate by naturally occurring microbial isolates. *FEMS Microbiol. Lett.* **155**, 155–159.

Thomas, R.A.P., Greated, A., Lawlor, K., Bailey, M. and Macaskie, L.E. (1997b) Stabilisation of tributyl phosphate biodegradative ability of naturally-occurring Pseudomonads using ampicillin. *Biotechnol. Tech.* **11**,781–785.

Thomas, R.A.P., Beswick, A.J., Basnakova, G., Moller, R. and Macaskie, L.E. (2000). Growth of naturally occcurring microbial isolates in metal-citrate medium and bioremediation of metal citrate wastes. *J. Chem. Technol. Biotechnol.* **75**, 187–195.

Tolley, M.R., Strachan, L.F. and Macaskie L.E. (1995) Lanthanum accumulation from acidic solution using a *Citrobacter* sp. immobilized in a flow-through bioreactor. *J. Ind. Microbiol.* **14**, 271–280.
Ullah, A.H.J., Cummins, B.J. and Dischinger, J.H.C. (1991) Cyclohexanedione modification of arginine at the active site of *Aspergillus ficuum* phytase. *Biochem Biophys Res. Comm.* **178**, 45–53.
Upton, J. (1998) Nutrient removal in the UK- now and in the future. *Proceedings, International Conference on Phosphorus Recovery*, 6/7 May 1998, Warwick, UK.
Van Groenestijn, J.W. and Deinema, M.H. (1985) Effects of cultural conditions on the accumulation and release of polyphosphate by *Acinetobacter* strain 210A In *Proceedings of the International Conference on Management Strategies for Phosphorus in the Environment* (eds Lester, J.N. and Kirk, P.W.), Selper, London, pp. 405–410.
Van Groenestijn, J.W., Zuidema, M., van de Worp, J.J.M., Deinema, M.H. and Zehnder, A.J.B. (1989) Influence of environmental parameters on polyphosphate accumulation in *Acinetobacter* sp. *Ant Van Leeuwenhoek* **55**, 67–82.
Volesky, B. (1994) Advances in biosorption of metals – selection of biomass types. *FEMS Microbiol. Rev.* **14**(4), 291–302.
Volesky, B. and Holan, Z. R. (1995) Biosorption of Heavy Metals. *Biotechnol. Prog.* **11**, 235–250.
Wagner, M., Amman, R., Lemmer, H., Manz, W. and Schleifer, K.H. (1994a) Probing activated sludge with fluorescently labelled rRNA-targeted oligonucleotides. *Water Sci. Technol.* **29**, 15–23.
Wagner, M., Erhart, R., Manz, W., Amann, R., Lemmer, H., Wedi, D. and Schleifer, K.H. (1994b) Development of an rRNA-targeted oligonucleotide probe specific for the genus *Acinetobacter* and its application for *in situ* monitoring in activated sludge *Appl. Environ. Microbiol.* **60**, 792–800.
Watson, J.H.P. and Ellwood, D.C. (1994) Biomagnetic separation and extraction process for heavy-metals from solution. *Minerals Eng.* **7**(8), 1017–1028.
White, C. and Gadd, G.M. (1996) Mixed sulphate-reducing bacterial cultures for bioprecipitation of toxic metals: Factorial and response-surface analysis of the effects of dilution rate, sulphate and substrate concentration. *Microbiology* **142**(8), 2197–2205.
White, C., Sharman, A.K. and Gadd, G.M. (1998) An integrated microbial process for the bioremediation of soil contaminated with toxic metals. *Nature Biotechnol.* **16**(6): 572–575.
Yakubu, N.A. and Dudeney, A.W.L. (1986) Biosorption of uranium with *Aspergillus niger*. In *Immobilisation of Ions by Biosorption (*eds Eccles, H. and Hunt, S.), Chichester, Sussex, UK pp. 183–200
Yong, P. and Macaskie, L.E. (1995a) Enhancement of uranium bioaccumulation by a *Citrobacter* sp. via enzymically-mediated growth of polycrystalline $NH_4UO_2PO_4$. *J. Chem. Technol. Biotechnol.* **63**, 101–108.
Yong, P. and Macaskie, L.E. (1995b) Removal of the tetravalent actinide thorium from solution by a biocatalytic system. *J. Chem. Technol. Biotechnol.* **64**, 87–95.
Yong, P. and Macaskie, L.E. (1997) The effect of substrate concentration and nitrate inhibition on product release and heavy metal removal by *Citrobacter* sp. *Biotechnol. Bioeng.* **55**, 821–830.
Yong, P. and Macaskie, L.E. (1998) Bioaccumulation of lanthanum, uranium and thorium, and use of a model system to develop a method for the biologically-mediated removal of plutonium from solution. *J. Chem. Technol. Biotechnol.* **71**, 15–26.
Yong, P. and Macaskie, L.E. (1999) The role of sulphate as a competitive inhibitor of enzymatically-mediated heavy metal uptake by *Citrobacter* spp. implications in the bioremediation of acid mine drainage water using biogenic phosphate precipitant. *J. Chem. Technol. Biotechnol.* **74**, 1149–1156.
Yong, P., Sammons, R.L., Marquis, P.M., Lugg, H. and Macaskie, L.E. (2003) Synthesis of nanophase hydroxyapatite by *Serratia* sp. N14. In *International Biohydrometallurgy Symposium*, Athens Greece, September 2003.

25
Developments in the use of calcium phosphates as biomaterials

R.L. Sammons, P.M. Marquis, L.E. Macaskie, P. Yong and C. Basner

25.1 INTRODUCTION

Calcium phosphates are used in biomaterials in four main forms: ceramics, bioactive glasses and glass-ceramics, cements and composite materials. The major clinical uses of these materials are as bone substitutes and as osteoconductive coatings for metal prostheses, such as hip and knee joint replacements. In addition, because they permit vascularisation and cellular ingrowth, some porous calcium phosphate-containing materials have been developed for use as ocular implants or to anchor medical devices in soft tissues (Table 25.1). This chapter will concentrate primarily on the use and development of synthetic calcium phosphate bone graft materials and coatings.

25.2 THE DEMAND FOR BONE SUBSTITUTE MATERIALS

When bone is lost due to trauma or disease a bone graft may be required in order to fully restore structure and function. Autologous bone from the same patient (usually from the iliac crest or the ribs) is theoretically the ideal graft material as it provides three desirable properties: scaffolding for osteoconduction, growth factors

© 2004 IWA Publishing. *Phosphorus in Environmental Technology: Principles and Applications.*
Edited by Eugenia Valsami-Jones. ISBN: 1 84339 001 9

Table 25.1 Applications of calcium phosphates in biomaterials and medical devices (Aoki, 1991; Denisson *et al.*, 1991; Le Geros, 1991, 2002; Hench and Wilson, 1996; Kundu *et al.*, 2002).

Nature of Application	Clinical area
Bone replacement	Removal of cysts and tumours, trauma, periodontal disease, ossicular replacement
Bone augmentation	Alveolar ridge, sinus lifts, cosmetic surgery, palatal reconstruction, hip and knee revision surgery
Bone repair	Spinal fusion, fracture union
Bone fixation (prosthetic coatings)	Hip, knee replacements, dental root implants
Tooth repair	Caries restoration
Prosthetic devices	Ocular implants
Soft tissue attachment	Implanted vascular access devices

for osteoinduction and progenitor cells for osteogenesis (Lidgren, 2002; Vaccaro 2002). However, it is sometimes neither desirable nor practical to obtain autologous bone: its provision requires a second operation, causing further trauma, blood loss and discomfort to the patient. There is also a risk of infection, nerve damage and in many cases acute or prolonged post-operative pain occurring at the donor site. Moreover, in some patients, especially the elderly or small children, there may be an insufficient supply of bone for grafting or the bone may not be of sufficiently good quality, for example, in patients with metastatic disease (Betz, 2002).

Allograft bone from a human donor is an alternative to autologous bone but apart from the absence of cells, there are potential disadvantages and hazards associated with this approach. These include its potential immunogenicity and the risk of transmission of tumour cells or infective agents of bacterial, viral or prion-related origin, which are difficult to eradicate despite costly sterilisation procedures. Sterilisation itself inactivates growth factors and may adversely affect the mechanical properties of the bone. Apart from cost and potential risk of infection there is a limited and insufficient supply of cadaver bone to meet an increasing need. Around 500,000–600,000 bone grafting procedures are carried out annually in the US (Bucholz, 2002) and an estimated 2.2 million procedures worldwide (Vaccaro, 2002).

Synthetic bone-substitute materials are a convenient alternative to autograft or allograft bone, since they can be generated in unlimited supply, pose no inherent risk of infection and are easy to store and to sterilise. However, they currently represent only about 10% of the bone-graft market (Bucholz, 2002; Vaccaro, 2002).

25.3 THE IDEAL BONE-SUBSTITUTE: WHAT IT HAS TO DO?

25.3.1 Why phosphates?

Calcium phosphate in mature bone, dentine, cementum and enamel is structurally and chemically similar to the mineral hydroxylapatite (HA, $Ca_5(PO_4)_3(OH)$), although containing a number of additional ions such as fluoride, carbonate and magnesium, as well as showing deficiencies, notably in calcium and hydroxyls (see also Valsami-Jones, this volume). To make this distinction, it will hereby be called biological apatite (BA). Bone is formed by ordered nucleation and deposition of Ca–P crystals in and around fibrils of collagen type I by a process that is controlled by the structure of the collagen fibrils themselves and by non-collagenous proteins, such as osteonectin and bone sialoproteins (Lian *et al.*, 1999). In immature bone the calcium phosphate crystals are mainly in the form of brushite ($CaHPO_4 \cdot 2H_2O$). These grow and transform to β-tricalcium phosphate (β-TCP; $Ca_3(PO_4)_2$), octa-calcium phosphate and finally biological apatite in mature bone, increasing in size and decreasing in solubility as they do so (Simmons and Grynpas, 1990). Because HA is sparingly soluble at physiological pH and occurs naturally in bone, it is the logical choice for use in non-resorbable bone-substitute materials, whilst β-TCP is used when it is desirable to have a resorbable implant. It may be advantageous to use biphasic calcium phosphate ceramics, consisting of mixtures of HA and β-TCP, since β-TCP is a more reactive and soluble material which stimulates new osteogenesis, whilst the HA provides long-term stability. These materials have been used clinically for 20–30 years and numerous *in vitro* studies and animal experiments have demonstrated their osteoconductive properties (Hench and Wilson 1996; Jarcho, 1981; Manley, 1993).

25.3.2 Mechanical properties of bone

Ideally a bone substitute material should mimic the morphology of natural bone and have similar mechanical properties. This is a difficult challenge: Bone is a natural composite of organic (collagen) and mineral (BA) components that combine to give it its elasticity and strength. The construction of bone further underlies its unique properties: It is made up of Haversian systems or osteons in which a central blood capillary-containing channel is surrounded by concentric cylinders of bone called lamellae. In adjacent lamellae collagen fibers are arranged in opposing directions, thus reinforcing the structure and contributing to the fracture toughness. The mechanical properties of synthetic HA and bone are compared in Table 25.2.

It can be seen from Table 25.2 that the compressive and tensile strengths of HA and bone are similar but the Young's modulus is 5–20 times that of cortical bone, meaning that it is more rigid. As a ceramic rather than a composite it is also brittle. This is the reason that most synthetic bone graft materials can only be used in non-load bearing sites and if they are used for fracture repair in, for example, metaphyseal defects, they must be supported during the resorption/integration process by internal and/or external fixation. Mechanical properties of HA and β-TCP are quite similar (Aoki, 1991). However, if the material resorbs and is replaced by bone or augmented by bone growing into the pores, as in the case of β-TCP or porous graft materials, the mechanical strength of the implant site may eventually increase to that of natural bone (Gazdag *et al.*, 1995; Shors, 1999). The initial lack of strength may therefore not be a problem so long as additional support is provided. The mechanical properties of the implant site should finally match that of the host bone,

Table 25.2 The mechanical properties of bone and synthetic hydroxylapatite. Data from Aoki (1991), Le Geros and Le Geros (1993); Rubin and Rubin (1999).

Property	Cortical bone	Trabecular bone	Hydroxylapatite (HA)
Compressive strength (MPa)	100–224	1–7	300–500
Tensile strength (MPa)	70–150		60–220
Young's Modulus (GPa)	18 (longitudinal) 12 (transverse) 3.3 (shear)	0.1–3.5	40–85
Ca/P Ratio	1.65–1.75 (varies with age/gender/type)		1.67
HA Crystal size (nm)	40 × 10 × 1 – 3		60–90

since if the material does not transfer stresses to the adjacent bone this may resorb due to stress-shielding, possibly also leading to loss of the implant (Williams, 1990). This is one of the reasons why dense HA is now no longer favoured as a bone-substitute material and resorbable materials such as β-TCP or composites, in which the ceramics are combined with another component which confers some elasticity, are being developed.

25.3.3 General properties of bone substitute materials

All bone substitute materials must be non-toxic, non-carcinogenic, non-allergenic, non-thrombogenic and non-pyrogenic. They should also ideally be osteoconductive. This means that they form a biological bond with bone, thus preventing fibrous encapsulation, and will permit, support and direct new bone growth. Some materials may also be osteoinductive, i.e. capable of inducing new bone growth in a non-osseous site via the promotion of mesenchymal stem cell recruitment and differentiation into chondroblasts and osteoblasts (Urist *et al.*, 1967).

25.3.3.1 Integration into host bone

Graft materials must be capable of attachment to the host bone, although the bonding mechanisms may be different. Bone substitute materials may be either non-resorbable or resorbable, depending upon whether they are required to function throughout the life-time of the recipient or to provide a temporary bone substitute which will eventually be replaced by natural bone. If resorbable, they should resorb in a predictable manner, in concert with bone growth. Non-resorbable materials must provide long-term stability but, as discussed above, they should not promote stress shielding.

25.3.3.2 Scaffold materials

Bone substitute materials can provide a scaffold to support the growing bone and to maintain space within the bone for new bone growth, thus preventing the ingress of soft tissue. Porous materials are preferred, since they allow ingrowth of bone cells, deposition of bone mineral and ingress of vascular tissue. They also permit wicking of blood by capillary action, carrying with it cells and plasma proteins which may accelerate osseointegration. It is generally agreed that the minimum macropore interconnection size of 50–100 µm is required to host blood capillaries (Klawitter *et al.*, 1976; Le Geros, 2002), whilst the optimum pore diameter for cellular ingrowth and osteon formation is in the range of 100–500 µm (Le Geros, 2002; White and Shors, 1986).

25.3.3.3 Sterilisation and handling properties

Materials used in surgery must be capable of being sterilized without significant alteration of composition or mechanical properties and they should be easily handled, malleable and ideally capable of being adapted in the operating theatre to conform to the dimensions of the defect they are filling. To this end, graft materials are usually supplied in a variety of forms including blocks, morsels and pastes.

25.4 SUCCESSFUL APPLICATIONS OF CALCIUM PHOSPHATES IN BIOMATERIALS

At the present time, all synthetic bone-graft materials are limited in their usage to non-load-bearing situations in the skeleton, due to their poor mechanical properties.

25.4.1 Orthopaedic applications

25.4.1.1 Spinal fusion

In the US and Europe, approximately half of all bone grafting procedures are for spinal fusions (Bucholz, 2002). These involve the insertion of metal rods and bone grafts to fuse two adjacent vertebrae into a single solid bone. Fusion can eliminate otherwise painful movement between the vertebrae and is sometimes used to treat common conditions such as slipped discs. Fusion halts the progress of deformities of the spine such as scoliosis, and may also be performed to support a weakened spine following removal of a tumour. Current success rates for this procedure are greater than 90% with autologous bone in the posterior cervical or lumber spine, but non-unions have been reported in 5–44% of cases in the posterolateral spine (Betz, 2002; Sandhu *et al.*, 1999). Calcium phosphate ceramics, cements and composites have all been used either alone or as a supplement to extend autograft bone in spinal fusions (Boden *et al.*, 1999; Passuti *et al.*, 1989; Sandhu *et al.*, 1999).

25.4.1.2 Fracture repair and/or bone augmentation

Around 40% of all bone graft procedures are for general orthopaedic applications (Bucholz, 2002), where they are used to reinforce or augment deficient host bone,

Developments in the use of calcium phosphates as biomaterials 587

in combination with metal rods for internal fracture fixation, especially in the case of osteoporotic fractures, non-unions and delayed unions. The material can be used alone or as a supplement and therefore reduce the amount of autograft bone required. Calcium phosphate ceramics have been used particularly successfully for fracture fixation in foot and ankle surgery (Rahimi *et al.*, 1997).

Osteoporotic bone is characterised by thin or absent trabeculae and presents a challenge not only for fracture fixation, but also for fixation of hip and knee replacements, especially in revision surgery. If, after removal of a hip replacement, the remaining bone stock is inadequate for fixation of the new prosthesis, it may be augmented by a graft of autologous or synthetic bone made from calcium phosphate (Larsson and Bauer, 2002).

25.4.1.3 *Calcium phosphate coatings for metal prostheses*

In addition to their use for bone grafts, calcium phosphates are used in orthopaedic surgery as coatings on metal prostheses, such as hip and knee replacements, to accelerate healing and promote fixation to bone without the use of cement (Capello *et al.*, 1998; Geesink, 2002, Figure 25.1).

Coated implants combine the strength of the metal (usually titanium or titanium alloy) and the osteoconductivity of the calcium phosphate. HA is most frequently

Figure 25.1 Hydroxylapatite (HA)-coated components of hip and knee prostheses. An HA-coated acetabular cup is seen bottom left, and a component of a knee prosthesis centre right. In this design the coating is plasma-sprayed onto titanium wire to create a porous surface. On the femoral component of the hip replacements this is applied to the proximal region to promote bone attachment in this region only. Photograph reproduced with permission from John Cresser Brown, Zimmer Inc. UK.

used for coatings, either alone or in combination with β-TCP, which is added to promote osseointegration. Porous coatings allow bone ingrowth, providing mechanical interlocking and thus avoiding the need for cement. This is an advantage because polymethylmethacrylate (PMMA) bone cement can deteriorate over time, contributing to aseptic loosening of implants (Schmalzried et al., 1992). Because HA forms a biological bond with bone, stresses in the femur are transferred via attached bone trabeculae to the metal shaft of the prosthesis. *In vivo* experiments on animals have shown that HA can bridge gaps of up to 2 mm between the implant and bone and stabilise micromotion. Therefore HA-coated hip prostheses are claimed to be potentially advantageous in the case of over-reaming in which too large a gap is formed between the stem of the prosthesis and the bone (Geesink, 2002; Søballe et al., 1990, 1993a, b).

25.4.2 Dental applications of bone graft materials

25.4.2.1 Treatment of periodontal disease

Periodontal disease is a very common condition, affecting over 50% of the adult population and if allowed to progress, the disease may cause sufficient damage to the periodontal ligaments and surrounding bone that the teeth may ultimately be lost. The disease may be treated by removal of infected soft and hard tissues by surgery and debridement followed by the use of a bone graft to restore damaged bone. This is a situation where it is clearly advantageous to use a synthetic graft rather than autologous bone from another site and in fact the first reported medical application of a calcium phosphate ceramic was for this purpose (Nery, 1975). A variety of calcium phosphate materials including HA and β-TCP, bioglasses and composite materials may be used (King, 2001).

25.4.2.2 Augmentation of the alveolar ridge and sinus lifts

If bone is not subjected to loading it resorbs. This can be seen in the alveolar ridge after teeth have been extracted or fall out. Over time the ridge reduces in height and may become narrow, preventing or limiting denture retention. To avoid this behaviour, bone graft materials may be used to pack the socket after tooth extraction and bioceramics are particularly suited to this application. Ceramics or composite materials can also be used to augment the height or alter the profile of the alveolar bone if it has already subsided, for example in long-term edentulous patients, to improve denture fitting and retention and to allow placement of dental root implants (Ashman, 2000; Denisson et al., 1985).

Bone in the upper jaw is sometimes too shallow to permit implant retention and in this case a sinus lift operation may be carried out, in which the floor of the sinus is raised by addition of bone graft particles introduced through an opening into the sinus. A variety of calcium phosphate containing materials have been successfully used for this purpose in the form of ceramic, bioactive glasses and composites. Osseointegration of the particles into the bone raises the floor of the sinus by as much as 10–15 mm, thus increasing the height of the bone available for attachment of the implant (Chanavaz, 2000). Another problem with placement of dental root implants in the upper jaw is that here the bone density is relatively low compared with the lower jaw and may be inadequate for fixation. In this case a hydroxylapatite-coated implant may accelerate healing and improve retention (Misch, 1990).

25.4.3 Applications in craniofacial and ear surgery

25.4.3.1 Craniofacial surgery

Many of the bones in the face and skull are suitable for repair using synthetic bone substitute materials because they are not load-bearing. Calcium phosphate biomaterials may be used, for example, for reconstruction of the eye orbit and other facial bones following trauma, the removal of tumours and in cosmetic facial surgery, for example to alter the profile of the nose (Aitosolo *et al.*, 2001). Calcium phosphate ceramic plates are also used successfully in the reconstruction of congenital defects such as cleft palate (Wolford and Stevao, 2002).

25.4.3.2 Ear

Synthetic bone graft materials have many applications in the ear. In congenital absence of the outer ear it is sometimes necessary to reconstruct the ear canal. It may also be necessary to repair the wall of the ear canal following trauma to the skull. In surgery to the middle ear, access is sometimes gained via the mastoid bone and in this case, and following removal of tumours, it is necessary to fill the cavity. The mastoid is a difficult region to repair because it is connected to the middle ear and hence particularly vulnerable to infection. Dense pure HA has been used for this purpose but more recently biphasic 60% HA, 40% β-TCP granules have been used successfully in combination with fibrin glue. The glue has osteogenic properties and aids implantation, ensuring cohesion of the granules for packing into deep cavities and improving stability (Daculsi *et al.*, 1992).

Another use of calcium phosphate biomaterials is for replacement of the ossicular bones of the ear: loss of the malleus, incus and stapes can result from infection and cholesteatoma, a benign epithelial cell tumour that can invade the bones of the ear, causing deafness. Hearing may be partly restored by insertion of a small prosthesis that is attached to the tympanic membrane and vibrates against the round window, thus simulating the function of the ossicular chain. However, construction and placement of these devices is particularly challenging, because they have to have the correct acoustic properties and because they are being placed in a site that is open to infection due to the Eustachian tube connecting the middle ear to the pharynx. Various designs and constructions of replacement ossicles are available, including ones made from hydroxylapatite, either as the ceramic or in composites such as Hapex™, described later.

25.5 CHEMISTRY AND MANUFACTURE OF BIOMEDICAL CALCIUM PHOSPHATES

25.5.1 Ceramics

25.5.1.1 Manufacture of HA and β-TCP

HA has the empirical formula $Ca_5(PO_4)_3OH$. It can be synthesised by several routes, some of which are listed in Table 25.3.

Unless the stoichiometry of the reactants is tightly controlled aqueous methods may give Ca-deficient or Ca-rich apatites which have Ca/P molar ratios less than or greater than 1.67, respectively. On sintering, these defective apatites with partially decompose resulting in phase impure HA where second phases such as α or β-TCP, tetracalcium phosphate and calcium oxide are common.

The first method will be briefly discussed here to illustrate the steps in the process. Further details of the other methods can be obtained from the references. In this, phosphoric acid is added to a stirring suspension of calcium hydroxide in water, forming a gelatinous precipitate of HA at alkaline pH (10–12).

$$6H_3PO_4 + 10Ca(OH)_2 \longrightarrow Ca_{10}(PO_4)_6(OH)_2 + 18H_2O$$

After several days of aging, in which amorphous calcium phosphate forms and the particles agglomerate, the material is dried, pressed, ground and sintered between 950°C and 1300°C. Electron microscopy of the sintered powder product reveals cohesive grains approximately 1–4 μm diameter, although the production of pure hydroxylapatite depends on the duration and temperature of sintering.

25.5.1.2 Effect of manufacturing conditions on mechanical properties

Mechanical properties of HA including compressive strength, hardness and fracture toughness are strongly influenced by sintering temperature and the nature of the initial powder compact. At temperatures between 1200 and 1300°C, α-TCP

Table 25.3 Hydroxylapatite methods of synthesis.

System/Type of Reaction	Reaction components
Aqueous precipitation	Phosphoric acid + calcium hydroxide.[1]
	Calcium nitrate + ammonium phosphate.[2]
Aqueous hydrolysis	Calcium phosphates + ammonium or sodium hydroxide.[3]
Hydrothermal conversion	Calcium carbonate ($CaCO_3$) (or β-TCP, or tetracalcium phosphate[4]) + calcium phosphate ($CaHPO_4$) or ammonium phosphate at 275°C under steam pressure.[3]
Solid-state conversion	Calcium carbonate + hydrated calcium phosphate ($CaHPO_4 \cdot 2H_2O$) fired above 900°C.[5]

[1] Jarcho (1981); [2] Hayek and Newesly (1963); [3] Le Geros and Le Geros (1993); [4] Muller-Mai et al. (1996); [5] Aoki (1991).

starts to form and the fracture toughness drops (de Groot, 1983). Fracture toughness is normally inversely proportional to grain size, because of the increased resistance to crack propagation, caused by increased crack deflection by smaller grains. As grain size reduces strength should also increase, particularly when grain size enters the nanoscale. However as initial powder size decreases problems arise as particles cluster together to form agglomerates, which, on sintering, lead to porosity and crack formation. By the addition of emulsion oil to precipitated HA during the aging process to prevent the formation of 'hard agglomerates' with large pores, in favour of 'soft agglomerates', Murray et al. (1995) showed that smaller grain sizes could be achieved in a low porosity denser product with high strength.

25.5.1.3 Transformation of phases of calcium phosphate and formation of β-TCP

Different phases of calcium phosphate can transform to HA, as is known to occur with biological apatite in bone. Thus HA can be formed from octacalcium phosphate ($Ca_8H_2(PO_4)_6 \cdot 5H_2O$), dicalcium phosphate dihydrate ($CaHPO_4 \cdot 2H_2O$; DCPD) or amorphous Ca–P (ACP) by hydrolysis in neutral or basic medium. In addition, HA can be converted to DCPD and ACP by dissolution and reprecipitation in acid media. These transformations may occur in the body but all phases are biocompatible. Pure β-TCP cannot easily be formed in aqueous solutions and is generally synthesised from solid-state reactions at temperatures typically between 800–1000°C, for example by heating HA (Le Geros, 1991).

25.5.1.4 Porous bone substitute materials

Porous calcium phosphate ceramics can be made by a number of different methods, some of which are shown in Table 25.4.

Table 25.4 Manufacturing methods for porous ceramics.

1.	By mixing the starting powder with organic powders or sponges which are burnt out on sintering leaving their replicas as voids.	De Groot et al. (1990a)
2.	By addition of chemicals, which give off gas, e.g., hydrogen peroxide, into HA slurry; dissociation of this at high temperatures produces pores in the ceramic body.	De Groot et al. (1990b)
3.	From biological structures (replaniform process): By embedding wax or metal into a biological structure e.g. coral, which is then removed by acid or heating. The porous mould is then filled with apatite slurries, and later sintered to remove the wax and yield porous apatite.	White et al. (1972); de Groot (1983); Ravaglioli and Krajewski (1992).
4.	By direct hydrothermal conversion of calcium carbonate of biological structure to calcium phosphate. Material is dipped into aqueous solution of NaCl, ammonium phosphate and tris buffer and a hydrothermal reaction occurs, producing HA.	Roy and Linnehan (1974).

Bone substitutes of marine origin. Many living organisms accumulate inorganic species from their biological habitat to build up their individual, three-dimensional skeleton. Some of these have an interconnected pore system which resembles that of human bone and their skeletons have been converted to HA for clinical use:

FRIOS® ALGIPORE® Marine red algae live one to five meters below sea level in certain areas of the Atlantic ocean (*Corallina officinalis*), or off the coast of South Africa (*Amphiroa ephedra*). They build up a cellular wall made of different marine minerals for increased mechanical stability. The plants are collected manually, cleaned and heated at over 700°C for many hours. During this first step of the patented manufacturing process, organic plant substances are eliminated. What remains is the cellular scaffold, mostly made of magnesium-containing calcite. In a second process, phosphate in an aqueous solution is added to hydrothermally transform the calcite into hydroxylapatite. What formerly were small branches of the algae, connected to each other by proteins, appear as natural granules, with a diameter of 0.3–2 mm (Figure 25.2a). Separated into different granule sizes by a mesh, packed, sealed and sterilised, the granules serve as a highly porous bone graft material in dentistry and orthopaedic surgery (Ewers *et al.*, 1992; Kasperk *et al.*, 1988; Kasperk and Ewers, 1986; 1987).

FRIOS® ALGIPORE® (Dentsply, FRIADENT, Mannheim, Germany) was developed by Ewers and Simons at the University of Kiel, Germany and the trade mark has been in clinical use for over 15 years. The material is well characterised in terms of porosity, specific surface and crystallographic data (Ewers *et al.*, 1992; Kasperk *et al.*,1988; Kasperk and Ewers, 1987).

In the clinic the granules are mixed with a little blood. Usually the amount of blood in the surgical site of the patient is sufficient. If a preparation of bone is performed in the same session, some collected small bone particles can be mixed with the graft material to improve the regeneration process. After placement in an area of a local bone volume deficiency, it is sealed and fixed with a surgical membrane and left to heal. The mineral scaffold is osseoconductive and acts as a guiding structure for the ingrowth of newly formed bone. The bone is not only formed on the surface of each granule, but also within the open porosity of each granule: a unique phenomenon that can only take place in highly porous materials. Over the years, the mineral gets resorbed and replaced by the patient's own bone (Figure 25.2c–d, Schopper *et al.*, 1998, 1999).

The porosity and biomimetic mineral structure of FRIOS®ALGIPORE® encourage further developments: it has already been used as a carrier for bone morphogenetic proteins with encouraging results (Ewers *et al.*, 1998) and tests with the material loaded with osteoblasts derived from cell cultures are also planned.

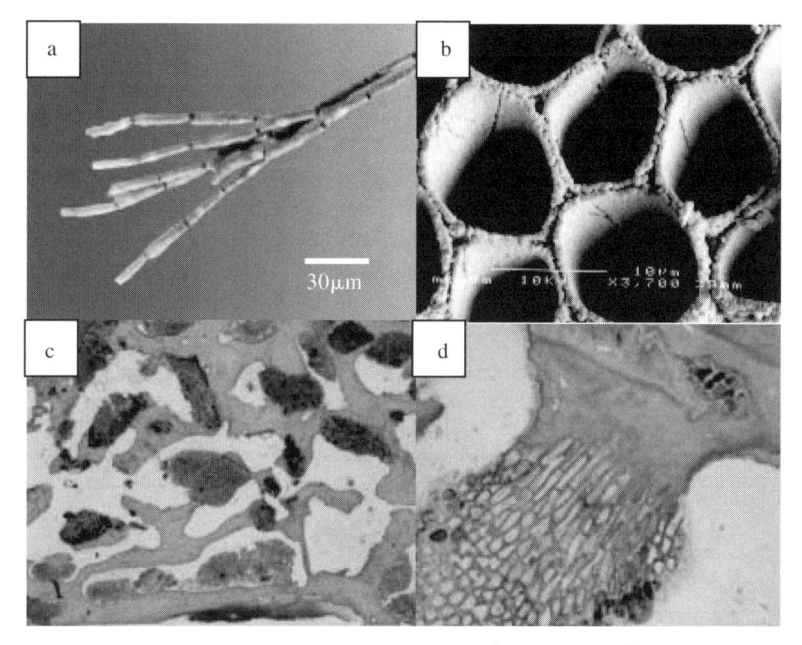

Figure 25.2 From the ocean to human bone – FRIOS® ALGIPORE®. (a) A single branch of the alga *Amphiroa ephedra*, which grows in the sea off the coast of South Africa. (b) SEM image of a granule, showing honeycomb-like structure. Pores are all interconnected and range from 10–15 μm in diameter. Lateral longitudinal pores form the outer part of the plant scaffold. Size bar = 10 μm. (c) Bone biopsy from a patient eleven months after a sinus lift operation and augmentation with FRIOS® ALGIPORE®. Newly formed mineralised bone (violet), can be seen bridging the ALGIPORE® granules (grey). Thionine staining; original magnification × 4 objective. (d) Higher magnification (×40) of the same biopsy showing ingrowth of mineralised bone (purple), and non-mineralised bone (blue) within pores. Photographs courtesy of DENTSPLY Friadent, Mannheim, Germany and of Prof. Rolf Ewers, University of Vienna, Austria. [This figure is also reproduced in colour in the plate section after page 336.]

Biocoral™, *Interpore*™ *and ProOsteon*™

> *'Full fathoms five my father lies:*
> *Of his bones are coral made;*
> *Those are pearls that were his eyes:*
> *Nothing of him that doth fade*
> *But doth suffer a sea-change*
> *Into something rich and strange'*

Shakespeare: The Tempest (quoted in company literature on Biocoral™).

Biocoral™ (Biocoral Inc. France) is a commercial calcium carbonate (aragonite) derived from natural coral whose pore structure fortuitously resembles that of natural bone. The diameter of the pores varies, depending on the coral species: *Porites* has interconnected pores of 140–160 μm; *Goniapora* pores range from 200–1000 μm. Biocoral™ is resorbed by the body and gradually replaced by natural bone over 6–12 months (Braye *et al.*, 1996).

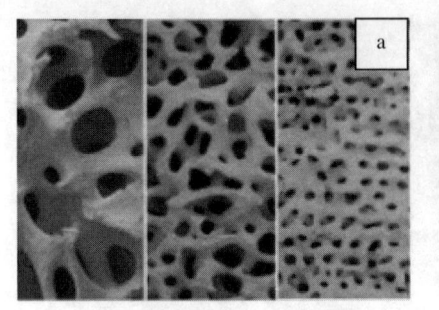

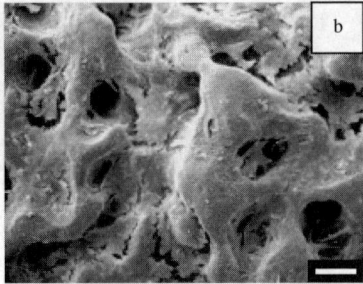

Figure 25.3 Porous bone substitutes made from natural coral. (a) Macroporosity of human cancellous bone (left), hydroxylapatite made from *Goniopora* coral (middle) and hydroxylapatite made from *Porites* coral (right). Figure courtesy of Edwin Shors (Interpore Cross Int. CA); (Shors, 1999, reproduced with permission). (b) Rat osteoblasts traversing and growing into pores of Interpore™; bar =200 μm. Image by R. Sammons, University of Birmingham, UK.

Interpore™ and ProOsteon™ (Interpore Inc. CA) are calcium phosphate ceramic materials consisting mainly of HA that are formed from the *Porites* or *Goniapora* corals by hydrothermal exchange, perfectly preserving the porosity (Figure 25.3). These materials have been used extensively for bone augmentation and repair where a bone substitute material is required (Shors, 1999). According to this publication, recently a hybrid coralline material, consisting mainly of calcium carbonate with a thin layer of calcium phosphate (mainly HA) over all the internal and external surfaces has been made, in which the thickness of the HA layer can be adjusted to give a range of materials with different resorption rates. Surface modification of a coralline hydroxylapatite for the controlled release of gentamycin antibiotic has recently been reported (Muragan and Panduranga Rao, 2002).

25.5.2 Coatings and coating procedures

Hydroxylapatite is the bioactive calcium phosphate ceramic of choice for coating metal prostheses because of its relatively low solubility at physiological pH. *In vivo* observations have indicated that coating thicknesses in the range 30–100 μm are optimal, since above this they are at risk of mechanical failure and delamination of the coating. There are a number of different coating techniques, of which the most common method is plasma-spraying. With this method it is claimed that dense tightly adhesive coatings of up to 100 μm thickness can be applied in two minutes. In plasma-spraying, particulate HA is fed into a high temperature plasma gas flame. As the particles pass through the flame at temperatures of up to 30,000°C, surface melting or full melting of the particles occurs as they accelerate towards the target, which has usually been grit blasted to aid adhesion. The temperature of the target is about 300°C. As the particles impact on the target they form a splat and quench into a crystalline or amorphous form. Although plasma

Table 25.5 Advantages and disadvantages of plasma spraying.

Advantages	Disadvantages
Tight adhesion between HA and metal. Creates porous surface conducive to bone cell attachment and mechanical interlocking. Quick.	High temperature to form plasma. Cannot be used to coat heat-labile substrata that degrade at target temperature (~300°C). Line-of-sight. Non-uniformity of coating density. Introduction of different Ca–P phases e.g. tetracalcium phosphate. Introduction of glassy and amorphous phases. Changes in crystallinity. High cost.

spraying is the most efficient method of coating a metal, it has a number of disadvantages listed in Table 25.5. Of particular concern are the compositional and phase differences in the final coating, which may be responsible for inconsistency of clinical performance. This has led to research being carried out into alternative coating procedures, as discussed in the final section.

25.5.3   Bioactive glass and glass ceramics

Bioglass® was first made in the early 1970's by Hench and co-workers (Kokubo, 1992). It consists of Na_2O-CaO-SiO_2-P_2O_5 but can vary in composition and phase and also be converted by crystallisation into machinable glass ceramics, some of which contain apatite e.g. Ceravital®. The composition of the glasses and glass-apatites renders them extremely surface reactive in aqueous environments such that they bond to bone, as described briefly in the next section. A very good account of the chemistry of these materials and the mechanism by which they bond to bone is given by Hench (1996). Bioactive glasses and glass ceramics have been successfully used in middle ear surgery, alveolar ridge maintenance, vertebral surgery and periodontal defects (Hench, 1996).

25.5.4   Composites

As mentioned above, a major disadvantage of calcium phosphate ceramics as bone-substitute materials is their brittle nature. Composites have been made in an attempt to overcome this problem by combining calcium phosphates (HA or β-TCP) with synthetic or natural polymers, which can be either resorbable or non-resorbable. Resorbable polymers include collagen, fibrin and synthetic polymers such as polylactic acid, glycolic acids and polypropylenefumarate, non-resorbable ones include polysulfone, polyethylene, and polymethylmethacrylate. Polylactic acid combined with β-TCP is used, for example, for fabricating biodegradable screws for fracture fixation. Composite materials combine the osteoconductivity of the calcium phosphate with the superior elasticity of the polymer (Burg et al., 2000; Marra et al., 1999; Zhang et al., 1999). It is possible to control the adsorption time of certain polymers such as poly (lactide-co-glycolide) and to modulate pore topography and size to suit a particular cell type and it is likely that this knowledge will be utilised in the future in the design of composite bone graft materials (Burg et al., 2000).

The composite Hapex™, which was developed in the 1980's as a bone-substitute material, combines 40% HA particles with polyethylene to give a material of greater toughness than HA. However, the compressive strength and stiffness are still too low to be used in load-bearing situations (Dornhoffer, 1998). Some improvement in the mechanical properties has been achieved by extrusion of the composite to align the polyethylene chains, increasing the potential clinical scope of this material (Bonner et al., 2002).

Composites of calcium phosphate ceramics and collagen have been developed. Collagen of animal origin is not highly immunogenic in humans and if necessary it can be enzymatically treated to further reduce its immunogenicity (Yannas, 1996). Aoki (1991) described precipitation of HA crystals onto insoluble collagen and Clarke et al. (1993) created a resorbable HA – collagen composite using an aminoacid ester unit as the binding agent, with a tensile strength of 45 MPa, i.e. similar to some types of bone. Du et al. (2000) formed a range of calcium phosphate/collagen composites by mineralisation of collagen matrix by presoaking type I collagen with a solution containing phosphate ions and then immersing it in a solution containing Ca^{2+} ions to allow mineral deposition at high ion concentrations and high pH. In tensile tests the material showed strength and Young's modulus values similar to the lower range of reported values for human bone. It is likely that composite materials of this kind will be further developed for clinical use.

25.5.5 Phosphate cements

Cements consist of a solid phase, which may be a calcium phosphate or another calcium salt, and a liquid component, which can be an inorganic phosphate or saline salt solution or an organic acid (LeGeros, 1991). When the two components are mixed, a cement is precipitated and eventually transforms into a hard bone-like apatite structure as the crystals grow and become entangled. Cements can be made into injectable pastes for filling cavities and have applications in restorative dentistry, repair of alveolar bone defects and as bone fillers in orthopaedic surgery. For example, injectable calcium phosphate cements are used in fracture fixation to fill voids in metaphyseal bone, thereby reducing the need for a bone graft, and to improve the holding strength around implants in osteoporotic bone (Larsson and Bauer, 2002). Their advantages are their malleability – they can be made to fill any shaped defect; their disadvantages include their low tensile strength, small pore size (1 μm), which precludes bone ingrowth, and initial liquidity that increases the risk of extrusion outside the cavity. Commercial examples of calcium phosphate cements are Norian SRS Cement (Norian Corp, Cupertino and Calif.) and α-BSM™ (Etex Corporation, Cambridge, MA). Composite materials with increased compressive strengths compared to α-BSM™ have been made by combining it with poly(ethyleneimine) and poly(allylamine hydrochloride) (Lee et al., 1999).

25.6 REACTIONS OF BIOMEDICAL PHOSPHATES IN THE BODY

25.6.1 Protein interactions

When a material is first implanted into the body proteins immediately start to adsorb to the surface from body fluids, forming a layer known as a conditioning film. This is a complex and dynamic process that has been extensively researched, since it is believed that protein adsorption is the key event that influences all subsequent cellular reactions and thus governs the fate of an implant in terms of integration or fibrous encapsulation. Which proteins adsorb first to an implant surface depends on the material surface properties including surface charge, hydrophobicity and the nature of chemical groups, and on protein concentration and affinity for the surface, which is in turn dependent on protein surface properties. Protein adsorption is dynamic – one protein may displace another over time, and proteins may change conformation on adsorption (Horbett, 1982; Kasemo and Gold, 1999; Schakenraad and Busscher, 1989).

25.6.2 Cellular interactions

Specific proteins in the conditioning film mediate cell adherence via attachment to integrins. These are heterodimeric transmembrane proteins in the cell membrane, linked intracellularly to the actin cytoskeleton. Integrins recognise and bind to specific 'adhesion-motif' sequences of amino acids, which are exposed on the surface of adsorbed proteins. *In vitro,* the plasma proteins fibronectin and vitronectin have been shown to be important in adhesion. Cell adherence is also mediated by interactions with extracellular matrix proteins such as collagen, osteonectin, and osteopontin (Anselme, 2000; Schakenraad, 1996). Attachment of integrins to the cell cytoskeleton occurs at focal contacts where it is mediated by the protein vinculin, a useful marker that can be located using immunocytochemistry and used to compare numbers of focal contacts on different types of surface. Cell attachment to surfaces and subsequent differentiation has been shown to be influenced by a variety of factors including surface chemistry, roughness, topography and energy (reviewed by Schwartz *et al.*, 1999).

25.6.3 Tissue interactions

In vivo, osseointegration occurs in several stages. In an osseous site the first cells to colonise an implant surface are blood cells and platelets. Macrophages will remove any dead cells and bone, and, as the blood clot is broken down, osteoprogenitor cells and undifferentiated mesenchymal stem cells migrate into the area and attach to the implant surface, modifying it by secreting their own matrix. Osteoblasts begin to differentiate within 3–6 days, calcification occurs within 6–14 days and remodelling after 21 days (Schwartz and Boyan, 1994).

The success of a skeletal tissue implant, however, depends on achieving a stable attachment to connective tissue. Three reactions are possible: fibrous encapsulation, bone bonding, dissolution/degradation, with subsequent replacement by host bone (Hench, 1996).

25.6.3.1 Fibrous encapsulation

If the material is non-bioactive a fibrous tissue forms around the implant, caused by fibroblasts adhering to their neighbours rather than the material surface (Brunette and Chehroudi, 1999). The collagen fibres which are present in the fibrous layer are orientated parallel to the surface of the implant, not bonded to it. A fibrous layer (0.1–1.5 mm thick) may be present adjacent to cemented (PMMA) and non-cemented press-fit hip prostheses, although bone may grow into pores in the cement in areas of less stress (Williams, 1990). The thickness of the capsule varies according to the tissue reactions to the implant: if it is nearly inert a thin layer develops; if the material elicits an inflammatory response (for example to particulates), a thicker fibrous synovium-like layer will develop whose thickness increases in proportion to the extent and duration of the inflammation (Schmalzried *et al.*, 1992; Williams, 1990). Once formed, fibroblasts in the layer release cytokines that inhibit osteogenesis, precluding osseointegration. The thickness of the fibrous layer is also increased by interfacial micromovement; the bigger the movement, the thicker the fibrous tissue becomes, increasing the risk of failure of the implant, by loosening or fracture (Hench, 1991). In addition, the layer will be relatively avascular and potentially vulnerable to infection, forming a 'fibro-inflammatory immunoincompetent zone' (Gristina, 1994). In theory, calcium phosphate biomaterials should not become encapsulated because they form a direct bond with bone.

25.6.3.2 Bone bonding to ceramics

If the material is bioactive an interfacial bond forms between the bone and the material surface. This occurs with HA, β-TCP, and bioglasses, although the interface is not the same in each case.

In the case of bioactive ceramics, there is no intervening fibrous layer between the bone and the implant but Transmission Electron Microscopy shows an amorphous or proteinaceous layer approximately 1 μm thick. At the HA-bone interface there is epitaxial alignment of the apatite crystals of bone and HA (Jarcho, 1981; Ravaglioli and Krajewski, 1992) and in pull-out tests fracture characteristically occurs in the bone, not at the interface (Miller *et al.* 1976). The mechanism by which bone-bonding occurs has been described by Ducheyne *et al.* (1992). It was shown that HA and β-TCP surface dissolution is followed by reprecipitation of calcium and phosphate ions from the supersaturated solution that forms at the interface, with the formation of carbonate apatite. Exchange of ions between the ceramic and the hydrated layer on the surface of the bone then occurs and this explains the bridging of the gap and subsequent bone-bonding, as the ions are deposited on collagenous matrix as the newly formed mineral inter-digitates with surface adsorbed matrix proteins. Materials that degrade faster may therefore favour early bone formation (de Bruijn *et al.* 1993). A second phenomenon involved in bone-bonding is the formation in the early stages of afibrillar mineralised globular accretions which contain glycosaminoglycan on the surface of the material. These are of cellular origin, and occur on both calcium phosphate and metal surfaces. Collagen fibres integrate into this matrix and subsequently mineralise (de Bruijn et al., 1993).

25.6.3.3 Bone-bonding to bioactive glasses

The interface between bone and bioactive glasses differs from ceramics in that interdigitation of collagen fibres with the surface is seen in the case of the latter (de Bruijn et al., 1993). The surface of bioglasses is highly reactive in an aqueous environment: CaO-, SiO_2- based glasses dissolve and release Ca^{2+} and silicate ions into the surrounding tissue fluids, thereby increasing their concentration to the point of supersaturation. The loss of silica ($Si(OH)_4$) and the formation of silanol groups, SiOH, at the glass-solution interface create a silica-rich hydrogel layer which in vitro provides favourable sites for nucleation of apatite crystals (Kokubo, 1992; Hench, 1996). Once these are formed, they grow spontaneously by taking up Ca^{2+} and PO_4^{3-} ions and simultaneously incorporating the OH^- and CO_3^{2-} ions from the surrounding fluids to make a hydroxyl-carbonate-apatite layer, which is similar in composition to bone apatite. It has been shown more recently that in the presence of proteins the calcium phosphate layer that forms on the surface is non-crystalline, amorphous HA, and it is thought that this layer promotes osteoblast differentiation from osteoprogenitor cells. Furthermore, the layer seemed to act synergistically with bone morphogenetic protein in promoting osteoblast differentiation (Ducheyne and Qiu, 1999; Santos et al., 1998).

The consequence of the corrosion of the surface, described above, is the production of 'protective pouches' into which undifferentiated mesenchymal cells migrate and differentiate into osteoblasts; there is no such differentiation on the outside of the pouches (Ducheyne et al., 1992). Depending on the particle size and thus the space between them, bone formed in adjacent pockets may eventually fuse together, filling the space occupied by the particles as they slowly dissolve (Ducheyne and Qiu, 1999). Bone formation in the presence of silica glasses and glass ceramics therefore differs from that with calcium phosphate ceramics in that bone forms from the inside out, rather than the other way round.

25.6.3.4 Bone bonding to coatings

Examination of retrieved HA-coated femoral hip replacements has shown that there is direct apposition of bone to the coating without any intervening fibrous capsule and that bone trabeculae are apposed to the HA surface at right angles rather than tangentially, in locations of significant stress (Furlong, 1993). It is clear that the coatings do not remain intact all over the implant but that some resorption occurs, particularly in non-stressed regions (Geesink, 2002).

Coating characteristics, such as percentage porosity and amount of amorphous phase, influence the rate of degradation of calcium phosphate ceramics, whilst implant geometry and surface roughness may affect delamination. Investigation of implants retrieved after autopsy showed that the amorphous phase had disappeared within one year of implantation, probably as a result of dissolution in tissue fluids. It is shown that if this happens the bone can still bind to the titanium surface, and it is possible that the coating serves its most useful function in promoting initial osseointegration and that subsequent resorption is not disadvantageous (de Groot et al., 1993; Geesink, 2002). Bloebaum et al. (1991) suggested that HA particles might migrate to the polyethylene acetabular cup causing damage and release of polyethylene particulates into the tissues, leading to osteolysis and aseptic loosening of the implant. However, Frayssinet et al. (1995) concluded from histological studies of 15 autopsied femurs containing hydroxylapatite-coated prostheses that the HA was resorbed by the normal cell-mediated remodelling processes. HA particles were released, phagocytosed, solubilised or osseointegrated into trabeculae with no fibrous encapsulation and there were no signs of osteolysis despite the presence of calcium phosphate particles inside bone tissues. Geesink (2002) also reports that there is no evidence that HA accelerates polyethylene wear.

HA coatings would appear to have many advantages but one problem with the use of coated prostheses is the fact that if, for any reason, the implant has to be removed, this may be difficult and involve extensive destruction of bone. For this reason, and possibly because of the inconsistency of performance attributed to HA-coated prostheses, many surgeons prefer to use PMMA-cemented prostheses.

25.6.3.5 Dissolution and degradation

Stability of an implanted material depends on material properties, including solubility product, crystallinity and crystal size, grain size (ceramics), and surface area. Solute (tissue fluid) properties affecting dissolution include ionic composition, pH, and temperature. HA is stable only above pH 5.5 and in the presence of inflammation the local pH may fall to 4.0 due to release of acids and oxygen depletion in the tissues. Bacterial infection may affect stability indirectly, because of the pH reduction due to release of acidic bacterial metabolites, or because of an ensuing inflammatory response (Geesink, 2002; Hench, 1996; Le Geros, 1991).

As discussed earlier, degradation of an implanted bone graft (or a coating) may also be brought about by cells such as macrophages, which ingest particulate material up to 2 μm in diameter, multinuclear giant cells, which may engulf larger particles, and osteoclasts. Multinuclear giant cells have been seen in histological sections surrounding dense HA particles six months after implantation (Ducheyne et al., 1992). Osteoclasts are responsible for the natural process of bone resorption in remodelling and are also capable of resorbing some calcium phosphate bone graft materials, including β-TCP. They are partially responsible for the rapid resorption rates of calcium sulphate and carbonate materials, which undergo chemical dissolution in the physiological environment and are used as temporary grafts. Calcium sulphate (gypsum) is used in infected sites because it can be used to deliver antibiotics as it gradually resorbs (Lidgren, 2002). As the ceramics resorb, particulates and new calcium phosphate phases may be formed (Hench, 1996). Dense stoichiometric HA is relatively insoluble and not readily resorbed by osteoclasts, macrophages or multinuclear giant cells. Consequently it has a resorption rate of only 5–15% per annum (Ducheyne *et al.*, 1992; Fleming *et al.*, 2000).

25.7 FUTURE DEVELOPMENTS

In the last 10 years considerable advances have been made in our understanding, at the molecular level, of the biological processes involved in wound healing, osteogenesis and bone-remodelling and the growth factors that control them. In parallel, materials research is leading to the development of materials whose mechanical and morphological properties more closely resemble those of the bone they are replacing. New developments aim to combine synthetic or biological components with signalling molecules and growth factors to stimulate osteogenesis.

25.7.1 Improving cell adhesion properties

The first stage in osseointegration of an implant involves the adsorption of proteins onto the surface. This is followed by migration of mesenchymal stem cells into the area and their differentiation into chondroblasts and osteoblasts. It may be possible to enhance materials properties by pre-adsorbing proteins that promote cell adhesion and differentiation onto the surface of scaffold materials. Qian and Bhatnagar (1996) identified a 15 amino-acid peptide, P-15, which resembles the biologically active domain of type I collagen, and binds avidly to cells. This peptide is now used in a commercial product (PepGen P-15™, DENTSPLY Ceramed, Lakewood, Colorado, USA) in which it is bound to a carrier of deproteinised bovine bone, and, according to Yukna *et al.* (2000), has been successfully used for the repair of periodontal defects.

25.7.2 Use of growth factors and cytokines to promote osteogenesis

Bone contains a large number of growth factors and cytokines involved in wound healing and the regulation of bone remodelling. These include TGF-β, platelet derived growth factor, bone morphogenetic proteins, insulin-like growth factors and fibroblast growth factors. BMP is the collective name for a family of osteo-inductive proteins discovered by Urist (Urist *et al.*, 1967; Urist and Strates, 1971) and involved in development of the skeleton in embryogenesis and in the postnatal skeleton. 15 BMPs have been characterized and cloned and rhBMP two and seven have been shown to promote osteogenesis in non-osseous sites and to aid fracture fixation in rats (Burg *et al.*, 2000). These are now being investigated in a number of human clinical trials including spinal fusion, alveolar ridge augmentation, sinus lifts and open tibial fracture repair with encouraging results (Li and Wozney, 2001; Vaccaro, 2002). The choice of delivery system is critical and research is concentrating on developing appropriate carriers, including porous calcium phosphate ceramics and composites, for controlled release of these and other factors to keep pace with osseogenesis (Arm et al., 1996; Kirker–Head, 2000; Li and Wozney, 2001; Winn et al., 1999).

25.7.3 Development of scaffold materials for cell delivery and gene therapy

Scaffold materials are also being developed as carriers for bone marrow and other types of cells (Blokhuis et al., 2000). The scaffold should have the correct porosity and resorption characteristics to keep the cells at the fracture site for long enough for them to differentiate, proliferate and produce growth factors to aid wound healing and bone formation (Vaccaro, 2002). Gene therapy, in which DNA encoding required growth factors are introduced into cells via plasmid transfer, is a realistic possibility (Boden et al., 1999; Wehling et al., 1997). Introducing recombinant DNA encoding BMP within cells, instead of in an inanimate carrier, could be an alternative and potentially more reliable way of delivering the protein at the right time and concentration (Betz, 2002).

25.7.4 Improving material fabrication

25.7.4.1 Coatings

The inherent mechanical deficiencies of current bone substitute materials and plasma-sprayed coatings have been highlighted in previous sections. A number of different methods are available which avoid the use of high temperatures and associated disadvantages of the plasma spray method. The laser ablation technique (Clèries et al., 1998) is capable of producing a range of different coatings of different crystallinity. Milev et al. (2001) used the sol-gel method to produce nanocrystalline thin films 100–1000 nm thick with a grain size of 500 nm after sintering, and Kannan et al. (2002) described the advantages of using the electrochemical deposition technique for production of coatings on any shape of material. Biomimetic coatings, in which a supersaturated calcium phosphate solution is precipitated onto a substrate at low temperature, mimicking the physiological process of bone apposition, show promising results, especially if they can be developed for delivery of growth factors or antibiotics (Geesink, 2002).

25.7.4.2 Nanoscale HA and β-TCP

As discussed above, the use of hydroxylapatite is limited by its poor mechanical properties. It is well known that as the grain size of materials reduces towards the nanoscale (<200 nm, ideally below 50 nm), properties such as strength are dramatically improved (Ozin, 1992; Siegel, 1997). To make nanophase HA would require precursor crystals of this nanoscale dimension, which could be subsequently manipulated and fabricated without loss of the fine grain size. Conventional aqueous chemical synthesis and drying approaches do not readily produce particles of the required dimensions and also invariably lead to agglomerates of particles, as discussed earlier, which lead to inherent weakness in the final implant. However, nanophase carbonated apatite crystals have been made by a hydrothermal conversion process. The crystals are acicular (needle-shaped) and 9–15 nm long (Li *et al.*, 1994), similar to bone crystals. Liu *et al.* (1997) incorporated these crystals into Polyactive™, a bone-bonding polymeric material and showed that they enhanced the mechanical properties in the dry but not the wet state, possibly because of their hygroscopic nature or poor dispersion within the polymer. Further research is needed but the approach is promising.

A novel route to the manufacture of nanoscale hydroxylapatite could be via the use of bacteria. *Serratia* species N14 expresses a high level of the enzyme acid phosphatase, present on the bacterial cell surface and enmeshed within the extracellular polymeric matrix of the cells. This enzyme cleaves organic phosphates such as glycerol-2-phosphate to yield phosphate ions, which combine with added calcium ions to form 15–25 nm crystals of hydroxylapatite, which can be seen in TEM and SEM surrounding the bacterial cells (Figure 25.4a–b). The crystals should be prevented from agglomerating by the polysaccharidic extracellular matrix surrounding the bacterial cells in a manner analogous to the oil emulsion method of Murray *et al.* (1995), described earlier, and thus retain their nanoscale dimensions. The bacteria form a crystal-encrusted biofilm on polyurethane reticulated foam, which, when sintered and the foam burnt away, yields a macroporous lattice structure of β-TCP. The material is highly microporous and supports the growth of osteoblasts in culture (Figure 25.4 c–e). With further refinements, this could become a potentially useful bone-substitute material (Thackray *et al.*, 2003, in press; Yong *et al.*, 2003, in press).

25.8 CONCLUDING REMARKS

Calcium phosphate biomaterials have been in use for over 20 years. Whilst the use of dense non-porous hydroxylapatite materials is now decreasing, because of its poor mechanical properties and lack of resorption in the body, porous HA and β-TCP scaffolds are likely to be used increasingly as vehicles for cell and drug delivery. These, and composite materials in which calcium phosphates are even more ingeniously combined with natural and synthetic polymers, are likely to replace autografts as the bone-substitute materials of choice as their mechanical properties are improved. Calcium phosphate ceramics will continue to be used as coatings to aid bone fixation to metal prostheses but it is likely that new methods will be developed to overcome the current inadequacies of plasma-spraying and that coatings will be further developed to promote initial osseointegration and improve consistency of performance.

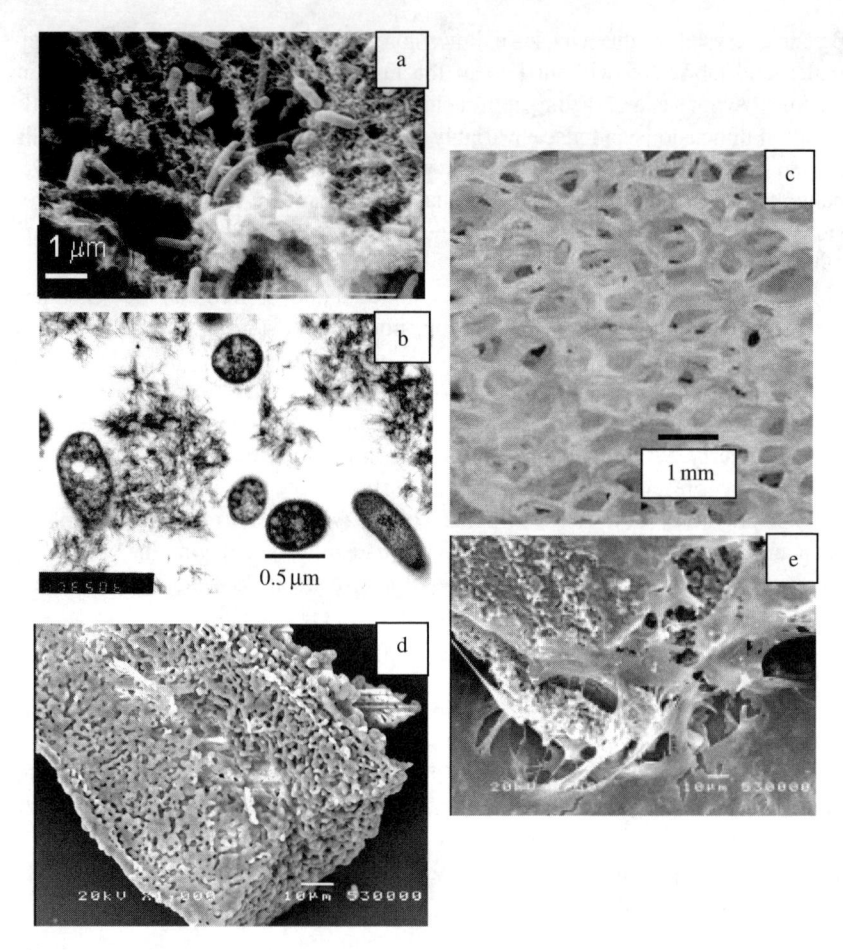

Figure 25.4 Calcium phosphate ceramic crystal lattice formed by *Serratia* N14. (a) SEM of cells and crystals on polyurethane foam (bar = 1 μm). (b) TEM of cells surrounded by crystals. (c) Sintered crystal lattice (bar = 1 mm). (d) Sintered strut of lattice showing microporosity (bar = 10 μm). (e) Rat osteoblasts migrating from underlying Petri dish onto lattice (bar = 10 μm). Images by H. Lugg, P. Yong, A. Thackray, University of Birmingham, UK.

REFERENCES

Aitosolo, K., Kinnunen, I., Palmgren, J. and Varpula, M. Repair of orbital floor fractures with bioactive glass implants. (2001) *J. Oral Maxillofac. Surg.* **59** (12) 1390–1395.

Anselme, K. (2000) Osteoblast adhesion on biomaterials. *Biomaterials* **21**, 667–681.

Aoki, H. (1991) *Science and Medical applications of hydroxyapatite*. Japanese Association of Apatite Science. Tokyo.

Arm, D.M., Tencer, A.F., Bain, S.D., and Celino, D. (1996) Effect of controlled release of platelet-derived growth factor from a porous hydroxyapatite implant on bone ingrowth. *Biomaterials* **17**, 703–709.

Ashman, A. (2000) Postextraction ridge preservation using a synthetic alloplast. *Implant Dent.* **9**(3), 168–176.
Balamurugan, A., Kannan, S. and Rajeswari, S. (2002) Bioactive sol-gel hydroxyapatite surface for biomedical applications – *in vitro* study. *Trends Biomater. Artif. Org.* **16**, 18–20.
Basle M., Chappard, D., Grizon, F., Filmon, R , Delectrin, J. Daculsi, G. and Rebel, A. (1993) Osteoclastic resorption of CaP biomaterials implanted in rabbit bone. *Calcif. Tissue Int.* **53**, 348–356.
Betz, R.R. (2002) Limitations of autograft and allograft – new synthetic solutions. *Orthopedics* **25S**, available at http://www.orthobluejournal.com/supp/0502/betz/
Blokhuis, T.J., Wippermann B.W., den Boer, F.C., van Lingen A., Patka, P., Bakker, F.C. and Haarman, H.J.T.M. (2000) Resorbable calcium phosphate particles as a carrier material for bone marrow in an ovine segmental defect. *J. Biomed. Mater. Res.* **51**, 369–375.
Bloebaum, R.D., Merrell, M., Gustke, K. and Simmons, M. (1991) Retrieval analysis of a hydroxyapatite-coated hip prosthesis. *Clin. Orthop.* **267**, 97–112.
Boden S.D., Martin G.J.J., Morone M., Ugbo J.L., Titus L. and Hutton W.C. (1999) The use of coralline hydroxyapatite with bone marrow, autogenous bone graft, or osteoinductive bone protein extract for posterolateral lumbar spine fusion. *Spine* **24**, 320–327.
Bonner, M., Saunders, L.S., Ward, I.M., Davies, G.W., Wang, M., Tanner, K.E. and Bonfield, W. (2002) Anistropic Mechanical Properties of Orientated HAPEX™. *J. Mater. Sci.* **37**, 325–334.
Braye, F., Irigaray, J.L., Jallot, E., Oudadesse, H., Weber, N., Deschamps, N., Deschamps, D. and Frayssinet P. (1996) Resorption kinetics of osseus substitute: natural coral and synthetic hydroxyapatite. *Biomaterials* **17**, 1345–1350.
Brunette, D.M. and Chehroudi, B. (1999) The effects of the surface topography of micro-machined titanium substrata on cell behavior *in vitro* and *in vivo*. *J. Biomech. Eng.-T ASME* **121**(1), 49–57.
Bucholz, R.W. (2002) Nonallograft osteoconductive bone graft substitutes. *Clin. Orthop.* **395**, 44–52.
Burg, K.J.L., Porter, S. and Kellam J. Biomaterial development for bone tissue engineering. *Biomaterials* **21**, 2347–2359.
Capello, W.N., D'Antonio J.A., Manley, M.T., and Feinberg, J.R. (1998) Hydroxyapatite in total hip arthroplasty. *Clin. Orthop.* **355**, 200–211.
Chanavaz, M. (2000) Sinus graft procedures and implant dentistry: a review of 21 years of surgical experience (1979–2000) *Implant Dent.* **9**(3), 197–206.
Clarke, K.I., Graves, S.E., Wong, A.T.C., Triffitt, J.T., Francis, M.J.O. and Czernuszka, J.T. (1993) Investigation into the formation and mechanical properties of a bioactive mater-ial based on collagen and calcium phosphate. *J. Mater. Sci. Mater. Med.* **4**, 107–110.
Clèries, L., Fernández-Pradas, J.M., Sardin, G. and Morenza, J.L. (1998) Dissolution behaviour of calcium phosphate coatings obtained by laser ablation. *Biomaterials* **19**, 1483–1487.
Colen, T.P., Paridaens, D.A., Lemij, H.G., Mourits, M.P. and Van der Bosch, W.A. (2000) Comparison of artificial eye amplitudes with acrylic and hydroxyapatite spherical enu-cleation implants. *Ophthalmology* **107**, 1889–1894.
Daculsi G., d'Arc, M.B., Corlieu, P. and Gersdorff, M. (1992) Macroporous biphasic cal-cium phosphates. Efficiency in mastoid cavity obliteration. Experimental and clin-ical findings. *Ann. Otol. Rhinol. Laryngol.* **101**(8), 669–674.
De Bruijn, J.D., Davies, J.E., Klein, C.P.A.T. de Groot K. and van Blitterswijk, C.A. (1993) Biological Responses to calcium phosphate ceramics. In *Bone Bonding Biomaterials* (eds Ducheyne, P., Kokubo T. and van Blitterswijk, C.A.), pp. 57–72, Reed Healthcare Publications, Leiderdorp, The Netherlands.
De Groot (1983) *Bioceramics of calcium phosphate*. CRC Press Inc. Boca Raton, Florida.

De Groot, K., Klein, C., Wolke. J. and de Blieck-Hogervorst, J. (1990a) Chemistry of calcium phosphate bioceramics. In *CRC Handbook of Bioactive Ceramics, vol 2* (eds Yamamuro, T., Hench, L. and Wilson, J.), CRC press, Boca Raton, Florida, pp. 3–16.

De Groot, K., Klein, C., Wolke, J, and de Blieck-Hogervorst, J. (1990b) Plasma-sprayed coatings of calcium phosphate. In *CRC Handbook of Bioactive Ceramics, vol 2* (eds Yamamuro, T., Hench, L. and Wilson, J.), CRC press, Boca Raton, Florida, pp. 133–142.

De Groot K., Jansen, J., Wolke, J., Klein, C. and van Blittrswijk, C. (1993) Developments in bioactive coatings. *Hydroxyapaite coatings in Orthopaedic surgery* (eds Geesink, R. and Manley, M.), pp. 49–62 Raven press Ltd., NY.

Denisson, H., Mangano, C. and Venini, G. (1985) *Hydroxylapatite implants*. Piccin, Padua, Italy.

Dornhoffer, J.L. (1998) Hearing Results with the Dornhoffer ossicular replacement prostheses. *Laryngoscope* **108**, 531–536.

Du, C., Cui, F.Z., Zhang, W., Feng, Q.L., Zhu, X.D. and de Groot, K. (2000) Formation of calcium phosphate/collagen composites through mineralization of collagen matrix *J. Biomed. Mater. Res*. **50**, 518–527.

Ducheyne, P., Bianco, S., Radin, S. and Schepers, E. (1992) Bioactive materials: mechanisms and bioengineering considerations. In *Bone-Bonding Biomaterials* (eds Ducheyne, P., Kokubo T. and van Blitterswijk C.A.), pp. 1–12, Reed Healthcare Communications, Leiderdorp, The Netherlands.

Ducheyne, P. and Qiu, Q. (1999) Bioactive ceramics: the effect of surface reactivity on bone formation and bone cell function. *Biomaterials* **20**, 2287–2303.

Ewers, R., Rasse, M. and Schumann, B. (1992) Development and clinical experience with the bone substitute Material Algipore®. *Osteologie* **1**, Suppl. 1, 18–19.

Ewers, R., Schopper, Ch., Gössweiner, S., Spassova, E. and Wild, K. (1998) Algipore: Carrier Material for Bone Morphogenic Protein. *J. Cranio. Max. Fac. Surg.* **26**, Suppl. 1, 46.

Fleming, J.E., Cornell, C.N. and Muschler, G.F. (2000) Bone cells and matrices in orthopaedic tissue engineering. *Orth. Clin. North Am*. **31**(3), 357–354.

Frayssinet, P., Hardy, D., Hanker, J.S. and Giammara, B.L. (1995) Natural history of bone response to hydroxyapatite-coated hip prostheses implanted in humans. *Cells and Materials* **5**, 125–138.

Furlong, R. (1993) A new concept of prosthetic fixation. In *Hydroxylapatite coatings in Orthopaedic Surgery* (eds Geesink R.G.T. and Manley M.T.), pp. 25–32, Raven Press, NY.

Gazdag A.R, Lane J.M., Glaser, D. and Forster R.A. (1995) Alternatives to autogenous bone graft: efficacy and indications. *J. Am. Acad. Orthop. Surg.* **3**(1), 1–8.

Geesink, R.G.T. (2002) Osteoconductive coatings for total joint arthroplasty. *Clin. Orthop.* **395**, 53–65.

Gristina, A.G., (1994) Implant failure and the immunoincompetent fibro-inflammatory zone. *Clin. Orthop.* **298**, 106–108.

Hayek, E. and Newesly, H. (1963) *Pentacalcium monohydroxy-orthophosphate (hydroxyapatite): inorganic synthesis*. Vol. 7. McGraw-Hill Book Co. Inc., pp 63–65.

Helm G.A. Dayoub, M.D. and Jane, J.A. (2001) Bone graft substitutes for the promotion of spinal arthrodesis. *Neurosurg. Focus* **10**(4) 1–5. Available at http://www.neurosurgery.org/focus/apr01/10-4-4.pdf

Hench, L.L. (1991) Bioceramics from concept to clinic. *J. Am. Cer. Soc*. **7**, 1487–1510.

Hench, L.L. (1996) Ceramics, glasses and glass-ceramics. In *Biomaterials Science. An introduction to materials in medicine* (eds Ratner, B.D., Hoffman, A.S., Schoen F.J. and Lemons, J.E.), pp. 73–83, Academic Press, San Diego, USA.

Hench, L.L. and Wilson, J.W. (1996) *Clinical performance of skeletal prostheses*. Chapman and Hall, London.

Hoen, M., Strittmatter, E.J., Labounty, G.L. Keller D.L. and Nespeca, J.A. (1989) Preserving the Maxillary Alveolar Ridge Contour Using Hydroxylapatite. *JADA* **118**, 739–741.

Horbett, T.A. (1982) Protein adsorption on biomaterials. *Adv. Chem. Ser*. **199**, 233–244.

Jarcho, M. (1981) Calcium phosphate ceramics as hard-tissue prosthetics. *Clin. Orthop*. **157**, 259–278.

Jiang, G. and Shi, D. (1998) Coating of Hydroxyapatite on highly porous Al_2O_3 Substrate for Bone Substitutes. *J. Biomed. Mater. Res.* **43**, 77–81.
Kannan, S., Balamurugan, A. and Rajeswari, S. (2002) Development of calcium phosphate coatings on type 316L SS and their *in vitro* response. *Trends Biomater. Artif. Organs* **16**, 8–11.
Karen, J. and Burg L. (2000) Biomaterial developments for bone tissue engineering. *Biomaterials* **21**, 2347–2359.
Kasemo, J. and Gold, J. (1999) Implant surfaces and interface processes. *Adv. Dent. Res.* **13**, 8–20.
Kasperk C. and Ewers R. (1987) Phycogen Hydroxyapatite – A New Interconnencting Porous Biomaterial. In *Advances in Biomaterials, 7 – Biomaterials and Clinical Applications* (eds Pizzoferrato, A., Marchetti, P.G., Ravaglioli and Lee, A.J.C.), Elsevier, Amsterdam, Oxford, New York,Tokyo.
Kasperk C., Ewers R., Simons B. and Kasperk R. (1988) Algae-derived (phycogene) Hydroxyapatite. A comparative histological study. *Int. J. Oral Maxillofac. Surg.* **17**, 319–324.
King, G.N. (2001) New regenerative technologies: rationale and potential for periodontal regeneration: 1. New advances in established regenerative strategies. Dent. Update, **28**(1), 7–12.
Kirker-Head, C.A. (2000) Potential applications and delivery strategies for bone morphogenetic proteins. *Adv. Drug Deliv. Rev.* **43**, 65–92.
Klawitter J.J., Bagwell J.G, Weinstein A.M., Sauer B.W., Pruitt J.R. (1976) An evaluation of Bone Growth into Porous High Density Polyethylene. *J. Biomed. Mater. Res.* **10**, 311–323.
Kundu, B. Sinha, M.K. and Basu, D. (2002) Development of a bio-active integrated ocular implant for anophthalmic human patients. *Trends Biomater. Artif. Organs* **16**, 1–4.
Kokubo, T. (1992) Bioactivity of glasses and glass ceramics. In *Bone-Bonding Biomaterials* (eds Ducheyne, P., Kokubo, T. and van Blitterswijk, C.A.), pp. 31–46, Reed Healthcare Communications, Leiderdorp, The Netherlands.
Larsson, S and Bauer, T.W. (2002) Use of injectable calcium phosphate cement for fracture fixation: A review. *Clin. Orthop.* **395**, 23–32.
Le Geros R.Z. (1991) *Calcium phosphates in oral biology and medicine. Monographs in Oral Science 15* (ed. Myers, H.M.), Karger, Basel.
Le Geros, R.Z. (2002) Properties of osteoconductive biomaterials: calcium phosphates. *Clin. Orthop.* **395**, 81–98.
Le Geros, R.Z. and Le Geros, J. (1993) *Dense Hydroxyapatite, An introduction to Bioceramics*, **1** pp. 139–180, (eds Hench, L. and Wilson, J.), World Science Publications, Singapore.
Lee, D.D., Tofighi, A., Aiolova, M., Chakravarthy, P., Catalano, A., Majahad, A. and Knaack, D. (1999) α-BSM: a biomimetic bone substitute and drug delivery vehicle. *Clin. Orthop.* **367S**, S396–S405.
Li, R.H. and Wozney J.M. (2001) Delivering on the promise of bone morphogenetic proteins. *Trends Biotechnol.* **19**, 255–265.
Li, Y., de Wijn, J.R., Klein, C.P.A.T., van der Meer, S. and de Groot, K. (1994) Preparation and characterisation of nanograde osteoapatite-like rod crystals. *J. Mater. Sci. Mater. Med.* **5**, 263–268.
Lian, J.B., Stein, G.S., Canalis, E., Gehron Robey, P. and Boskey, A.L. (1999) Bone formation: osteoblast lineage cells gowth factors, matrix proteins and the mineralisation process. In *Primer on the metabolic diseases and disorders of mineral metabolism* (ed. Favus, M.J.), 4th edn. pp. 14–29. American Society for Bone and Mineral Research, Lippincott Williams and Wilkins, Philadelphia PA.
Lidgren, L. (2002) Bone substitutes. *Karger Gazette* **65**, 1–5. Available at http://www.karger.com/gazette/65/lidgren/art_5_1.htm
Liu, Q., de Wijn, J.R. and van Blitterswijk, C.A (1997) Nano-apatite/polymer composites: mechanical and physicochemical characteristics. *Biomaterials* **18**, 1263–1270.
Manley, M.T. (1993) Calcium phosphate Biomaterials: a review of the Literature. In *Hydroxylapatite coatings in Orthopaedic Surgery.* (eds Geesink, R.G.T. and Manley, M.T.), pp. 1–24, Raven Press, NY.

Marra, K.G., Szem, J.W., Kumta, P.N., DiMilla, P.A. and Weiss, L.E. (1999) *In vitro* analysis of biodegradable polymer blend/hydroxyapatite composites for bone tissue engineering. *J. Biomed. Mat. Res.* **47**, 324–335.

Milev, A., Green, D., Chai, C.S. and Ben-Nissan, B. (2001) coating of orthopaedic implants with sol-gel derived hydroxyapatite. Available at http://www.ansto.gov.au/ainse/nta/a26.pdf

Miller, G.J., Greenspan, D.C., Piottrowski, G. and Hench, L.L. (1976) Mechanical evaluation of bone-bioglass bonding. *8th Am. Int. Biomat. Symp.* Philadelphia, PA.

Misch, C. (1990) Density of bone: effect on treatment plans, surgical approach, healing and progressive loading. *Int. J. Oral Implant.* **6**, 23–31.

Muller-Mai, C., Voigt, C., de Almeida Reiss, S., Herbst H. and Gross, U. (1996) Substitution of natural coral by cortical bone and bone marrow in the rat femur. *J. Mater. Sci. Mat. Med.* **7**, 479–488.

Muragan, R. and Panduranga Rao, K. (2002) Controlled release of antibiotic from surface modified coralline hydroxyapatite. *Trends Biomater. Art. Organs* **16**, 43–45.

Murray, M.G.S., Wang, J., Ponton, C.B. and Marquis, P.M. (1995) An improvement in processing of hydroxyapatite ceramics, *J. Mater. Sci.* **30**(12), 3061–3074.

Nery, E.B., Lynch, K.L, Hirthe, W.M. and Mueller K.H. (1975) Bioceramic implants in periodontal osseous defects. *J. Periodontol.* **46**, 328–339.

Ozin, G.A. (1992) Nanochemistry: synthesis in diminishing dimensions. *Adv. Mater.* **4**, 612–649.

Passuti, N., Daculsi, G., Rogez, J.M., Martin, S. and Bainvel J.V. (1989) Macroporous calcium phosphate ceramic performance in human spine fusion. *Clin. Orthop.* **248**, 169–176.

Qian, J.J. and Bhatnagar, R.S. (1996) Enhanced cell attachment to anorganic bone mineral in the presence of a synthetic peptide related to collagen. *J. Biomed. Mat. Res.* **31**, 545–554.

Rahimi, F., Maurer, B.T., Enzweiler, M.G. (1997) Coralline hydroxyapatite: a bone graft alternative in foot and ankle surgery. *J. Foot Ankle Surg.* **36**(3), 192–203.

Ravaglioli A. and Krajewski, A. (1992) *Bioceramics* Chapman and Hall, London, UK.

Roy, D.M. and Linnehan, S.K. (1974) Hydroxyapatite formed from coral skeletal carbonate by hydrothermal exchange. *Nature* **247**, 220–222.

Rubin, C.T. and Rubin J. (1999) Biomechanics of Bone. In *Primer on the metabolic diseases and disorders of mineral metabolism* (ed. Favus, M.J.), 4th edn. pp. 39–44, American society for Bone and Mineral Research, Lippincott Williams and Wilkins, Philadelphia PA.

Santos, E.M., Radin, S., Shenker, B.J. Shapiro, I.M. and Ducheyne, P. (1998) Si-Ca-P xerogels and bone morphogenetic protein act synergistically on rat stromal marrow cell differentialtion *in vitro*. *J. Biomed. Mater. Res.* **41**, 87–94.

Sandhu, H.S., Grewal, H.S. and Parvataneni, H. (1999) Bone grafting for spinal fusion. *Orthop. Clin. North Am.* **30**, 685–698.

Schakenraad, J.M. and Busscher, H.J. (1989) Cell-polymer interactions: the influence of protein adsorption. *Colloid Surf.* **42**, 331–343.

Schakenraad, J.M. (1996) In *Biomaterials Science. An introduction to materials in medicine* (eds Ratner, B.D., Hoffman, A.S., Schoen, F.J. and Lemons, J.E.), pp. 141–147, Academic Press, San Diego, USA.

Schopper Ch., Moser D., Wanschitz F., Watzinger F. Lagogiannis G., Spassova E. and Ewers R. (1999) Histomorphologic Findings on Human Bone Samples Six Months after Bone Augmentation of the Maxillary Sinus with Algipore®. *J. Long-Term Effects Med. Implants* **9**, 203–213.

Schopper Ch., Ewers R. and Moser D. (1998) Bioresorption of Algipore® at Human Recipient Sites. *J. Cranio. Max. Fac. Surg.* **26** Suppl. 1, 172–173.

Simmons, D.J. and Grynpas, M.D. (1990) The osteoblast and osteocyte. In *Bone* **1** (ed. Hall, B.), pp. 193–302. Telford Press NJ, USA.

Shors, E.C. (1999) Coralline bone graft substitutes. *Orth. Clin. North Am.* **30** (4), 599–613.

Siegel, R.W. (1997) Mechanical properties of nanophase materials. *Mater. Sci. Forum* **235**, 851–859.

Søballe K., Hansen, E.S., Brockstedt-Rasmussen H., Pedersen, C.M. and Bünger, C. (1990) Hydroxyapatite coating enhances fixation of porous coated implants. A comparison in dogs between press fit and non-interference fit. *Acta Orthop. Scand.* **61**, 299–306.

Søballe K., Hansen, E.S., Brockstedt-Rasmussen H. and Bünger C. (1993a) Hydroxyapatite coating converts fibrous tissue to bone around loaded implants. *J. Bone Joint Surg.* **75B**, 270–278.

Søballe K., Hansen, E.S., Brockstedt-Rasmussen H. and Bünger C. (1993b) The effect of osteoporosis, bone deficiency, bone grafting and micromotion on fixation of porous-coated versus hydroxlapatite-coated implants. In *Hydroxylapatite coatings in Orthopaedic Surgery* (eds Geesink, R.G.T. and Manley, M.T.), pp. 107–136, Raven Press, NY.

Schmalzried, T.K., Kwong, I.M., Jasty, M, Sedlacek, R.C., Haire, T.C., O'Connor, D.O. Bragdon, C.R., Kabo, M., Malcolm, A.J. and Harris, W.H. (1992) The mechanisms of loosening of cemented acetabular components in total hip arthroplasty: analysis of specimens retrieved at autopsy. *Clin. Orthop.* **274**, 60–68.

Schwartz, Z. and Boyan, B.D. (1994) Underlying mechanisms at the bone-biomaterial interface. *J. Cell Biochem.* **56**, 340–347.

Schwartz, Z., Lohmann, C.H. Oefinger, J. Bonewald, L.F., Dean, D.D. and Boyan, B.D. (1999) Implant surface characteristics modulate differentiation behaviour of cells in the osteoblastic lineage. *Adv. Den. Res.* **13**, 38–48.

Thackray, A., Sammons, R.L., Macaskie, L.E., Yong, P., Lugg, H. and Marquis, P.M. (2003) Bacterial biosynthesis of a calcium phosphate bone substitute material. *J. Mater. Sci. Mater. Med.* (in press).

Urist, M.R., Silverman, M.F., Buring, K., Dubek, F.L. and Rosenburg, J.M. (1967) The bone induction principle. *Clin. Orthop.* **53**, 243–283.

Urist, MR, Strates, B.S. (1971) Bone morphogenetic protein. *J. Dent Res Suppl* **50**, 1392–1406.

Vaccaro, A.R. (2002) The Role of the Osteoconductive Scaffold in Synthetic Bone Graft. *Orthopedics* **25S**. Available at: http://www.orthobluejournal.com/supp/0502/vaccaro/

Wehling, P., Schulitz K.P., Robbins P.D., Evans C.H. and Reinecke J.A. (1997) Transfer of genes to chondrocytic cells of the lumbar spine. Proposal for a treatment strategy of spinal disorders by local gene therapy. *Spine* **22**, 1092–1097.

Williams, D. (1990) *Concise Encyclopaedia of Medical and Dental Materials*. Pergamon Press, Oxford, UK.

Wolford L.M. and Stevao L.L. (2002) Correction of jaw deformities in patients with cleft lip and palate. *BUMC Proc.* **15**, 250–254.

White, E. and Shors, E.C. (1986) Biomaterial Aspects of Interpore-200 Porous Hydroxyapatite. *Dental Clinic North Am.* **30**(1), 49–67.

White, R., Weber, J. and White, E. (1972) Replaniform: A new process for preparing porous ceramic, metal and polymer prosthetic matrials. *Science* **176**, 922–924.

Winn, S.R., Uladag, H. and Hollinger, J.O. (1999) Carrier systems for bone morphogenetic proteins. *Clin. Orthop.* **367S**, S95–S106.

Yong, P., Sammons, R.L., Marquis, P.M., Lugg, H. and Macaskie, L.E. (2003) Synthesis of nanophase hydroxyapatite by *Serratia* sp. N14. In *International Biohydrometallurgy Symposium*, Athens Greece, September 2003 (in press).

Yannas, I.A. (1996) Classes of materials used in medicine: Natural Materials. In *Biomaterials Science. An introduction to materials in medicine* (eds Ratner, B.D., Hoffman, A.S., Schoen, F.J. and Lemons, J.E.), pp. 84–93, Academic Press, San Diego, USA.

Yukna, R.A. Krauser, J.T., Callan, D.P. (2000) Multi-center clinical comparison of combination anorganic bovine-derived hydroxyapatite matrix (ABM)/cell-binding peptide (P-15) as a bone-replacement graft material in human periodontal osseus defects: 6 month results. *J. Periodontol.* **71**, 1671–1679.

Zhang, R.Y. and Ma, P.X. (1999) Porous poly(L-lactic acid)/apatite composites created by biomimetic process. *J. Biomed. Mater. Res.* **45**, 285–293.

26
Agronomic-based technologies towards more ecological use of phosphorus in agriculture

W.J. Horst and M. Kamh

26.1 INTRODUCTION

The optimum amount of plant-available phosphorus (P) in agricultural soils, usually called *optimum soil-P level*, can be defined as the amount sufficient to achieve an economically optimal yield, through the application of P to replace the amount removed with the harvested crop. Whether practicing low or high input agriculture, there is a major economic and environmental drive towards adjusting the necessary soil-P level as low as possible. In this chapter we explore the management practices, which could allow to successfully crop soils, whilst keeping soil-P levels low.

Acquisition of soil and fertiliser P by crops depends on soil and plant properties. Soil processes determining P availability to plants are P solubility/sorption, P transport, root/soil contact and mineralisation/immobilisation. Plants have evolved strategies contributing to a more efficient use of plant-available soil-P and to the mobilisation of P from less available soil-P fractions. Agronomic practices may affect P availability to crops through the modification of soil properties or through direct quantitative and qualitative crop impact on soil-P dynamics. Among the agronomic practices, the method of placement of P fertiliser can reduce the fertiliser application rate and P fixation in the soil and thus increase fertiliser use-efficiency. Application of organic matter, such as green manure and crop residues, to maintain or increase soil organic matter content and to enhance soil biological activity, and incorporation into the cropping system of P-mobilising plant species are particularly beneficial. There is increasing evidence that the growth of P-efficient leguminous cover crops in mixed cropping or in rotation positively affects growth and yield of crops on soils with low P availability. This can be attributed to the transfer of P from sparingly soluble to more readily available soil P fractions either directly or indirectly through return of the crop residues.

26.2 DEFINING OPTIMUM SOIL-P LEVELS

Phosphorus deficiency is a major factor limiting crop production particularly in tropical and sub-tropical soils (Fairhust *et al.*, 1999; Mokwuny *et al.*, 1986; Sanchez and Salinas, 1981). Correcting P deficiency with applications of P fertiliser is not possible for the mostly resource-poor farmers in the tropics and subtropics, especially on soils with high P-fixing capacity. Independent of the P availability in soils, crops cannot utilise more than 20–30% of the fertiliser P applied to that crop, due to spatial limitations of the root system together with the low mobility of P in most soils meaning that crops have to meet their P requirement predominantly from soil P. This has led to the recommendation to substantially enrich soils in P through the use of inorganic and/or organic fertilisers in intensive land-use systems. High P contents in surface soils increase the risk of transfer of P to surface and ground waters via surface runoff, erosion, drainage and preferential flow (Andraski *et al.*, 1985; Haygarth and Jarvis, 1999; Haygarth and Condron, this volume; Heckrath *et al.*, 1995; McDowell and Sharpley, 2001). Optimum and thus target soil-P levels are, therefore, defined as sufficient to achieve economically optimum yields through the application of P amounts sufficient to replace the P removed with the harvested product (Helyar, 1998). In both low and high input agriculture, there is a major economic and environmental interest to adjust the necessary soil-P levels as low as possible. In this chapter, we explore the management practices, which could allow to successfully crop soils at lower soil-P levels.

26.3 PHOSPHORUS DYNAMICS IN SOILS

The main interrelated processes governing the acquisition of soil and fertiliser P by crops are dissolution/precipitation and desorption/sorption (buffer capacity), transport primarily by diffusion, soil/root contact and biological P transformations (Figure 26.1). Dissolution/precipitation and desorption/sorption reactions are

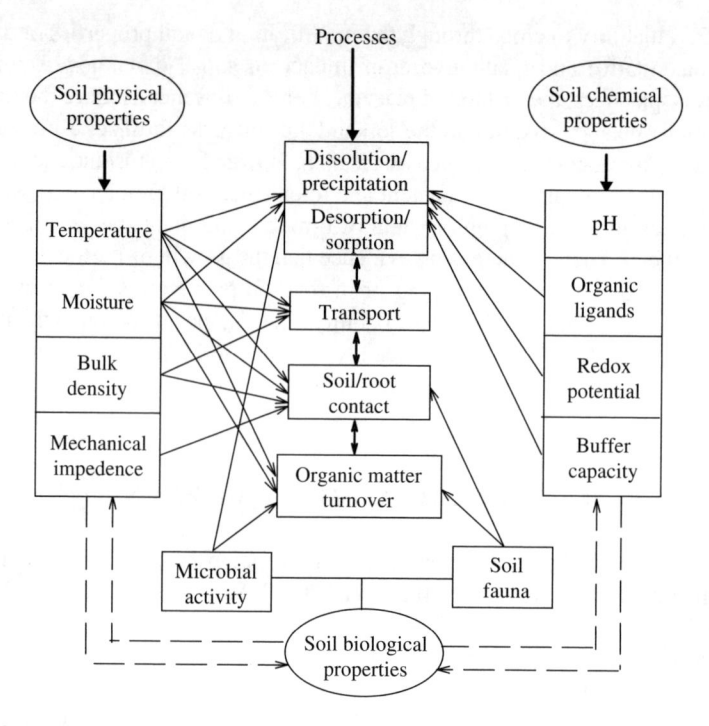

Figure 26.1 Processes governing acquisition of soil and fertiliser P by crops as affected by soil properties.

affected by chemical soil properties, and by temperature and soil moisture directly, or indirectly through microbial activity. Transport of P to the plant primarily depends on the P diffusion coefficient (Barber, 1995; Jungk and Claassen, 1997), which is determined by soil temperature, moisture, tortuosity (bulk density) and buffer capacity. Root growth is decisive for establishing soil/root contact, and is primarily affected by soil physical properties (Donald *et al.*, 1987; Faiz and Weatherley, 1982), although chemical factors such as Al toxicity and Ca deficiency in acid soils may be even more important (Carvalho and van Raij, 1997; Hardy *et al.*, 1990). Turnover of organic matter may lead to release or immobilisation of P depending on soil microbial and faunal activity.

26.4 PHOSPHORUS USE EFFICIENCY OF PLANTS

Plants may interfere with the above mentioned processes either directly or indirectly through the modification of soil properties, thus enhancing P availability and uptake (Figure 26.2). It is especially important to clearly differentiate between those plant properties that contribute to a more efficient use of the plant-available soil-P and those that mobilise P from less available soil-P fractions.

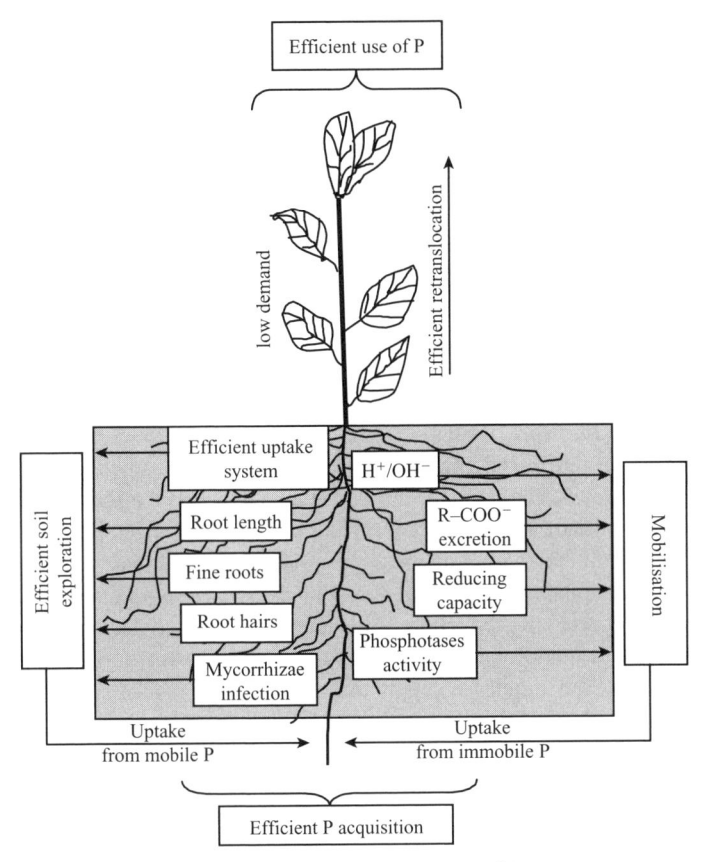

Ecological use of phosphorus in agriculture 613

Figure 26.2 Plant properties contributing to efficient use of P.

An efficient uptake system may enhance diffusion of P to the root through creating a steeper concentration gradient at the soil/root interface. However, in most soils, the transport of P to the root is the main limiting factor for P acquisition rather than root uptake of P (Barber, 1995; Jungk and Claassen, 1997). Therefore, enhancing P transport and soil/root contact through lateral root formation (Lynch, 1997; Manske et al., 2000), root-hair length (Gahoonia et al., 1997, 1999; Itoh and Barber, 1983) and reduced root diameter (Föhse et al., 1991; Silberbush and Barber, 1983), or establishing a symbiosis with arbuscular mycorrhizal fungi, which allows plant access to soil P up to several cm away from the root (George et al., 1995; Li et al., 1997), will enhance plant P nutrition more than a highly efficient P transport across the plasma membrane.

The release of H^+ or OH^- (Gahoonia et al., 1992), organic acid anions (Kirk et al., 1999; Gerke et al., 2000), the increase of reduction capacity (Holford and Patrick, 1979) and rhizosphere phosphatase activity (Tarafdar and Jungk, 1987) will allow the plant to access poorly available inorganic and organic soil-P fractions and thus increase the pool of soil/fertiliser P which contributes to plant P nutrition.

26.5 AGRONOMIC PRACTICES

26.5.1 Fertiliser P placement

The term *P placement* is used to describe a P fertiliser application technique that differs from the traditional broadcast application in that the fertiliser P is mixed with only a small soil volume. With the most common band application, P is placed in a fertiliser band into the soil at planting.

Due to the high chemical reactivity of phosphorus in soil, soluble P fertiliser reacts quickly with soil constituents and, therefore, is transformed to immobile forms, at a rate which depends on soil conditions, particularly P sorption/precipitation capacity, and reaction time (Barrow, 1974; Lindsay, 1979). The homogeneous incorporation of P fertiliser into the soil prior to planting, results in high intimate mixing with the soil and, therefore, formation of reaction products of low solubility too early. A prerequisite for fertiliser P uptake is P transport to the root. As the diffusion is the main transport process for P, and the diffusion rate of P is low, it is the transport which determines the rates of P uptake by roots (Jungk, 1991). Localized placement of soluble P fertilisers in the zone of maximum root-length density may reduce P fixation in the soil, via reduction of soil/fertiliser contact (Aitken and Hughes, 1980), and increase the P uptake, via a reduction of the transport distance. Also, under conditions of insufficient P availability for optimum plant growth, placement of P may enhance root branching in the zone of P supply thus enhancing P uptake (Drew and Saker, 1978; Mullinus, 1993). This effect of P placement on root branching is particularly well expressed in combination with nitrogen placement (Grunes, 1959).

A mechanistic model of P uptake predicted that the beneficial effect of P placement is particularly pronounced when the quantity of applied P is low compared to the P sorption capacity of the soil (Anghinoni and Barber, 1980; Borkert and Barber, 1985). This is in agreement with a range of field studies showing that the amount of fertiliser P for optimum yield, and/or the soil-P level necessary to achieve optimum yields are lower with P placement than with P broadcast application (Aitken and Hughes, 1980; Fiedler *et al.*, 1989; Fox and Kang, 1978; Peterson *et al.*, 1981; Ron and Loewy, 2000; Sinaj *et al.*, 2001). The relative efficiency of placed versus broadcast fertiliser P was greater than 3 : 1 and was reduced to 1 : 1 when the soil P content was <0.2 and $1.2\,\mathrm{g\,m^{-3}}$, respectively (Sanchez *et al.*, 1991). Buerkert *et al.* (2001) presented overwhelming evidence of the effectiveness of very small amounts of P placed to millet on P-deficient soils in Sub-Saharan Africa.

Phosphorus placement is particularly advantageous for crops grown in wider spaced rows which do not exploit with their roots the soil volume homogeneously, particularly at early growth stages (Eghball and Sander, 1989a, b; Mallarino *et al.*, 1999; Sweeney, 1993), and for plant species with small seeds, and thus low seed P reserves, which require their P demand to be met from P uptake at very early growth stages (Costigan 1984, 1987).

The advantage of enhanced plant growth at early growth stages through P placement, however, may not always translate into yield increases (Buah *et al.*, 2000; Mallarino *et al.*, 1999). This may be due to optimum conditions for P uptake from broadcast fertiliser and soil P at later growth stages, or other factors, rather than P limiting yield. Also, according to the model calculation by Anghinoni and Barber (1980), maximum P uptake and biomass production can only be achieved if 50% of the rooting zone is fertilised and thus 50% of the roots contribute to P uptake. Particularly in highly P-fixing soils, maximum yield could not be achieved by banding P fertiliser (Borkert and Barber, 1985). With an adequate rate of P fertilisation, incorporating P into the entire cultivated soil volume gave better results (Fox and Kang, 1978; Kamprath, 1967), due to building up the level of residual soil P in the whole soil volume. However, such recommendation is often economically prohibitive (Fiedler et al., 1989).

26.5.2 Organic matter application

The application of organic manure and crop residues have important roles in maintaining soil structure, water holding capacity, microbial biomass and soil fauna, and in nutrient availability and cycling (Kumar and Goh, 2000). Organic inputs, such as animal manures, cover crops, and green manures, have usually been evaluated in terms of nitrogen and carbon while little attention has been paid to phosphorus or other nutrients. The return of crop residues might be especially beneficial for phosphorus, because the slower decomposition and release of P from organic matter prevents rapid fixation as inorganic P (Blevins et al., 1983; Friessen et al., 1997), and may better match the P requirement of the subsequently grown crop. The return of crop residues will reduce the P export from the soil (Figure 26.3 Roland et al., 1997), but the amount of P returned in crop residues is generally low due to very efficient retranslocation of P from vegetative to reproductive plant organs especially

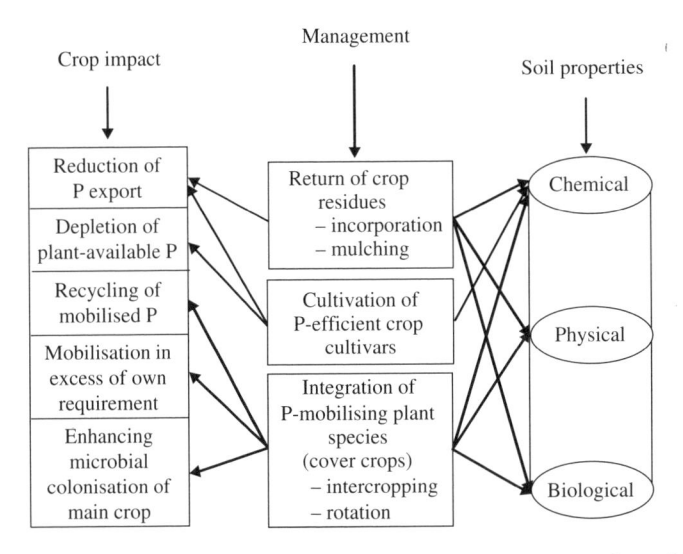

Figure 26.3 Agronomic practices affecting P availability to crops through modification of soil properties and related crop impact. Modified after Horst et al. 2001.

Table 26.1 Effect of 40 years organic and single superphosphate (SSP) fertilisation on soil P content, water (H_2O), and double lactate (DL)-extractable soil P. Modified from Amann and Amberger (1984).

P fraction	P content [mg P kg^{-1}] Fertilisation			
	No fertiliser P		Manure	SSP
	−Straw	+Straw		
P (H_2O)	3.9	9.5	7.0	25.7
P (DL)	10.8	21.3	17.0	73.6
P (total)	661	679	638	823

at limiting P supply (Palm et al., 1997). The main advantage of crop residues, therefore, relies on their positive effects on soil properties (Rasse et al., 2000).

Soil processes can be shifted to dissolution/desorption of P thus increasing the more readily available P fractions (Amann and Amberger, 1984; Iyamuremye and Dick, 1996; Hafner et al., 1993). Adsorption of organic acid anions produced from the decomposition of organic matter on P-fixing surfaces (Bhatti et al., 1998; Lyamuremye and Dick, 1996; Palm et al., 1997) and neutralization of the pH (Kretzschmar et al., 1991; Tang and Yu, 1999) will enhance the availability of soil and fertiliser P due to the reduction of the P buffer/sorption capacity. In addition, improving the conditions for P transport (Amberger and Amann, 1984), root growth (Hafner et al., 1993), and soil/root contact will enhance the P acquisition capacity of the crops. The positive effect of long-term return of straw as crop residues on soil-P availability is particularly well documented by Amann and Amberger (1984, Table 26.1). Also, in this context, the accumulation of organic P fractions seems to be of special importance for the maintenance of soil-P availability (Lee et al., 1990; Oberson et al., 1999; Seeling and Zasoski, 1993; Sharpley, 1985). Acid and alkaline phosphatase activity in the soil, which is related to organic P mineralization, may increase as a result of crop-residues application (Aggarwal et al., 1997).

Application of crop residues/organic matter as mulch not only protects the soil surface against erosion but is also especially important for the maintenance and improvement of soil physical properties and the activity of soil organisms (Hauser, 1993). These reduce mechanical impedance (Logsdon and Linden, 1992) and contribute to P mobilisation (Atlavinyte and Vanagas, 1973; Kang et al., 1994; Sinaj et al., 2001).

Although the contribution of crop-residue P to the soil inorganic and organic P pools is clearly established (McLaughlin et al., 1988a), the contribution to the P nutrition of the subsequent crop is highly variable and not yet well understood. McLaughlin et al. (1988b) found that only a small proportion of the P accumulated by subsequent wheat was derived from medic residues, and wheat yield was negatively affected (McLaughlin and Alston, 1986). Application of low-P crop residues may not only reduce N mineralisation (Nguluu et al., 1996; Tiessen and Shang, 1998),

Table 26.2 Effect of organic matter source and triple superphosphate (TSP, 15 kg P) on soil-P availability. P_i = P inorganic, P_o = P organic. Modified from Nziguheba et al. (2000).

Treatment	Organic matter quality indicator			Soil-P fractions [mg P kg^{-1}]				
	P [%]	N [%]	Lignin [%]	Resin	NaHCO$_3$		NaOH	
					P_i	P_o	P_i	P_o
Control				2.53	3.46	24.40	36.00	159
TSP				3.63	4.90	24.50	39.20	160
Tithonia leaves	0.27	3.80	12.00	8.09	5.19	26.40	43.60	166
Maize straw	0.07	0.64	6.00	3.39	3.86	24.80	37.10	165

but also lead to P immobilisation (Schomberg and Steiner, 1999). This may be a contributing factor to the observed negative effect of low-P maize residues on subsequent maize yields (Horst et al., 2001). It is not the total amount of P returned to the soil in crop residues, but rather their P concentration, C–P ratio (Singh and Jones, 1976), and the biodegradability of the organic matter (Friesen et al., 1997) that appear to be decisive for possible fertilisation effects of crop-residue P. This can be clearly shown for the high quality organic source *Tithonia diversifolia* (Table 26.2). Application of tithonia leaves increased the resin, bicarbonate and sodium hydroxide inorganic P (P_i) fractions more than the application of maize stover and triple superphosphate (TSP) (Nziguheba et al., 1998). The increase in measured P_i fractions was accompanied by a reduction in P adsorption.

Another plant quality-factor might be the ash alkalinity of the plant material which determines the amount of organic acid anions released during decomposition, not only with regard to neutralisation of soil acidity and complexation of phytotoxic Al (Kretzschmar et al., 1991; Noble et al., 1996; Yan and Schubert, 2000), but also to the decrease of P sorption-capacity (Nziguheba et al., 1998, see above). There is a need to study the complex interactions of properties of plant residues and soil properties in a more detailed and quantitative way not only with regard to the C and N, but also to the P cycle.

26.5.3 Integration of P-efficient plants

Under conditions of low/limiting P supply in the soil, the integration of crop species and/or crop varieties that can make most efficient use of the P supplied by the soil and maintenance fertiliser applications represent a key element of sustainable cropping systems (Ae et al., 1990; Lynch, 1998). Although knowledge about the principle mechanisms involved in efficient P acquisition by plants has evolved substantially during recent years (Miyasaka and Habte, 2001; Raghothama, 1999; Rao et al., 1999; Sattelmacher et al., 1994; see Figure 26.2), their possible contribution to the overall P efficiency of the cropping system is still not well understood.

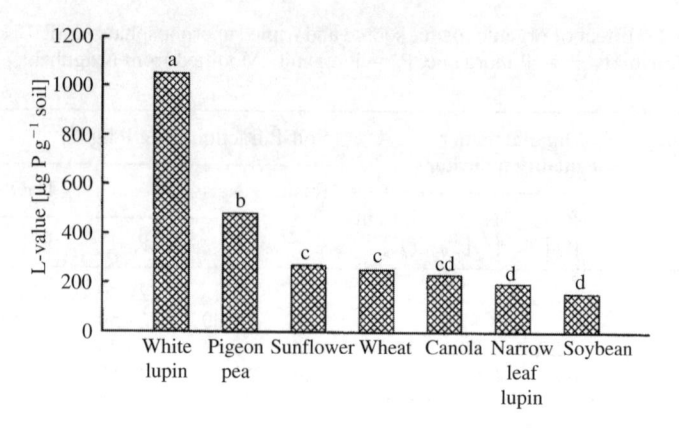

Figure 26.4 Exchangeable P (L-value) for different crop species grown in an Oxisol (from Hocking, 2001). Values with the same letters (a, b, c, cd, d) are not significantly different at the $P > 0.05$ level (Tukey).

The selection and cultivation of crop varieties with high P-uptake efficiency will enhance the depletion of plant-available P (see Figure 26.3). These may be viable options only in combination with maintenance P application on soils with low P-sorption capacity. Phosphorus efficiency based on P-utilisation efficiency will reduce the export of P from the soil in harvested products (Lynch, 1998). However, this may lead to an undesirable low nutritional quality of the harvested product (Batten, 1986) and of crop residues (Buresh, 1999). If P efficiency is due to mobilisation of P from less available soil-P fractions (Helal and Dressler, 1989; Braum and Helmke, 1995), then the depletion of readily available soil P might be reduced.

The most promising agronomic approach appears to be the integration into the cropping system of P-mobilising plant species as inter-crops or in rotation. This not only positively affects soil properties, but also may make mobilised P available to the main crop. This requires that the cover crop has access to soil P fractions that are not available to the main crop, thus minimising competition for plant-available P. There is good evidence showing that some plant species such as white lupin (*Lupinus albus*) possess a remarkable ability to mobilise sparingly soluble soil P (Braum and Helmke, 1995; Hocking, 2001). Hocking (2001) used the L-value (a measure of P availability to the plant using ^{32}P isotopic dilution technique) to compare different species in their ability to access P from different soil-P pools. The L-values indicated that the pool of soil P available for white lupin but also for pigeon pea (*Cajanus cajan*) were larger than that for other species (Figure 26.4). Horst et al. (2001) showed that some leguminous cover crops, well adapted to the humid savannah of West Africa, were also capable of acquiring soil-P fractions not accessible to maize. Comparable to white lupin (Gardner et al., 1983; Keerthisinghe et al., 1998; Neumann and Römheld, 1999), this capability seems to be related to an enhanced release of organic acid anions at limiting P supply (Kamh et al., 2002), and an enhanced root-surface phosphatase activity (Kamh et al., 1999).

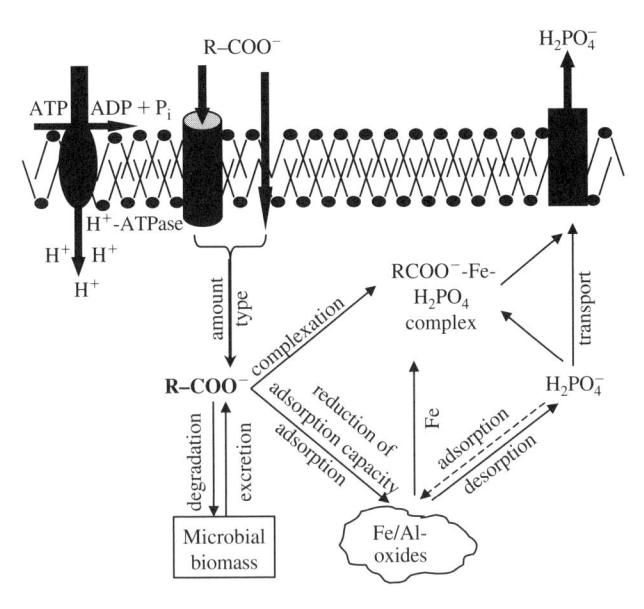

Figure 26.5 Effects of organic acid anions exuded from plant roots on P availability in the rhizosphere.

The effect of P deficiency-induced synthesis and release of organic acid anions to soil P availability and P uptake by plants is complex as shown in Figure 26.5. It is primarily aimed at mobilising P to meet the plant's own P requirement. Given the energy cost of this mechanism and its sensitive regulation by the P nutritional status of the plant (Jones, 1998), it appears unlikely that P is mobilised in excess of the plant's P requirement. This assumption is supported by the demonstration of a strong depletion not only of less soluble but also of readily plant-available P fractions in the rhizosphere soil of P-efficient plant species (Braum and Helmke, 1995; Horst *et al.*, 2001; Kamh *et al.*, 1999). Large amounts of organic acid anions have to be excreted in order to mobilise sparingly soluble P (Gerke, 1992, 1993; Jones, 1998). Therefore, concentrating the root exudates on a small soil volume by confining the exudation to the root apices (Hoffland *et al.*, 1989) or proteoid roots (Dinkelaker *et al.*, 1995) is a particularly efficient adaptation mechanism. High exudation rates are necessary because the organic acid anions are rapidly adsorbed by Fe/Al hydroxides in acid soils (Gerke *et al.*, 2000; Jones and Brassington, 1998), precipitated as Ca citrate in calcareous soils (Dinkelaker *et al.*, 1989), or decomposed by soil micro-organisms (Bowen and Rovira, 1991; Jones *et al.*, 1996) followed by re-adsorption of the released P (Hinsinger and Gilkes, 1996). However, in spite of these immobilisation processes, the exudation of organic acid anions may contribute to enhanced soil-P availability. Sorption of organic acid anions on Fe/Al hydroxides may desorb P (Geelhoed *et al.*, 1999). Since P is sorbed more strongly on Fe hydroxides than organic acid anions (Jones and Brassington, 1998; Kirk, 1999) the main positive effect will rely on the blockage of potential P-sorption sites by the organic anions, thus decreasing the soil P buffer capacity and increasing the equilibrium soil-solution P concentration when P is applied as fertiliser or released from organic matter decomposition (Iyamuremye and Dick, 1996; Nziguheba et al., 1998). The biodegradation of organic acid anions adsorbed on Fe/Al hydroxides is greatly reduced (Jones and Edwards, 1998), which may lead to a sustained reduction in soil P-sorption capacity. Enhanced microbial colonisation of the rhizosphere may lead to a build-up of organic P in the rhizosphere (Kamh et al., 1999). While this may lead to competition between plant and microbes for available P (P immobilisation), it may also add to the soil organic P pool which has been shown to be of special importance for the maintenance of soil-P availability, especially in highly P-fixing soils (Oberson et al., 1999). The role of rhizosphere microbes may be particularly beneficial for the plant, if non-P-mobilising root exudates, such as sugars, are transformed to P-mobilising microbial exudates (Schilling et al., 1998).

Not only the turnover of the organic root exudates and organic P fractions, but also the equilibrium between readily and sparingly plant-available inorganic P fractions is time-dependent (Morel and Hinsinger, 1999). The main driving force for P mobilisation is the release of root and microbial exudates, and the uptake and depletion of the soil-solution P by plants and soil organisms.

A transfer of mobilised P to the main crop grown in rotation may also occur. In a glasshouse experiment Hocking (2001) showed that wheat following the P-mobilising plant species white lupin or pigeon pea had better growth and P nutrition as well as higher L-values than wheat grown after wheat or soybean (Table 26.3). Also Kamh et al. (1999) showed that white lupin improved the P nutrition and growth of subsequently grown wheat in a pot experiment. The mobilised P could also be made available to the main crop grown in rotation, after decomposition of plant residues. This is shown by a significant positive relationship between P applied with the crop residues of different cover crops and yield of subsequently grown maize in a field experiment on P-deficient and highly P-fixing soil (Horst et al., 2001).

However, during the vegetation-free period between cropping seasons, plant-available/mobilised P may be immobilised thus limiting the beneficial effect of

Table 26.3 Effect of preceding crops on growth, P content, and L-value of following wheat plants grown in pots with an Oxisol. Mean values with the same letters (a or b) are not significantly different at the $P > 0.05$ level (Tukey). Modified from Hocking (2001).

Preceding crop	Dry matter [mg plant^{-1}]	P content [mg plant^{-1}]	L-value
White lupin	0.71 a	2.70 a	681 a
Pigeon pea	0.67 a	3.17 a	529 a
Wheat	0.48 b	1.53 b	289 b
Soybean	0.38 b	1.88 b	215 b

Ecological use of phosphorus in agriculture 621

P mobilisation by one crop to the next crop in the rotation. Therefore, the transfer of mobilised P from a P-efficient crop to an inefficient crop is more likely to occur in a mixed cropping system. This is indicated by a wheat/white lupin intercropping experiment in the field (Gardner and Boundy, 1983), and more clearly under controlled conditions in a pot experiment (Horst and Waschkies, 1987; Kamh et al., 1999). However, the positive effect of intercropping a P-mobilising crop may not occur if the component crops compete for other growth factors than P. Competition for water may lead to lower soil moisture and thus reduced P transport in the soil (Härdter and Horst, 1991).

The integration of P-efficient species into the cropping system could also improve the P acquisition of the main crop through enhancing root colonisation by

Table 26.4 Root infection by mycorrhizae (AM) in millet assessed 35 to 75 days after seeding (DAS) as affected by cropping system at three locations in West Africa. From Bagayoko et al. (2000).

Cropping systems	Experimental locations				
	Sadore		Gaberi		Gaya
	35 DAS	75 DAS	45 DAS	75 DAS	50 DAS
	AM infection (% of roots)				
Continuous millet	23	44	27	48	11
Millet after cowpea	32	50	48	64	31

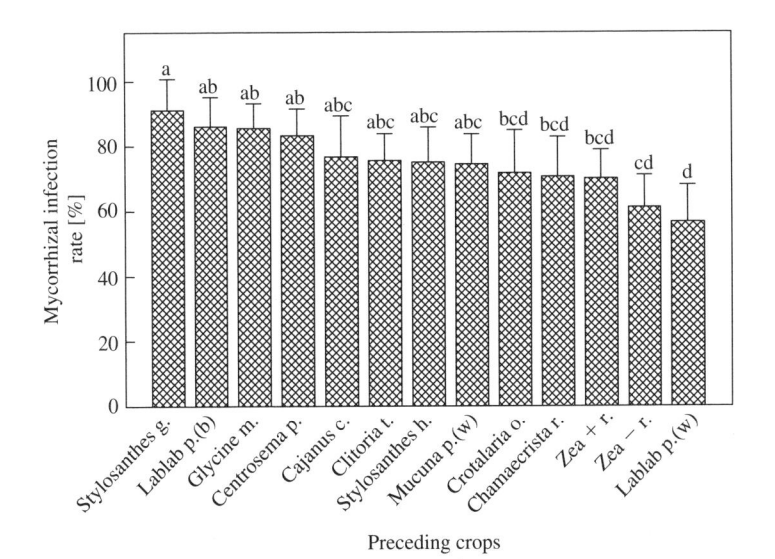

Figure 26.6 Effect of preceding crops on mycorrhizal infection (five weeks after planting) on the subsequently grown maize in rotation (field experiment in northern Nigeria). Means with the same letters (a, ab, abc, bcd, cd or d) are not significantly different at the $P > 0.05$ level (Tukey). Modified from Horst et al. (2001).

mycorrhizal fungi (Bagayoko *et al.*, 1999), P-mobilising (Schilling *et al.*, 1998) and plant growth-promoting rhizobacteria (de Freitas *et al.*, 1997). Bagayoko *et al.* (2000) demonstrated that enhanced mycorrhization greatly contributed to the positive rotational effect of grain legumes on millet yields (Table 26.4). Additional evidence for enhancing mycorrhizal infection comes from a field experiment of Horst *et al.* (2001, Figure 26.6). Most of the cover-crop species significantly enhanced mycorrhizal colonisation of subsequent maize. On the other hand, the non-host crop canola, clearly delayed mycorrhizal infection and P uptake of subsequently grown maize (Gavito and Miller, 1998a, b).

26.6 CONCLUSIONS

The available knowledge shows that agronomic measures can greatly contribute to increased P availability to crops and thus reduce the necessity to enrich soils in plant available P for optimum crop yields. Among the different contributing factors, fertiliser-P placement, the application of organic matter, such as green manure and crop residues, to maintain or increase soil organic matter content and to enhance soil biological activity, and the incorporation into the cropping system of P-mobilising plant species are particularly beneficial. Such agronomic management practices can contribute to reduce the release of P to the environment and the P fertiliser requirement for optimum yield. However, especially on strongly P-fixing soils these agronomic measures cannot substitute for maintenance of P fertiliser applications.

REFERENCES

Ae, N.A.J., Okada, K., Yoshihara, T. and Johansen, C. (1990) Phosphorus uptake by pigeon pea and its role in cropping systems of the Indian subcontinent. *Science* **248**, 477–480.

Aggarwal, R.K., Kumar, P. and Power, J.F. (1997) Use of crop residues and manure to conserve water and enhance nutrient availability and pearl millet yields in an arid tropical region. *Soil Tillage Res.* **41**, 43–51.

Aitken, R.L. and Hughes, J.D. (1980) The effect of method of application and presence of sulphate on phosphate fixation from three phosphate fertilizers applied to a krsnozem soil. *Aust. J. Exp. Agric. Anim. Husb.* **20**, 486–491.

Amberger, A. and Amann, Ch. (1984) Wirkung organischer Substanzen auf Boden- und Düngerphosphat. II. Einfluß verschiedener organischer Stoffe auf die Mobilität von Dünger-P. *Z. Pflanzenernähr. Bodenk.* **147**, 60–70.

Amann, Ch. and Amberger, A. (1984) Wirkung organischer Substanzen auf Boden- und Düngerphosphat. I. Einfluß von Stroh- und Maiswurzelextakten auf die Löslichkeit von Boden und Dünger-P. *Z. Pflanzenernähr. Bodenk.* **147**, 49–59.

Andraski, B.J., Mueller, D.H. and Daniel, T.C. (1985) Phosphorus losses in runoff as affected by tillage. *Soil Sci. Soc. Am. J.* **49**, 1523–1527.

Anghinoni, I. and Barber, S.A. (1980) Predicting the most efficient phosphorus placement for corn. *Soil Sci. Soc. Am. J.* **44**, 1016–1020.

Atlavinyte, O. and Vanagas, J. (1973) Mobility of nutritive substances in relation to earthworm members in the soil. *Pedobiologia* **13**, 344–352.

Bagayoko, M., Buerkert, A., Lung, G., Bationo, A. and Römheld, V. (2000) Cereal/legume rotation effects on cereal growth in Sudano-Sahelian West Africa: soil mineral nitrogen, mycorrhizae and nematodes. *Plant Soil* **218**, 103–116.

Barber, S.A. (1995) Soil nutrient bioavailability: a mechanistic approach, 2nd edn. John Wiley & Sons, New York.

Barrow, N.J. (1974) The slow reactions between soil and anions: I. Effects of time, temperature, and water content of a soil on the decrease in effectiveness of phosphate for plant growth. *Soil Sci.* **118**, 380–386.

Batten, G.D. (1986) Phosphorus fractions in the grain of diploid, tetraploid, and hexaploid wheat grown with contrasting phosphorus supplies. *Cereal Chem.* **63**, 384–387.

Bhatti, J.S., Comerford, N.B. and Johnston, C.T. (1998) Influence of oxalate and soil organic matter on sorption and desorption of phosphate onto a spodic horizon. *Soil Sci. Soc. Am. J.* **62**, 1089–1095.

Blevins, R.L., Thomas, G.W., Frye, W.W. and Corneliues, P.L. (1983) Changes in soil properties after 10 years continuous non-tilled and conventionally tilled corn. *Soil Tillage Res.* **3**, 135–146.

Borkert, C.M. and Barber, S.A. (1985) Predicting the most efficient phosphorus placement for soybeans. *Soil Sci. Soc. Am. J.* **49**, 901–904.

Bowen, G.D. and Rovira, A.D. (1991) The rhizosphere: the hidden half of the hidden half. In *Plant Roots: The Hidden Half* (eds Waisel, Y., Eshel, A. and Kafkaki, U.), pp. 641–669, Marcel Dekker, New York.

Braum, S.M. and Helmke, P.A. (1995) White lupin utilise soil phosphorus that is unavailable to soybean. *Plant Soil* **176**, 95–100.

Buah, S., Polito, T.A. and Killorn, R. (2000) No-tillage corn response to placement of fertilizer nitrogen, phosphorus, and potassium. *Commun. Soil Sci. Plant Anal.* **31**, 3121–3133.

Buerkert, A., Bationo, A. and Piepho, H-P. (2001) Efficient phosphorus application strategies for increased crop production in sub-Saharan West Africa. *Field Crop Res.* **72**, 1–15.

Buresh, R.J. (1999) Phosphorus management in tropical agroforestry: current knowledge and research challenges. *Agroforesty Forum* **9**, 61–66.

Carvalho, M.S.C. and van Raij, B. (1997) Calcium sulphate, phosphogypsum and calcium carbonate in the amelioration of acid subsoils for root growth. *Plant Soil* **192**, 37–48.

Costigan, P.A. (1984) The effect of placing small amounts of phophate fertilizer close to the seed on growth and nutrient concentration of lettuce. *Plant Soil* **79**, 191–201.

Costigan, P.A. (1987) The importance of seedling nutrient stress in producing site to site variations in yield. *J. Plant Nutr.* **10**, 1523–1530.

De Freitas, J.R., Banerjee, M.R. and Germida, J.J. (1997) Phosphate-solubilising rhizobacteria enhance the growth and yield but not phosphorus uptake of canola (*Brassica napus* L.). *Biol. Fertil. Soils* **24**, 358–364.

Dinkelaker, B., Hengeler, Ch. and Marschner, H. (1995) Distribution and function of proteoid roots and other root clusters. *Bot. Acta.* **108**, 183–200.

Dinkelaker, B., Römheld, V. and Marschner, H. (1989) Citric acid excretion and precipitation of calcium citrate in the rhizosphere of white lupin (*Lupinus albus* L.). *Plant Cell Environ.* **12**, 285–292.

Drew, M.C. and Saker, L.R. (1978) Nutrient supply and the growth of the seminal root system in barley. III. Compensatory increase in growth of lateral roots and in rate of P uptake in response to a localized supply of phosphate. *J. Exp. Bot.* **29**, 435–451.

Donald, R.G., Kay, B.D. and Miller, M.H. (1987) The effect of soil aggregate size on early shoot and root growth of maize (*Zea mays* L.). *Plant Soil* **103**, 251–259.

Eghball, B. and Sander, D.H. (1989a) Band spacing effects of dual-placed nitrogen and phosphorus fertilizers on corn. *Agron. J.* **81**, 178–184.

Eghball, B. and Sander, D.H. (1989b) Distance and distribution effects of phosphorus fertilizer on corn. *Soil Sci. Soc. Am. J.* **53**, 282–287.

Fairhust, T., Lefroy, R., Mutert, E. and Batjes, N. (1999) The importance, distribution and causes of phosphorus deficiency as a constraint to crop production in the tropics. *Agroforestry Forum* **9**, 2–8.
Faiz, S.M.A. and Weatherley, P.E. (1982) Root contraction in transpiring plants. *New Phytol.* **92**, 333–343.
Fiedler, R.J., Sander, D.H. and Peterson, G.A. (1989) Fertilizer phosphorus recommendation for winter wheat in terms of method of phosphorus application, soil pH, and yield goal. *Soil Sci. Soc. Am. J.* **53**, 1282–1287.
Föhse, D., Claassen, N. and Jungk, A. (1991) Phosphorus efficiency of plants. II. Significance of root radius, root hairs and cation-anion balance for phosphorus influx in seven plant species. *Plant Soil* **132**, 261–272.
Fox, R.L. and Kang, B.T. (1978) Influence of phosphorus fertilizer placement and fertilization rate on maize nutrition. *Soil Sci.* **125**, 34–40.
Friesen, D.K., Rao, I.M., Thomas, R.J., Oberson, A. and Sanz, J.I. (1997) Phosphorus acquisition and cycling in crop and pasture system in low fertility soils. In *Plant nutrition – for sustainable food production and environment* (eds Ando, T., Fujiter, K., Mae, T., Matsumoto, H., Mori, S. and Sekiya, J.), pp. 493–498, Kluwer Academic Publishers Japan.
Gahoonia, T.S., Claassen, N. and Jungk, A. (1992) Mobilization of phosphate in different soils by ryegrass supplied with ammonium and nitrate. *Plant Soil* **143**, 241–248.
Gahoonia, T.S., Care, D. and Nielsen, N.E. (1997) Root hairs and phosphorus acquisition of wheat and barley cultivars. *Plant Soil* **191**, 181–188.
Gahoonia, T.S., Nielsen, N.E. and Lyshede, O.B. (1999) Phosphorus (P) acquisition of cereal cultivars in the field at three levels of P fertilisation. *Plant Soil* **211**, 265–281.
Gardner, W.K. and Boundy, K.A. (1983) The acquisition of phosphorus by *Lupinus albus* L. IV. the effect of intercropping wheat and white lupin on the growth and mineral composition of the two species. *Plant Soil* **70**, 391–402.
Gardner, W.K., Barber, D.A. and Parbery, K.G. (1983) The acquisition of phosphorus by *Lupinus albus* L. III. The probable mechanism by which phosphorus movement in the soil/root interface is enhanced. *Plant Soil* **70**, 107–124.
Gavito, M.E. and Miller, M.H. (1998a) Early phosphorus nutrition, mycorrhizae development, dry matter partitioning and yield of maize. *Plant Soil* **199**, 177–186.
Gavito, M.E. and Miller, M.H. (1998b) Changes in mycorrhizae development in maize induced by crop management practices. *Plant Soil* **198**, 185–192.
Geelhoed, J.S., van Riemsdijk, W.H. and Findenegg, G.R. (1999) Simulation of the effect of citrate exudation from roots on the plant availability of phosphate adsorbed on goethite. *Eur. J. Soil Sci.* **50**, 379–390.
George, E., Marschner, H. and Jakobsen, I. (1995) Role of arbuscular mycorrhizal fungi in uptake of phosphorus and nitrogen from soil. *Crit. Rev. Biotechnol.* **15**, 257–270.
Gerke, J. (1992) Phosphate, iron, and aluminium in soil solution of three different soils in relation to varying concentration of citric acid. *Z. Pflanzenernäh. Bodenk.* **155**, 339–343.
Gerke, J. (1993) Solubilization of Fe (III) from humic-Fe complexes, humic/Fe oxide mixtures and from poorly ordered Fe-oxides by organic acids – consequences for P adsorption. *Z. Pflanzenernähr. Bodenk.* **156**, 253–257.
Gerke, J., Beissner, L. and Römer, W. (2000) The quantitative effect of chemical phosphate mobilisation by carboxylate anions on P uptake by single root. I. The basic concept and determination of soil parameters. *J. Plant Nutr. Soil Sci.* **163**, 207–212.
Grunes, D.L. (1959) Effect of nitrogen on the availability of soil and fertilizer phosphorus to plants. *Adv. Agron.* **11**, 369–396.
Hafner, H., George, E., Bationo, A. and Marschner H. (1993) Effect of crop residues on root growth and phosphorus acquisition of pearl millet in an acid sandy soil in Niger. *Plant Soil* **150**, 117–127.

Härdter, R. and Horst, W.J. (1991) Nitrogen and phosphorus use in maize sole cropping and maize/cowpea mixed cropping systems on an Alfisol in the Northern Guinea Savanna of Ghana. *Biol. Fertil. Soils* **10**, 267–275.

Hardy, D.H., Barber, C.D. and Miner, G.S. (1990) Chemical restriction of roots in Ultisol subsoil by long-term management. *Soil Sci. Soc. Am. J.* **54**, 1657–1660.

Hauser, S. (1993) Distribution and activity of earthworms and contribution to nutrient recycling in ally cropping. *Biol. Fertil. Soil* **15**, 16–20.

Haygarth, P.M. and Jarvis, S.C. (1999) Transfer of phosphorus from agricultural soils. *Adv. Agron.* **66**, 195–249.

Heckrath, G., Brookes, P.C., Poulton, P.R. and Goulding, K.W.T. (1995) Phosphorus leaching from soils containing different phosphorus concentrations in the broadbalk experiment. *J. Environ. Qual.* **24**, 904–910.

Hedley, M.J., White, R.E. and Nye, P.H. (1982) Plant-induced changes in the rhizosphere of rape (*Brassica napus* var. Emerald) seedlings. III. Changes in L value, soil phosphate fractions and phosphatase activity. *New Phytol.* **91**, 45–56.

Helal, H.M. and Dressler, A. (1989) Mobilisation and turnover of soil phosphorus in the rhizosphere. *Z. Pflanzenernähr. Bodenk.* **152**, 175–180.

Helyar, K.R. (1998) Efficiency of nutrient utilization and sustaining soil fertility with particular reference to phosphorus. *Field Crops Res.* **56**, 187–195.

Hinsinger, P. and Gilkes, R.J. (1996) Mobilisation of phosphate from phosphate rock and alumina-sorbed phosphate by roots of ryegrass and clover as related to rhizosphere pH. *Eur. J. Soil Sci.* **47**, 533–544.

Hocking, J. (2001) Organic acids exuded from roots in phosphorus uptake and aluminium tolerance of plants in acid soils. *Adv. Agron.* **74**, 63–97.

Hoffland, E., Finddenegg, G.R. and Nelemans, J.A. (1989) Solubilization of rock phosphate by rape. II. Local root exudation of organic acids as response to P starvation. *Plant Soil* **113**, 161–165.

Holford, I.C.R. and Patrick, W.H. (1979) Effects of reduction and pH change on phosphate sorption and mobility in an acid soil. *Soil Sci. Soc. Am. J.* **43**, 292–297.

Horst, W.J. and Waschkies, C. (1987) Phosphorus nutrition of spring wheat in mixed culture with white lupin. *Z. Pflanzenernähr. Bodenk.* **150**, 1–8.

Horst, W.J. Kamh, M., Jibrin, J.M. and Chude, V.O. (2001) Agronomic measure for increasing P availability to crops. *Plant Soil* **237**, 211–223.

Itoh, S. and Barber, S.A. (1983) Phosphorus uptake by six plant species as related to root hairs. *Agron. J.* **75**, 457–461.

Iyamuremye, F. and Dick, R.P. (1996) Organic amendments and phosphorus sorption by soils. *Adv. Agron.* **56**, 139–185.

Jones, D.L. (1998) Organic acids in the rhizosphere – a critical review. *Plant Soil* **205**, 25–44.

Jones, D.L. and Brassington, D.S. (1998) Sorption of organic acids in acid soils and its implications in the rhizosphere. *Eur. J. Soil Sci.* 49, 447–455.

Jones, D.L. and Edwards, A.C. (1998) Influence of sorption on the biological utilisation of two simple carbon substances. *Soil Biol. Biochem.* **30**, 1895–1902.

Jones, D.L., Prabowo A.M. and Kochain, L.V. (1996) Kinetics of malate transport and decomposition in acid soils and isolated bacterial-populations. The effect of microorganisms on root exudation of malate under Al stress. *Plant Soil* **182**, 239–247.

Jungk, A. (1991) Dynamics of nutrient movement at the soil-root interface. In *Plant roots: The Hidden Half* (eds Waisel, Y., Eshel, A. and Kafkafi, U.), pp. 455–481, Dekker, New York.

Jungk, A. and Claassen, N. (1989) Availability in soil and acquisition by plants as the bases for potassium and phosphorus supply to plants. *Z. Pflanzenernähr. Bodenk.* **152**, 151–158.

Jungk, A. and Claassen, N. (1997) Ion diffusion in the soil-root system. *Adv. Agron.* **61**, 53–110.

Kamh, M., Horst, W.J., Amer, F., Mostafa, H. and Maier, P. (1999) Mobilisation of soil and fertilizer phosphate by cover crops. *Plant Soil* **211**, 19–27.

Kamh, M., Abdou, M. Chude, V., Wiesler, F. and Horst, W.J. (2002) Mobilisation of phosphorus contributes to positive rotational effects of leguminous cover crops on maize grown on soils from northern Nigeria. *Plant Nutr. Soil Sci.* **165**(5), 557–660.

Kamprath, D.J. (1967) Residual effect of large application of phosphate on high phosphorus fixing soils. *Agron. J.* **59**, 25–27.

Kang, B.T., Akinnifesi, F.K. and Pleysier, J.L. (1994) Effect of agroforestry woody species on earthworm activity and physicochemical properties of worm casts. *Biol. Fertil. Soils* **18**, 193–199.

Keerthisinghe, G., Hocking, P.J., Ryan, P.R. and Delhaize, E. (1998) Effect of phosphorus supply on the formation and function of proteoid roots of white lupin (*Lupinus albus* L.). *Plant Cell Environ.* **21**, 467–478.

Kirk, G.J.D. (1999) A model of phosphate solubilisation by organic anion excretion from plant roots. *Eur. J. Soil Sci.* **50**, 369–378.

Kirk, G.J.D, Santos, E.E. and Findenegg G.R. (1999) Phosphate solubilization by organic anion excretion from rice (*Oryza sativa* L.) growing in aerobic soil. *Plant Soil* **211**, 11–18.

Kretzschmar, R.M., Hanfner, H., Bationo, A. and Marschner, H. (1991) Long and short-term effects of crop residues on aluminium toxicity, phosphorus availability and growth of pearl millet in an acid sandy soil. *Plant Soil* **136**, 215–223.

Kumar, K. and Goh, K.M. (2000) Crop residues and management practices: Effects on soil quality, soil nitrogen dynamic, crop yield, and nitrogen recovery. *Adv. Agron.* **68**, 197–318.

Lee, D. Han, X.G. and Jordan, C.E. (1990) Soil phosphorus fractions, aluminium, and water retention as affected by microbial activity in an Ultisol. *Plant Soil* **121**, 125–136.

Lindsay, W.L. (1979) *Chemical Equilibria in Soils*. John Wiley & sons, New York.

Li, X.L., George, E., Marschner, H. and Zhang, J.L. (1997) Phosphorus acquisition from compacted soil by hyphae of a mycorrhizal fungus associated with red clover (Trifolium pratense). *Can. J. Bot.* **75**, 723–729.

Logsdon, S.D. and Linden D.R. (1992) Interactions of earthworms with soil physical conditions influencing plant growth. *Soil Sci.* **154**, 330–337.

Lynch, J. (1998) The role of nutrient-efficient crops in modern agriculture. *J. Crop Production* **1**, 241–264.

Lynch, J. (1997) Root architecture and phosphorus acquisition efficiency in common bean. In *Radical Biology: Advances and perspectives on the function of plant roots* (eds Flores, H.E., Lynch, J.P. and Eissenstadt, D.), Current topics in Plant Physiology. *Ann. Am. Soc. Plant Phys. Series* **18**, 81–91.

Lyamuremye, F. and Dirck, R.P. (1996) Organic amendments and phosphorus sorption by soils. *Adv. Agron.* **56**, 139–185.

Manske, G.G.B., Ortiz-Monasterio, J.I., Van Ginkel, M., Gonzalez, R.M., Rajaram, S., Molina, E. and Vlek, P.L.G. (2000) Traits associated with improved P-uptake efficiency in CIMMYT's semidwarf spring bread wheat grown on an acid andisol in Mexico. *Plant Soil* **221**, 189–204.

Mallarino, A.P., Bordoli, J.M. and Borges, R. (1999) Phosphorus and potassium placement effect on early growth and nutrient uptake of no-till corn and relationships with grain yield. *Agron. J.* **91**, 37–45.

Marschner, H. (1991) Mechanisms of adaptation of plants to acid soils. In *Plant-Soil Interaction at Low pH* (eds Wright, R.J., Baligar, V.C. and Murrmann, R.P.), pp. 683–720, Kluwer Acadimic Publishers, Netherlands.

McDowell, R.W. and Sharpley, A.N. (2001) Approximating phosphorus release from soils to surface runoff and subsurface drainage. *J. Environ. Qual.* **30**, 508–520.

McLaughlin, M.J. and Alston, A.M. (1986) The relative contribution of plant residues and fertilizer to the phosphorus nutrition of wheat in a pasture/cereal rotation. *Aust. J. Soil Res.* **24**, 517–526.

McLaughlin, M.J., Alston, A.M. and Martin, J.K. (1988a) Phosphorus cycling in wheat-pasture rotations. III. Organic phosphorus turnover and phosphorus cycling. *Aust. J. Soil Res.* **26**, 343–353.

McLaughlin, M.J., Alston, A.M. and Martin, J.K. (1988b) Phosphorus cycling in wheat-pasture rotations. I. The source of phosphorus taken up by wheat. *Aust. J. Soil Res.* **26**, 323–331.

Miyasaka, S.C. and Habte, M. (2001) Plant mechanisms and mycorrhizal symbioses to increase phosphorus uptake efficiency. *Commun. Soil Sci. Plant Anal.* **32**, 1101–1147.

Mokwuny, A.S.H., Chien, S.H. and Rhodes, E.R. (1986) Reactions of phosphate with tropical African soils. In *Management of Nitrogen and Phosphorus Fertilizers in Sub-Saharan Africa* (eds Mokwuny, A. and Vlek, P.L.G.), pp. 253–282, Martinus Nijhoff Publishers, Dordrecht, Netherlands.

Morel, C. and Hinsinger, P. (1999) Root-induced modification of the exchange of phosphate ion between soil solution and soil solid phase. *Plant Soil* **211**, 103–110.

Mullinus, G.L. (1993) Cotton root growth as affected by P fertilizer placement. *Fertl. Res.* **34**, 23–26.

Neumann, G. and Römheld, V. (1999) Root excretion of carboxylic acids and protons in phosphorus-deficient plants. *Plant Soil* **211**, 121–130.

Nguluu, S.N., Probert, M.E., Myers, R.J.K. and Waring, S.A. (1996) Effect of tissue phosphorus concentration on the mineralization of nitrogen from stylo and cowpea residues. *Plant Soil* **191**, 139–146.

Noble, A.D., Zenneck, I. and Randall, P.J. (1996) Leaf litter ash alkalinity and neutralisation of soil acidity. *Plant Soil* **179**, 293–302.

Nziguheba, G., Palm, C.A., Buresh, R.J. and Smithson, P.C. (1998) Soil phosphorus fractions and adsorption as affected by organic and inorganic sources. *Plant Soil* **198**, 159–168.

Oberson, A., Friesen, D.K., Tiessen, H., Morel, C. and Stahel, W. (1999) Phosphorus status and cycling in native savanna and improved pastures on an acid low-P Colombian Oxisol. *Nut. Cycling Agroecosyst.* **55**, 77–88

Palm, C.A., Myers, R.J.K. and Nandwa, S.M. (1997) Combined use of organic and inorganic nutrient sources for soil fertility maintenance and replenishment. In *Replenishing Soil Fertility in Africa* (eds Roland, J.B., Sanchez, P.A. and Calhoun, F.), pp. 193–217, SSSA special Publication No. 51 Madison, Wisconsin, USA.

Peterson, G.A., Sander, D.H., Grabouski, P.H. and Hooker, M.L. (1981) A new look at row and broadcast phosphate recommendations for winter wheat. *Agon. J.* **73**, 13–17.

Raghothama, K.G. (1999) Phosphate acquisition. *Ann. Rev. Plant Physiol. Plant Mol. Biol.* **50**, 665–693.

Rao, I.M., Friesen, D.K. and Horst, W.J. (1999) Opportunities for germplasm selection to influence phosphorus acquisition from low-phosphorus soils. *Agroforesty Forum* **9**, 13–17.

Rasse, D.P., Smucker, A.J.M. and Santos, D. (2000) Alfalfa root and shoot mulching effects on soil hydraulic properties and aggregation. *Soil Sci. Soc. Am. J.* **64**, 725–731.

Roland, J.B., Paul, C.S. and Deborah, T.H. (1997) Building soil fertility capital in Africa. In *Replenishing Soil Fertility in Africa* (eds Roland, J.B., Sanchez, P.A. and Calhoun, F.), pp. 111–149, SSSA special publication No. 51 Madison, Wisconsin, USA.

Ron, M.M. and Loewy, T. (2000) Effect of phosphorus placement on wheat yield and quality in southwestern Buenos Aires (Argentina). *Commun. Soil Sci. Plant Anal.* **31**, 2891–2900.

Sanchez, P. and Salinas, J.G. (1981) Low input technology for managing oxisols and ultisols in tropical america. *Adv. Agron.* **34**, 280–406.

Sanchez, C.A., Porter, P.S. and Ulloa, M.F. (1991) Relative efficiency of broadcast and banded phosphorus for sweet corn produced on histosol. *Soil Sci. Soc. Am. J.* **55**, 871–875.

Sattelmacher, B., Horst, W.J. and Becker, H.C. (1994) Factors that contribute to genetic variation for nutrient efficiency of crop plants. *Z. Pflanzenernähr. Bodenk.* **157**, 215–224.

Schilling, G., Gransee, A., Deubel, A., Lezovic, G. and Ruppel, S. (1998) Phosphorus availability, root exudates, and microbial activity in the rhizosphere. *Z. Pflanzenernähr. Bodenk.* **161**, 465–478.

Schomberg, H.H. and Steiner, J.L. (1999) Nutrient dynamic of crop residues decomposing on a fallow no-till soil surface. *Soil Sci. Soc. Am. J.* **63**, 607–613.

Seeling, B. and Zasoski, R.J. (1993) Microbial effects in maintaining organic and inorganic solution phosphorus concentrations in a grassland tropical. *Plant Soil* **148**, 277–284.

Sharpley, A.N. (1985) Phosphorus cycling in unfertilised and fertilised agricultural soils. *Soil Sci. Soc. Am. J.* **49**, 905–911.

Silberbuch, M. and Barber, S.A. (1983) Sensitivity of simulated phosphorus uptake to parameters used by mechanistic-mathematical model. *Plant Soil* **74**, 93–100.

Sinaj, S., Buerkert A., El-Hajj, G., Bationo, A., Traore, H. and Frossard, E. (2001) Effect of fertility management strategies on phosphorus bioavailability in four West African soils. *Plant Soil* **233**, 71–83.

Singh, B.B. and Jones, J.P. (1976) Phosphorus sorption and desorption characteristic of soil as affected by organic residues. *Soil Sci. Soc. Am. J.* **40**, 389–394

Sweeney, D.W. (1993) Fertilizer placement and tillage effects on grain sorghum growth and nutrient uptake. *Soil Sci. Soc. Am. J.* **57**, 532–537.

Tang, C. and Yu, Q. (1999) Impact of chemical composition of legume residues and initial soil pH on pH change of a soil after residue incorporation. *Plant Soil* **215**, 29–38.

Tarafdar, J.C. and Jungk, A. (1987) Phosphatase activity in the rhizosphere and its relation to the depletion of soil organic phosphorus. *Biol. Fertil. Soils* **3**, 199–204.

Tiessen, H. and Shang, C. (1998) Organic-matter turnover in tropical land-use systems. In *Carbon and nutrient dynamics in natural and agricultural tropical systems* (eds Bergström, L. and Kirchmann, H.), CAB International, Wallingford, Oxon, UK.

Yan, F. and Schubert, S. (2000) Soil pH changes after application of plant shoot materials of faba bean and wheat. *Plant Soil* **220**, 279–287.

27
Biodegradation of organophosphate nerve agents

*M. Shimazu, W. Chen and
A. Mulchandani*

27.1 INTRODUCTION

The growing population in the twentieth century has necessitated higher yields in agriculture. To this end, pesticides have become an integral part of modern agriculture. The use of pesticides has resulted in higher crop yields, but at the same time released hazardous chemicals into the environment. Due to their low cost, efficacy, and wide spread availability, organophosphate (OP) compounds have been one of the most widely used classes of pesticides throughout the world. In the United States alone, over 40 million kg of organophosphate pesticides are applied (Mulchandani *et al*., 1999). OP compounds are available for industrial, agricultural as well as home use. Commonly used OP pesticides include parathion, coumaphos, dursban, diazinon and malathion. Recent concerns about the toxicity and environmental impact of OP compounds has lead to the phasing out or banning of this class of compounds for home and agricultural operations, requiring manual picking in more industrialized nations such as the United States and Western Europe. Despite the health and environmental risks, many developing nations continue to rely on OP compounds in pest management and industrialised nations are obliged to use them under certain circumstances.

More specifically, the United States Department of Agriculture (USDA) operates a large scale tick (*Boophilus annulatus*) eradication program along the United States–Mexico border. The Tick Eradication Program effectively prevents the re-introduction of Cattle Fever into the United States from imported cattle from Mexico. The treatment process includes the dipping of cattle in large vats containing approximately 15,000 litres of coumaphos. Coumaphos waste from these vats is pumped into evaporation pits. Leaching, groundwater and soil contamination become major concerns because many of these pits are unlined (Mulbry et al., 1996). In addition, diazinon is used in Australia, New Zealand and the United Kingdom to control flystrike and blowfly larvae (Rammell and Bentley, 1988).

The effectiveness of OP compounds as pesticides and insecticides also makes them hazardous to humans and to the environment. Organophosphates and their family of compounds are potent neurotoxins that share structural similarities to chemical warfare agents such as sarin, soman and VX. Organophosphates act as cholinesterase inhibitors and in turn disrupt neurotransmission in insects as well as mammals. Classical symptoms of OP exposure include salivation, lacrimination, urination and defecation. Exposure to OP compounds can cause fatigue, dizziness, vomiting, paralysis and even death (Casarett and Doull, 1996).

As a result of their toxicity and wide spread usage, there is a need to treat the large amounts of wastes from unused pesticide concentrates, agricultural runoff, accidental spillage, and from cleaning of spray equipment and storage tanks. In addition to OP pesticides, disposal of huge amounts of OP compounds in the form of chemical weapons must be handled. The Chemical Weapons Convention (CWC) disarmament agreement required the immediate ceasing of development, production and stockpiling of chemical weapons. Further, the ratification by the United States of the CWC on April 30, 1997 required the destruction of all chemical weapons stockpiles by April 30, 2007. Clearly there is a need for rapid and efficient methods by which to dispose of these compounds.

Current disposal methods involve the use of landfills, chemical hydrolysis or incineration. With landfills, the threat of breaching of the containment vessel while handling or in transit and groundwater leaching are always a concern. Chemical hydrolysis relies on the use of harsh acids and bases whose by-product must in turn be treated and disposed of. Currently the only EPA approved method, incineration, generates strong public outcry over the potential release of toxic by-products into the atmosphere (Munnecke, 1979; Munnecke et al., 1976). With the drawbacks of traditional approaches and the need to find environmentally friendly methods towards pesticide decontamination, biological approaches have gained increasing attention (Grimsley et al., 1997; National Research Council, 1993).

27.2 MICROBIAL DEGRADATION

The study of the biodegradation of OP compounds has dated back more than 20 years and started with the identification and isolation of organisms in the environment capable of degrading OP compounds. In 1964, Lichtenstein and Schultz described a mixed population of soil microorganisms capable of degrading parathion and malathion. In 1976, *Pseudomonas diminuta* MG, capable of hydrolyzing parathion, was isolated and was shown to decrease the toxicity of OP compounds two to 120-fold (Munnecke *et al*., 1976). Since that time, many soil microorganisms with the ability to hydrolyze OP compounds have been isolated (Mulbry and Karns, 1998; Serdar *et al*., 1989; Zhongli *et al*., 2001). Microorganisms capable of using OP compounds as sole carbon sources have also been isolated (Rani and Lalithakumari, 1994; Siddaramappa *et al*., 1973), along with microorganisms capable of degrading by-products of the hydrolysis such as p-nitrophenol, a by-product of parathion and paraoxon (Leung *et al*., 1997; Nishino and Spain, 1993; Spain *et al*., 1979). A *Burkholderia* sp. strain NF100, capable of using fenitrothion as a sole carbon source, has also been isolated. Interestingly, the genes responsible for degradation of fenitrothion and its hydrolysis product, 3-methyl-4-nitrophenol, were found on separate 105 kb and 33 kb plasmids respectively (Hayatsu *et al*., 2000).

Two parathion degrading bacteria, *Pseudomonas diminuta* MG and *Flavobacterium* sp., have been well characterized and the organophosphate degrading (*opd*) genes coding for hydrolase activity have been cloned and sequenced. The two genes were shown to be identical on the amino acid level (Mulbury *et al*., 1987). The hydrolases from both bacteria have also been purified and extensive kinetic characterizations have been performed (Dumas *et al*., 1989; Mulbry and Karns, 1989).

Organophosphorus hydrolase (OPH, EC 3.1.8.1) is a homodimer with a binuclear metal center. Although the natural substrate is not known, OPH can hydrolyze the P—O and P—S bonds used in a variety of commercial pesticides, as well as P—F and P—CN bonds found in chemical warfare agents (Table 27.1) (Dumas *et al*., 1989; Karns *et al*., 1987; Lai *et al*., 1994; Mason *et al*., 1997). However,

Table 27.1 OPH kinetic constants; k_{cat}: hydrolysis rate, K_m: substrate concentration.

Compound	k_{cat} (s^{-1})	K_m (mM)	k_{cat}/K_m (M^{-1}s^{-1})
Paraoxon	3180	0.058	5.5×10^7
Coumaphos	800	0.39	2.1×10^3
Parathion	630	0.24	2.6×10^6
Methyl parathion	189	0.08	2.4×10^4
Diazinon	176	0.45	3.9×10^6
DFP	465	0.048	9.7×10^4
Sarin	56	0.7	80×10^3
Soman	5	0.5	10×10^3
Demeton-S	1.3	0.78	1.6×10^3

hydrolysis rates varied from very fast for phosphotriesters and phosphothiolester pesticides (P—O bond), such as paraoxon ($k_{cat} > 3100\,\text{s}^{-1}$) and coumaphos ($k_{cat} = 800\,\text{s}^{-1}$), to limited hydrolysis for chemical warfare agents such as sarin (P—F bond) ($k_{cat} = 56\,\text{s}^{-1}$) and VX (P—CN bond) ($k_{cat} = 0.3\,\text{s}^{-1}$) (Kolakowski et al., 1997). The presence of divalent metal ions, such as Co^{+2} or Zn^{+2}, is essential for enzyme activity.

With the current advent of biotechnology, several avenues for bioremediation of OP compounds using OPH are available. Due to the slow specific growth rates of native OPH producers *Pseudomonas diminuta* MG and *Flavobacterium* sp., the organophosphate degrading (*opd*) gene has been expressed in a wide variety of heterogeneous hosts, including *E. coli* (Dave et al., 1993), insect cell (fall armyworm) (Phillips et al., 1990), *Streptomyces* (Steiert et al., 1989), and a soil fungus (Xu et al., 1996). Of these, *E. coli* has been the most widely used and studied, in large part due to ease of growth and maintenance and the high cell densities achieved in fermentation. Initial attempts at expression using the native promoter resulted in very poor expression levels (Serdar et al., 1989; Serdar and Gibson, 1985). However, increased levels of expression was achieved with the use of a strong *lac* promoter (Mulbry et al., 1987).

27.3 ENZYMATIC DETOXIFICATION OF OP NEUROTOXINS

27.3.1 Immobilized OPH

Both native and recombinant OPH has been immobilized onto various support materials, such as nylon membranes, porous glass and nanometer size silica beads that were subsequently incorporated into silicone polymers and applied for the detoxification of OP compounds (Caldwell and Raushel, 1991; Gill and Ballesteros, 2000; Munnecke, 1979).

Partially purified OPH was immobilized onto nylon powders and membranes (Caldwell and Raushel, 1991). Through the reaction of glutaraldehyde with the free amino groups on the nylon polymers, followed by coupling of the enzymes to the aldehyde functionality, partially purified OPH was cross-linked onto nylon 6 and 66 membranes and nylon 11 powder and nylon 6 tubing. Kinetic measurements were performed for OPH immobilized onto the nylon 6 membrane and were found to increase the K_m by five to six fold. More dramatic effects were seen in the overall hydrolysis efficiency as the nylon-immobilized OPH showed less than 10% of the original hydrolysis compared to soluble enzyme (Caldwell and Raushel, 1991).

Immobilization of OPH within polyurethane foams has also been explored. Enzyme-modified polyurethane sponges offer a quick and effective response for cleaning of minor residues or emergency response to accidental spillage. The use of sponges combines a large surface area for immobilization with high liquid retention capacity. Havens and Rase (1993) and LeJeune et al. (1997) immobilized OPH into polyurethane foam for the simultaneous cleanup and degradation of OP spills. This strategy resulted in the retention of approximately 50% of the enzyme activity and increased storage stability of the enzyme.

Purified OPH has also been used to generate cross-linked enzyme crystals (CLEC) (St. Claire and Navia, 1992; Hoskin et al., 1999). The CLEC OPH hydrolyzed Tetriso (*O,O*-diisopropyl-*S*-(2-diisopropylaminoethyl)) phosphorothiolate and Demeton-S (*O,O*-diethyl-*S*-(2-ethylthioethyl)) phosphorothiolate corresponding closely with the K_m of soluble OPH.

OPH immobilized by covalent attachment to nano silica particles have been incorporated into silicone polymers to produce highly active and stable biocatalyst, capable of both liquid and gas phase degradation of OP compounds. In addition these nanocomposites have the ability to be formed into various sheets, films, granulates and foams. When immobilized, OPH showed up to a 90% retention of activity as compared to soluble OPH, with immobilization efficiencies up to 80% with 400 mg OPH applied per gram material and minimum activity loss when stored at 5°C over a period of 6 months (Gill and Ballesteros, 2000).

27.3.2 OPH immobilization by affinity tags

Unfortunately, physical adsorption offers poor and nonspecific binding, while covalent modifications to OPH often results in reductions in enzyme activity and kinetic properties (Caldwell and Raushel, 1991; LeJeune and Russell, 1996). In addition, such methods can result in improper orientation of the enzyme leading to inaccessibility of the substrate to the enzyme active site. One solution is the use of an affinity tag fused to the OPH moiety enabling one-step purification and oriented immobilization. Affinity tags offer strong reversible binding under mild non-denaturing conditions, as well as proper orientation of the enzyme for full substrate accessibility to the enzyme active site.

Recently, a fusion protein between a cellulose binding domain (CBD_{clos}), isolated from *Clostridium cellulovorans*, and OPH was capable of immobilization onto various cellulose materials (Richins et al., 2000). The use of cellulose as an immobilization matrix is advantageous due to its low cost, wide spread availability and non-toxic nature. The kinetic parameters of OPH fused to the CBD domain were essentially identical to the soluble protein, with a slight increase in K_m from 0.058 to 0.220 and a modest decrease in k_{cat} from 3170 to 2840. Additionally, the immobilized fusion protein offered superior stability over that of soluble OPH, retaining over 85% activity over the period of 45 days (Richins et al., 2000).

In another strategy to orient OPH correctly onto a solid support, an octapeptide, Asp-Tyr-Lys-Asp-Asp-Asp-Asp-Lys (FLAG), was fused to organophosphorus hydrolase. The OPH-FLAG fusion protein was then immobilized onto magnetic beads coated with a protein A-anti-FLAG monoclonal antibody. The fusion of the OPH to the FLAG showed no detrimental affect on enzyme activity. Free soluble OPH was measured with a catalytic efficiency (k_{cat}/K_m) of 2.9 ($\pm$0.3) $\times$ $10^7 M^{-1} s^{-1}$ while the fusion exhibited similar efficiency of 2.5 ($\pm$0.3) $\times$ $10^7 M^{-1} s^{-1}$. However, when immobilized onto the magnetic beads, there was a 2-fold increase in K_m

and a 90% decrease in k_{cat}. Stability of the enzyme improved with a reported half-life of 23 days when immobilized and stored at 4°C (Wang et al., 2002).

Shimazu et al. (2003) demonstrated the use of a thermally responsive peptide for the purification and immobilization of OPH. Elastin-like proteins (ELPs) are composed of repeats of the pentapeptide sequence Val-Pro-Gly-Xaa-Gly (where Xaa is any residue except for proline) and undergo a sharp temperature transition from soluble to insoluble with an increase of ionic strength of the buffer or increase in temperature (Urry, 1992). The phase transition is a hydrophobically driven process causing the side chains of hydrophobic amino acid residues aggregating to a thermodynamically favourable state with the increase in temperature of ionic strength (Urry, 1997). Using this property an ELP-OPH fusion protein was purified by inverse temperature transitioning. ELP-OPH was purified to homogeneity and the fusion protein exhibited only modest changes in the K_m and k_{cat} over wild-type OPH. In addition, the ELP portion of the protein served as an affinity tag for immobilization of OPH onto hydrophobic surfaces (Shimazu et al., 2003).

27.4 WHOLE CELL DETOXIFICATION OF OP NEUROTOXINS

A major economic obstacle in using enzymes to degrade OP compounds is the need to purify large amounts of enzymes. Traditionally, gel filtration and ion exchange chromatography are used. However these methods are costly, tedious and are not amenable to scale up. A strain of *Streptomyces lividans* has been developed for the secretion of OPH into the medium (Steiert et al., 1989). This method allows for minimal cell wall disruption and handling of the cells, but still required additional purification of the end product. To alleviate the need for any steps in purifying protein, whole cells were used as live biocatalyst instead of purified proteins.

The use of whole cells as live biocatalysts for the detoxification of OP compounds has been demonstrated using *Pseudomonas putida* (Rani and Lalithakumari, 1994) and *Pseudomonas* sp. A3 (Ramanathan and Lalithakumari, 1996). *P. putida* immobilized using 3% sodium alginate beads (w/v) and a wet cell mass of 5% (w/v) was capable of degrading 99% of 1 mM methyl parathion during the course of a 48 hour incubation time. Further, the reactor was re-usable for numerous batch runs (Ramanathan and Lalithakumari, 1996).

A microbial consortium was isolated from soil contaminated with coumaphos and used in a biofilter. The consortium was cultured onto sand, gravel and Celite and packed into a bio-trickling filter. After 7–10 days at 25°C, the coumaphos concentrations dropped from 1200 mg/l to between 0.010 mg/l and 0.050 mg/l (Mulbry et al., 1996). Subsequently, a field-scale bioreactor capable of treating 15,000 l batches was constructed. In large scale runs, the bioreactor was capable of reducing the coumaphos concentrations from 2000 mg/l to 10 mg/ml. The reactor was capable of a 200-fold reduction in coumaphos in two successive 11,000 l batch runs. However, fouling became a problem with prolonged use (Mulbry et al., 1998).

27.4.1 Surface expression of OPH

Unfortunately, the cell walls of the bacteria can act as a diffusional barrier for substrate to access the OPH, which are traditionally expressed intercellularly. In terms of parathion and paraoxon, it has been previously shown that uptake was the rate-limiting step in degradation of these pesticides by recombinant *E. coli* expressing OPH intercellularly (Elashvili *et al.*, 1998; Hung and Liao, 1996; Richins *et al.*, 1997). One solution is to express the enzymes onto the cell surface thereby bypassing any diffusional barrier caused by the cell membrane (Figure 27.1).

Using the Lpp-OmpA anchor system, a Lpp-OmpA-OPH fusion was targeted successfully onto the surface of *E. coli*. The fusion was targeted with over 90% efficiency and afforded a seven-fold increase in hydrolysis of parathion when compared to cells expressing similar amounts of OPH intercellularly. In addition to being more efficient, the live biocatalysts were more stable than soluble OPH and remained active over the span of 30 days (Richins *et al.*, 1997).

More recently, OPH was targeted onto the surface of an *E. coli* and *Moraxella* sp. using an ice-nucleation protein (INP) anchor derived from *Pseudomonas syringe* (Shimazu *et al.*, 2001a, b). Combined with the native p-nitrophenol degrading pathway, the genetically engineered *Moraxella* sp. with surface expressed OPH was capable of completely degrading 0.4 mM paraoxon and its hydrolysis product within 10 hours. More impressively, *Moraxella* sp. was capable of a 80-fold increase in OPH activity over *E. coli* (Shimazu *et al.*, 2001b) providing a potentially powerful platform for microbial degradation of various toxic compounds using this robust bacterium.

Cells with surface-expressed OPH were immobilized on a nonwoven polypropylene fabric support and effectively degraded 90–100% of diazinon, methyl parathion, paraoxon, and coumaphos (Mulchandani *et al.*, 1999) in less than 3 h. The ability of immobilized cells to carryout repeated cycles of degradation in sequence batch reactor was demonstrated, and only a small decline in degradation performance was detected during 12 consecutive repeated sequence batch degradation of paraoxon cycles, over a period of 19 days. In addition to polypropylene

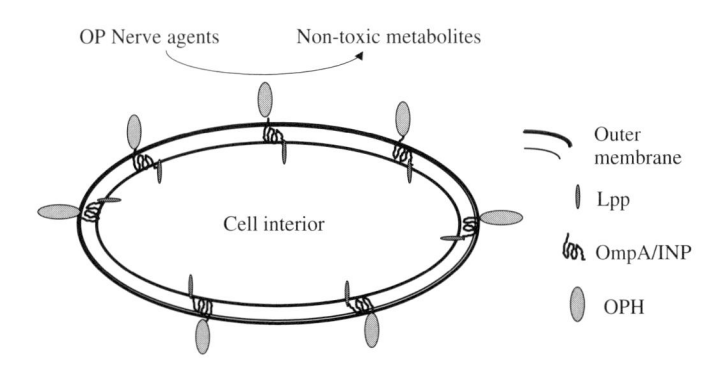

Figure 27.1 Cell surface expressed OPH.

supports, cells were also immobilized onto Siran™ porous glass beads. The reactor degraded 100% of 0.2 mM coumaphos at the flow rate of 23 ml/h within 2 hours (Mansee *et al.*, 2000).

Immobilization of CBD as a fusion with OPH has been taken a step further with the use of surface expressed CBD using the Lpp-OmpA anchor to immobilize whole cells onto cellulose materials (Wang *et al.*, 2002). With surface expressed CBD on the surface of bacteria, loss of whole cell bioreactor activity from gradual cell detachment was overcome. The surface anchored CBD immobilized 250 mg/cm^2 (dry weight) of cells forming a monolayer when observed under an electron microscope. Combined with surface expressed OPH, the coexpression system was capable of degrading paraoxon very rapidly with the initial rate of 0.65 mM/min/g (dry weight) of cell. The bioreactor also retained nearly 100% efficiency over the period of 45 days (Wang *et al.*, 2002).

27.5 MODIFICATIONS OF SPECIFICITY AND ACTIVITY

27.5.1 Site-directed mutagenesis

OPH possesses the ability to catalyze the hydrolysis of a wide range of OP compounds. Unfortunately, the catalytic efficiencies towards some substrates do not occur at an industrial or economically feasible time scale. Widely used pesticides such as methyl parathion, chlorpyrifos and diazinon are hydrolyzed 30 to 1000 times slower than the preferred substrate paraoxon, and OP compounds used as chemical warfare agents degrade at an even slower rate (Dumas *et al.*, 1989). Although OP degradation remains an attractive technology, catalytic efficiencies for a number of OP compounds leave room for improvement.

Fortunately, the crystal structure of OPH has been elucidated (Benning *et al.*, 1995). OPH is composed of a distorted α/β barrel with eight parallel β-pleated sheets making up the barrel structure linked to the outer surface by 14 α-helices (Benning *et al.*, 1995; 2000; Vanhooke *et al.*, 1996). The active site for this enzyme has also been described (Vanhooke *et al.*, 1996). The active site requires the presence of a divalent metal, with preference for Co^{+2}, and is complexed through three histidine residues (Dumas *et al.*, 1989; Omburo *et al.*, 1992). Extensive knowledge of structural detail allows a reasonable redesign of OPH, using site directed mutagenesis, thus improving efficiencies towards a number of OP compounds (Table 27.2).

Initial efforts focused on the substitution of amino acids directly responsible for binding the divalent metals. Site directed mutagenesis was used to substitute the original histidine residues at positions 254 and 257 (diSioudi *et al.*, 1999a; 1999b). Three of the variants (H254R, H254S and H257L) caused the enzyme to retain only one of the original two divalent metal ions at the active site. The mutants retained their ability to hydrolyze the preferred substrate paraoxon and simultaneously caused a four to fivefold increase in catalytic activity towards VX (*O*-ethyl-S-(2-diisopropyl aminoethyl) methylphosphonothioate) and insecticide

Table 27.2 Improved OPH activity through mutagenesis.

Amino acid substitution	Specificity towards	Method[e,f]/ type	Reference
Binuclear metal centre mutations			
H257L	VX[a], Demeton-S[b]	SD	diSioudi *et al.*, 1999a
H254R	VX[a], Demeton-S[b]	SD	diSioudi *et al.*, 1999a
H254R/H257L	Demeton-S[b], NPPMP[c]	SD	diSioudi *et al.*, 1999a
Active-site mutations			
F132H/F306H	DFP[d]	SD	Watkins *et al.*, 1997
F132H/F306Y	DFP[d]	SD	Watkins *et al.*, 1997
I274N/H257Y/K185R/A80V	Methyl parathion	DE	Cho *et al.*, 2002

[a](*O*-ethyl-S-(2-diisopropyl aminoethyl) methylphosphonothioate); [b](*O,O*-diethyl-*S*-(2-ethylthioethyl) phosphorothiolate); [c]soman analogue; [d]sarin analogue (diisopropyl fluorophosphate); [e]SD – site-directed, [f]DE – directed evolution.

analogue demeton-S. A double mutant of H254R/H257 had a 20-fold increase activity towards demeton-S over wild type (diSioudi *et al.*, 1999a; 1999b).

Another approach was the mutation of residues of OPH surrounding the leaving group. Here, improvement of OPH activity towards a relatively small substrate, DFP (diisopropyl fluorophosphate – a sarin analogue) was targeted. Residues lining the leaving group were substituted with hydrophobic amino acids (Trp131, Phe132, Leu271, Phe306, and Try309). Because DFP has a fluoride leaving group, replacement of one side chain with a residue capable of hydrogen bond formation and proton donation (His, Tyr, or Lys) was predicted to enhance catalysis. Values for k_{cat} of Phe132 and Phe306 and some double mutants showed an increase of ten-fold (Watkins *et al.*, 1997). When attempts at mutagenesis, using the crystal structure of OPH and the similarities it shares with acetylcholinesterase, were used in a rational design approach, a 33% increase in the relative VX hydrolysis rate compared to wild type enzyme in a L136Y mutant was achieved (Sriram *et al.*, 2000).

27.5.2 Directed evolution

DNA shuffling has received considerable attention in the past decade. The technique mimics natural homologous recombination through the fragmentation of a population of related genes followed by denaturation and hybridization. The recombination occurs when a fragment derived from one template primes a template with different sequences in a polymerase chain reaction (Stemmer 1994a; 1994b). DNA shuffling is a powerful technique that allows for the generation of mutants without prior knowledge of structure. The technique also allows for the generation of mutants distal to the active site.

Recently this technique was applied to the evolution of OPH for increased methyl parathion hydrolysis. After two rounds of DNA shuffling, a variant that could hydrolyze methyl parathion 25-fold faster than wild type was isolated. The mutations were not directly located in the active site and could not be otherwise predicted *a priori* (Cho *et al.*, 2002). This technique could be used to target other slow degrading pesticides such as chlorpyrifos and diazinon and against chemical warfare agents VX and sarin.

27.6 ORGANOPHOSPHORUS ACID ANHYDROLASE

Another enzyme capable of hydrolyzing the P—CN and P—F bonds of OP compounds is organophosphorus acid anhydrolase (OPAA: EC 3.1.8.2). Like OPH, OPAA can hydrolyze a variety of chemical warfare agents. Bacteria capable of producing OPAA were first isolated from a halophilic *Alteromonosas* strain JD6.5 (Cheng *et al.*, 1993) and subsequently identified in *Alteromonas haloplanktis* C (Cheng *et al.*, 1997). The genes coding for OPAA have been cloned from *A.* sp. Strain JD6.5 (Cheng *et al.*, 1996) and *A. haloplanktis* (Cheng *et al.*, 1997) and sequences were found to have over 80% sequence homology (Cheng *et al.*, 1997). The OPAA genes from both bacteria have also been cloned in the heterogeneous host *E. coli* (Cheng and Calomiris, 1996). The enzyme is a single polypeptide with a molecular weight of 60,000 that requires a divalent metal (DeFrank and Cheng, 1991). Through amino acid sequence analysis, it has been determined that the natural role of OPAA is as a prolidase and more specifically a X-Pro dipeptidase hydrolyzing dipeptides with a prolyl residue in the carboxyl-terminal position (Cheng *et al.*, 1996, 1997).

For increased production of OPAA at a practically useful scale, the *opaA* gene was cloned into *E. coli*. The recombinant OPAA was purified to homogeneity and had the same apparent molecular weight as the native enzyme. Small-scale batch operations gave yields of up to 200 mg/l of culture (Cheng *et al.*, 1996).

The recombinant enzyme showed remarkable stability when lyophilized with trehalose. Lyophilized OPAA was stored for 12 months then reconstituted in ammonium carbonate without showing any decrease in activity. The ability to retain 100% activity when stored under these conditions allowed for the reconstitution of the OPAA with any solvent and application to the contaminated surface with the advantages of ease of storage and long term stability. The ability to lyophilize the enzyme also allows for the enzyme to be reconstituted in buffer and used in conjunction with a foam based system for decontamination. Combined with the reconstitution in the innocuous ammonium carbonate buffer, such foam could be used to coat personnel, equipment, as well as applied to contain or clean up discharge (Cheng *et al.*, 1999).

OPAA activity in a variety of wetting foam agents has also been investigated. Two such agents, 'Cold Fire', a fire suppressing agent, and 'Odor Seal', an odor removing agent, increased enzyme activity. Both are based on biosurfactants derived from plants. The foam allowed for the encapsulation of the OP compound, while at the same time the biosurfactants' nature of the foam allowed for increase solubility of the substrate and thus greater enzyme accessibility. The use of these foams is also very attractive as a fully biodegradable enzyme decontamination system (Cheng *et al.*, 1999).

27.7 CONCLUSIONS

Despite the recent advances in biotechnology, biocatalyst based treatment of organophosphate compounds has yet to be implemented in a large-scale cell or enzyme based detoxification. Yet, inevitably the future of remediation will rely heavily on 'green/clean' technologies such as bioremediation, due, in large part, to the public's consciousness regarding environmental contamination and its desire for environmentally friendly cleanup strategies. In addition, as the technology matures, the cost of biocatalysis will decrease and be more economical compared to traditional approaches. For these reasons, in coming years the area of biocatalysis will receive increased attention.

ACKNOWLEDGEMENTS

This work was partially supported by grants from the National Science Foundation, the U.S. Environmental Protection Agency and the U.S. Department of Agriculture.

REFERENCES

Benning, M.M., Kuo, J.M., Raushel, F.M. and Holden, H.M. (1995) Three-dimensional structure of the binuclear metal center of phosphotriesterase. *Biochemistry* **34**, 7973–7978.

Benning, M.M., Hong, S.B., Raushel, F.M. and Holden, H.M. (2000) The binding of substrate analogs to phosphotriesterase. *J. Biol. Chem.* **275**, 30,556–30,560.

Caldwell, S.R. and Raushel, F.M. (1991) Detoxification of organophosphate pesticides using a nylon based immobilized phosphotriesterase from *Pseudomonas diminuta*. *Appl. Biochem. Biotechnol.* **31**, 59–74.

Casarett and Doull's (1996) *Toxicology: the basic science of poisons* (ed. Curtis D. Klaassen), (eds emeriti, Mary, O., Amdur, John Doull), 5th edn. New York: McGraw-Hill, Health Professions Division.

Cheng, T.C., Harvey, H.P. and Chen, GL. (1996) Cloning and expression of a gene encoding a bacterial enzyme for decontamination of organophosphorus nerve agents and nucleotide sequence of the enzyme. *Appl. Environ. Microbiol.* **62**, 1636–1641.

Cheng, T.C. and Calomiris, J.J. (1996) A cloned bacterial enzyme for nerve agent decontamination. *Enzyme Microb. Technol.* **18**, 597–601.

Cheng, T.C., Liu, L., Wang, B., DeFrank, J.J., Anderson, D.M., Rastogi, V.K. and Hamilton, A.B. (1997) Nucleotide sequence of a gene encoding an organophosphorus nerve agent degrading enzyme from *Alteromonas haloplanktis*. *J. Ind. Microbiol. Biot.* **18**, 49–55.

Cheng, T.C., Steven, H.O. and Stroup, A.N. (1993) Purification and properties of a highly active organophosphorus acid anhydolase from *Alteromonas undina*. *Appl. Environ. Microb.* **59**, 3138–3140.

Cheng, T.C., DeFrank, J.J. and Rastogi, VK. (1999) *Alteromonas* prolidase for organophosphorus G-agent decontamination. *Chem.-Biol. Interact.* 119–120, 455–462.

Cho, C.M.H., Mulchandani, A. and Chen, W. (2002) Bacterial cell surface display of organophosphorus hydrolase for selective screening if improved hydrolysis of organophosphate nerve agents. *Appl. Environ. Microb.* **68**, 2026–2030.

Dave, K.I., Miller, C.E. and Wild, J.R. (1993) Characterization of organophosphorus hydrolases and the genetic manipulation of the phosphotriesterase from *Pseudomonas diminuta*. *Chem.-Biol. Interact.* **87**, 55–68.
DeFrank, J.J., Cheng, T.C. (1991) Purification and properties of an organophosphorus acid anhydrase from a halophilic bacterial isolate. *J. Bacteriol.* **173**, 1938–1943.
diSioudi, B., Grimsley, J.K., Lai, K., Wild, J.R. (1999a) Modification of near active site residues in organophosphorus hydrolase reduces metal stoichiometry and alters substrate specificity. *Biochemistry* **38**, 2866–2872.
diSioudi, B., Miller, C.E., Lai, K., Grimsley, J.K. and Wild J.R. (1999b) Rational design of organophosphorus hydrolase for altered substrate specificities. *Chem.-Biol. Interact.* 119–120, 211–223.
Dumas, D.P., Caldwell, S.R., Wild, J.R. and Raushel, F.M. (1989) Purification and properties of the phosphotriesterase from *Pseudomonas diminuta*. *J. Biol. Chem.* **261**, 19,659–19,665.
Elashvili, I., Defrank, J.J. and Culotta, V.C. (1998) phnE and glpT genes enhance utilization of organophosphates in *Escherichia coli* K-12. *Appl. Environ. Microb.* **64**, 2601–2608.
Gill, I. and Ballesteros, A. (2000) Degradation of organophosphorus nerve agens by enzyme-polymer nanocomposites: efficient biocatalytic materials for personal protection and large-scale detoxification. *Biotechnol. Bioeng.* **70**, 400–410.
Grimsley, J.K., Scholtz, J.M., Pace, C.N. and Wild, J.R. (1997) Organophosphorus hydrolase is a remarkably stable enzyme that unfolds through a homodimeric intermediate. *Biochemistry* **36**, 14,366–14,374.
Havens, P.L., Rase, H.F. (1993) Reusable immobilized enzyme/polyurethane sponge for removal and detoxification of localized organophosphate pesticide spills. *Ind. Eng. Chem. Res.* **32**, 2254–2258.
Hayatsu, M., Hirano, M. and Tokuda, S. (2000) Involvement of two plasmids in fenitrothion degradation by *Burkolderia* sp. Strain NF100. *Appl. Environ. Microb.* **66**, 1737–1740.
Hoskin, F.C.G., Walker, J.E. and Stote, R. (1999) Degradation of nerve gases by CLECS and cells: kinetics of heterogenous systems. *Chem.-Biol. Interact.* **120**, 439–444.
Hung, S.C. and Liao, J.C. (1996) Effects of ultraviolet light irradiation in biotreatment of organophosphates. *Appl. Biochem. Biotechnol.* **56**, 37–47.
Karns, J.S., Muldoon, M.T., Derbyshire, M.K. and Kearney, P.C. (1987) Use of microorganisms and microbial systems in the degradation of pesticides. In *Biotechnology in Agricultural Chemistry* (eds LeBaron, H.M., Mumma, R.O., Honeycutt, RC and Duesing, J.S.), ACS Symposium Series. No. 334. Washington, DC: American Chemical Society, pp. 156–170.
Kolakowski, J.E., DeFrank, J.J., Harvey, S.P., Szafraniec, L.L., Beaudry, W.T., Lai, K.H. and Wild, J.R. (1997) Enzymatic hydrolysis of the chemical warfare agent VX and its neurotoxic analogues by organophosphorus hydrolase. *Biocatal. Biotransfor.* **15**, 297–312.
Lai, K.H., Dave, K.I. and Wild, J.R. (1994) Bimetallic binding motifs in organophosphorus hydrolase are important for catalysis and structure organization. *J. Biol. Chem.* **269**, 16,579–16,584.
LeJeune, K.E., Mesiano, A.J., Bower, S.B., Grimsley, J.K., Wild, J.R. and Russell, A.J. (1997) Dramatic stabilized phosphotriesterase-polymers for nerve agent degradation. *Biotechnol. Bioeng.* **54**, 105–114.
LeJeune, K.E. and Russell, A.J. (1996) Covalent binding of a nerve agent hydrolyzing enzyme within polystyrene foams. *Biotechnol. Bioeng.* **51**, 450–457.
Leung, K.T., Tresse, O., Errampalli, D., Lee, H. and Trevors, J.T. (1997) Mineralization of *p*-nitrophenol by pentachlorophenol-degrading *Sphingomonas* spp. *FEMS Microbiol. Lett.* **155**, 107–114.
Lichtenstein, E.P. and Schultz, K.R. (1964) The effects of moisture and microorganisms on the persistence and metabolism of some organophosphate insecticides in soils, with special emphasis on parathion. *J. Econ. Entomol.* **57**, 618.
Mansee, A., Chen, W. and Mulchandani, A. (2000) Biodetoxification of coumaphos insecticide using immobilized Escherichia coli expressing organophosphorus hydrolase enzyme on cell surface. *Biotechnol. Bioproc. Eng.* **5**, 436–440.

Mason, J.R., Briganti, F. and Wild, J.R. (1997) Protein engineering for improved biodegradation of recalcitrant pollutants. In *Perspectives in Bioremediation: Technologies for Environmental Improvement*, Proceedings of the NATO Advanced Research Workshop on Biotechnical Remediation of Contaminated Sites, Lviv, Ukraine, March 5–9, 1996 (NATO ASI Partnership Sub-series) (eds. Wild, J.R., Varfolomeyev, S.D. and Scozzafava, A.), pp. 107–118.

Mulbry, W.W., Kearney, P.C., Nelson, J.O. and Karns, J.S. (1987) Physical comparison of parathion hydrolase plasmids from *Pseudomonas diminuta* and *Flavobacterium sp*. *Plasmid* **8**, 173–177.

Mulbry, W.W. and Karns, J.S. (1989) Parathion hydrolase specified by the *Flavobacterium* opd gene: Relationship between the gene and protein. *J. Bacteriol.* **171**, 6740–6746.

Mulbry, W.W., Del Valle, P.L. and Karns, J.S. (1996) Biodegradation of the organophosphate insecticide coumaphos in highly contaminated soils and in liquid wastes. *Pesticide Sci.* **48**, 149–155.

Mulbry, W.W., Ahrens, E. and Karns, J. (1998) Use of a field-scale biofilter for the degradation of the organophosphate insecticide coumaphos in cattle dip wastes. *Pesticide Sci.* **52**, 268–274.

Mulchandani, A., Kaneva, I. and Chen, W. (1999) Detoxification of organophosphate nerve agents by immobilized *Escherichia coli* with surface-expressed organophosphorus hydrolase. *Biotechnol. Bioeng.* **63**, 216–223.

Munnecke, D.M. (1979) Hydrolysis of organophosphate insecticides by an immobilized-enzyme system. *Biotechnol. Bioeng.* **21**, 2247–2261.

Munnecke, D.M. and Hsieh, D.P.H. (1976) Pathways of microbial metabolism of parathion. *Appl. Environ. Microb.* **31**, 63–69.

Munnecke, D.M., Day, R. and Trask, H.W. (1976) *Review of Pesticide Disposal Research*. United States Environmental Protection Agency, Washington, DC.

National Research Council (1993) *Alternative Technologies for the Destruction of Chemical Agents and Munitions*. National Academy of Sciences, Washington, D.C.

Nishino, S.F. and Spain, J.C. (1993) Cell density-dependent adaptation of *Pseudomonas putida* to biodegradation of *p*-nitrophenol. *Environ. Sci. Technol.* **27**, 489–494.

Omburo, G.A., Kuo, J.M., Mullins, L.S. and Raushel, F.M. (1992) Characterization of the zinc binding site of bacterial phosphotriesterase. *J. Biol. Chem.* **267**, 13278–13283.

Phillips, J.P., Xin, J.H., Kirby, K., Milne, C.P., Krell, P. and Wild, J.R. (1990) Transfer and expression of an organophosphate insecticide-degrading gene from *Pseudomonas* in *Drosophila melanogaster*. *Procedings for the National Academy Sciences (USA)* **87**, 8155–8159.

Ramanathan, M.P. and Lalithakumari, D. (1996) Short Communication: Methylparathion degradation by *Pseudomonas* sp. A3 immobilized in sodium alginate beads. *World J. Microb. Biotechnol.* **12**, 107–108.

Rammell, C.R. and Bentley, G.R. (1988) Organophosphate residues in the wool of sheep dipped for flystrike control. *NZJ. Agr.* **31**, 151–154.

Rani, N.L. and Lalithakumari, D. (1994) Degradation of methyl parathion by *Pseudomonas putida*. *Can. J. Microbiol.* **40**, 1000–1006.

Richins, R., Mulchandani, A. and Chen, W. (2000) Expression, immobilization, and enzymatic characterization of cellulose-binding domain-organophosphorus hydrolase fusion proteins. *Biotechnol Bioeng.* **69**, 591–596.

Richins, R., Kaneva, I., Mulchandani, A. and Chen, W. (1997) Biodegradation of organophosphorus pesticides by surface-expressed organophosphorus hydrolase. *Nat. Biotechnol.* **15**, 984–987.

Serdar, C.M. and Gibson, D.T. (1985) Enzymatic hydrolysis of organophosphates cloning and expression of a parathion hydrolase gene from *Pseudomonas diminuta*. *Bio-Technology* **3**, 567–571.

Serdar, C.M., Murdock, D.C. and Rohde, M.F. (1989) Parathion hydrolase gene from *Pseudomonas diminuta* MG: subcloning, complete nucleotide sequence, and expression of the mature portion of the enzyme in *Escherichia coli*. *Bio-Technology* **7**, 1151–1155.

Shimazu, M., Mulchandani, A. and Chen, W. (2003) Thermally triggered purification and immobilization of elastin-OPH Fusion. *Biotechnol. Bioeng.* **81**, 74–79.

Shimazu, M., Mulchandani, A. and Chen, W. (2001a) Cell surface display of organophosphorus hydrolase using ice nucleation protein. *Biotechnol. Progr.* **17**, 76–80.

Shimazu, M., Mulchandani, A. and Chen, W. (2001b) Simultaneous degradation of organophosphorus pesticides and *p*-nitrophenol by a genetically engineered *Moraxella* sp. with surface-expressed organophosphorus hydrolase. *Biotechnol. Bioeng.* **76**, 318–324.

Siddaramappa, R., Rajaram, K.P. and Sethunatha, N. (1973) Degradation of parathion by bacteria isolated from flooded soil. *Appl. Environ.l Microb.* **26**, 846–849.

Spain, J.C., Wyss, O. and Gibson, D.T. (1979) Enzymatic oxidation of *p*-nitrophenol. *Biochem. Bioph. Res. Co.* **88**, 634–641.

Sriram, G., Rastogi, V., Ashman, W. and Mulbry, W. (2000) Mutagenesis of organophosphorus hydrolase to enhance hydrolysis of the nerve agent VX. *Biochem. Bioph. Res. Co.* **279**, 516–519.

St. Claire, N.L. and Navia, M.A. (1992) Cross-linked enzyme crystals as robust catalysts. *J. Am. Chem. Soc.* **114**, 7314–7316.

Steiert, J.G., Pogell, B.M., Speedie, M.K. and Laredo, J. (1989) A Gene coding for a membrane-bound hydrolase is expressed as a secreted soluble enzyme in *Streptomyces lividans*. *Bio-Technology* **7**, 65–68.

Stemmer, W.P.C. (1994a) Rapid evolution of a protein in vitro by DNA shuffling. *Nature* **370**, 389–391.

Stemmer, W.P.C. (1994b) DNA shuffling by random fragmentation and reassembly: In vitro recombination for molecular evolution. *P. Natl. Acad. Sci. USA* **91**, 10, 747–10,751.

Urry, D.W. (1992) Free energy transduction in polypeptides and proteins based on inverse temperature transitions. *Prog. Biophys. Mol. Bio.* **57**, 23–57.

Urry, D.W. (1997) Physical chemistry of biological free energy transductions as demonstrated by elastic protein-based polymers. *J. Phys. Chem.* **B101**, 11,007–11,028.

Vanhooke, J.L., Benning, M.M., Rausel, F.M. and Holden, H.M. (1996) Three-dimensional structure of the zinc-containing phosphotriesterase with the bound substrate analog diethyl 4-methylbenzylphosphonate. *Biochemistry* **35**, 6020–6025.

Wang, J., Bhattacharyya, D. and Bachas, L.G. (2001) Orientation specific immobilization of organophosphorus hydrolase on magnetic particles through gene fusion. *Biomacromolecules* **3**, 700–705.

Wang, A., Mulchandani, A. and Chen, W. (2002) Specific adhesion to cellulose and hydroysis of organophosphate nerve agents by a genetically engineered *E. coli* with surface-expressed cellulose-binding domain and organophosphorus hydrolase. *Appl. Environ. Microb.* **68**, 1684–1689.

Watkins, L.M., Mahoney, J.J., McCulloch, J.K. and Raushel, F.M. (1997) Augmented hydrolysis of diisopropyl fluorophosphates in engineered mutants of phosphotriesterase. *J. Biol. Chem.* **272**, 25,596–25,601.

Xu, B., Wild, J.R. and Kenerley, C.M. (1996) Enhanced expression of a bacterial gene for pesticide degradation in a common soil fungus. *J. Fermenat. Bioeng.* **81**, 473–481.

Zhongli, C., Shunpeng, L. and Guoping, F. (2001) Isolation of methyl-degrading strain M6 and cloning of the methyl parathion hydrolase gene. *Appl. Environ. Microb.* **67**, 4922–4925.

Index

abattoir waste 343
Acinetobacter 278, 280–3, 570–2
ACP *see* amorphous calcium phosphate
activated alumina 294, 298–9
adenosine diphosphate (ADP) 17
adenosine diphosphate phosphotransferase 281–2
adenosine monophosphate (AMP) 63, 64, 71
adenosine monophosphate phosphotransferase 280
adenosine triphosphate (ATP) 17
  biological phosphorus removal 280–1
  biosynthesis 62–8, 72–4
  breakdown 68–70
  energy transduction 68–70
  metabolic regulation 70–2
  transport 53, 55, 58
adenylate kinase 280
ADP *see* adenosine diphosphate
adsorption 151–3, 201, 202, 213–21
  see also ochre; solid phase adsorbents
adsorption–desorption reactions 98–9
AEP *see* 2-aminoethylphosphonic acid
aeration 417, 424–5, 471–2
aerobic digestion 273, 511–12, 525–6
affinity tags 633–4
AFM *see* atomic force microscopy
Africa 189–90
ageing of supernatant 440, 441
agglomeration 377–9
aggregation 196
Agricultural Pollution Control Programme 179
agriculture
  diffuse phosphate pollution 179–81

ecological phosphorus use 610–28
recovery 342–3
runoff 332–3
agronomy 614–22
air lift 393–4
air mixing 392–4
air stripping 440–2, 443–55, 459, 466
Alcaligenes eutrophus 552–3
algae 254
  see also eutrophication
allograft bone 583
allophanes 294, 295–6, 306
Alteromonosas 638
aluminium oxides 297–8, 306
aluminium salts
  chemical phosphorus removal 261, 266–7
  phosphorus recovery 526
  sludge production 433–4
  struvite recovery 500–1
alveolar ridge augmentation 588–9
2-aminoethylphosphonic acid (2 AEP) 61
aminophosphines 15–16
aminopolycarboxylates 147, 148
aminopolyphosphonates 148, 149
amorphous calcium phosphate (ACP) 210
AMP *see* adenosine monophosphate
anaerobic digestion
  biological phosphorus removal 273
  Geestmerambacht case study 480–1, 482–3
  phosphorus recovery 525–7
  sludge treatment/disposal 257–8
  unprocessed manure 512
anchor impellers 386
anion exchange resins 107

643

Index

anionic polymers 237
Antarctica 190
anthocyanins 103
apatites 9
  see also calcium phosphates
  bioapatites 39–41
  ceramic implants 229
  dissolution 207
  geochemistry 23–4
  mineralogy 33–7
  precipitation 196
  structure 34
  transport in animals 55–6
apparent secondary nucleation 375
aragonite 304, 593
Asia 174, 186–9
Aspergillus 551
atomic clusters 212
atomic force microscopy (AFM) 235, 239, 240
atomic properties 4
ATP *see* adenosine triphosphate
ATPases 69–70
attrition 374–5
Australia 137, 174, 185–6
axial flow impellers 384, 385–6

BA *see* biological apatite
Bacillus 57, 551, 564, 568
background release 79, 80, 81
bacteria *see* biomineralisation; *individual genera*; microorganisms
base metal phosphates 41–2
basic oxygen furnace (BOF) slag 302
batch precipitation 368–70
batch stirred vessels 360–1
bauxite residue 292–3, 299–300
BCF *see* Burton, Cabrera and Frank
bed expansion 391
bench scale work 411–15
BET *see* Brunauer, Emmet and Teller
BFS *see* blast furnace slag
BINAP 12–13
bioaccumulation 21
bioactive glass 595, 599
bioapatites 39–41
bioavailability
  cadmium 115
  chelating agents 147
  crop nutrition 94, 98
  elevated release 85
  eutrophication 122
  geochemistry 24–5, 26
  pH 24–5
  phosphates 195–6
biodegradation
  biomaterials 600–1

organophosphate nerve agents 629–41
phosphonates 59
phosphonic acids 157–8
biodiversity 121
biogas 403
biogenic phosphates 207
biological apatite (BA) 584
biological nutrient removal (BNR)
  bench scale work 411–15
  mass balances 407–11
  performance improvement 437–9
  production rates 422
  reactor specifications 415–16
  Slough STW case study 402–27
  struvite 402–27
  Treviso case study 429–42, 455–7, 461
biological oxygen demand (BOD) 189, 249
biological phosphorus pumps 516–18
biological phosphorus removal 272–90
  advances 283–5
  enzymes, polyphosphate-metabolising 277–82
  industrial recycling 522–4
  microorganisms 277–85
  Phostrip process 274–5
  polyphosphates 273, 276–83
  process design 274–6
  recovery 346
  University of Capetown process 275–6
biological pre-concentration 476
bioluminescence 7
biomass 253
biomembranes 16–17
biomineralisation 21, 32, 549–81
biophosphates 37–40
biotechnologies
  bacterial precipitation 549–81
  calcium phosphates 582–609
  ecological uses 610–28
  organophosphates 629–41
'birth and spread' model 203
1,3-bisphosphoglycerate 62–3
bisphosphonates 149
black phosphorus 6
blast furnace slag (BFS) 292–3, 294, 302–3
BOD *see* biological oxygen demand
BOF *see* basic oxygen furnace
bonding 4–5
bone
  apatite 39–41
  augmentation 586–7
  biotechnologies 583–9
  heterogeneous nucleation-epitaxial growth 229–30
  integration of graft materials 585, 598–600, 601
  interactions with biomaterials 598–600

mechanical properties 584–5
transport 55–6
Boophilus annulatus 630
bovine spongiform encephalopathy (BSE) 343, 353
Brandt, Hennig 3
British Code of Practice for Small Sewage Works 256
Brunauer, Emmet and Teller (BET) surface area 207
BSE *see* bovine spongiform encephalopathy
bubble aeration 472
buffering capacity 86
buffer zones 140
bulk precipitation 208–12
Burkholderia 285, 631
Burton, Cabrera and Frank (BCF) theory 202–3

cadmium 113–15, 116
Cajanus cajan 618
calcite 231–2, 304
calcitonin 56
calcium fluoride 229
calcium phosphates 33–7
  see also apatites
  adsorption mechanism 217–21
  biotechnologies 582–609
  bulk precipitation 208–12
  competition/co-precipitation with carbonates 231–3
  crystal growth 211–12, 213–17
  dissolution 234–8
  Geestmerambacht case study 475–6, 489–90
  heterogeneous nucleation-epitaxial growth 221–31
  inhibition 213–17
  nucleation 212
  precipitation 196, 208–33
cAMP *see* cyclic AMP
Canada 121, 263–4
Candida 277, 285
capillary suction time (CST) 268
carbonate fluorapatite 29, 36
carbonate hydroxylapatite 36
carbonates
  competition/co-precipitation 231–3
  Geestmerambacht case study 495
  solid phase adsorbents 294, 304–5
  stripping cascade 484–5
carbon cycle 21
cell adhesion 601
cellular interactions 597
cements 596
ceramics 229, 590–4, 595, 598–9
cGMP *see* cyclic GMP
change point 83
chelating agents 147–9

chemical degradation 158–60
chemical oxygen demand (COD) 430–1, 438–9, 513–14
chemical phosphorus removal 260–71, 346
chemical potential 199
chemical warfare 3
Chemical Weapons Convention (CWC) 630
chemiluminescence 7
China 188
chiral phosphines 12–14
chlorapatite 33–4, 35–7, 234, 236–7
Citrobacter 553
clay minerals 25–6
Clean Water Act 533, 534
clinker 305
Clostridium cellulovorans 633
coal-fired power stations 300
coal mines 321
coating biomaterials 594–5, 599–600, 602
coherent interfaces 224
collagen type I 230
colloids 132–4
Comamonas 566–7
competitive precipitation 231–3
composite biomaterials 595–6
composite samples 407
constructed wetland systems (CWS)
  ochre 321, 323, 332
  operational lifetimes 332
  solid phase adsorbents 292–3, 301–2, 305, 306
continuous precipitation 370–2
continuous reactors 362–3
Convention on Biological Diversity 539
co-precipitation 231–3, 304–5, 447–8, 562–4
coral biomaterials 592–4
coral reefs 24, 189
corrosion 268
cosmic abundance 22
CPK *see* creatine phosphokinase
craniofacial surgery 589
creatine phosphokinase (CPK) 71
critical levels 105–6
critical source areas (CSAs) 125–6, 129, 131, 140
crop cover 128
crop nutrition 93–119
  critical levels 105–6
  deficiency symptoms 103
  fertilisers
    development 96–8
    environmental issues 113–16
    practices 110–13
    recommendations 106–10
  resources 95–6
  soil P tests 106–10
  soils 98–102

Crystalactor® 476–7, 479–80, 484–90, 493, 526–8
crystal growth
  fluid dynamics 375–7
  models 450
  precipitation 196, 199, 200–4, 211–12, 213–17
  recovery 346–7
CSAs *see* critical source areas
CST *see* capillary suction time
CWC *see* Chemical Weapons Convention
CWS *see* constructed wetland systems
cyclic AMP (cAMP) 71
cyclic GMP (cGMP) 72
cyclic phosphorylation 67
cytokines 601–2
cytoplasmic alkalisation 53
cytosolic phosphate 54, 55

DAP *see* diammonium phosphate
DCPA *see* dicalcium phosphate anhydrous
DCPD *see* dicalcium phosphate dihydrate
deep coal mines 321
deficiency 103–5, 611, 619
degradation 158–60
  see also biodegradation
degree of soil P saturation (DPS) 132
denitrification
  Geestmerambacht case study 491–3
  sequencing batch reactors 514
  Treviso case study 429, 433
  waste water treatment plants 254, 257
Denmark 179
dental applications 229–30, 237, 588–9
dentine 39, 56
deoxyribonucleic acid (DNA) 17, 73–4, 637–8
desolvation 213
desorption 325, 330
detachment 85–6
detergents 10, 353, 530–1
dewatering 258–9
diagenetic phosphate 29
dialkyl monofluorophosphate esters 16
1,2-diaminoethanetetrakis(methylene phosphonic acid) (EDTMP) 148, 149, 152, 158, 164–5
diammonium phosphate (DAP) 96, 111, 114
diazinon 630
dicalcium phosphate anhydrous (DCPA) 234
dicalcium phosphate dihydrate (DCPD) 209, 210, 222, 227–9, 234
diethylenetriaminepentakis(methylene phosphonic acid) (DTPMP) 148, 149, 151, 162–4
diffuse sources 122–3, 179–82, 186
diffusion-reaction 202

digestion *see* aerobic digestion; anaerobic digestion
Diminishing Returns, Law of 108
2,4-dinitrophenol 272
DIOP 12–13
diphosphine 11
2,3-diphosphoglycerate (2,3-DPG) 56
direct flow measurement 407
dissolution
  applications 234–40
  aqueous transport 30–2
  biomaterials 600–1
  calcium phosphates 234–8
  equations 205–8
  inhibition 156
  mechanisms 204–5
  metal phosphates 238–9
  minerals 153–4
  phosphonic acids 155–6
dissolution pits 201, 214–15, 235
DNA *see* deoxyribonucleic acid
dolomite 305
double jet semi-batch precipitation 370–2
double saturation model 450–1
downstream impacts 267–9
2,3-DPG *see* 2,3-diphosphoglycerate
DPS *see* degree of soil P saturation
drainage 86, 128
drinking water 147, 308
DTPA 147
DTPMP *see* diethylenetriaminepentakis(methylene phosphonic acid)
Dunaliella salina 284

ear surgery 589
East Asia 188–9
EBPR *see* enhanced biological phosphorus removal
ecological phosphorus use 610–28
ectomycorrhizae 237
EDTA 147, 148, 152–4, 158, 161
EDTMP *see* 1,2-diaminoethanetetrakis(methylene phosphonic acid)
EDX *see* energy dispersive X-ray analysis
effluent recycling 425
elastin 230–1
electric arc furnace (EAF) steel slag 292–3, 294, 302
electric furnace process 349–50
electrodialysis 261
electronic structure 4–5, 20
elemental phosphorus 6–7
elevated release
  detachment 85–6
  mobilisation 82–6

sources 82
  terrestrial 79–92
  transport 86–7
emission limits 188
enamel 39
endopolyphosphatases 279
energy dispersive X-ray analysis (EDX) 102
energy transduction 68–70
enhanced biological phosphorus removal
    (EBPR) 273–85, 514–16, 536–7, 569–73
Environment Action Programmes (EAPs) 341
Environment Agency (England and Wales) 79,
    122, 249
Environment Agency (Japan) 187
Environmental Protection Agency (Sweden) 354
Environmental Protection Agency (USEPA)
    184–5, 263–4, 533, 534
enzymes 277–82, 632–4
Ephydatia muelleri 285
epitaxial growth 221–31
EPS *see* extracellular polymeric substances
erosion 139–40, 175
Escherichia coli 280, 284
  bacterial precipitation 554, 568
  organophosphates 632, 635
etch pits 201, 214–15, 235
Europe
  chemical phosphorus removal 260
  fertilisation practices 111–12, 114
  phosphates 174, 176–83
  recovery 341–3, 349
  recycling 529–33, 538
  regulations 532–3, 538
  terrestrial release 79–80
  waste water treatment plants 249
European Commission 178
eutrophication
  definition 120–1
  detergents 10, 530–1
  geochemistry 26
  husbanding soil 116
  hydrology 127–30, 134–8
  Japan 496, 497
  management minimisation 138–40
  sources and pathways 122–31
  terrestrial release 79–90
  transfer 120–46
exchange reactions 151–6
exopolyphosphatases 279
expanded clay aggregates 292, 294, 301–2
extended aeration processes 471–2
external carbon sources 438–9
extracellular polymeric substances (EPS) 558–9
extraction 106–7

F_1F_0-ATPases 69–70
farmyard manure (FYM) 111, 509

FBRs *see* fluidised bed reactors
Federal Environment Agency 354
Federal Water Pollution Control Act 184
feed concentrates 81
feed supplements 352–3
ferrihydrite 297–8, 324
Fertiliser Manufacturers' Association 111
fertilisers
  aquatic environment 113
  Asia 188
  cadmium 113–15, 116
  consumption 97
  development 96–8
  economic analysis 535
  elevated release 81
  optimum soil-phosphorus levels 612
  phosphates 176
  phosphorus placement 614–15
  practices 110–13
  radioactivity 115
  recommendations 106–10
  recycled phosphorus 351–2
  slow-release 140
  struvite 425–6, 499–500, 506
Ferti Mieux 179
fibrous encapsulation 598
FIDMP *see* formyl-
    iminodimethylenephosphonic acid
FISH *see* fluorescence *in situ* hybridisation
fixed film *see* trickling filters
FLAG fusion proteins 633–4
Flavobacterium 631–2
flocculation 156, 261
fluid dynamics
  agglomeration 377–9
  air mixing 392–4
  crystal growth 375–7
  fragmentation 379
  mixing 359–73
  modelling 394–7
  precipitation 368–72
  primary nucleation 366–73
  recovery 358–401
  secondary nucleation 373–5
  stirred vessels 379–88
fluidised bed reactors (FBRs)
  air injection 394
  air stripping 443–55, 466
  Crystalactor® 486–90
  design guidelines 454–5
  economic balances 453
  fluid dynamics 372–3, 388–92, 445
  mass balances 453
  mathematical models 448–52
  performance 446–8
  supersaturation curves 442–3
  Treviso case study 429

fluorapatite (FAP) 33–4, 35–7, 236–7
fluorescence *in situ* hybridisation (FISH) 282–3
fluorine 40
fly ash 292–3, 294, 300–1
foreign substrates *see* impurities; seed crystals
formyl-iminodimethylenephosphonic acid (FIDMP) 150–1, 159–60
fracture repair 586–7
fragmentation 379
France 179, 533
Freundlich equation 326–8
FRIOS® ALGIPORE® 592–4
Fukuoka City West Waste Water Treatment Centre 497–500
fulvic acids 26
fungi 254
  see also mycorrhizae
fungicides 14
FYM *see* farmyard manure

Geestmerambacht case study 470–95
gene therapy 602
geochemistry 20–32
Germany 179–80, 181–3, 354
glass ceramics 595, 599
Global Positioning Satellites (GPS) 112
glucokinase 64
glucose 6-phosphate 63
glucose 6-phosphate/phosphate translocator (GPT) 54
glucose phosphotransferase 280–1
glutaraldehyde 230
glycerol 3-phosphate 63
glycerolipids 73
glycolysis 63–5
glyphosate 160
goethite 297–8, 324
Goniapora 593–4
GPS *see* Global Positioning Satellites
GPT *see* glucose 6-phosphate/phosphate translocator
grab samples 408, 411
graft materials 585, 598–600, 601
granular magnesia clinker 305
granulation 331
gravels 294, 305–6
grit removal 251
groundwaters 86, 129, 147
growth factors 601–2
guanosine triphosphate (GTP) 62
gypsum 299–300

Habitats Directive 538
halides 10–11
HAP *see* hydroxylapatite
'haze' effect 266
heavy metals 550, 551–3, 562–4

HEDP *see* 1-hydroxyethane(1,1-diylbis phosphonic acid)
Helicobacter pylori 59
heterogeneous nucleation-epitaxial growth 221–31
high-affinity phosphorus transporters 53
Holland 180–1, 470–95
homeostasis 56, 57
Hortonian flow 136
humic acids 26
HUP *see* hydrogen uranyl phosphates
husbandry, soil 116
HYDRAQL model 215–17, 218–20
hydraulic retention time 425
hydrides 11
hydrodynamics *see* fluid dynamics
hydrogen uranyl phosphates (HUP) 564–6
hydrology 126, 127–30, 134–8
hydrolysis 158–9
hydrotalcites 293–5, 300
1-hydroxyethane(1,1-diylbis phosphonic acid) (HEDP) 148, 149, 157, 161–5
hydroxylapatite (HAP) 33–4, 35–7
  biotechnologies 583–4, 587–96, 602–3
  co-precipitation 231–3
  dissolution 234, 236–7
  heterogeneous nucleation-epitaxial growth 222, 227–31
  precipitation 209, 210–11, 215–21
  solid phase adsorbents 293
  Treviso case study 436–7, 440–3, 446
hypertrophic conditions 121, 122
hypocalcemia 52
hypophosphatemia 52
hysteresis, ochre 325

iminodimethylenephosphonic acid (IDMP) 150–1, 159
immobilised organophosphate hydrolase 632–4
immunogold labelling 556–7
imogolites 294, 295–6
IMPHOS *see* World Phosphate Institute
impounded rivers 308, 309
improved aerobic digestion 512
impurities 196, 213–17, 221–31
incidental transfers 86, 135
incineration 345, 347, 524, 527, 532, 630
incoherent interfaces 224
India 188–9
induction times 200, 226–7, 230–3
industrial byproducts 292–3
industrial recycling 339, 521–8
infiltration-excess overland flow 136
inhibition 156, 213–17, 237
inorganic phosphate 51, 131–2, 331, 510, 569–73

Index

inorganic polyphosphate 60–1
inositol hexaphosphate 84, 567–8
inositol triphosphate 72
integration of graft materials 585, 598–600, 601
internal carbon sources 438
International Commission for the Protection of the Rhine 181
ion exchange 564–6
ionic phosphides 7–8
IPPC Directive 342
Ireland 84
iron oxides 294, 297–8, 306
iron phosphonates 158
iron salts 261–2, 263–7, 501–2
isotopes 4
Italy 178, 428–69

Japan 186–8, 496–506, 533

kinks 201, 214–15
Kossel model 213

lakes *see* surface waters
lake sediments 30
lamella clarifiers 415–16
Langmuir isotherm 215–17, 326–8
lanthanum 303–4, 561–2, 572
lanthanum carbonates 305
Law of Diminishing Returns 108
Law of the Minimum 93–4
'leached layers' 205
lime 261–2, 489, 526
limestone 304
limiting criteria 23
long-term immobilisation 311
long-term perfomance 463–5
long-term removal, ochre 329–30
Lupinus albus 618
'luxury uptake' 272, 479, 480
lysimetry 134

macromixing 359, 363–5, 368–9
macropores 138
magnesium ammonium phosphate (MAP) *see* struvite
magnesium chloride dosing 414–15, 418, 424, 498
magnesium phosphate 495
major facilitator superfamily (MFS) 53
malathion 16
manures 111, 112–13, 180, 507–20, 615–17
MAP *see* monoammonium phosphate; struvite
marine environment 31
marine red algae 592–4

marine sediments 24, 28–9
mass balances
  Geestmerambacht case study 477, 478–82
  Slough STW case study 407–11
  struvite 407–11
  Treviso case study 432–5, 439–40, 453
matrix flow 135
MBBR *see* moving bed biofilm reactors
membrane bioreactors (MBR) 255
mesomixing 365–6
mesophilic anaerobic digestion 258
metabolism 51, 70–2
metal phosphates
  bacterial precipitation 549–81
  dissolution 238–9
  ion exchange 564–6
  mineralogy 41–2
  precipitation 238–9
metal-rich phosphides 7–8
metals
  bacterial precipitation 550, 551–3
  chelating agents 147–9
  co-precipitation 562–4
  phosphines 5
  prostheses 587–8
  remobilisation 154–5
  struvite 419–21
  transuranic elements 561–2
metal salts 261–9
metamorphic rocks 32
metaphosphoric acid 9
methane gas 253, 258
Methanobacterium thermoautotrophicum 279
metolachlor 12–13
MFS *see* major facilitator superfamily
micromixing 359, 360–3, 368–9
microorganisms
  see also biodegradation; biological phosphorus removal; mineralisation
  apatite dissolution 237
  bacterial precipitation 549–81
  photosynthesis 61
  secondary treatment 252–6
  struvite 421
  transport 57–61
  unprocessed manure 510
  waste water treatment plants 253–4
migration 201
mineral dissolution 153–4
mineralisation 27, 32
  heterogeneous nucleation-epitaxial growth 229–30
  microorganisms 57
  soils 84–5, 98

mineralogy 32–44
  base metal phosphates 41–2
  biophosphates 37–40
  calcium phosphates 33–7
  monazites 32, 43–4
  solid phase adsorbents 292
  struvite 43
  vivianite 42–3
  xenotime 32, 43–4
mine water treatment plants (MWTPs) 322–5
minimum fluidising velocity 389–90
mixing 359–75
mobilisation 82–6
monazites 32, 43–4
monoammonium phosphate (MAP) 96, 111, 114
mononuclear mechanism 203
monophosphides 7–8
monosaccharides 72–3
Moraxella 635
moving bed biofilm reactors (MBBR) 255
Mozambique 189
municipal waste water treatment plants (MWWTPs) 470–95, 521–8
muscle 56
mutagenesis 636–7
MWTPs *see* mine water treatment plants
MWWTPs *see* municipal waste water treatment plants
Mycobacterium 280
mycorrhizae 27, 58, 105, 237, 621–2
Myxococcus xanthus 280

NAD/NADH/NADP/NADPH *see* nicotinamide adenine dinucleotides
nanoparticles 207
National Eutrophication Survey 184
naturally occurring radioactive materials (NORM) 115
NDBEPR *see* nitrification–denitrification biological enhanced phosphate removal
nerve agents 16–17, 629–41
Netherlands, The 180–1, 470–95
neutralised red mud 299–300
New Zealand 80, 115
nicotinamide adenine dinucleotides (NAD/NADH/NADP/NADPH) 64–5
Nitrates Directive 344
nitrification
  Geestmerambacht case study 491–3
  sequencing batch reactors 514
  Treviso case study 429
  waste water treatment plants 254, 256, 257
nitrification–denitrification biological enhanced phosphate removal (NDBEPR) 284

nitrilotris(methylene phosphonic acid) (NTMP) 148, 149, 157, 159–62
'nitrogen break point' 515
nitrogen fixation 122
nitrogen–phosphorus compounds 15–16
non-point sources 122–3, 179–82, 186
non-stoichiometric phases 196, 205
NORM *see* naturally occurring radioactive materials
NTA 147, 148
NTMP *see* nitrilotris(methylene phosphonic acid)
nucleation 196, 200–4, 212, 446–7
nucleation-epitaxial growth 221–31
nucleic acids 73–4
nucleotide phosphates 55
nutrition *see* crop nutrition

oceanic *see* marine
ochre 321–35
  agricultural runoff 332–3
  batch experiments 326–8
  desorption 325, 330
  engineering issues 330–1
  formation 322–3
  kinetics 329
  long-term removal 329–30
  mechanisms 325
  pH effects 328
  properties 323–5
  recycling 330, 333
  sewage effluent 331–2
octacalcium phosphate (OCP) 230, 234
OFMSW *see* organic fraction of municipal solid waste
Olsen's reagent 99–101, 104, 108–10
'one-step' removal 285
OPAA *see* organophosphorus acid anhydrolase
OPHY *see* organophosphate hydrolase
optimum soil-phosphorus levels 610, 611
organic acids 237, 619–20
organic content 421
organic farming 343
organic fraction of municipal solid waste (OFMSW) 438–9, 455–7, 460–3
organic manures 111, 615–17
organic phosphorus 124, 132, 262, 510
organophosphate hydrolase (OPH) 631–7
organophosphate nerve agents 16–17, 629–41
organophosphorus acid anhydrolase (OPAA) 638
orthopaedic applications 586–8
orthophosphates 52–3, 98, 104, 262, 265
orthophosphoric acid 8, 9
OSPAR Convention 539
osteoclasts 237–8
osteonectin 56

Ostwald's rule of stages 210
overgrowth 223–5
oxidative degradation 160
oxidative phosphorylation 65–6
oxides of phosphorus 8
oxo-acids 8–9
oxyhydroxides 24–6

PACl *see* poly-aluminium chloride
PAOs *see* polyphosphate accumulating organisms
Paracoccus denitrificans 284
parathion 631
parathyroid hormone (PTH) 56
partially calcined dolomite 305
particle segregation 391–2
particulates 30–2, 132–4
pathologic minerals 38
P-ATPases 69–70
PBTC *see* phosphonobutane-tricarboxylic acid
P-efflux 53
pentahalides 10–11
PEP *see* phosphoenolpyruvate
peptide bonds 62
percolating filters *see* trickling filters
percolation 130, 136
periodontal disease 588
pesticides 629–41
Pfiesteria piscicida 80, 121
PFK I *see* phosphofructokinase I
pH
  adsorption of calcium phosphates 217–19
  bioavailability 24–5
  dissolution rate 204–5, 236–7
  ochre 328
  soils 99
  solid phase adsorbents 301, 302, 309–10
  speciation 31
  struvite 417–18, 422–3
  transfer 133
  transport in plants 53
PHB *see* polyhydroxybutyrate
phosphabenzene 16
phosphatases 27, 55, 58
phosphates 9–10
  see also calcium phosphate
  acquisition 51–2
  Africa 189–90
  agricultural diffuse pollution 179–81
  Antarctica 190
  Asia 174, 186–9
  Australia 174, 185–6
  bacterial precipitation 549–81
  cements 596
  European Union 174, 176–83
  global pollution overview 174–91
  mineralogy 21

  solid phase adsorbents 296–9
  United States 174, 183–5
phosphatidylcholine 17
phosphatidylinositol (PI) 72
phosphazenes 15
phosphides 7–8
phosphines 5, 11–14
phosphinic acid 9
phosphobacterin 57
phosphocreatine 62–3
phosphodiester linkages 74
phosphoenolpyruvate (PEP) 54, 62–3
phosphoenolpyruvate/phosphate translocator (PPT) 54
phosphofructokinase I (PFK I) 64
phosphoinositide specific phospholipase (PLC) 72
phospholipids 16–17, 73
phosphonates 14, 15
  see also phosphonic acids
  analysis 150–1, 162–3
  biodegradation 59, 157–8
  chemical degradation 158–60
  microorganisms 58–9
  oxidative degradation 160
  photodegradation 158
  preconcentration 151
  properties 149
  toxicology 164–5
  uses 149
  waste water treatment 163–4
phosphonic acids
  adsorption 151–3
  analysis 150–1, 162–3
  biodegradation 157–8
  chemical degradation 158–60
  dissolution 155–6
  environmental chemistry 147–73
  exchange reactions 151–6
  mineral dissolution 153–4
  oxidative degradation 160
  photodegradation 158
  precipitation 155–6
  properties 149
  remobilisation of metals 154–5
  speciation 161–3
  tautomerism 9
  toxicology 164–5
  uses 149
  waste water treatment 163–4
phosphonium salts 11
phosphonoacetate 59
phosphonobutane-tricarboxylic acid (PBTC) 157
phosphonopyruvate 59
phosphorescence 7
phosphoric acid 348–9, 353

phosphorus
 atomic properties 4
 cycle 21, 22–4, 99
 electronic structure 4–5
 elemental 6–7
 placement 614–15
phosphorus accumulating organisms (PAOs) 476, 482, 513–14, 516–19
phosphorus nuclear magnetic resonance (^{31}PNMR) spectroscopy 4
phosphorus oxides 8
phosphorus recovery process (PRP) 453, 457–63, 474–90
phosphorus-rich phosphides 7–8
phosphorylation 63–8, 70, 73
phosphotransferases 280–2
Phostrip process 274–5, 571
photodegradation 158
photophosphorylation 66–8
photosynthesis 54, 61
PHV *see* polyhydroxyvalerate
phytic acid *see* inositol phosphate
phytol 508
PI *see* phosphatidylinositol
pig farm effluents 299
pig slurry 508–10, 511
pilot plants 411–15, 477, 487–8
pitched-blade turbines 386
placement, phosphorus 614–15
plants
 acquisition 51–2
 biological phosphorus removal 272
 crop nutrition 93–119
 ecological phosphorus use 610–28
 metabolic regulation 71
 optimum soil-phosphorus levels 611, 612
 phosphates 175
 phosphorus use efficiency 612–13, 617–22
 transport/translocation 52–5
PLC *see* phosphoinositide specific phospholipase
PMMA *see* polymethylmethacrylate
^{31}PNMR *see* phosphorus nuclear magnetic resonance
point sources 122–3, 177, 181–2, 184–5
polluter pays principle 534
poly-aluminium chloride (PACl) 500–1
polyferric sulfate 501–2
polyhydroxybutyrate (PHB) 273, 276, 570
polyhydroxyvalerate (PHV) 273, 276
polymethylmethacrylate (PMMA) 588
polymorphic precipitation 232
polynucleation mechanism 203
polyphosphate accumulating organisms (PAOs) 405–6
polyphosphate glucokinase 280–1
polyphosphatekinase (PPK) 278–9, 281–2

polyphosphate-metabolising enzymes 277–82
polyphosphates 9–10, 60–1, 262, 273, 276–83
polyphosphonates 148
polyphosphoric acid 9
polysaccharide phosphate esters 73
Porites 593–4
porous bone substitutes 591–2
Posner's cluster 212
post-precipitation 263
PPK *see* polyphosphatekinase
PPT *see* phosphoenolpyruvate/phosphate translocator
precipitation
 applications 208–33, 238–9
 bacterial 549–81
 calcium phosphates 208–33
 chemical phosphorus removal 261–9
 competition/co-precipitation with carbonates 231–3
 equations 198–204
 fluid dynamics 368–72
 heterogeneous nucleation-epitaxial growth 221–31
 inhibition 156, 213–17
 mechanisms 196–8
 metal phosphates 238–9
 ochre 325
 phosphonic acids 155–6
 reactor design 358–401
 recovery 344–7
 struvite 402–27, 511
 unprocessed manure 511–12
precision farming 112
preconcentration 151, 476
preliminary treatment 250, 251–2
pre-precipitation 262
primary nucleation 200, 366–73, 448–9
primary sludge 258, 525–7
primary treatment 250, 251–2
privatisation 534
prostheses 587–8
protein interactions 597
protein kinases 70
protein phosphorylation 70, 73
Proteobacteria 282, 283
protozoa 254
PRP *see* phosphorus recovery process
Pseudomonas 57, 281, 284, 566–8, 631–2, 634
PTH *see* parathyroid hormone
public health 310
pyrophosphate 60
pyruvate 66

quadramet 14

radial flow impellers 383–5
radioactivity 115

Index

radiotherapy 4
rainfall 127, 136
Ralstonia metallidurans 552–3
rare earth elements (REE) 43–4, 305
rare earth modified clays 294, 303–4
raw sewage 249, 250
reactive phosphorus 28, 31, 331
recovered ochre *see* ochre
recovery
   see also recycling
   air stripping 440–2, 443–55, 459, 466
   fluid dynamics 358–401
   Geestmerambacht case study 470–95
   industrial uses 339–57
   precipitation 195, 344–6
   precipitation reactor design 358–401
   Slough STW case study 402–27
   solid phase adsorbents 293
   struvite 402–27, 496–506
   Treviso case study 428–69
   unprocessed manure 507–20
recycling 348–54
   economic analysis 533–7
   industrial 339, 521–8
   ochre 330, 333
   regulation 532–3
   sludge spreading 537–40
   solid phase adsorbents 293
   system approach 540–2
Redfield ratio 22–3
red mud 292–3, 294, 299–300
redox 133, 265, 308–9, 513–18
red phosphorus 6–7
red tides 187
REE *see* rare earth elements
remobilisation of metals 154–5
remote sensing 112
removal
   biological 272–90
   chemical 260–71
   fluidised bed reactors 392
   ochre 321–35
   precipitation 195
   solid phase adsorbents 291–320
   stirred vessels 387–8
residence time distribution (RTD) 360–4
return period 127
reverse PPK activity 281–2
Rhine Action Programme 182
Rhine case study 181–3
rhizosphere organisms 58
rhodium, phosphines 12
ribonucleic acid (RNA) 17, 73–4
rice 188, 189, 272, 506
rivers *see* surface waters
river sediments 30
RNA *see* ribonucleic acid

rock phosphates 124
root infection 621–2
Rothamsted Experimental Station 96, 97, 101
RTD *see* residence time distribution
runoff
   agricultural 332–3
   eutrophication 123, 125–6, 129, 131, 139–40
   transfer mechanisms 133–4, 136

Saccharomyces cerevisiae 279, 284
Salmonella 554
sands 294, 305–6
saturation-excess overland flow 136
saturation phenomena 83
SBR *see* sequencing batch reactors
scaffold materials 585–6, 602
scale formation 156, 231, 465
scanning electron microscopy (SEM) 102, 239, 240
screens 251, 252
scum 466
seasonality 127
seawater-neutralised red mud 300
SECC *see* Shimane Prefecture Lake Shinji East Clean Centre
secondary nucleation 196, 200, 373–5, 448, 449–50
secondary sludge ash 524, 527
secondary treatment 250, 252–6
second messengers 72
sedimentary rocks 33
sedimentation 250, 251–2, 257
sediments 24, 28–30, 301, 304
seed crystals 211, 222, 227, 231–3, 489
SEM *see* scanning electron microscopy
'sensitive areas' 260
septic tanks 190, 301
sequencing batch reactors (SBR) 513–16
L-serine 215–17, 218–21
Serratia
   bacterial precipitation 553–7, 558, 562, 565, 572–3
   biotechnologies 603, 604
settlement ponds 321
Severn Trent Water Ltd 275–6
Sewage Sludge Directive 342
sewage treatment
   industrial recycling 521–8
   ochre 331–2
   phosphates 175–6, 177–8
   recovery 470–95
   solid phase adsorbents 307, 309
   struvite 402–27
shikimate pathway 54
Shimane Prefecture Lake Shinji East Clean Centre 500–6

side-stream phosphorus recovery process 474–90
signalling pathways 71–2
simultaneous precipitation 262–3
single jet semi-batch precipitation 368–70
sintered ceramic implants 229
sinus lifts 588–9
site-directed mutagenesis 636–7
slags 97, 292–3, 294, 302–3
Slough STW case study 402–27
slow-release fertilisers 140, 322, 333
sludge
  aluminium salts 433–4
  chemical phosphorus removal 268–9
  industrial recycling 522–4
  phosphorus rich 537–8
  primary 525–7, 528
  recovery 344–5
  recycling 531–2, 537–40
  secondary sludge ash 524, 527
  settling 490–1
  spreading 537–40
  sulphur 406–7
  thickening 477, 480, 483–4
  treatment/disposal 257–9
  waste activated 435
sludge volume index (SVI) 266–7
sodium tripolyphosphate (STPP) 530–1
soil organic matter (SOM) 110
soil P tests 83–4, 106–10, 124–5, 132, 139–40
soils
  background release 79, 80, 81
  classification 109–10
  crop nutrition 98–102
  detachment 85–6
  ecological phosphorus use 610–28
  elevated release 79–92
  erosion 85–6
  fertilisers 113–15
  geochemistry 24–8
  husbandry 116
  mobilisation 82–6
  optimum soil-phosphorus levels 611, 612
  phosphorus dynamics 611–12
  recycled phosphorus 351–2
  solid phase adsorbents 292, 294, 305–6
  sources 82
  transfer 127–30
  transport 86–7
solid phase adsorbents 291–320
  activated alumina 294, 298–9
  allophanes 294, 295–6, 306
  aluminium oxides 297–8, 306
  carbonates 294, 304–5
  evaluation 310–11
  expanded clay aggregates 292, 294, 301–2
  fly ash 292–3, 294, 300–1
  hydrotalcites 293–5, 300
  imogolites 294, 295–6
  iron oxides 294, 297–8, 306
  pH sensitivity 301, 302, 309–10
  public health 310
  rare earth modified clays 294, 303–4
  red mud 292–3, 294, 299–300
  redox sensitivity 308–9
  selection criteria 307–11
  slags 292–3, 294, 302–3
  soils, sands and gravels 292, 294, 305–6
solubilisation 57–8, 82–5, 132, 133–5
solubility 197–205, 207, 209
SOM *see* soil organic matter
sorption 25–6
South Asia 188–9
sparingly soluble salts 198–9, 204
speciation 161–3, 199
spinal fusion 586
spiral growth mechanism 202–3
spontaneous precipitation 210–11
SPP *see* struvite precipitation potential
SPS *see* sucrose phosphate synthase
starch 53–4
steps 201, 214–15
sterilisation 586
sterrettite 25
stirred vessels 379–88
stocking density 128
Stokes Law 251
stormflow 136, 137–8
STPP *see* sodium tripolyphosphate
strengite 25
struvite 43
  bench scale work 411–15
  economic analysis 535
  economic viability 422–4
  fertilisers 499–500, 506
  Japan 496–506
  mass balances 407–11
  precipitation 511
  production rates 422
  quality assessment 419–25
  reactor specifications 415–16
  recovery 402–27
  Treviso case study 436–7, 440–3, 446
  uses 425–6
struvite precipitation potential (SPP) 408, 411
substrate level phosphorylation 63–5
sucrose 53–4
sucrose phosphate synthase (SPS) 71
sudden oak death 14
sulphate-reducing bacteria (SRB) 553
sulphides 10
sulphur 94, 406–7
supercritical nuclei 200

supersaturation
    co-precipitation 232–3
    curves 442–3
    heterogeneous nucleation-epitaxial growth 222–3, 225–7, 229–31
    model 448, 451
    precipitation 197–204, 209–11
surface area 207
surface expression of OPH 635–6
surface secondary nucleation 373–4
surface waters 31–2, 120–46, 174–5, 291–320
sustainability 95–6
SVI *see* sludge volume index
Sweden, recycling 354
swimming pools 308, 309
symbiosis 105

Tanzania 189
tautomerism 9
TBP *see* tributylphosphate
TCA *see* tricarboxylic acid cycle
TCLP *see* Toxicity Characteristic Leaching Protocol
TCP *see* tricalcium phosphate
teeth *see* dental; dentine
terminal settling velocity 390–1
terraces 201, 214–15
terrestrial *see* soils
tertiary treatment 250, 256–7, 537–8
tetrahalides 10–11
thickening sludge 477, 480, 483–4
thiobacilli 57–8
threshold inhibitors 156
thyalkaloid membrane 67
Tick Eradication Program 630
tinticite 25
tissue interactions 597–601
Tolman's cone angle 5
topography 86, 125–6, 128, 131
toxicity 3, 164–5, 253, 301, 303–4
Toxicity Characteristic Leaching Protocol (TCLP) 303
TPT *see* triose phosphate/phosphate translocator
transfer 120–46
transition metals 5
transition state theory 204
translocation 54–5
transport
    animals 55–7
    elevated release 86–7
    geochemistry 30–2
    microorganisms 57–61
    plants 52–5
    precipitation 201
transuranic elements 561–2
travelling band screens 252

Treviso case study 428–69
tributylphosphate (TBP) 16, 566–7, 572–3
tricalcium phosphate (TCP) 234, 236–7, 584–5, 588, 590–1, 602–3
tricarboxylic acid cycle (TCA) 66
trickling filters 254
trihalides 10–11
triose phosphate/phosphate translocator (TPT) 54
triphenylphosphine 11
triphosphoric acid 10
triple superphosphate (TSP) 96, 111, 114, 617
Trypanosoma cruzi 60
TSP *see* triple superphosphate
two-stage sequencing batch reactors (TSSBR) 514–16

UCT *see* University of Capetown
United Kingdom
    eutrophication 121–2, 137
    fertilisers 111, 114, 531, 539
    ochre 321–2, 324
    phosphates 178
    struvite 402–27
    unprocessed manure 508
    waste water treatment plants 249
United States
    chemical phosphorus removal 263–4
    crop nutrition 96
    elevated release 83
    eutrophication 121, 132
    fly ash production 300
    phosphates 174, 183–5
    slags 302
United States Department of Agriculture (USDA) 630
United States Environmental Protection Agency (USEPA) 184–5, 263–4, 533, 534
University of Capetown (UCT) process 275–6
unprocessed manure 507–20
uranium
    bacterial precipitation 550, 552, 555, 557–9, 564–7, 571–2
    fertilisers 115
    phosphate fixation 239
uranium phosphates 41
Urban Waste Water Treatment Directive (UWWTD) 138, 249, 260, 342–3, 538
USDA *see* United States Department of Agriculture
USEPA *see* United States Environmental Protection Agency
UWWTD *see* Urban Waste Water Treatment Directive

vacancies 35–6, 196, 235
vacuolar phosphate 53, 55

VAM *see* vesicular arbuscular mycorrhizal
variable source areas (VSAs) 125, 128, 130, 131, 137
variscite 25
V-ATPases 69–70
vertical flows 136
vesicular arbuscular mycorrhizal (VAM) fungi 105
VFAs *see* volatile fatty acids
vitamins 74
vivianite 42–3, 293
volatile fatty acids (VFAs) 435–6
Volmer–Weber–Becker theory 226
voluntary national initiatives 179
VSAs *see* variable source areas

waste activated sludge (WAS) 435
Waste Basis Directive 538
waste water treatment plants (WWTPs) 249–59
   economic analysis 533–7
   eutrophication 123
   Geestmerambacht case study 470–95
   phosphates 177–8, 184–5, 188–9
   phosphonates 151, 163–4
   preliminary treatment 250, 251–2
   primary treatment 250, 251–2
   recovery 344–5
   recycling 521–8, 529–46
   regulation 532–3
   secondary treatment 250, 252–6
   sludge treatment/disposal 257–9
   solid phase adsorbents 291–320
   tertiary treatment 250, 256–7
   Treviso case study 429–37
Water Framework Directive (WFD) 80, 341–2, 538
Water Research Council (WRC) 411
weathering 21, 23
wet recycling processes 348, 350
WFD *see* Water Framework Directive
white phosphorus 6–7
whitlockite 32, 34
WHO *see* World Health Organization
whole cell detoxification 634–6
Wilkinson's catalyst 12
Wittig reaction 11
World Health Organization (WHO) 16
World Phosphate Institute (IMPHOS) 101
WRC *see* Water Research Council
WWTPs *see* waste water treatment plants

xenotime 32, 43–4

yeasts 58
ylides 11
Young's modulus 584–5
yttrium carbonate 305

Zanzibar 189
zeolites 301, 530, 550